T0314036

ENGINEERING BIOSTATISTICS

ENGINEERING BIOSTATISTICS

An Introduction using MATLAB® and WinBUGS

BRANI VIDAKOVIC

WILEY

Registered Office
John Wiley & Sons, Inc., 111 River Street, Hoboken, NJ 07030, USA

Editorial Office
111 River Street, Hoboken, NJ 07030, USA

For details of our global editorial offices, customer services, and more information about Wiley products visit us at www.wiley.com.

Wiley also publishes its books in a variety of electronic formats and by print-on-demand. Some content that appears in standard print versions of this book may not be available in other formats.

Library of Congress Cataloging-in-Publication Data:

Names: Vidakovic, Brani, 1955- author.
Title: Engineering biostatistics : an introduction using MATLAB and WinBUGS / Brani Vidakovic.
Description: Hoboken, New Jersey : John Wiley & Sons, 2017. | Includes bibliographical references and index.
Identifiers: LCCN 2016042853| ISBN 9781119168966 (cloth) | ISBN 9781119168980 (epub)
Subjects: LCSH: MATLAB. | WinBUGS. | Biometry--Statistical methods. | Engineering--Statistical methods.
Classification: LCC QH323.5 .V53 2017 | DDC 570.1/5195--dc23 LC record available at https://lccn.loc.gov/2016042853

Cover images: (Background) © John Lund/Gettyimages; (Figures) Courtesy of author
Cover design by Wiley

10 9 8 7 6 5 4 3 2 1

Contents

Preface

Let's say you commit to 2 hours per day and you're able to write 3 pages per hour. To write an average length book of 300 pages will take 50 days. (300 pages per book/6 pages per day = 50 days).

A posting on *Life Learning Today*: How to Write a Book in 60 Days or Less.

By not following the recommendations from the above quote the writing of this book took much longer. The book is a result of many semesters of teaching various statistical courses to engineering students at Duke University and the Georgia Institute of Technology. Through its scope and depth of coverage, the book addresses the needs of the vibrant and rapidly growing engineering fields while implementing software that engineers are familiar with.

This book is substantially revised version of the text originally published by Springer in 2011 (ISBN 978-1-4614-0394-4). In addition to providing many new examples and exercises, a number of new sections is added. The original edition served as a primary textbook for the course *Introduction to Bioengineering Statistics*, at The Wallace H. Coulter Department of Biomedical Engineering at Georgia Tech for 6 consecutive semesters. I noticed that it was used not only as a textbook but students found it useful as a repository of techniques for data analysis in other courses, class projects, senior design projects, and day-to-day laboratory data analysis. So by its scope this book is both: a textbook for introductory-to-intermediate applied biostatistics courses and a reference book with a coverage of a number of rather specialized techniques.

This book is heavily oriented to computation and hands-on approaches. The approach enforced avoids the use of mainstream statistical packages in which the procedures are often black-boxed. Rather, the students are expected to code the procedures on their own. The results may not be as flashy as they would be if the specialized packages were used, but the student will go through the process and understand each step of the program.

The computational support for this text is the MATLAB© programming environment since this software is predominant in the engineering communities.

Another dimension of this book is in the substantial coverage of Bayesian approaches to statistical inference. I avoided taking sides on the traditional (classical, frequentist) vs. Bayesian approach; it was my goal to expose students to both approaches. It is undeniable that classical statistics is overwhelmingly used in conducting and reporting inference among practitioners, and that Bayesian statistics is gaining in popularity, acceptance, and usage (FDA, Guidance for the Use of Bayesian Statistics in Medical Device Clinical Trials, 5 February 2010). Many examples in this book are solved using both the traditional and Bayesian methods, and the results are compared and commented upon.

This diversification is made possible by advances in Bayesian computation and the availability of the free software WinBUGS/OpenBUGS that provides painless computational support for Bayesian solutions. WinBUGS and MATLAB communicate well due to the interface software MATBUGS, written by Kevin P. Murphy and coauthors. The book also relies on stat toolbox within MATLAB.

The World Wide Web (WWW) facilitates the book. All custom-made MATLAB and WinBUGS programs (compatible with MATLAB R2017a and WinBUGS 1.4.3 or OpenBUGS 3.2.3) as well as data sets used in this book are available on the Web:

```
http://statbook.gatech.edu/
```

With the size of this book in mind the solutions and hints to some exercises can be found on the book's Web site. The computer scripts and examples are an integral part of the book, and all MATLAB codes and outputs are shown in blue typewriter font while all WinBUGS programs are given in red-brown typewriter font. The comments in MATLAB and WinBUGS codes are presented in green typewriter font.

The three icons 📖, ◢, and ✳ are used to point to data sets, MATLAB codes, and WinBUGS codes, respectively.

The difficulty of the material covered necessarily varies. More difficult sections that may be omitted in the basic coverage are denoted by a star, *. However, it is my experience that advanced undergraduate bioengineering students affiliated with school research labs need and use the "starred" material, such as functional ANOVA, variance stabilizing transforms, and nested experimental designs, to name just a few. Tricky or difficult places are marked with Donald Knut's "bend" ✆.

Each chapter starts with a box titled WHAT IS COVERED IN THIS CHAPTER and ends with chapter exercises, a box called MATLAB AND WINBUGS FILES AND DATA SETS USED IN THIS CHAPTER, and chapter references.

The examples are numbered and the end of each example is marked with ✐.

I am aware that this work could be improved with respect to both exposition and coverage. Thus, I would welcome any criticism and pointers from readers as to how this book could be improved.

Acknowledgments. I am indebted to many students and colleagues who commented on various drafts of the book. In particular I am grateful to colleagues from the Department of Biomedical Engineering at the Georgia Institute of Technology and Emory University and their undergraduate and graduate advisees/researchers who contributed many examples and data sets from their research labs.

Colleagues Tom Bylander of the University of Texas at San Antonio, John H. McDonald of the University of Delaware, and Roger W. Johnson of the South Dakota School of Mines & Technology kindly gave permission to use their data and examples. Special thanks go to Dr. Gary M. Raymond from University of Washington who provided useful feedback on several occasions. Several MATLAB codes used in this book come from the MATLAB Central File Exchange forum. In particular, I am grateful to Antonio Truillo-Ortiz and his team (Universidad Autonoma de Baja California) and to Giuseppe Cardillo (MeriGen Research) for their excellent contributions.

The book benefited from the input of many diligent students when it was used either as a supplemental reading or later as a draft textbook for a semester-long course at Georgia Tech: BMED2400 Introduction to Bioengineering Statistics. A complete list of students who provided useful feedback would be quite long, but the most diligent ones were Erin Hamilton, Kiersten Petersen, David Dreyfus, Jessica Kanter, Radu Reit, Amoreth Gozo, Nader Aboujamous, and Allison Chan. Special thanks go to Brett Jordan, who, as multiple-time Teaching Assistant for Bioengineering Statistics course, pointed out numerous places for text improvement.

Wiley's team kindly helped along the way. I am grateful to Jon Gurstelle, Allison McGinniss, Kathleen Pagliaro, and Melissa Yanuzzi for their encouragement and support. Finally, it hardly needs stating that the book would have been considerably less fun to write without the unconditional support of my family.

BRANI VIDAKOVIC
School of Industrial and Systems Engineering and
School of Biomedical Engineering
Georgia Institute of Technology
brani@gatech.edu

Chapter 1
Introduction

Many people were at first surprised at my using the new words "Statistics" and "Statistical," as it was supposed that some term in our own language might have expressed the same meaning. But in the course of a very extensive tour through the northern parts of Europe, which I happened to take in 1786, I found that in Germany they were engaged in a species of political inquiry to which they had given the name of "Statistics". . . . I resolved on adopting it, and I hope that it is now completely naturalised and incorporated with our language.

– Sinclair, 1791; Vol XX

WHAT IS COVERED IN THIS CHAPTER

- What is the subject of statistics?
- Population, sample, data
- Appetizer examples

The problems confronting health professionals today often involve fundamental aspects of device and system analysis, and their design and application. As such they are of extreme importance to engineers and scientists.

Due to many aspects of engineering and scientific practice involving nondeterministic outcomes, understanding and knowledge of statistics is important to any engineer and scientist. Statistics is a *guide to the unknown*. It is a science that deals with designing experimental protocols; collecting, summarizing, and presenting data; and, most important, making inferences and aiding decisions in the presence of variability and uncertainty.

For example, R. A. Fisher's 1943 elucidation of the human blood-group system Rhesus in terms of the three linked loci C, D, and E, as described in Fisher (1947) or Edwards (2007), is a brilliant example of building a coherent structure of new knowledge guided by a statistical analysis of available experimental data.

The uncertainty that statistical science addresses derives mainly from two sources: (1) from observing only a part of an existing, fixed, but large population or (2) from having a process that results in nondeterministic outcomes. At least a part of the process needs to be either a *black box* or inherently stochastic, so the outcomes cannot be predicted with certainty.

A *population* is a statistical universe. It is defined as a collection of existing attributes of some natural phenomenon or a collection of potential attributes when a process is involved. In the case of a process, the underlying population is called hypothetical, for obvious reasons. Thus, populations can be either finite or infinite. A subset of a population selected by some relevant criteria is called a subpopulation.

Often we think about a population as an assembly of people, animals, items, events, times, etc., in which the attribute of interest is measurable. For example, the population of all US citizens older than 21 is an example of a population for which many attributes can be assessed. Attributes might be *a history of heart disease, weight, political affiliation, level of blood sugar*, etc.

A sample is an observed part of a population. Selection of a sample is a rich methodology in itself, but, unless otherwise specified, it is assumed that the sample is selected at random. The randomness ensures that the sample is representative of its population.

The sampling process depends on the nature of the problem and the population. For example, a sample may be obtained via a retrospective study (usually existing historical outcomes over some period of time), an observational study (an observer monitors the process or population in real time), a sample survey (a researcher administers a questionnaire to measure the characteristics and/or attitudes of subjects), or a designed study (a researcher makes deliberate changes in controllable variables to induce a cause/effect relationship), to name just a few.

Example 1.1. **Ohm's Law Measurements.** A student constructed a simple electric circuit in which the resistance R and voltage E were controllable. The output of interest is current I, and according to Ohm's law it is

$$I = \frac{E}{R}.$$

This is a mechanistic, theoretical model. In a finite number of measurements under an identical R, E setting, the measured current varies. The population here is hypothetical – an infinite collection of all potentially obtainable measurements of its attribute, current I. The observed sample is finite. In the presence of sample variability, one establishes an empirical (statistical) model for currents from the population as either

$$I = \frac{E}{R} + \epsilon \quad \text{or} \quad I = \epsilon \frac{E}{R}.$$

On the basis of a sample, one may first select the model and then proceed with the inference about the nature of the discrepancy, ϵ.

Example 1.2. **Cell Counts.** In a quantitative engineering physiology laboratory, a team of four students was asked to make a LabVIEW© program to automatically count MC3T3-E1 cells in a hemocytometer (Fig. 1.1). This automatic count was to be compared with the manual count collected through an inverted bright field microscope. The manual count is considered the gold standard.

The experiment consisted of placing 10 μL of cell solutions at two levels of cell confluency: 20% and 70%. There were $n_1 = 12$ pairs of measurements (automatic and manual counts) at 20% and $n_2 = 10$ pairs at 70%, as in the table below.

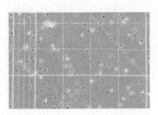

Fig. 1.1 Cells on a hemocytometer plate.

	20% confluency											
Automated	34	44	40	62	53	51	30	33	38	51	26	48
Manual	30	43	34	53	49	39	37	42	30	50	35	54
	70% confluency											
Automated	72	82	100	94	83	94	73	87	107	102		
Manual	76	51	92	77	74	81	72	87	100	104		

The students wish to answer the following questions:

(a) Are the automated and manual counts significantly different for a fixed confluency level? What are the confidence intervals for the population differences if normality of the measurements is assumed?

(b) If the difference between automated and manual counts constitutes an error, are the errors comparable for the two confluency levels?

We will revisit this example later in the book (Exercise 10.20) and see that for the 20% confluency level there is no significant difference between the automated and manual counts, whereas for the 70% level the difference is significant. We will also see that the errors for the two confluency levels significantly differ. The statistical design for comparison of errors is called a difference in differences (DiD) and is quite common in biomedical data analysis.

✐

Example 1.3. **Rana Pipiens.** Students in a quantitative engineering physiology laboratory were asked to expose the gastrocnemius muscle of the northern leopard frog (*Rana pipiens,* and stimulate the sciatic nerve to observe contractions in the skeletal muscle. Students were interested in modeling the length–tension relationship. The force used was the active force, calculated by subtracting the measured passive force (no stimulation) from the total force (with stimulation).

The active force represents the dependent variable. The length of the muscle begins at 35 mm and stretches in increments of 0.5 mm, until a maximum length of 42.5 mm is achieved. The velocity at which the muscle was stretched was held constant at 0.5 mm/s.

Reading	Change in Length (in %)	Passive force	Total force
1	1.4	0.012	0.366
2	2.9	0.031	0.498
3	4.3	0.040	0.560
4	5.7	0.050	0.653
5	7.1	0.061	0.656
6	8.6	0.072	0.740
7	10.0	0.085	0.865
8	11.4	0.100	0.898
9	12.9	0.128	0.959
10	14.3	0.164	0.994
11	15.7	0.223	0.955
12	17.1	0.315	1.019
13	18.6	0.411	0.895
14	20.0	0.569	0.900
15	21.4	0.751	0.905

The correlation between the active force and the percent change in length from 35 mm is –0.0941. Why is this correlation so low?

For example, one possible model can be found using linear regression (least squares):

$$\hat{F} = 0.0618 + 0.2084 \cdot \delta - 0.0163 \cdot \delta^2 + 0.0003 \cdot \delta^3$$
$$- 0.1732 \cdot \sin\left(\frac{\delta}{3}\right) + 0.1242 \cdot \cos\left(\frac{\delta}{3}\right),$$

where $\hat{F}$ is the fitted active force and δ is the percent change. This model is nonlinear in variables but linear in coefficients, and standard linear regression methodology is applicable (Chapter 14). The model achieves a coefficient of determination of $R^2 = 87.16\%$.

A plot of the original data with superimposed model fit is shown in Figure 1.2a. Figure 1.2b shows the residuals $F - \hat{F}$ plotted against δ.

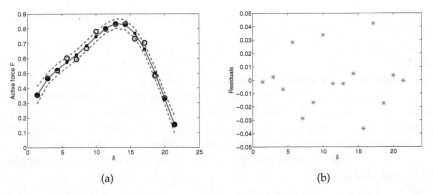

(a)   (b)

Fig. 1.2 (a) Regression fit for active force. Observations are shown as *yellow* circles, while the smaller *blue* circles represent the model fits. *Dotted (blue) lines* are 95% model confidence bounds. (b) Model residuals plotted against the percent change in length δ.

Suppose that students are interested in estimating the active force for a change of 12%. The model prediction for $\delta = 12$ is 0.8183, with a 95% confidence interval of $[0.7867, 0.8498]$.

Example 1.4. **The 1954 Polio Vaccine Trial.** One of the largest and most publicized public health experiments was performed in 1954 when the benefits of the Salk vaccine for preventing paralytic poliomyelitis was assessed. To ensure that there was no bias in conducting and reporting, the trial was blind to doctors and patients. In boxes of 50 vials, 25 had active vaccines and 25 were placebo. Only the numerical code known to researchers distinguished the well-mixed vials in the box. The clinical trial involved a large number of first-, second-, and third-graders in the United States.

The results were convincing. While the numbers of children assigned to active vaccine and placebo were approximately equal, the incidence of polio in the active group was almost four times lower than that in the placebo group.

	Inoculated with vaccine	Inoculated with placebo
Total number of children inoculated	200,745	201,229
Number of cases of paralytic polio	33	115

On the basis of this trial, health officials recommended that every child be vaccinated. Since the time of this clinical trial, the vaccine has improved; Salk's vaccine was replaced by the superior Sabin preparation and polio is now virtually unknown in the United States. A complete account of this clinical trial can be found in Francis et al.'s (1955) article or Paul Meier's essay in a popular book by Tanur et al. (1972).

The numbers are convincing, but was it possible that an ineffective vaccine produced such a result by chance?

In this example there are two hypothetical populations. The first consists of all first-, second-, and third-graders in the United States who would be inoculated with the active vaccine. The second population consists of US children of the same age who would receive the placebo. The attribute of interest is the presence/absence of paralytic polio. There are two samples from the two populations. Randomness of the samples was ensured by randomization of vials in the boxes and random selection of geographic regions for schools. Further analysis of this data can be found in Examples 10.17 and 18.11.

✐

The term *statistics* has a plural form but is used in the singular when it relates to methodology. To avoid confusion, we note that *statistics* has another meaning and use. Any sample summary will be called a *statistic*. For example, a sample mean is a statistic, and sample mean and sample range are statistics. In this context, statistics is used in the plural.

The ultimate summary for quantifying a population attribute is a statistical model. The statistical model term is used in a broad sense here, but a component quantifying inherent uncertainty is always present. For example, random variables, discussed in Chapter 5, can be interpreted as basic statistical models when they model realizations of the attributes in a sample. The model is often indexed by one, several, or sometimes even an infinite number of unknown parameters. An inference about the model translates to an inference about its parameters.

Data are the specific values pertaining to a population attribute recorded from a sample. Often, the terms sample and data are used interchangeably. The term *data* is used as both singular and plural. The singular mode relates to a set, a collection of observations, while the plural is used when referring to the observations. A single observation is called a *datum*.

The following table summarizes the fundamental statistical notions that we discussed:

attribute	Quantitative or qualitative property, feature(s) of interest
population	Statistical universe; an existing or hypothetical totality of attributes
sample	A subset of a population
data	Recorded values/realizations of an attribute in a sample
statistical model	Mathematical description of a population attribute that incorporates incomplete information, variability, and the nondeterministic nature of the population
population parameter	A component (possibly multidimensional) in a statistical model; the models are typically specified up to a parameter that is left unknown

CHAPTER REFERENCES

Edwards, A. W. F. (2007). R. A. Fisher's 1943 unravelling of the Rhesus blood-group system. *Genetics*, **175**, 471–476.

Fisher, R. A. (1947). The Rhesus factor: A study in scientific method. *American Scientist*, **35**, 95–102.

Francis, T. Jr., Korns, R., Voight, R., Boisen, M., Hemphill, F., Napier, J., and Tolchinsky, E. (1955). An evaluation of the 1954 poliomyelitis vaccine trials: Summary report. *American Journal of Public Health*, **45**, 5, 1–63.

Sinclair, Sir John. (1791). The Statistical Account of Scotland. Drawn up from the communications of the ministers of the different parishes. Volume first. Edinburgh: printed and sold by William Creech, Nha. V27.

Tanur, J. M., Mosteller, F., Kruskal, W. H., Link, R. F., Pieters, R. S. and Rising, G. R., eds. (1989). *Statistics: A Guide to the Unknown*, 3rd Edition. Wadsworth, Belmont, CA.

Chapter 2
The Sample and Its Properties

When you're dealing with data, you have to look past the numbers.

– Nathan Yau

WHAT IS COVERED IN THIS CHAPTER

- MATLAB Session with Basic Univariate Statistics
- Numerical Characteristics of a Sample
- Multivariate Numerical and Graphical Sample Summaries
- Exploratory Principal Components
- Typology of Data

2.1 Introduction

The famous American statistician John Tukey once said, "Exploratory data analysis can never be the whole story, but nothing else can serve as the foundation stone – as the first step." The term *exploratory data analysis* is self-defining. Its simplest branch, *descriptive statistics*, is the methodology behind approaching and summarizing experimental data. No formal statistical training is needed for its use. Basic data manipulations such as calculating averages of experimental responses, translating data to pie charts or histograms, or assessing the variability and inspection for unusual measurements are all examples of descriptive statistics. Rather than

9

focusing on the population using information from a sample, which is a staple of statistics, descriptive statistics is concerned with the description, summary, and presentation of the sample itself. For example, numerical summaries of a sample could be measures of location (mean, median, percentiles, mode, extrema), measures of variability (sample standard deviation/variance, robust versions of the variance, range of data, interquartile range, etc.), higher-order statistics (kth moments, kth central moments, skewness, kurtosis), and functions of descriptors (coefficient of variation). Graphical summaries of samples involve various visual presentations such as box-and-whisker plots, pie charts, histograms, empirical cumulative distribution functions, etc. Many basic data descriptors are used in everyday data manipulation.

Ultimately, exploratory data analysis and descriptive statistics contribute to the principal goal of statistics – inference about population descriptors – by guiding how the statistical models should be set.

It is important to note that descriptive statistics and exploratory data analysis have recently regained importance due to ever increasing sizes of data sets. Some complex data structures require several terabytes of memory just to be stored. Thus, preprocessing, summarizing, and dimension-reduction steps are needed to prepare such data for inferential tasks such as classification, estimation, and testing. Consequently, the inference is placed on data summaries (descriptors, features) rather than the raw data themselves.

Many data-managing software programs have elaborate numerical and graphical capabilities. MATLAB provides an excellent environment for data manipulation and presentation with superb handling of data structures and graphics. In this chapter we intertwine some basic descriptive statistics with MATLAB programming using data obtained from real-life research laboratories. Most of the statistics are already built in; for some we will make a custom code in the form of m-functions or m-scripts.

This chapter has two goals: (i) to help you gently relearn and refresh your MATLAB programming skills through annotated sessions, and (ii) introduce some basic statistical measures, many of which should already be familiar to you. Many of the statistical summaries will be revisited later in the book in the context of inference. You are encouraged to continuously consult MATLAB's online help pages since details and command options are not fully covered in this text.

2.2 A MATLAB Session on Univariate Descriptive Statistics

In this section we will analyze data derived from an experiment, step by step with a brief explanation of the MATLAB commands used. The whole session can be found in a single annotated file ◀ cellarea.m available

at the book's Web page, http://statbook.gatech.edu/Ch2.Descriptive/.
The data can be found in the file 🖳 cellarea.dat|mat|xlsx, which features
measurements from the lab of Professor Todd McDevitt at Georgia Tech.

This experiment on cell growth involved several time durations and two
motion conditions. Here is a brief description:

> Embryonic stem cells (ESCs) have the ability to differentiate into all somatic cell
> types, making ESCs useful for studying developmental biology, *in vitro* drug
> screening, and as a cell source for regenerative medicine and cell-based therapies.
> A common method to induce differentiation of ESCs is through the formation of
> multicellular spheroids termed embryoid bodies (EBs). ESCs spontaneously ag-
> gregate into EBs when cultured on a nonadherent substrate; however, under static
> conditions, this aggregation is uncontrolled and EBs form in various sizes and
> shapes, which may lead to variability in cell differentiation patterns. When rotary
> motion is applied during EB formation, the resulting population of EBs appears
> more uniform in size and shape.

> After 2, 4, and 7 days of culture, images of EBs were acquired using phase-contrast
> microscopy. Image analysis software was used to determine the area of each EB
> imaged. At least 100 EBs were analyzed from three separate plates for both static
> and rotary cultures at the three time points studied.

Here we focus only on the measurements of visible surface areas of cells
(in μm^2) after growth time of 2 days, $t = 2$, under the static condition. The
data are recorded as an ASCII file 🖳 cellarea.dat. Importing the data set
into MATLAB is done using the command

```
load('cellarea.dat');
```

provided that the data set is on the MATLAB path. If this is not the case,
use addpath('foldername') to add the path foldername in which the file resides.
A glimpse at the data is obtained by histogram command, hist:

```
hist(cellarea, 100)
```

On inspecting the histogram (Fig. 2.1), we find that one quite unusual
observation that is inconsistent with the remaining experimental measurements. We can assume that the unusual observation is an outlier and omit
it from the data set:

```
car = cellarea(cellarea ~= max(cellarea));
```

(Some formal diagnostic tests for outliers will be discussed later in the text.)

Next, the data are rescaled to more moderate values, so that the area is
expressed in thousands of μm^2 and the measurements have a convenient
order of magnitude:

```
car = car/1000;
n = length(car); %n is sample size
%n=462
```

Thus, we obtain a sample of size $n = 462$ that we can further explore by
descriptive statistics. The histogram we have plotted has already given us

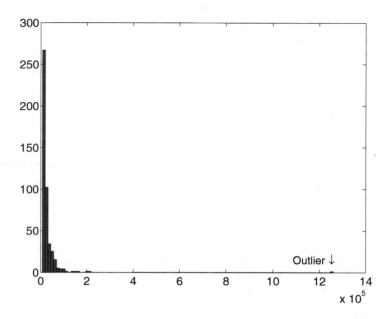

Fig. 2.1 Histogram of the raw data. Notice the unusual measurement beyond the point 12×10^5.

a sense of the distribution within the sample, and we have an idea of the shape, location, spread, symmetry, etc., of the observations.

Yet we need to find the numerical characteristics of the sample. First we will discuss its location measures, which, as the name indicates, evaluate the relative location of the sample.

2.3 Location Measures

Means. The three averages – arithmetic, geometric, and harmonic – are known as Pythagorean means.

The arithmetic mean (mean),

$$\overline{X} = \frac{X_1 + \cdots + X_n}{n} = \frac{1}{n} \sum_{i=1}^{n} X_i,$$

is a fundamental summary statistic. The geometric mean (geomean) is

$$\sqrt[n]{X_1 \times X_2 \times \cdots \times X_n} = \left(\prod_{i=1}^{n} X_i\right)^{1/n},$$

and the harmonic mean (harmmean) is

$$\frac{n}{1/X_1 + 1/X_2 + \cdots 1/X_n} = \frac{n}{\sum_{i=1}^{n} 1/X_i}.$$

For the data set $\{1,2,3\}$ the mean is 2, the geometric mean is $\sqrt[3]{6} = 1.8171$, and the harmonic mean is $3/(1/1 + 1/2 + 1/3) = 1.6364$. In standard statistical practice geometric and harmonic means are not used as often as arithmetic means. To illustrate the contexts in which they should be used, consider several simple examples.

Example 2.1. **Use of Geometric Mean.** You visit the bank to deposit a long-term monetary investment in the hope that it will accumulate interest over a three year span. Suppose that the investment earns 10% the first year, 50% the second year, and 30% the third year. What is its average rate of return? In this instance it is not the arithmetic mean, because in the first year the investment was multiplied by 1.10, in the second year it was multiplied by 1.50, and in the third year it was multiplied by 1.30. The correct measure is the geometric mean of these three numbers, which is about 1.29, or 29% of the annual interest. If, for example, the ratios are averaged (i.e., ratio = new method/old method) over many experiments, the geometric mean should be used. This is evident by considering an example. If one experiment yields a ratio of 10 and the next yields a ratio of 0.1, an arithmetic mean would misleadingly report that the average ratio was near 5. Taking a geometric mean will report a more meaningful average ratio of 1.

Example 2.2. **Use of Harmonic Mean.** Consider now two scenarios in which the harmonic mean should be used.

(i) If for half the distance of a trip one travels at 40 miles per hour and for the other half of the distance one travels at 60 miles per hour, then the average speed of the trip is given by the harmonic mean of 40 and 60, which is 48; that is, the total amount of time for the trip is the same as if one traveled the entire trip at 48 miles per hour. Note, however, that if one had traveled for half the time at one speed and then half time at another speed, the arithmetic mean, in this case 50 miles per hour, would be the correct average.

(ii) In financial calculations, the harmonic mean is used to express the average cost of shares purchased over a period of time. For example, an investor purchases $1000 worth of stock every month for 3 months. If the three spot prices at execution time are $8, $9, and $10, then the average price the investor paid is $8.926 per share. However, if the investor purchased 1000 shares per month, then the arithmetic mean should be used.

Order Statistic. If the sample $X_1, \ldots, X_n$ is ordered as $X_{(1)} \le X_{(2)} \le \cdots \le X_{(n)}$ so that $X_{(1)}$ is the minimum and $X_{(n)}$ is the maximum, then $X_{(1)}, X_{(2)}, \ldots, X_{(n)}$ is called the *order statistic*. For example, if $X_1 = 2$, $X_2 = -1$, $X_3 = 10$, $X_4 = 0$, and $X_5 = 4$, then the order statistic is $X_{(1)} = -1$, $X_{(2)} = 0$, $X_{(3)} = 2$, $X_{(4)} = 4$, and $X_{(5)} = 10$.

Median. The median[1] is the middle of the sample sorted in numerical order. In terms of the order statistic, the median is defined as

$$
Me = \begin{cases} X_{((n+1)/2)}, & \text{if } n \text{ is odd,} \\ (X_{(n/2)} + X_{(n/2+1)})/2, & \text{if } n \text{ is even.} \end{cases}
$$

If the sample size is odd, then there is a single observation in the middle of the ordered sample at the position $(n+1)/2$. For even sample sizes, the ordered sample has two elements in the middle at positions $n/2$ and $n/2 + 1$, and the median is their average. The median is a robust estimator of location, that is, not easily affected by extremes and outliers. For instance, in both data sets, $\{-1, 0, 4, 7, 20\}$ and $\{-1, 0, 4, 7, 200\}$, the median is 4. However, the means are 6 and 42, respectively.

Mode. The most frequent (fashionable[2]) observation in the sample (if such exists) is the mode of the sample. If the sample is composite, the observation x_i corresponding to the largest frequency f_i is the mode. Composite samples consist of realizations x_i and their frequencies f_i, as in

$$
\begin{pmatrix} x_1 \ x_2 \ \ldots \ x_k \\ f_1 \ f_2 \ \ldots \ f_k \end{pmatrix}.
$$

Mode may not be unique. If there are two modes, the sample is bimodal, three modes make it trimodal, etc.

Trimmed Mean. As mentioned earlier, the mean is a location measure sensitive to extreme observations and possible outliers. To make this measure more robust, one may trim $\alpha \cdot 100\%$ of the data symmetrically from both sides of the ordered sample (trim $\alpha/2 \cdot 100\%$ smallest and $\alpha/2 \cdot 100\%$ largest observations, Figure 2.2b).

If your sample, for instance, is $\{1, 2, 3, 4, 5, 6, 7, 8, 9, 100\}$, then a 20% trimmed mean is a mean of $\{2, 3, 4, 5, 6, 7, 8, 9\}$.

Here is the command in MATLAB that determines the discussed locations for the cell data:

```
location = [geomean(car) harmmean(car) mean(car) ...
    median(car) mode(car) trimmean(car,20)]
%location = 18.8485 15.4211  24.8701 17 10 20.0892
```

---

[1] Latin: *medianus* = middle

[2] *Mode* (fr) = fashion

By applying $\alpha \times 100\%$ trimming, we end up with a sample of reduced size $[(1 - \alpha) \times 100\%]$. Sometimes the sample size is important to preserve.

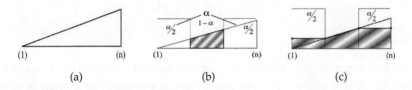

Fig. 2.2 (a) Schematic graph of an ordered sample; (b) Part of the sample from which an α-trimmed mean is calculated; (c) Modified sample for the winsorized mean.

Winsorized Mean. A robust location measure that preserves sample size is the winsorized mean. Similar to a trimmed mean, a winsorized mean identifies outlying observations, but instead of being excluded, these outlying observations are replaced by either the minimum or maximum of the trimmed sample, depending on whether the trimming was done from below or above (Fig. 2.2c).

The winsorized mean is not a built-in MATLAB function. However, it can be calculated easily by the following code:

```
alpha=20;
sa = sort(car);
sa(1:floor( n*alpha/200 )) = sa(floor( n*alpha/200 ) + 1);
sa(end-floor( n*alpha/200 ):end) = ...
        sa(end-floor( n*alpha/200 ) - 1);
winsmean = mean(sa) % winsmean = 21.9632
```

Figure 2.2 shows schematic graphs of an ordered sample, part of the sample from which an α-trimmed mean is calculated, and the modified sample for the winsorized mean.

2.4 Variability Measures

Location measures are intuitive, but they give a minimal glimpse at the nature of a sample. An important set of sample descriptors are dispersion measures, or measures of spread. There are many measures of variability in a sample. In the early nineteenth century, Karl Friedrich Gauss already used several variability measures on a set of 48 astronomical measurements concerning relative positions of Jupiter and its satellite Pallas (Gauss, 1816).

Sample Variance and Sample Standard Deviation. The variance of a sample, or sample variance, is defined as

$$s^2 = \frac{1}{n-1} \sum_{i=1}^{n} (X_i - \overline{X})^2.$$

Note that we use $\frac{1}{n-1}$ instead of $\frac{1}{n}$ as one would expect. The reasons for this will be discussed later. An alternative expression for s^2 that is more suitable for calculation (by hand) is

$$s^2 = \frac{1}{n-1} \left(\sum_{i=1}^{n} (X_i^2) - n(\overline{X})^2 \right);$$

see Exercises 2.5 and 2.7.

In MATLAB, the sample variance of a data vector x is var(x) or var(x,0) Flag 0 in the argument list indicates that the ratio $1/(n-1)$ is used to calculate the sample variance. If the flag is 1, then var(x,1) stands for

$$s_*^2 = \frac{1}{n} \sum_{i=1}^{n} (X_i - \overline{X})^2,$$

which is sometimes used instead of s^2. We will see later that both estimators have good properties: s^2 is an unbiased estimator of the population variance while s_*^2 is the maximum likelihood estimator. The square root of the sample variance is the *sample standard deviation*:

$$s = \sqrt{\frac{1}{n-1} \sum_{i=1}^{n} (X_i - \overline{X})^2}.$$

In MATLAB the standard deviation can be calculated by std(x)=std(x,0) or std(x,1), depending on whether the sum of squares is divided by $n-1$ or by n.

```
%Variability Measures
var(car)    % standard sample variance, also var(car,0)
            %ans = 588.9592
var(car,1) % sample variance with sum of squares
            % divided by n
   %ans =  587.6844
std(car)    % sample standard deviation, sum of squares
            % divided by (n-1), also std(car,0)
   %ans = 24.2685
```

```
std(car,1) % sample standard deviation, sum of squares
           % divided by n
   %ans = 24.2422
sqrt(var(car))   %should be equal to std(car)
   %ans = 24.2685
sqrt(var(car,1)) %should be equal to std(car,1)
   %ans = 24.2422
```

Remark. When a new observation is obtained, one can update the sample variance without having to recalculate it. If $\bar{x}_n$ and s_n^2 are the sample mean and variance based on $x_1, x_2, \ldots, x_n$ and a new observation x_{n+1} is obtained, then

$$s_{n+1}^2 = \frac{(n-1)s_n^2 + (x_{n+1} - \bar{x}_n)(x_{n+1} - \bar{x}_{n+1})}{n},$$

where $\bar{x}_{n+1} = (n\bar{x}_n + x_{n+1})/(n+1)$.

MAD-Type Estimators. Another group of estimators of variability involves absolute values of deviations from the center of a sample. These estimators, known as MAD estimators, are less sensitive to extreme observations and outliers compared to the sample standard deviation. They belong to the class of so-called robust estimators. The acronym MAD stands for either *mean absolute difference from the mean* or, more commonly, *median absolute difference from the median*. According to statistics historians (David, 1998), both MADs were already used by Gauss at the beginning of the nineteenth century.

MATLAB uses `mad(car)` or `mad(a,0)` for the first and `mad(car,1)` for the second definition:

$$\mathrm{MAD}_0 = \frac{1}{n}\sum_{i=1}^{n}|X_i - \bar{X}|, \quad \mathrm{MAD}_1 = \mathrm{median}\{|X_i - \mathrm{median}\{X_i\}|\}.$$

A typical convention is to multiply the MAD_1 estimator `mad(car,1)` by `1/norminv(3/4)=1.4826`, to make it comparable in magnitude to the sample standard deviation.

```
mad(car) % mean absolute deviation from the mean;
         % MAD is usually referring to
         % median absolute deviation from the median
   %ans = 15.3328
realmad = 1.4826 * median( abs(car - median(car)))
         %real mad in MATLAB is 1.4826 * mad(car,1)
   %realmad = 10.3781
```

Sample Range and IQR. Two simple measures of variability, or rather the spread of a sample, are the range R and interquartile range (IQR). In MATLAB these are range and iqr respectively. They are defined by the order statistic of the sample. The range is the maximum minus the minimum of the sample, $R = X_{(n)} - X_{(1)}$, while IQR is defined by sample quantiles, to be explained later.

```
range(car)  %Range, span of data, Max - Min
    %ans = 212
iqr(car)        %inter-quartile range, Q3-Q1
    %ans = 19
```

If the sample is bell-shape distributed, a robust estimator of variance is $\hat{\sigma}^2 = (IQR/1.349)^2$, and this summary was used by Adolphe Quetelet in the first part of the nineteenth century. It is a simple estimator, not affected by outliers (it ignores 25% of observations in each tail), but its variability is large.

Sample Quantiles/Percentiles. Sample quantiles (in units between 0 and 1) or sample percentiles (in units between 0 and 100) are very important summaries that reveal both the location and the spread of a sample. For example, we may be interested in a point x_p that partitions the ordered sample into two parts, one with $p \cdot 100\%$ of observations smaller than x_p and another with $(1 - p)100\%$ observations greater than x_p. In MATLAB, we use the commands quantile or prctile, depending on how we express the proportion of the sample. For example, for the 5, 10, 25, 50, 75, 90, and 95 percentiles we have

```
%5%, 10%, 25%, 50%, 75%, 90%, 95% percentiles are:
prctile(car, 100*[0.05, 0.10, 0.25, 0.50, 0.75, 0.90, 0.95] )
%ans =  7    8    11    17    30    51    67
```

The same results can be obtained using the command

```
qts = quantile(car,[0.05 0.1 0.25 0.5 0.75 0.9 0.95])
%qts =   7    8    11    17    30    51    67
```

In our data set, 5% of the observations are less than 7, and 90% of the observations are less than 51.

Some percentiles/quantiles are special. For example, the median of the sample is the 50th percentile. Quartiles divide an ordered sample into four parts; the 25th percentile is known as the first quartile, Q_1, and the 75th percentile is known as the third quartile, Q_3. The median is Q_2, of course.[3] In MATLAB, Q1=prctile(car,25); Q3=prctile(car,75). Now we can define the IQR as $Q_3 - Q_1$:

---

[3] The range is equipartitioned by a single median, two terciles, three quartiles, four quintiles, five sextiles, six septiles, seven octiles, eight naniles, or nine deciles.

```
prctile(car, 75)- prctile(car, 25) %should be equal to iqr(car).
%ans = 19
```

The five-number summary for univariate data is defined as (*Min*, Q_1, *Me*, Q_3, *Max*).

z-Scores. For a sample $x_1, x_2, \ldots, x_n$ the z-score is the standardized sample $z_1, z_2, \ldots, z_n$, where $z_i = (x_i - \overline{x})/s$. In the standardized sample, the mean is 0 and the sample variance (and standard deviation) is 1. The basic reason why standardization may be needed is to assess extreme values, or compare samples taken at different scales. Some other reasons will be discussed in subsequent chapters.

```
zcar = zscore(car);
mean(zcar)
%ans = -5.8155e-017
var(zcar)
%ans = 1
```

Moments of Higher Order. The term *sample moments* is drawn from mechanics. If the observations are interpreted as unit masses at positions $X_1, \ldots, X_n$, then the sample mean is the first moment in the mechanical sense – it represents the balance point for the system of all points. The moments of higher order have their corresponding mechanical interpretation. The formula for the kth moment is

$$m_k = \frac{1}{n}(X_1^k + \cdots + X_n^k) = \frac{1}{n}\sum_{i=1}^{n} X_i^k.$$

The moments m_k are sometimes called *raw* sample moments. The *power k mean* is $(m_k)^{1/k}$, that is,

$$\left(\frac{1}{n}\sum_{i=1}^{n} X_i^k\right)^{1/k}.$$

For example, the sample mean is the first moment and power 1 mean, $m_1 = \overline{X}$. The *central* moments of order k are defined as

$$\mu_k = \frac{1}{n}\sum_{i=1}^{n}(X_i - m_1)^k.$$

Notice that $\mu_1 = 0$ and that μ_2 is the sample variance (calculated by var(.,1) with the sum of squares divided by n). MATLAB has a built-in function moment for calculating the central moments.

```
%Moments of Higher Orders
    %kth (row) moment: mean(car.^k)
mean(car.^3) %third
    %ans = 1.1161e+005
    %kth central moment mean((car-mean(car)).^k)
mean( (car-mean(car)).^3 ) %ans=5.2383e+004
    %is the same as
moment(car,3) %ans=5.2383e+004
```

Skewness and Kurtosis. There are many uses of higher moments in describing a sample. Two important sample measures involving higher-order moments are *skewness* and *kurtosis*.

Skewness is defined as

$$\gamma_n = \mu_3 / \mu_2^{3/2} = \mu_3 / s_*^3$$

and measures the degree of asymmetry in a sample distribution. Positively skewed distributions have longer right tails, and their sample mean is larger than the median. Negatively skewed sample distributions have longer left tails, and their mean is smaller than the median.

Kurtosis is defined as

$$\kappa_n = \mu_4 / \mu_2^2 = \mu_4 / s_*^4.$$

It represents the measure of "peakedness" or flatness of a sample distribution. In fact, there is no consensus on the precise interpretation of kurtosis since flat but fat-tailed distributions would also have high kurtosis. Distributions that have a kurtosis of <3 are called *platykurtic* and those with a kurtosis of >3 are called *leptokurtic*. Kurtosis is sometimes defined as $\mu_4 / \mu_2^2 - 3$ and termed *excess kurtosis*, or simply *excess*.

```
%sample skewness mean(car.^3)/std(car,1)^3
mean( (car-mean(car)).^3 )/std(car,1)^3 %ans = 3.6769
skewness(car) %ans = 3.6769
%sample kurtosis
mean( (car-mean(car)).^4 )/std(car,1)^4 % ans = 22.8297
kurtosis(car)% ans = 22.8297
```

A robust version of the skewness measure was proposed by Bowley (1920) as

$$\gamma_n{}^* = \frac{(Q_3 - Me) - (Me - Q_1)}{Q_3 - Q_1},$$

and ranges between –1 and 1. Moors (1988) proposed a robust measure of kurtosis based on sample octiles:

$$\kappa_n{}^* = \frac{(O_7 - O_5) + (O_3 - O_1)}{O_6 - O_2},$$

where O_i is the $i/8 \times 100$ percentile (*i*th octile) of the sample for $i = 1, 2, \ldots, 7$. If the sample is large, one can take O_i as $X_{(\lfloor i/8 \times n \rfloor)}$. The constant 1.766 is sometimes added to κ_n^* as a calibration adjustment so that it becomes comparable with the traditional measure of kurtosis for samples from Gaussian populations, which will be discussed extensively later in Chapter 6.

```
%robust skewness
(prctile(car, 75)+prctile(car, 25) - ...
2 * median(car))/(prctile(car, 75) - prctile(car, 25))
%0.3684

%robust kurtosis
(prctile(car,7/8*100)-prctile(car,5/8*100)+prctile(car,3/8*100)- ...
  prctile(car,1/8*100))/(prctile(car,6/8*100)-prctile(car,2/8*100))
%1.4211
```

Coefficient of Variation. The coefficient of variation, CV, is the ratio

$$CV = \frac{s}{\overline{X}}.$$

The CV expresses the variability of a sample in the units of its mean. In other words, a CV equal to 2 would mean that the standard deviation is equal to $2\overline{X}$. The assumption is that the mean is positive. The CV is used when comparing the variability of data reported on different scales. For example, instructors A and B teach different sections of the same class but design their final exams individually. To compare the effectiveness of their respective exam designs at creating a maximum variance in exam scores (a tacit goal of exam designs), they calculate CVs. It is important to note that the CVs would not be related to the exam grading scale, to the relative performance of the students, or to the difficulty of the exam.

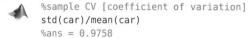

```
%sample CV [coefficient of variation]
std(car)/mean(car)
%ans = 0.9758
```

The reciprocal of CV, $\overline{X}/s$, is sometimes called the signal-to-noise ratio, and it is often used in engineering quality control.

Composite Sample. When a sample is large and many observations are repetitive, data are often recorded as grouped. For example, the data set

$$
\begin{array}{l}
4\ 5\ 6\ 3\ 4\ 3\ 6\ 4\ 5\ 4\ 3 \\
7\ 3\ 5\ 2\ 5\ 6\ 4\ 2\ 4\ 3\ 4 \\
7\ 7\ 4\ 2\ 2\ 5\ 4\ 2\ 5\ 3\ 8
\end{array}
$$

is called a simple sample, or raw sample, as it lists explicitly all observations. It can be presented in a more compact form, as grouped or composite sample:

$$
\begin{array}{c|ccccccc}
X_i & 2 & 3 & 4 & 5 & 6 & 7 & 8 \\
\hline
f_i & 5 & 6 & 9 & 6 & 3 & 3 & 1
\end{array}'
$$

where X_i are distinctive values in the data set with frequencies f_i, and the number of groups is $k = 7$. Notice that $X_i = 5$ appears six times in the simple sample, so its frequency is $f_i = 6$.

The function ◢ [xi fi]=simple2comp(a) provides frequencies fi for a list xi of distinctive values in a.

```
a=[  4  5  6  3  4  3  6  4  5  4  3 ...
     7  3  5  2  5  6  4  2  4  3  4 ...
     7  7  4  2  2  5  4  2  5  3  8];
[xi fi]  = simple2comp( a )
% xi =
%      2     3     4     5     6     7     8
% fi =
%      5     6     9     6     3     3     1
```

Here, $n = \sum_i f_i = 33$.

When a sample is composite, the sample mean and variance are calculated as

$$
\overline{X} = \frac{\sum_{i=1}^{k} f_i X_i}{n}, \quad s^2 = \frac{\sum_{i=1}^{k} f_i (X_i - \overline{X})^2}{n-1}
$$

for $n = \sum_i f_i$. By defining the mth raw and central sample moments as

$$
\overline{X^m} = \frac{\sum_{i=1}^{k} f_i X_i^m}{n} \quad \text{and} \quad \mu_m = \frac{\sum_{i=1}^{k} f_i (X_i - \overline{X})^m}{n-1},
$$

one can express skewness, kurtosis, CV, and other sample statistics that are functions of moments.

Diversity Indices for Categorical Data. If the data are categorical and numerical characteristics such as moments and percentiles cannot be defined, but the frequencies f_i of classes/categories are given, one can define Shannon's diversity index:

$$H = \frac{n \log n - \sum_{i=1}^{k} f_i \log f_i}{n},$$ (2.1)

where n is total sample size and k is the number of categories.

If some frequency is 0, then $0 \times \log 0 = 0$. The maximum of H is $\log k$; it is achieved when all f_i are equal. The normalized diversity index, $E_H = H/\log k$, is called Shannon's homogeneity (equitability) index of the sample.

Neither H nor E_H depends directly on the sample size but on relative class-frequencies f_i/n since H can be expressed as $-\sum_{i=1}^{k}(f_i/n)\log(f_i/n)$.

Example 2.3. **Homogeneity of Blood Types.** Suppose that samples from Brazilian, Indian, Norwegian, and US populations are taken and the frequencies of blood types (ABO/Rh) are obtained.

Population	O+	A+	B+	AB+	O–	A–	B–	AB–	Total
Brazil	115	108	25	6	28	25	6	1	314
India	220	134	183	39	12	6	6	12	612
Norway	83	104	16	8	14	18	2	1	246
US	99	94	21	8	18	18	5	2	265

Which country's sample is most homogeneous with respect to the blood type attribute?

```
br = [115    108     25      6     28     25      6      1];
in = [220    134    183     39     12      6      6     12];
no = [ 83    104     16      8     14     18      2      1];
us = [ 99     94     21      8     18     18      5      2];

Eh = @(f) (sum(f)*log(sum(f)) - ...
       sum( f.*log(f)))/(sum(f)*log(length(f)))

Eh(br)   % 0.7324
Eh(in)   % 0.7125
Eh(no)   % 0.6904
Eh(us)   % 0.7306
```

Among the four samples, the sample from Brazil is the most homogeneous with respect to the blood types of its population, as it maximizes the statistic E_H. See also Exercise 2.14 for an alternative definition of diversity/homogeneity indexes.

2.5 Ranks

Let $X_1, X_2, \ldots, X_n$ be a sample. The ranks of a sample $X_1, X_2, \ldots, X_n$ are defined as indices of ordered sample

$$r(X_1), r(X_2), \ldots, r(X_n).$$

For example,

```
ranks([10 20 25 7])
%ans = 2      3      4      1
```

The function ranks.m is

```
function  r = ranks(data, glob)
%----------------------------------------
if nargin < 2
    glob = 1;
end
    shape = size(data);
if glob == 1
    data=data(:);
 end
% Ties ranked from UptoDown
  [ irrelevant , indud ]  =  sort(data);
  [ irrelevant , rUD ]    =  sort(indud);
% Ties ranked from RtoL
  [ irrelevant , inddu ]  =  sort(flipud(data));
  [ irrelevant , rDU ]    =  sort(inddu);
% Averages ranks of ties, keeping ranks
% of no-tie-observations the same
r = (rUD + flipud(rDU))./2;
r = reshape(r,shape);
```

For example, when the input is a matrix, the optional parameter `glob = 1` produces global ranking, while for `glob` not equal to 1, columnwise ranking is performed.

```
%a =
%     0.8147     0.9134     0.2785     0.9649
%     0.9058     0.6324     0.5469     0.1576
%     0.1270     0.0975     0.9575     0.9706
```

```
   ranks(a)
% ans =
%      7       9       4      11
%      8       6       5       3
%      2       1      10      12

   ranks(a,2)
% ans =
%      2       3       1       2
%      3       2       2       1
%      1       1       3       3
```

In the case of ties, it is customary to average the tied rank values. The script ◀ ranks.m does just that:

```
ranks([2 1 7 1 15 9])
%ans = 3.0000   1.5000   4.0000   1.5000   6.0000   5.0000
```

Here $r(2) = 3$, $r(1) = 1.5$, $r(7) = 4$, and so on. Note that 1 appears twice and ranks 1 and 2 are averaged. In the case

```
ranks([9 1 7 1 9 9])
%ans = 5.0000   1.5000   3.0000   1.5000   5.0000   5.0000
```

the ranks of three 9s are 4, 5, and 6, which are averaged to 5.

2.6 Displaying Data

Besides being represented by their numerical descriptors, samples are often presented in a graphical manner. In this section, we discuss some basic graphical summaries.

Box-and-Whiskers Plot. The top and bottom of the "box" are the 25th and 75th percentile of the data, respectively, with the distances between them representing the IQR. The line inside the box represents the sample median. If the median is not centered in the box, it indicates sample skewness. Whiskers extend from the lower and upper sides of the box to the data's most extreme values within 1.5 times the IQR. Potential outliers are displayed with red "+" beyond the endpoints of the whiskers.

The MATLAB command boxplot(X) produces a box-and-whisker plot for X. If X is a matrix, the boxes are calculated and plotted for each column. Figure 2.3a is produced by

```
%Some Graphical Summaries of the Sample
figure;
boxplot(car)
```

Histogram. As illustrated previously in this chapter, the histogram (Greek: *histos*, a web or tissue; *gramma*, a thing written or drawn) is a rough

approximation of the population distribution based on a sample. Plotted in a histogram are frequencies (or relative frequencies for normalized histograms) for interval-grouped data. Graphically, the histogram is a barplot over contiguous intervals or bins spanning the range of data (Fig. 2.3b). In MATLAB, the typical command for a histogram is `[fre,xout]
= hist(data,nbins)`, where `nbins` is the number of bins and the outputs `fre` and `xout` are the frequency counts and the bin locations, respectively. Given the output, one can use `bar(xout,n)` to plot the histogram. When the output is not requested, MATLAB produces the plot by default.

```
figure;
hist(car, 80)
```

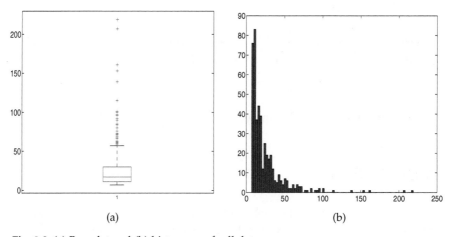

Fig. 2.3 (a) Box plot and (b) histogram of cell data `car`.

Histogram is only an approximation of the distribution of measurements in the population from which the sample is obtained.

There are several rules on how to automatically determine the number of bins or, equivalently, bin sizes, none of them superior to the others on all possible data sets. A commonly used proposal is Sturges' rule (Sturges, 1926), where the number of bins k is given as

$$k = 1 + \log_2 n,$$

where n is the size of the sample. Sturges' rule was intended for bell-shaped distributions of data and may oversmooth data that are skewed, multimodal, or have some other features. Other suggestions specify the bin size as $h = 2 \cdot \text{IQR}/n^{1/3}$ (Diaconis–Freedman rule) or, alternatively, $h = (7s)/(2n^{1/3})$ (Scott's rule; s is the sample standard deviation). The number of bins is found by dividing the range of the data by h.

For example, for cell-area data `car`, Sturges' rule suggests 10 bins, Scott's 19 bins, and the Diaconis–Freedman rule 43 bins. The default `nbins` in MAT-LAB is 10 for any sample size.

The histogram is a crude estimator of probability densities discussed in detail later in Chapter 5. A more esthetic estimator of the population distribution is given by the *kernel smoother density* estimate, or `ksdensity`. We will not go into the details of kernel smoothing at this point in the text; however, note that the spread of a kernel function (such as a Gaussian kernel) regulates the degree of smoothing and in some sense is equivalent to the choice of bin size in histograms.

Command `[f,xi,u]=ksdensity(x)` computes a density estimate based on data `x`. Output `f` is the vector of density values evaluated at the points in `xi`. The estimate is based on a normal kernel function, using a window parameter `width` that depends on the number of points in `x`. The default width `u` is returned as an output and can be used to tune the smoothness of the estimate, as is done in the example below. The density is evaluated at 100 equally spaced points that cover the range of the data in `x`:

```
figure;
[f,x,u] = ksdensity(car);
plot(x,f)
hold on
[f,x] = ksdensity(car,'width',u/3);
plot(x,f,'r');
[f,x] = ksdensity(car,'width',u*3);
plot(x,f,'g');
legend('default width','default/3','3 * default')
hold off
```

Empirical Cumulative Distribution Function. The empirical cumulative distribution function (ECDF) $F_n(x)$ for a sample $X_1, \ldots, X_n$ is defined as

$$F_n(x) = \frac{1}{n} \sum_{i=1}^{n} \mathbf{1}(X_i \leq x) \tag{2.2}$$

and represents the proportion of sample values smaller than x. Here $\mathbf{1}(X_i \leq x)$ is either 0 or 1. It is equal to 1 if $\{X_i \leq x\}$ is true, 0 otherwise.

The function ◀ `empiricalcdf(x,sample)` will calculate the ECDF based on the observations in `sample` at a value `x`.

```
xx = min(car)-1:0.01:max(car)+1;
yy = empiricalcdf(xx, car);
plot(xx, yy, 'k-','linewidth',2)
xlabel('x');  ylabel('F_n(x)')
```

In MATLAB, `[f xf]=ecdf(x)` is used to calculate the proportion `f` of the sample `x` that is smaller than `xf`. Figure 2.4b shows the ECDF for the cell area data, `car`.

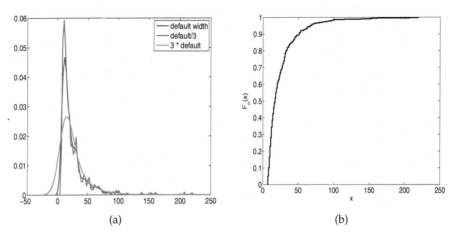

(a)   (b)

Fig. 2.4 (a) Smoothed histogram (density estimator) for different widths of smoothing kernel; (b) Empirical CDF.

Q–Q Plots. Q–Q plots, short for quantile–quantile plots, compare the distribution of a sample with some standard theoretical distribution, such as normal distribution, or with a distribution of another sample. This is done by plotting the sample quantiles of one distribution against the corresponding quantiles of the other. If the plot is approximately linear, then the distributions are close (up to a scale and shift). If the plot is close to the 45° line, then the compared distributions are approximately equal. In MATLAB the command `qqplot(X,Y)` produces an empirical Q–Q plot of the quantiles of the data set X versus the quantiles of the data set Y. If the data set Y is omitted, then `qqplot(X)` plots the quantiles of X against standard normal quantiles and essentially checks the normality of the sample.

Figure 2.5 gives us the Q–Q plot of the cell area data set against the normal distribution. Note the deviation from linearity suggesting that the distribution is skewed. A line joining the first and third sample quartiles is superimposed in the plot. This line is extrapolated out to the ends of the sample to help visually assess the linearity of the Q–Q display. Q–Q plots will be discussed in more detail in Chapter 17.

Pie Charts. If we are interested in visualizing proportions or frequencies, a pie chart is appropriate. A pie chart (`pie` in MATLAB) is a graphical display in the form of a circle in which particular sample proportions are assigned segments.

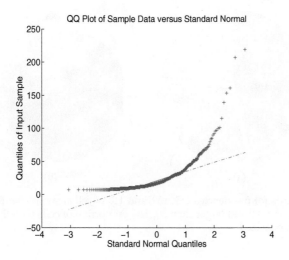

Fig. 2.5 Quantiles of data plotted against corresponding normal quantiles, via `qqplot`.

Suppose that in the cell area data set we are interested in comparing proportions of cells with areas in three regions: smaller than or equal to 15, between 15 and 30, and larger than 30. We would like to emphasize the proportion of cells with areas between 15 and 30. The following MATLAB code plots the pie charts (Fig. 2.6).

```
n1 = sum( car <= 15 ); %n1=213
n2 = sum( (car > 15 ) & (car <= 30) ); %n2=139
n3 = sum( car > 30 ); %n3=110
% n=n1+n2+n3 = 462
% proportions n1/n, n2/n, and n3/n are
%              0.4610,0.3009 and 0.2381
explode = [0 1 0]
pie([n1, n2, n3], explode)
pie3([n1, n2, n3], explode)
```

Note that option `explode=[0 1 0]` separates the second segment from the circle. The command `pie3` plots a 3D version of a pie chart (Fig. 2.6b).

2.7 Multidimensional Samples: Fisher's Iris Data and Body Fat Data

In the cell area example, the sample was univariate, that is, each measurement was a scalar. If a measurement is a vector of data, then descriptive statistics and graphical methods increase in importance, but they are much

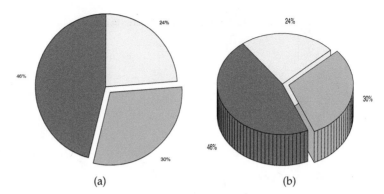

(a)   (b)

Fig. 2.6 Pie charts for frequencies 213, 139, and 110 of cell areas smaller than or equal to 15, between 15 and 30, and larger than 30. The proportion of cells with the area between 15 and 30 is emphasized.

more complex than in the univariate case. The methods for understanding multivariate data range from the simple rearrangements of tables in which raw data are tabulated to quite sophisticated computer-intensive methods in which exploration of the data is reminiscent of futuristic movies of space explorations.

Multivariate data from an experiment are first recorded in the form of tables, by either a researcher or a computer. In some cases, such tables may appear uninformative simply because of their format of presentation. By simple rules, such tables can be rearranged in more useful formats. There are several guidelines for successful presentation of multivariate data in the form of tables. (i) Numbers should be maximally simplified by rounding as long as it does not affect the analysis. For example, the vector (2.1314757, 4.9956301, 6.1912772) could probably be simplified to (2.14, 5, 6.19); (ii) Organize the numbers to compare columns rather than rows; and (iii) The user's cognitive load should be minimized by spacing and table layout so that the eye does not travel long in making comparisons.

Fisher's Iris Data. An example of multivariate data is provided by the celebrated Fisher's iris data. Plants of the family *Iridaceae* grow on every continent except Antarctica. With a wealth of species, identification is not simple. Even iris experts sometimes disagree about how some flowers should be classified. Fisher's (Anderson, 1935; Fisher, 1936) data set contains measurements on three North American species of iris: *Iris setosa canadensis, Iris versicolor,* and *Iris virginica.* The 4-dimensional measurements on each of the species consist of sepal length, sepal width, petal length, and petal width.

The data set fisheriris is part of the MATLAB distribution and contains two fields: meas and species. The meas field, shown in Figure 2.7a, is a 150 × 4 matrix and contains 150 entries, 50 for each species. Each row in the matrix meas contains four elements: sepal length, sepal width, petal length, and petal width. Note that the convention in MATLAB is to store variables as columns and observations as rows.

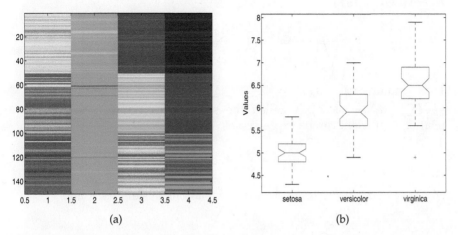

(a)   (b)

Fig. 2.7 (a) Matrix meas in fisheriris; (b) Box plots of sepal length (the first column in matrix meas) versus species.

The data set species contains names of species for the 150 measurements. The following MATLAB commands plot the data and compare sepal lengths among the three species.

```
load fisheriris
s1 = meas(1:50, 1);    %setosa,    sepal length
s2 = meas(51:100, 1);  %versicolor, sepal length
s3 = meas(101:150, 1); %virginica,  sepal length
s = [s1 s2 s3];
figure;
imagesc(meas)
figure;
boxplot(s,'notch','on',...
        'labels',{'setosa','versicolor','virginica'})
```

Correlation in Paired Samples. We will briefly describe how to find the correlation between two aligned vectors, leaving detailed coverage of correlation theory to Chapter 13.

Sample correlation coefficient r measures the strength and direction of the linear relationship between two paired samples $X = (X_1, X_2, \ldots, X_n)$ and $Y = (Y_1, Y_2, \ldots, Y_n)$. Note that the order of components is important and the samples cannot be independently permuted if the correlation is of interest. Thus the two samples can be thought of as a single bivariate sample (X_i, Y_i), $i = 1, \ldots, n$.

The correlation coefficient between samples $X = (X_1, X_2, \ldots, X_n)$ and $Y = (Y_1, Y_2, \ldots, Y_n)$ is

$$r = \frac{\sum_{i=1}^{n}(X_i - \overline{X})(Y_i - \overline{Y})}{\sqrt{\sum_{i=1}^{n}(X_i - \overline{X})^2 \cdot \sum_{i=1}^{n}(Y_i - \overline{Y})^2}}.$$

The summary $s_{XY} = \frac{1}{n-1} \sum_{i=1}^{n}(X_i - \overline{X})(Y_i - \overline{Y}) = \frac{1}{n-1}\left(\sum_{i=1}^{n} X_i Y_i - n\overline{X}\,\overline{Y}\right)$ is called the sample covariance. The correlation coefficient can be expressed as a ratio:

$$r = \frac{s_{XY}}{s_X \, s_Y},$$

where s_X and s_Y are sample standard deviations of samples X and Y.

Covariances and correlations are basic exploratory summaries for paired samples and multivariate data. Typically, they are assessed in data screening before building a statistical model and conducting an inference. The correlation ranges between −1 and 1, which are the two ideal cases of decreasing and increasing linear trends. Zero correlation does not, in general, imply independence but signifies the lack of any linear relationship between samples.

To illustrate the preceding principles, we find covariance and correlation between sepal and petal lengths in Fisher's iris data. These two variables correspond to the first and third columns in the data matrix. The conclusion is that these two lengths exhibit a high degree of linear dependence as evident in Figure 2.8. The covariance of 1.2743 by itself is not a good indicator of this relationship since it is scale (magnitude) dependent. However, the correlation coefficient is scale independent and, in this case, shows a strong positive relationship between the variables:

```
load fisheriris
X = meas(:, 1);    %sepal length
Y = meas(:, 3);    %petal length
cv = cov(X, Y); cv(1,2)   %1.2743
r = corr(X, Y)            %0.8718
```

In the next section we will describe an interesting multivariate data set and, using MATLAB, find some numerical and graphical summaries.

Example 2.4. **Body Fat Data.** We now discuss a multivariate data set analyzed in Johnson (1996) that was submitted to http://www.amstat.org/publications/jse/datasets/fat.txt and featured in Penrose et al. (1985). This data set can be found on the book's Web page as well, as fat.dat.

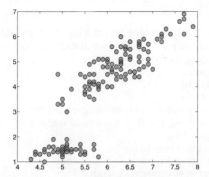

Fig. 2.8 Correlation between petal and sepal lengths (columns 1 and 3) in iris data set. Note the strong linear dependence with a positive trend. This is reflected by a covariance of 1.2743 and a correlation coefficient of 0.8718.

Fig. 2.9 Water test to determine body density. It is based on underwater weighing (Archimedes' principle) and is regarded as the gold standard for body composition assessment.

Percentage of body fat, age, weight, height, and ten body circumference measurements (e.g., abdomen) were recorded for 252 men. The percentage of body fat was estimated through an underwater weighing technique (Fig. 2.9).

The data set has 252 observations and 19 variables. The Brozek and Siri indexes (Brozek et al., 1963; Siri, 1961) and fat-free weight are obtained by underwater weighing, while other anthropometric variables are obtained using scales and a measuring tape. The anthropometric variables are less intrusive but also less reliable in assessing the body fat index.

Remark. There are a few erroneous recordings. The body densities for cases 48, 76, and 96, for instance, each seem to have one digit in error as seen from the two body fat percentage values. You will also note the presence of a man (case 42) over 200 lb. in weight who is less than 3 ft. tall (the height should presumably be 69.5 in., not 29.5 in.)! The percent body fat estimates are truncated to zero when negative (case 182).

```
load('\your path\fat.dat')
```

Column	Name	Variable description
1	casen	Case number
2	broz	Percent body fat using Brozek's equation: 457/density – 414.2
3	siri	Percent body fat using Siri's equation: 495/density – 450
4	densi	Density (g/cm^3)
5	age	Age (years)
6	weight	Weight (lb)
7	height	Height (in)
8	adiposi	Adiposity index = weight/(height^2) (kg/m^2)
9	ffwei	Fat-free weight = (1 – fraction of body fat) × weight, using Brozek's formula (lb)
10	neck	Neck circumference (cm)
11	chest	Chest circumference (cm)
12	abdomen	Abdomen circumference (cm)
13	hip	Hip circumference (cm)
14	thigh	Thigh circumference (cm)
15	knee	Knee circumference (cm)
16	ankle	Ankle circumference (cm)
17	biceps	Extended biceps circumference (cm)
18	forearm	Forearm circumference (cm)
19	wrist	Wrist circumference (cm) "distal to the styloid processes"

Table 2.1 Structure of file `fat.dat`

```
casen = fat(:,1);
broz = fat(:,2);
siri = fat(:,3);
densi = fat(:,4);
age = fat(:,5);
weight = fat(:,6);
height = fat(:,7);
adiposi = fat(:,8);
ffwei = fat(:,9);
neck = fat(:,10);
chest = fat(:,11);
abdomen = fat(:,12);
hip = fat(:,13);
thigh = fat(:,14);
knee = fat(:,15);
ankle = fat(:,16);
biceps = fat(:,17);
forearm = fat(:,18);
wrist = fat(:,19);
```

We will analyze this data set again in this chapter, and later in Chapter 14, in the context of multiple regression.

2.8 Multivariate Samples and Their Summaries*

Multivariate samples are organized as a data matrix, where the rows are observations and the columns are variables or components. One such data matrix of size $n \times p$ is shown in Figure 2.10.

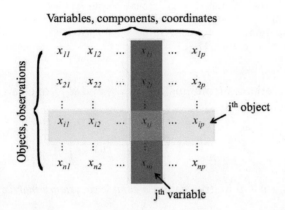

Fig. 2.10 Data matrix X. In the multivariate sample the rows are observations and the columns are variables.

The measurement x_{ij} denotes the jth component of the ith observation. There are n row vectors $x_1', x_2', \ldots, x_n'$ and p columns $x_{(1)}, x_{(2)}, \ldots, x_{(p)}$, so that

$$X = \begin{bmatrix} x_1' \\ x_2' \\ \vdots \\ x_n' \end{bmatrix} = \begin{bmatrix} x_{(1)}, x_{(2)}, \ldots, x_{(p)} \end{bmatrix}.$$

Note that $x_i = (x_{i1}, x_{i2}, \ldots, x_{ip})'$ is a p-vector denoting the ith observation, while $x_{(j)} = (x_{1j}, x_{2j}, \ldots, x_{nj})'$ is an n-vector denoting values of the jth variable/component.

The mean of data matrix X is a vector $\bar{x}$, which is a p-vector of column means

$$\bar{x} = \begin{bmatrix} \frac{1}{n}\sum_{i=1}^{n} x_{i1} \\ \frac{1}{n}\sum_{i=1}^{n} x_{i2} \\ \vdots \\ \frac{1}{n}\sum_{i=1}^{n} x_{ip} \end{bmatrix} = \begin{bmatrix} \bar{x}_1 \\ \bar{x}_2 \\ \vdots \\ \bar{x}_p \end{bmatrix}.$$

By denoting a vector of ones of size $n \times 1$ as $\mathbf{1}$, the mean can be written as $\bar{x} = \frac{1}{n}X' \cdot \mathbf{1}$, where X' is the transpose of X.

Note that $\bar{x}$ is a column vector, but MATLAB's command `mean(X)` will produce a row vector. It is instructive to take a simple data matrix and inspect step by step how MATLAB calculates the multivariate summaries. For instance,

```
X = [1 2 3; 4 0 1; -1 1 2; 3 6 9];
[n p]=size(X)  %[4 3]: four 3-dimensional  observations
meanX = mean(X)'       %or mean(X,1), along dimension 1
       %transpose of meanX needed to be a column vector
meanX = 1/n * X' * ones(n,1)
```

For any two variables (columns) in X, $x_{(i)}$ and $x_{(j)}$, one can find the sample covariance:

$$s_{ij} = \frac{1}{n-1}\left(\sum_{k=1}^{n} x_{ki}x_{kj} - n\bar{x}_i\bar{x}_j\right).$$

All s_{ij}s form a $p \times p$ matrix, called a *sample covariance matrix* and denoted by S.

A simple representation for S uses matrix notation: .

$$S = \frac{1}{n-1}\left(X'X - \frac{1}{n}X'JX\right).$$

Here J is a standard notation for a matrix consisting of ones. If one defines a *centering matrix* H as $H = I - \frac{1}{n}J$, then $S = \frac{1}{n-1}X'HX$. Here I is the identity matrix.

```
X = [1 2 3; 4 0 1; -1 1 2; 3 6 9];
[n p]=size(X);
H = eye(n) - 1/n * ones(n);  %centering matrix
S = 1/(n-1) * X' * H * X      %the same as
S = cov(X)                    %built-in command
```

An alternative definition of the covariance matrix, $S^* = \frac{1}{n}X'HX$, is coded in MATLAB as `cov(X,1)`. Note that the diagonal of S contains sample variances of variables, since $s_{ii} = \frac{1}{n-1}\left(\sum_{k=1}^{n} x_{ki}^2 - n\bar{x}_i^2\right) = s_i^2$.

Matrix S describes scattering in data matrix X. Sometimes it is convenient to have scalars as measures of scatter, and for that purpose two summaries of S are typically used: (i) the determinant of S, $|S|$, as a generalized variance and (ii) the trace of S, $\text{tr}S$, as the total variation.

The sample correlation coefficient between the ith and jth variables is

$$r_{ij} = \frac{s_{ij}}{s_i\, s_j},$$

where $s_i = \sqrt{s_i^2} = \sqrt{s_{ii}}$ is the sample standard deviation. Matrix R with elements r_{ij} is called a sample correlation matrix. If $R = I$, the variables are uncorrelated. If $D = diag(s_i)$ is a diagonal matrix with $(s_1, s_2, \ldots, s_p)$ on its diagonal, then

$$S = DRD, \quad R = D^{-1}SD^{-1}.$$

Next we show how to standardize multivariate data. Data matrix Y is a standardized version of X if its rows y_i' are standardized rows of X,

$$Y = \begin{bmatrix} y_1' \\ y_2' \\ \vdots \\ y_n' \end{bmatrix}, \quad \text{where } y_i = D^{-1}(x_i - \bar{x}), \ i = 1, \ldots, n.$$

Y has a covariance matrix equal to the correlation matrix. This is a multivariate version of the z-score. For the two-column vectors from Y, $y_{(i)}$ and $y_{(j)}$, the correlation r_{ij} can be interpreted geometrically as the cosine of the angle φ_{ij} between the vectors. This shows that correlation is a measure of similarity because close vectors (with a small angle between them) will be strongly positively correlated, whereas the vectors orthogonal in the geometric sense will be uncorrelated. This is why uncorrelated vectors are sometimes called orthogonal.

Another useful transformation of multivariate data is the Mahalanobis transformation. When data vector is subjected to the Mahalanobis transformation, its components become decorrelated. For this reason, such transformed data are sometimes called "sphericized."

$$Z = \begin{bmatrix} z_1' \\ z_2' \\ \vdots \\ z_n' \end{bmatrix}, \quad \text{where } z_i = S^{-1/2}(x_i - \bar{x}), \ i = 1, \ldots, n.$$

Because the Mahalanobis transform decorrelates the components, the covariance matrix $\text{Cov}(Z)$ is an identity matrix. The Mahalanobis transformation is useful in defining the distances between multivariate observations. For further discussion on the multivariate aspects of statistics, we direct the student to the excellent classical book by Morrison (2004).

Example 2.5. The iris data set was a data matrix of size 150×4, while the size of the body fat data was 252×19. To illustrate some of the multivariate summaries just discussed, we construct a new, 5-dimensional data matrix from the body fat data set. The selected columns are broz, densi, weight, adiposi, and biceps. All 252 rows are retained.

```
% From multifatstat.m
   X = [broz densi weight adiposi biceps];
   varNames = {'broz'; 'densi'; 'weight'; 'adiposi'; 'biceps'};

varNames =
     'broz'    'densi'   'weight'   'adiposi'   'biceps'

Xbar = mean(X)

Xbar = 18.9385 1.0556 178.9244 25.4369 32.2734

S = cov(X)

 S =
    60.0758    -0.1458   139.6715    20.5847    11.5455
    -0.1458     0.0004    -0.3323    -0.0496    -0.0280
   139.6715    -0.3323   863.7227    95.1374    71.0711
    20.5847    -0.0496    95.1374    13.3087     8.2266
    11.5455    -0.0280    71.0711     8.2266     9.1281

R = corr(X)

 R =
     1.0000    -0.9881     0.6132     0.7280     0.4930
    -0.9881     1.0000    -0.5941    -0.7147    -0.4871
     0.6132    -0.5941     1.0000     0.8874     0.8004
     0.7280    -0.7147     0.8874     1.0000     0.7464
     0.4930    -0.4871     0.8004     0.7464     1.0000

% By ''hand''
[n p]=size(X);
H = eye(n) - 1/n * ones(n,1)*ones(1,n);
S = 1/(n-1) * X' * H * X;
stds = sqrt(diag(S));
D = diag(stds);
R = inv(D) * S * inv(D);
%S and R here coincide with S and R
%calculated by built-in functions cov and cor.

Xc= X - repmat(mean(X),n,1);  %center X
%subtract component means
%from variables in each observation.

%standardization
Y =  Xc * inv(D);  %for Y, S=R

%Mahalanobis transformation
M = sqrtm(inv(S))  %sqrtm is a square root of matrix

 %M =
```

```
%     0.1739     0.8423    -0.0151    -0.0788     0.0046
%     0.8423   345.2191    -0.0114     0.0329     0.0527
%    -0.0151    -0.0114     0.0452    -0.0557    -0.0385
%    -0.0788     0.0329    -0.0557     0.6881    -0.0480
%     0.0046     0.0527    -0.0385    -0.0480     0.5550

Z = Xc * M;   %Z has uncorrelated components
cov(Z)        %should be identity matrix
```

Figure 2.11 shows data plots for a subset of five variables and the two transformations, standardizing and Mahalanobis. Panel (a) shows components broz, densi, weight, adiposi, and biceps over all 252 measurements. Note that the scales are different and that weight has much larger magnitudes than the other variables.

Panel (b) shows the standardized data. All column vectors are centered and divided by their respective standard deviations. Note that the data plot here shows the correlation across the variables. The variable density is negatively correlated with the other variables.

Panel (c) shows the decorrelated data. Decorrelation is done by centering and multiplying by the Mahalanobis matrix, which is the matrix square root of the inverse of the covariance matrix. The correlations visible in panel (b) disappeared.

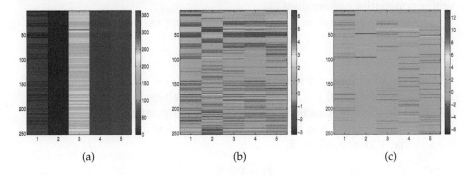

(a)   (b)   (c)

Fig. 2.11 Data plots for (a) 252 five-dimensional observations from Body Fat data where the variables are broz, densi, weight, adiposi, and biceps. (b) Y is standardized X, and (c) Z is a decorrelated X.

2.9 Principal Components of Data

Principal components were introduced by Pearson (1901) and fully developed by Hotelling (1933). Without insisting on their theoretical properties, we will apply this methodology as algebraic transformations of multivariate data sets. Our goal is to better understand the multivariate data and reduce its dimensionality.

Let $X_1, \ldots, X_n$ be a sample consisting of n of p-dimensional observations $X_i = (X_{i1}, \ldots, X_{ip})$. The sample is organized as a $(n \times p)$ data matrix X, in which, as before, the observations are rows while the components within the observations form columns:

$$X = \begin{bmatrix} X_{11} & X_{12} & \ldots & X_{1p} \\ X_{21} & X_{22} & \ldots & X_{2p} \\ \vdots & & & \\ X_{n1} & X_{n2} & \ldots & X_{np} \end{bmatrix}_{n \times p} .$$

Operationally, the principal component analysis is an algebraic decomposition on the sample covariance matrix S. As a background, we will briefly review notions of *eigenvectors* and *eigenvalues*.

When multiplied by a matrix A, vectors generally change their direction. However, certain vectors Ax remain in the same direction as the original x. Such vectors x are called *eigenvectors*. Thus, the vector Ax is λ times the original x. In this case, λ is an *eigenvalue* corresponding to eigenvector x. When A is a full rank covariance matrix of size $p \times p$, it has p eigenvectors with nonnegative eigenvalues.

The sample covariance matrix S can be decomposed as

$$S = VDV',$$

where V is an orthogonal matrix consisting of eigenvectors (as columns) of S, and D is a diagonal $p \times p$ matrix with eigenvalues $\lambda_1 \geq \ldots \lambda_p \geq 0$ on the diagonal. In algebra, this is called spectral decomposition.

After multiplying $S = VDV'$ by V' from the left and V from the right, we see that V diagonalizes S,

$$D = V'SV.$$

In MATLAB,

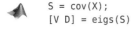

```
S = cov(X);
[V D] = eigs(S)
```

Matrix V in the output consists of coefficients defining the principal axes. The columns of V are called principal components. The principal components form an orthonormal basis of p-dimensional space to which the data form X are transformed.

The *scores*, organized as data matrix Y,

$$Y = XV, \tag{2.3}$$

consist of p-tuples, which are observations from X represented in this new coordinate system. Geometrically, each row of X represents a point in a p-dimensional space. The space is rotated (V is an orthogonal matrix) and the scores Y represent coordinates of the points in this rotated coordinate system. Note that transformation in (2.3) is linear; each component of ith score Y_i (ith row of Y) is a linear combination of all components of ith observation X_i where coefficients are the elements of the corresponding eigenvectors.

The diagonal matrix D contains eigenvalues of S ordered from the largest to the smallest. This matrix is a covariance matrix of the transformed data, and it is diagonal. This means that principal components are uncorrelated and the total variance trace(S) = trace(D) is preserved.

When the coordinates have different magnitudes and scales, it is advisable to use sample correlation matrix R instead of covariance matrix S. This is equivalent to applying the previous transformations on the z-scores of X. In this case trace(R) = trace(D) = p, since we have p components, each with variance one.

MATLAB's built-in function is `[V, Y, d] = pca(X);` where V, Y are as above and diag(d) = D.

In its most rudimentary form, the principal components can be found by applying the *singular value decomposition* directly on the centered data matrix:

```
cX = X - repmat(mean(X), n,1); %centered data matrix
[U lam V]=svd(cX);
                              %v=coeffs
D =  lam.^2/(n-1)             %variances
Y = cX * V;                   %scores
```

Example 2.6. **Principal Components of Fisher's Iris Data.** Fisher's iris data matrix (150×4) was described on page 30. We will find principal components for this data set and explore how they behave for different iris species. The following MATLAB script imports the data, finds principal components, scores, and variances, and plots the results:

```
load fisheriris % meas (150 x 4), species cell(150)

[coefs,vars] = eigs(cov(meas));
scores=meas*coefs;
variances=diag(vars);

% species is cell array
gse = ismember(species,'setosa');
gvi = ismember(species,'virginica');
gve = ismember(species,'versicolor');
```

```
% gse, gvi, gve are logical arrays

figure;
plot( meas(find(gse),1), meas(find(gse),2), 'ro')
hold on
plot( meas(find(gvi),1), meas(find(gvi),2), 'go')
plot( meas(find(gve),1), meas(find(gve),2), 'ko')
legend('Setosa','Virginica','Versicolor',0)
xlabel('1st Coordinate')
ylabel('2nd Coordinate')

figure;
plot( scores(find(gse),1), scores(find(gse),2), 'ro')
hold on
plot( scores(find(gvi),1), scores(find(gvi),2), 'go')
plot( scores(find(gve),1), scores(find(gve),2), 'ko')
legend('Setosa','Virginica','Versicolor',0)
xlabel('1st Principal Component')
ylabel('2nd Principal Component')

variances(1)/sum(variances)   % 92.46% (explained by 1st PC)
```

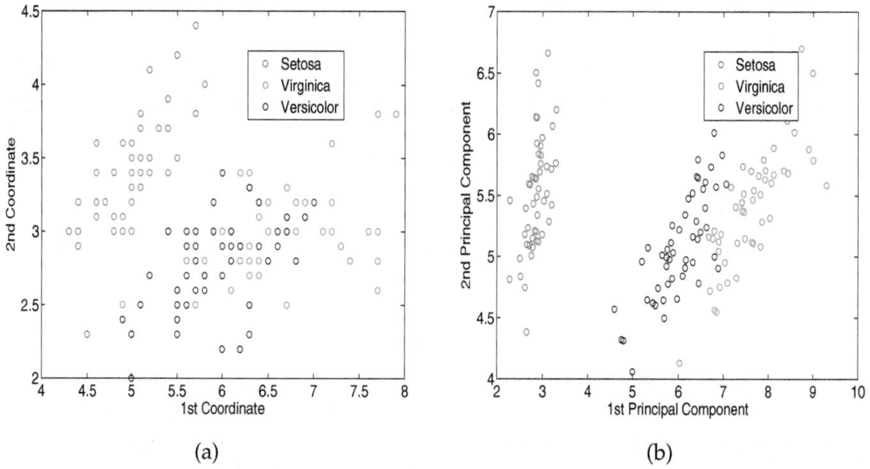

(a)	(b)

Fig. 2.12 (a) First vs second component in the original data; (b) First vs second principal component in scores.

Note that in the domain of original measurements the first two components do not separate species as well as in the domain of principal components (Fig 2.12). Here the principal components help identify features in the data that discriminate between the species. Also, the first principal component is responsible for 92.46% of the variability in this data set.

Example 2.7. **Wisconsin Diagnostic Breast Cancer (WDBC).** Wolberg, Street, and Mangasarian (1994) were interested in applying machine learning to

diagnosing breast cancer from fine-needle aspirates (FNA). The data set
wdbc.mat contains a matrix wdbc with 569 rows (subjects) of which 357
correspond to controls and 212 to cancer. The matrix has 31 columns: column 1 is diagnosis (0 = control, 1 = cancer), and columns 2–31 contain 30
features. The features are computed from a digitized image of a FNA of a
breast mass, as shown in Figure 2.13. The characteristics of the cell nuclei
present in the image are listed as follows:

Variable	Mean	S.Error	Extreme
Radius (average distance from the center)	Col 2	Col 12	Col 22
Texture (standard deviation of gray-scale values)	Col 3	Col 13	Col 23
Perimeter	Col 4	Col 14	Col 24
Area	Col 5	Col 15	Col 25
Smoothness (local variation in radius lengths)	Col 6	Col 16	Col 26
Compactness (perimeter^2 / area - 1.0)	Col 7	Col 17	Col 27
Concavity (severity of concave portions of the contour)	Col 8	Col 18	Col 28
Concave points (number of concave portions of the contour)	Col 9	Col 19	Col 29
Symmetry	Col 10	Col 20	Col 30
Fractal dimension ("coastline approximation" - 1)	Col 11	Col 21	Col 31

The mean, standard error, and extreme (largest) of nuclei measures were
computed for each image, resulting in 30 features. For instance, column 2
is Mean Radius, column 12 is Radius Standard Error, column 22 is Extreme
Radius.

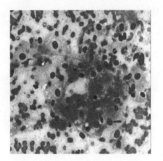

Fig. 2.13 FNA: A digitized image of a fine-needle aspirate of a breast mass.

Below is MATLAB code that finds PCs for this data set and plots the
Pareto graph.

```
load('wdbc.mat')
Y = wdbc(:,1);
X = wdbc(:,2:31);
[n p]=size(X);
%
Z=zscore(X); % components of matrix X are on very different scales,
```

```
              % so a zscored matrix is used.
C=cov(Z)      % cov(Z) is the same as corr(X)
       % pca is MATLAB built in. Gives PC (coeffs), transformed data (scores),
       % and vecyor of variances (latent). When Z is used sum(variances)= p= 14.
[coeffs1, scores1, latent1] = pca(Z);
       %
       %Equivalent task using eig or eigs. Note that eigenvectors
       %(columns of V) are PCs, and variances are on the diagonal of D.
       % Scores are recovered as Z*V
[V D]=eigs(C,30);
coeffs2=V;
scores2=Z*V;
latent2=diag(D);
%
figure;
  pareto(latent2)  % Pareto plot of variance balance
  xlabel('PC')
  ylabel('Variance')
%
figure;
plot(scores1(Y==0, 1), scores1(Y==0, 2),'b.')
hold on
plot(scores1(Y==1, 1), scores1(Y==1, 2),'r.')
legend('No cancer','Cancer',2)
xlabel('1st PC'); ylabel('2nd PC')
```

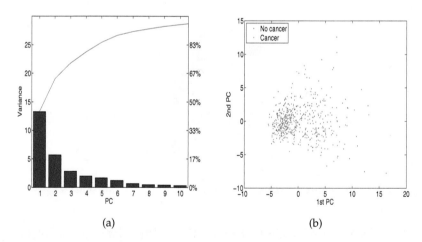

(a)   (b)

Fig. 2.14 (a) Pareto plot for principal components of wdbc data; (b) Scatterplot of scores for the first two principal components.

The dimension reduction in this context means that most of the variance in data is explained by only a few components. Total variance is 30, equal to p, since we used z-scores, or equivalently, the correlation matrix. The first two components explain 63.24% of variance, (sum(latent1(1:2))/30 = 0.6324).

The first 5 components contain roughly 85% of information (variability) and the applicability of this methodology in data compression is obvious.

Often a scatterplot of the leading components of scores may reveal patterns (as in Example 2.6) that are useful in data mining. In this example the first two components are not discriminatory of cancer because their scatterplot is well mixed (Fig. 2.14(b)).

2.10 Visualizing Multivariate Data

The need for graphical representation is much greater for multivariate data than for univariate data, especially if the number of dimensions exceeds three.

For data given in matrix form (observations in rows, components in columns), we have already seen an illuminating graphical representation, which we called a data matrix.

It is straightforward to extend the histogram to bivariate data. An example of a 2D histogram obtained by m-file hist2d is given in Figure 2.15a. The histogram (in the form of an image) shows the sepal and petal lengths from the fisheriris data set. A scatterplot of the 2D measurements is superimposed.

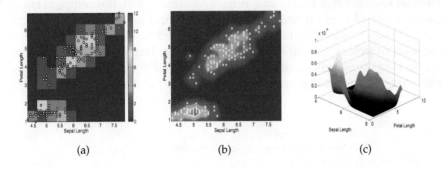

(a)   (b)   (c)

Fig. 2.15 (a) Two-dimensional histogram of Fisher's iris sepal (X) and petal (Y) lengths. The plot is obtained by hist2d.m; (b) Scattercloud plot – smoothed histogram with superimposed scatterplot, obtained by scattercloud.m; (c) Kernel-smoothed and normalized histogram obtained by smoothhist2d.m.

Panels (b) and (c) in Figures 2.15 show the smoothed histograms. The histogram in panel (c) is normalized so that the area below the surface is 1. The smoothed histograms are plotted by ◄ scattercloud.m and ◄ smoothhist2d.m (S. Simon and E. Ronchi, MATLAB Central).

If the dimension of the data is three or more, one can gain additional insight by plotting pairwise scatterplots. This is accomplished by the MAT-LAB command `gplotmatrix(X,Y,group)`, which creates a matrix arrangement of scatterplots. Each subplot in the graphical output contains a scatterplot of one column from data set X against a column from data set Y. For a single data set (as in body fat and Fisher iris examples), Y is omitted or set at `Y=[ ]`, and the scatterplots contrast the columns of X. The plots can be grouped by the grouping variable `group`. This variable can be a categorical variable, vector, string array, or cell array of strings.

The variable `group` must have the same number of rows as X. Points with the same value of `group` appear on the scatterplot with the same marker and color. Other arguments in `gplotmatrix(x,y,group,clr,sym,siz)` specify the color, marker type, and size for each group. An example of the `gplotmatrix` command is given in the code below. The output is shown in Figure 2.16a.

```
% From multifat.m
X = [broz densi weight adiposi biceps];
varNames = {'broz'; 'densi'; 'weight'; 'adiposi'; 'biceps'};
agegr = age > 55;
gplotmatrix(X,[],agegr,['b','r'],['x','o'],[],'false');
text([.08 .24 .43 .66 .83],  repmat(-.1,1,5), varNames, ...
    'FontSize',8);
text(repmat(-.12,1,5), [.86 .62 .41 .25 .02], varNames, ...
    'FontSize',8, 'Rotation',90);
```

Parallel Coordinates Plots. In a *parallel coordinates plot*, the components of the data are plotted on uniformly spaced vertical lines called component axes. A p-dimensional data vector is represented as a broken line connecting a set of points, one on each component axis. Data represented as lines create readily perceived structures. A command for parallel coordinates plot `parallelcoords` is given below with the output shown in Figure 2.16b.

```
parallelcoords(X, 'group', age>55, ...
                'standardize','on', 'labels',varNames)
set(gcf,'color','white');
```

Figure 2.17a shows parallel coords for the groups age > 55 and age <= 55 with 0.25 and 0.75 quantiles.

```
parallelcoords(X, 'group', age>55, ...
    'standardize','on', 'labels',varNames,'quantile',0.25)
set(gcf,'color','white');
```

Andrews' Plots. An *Andrews plot* (Andrews, 1972) is a graphical representation that utilizes Fourier series to visualize multivariate data. With an observation $(X_1,\ldots,X_p)$, one associates the function

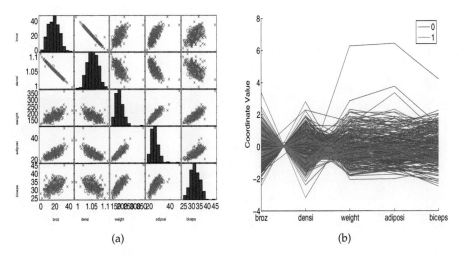

Fig. 2.16 (a) `gplotmatrix` for `broz`, `densi`, `weight`, `adiposi`, and `biceps`; (b) `parallelcoords` plot for X, by age>55.

$$F(t) = X_1/\sqrt{2} + X_2 \sin(2\pi t) + X_3 \cos(2\pi t)$$
$$+ X_4 \sin(2 \cdot 2\pi t) + X_5 \cos(2 \cdot 2\pi t) + \ldots,$$

where t ranges from -1 to 1. One Andrews' curve is generated for each multivariate datum – a row of the data matrix. Andrews' curves preserve the distances between observations. Observations close in the Euclidian distance sense are represented by close Andrews' curves. Hence, it is easy to determine which observations (i.e., rows when multivariate data are represented as a matrix) are most alike by using these curves. Due to the definition, this representation is not robust with respect to the permutation of coordinates. The first few variables tend to dominate, so it is a good idea when using Andrews' plots to put the most important variables first. Some analysts recommend running a principal components analysis first, then generating Andrews' curves for principal components.

An example of Andrews' plots is given in the code below with the output in Figure 2.17b.

```
andrewsplot(X, 'group', age>55, 'standardize','on')
set(gcf,'color','white');
```

Star Plots. The star plot is one of the earliest multivariate visualization objects. Its rudiments can be found in the literature from the early nineteenth century. Similar plots (rose diagrams) are used in Florence Nightingale's Notes on Matters Affecting the Health, Efficiency and Hospital Administration of the British Army (Nightingale, 1858).

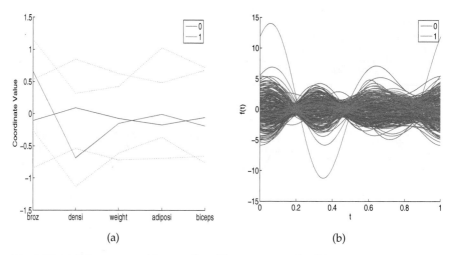

Fig. 2.17 (a) X by age>55 with quantiles; (b) andrewsplot for X by age>55.

The star glyph consists of a number of spokes (rays) emanating from the center of the star plot and connected at the ends. The number of spokes in the star plot is equal to the number of variables (components) in the corresponding multivariate datum. The length of each spoke is proportional to the magnitude of the component it represents. The angle between two neighboring spokes is $2\pi/p$, where p is the number of components. The star glyph connects the ends of the spokes.

An example of the use of star plots is given in the code below with the output in Figure 2.18a.

```
ind = find(age>67);
strind = num2str(ind);
h = glyphplot(X(ind,:), 'glyph','star', 'varLabels',...
       varNames,'obslabels', strind);
set(h(:,3),'FontSize',8); set(gcf,'color','white');
```

Chernoff Faces. People grow up continuously studying faces. Minute and barely measurable differences are easily detected and linked to a vast catalog stored in memory. The human mind subconsciously operates as a super computer, filtering out insignificant phenomena and focusing on the potentially important. Such mundane characters as :), :(, :0, and >:p are readily linked in our minds to joy, dissatisfaction, shock, or affection.

Face representation is an interesting approach to taking a first look at multivariate data and is effective in revealing complex relations that are not visible in simple displays that use the magnitudes of components. It can be used to aid in cluster analysis and discrimination analysis and to detect substantial changes in time series.

Each variable in a multivariate datum is connected to a feature of a face. The variable-feature links in MATLAB are as follows: variable 1 – size of face; variable 2 – forehead/jaw relative arc length; variable 3 – shape of forehead; variable 4 – shape of jaw; variable 5 – width between eyes; variable 6 – vertical position of eyes; variables 7–13 – features connected with location, separation, angle, shape, and width of eyes and eyebrows; and so on. An example of the use of Chernoff faces is given in the code below with the output in Figure 2.18b.

```
ind = find(height > 74.5);
strind = num2str(ind);
h = glyphplot(X(ind,:), 'glyph','face', 'varLabels',...
varNames,'obslabels', strind);
set(h(:,3),'FontSize',10);  set(gcf,'color','white');
```

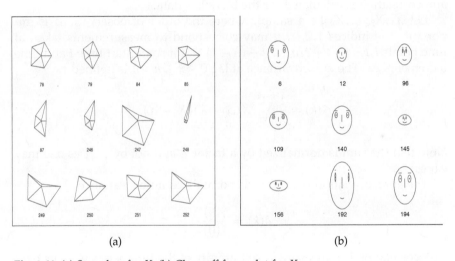

(a)   (b)

Fig. 2.18 (a) Star plots for X; (b) Chernoff faces plot for X.

2.11 Observations as Time Series

Observations that have a time index, that is, if they are taken at equally spaced instances in time, are called *time series*. EKG and EEG signals, high-frequency bioresponses, sound signals, economic indexes, and astronomic and geophysical measurements are all examples of time series. The following example illustrates a time series.

Example 2.8. **Blowflies Time Series.** The data set ⬛blowflies.dat consists of the total number of blowflies (*Lucilia cuprina*) in a population under controlled laboratory conditions. The data represent counts for every other day. The developmental delay (from egg to adult) is between 14 and 15 days for insects under the conditions employed. Nicholson (1954) made 361 bi-daily recordings over a 2-year period (722 days), see Figure 2.19a.
✐

In addition to analyzing basic location, spread, and graphical summaries, we are also interested in evaluating the degree of autocorrelation in time series. Autocorrelation measures the level of correlation of the time series with a time-shifted version of itself. For example, autocorrelation at lag 2 would be a correlation between $X_1, X_2, X_3, \ldots, X_{n-3}, X_{n-2}$ and $X_3, X_4,$ $\ldots, X_{n-1}, X_n$. When the shift (lag) is 0, the autocorrelation is just a correlation. The concept of autocorrelation is introduced next, and then the autocorrelation is calculated for the blowflies data.

Let $X_1, X_2, \ldots, X_n$ be a sample where the order of observations is important. The indices $1, 2, \ldots, n$ may correspond to measurements taken at time points $t, t + \Delta t, t + 2\Delta t, \ldots, t + (n-1)\Delta t$, for some start time t and time increments Δt. The autocovariance at lag $0 \leq k \leq n - 1$ is defined as

$$\hat{\gamma}(k) = \frac{1}{n} \sum_{i=1}^{n-k} (X_{i+k} - \overline{X})(X_i - \overline{X}).$$

Note that the sum is normalized by a factor $\frac{1}{n}$ and not by $\frac{1}{n-k}$, as one may expect.

The autocorrelation is defined as normalized autocovariance,

$$\hat{\rho}(k) = \frac{\hat{\gamma}(k)}{\hat{\gamma}(0)}.$$

Autocorrelation is a measure of self-affinity of the time series with its own shifts and is an important summary statistic. MATLAB has the built-in functions autocov and autocorr. The following two functions are simplified versions illustrating how the autocovariances and autocorrelations are calculated:

```
function acv = acov(ts, maxlag)
%acov.m: computes the sample autocovariance function
%            ts    = 1-D time series
%            maxlag = maximum lag ( < length(ts))
%usage: z = autocov (a,maxlag);
n = length(ts);
ts = ts(:) - mean(ts); %note overall mean
suma = zeros(n,maxlag+1);
suma(:,1) = ts.^2;
for h = 2:maxlag+1
```

```
    suma(1:(n-h+1), h) = ts(h:n);
    suma(:,h) = suma(:,h) .* ts;
end
acv = sum(suma)/n; %note the division by n
                   %and not by expected (n-h)


function [acrr] = acorr(ts , maxlag)
  acr =  acov(ts, maxlag);
  acrr = acr ./ acr(1);
```

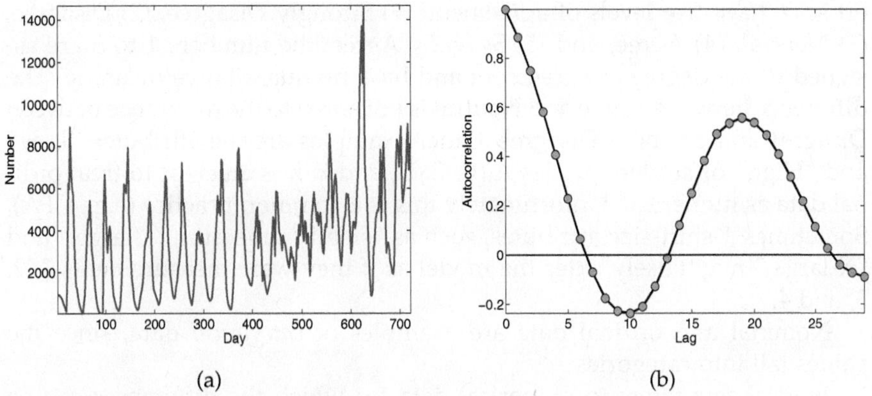

(a)   (b)

Fig. 2.19 (a) Bi-daily measures of size of the blowfly population over a 722-day period, (b) The autocorrelation function of the time series. Note the peak at lag 19 corresponds to the periodicity of 38 days.

Figure 2.19a shows the time series illustrating the size of the population of blowflies over 722 days. Note the periodicity in the time series. In the autocorrelation plot (Fig. 2.19b) the peak at lag 19 corresponds to a time shift of 38 days. This indicates a periodicity with an approximate length of 38 days in the dynamic of this population. A more precise assessment of the periodicity and related inference can be done in the frequency domain of a time series, but this theory is beyond the scope of this course. Good follow-up references are Brillinger (2001), Brockwell and Davis (2009), and Shumway and Stoffer (2005). Also, see Exercises 2.30 and 2.31.

2.12 About Data Types

The cell data elaborated in this chapter are *numerical*. When measurements are involved, the observations are typically *numerical*. Other types of data encountered in statistical analysis are categorical. Stevens (1946), who was influenced by his background in psychology, classified data as nominal, ordinal, interval, and ratio. This typology is loosely accepted in other scientific circles. However, there are vibrant and ongoing discussions and disagreements (e.g., Veleman and Wilkinson, 1993). *Nominal data*, such as race, gender, political affiliation, and names, cannot be sensibly ordered. For example, the counties in northern Georgia, Cherokee, Clayton, Cobb, DeCalb, Douglas, Fulton, and Gwinnett, cannot be ordered, though there is a nonessential alphabetical order of their names. Of course, numerical attributes of these counties, such as size, area, and revenue, can be ordered.

Ordinal data could be ordered and sometimes assigned numbers that do not convey their relative standing. For example, data on the five-point Likert scale have five levels of agreement: (1) Strongly Disagree, (2) Disagree, (3) Neutral, (4) Agree, and (5) Strongly Agree; the numbers 1 to 5 are assigned to the degree of agreement and have no quantitative meaning. The difference between Agree and Neutral is not equal to the difference between Disagree and Strongly Disagree. Other examples are the attributes "Low" and "High" or student grades A, B, C, D, and F. It is an error to treat ordinal data as numerical. Unfortunately this is a common practice (e.g., GPA). Sometimes T-shirt-size attributes, such as "small," "medium," "large," and "X-large," may falsely enter the model as if they were measurements 1, 2, 3, and 4.

Nominal and ordinal data are examples of *categorical* data, since the values fall into categories.

Interval data refers to numerical data for which the differences can be well interpreted. However, for this type of data, the origin is not defined in a natural way, so the ratios would not make sense. Temperature is a good example. We cannot say that a day in July with a temperature of $100°F$ is twice as hot as a day in November with a temperature of $50°F$. Test scores are another example of interval data as a student who scores 100 on a midterm may not be twice as good as a student who scores 50.

Ratio data are at the highest level; these are usually standard numerical values for which ratios make sense and the origin is absolute. Length, weight, and age are all examples of ratio data.

Interval and ratio data are examples of *numerical* data.

MATLAB provides a way to keep such heterogeneous data in a single structure array with a syntax resembling C language.

Structures are arrays comprised of structure elements and are accessed by named fields. The fields (data containers) can contain any type of data. Storage in the structure is allocated dynamically. The general syntax for a structure format in MATLAB is `structurename(recordnumber).fieldname=data`

For example,

```
patient.name = 'John Doe';
patient.agegroup = 3;
patient.billing = 127.00;
patient.test = [79 75 73; 180 178 177.5; 220 210 205];
```

```
patient
%To expand the structure array, add subscripts.
patient(2).name = 'Ann Lane';
patient(2).agegroup =  2;
patient(2).billing = 208.50;
patient(2).test = [68 70 68; 118 118 119; 172 170 169];
patient
```

2.13 Big Data Paradigm

The phrase *Big Data* usually refers to massive, heterogeneous, longitudinal, complex, and/or distributed data sets generated by devices, sensors, scanners, Internet, or other sources of digital information.

The bioengineering research community is undergoing a profound transformation with the use of large-scale and diverse data sets that allow for data-guided decision-making. New statistical models, prediction procedures, and multiscale domains for data analysis are enabling this paradigm shift in biomedical research.

In simplistic terms, Big Data initiative aims to accelerate the progress of scientific discovery and innovation. Under the umbrella of Big Data new fields of inquiry that would not otherwise be possible to discover can be formulated, analyzed, and utilized. The development of new data analytic tools would lead to more efficient and less expensive healthcare. In addition, it would lead to increased quality of life by enabling breakthrough discoveries and innovations in health and medical sciences. Big Data provides a platform to support cross-disciplinary collaborations necessary to make advances in complex grand challenges in bioengineering.

In data science, the term meta analysis describes the methodology that puts together isolated studies in order to improve overall inferential power. In a simplified way the Big Data paradigm can be thought as meta-analytic approach to fusion of distributed and massive data sets. Typically, for a data to be classified as "big," the conditions from popular "Four Vee" definition need to be satisfied: volume, velocity, variety, and veracity.

Volume. The data is massive, often measured in tera-, peta-, even exa-byte units. A commonsense understanding is that if storage and manipulation of data are not routine due to their size, such data can be classified as massive.

Size of data needs to be understood in relative terms, since for some complex inferential models, even moderate-sized data sets are "big." So in addition to their sheer volume posing storage and handling challenges, the data become big when the scalability of methodologies traditionally used for their processing breaks.

Velocity. Speed at which the data is created, recorded, stored, transmitted, and analyzed is data's Velocity. The speed at which data is created and processed nowadays is unimaginable. For example, every minute millions of email and Google quarries are conducted. Unlike the batch-processing approach, where data are static and processed in batches, in the Big Data era, information needs to be stored and analyzed dynamically, that is in real time or near real time.

Variety. In addition to classical text book data-type classes such as numeric, ordinal, nominal, etc., Big Data encompasses a range of heterogenous formats and structures. In fact, most of the data obtained today is nontraditional, unstructured, and distributed (video clips, images, data generated by biomedical devices, sensory data, incomplete data, preferences and sentiments, satellite data, click streams, etc). Such a variety of types and sources poses new challenges for data analysis and fusion. The concept of *variability* often attributed to Big Data differs form the concept of variety. It represents the a degree of variability for a repeated single attribute.

Veracity. Increasing the volume, speed, and variety of data is worthless if data are incorrect or irrelevant. Biased data can cause a lot of problems in decision making. Therefore, it is important to have safeguards for data quality, to eliminate or minimize the human-error factors, and to have robust data analytic procedures that are less sensitive to variation form the postulated data models.

Although this text is not about Big Data methodology, the concepts and procedures covered in this and subsequent chapters are critical for understanding and utilizing the Big Data. Many of the procedures covered here are directly scalable to massive data sets; however, most are prohibitively computationally expensive and require the interplay of statistics, computer science, and problem content science, to be tackled. To this end, this book may provide the first inferential step.

2.14 Exercises

2.1. **Auditory Cortex Spikes.** This data set comes from experiments in the lab of Dr. Robert Liu of Emory University[4] and concerns single-unit electrophysiological recordings in the auditory cortex of nonanesthetized female mice. The motivating question is the exploration of auditory neural

---

[4] http://www.biology.emory.edu/research/Liu/index.html

differences between female parents and female virgins and their relationship to cortical response.

Researchers in Liu's lab developed a restrained awake setup to collect
single neuron activity from both female types. Multiple trials are performed on the neurons from one maternal and one naïve animal.

The recordings are made from a region in the auditory cortex of the
mouse with a single tungsten electrode. A sound stimulus is presented
at a time of 200 ms during each sweep (time shown is 0–611 and 200 is
the point at which a stimulus is presented). Each sweep is 611 ms long
and the duration of the stimulus tone is 10 to 70 ms. The firing times for
maternal and naïve mice are provided in the data set ▦ spikes.dat, in
columns 2 and 3. Column 1 is the numbering from 1 to 611.

(a) Using MATLAB's diff command, find the inter-firing times. Plot a
histogram for both sets of inter-firing times. Use biplot.m to plot the histograms back to back.

(b) For inter-firing times in the maternal mouse's response find descriptive statistics similar to those in the cell area example.

2.2. **On Average.** It is an anecdotal truth that an average Australian has
less than two legs! Because some Australians have lost their leg(s), the
number of legs is less than twice the number of people.

Here is the exercise in which several averages are calculated and compared.. A small company reports the following salaries: 4 employees at
20K, 3 employees at 30K, the vice-president at 200K, and the president at
400K. Calculate the arithmetic mean, geometric mean, median, harmonic
mean, and mode. If the company is now hiring, would an advertising
strategy in which the mean salary is quoted be fair? If not, suggest an
alternative.

2.3. **Contraharmonic Mean and f-Mean.** The contraharmonic mean for
$X_1, X_2, \ldots, X_n$ is defined as

$$C(X_1, \ldots, X_n) = \frac{\sum_{i=1}^{n} X_i^2}{\sum_{i=1}^{n} X_i}.$$

(a) Show that $C(X_1, X_2)$ is twice the sample mean minus the harmonic
mean of X_1, X_2.

(b) Show that $C(x, x, x, \ldots, x) = x$.

The generalized f-mean of $X_1, \ldots, X_n$ is defined as

$$X_f = f^{-1}\left(\frac{1}{n}\sum_{i=1}^{n} f(X_i)\right),$$

where f is suitably chosen such that $f(X_i)$ and f^{-1} are well defined.

(c) Show that $f(x) = x, \frac{1}{x}, x^k, \log x$ gives the mean, harmonic mean, power k mean, and geometric mean.

2.4. **Mushrooms.** The unhappy outcome of uninformed mushroom picking is poisoning. In many such cases, the poisoning is due to ignorance or a superficial approach to identification. The most dangerous fungi are Death Cap (*Amanita phalloides*) and two species akin to it, *A. verna* and Destroying Angel (*A. virosa*). These three toadstools cause the majority of fatal poisoning.

One of the keys to mushroom identification is the spore deposit. Spores of *Amanita phalloides* are colorless, nearly spherical, and smooth. Measurements in microns of 28 spores are given below:

9.2 8.8 9.1 10.1 8.5 8.4 9.3
8.7 9.7 9.9 8.4 8.6 8.0 9.5
8.8 8.1 8.3 9.0 8.2 8.6 9.0
8.7 9.1 9.2 7.9 8.6 9.0 9.1

(a) Find the *five-number summary* (Min, Q_1, Me, Q_3, Max) for the spore measurement data.
(b) Find the mean and the mode.
(c) Find and plot the histogram of z-scores, $z_i = (X_i - \overline{X})/s$.

2.5. **Manipulations with Sums.** Prove the following algebraic identities involving sums, which are useful in demonstrating properties of some sample summaries:

(a) $\sum_{i=1}^{n}(x_i - \overline{x}) = 0$	(b) If $y_1 = x_1 + a, y_2 = x_2 + a, \ldots, y_n = x_n + a$, then $\sum_{i=1}^{n}(y_i - \overline{y})^2 = \sum_{i=1}^{n}(x_i - \overline{x})^2$
(c) If $y_1 = c \cdot x_1, y_2 = c \cdot x_2, \ldots, y_n = c \cdot x_n$, then $\sum_{i=1}^{n}(y_i - \overline{y})^2 = c^2 \sum_{i=1}^{n}(x_i - \overline{x})^2$	(d) If $y_1 = c \cdot x_1 + a, y_2 = c \cdot x_2 + a, \ldots, y_n = c \cdot x_n + a$, then $\sum_{i=1}^{n}(y_i - \overline{y})^2 = c^2 \sum_{i=1}^{n}(x_i - \overline{x})^2$.
(e) $\sum_{i=1}^{n}(x_i - \overline{x})^2 = \sum_{i=1}^{n} x_i^2 - n(\overline{x})^2$	(f) $\sum_{i=1}^{n}(x_i - \overline{x})(y_i - \overline{y}) = \sum_{i=1}^{n} x_i y_i - n(\overline{x})(\overline{y})$
(g) $\sum_{i=1}^{n}(x_i - a)^2 = \sum_{i=1}^{n}(x_i - \overline{x})^2 + n(\overline{x} - a)^2$	(h) For any constant a, $\sum_{i=1}^{n}(x_i - \overline{x})^2 \leq \sum_{i=1}^{n}(x_i - a)^2$

2.6. **Emergency Calculation.** Graduate student Rosa Juliusdottir reported the results of an experiment to her advisor who wanted to include these results in his grant proposal. Before leaving to Reykjavik for a short vacation, she left the following data in her advisor's mailbox: sample size $n = 12$, sample mean $\overline{X} = 15$, and sample variance $s^2 = 34$.

The advisor noted with horror that the last measurement X_{12} was wrongly recorded. It should have been 16 instead of 4. It would be easy to fix $\overline{X}$ and s^2, but the advisor did not have the previous 11 measurements nor the statistics training necessary to make the correction. Rosa was in Iceland, and the grant proposal was due the next day. The advisor was desperate, but luckily you came along. Can you update $\overline{X}$ and s^2?

2.7. **Sample Mean and Standard Deviation after a Change.** It is known that $\overline{y} = 11.6$, $s_y = 4.4045$, and $n = 15$. The observation $y_{12} = 7$ is removed and observation y_{13} was misreported; it was not 10, but 20. Find $\overline{y}_{new}$ and $s_{y(new)}$ after the changes.

2.8. **Aspirin Weights.** Stoodley (1984) provides 100 weights of aspirin tablets determined using a laboratory balance and rounded to the nearest mg. The data in 🖳 aspirin.dat are given as a simple sample.
(a) Form a composite sample using frequencies of the measurements.
(b) From the composite sample find location and spread measures, skewness, and kurtosis.

2.9. **Surveys on Different Scales.** We are interested in determining whether UK voters (whose parties have somewhat more distinct policy positions than those in the United States) have a wider variation in their evaluations of the parties than US voters. The problem is that the British election survey takes evaluations scored 0–10, whereas the US National Election Survey gets evaluations scored 0–100. Here are two surveys:

$$\begin{array}{l|l} \text{UK} & 6\ \ 7\ \ 5\ \ 10\ \ 3\ \ \ 9\ \ \ 9\ \ 6\ \ 8\ \ 2\ \ 7\ \ 5 \\ \text{US} & 67\ \ 65\ \ 95\ \ 86\ \ 44\ \ 100\ \ 85\ \ 92\ \ 91\ \ 65 \end{array}$$

Using CV, compare the degree of variation without worrying about the different scales.

2.10. **Merging Two Samples.** Suppose that $\overline{X}$ and s_X^2 are the mean and variance of the sample $X_1, \ldots, X_m$ and $\overline{Y}$ and s_Y^2 of the sample $Y_1, \ldots, Y_n$. If the two samples are merged into a single sample, show that its mean and variance are

$$\frac{m\overline{X} + n\overline{Y}}{m+n} \quad \text{and} \quad \frac{1}{m+n-1}\left[(m-1)s_X^2 + (n-1)s_Y^2 + \frac{mn}{m+n}(\overline{X}-\overline{Y})^2\right].$$

2.11. **Fitting the Histogram.** The following is a demonstration of MATLAB's built-in function histfit on a simulated data set:

```
dat = normrnd(4, 1,[1 500]) + normrnd(2, 3,[1 500]);
figure; histfit(dat(:));
```

The function histfit plots the histogram of data and overlays it with the best-fitting Gaussian curve. As an exercise, take Brozek index broz from the data set 🖳 fat.dat (second column) and apply the histfit command. Comment on how the Gaussian curve fits the histogram.

2.12. **Orientation of Stem Cells.** Human mesenchymal stem cells were seeded into synthetic poly(ethylene glycol)-based hydrogels. Two types of hydrogels were used – one hydrogel type that would degrade slowly, and one hydrogel type that would degrade quickly. Then, hydrogels were stretched repeatedly in a single direction. Cells in slow-degrading gels would be limited in changing their orientation, while cells in fast-degrading gels would be much more able to change their orientation. It was hypothesized that cells in fast-degrading gels would reorient in the direction of strain after 14 days of culture.

To measure orientation, pictures of gels were taken using a confocal microscope and calcein staining of cells. The data provided (courtesy of Dr. Peter Young from Temenoff Lab at Georgia Tech) describe the distribution of particle orientation angles of the longer axis of the ellipse relative to the positive direction of the x-axis. This angle is measured in absolute value and ranges between 0 and 90 degrees.

The data set `gel.mat` reads in as a structure `gel`, where the fields `gel.static14` are angles for slow degrading gel, and `gel.dynamic14` are angles for fast degrading gel.

For both static and dynamic data

(a) Plot histograms with 30 bins. Use back-to-back histogram code `bihist.m`.

(b) Plot box-and-whiskers summaries of the samples.

(c) Calculate location measures (mean, median, mode, 20% trimmed mean, 20% winsorized mean).

(d) Calculate measures of spread (variance, standard deviation, real-MAD).

(e) Find skewness and kurtosis.

(f) Find 20th percentile or 0.2-quantile of the sample.

Organize all tasks in a single m-file. MATLAB-publish the file as PDF report.

2.13. **QT Syndrome.** The QT interval is a time interval between the start of the Q wave and the end of the T wave in a heart's electrical cycle (Fig. 2.20). It measures the time required for depolarization and repolarization to occur. In a long QT syndrome, the duration of repolarization is longer than normal, which results in an extended QT interval. An interval above 440 ms is considered prolonged. Although the mechanical function of the heart could be normal, the electrical defects predispose affected subjects to arrhythmia, which may lead to sudden loss of consciousness (syncope) and, in some cases, to a sudden cardiac death.

The data set `QT.dat|mat` was compiled by Christov et al. (2006) and is described in `http://www.biomedical-engineering-online.com/content/5/1/31`. It provides 548 QT times taken from 293 subjects. The subjects

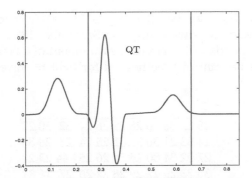

Fig. 2.20 Schematic plot of ECG, with QT time between the red bars.

include healthy controls (about 20%) and patients with various diagnoses, such as myocardial infarction, cardiomyopathy/heart failure, bundle branch block, dysrhythmia, and myocardial hypertrophy. Prolonged QT (> 440 ms) is a risk factor for abnormality of heart's electric system. The Q-onsets and T-wave ends are evaluated by five independent experts, and medians of their estimates are used in calculations of the QT for a subject.

Plot the histogram of this data set and argue that the data are reason-ably "bell-shaped." Find the location (mean, median, mode) and spread measures (s2, MAD, iqr) of the sample. What proportion of this sample has prolonged QT?

2.14. **Simpson's Diversity Index.** An alternative diversity measure to Shannon's in (2.1) is the Simpson diversity index defined as

$$D = \frac{n^2}{\sum_{i=1}^{k} f_i^2}.$$

This measure achieves its maximum k when all frequencies are equal; thus Simpson's homogeneity (equitability) index is defined as $E_D = D/k$. Repeat the calculations from Example 2.3 with Simpson's diversity and homogeneity indexes in place of Shannon's. Is the Brazilian sample still the most homogeneous, as it was according to Shannon's E_H index?

2.15. **Speed of Light.** Light travels very fast. It takes about 8 minutes to reach Earth from the Sun and over 4 years to reach Earth from the closest star outside the solar system. Radio and radar waves also travel at the speed of light, and an accurate value of that speed is important to communicate with astronauts and orbiting satellites. Because of the nature of light, it is very hard to measure its speed. The first reasonably accurate measurements of the speed of light were made by A. Michelson and

S. Newcomb. The table below contains 66 transformed measurements made by Newcomb between July and September 1882. Entry 28, for instance, corresponds to the actual measurement of 0.000024828 seconds. This was the amount of time needed for light to travel approximately 4.65 miles.

28	22	36	26	28	28	26	24	32	30 27
24	33	21	36	32	31	25	24	25	28 36
27	32	34	30	25	26	26	25	−44	23 21
30	33	29	27	29	28	22	26	27	16 31
29	36	32	28	40	19	37	23	32	29 −2
24	25	27	24	16	29	20	28	27	39 23

You can download 🖳 `light.data|mat` and read it in MATLAB.

If we agree that outlier measurements are outside the interval $[Q_1 - 2.5\ IQR, Q_3 + 2.5\ IQR]$, what observations qualify as outliers? Make the data "clean" by excluding outlier(s). For the cleaned data, find the mean, 20% trimmed mean, real MAD, std, and variance.

Plot the histogram and kernel density estimator for an appropriately selected bandwidth.

2.16. **Spatial Distribution of Weed.** Collecting exact counts of weed in an agricultural field is trivial but extremely time-consuming task. Instead, image analysis algorithms for object extraction applied to pictures of agricultural fields are used to estimate the weed content. High resolution (about 1 m^2), pictures that are acquired at a large number of sites can be used to obtain maps of weed content over a whole field at a reasonably low cost. However, these image-based estimates are not perfect, and acquiring exact weed counts is in fact highly useful both for assessing the accuracy of the image-based algorithms and for improving the estimates by use of the combined data.

The data file 🖳 `weed.dat|xlsx|mat` has 100 rows, where each row consists of two spatial coordinates, exact weed counts, and image estimate of weed counts.

(a) Using function `bihist.m` plot back-to-back histograms of exact and image counts.

(b) Find correlation between the two types of counts.

(c) Using MATLAB's `scatter`, plot a scatterplot of circles with centers at location coordinates, of size proportional to the exact count, and with color mapped to the difference between exact and image counts. Consult the help for `scatter`.

2.17. **AFM.** The AFM is a type of scanned probe microscopy (SPM) that can measure the adhesion strength between two materials at the nanonewton scale. In AFM, a cantilever beam is adjusted until it bonds with the surface of a sample, and then the force required to separate the beam and sample is measured from the beam deflection. Beam vibration can be caused by factors such as thermal energy of the surrounding air or the footsteps of someone outside the laboratory. The vibration of a beam acts as noise on the deflection signal.

The AFM data from the adhesion measurements between carbohydrate and the cell adhesion molecule (CAM) E-Selectin was collected by Bryan Marshal at Georgia Institute of Technology. The technical description is provided in Marshall et al. (2003)

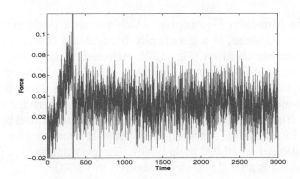

Fig. 2.21 The AFM measurements.

(a) Read data set ⬚ afm.dat into MATLAB. The array afm has 3000 measurements. Form data vector force by taking measurements with index greater or equal to 335, (force=afm(335:3000);), thus avoiding the "ramp" artifact, Figure 2.21.

(b) For vector force find mean, standard deviation, median, IQR, 0.95-quantile, skewness, and kurtosis.

(c) To visually check for normality of force, find qqplot. Is it linear?

(d) Use function acorr to find autocorrelations up to 20 lags to check whether the observations are autocorrelated. Plot the autocorrelations by MATLAB's stem plot. The autocorrelation at lag 0 is always 1, but what about other lags?

2.18. **Limestone Formations in Jamaica.** This data set contains 18 observations of nummulite specimens from the Eocene yellow limestone formation in northwestern Jamaica (⬚ limestone.dat). The use of faces to represent points in k-dimensional space graphically was originally illustrated on this data set (Chernoff, 1973). Represent this data set graphically using Chernoff faces.

ID	Z_1	Z_2	Z_3	Z_4	Z_5	Z_6	ID	Z_1	Z_2	Z_3	Z_4	Z_5	Z_6
1	160	51	10	28	70	450	45	195	32	9	19	110	1010
2	155	52	8	27	85	400	46	220	33	10	24	95	1205
3	141	49	11	25	72	380	81	55	50	10	27	128	205
4	130	50	10	26	75	560	82	70	53	7	28	118	204
6	135	50	12	27	88	570	83	85	49	11	19	117	206
41	85	55	13	33	81	355	84	115	50	10	21	112	198
42	200	34	10	24	98	1210	85	110	57	9	26	125	230
43	260	31	8	21	110	1220	86	95	48	8	27	114	228
44	195	30	9	20	105	1130	87	95	49	8	29	118	240

2.19. **Duchenne Muscular Dystrophy.** *Duchenne muscular dystrophy* (DMD), or Meryon's disease, is a genetically transmitted disease, passed from a mother to her children (Fig. 2.22). Affected female offspring usually suffer no apparent symptoms and may unknowingly carry the disease. Male offspring with the disease die at a young age. Not all cases of the disease come from an affected mother. A fraction, perhaps one-third, of the cases arise spontaneously, to be genetically transmitted by an affected female. This is the most widely held view at present. The incidence of DMD is about 1 in 10,000 male births. The population risk (prevalence) that a woman is a DMD carrier is about 3 in 10,000.

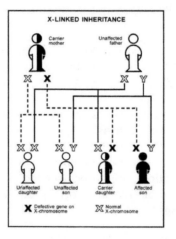

Fig. 2.22 Each son of a carrier has a 50% chance of having DMD and each daughter has a 50% chance of being a carrier.

Download data set 📥 dmd.dat|mat|xls from the text page. This data set is modified data from Percy et al. (1981) (entries containing missing values excluded). It consists of 194 observations corresponding to blood samples collected in a project to develop a screening program for female relatives of boys with DMD. The program was implemented in Canada, and its goal was to inform a woman of her chances of being a carrier based on serum markers as well as her family pedigree. Another question of interest was whether age should be taken into account. Enzyme levels were measured in known carriers (67 samples) and in a group of noncarriers (127 samples).

The first two serum markers, creatine kinase and hemopexin (ck,h), are inexpensive to obtain, while the last two, pyruvate kinase and lactate dehydroginase (pk,ld), are expensive.

The variables (columns) in the data set are:

Column	Variable	Description
1	age	Age of a woman in the study
2	ck	Creatine kinase level
3	h	Hemopexin
4	pk	Pyruvate kinase
5	ld	Lactate dehydroginase
6	carrier	Indicator if a woman is a DMD carrier

(a) Find the mean, median, standard deviation, and *real MAD* of pyruvate kinase level, pk, for all cases (carrier=1).

(b) Find the mean, median, standard deviation, and *real MAD* of pyruvate kinase level, pk, for all controls (carrier=0).

(c) Find the correlation between variables pk and carrier.

(d) Use MATLAB's gplotmatrix to visualize pairwise dependencies between the six variables.

(e) Plot the histogram with 30 bins and smoothed normalized histogram (density estimator) for pk. Use ksdensity.

2.20. **Ashton's Dental Data.** The evolutionary status of fossils (Australopithecinae, Proconsul, etc.) stimulated considerable discussion in the 1950s. Particular attention was paid to the teeth of the fossils, comparing their overall dimensions with those of human beings and of the extant great apes. As "controls," measurements were taken on the teeth of three types of the modern man (British, West African native, Australian aboriginal) and of the three living great apes (gorilla, orangutan, and chimpanzee).

The data in the table below are taken from Ashton et al. (1957, p. 565), who used 2D projections to compare the measurements. Andrews (1972) used an excerpt of these data to illustrate his methodology. The values in the table are not the original measurements, but the first eight *canonical variables* produced from the data in order to maximize the sum of distances between different pairs of populations.

A. West African	−8.09	0.49	0.18	0.75	−0.06	−0.04	0.04	0.03
B. British	−9.37	−0.68	−0.44	−0.37	0.37	0.02	−0.01	0.05
C. Au. aboriginal	−8.87	1.44	0.36	−0.34	−0.29	−0.02	−0.01	−0.05
D. Gorilla: male	6.28	2.89	0.43	−0.03	0.10	−0.14	0.07	0.08
E. Female	4.82	1.52	0.71	−0.06	0.25	0.15	−0.07	−0.10
F. Orangutan: Male	5.11	1.61	−0.72	0.04	−0.17	0.13	0.03	0.05
G. Female	3.60	0.28	−1.05	0.01	−0.03	−0.11	−0.11	−0.08
H. Chimpanzee: male	3.46	−3.37	0.33	−0.32	−0.19	−0.04	0.09	0.09
I. Female	3.05	−4.21	0.17	0.28	0.04	0.02	−0.06	−0.06
J. *Pithecanthropus*	−6.73	3.63	1.14	2.11	−1.90	0.24	1.23	−0.55
K. *pekinensis*	−5.90	3.95	0.89	1.58	−1.56	1.10	1.53	0.58
L. *Paranthropus robustus*	−7.56	6.34	1.66	0.10	−2.23	−1.01	0.68	−0.23
M. *Paranthropus crassidens*	−7.79	4.33	1.42	0.01	−1.80	−0.25	0.04	−0.87
N. *Meganthropus paleojavanicus*	−8.23	5.03	1.13	−0.02	−1.41	−0.13	−0.28	−0.13
O. *Proconsul africanus*	1.86	−4.28	−2.14	−1.73	2.06	1.80	2.61	2.48

Andrews (1972) plotted curves over the range $-\pi < t < \pi$ and concluded that the graphs clearly distinguished humans, the gorillas and orangutans, the chimpanzees, and the fossils. Andrews noted, for example, that the curve for the fossil *Proconsul africanus* corresponds to a plot inconsistent with that of all other fossils as well as those of humans and apes.

Graphically present this data using (a) star plots, (b) Andrews plots, and (c) Chernoff faces.

2.21. **Andrews Plots of Iris Data.** Fisher iris data are 4D, and Andrews plots can be used to explore clustering of the three species (*Setosa, Versicolor,* and *Virginica*). Discuss the output from the code below:

```
load fisheriris
andrewsplot(meas,'group',species);
```

What species clearly separate? What species are more difficult to separate?

2.22. **Leptoconops – Biting Flies.** Atchley (1974) collected morphological characteristics of two species of biting flies *Leptoconops torrens* and *Leptoconops carteri*. They are morphologically so similar that for many years they have been considered to be the same species.

The data set, as reported by Johnson and Whichern (1988), contains 35 multivariate observations of each species. An observation is composed of 7 dependent taxonomic responses: wing length, wing width, third palp length, third palp width, fourth palp length, length of antennal segment 12, and length of antennal segment 14.

This data set is given as a structure field `leptoconops.morpho` in MATLAB's structure file ⊞ `leptoconops.mat`. The field `leptoconops.names` contains names of seven recorded morphological measures: `winglen`, `wingwid`, `papl3len`, `palp3wid`, `palp4len`, `ant12len`, and `ant14len`. The two species are identified in the field leptoconops.spec where 0's correspond to *L.torrens* and 1's to *L.carteri*.

(a) Find the sample means and covariance matrices for *L.torrens* and *L.carteri*.

(b) Using MATLAB's built-in command gplotmatrix create a matrix of scatter plots where each figure (i, j) contains a scatter plot of a column i against a column j of leptoconops.morpho. The plots should use markers defined by the grouping variable leptoconops.spec.

(c) Explore whether it is possible to visually delimit 2 phenetic clusters within the torrens-carteri complex?

(d) Experiment with MATLAB's graphical tools imagesc, parallelcoords, andrewsplot, glyphplot with options star and face. Submit only your recommended method with discussion (an open-ended question). Check the MATLAB Help for the syntax options for the commands above.

2.23. **Cork Boring Data.** Cork is the bark of the cork oak (*Quercus suber L*), a noble tree with very special characteristics that grows in the Mediterranean. This natural tissue has unique qualities: light weight, elasticity, insulation and impermeability, fire retardancy, resistance to abrasion, etc. The data measuring cork boring of trees given in Rao (1948) consist of the weights (in centigrams) of cork boring in four directions (north, east, south, and west) for 28 trees. Data given in Table 2.2 can also be found in cork.dat|mat.

Table 2.2 Rao's cork data. Weights of cork boring in four directions (north, east, south, west) for 28 trees.

Tree	N E S W	Tree	N E S W
1	72 66 76 77	15	91 79 100 75
2	60 53 66 63	16	56 68 47 50
3	56 57 64 58	17	79 65 70 61
4	41 29 36 38	18	81 80 68 58
5	32 32 35 36	19	78 55 67 60
6	30 35 34 26	20	46 38 37 38
7	39 39 31 27	21	39 35 34 37
8	42 43 31 25	22	32 30 30 32
9	37 40 31 25	23	60 50 67 54
10	33 29 27 36	24	35 37 48 39
11	32 30 34 28	25	39 36 39 31
12	63 45 74 63	26	50 34 37 40
13	54 46 60 52	27	43 37 39 50
14	47 51 52 43	28	48 54 57 43

(a) Graphically display the data as a data plot, pairwise scatterplots, an Andrews plot, and Chernoff faces.

(b) Find the mean $\bar{x}$ and covariance matrix S for this data set. Find the trace and determinant of S.

(c) Find the Mahalanobis transformation for these data. Check that the covariance matrix for the transformed data is the identity matrix.

2.24. **Balance.** When a human experiences a balance disturbance, muscles throughout the body are activated in a coordinated fashion to maintain an upright stance. Researchers at Lena Ting Laboratory for Neuroengineering at Georgia Tech are interested in uncovering the sensorimotor mechanisms responsible for coordinating this automatic postural response (APR). Their approach was to perturb the balance of a human subject standing upon a customized perturbation platform that translates in the horizontal plane.

Platform motion characteristics spanned a range of peak velocities (5 cm/s steps between 25 and 40 cm/s) and accelerations (0.1 g steps between 0.2 and 0.4 g). Five replicates of each perturbation type were collected during the experimental sessions. Surface electromyogram (EMG) signals, which indicate the level of muscle activation, were collected at 1080 Hz from 11 muscles in the legs and trunk.

The data in ⬛ balance2.mat are processed EMG responses to backward-directed perturbations in the medial gastrocnemius muscle (an ankle plantar flexor located on the calf) for all experimental conditions. There is 1 s of data, beginning at platform motion onset. There are 5 replicates of length 1024, each collected at 12 experimental conditions (4 velocities crossed with 3 accelerations), so the data set is 3D, $1024 \times 5 \times 12$.

For example, data(:,1,4) is an array of 1024 observations corresponding to first replicate, under the fourth experimental condition (30 cm/s, 0.2 g).

Consider a fixed acceleration of 0.2g and only the first replicate. Form 1024 4D observations (velocities 25, 30, 35, and 40 as variables) as a data matrix. For the first 16 observations find multivariate graphical summaries using MATLAB's gplotmatrix, parallelcoords, andrewsplot, and glyphplot.

2.25. **Cats.** Cats are often used in studies about locomotion and injury recovery. In one such study, a bundle of nerves in a cat's legs were cut and then surgically repaired. This mimics the surgical correction of injury in people. The recovery process of these cats was then monitored. It was monitored quantitatively by walking a cat across a plank that has force plates, as well as by monitoring various markers inside the leg. These markers provided data for measures such as joint lengths and joint moments. A variety of data was collected from three different cats: Natasha, Riga, and Korina. Natasha (cat = 1) has 47 data entries, Riga (cat = 2) has 39 entries, and Korina (cat = 3) has 35 entries.

The measurements taken are the number of steps for each trial, the length of the stance phase (in milliseconds), the hip height (in meters), and the velocity (in meters/second). The researchers observe these variables for different reasons. They want uniformity both within and between samples (to prevent confounding variables) for steps and velocity. The hip height helps monitor the recovery process. A detailed description can be found in Farrell et al. (2009).

The data set, courtesy of Dr. Boris Prilutsky, School of Applied Physiology at, Georgia Tech, is given as the MATLAB structure file 🖫 cats.mat. Form a data matrix

```
X = [cat.nsteps cat.stancedur cat.hipheight cat.velocity cat.cat];
```

and find its mean and correlation matrix. Form matrix Z by standardizing the columns of X (use zscore). Plot the image of the standardized data matrix.

2.26. **BUPA Liver Data.** The BUPA liver disorders database (courtesy of Richard Forsyth, BUPA Medical Research Ltd.) consists of 345 records of male individuals. Each record has 7 attributes:

Attribute	Name	Meaning
1	mcv	Mean corpuscular volume
2	alkphos	Alkaline phosphotase
3	sgpt	Alamine aminotransferase
4	sgot	Aspartate aminotransferase
5	gammagt	Gamma-glutamyl transpeptidase
6	drinks	Number of half-pint equivalents of alcoholic beverages drunk per day
7	selector	Field to split the database

The first five variables are all blood tests that are thought to be sensitive to liver disorders that might arise from excessive alcohol consumption. The variable selector was used to partition the data into two sets, very likely into a training and validation part.

Using gplotmatrix, explore the relationship among variables 1 through 6 (exclude the selector).

2.27. **Triazines.** A common step in pharmaceutical development is the formation of a quantitative structure-activity relationship (QSAR) to model an exploratory series of compounds. A QSAR generalizes how the structure (shape) of a compound relates to its biological activity. The data set 🖫 triazines.mat involves variables/attributes potentially important for the inhibition of rat1mouse tumor DHFR by triazines. This data set is fully explained in Hirst et al. (1994), but here is the basic summary: Number of instances: 186; Number of attributes: 60; Attribute names as in the table below; Number of responses: 1 (activity).

p1_polar	p1_size	p1_flex	p1_h_doner
p1_h_acceptor	p1_pi_doner	p1_pi_acceptor	p1_polarisable
p1_sigma	p1_branch	p2_polar	p2_size
p2_flex	p2_h_doner	p2_h_acceptor	p2_pi_doner
p2_pi_acceptor	p2_polarisable	p2_sigma	p2_branch
p3_polar	p3_size	p3_flex	p3_h_doner
p3_h_acceptor	p3_pi_doner	p3_pi_acceptor	p3_polarisable
p3_sigma	p3_branch	p4_polar	p4_size
p4_flex	p4_h_doner	p4_h_acceptor	p4_pi_doner
p4_pi_acceptor	p4_polarisable	p4_sigma	p4_branch
p5_polar	p5_size	p5_flex	p5_h_doner
p5_h_acceptor	p5_pi_doner	p5_pi_acceptor	p5_polarisable
p5_sigma	p5_branch	p6_polar	p6_size
p6_flex	p6_h_doner	p6_h_acceptor	p6_pi_doner
p6_pi_acceptor	p6_polarisable	p6_sigma	p6_branch
activity			

Read the data set into MATLAB. Form a vector `activity` from the 61st column. Transform this vector as `y = activity.^3;`

(a) For `y`, find descriptive statistics: sample mean and variance, median, interquartile range, 0.05- and 0.9-sample quantiles. Plot the histogram with 20 bins for `y`.

(b) Plot data matrix `triazines(:,1:60)` using command `imagesc`.

(c) Conduct principal component analysis on `triazines(:,1:60)`. How much variability is contained in the first 5 principal components?

(d) Form vector `x` as the fourth coordinate of scores for triazines. Find the correlation between `x` and `y`.

2.28. **Principal Components for BUPA Liver Data.** The BUPA liver disorders database was discussed in Exercise 2.26.

(a) For the variables 1–6 conduct principal component analysis.

(b) What proportion of variance is contained in the first two principal components? Show the Pareto plot.

(d) Plot the scatterplot of scores for the first two principal components.

2.29. **Cell Circularity Data.** In the lab of Dr. Todd McDevitt at Georgia Tech, researchers wanted to elucidate differences between the "static" and "rotary" culture of embrionic bodies (EBs) that were formed under both conditions with equal starting cell densities. After 2, 4, and 7 days of culture, images of EBs were acquired using phase-contrast microscopy. Image analysis software was used to determine the circularity (defined as $4\pi(Area/Perimeter^2)$) of each EB imaged. A total of $n = 325$ EBs were analyzed from three separate plates for both static and rotary cultures at the three time points studied. The circularity measures were used to examine differences in the shape of EBs formed under the two conditions as well as differences in their variability.

The data set circ.dat|mat consists of six columns corresponding to six treatments (2d, rotary), (4d, rotary), (7d, rotary), (2d, static), (4d, static), and (7d, static). Note that this is not an example of multivariate data since the columns are freely permutable, but rather six univariate data sets.

(a) For rotation and static 2d measurements, plot back-to-back histograms (bihist.m) as well as boxplots.

(b) For static 7d measurements, graph by pie chart (pie) the proportion of EBs with circularity smaller than 0.75.

2.30. **Blowfly Count Time Series.** For the data in Example 2.8, it was postulated that a major transition in the dynamics of blowfly population size appeared to have occurred around day 400. This was attributed to biological evolution, and the whole series cannot be considered as representative of the same system. Divide the time series into two data segments with indices 1–200 and 201–361. Calculate and compare the autocorrelation functions for the two segments.

2.31. **Canadian Lynx Time Series.** The Canadian lynx data set is popular in time series modeling. The data set ⬛ lynx.mat|xlsx contains the annual record of the number of the Canadian lynx trapped in the Mackenzie River district of northwest Canada for the period 1821–1934.

(a) Plot this data. Find and plot autocorrelation function up to lag 40.

(b) Notice that autocorrelation function has local maximum every 9–10 years. What does this imply for the original time series?

MATLAB FILES AND DATA SETS USED IN THIS CHAPTER
http://statbook.gatech.edu/Ch2.Descriptive/

acorr.m, acov.m, ashton.m, balances.m, balancespca.m, bat.m, bihist.m, biomed.m, blowfliesTS.m, BUPAliver.m, carea.m, cats.m, cats1.m, circular.m, corkrao.m, corkraopca.m, crouxrouss.m, crouxrouss2.m, diversity.m, ecg.m, empiricalcdf.m, fisher1.m, fisheriris.m, grubbs.m, hist2d.m, histn.m, lightrev.m, limestone.m, lynx.m, mahalanobis.m, meanvarchange.m, multifat.m, multifatstat.m, mushrooms.m, myquantile.m, mytrimmean.m, piecharts.m, scattercloud.m, simple2comp.m, smoothhist2D.m, spikes.m, surveysUKUS.m, wdbcpca.m, weed.m

afm.dat|mat, amanita28.dat, ashton.dat, aspirin.dat, balance2.mat, bat.dat, blowflies.dat|mat, BUPA.dat|mat|xlsx, cats.mat, cellarea.dat|mat, circ.dat|mat, coburn.mat, cork.dat|mat, diabetes.xls, dmd.dat|mat|xls, fat.dat, leptoconops.mat|xlsx, light.dat, limestone.dat, lynx.mat|xlsx, QT.dat|mat, raman.dat|mat, spikes.dat, triazines.dat|mat|xlsx, tsdata.mat, wdbc.mat, weed.dat|mat|xlsx

CHAPTER REFERENCES

Anderson, E. (1935). The irises of the Gaspe Peninsula. *Bull. Am. Iris Soc.*, **59**, 2–5.

Andrews, F. D. (1972). Plots of high dimensional data. *Biometrics*, **28**, 125–136.

Atchley, W. R. (1974). A Quantitative Taxonomic Analysis of Leptoconops Torrens and L. Carteri (Diptera: Ceratopogonidae). *J. Med. Entomol.*, **11**, 4, 467–470.

Bowley, A. L. (1920). *Elements of Statistics*. Scribner, New York.

Brillinger, D. R. (2001). *Time Series: Data Analysis and Theory*. Classics Appl. Math. **36**, SIAM, pp 540.

Brockwell, P. J. and Davis, R. A. (2009). *Introduction to Time Series and Forecasting*. Springer, New York.

Brozek, J., Grande, F., Anderson, J., and Keys, A. (1963). Densitometric analysis of body composition: revision of some quantitative assumptions. *Ann. New York Acad. Sci.*, **110**, 113–140.

Chernoff, H. (1973). The use of faces to represent points in k-dimensional space graphically. *J. Am. Stat. Assoc.*, **68**, 361–366.

Christov, I., Dotsinsky, I. , Simova, I. , Prokopova, R., Trendafilova, E., and Naydenov, S. (2006). Dataset of manually measured QT intervals in the electrocardiogram. *BioMed. Eng. OnLine*, **5**, 31 doi:10.1186/1475-925X-5-31. The electronic version of this article can be found online at:
http://www.biomedical-engineering-online.com/content/5/1/31

David, H. A. (1998). Early sample measures of variability. *Stat. Sci.*, **13**, 4, 368–377.

Farrell B., Bulgakova M., Hodson-Tole E. F., Shah S., Gregor R. J., Prilutsky B. I. (2009). Short-term locomotor adaptations to denervation of lateral gastrocnemius and soleus muscles in the cat. In: Proceedings of the Society for Neuroscience meeting, 17–21 October 2009, Chicago.

Fisher, R. A. (1936). The use of multiple measurements in taxonomic problems. *Ann. Eugen.*, **7**, Pt. II, 179–188.

Gauss, C. F. (1816). Bestimmung der Genauigkeit der Beobachtungen. *Zeitschrift Astronomie*, **1**, 185–197.

Guillot, G., Loren N., and Rudemo, M. (2009). Spatial prediction of weed intensities from exact count data and image-based estimates, *Appl. Stat.*, **58**, 525–542.

Hirst, J. D., King, R. D., and Sternberg, M. J. (1994). Quantitative structure-activity relationships by neural networks and inductive logic programming. II. The inhibition of dihydrofolate reductase by triazines. *J. Comput. Aided Mol. Des.*, **8**, 4, 421–432.

Johnson, R. A. and Wichern, D. W. (1988). *Applied Multivariate Statistical Analysis*. Prentice-Hall, Upper Saddle River, NJ.

Johnson, R. W. (1996). Fitting percentage of body fat to simple body measurements. *J. Stat. Educ.*, **4**, 1.
http://www.amstat.org/publications/jse/v4n1/datasets.johnson.html

Kaufman, L. and Rock, I. (1962). The moon illusion. *Science*, **136**, 953–961.

Marshall, B. T., Long, M., Piper, J. W., Yago, T., McEver, R. P., and Zhu, C. (2003). Direct observation of catch bonds involving cell-adhesion molecules. *Nature*, **423**, 190–193.

Moors, J. J. A. (1988). A quantile alternative for kurtosis. *Statistician*, **37**, 25–32.

Morrison, D. F. (2004). *Multivariate Statistical Methods*, 4th ed. Duxbury Press, Pacific Grove, CA.

Nicholson, A. J. (1954). An outline of the dynamics of animal populations. *Aust. J. Zool.*, **2**, 1, 9–65.

Nightingale, F. (1858). Notes on matters affecting the health, efficiency, and hospital administration of the British army. Founded chiefly on the experience of the late war. Presented by request to the Secretary of State for War. Privately printed for Miss Nightingale, Harrison and Sons.

Penrose, K., Nelson, A., and Fisher, A. (1985). Generalized body composition prediction equation for men using simple measurement techniques (abstract). *Med. Sc. Sports Exerc.*, **17**, 2, 189.

Percy, M. E., Andrews, D. F., Thompson, M. W., and Opitz, J. M. (1981). Duchenne muscular dystrophy carrier detection using logistic discrimination: Serum creatine kinase and hemopexin in combination. *Am. J. Med. Genet.*, **8**, 4, 397–409.

Rao, C. R. (1948). Tests of significance in multivariate analysis. *Biometrika*, **35**, 58–79.

Shumway, R. H. and Stoffer, D. S. (2005). *Time Series Analysis and Its Applications.* Springer Texts in Statistics, Springer, New York.

Siri, W. E. (1961). Body composition from fluid spaces and density: Analysis of methods. In *Techniques for Measuring Body Composition*, eds. J. Brozek and A. Henzchel. National Academy of Sciences, Washington, 224–244.

Stevens, S. S. (1946). On the theory of scales of measurement. *Science*, **103**, 2684, 677–680. PMID 17750512.

Stoodley, K. (1984). *Applied and Computational Statistics, A First Course.* Ellis Horwood LTD, Chichester, UK, 229 pp.

Sturges, H. (1926). The choice of a class-interval. *J. Am. Stat. Assoc.*, **21**, 65–66.

Velleman, P. F. and Wilkinson, L. (1993). Nominal, ordinal, interval, and ratio typologies are misleading. *Am. Stat.*, **47**, 1, 65–72.

Chapter 3
Probability, Conditional Probability, and Bayes' Rule

Ultimately, in my extreme view, all reasoning reduces to probability calculations.

– Dennis Victor Lindley

3.1 Introduction

If statistics can be defined as the science that studies uncertainty, then probability is the branch of mathematics that quantifies it. One's intuition of chance and probability develops at a very early age (Piaget and Inhelder, 1976). However, the formal, precise definition of probability is elusive. There are several competing definitions for the probability of an event, but the most practical one uses its relative frequency in a potentially infinite series of experiments.

Probability is a part of all introductory statistics programs for a good reason: it is the theoretical foundation of statistics. The basic statistical concepts of random sample, sampling distributions, statistic, etc., require familiarity with probability to be understood, explained, and applied.

Probability is critical for the development of statistical concepts. Despite this fact, it will not be a focal point of this text. There is a dangerous temptation to dwell on urns, black and white balls, and combinatorics for so long that more important statistical concepts such as regression or ANOVA, fall into a zeitnot (a term used in chess to describe the pressure felt from having little remaining time).

Many students taking a university-level introductory statistics course have already been exposed to probability and statistics in their previous education. With this in mind, we will use this chapter as a survey of probability using a range of examples. The more important concepts of independence, conditioning, and Bayes' rule will be covered in more detail and repeatedly used later in various contexts. Ross (2009) is recommended for a review and comprehensive coverage.

3.2 Events and Probability

If an experiment has the potential to be repeated an infinite number of times, then the probability of an outcome can be defined through its relative frequency of appearing. For instance, if we rolled a die a number of times, we could construct a table showing how many times each face came up. These individual frequencies (n_i) can be transformed into proportions or relative frequencies, by dividing them by the total number of tosses $n : f_i = n_i/n$. If we were to see the outcome ⚅ in 53 out of 300 tosses, then that face's proportion, or relative frequency, would be $f_6 = 53/300 = 0.1767$. As more tosses are made, we would "expect" the proportion of ⚅ to stabilize around $\frac{1}{6}$. The "experiments" in the next example are often quoted in the literature on elementary probability.

Example 3.1. **Famous Coin Tosses.** Buffon tossed a coin 4,040 times. Heads appeared 2,048 times. K. Pearson tossed a coin 12,000 times and 24,000 times. The heads appeared 6,019 times and 12,012, respectively. For these three tosses the relative frequencies of heads are $2048/4040 \approx 0.5049$, $6019/12000 \approx 0.5016$, and $12012/24000 \approx 0.5005$.

What if the experiments cannot be repeated? For example, what is the probability that "Squiki" the guinea pig survives its first treatment by a particular drug? Or in the "experiment" of taking a statistics course this semester, what is the probability of getting an A? In such cases we can define probability *subjectively* as a measure of strength of belief. Here is another example.

Example 3.2. **Tutubalin's Problem.** In a desk drawer in the office of numismatist Mr. Jay Parrino there is a coin, a 1913 Liberty Head nickel, one of only five known. What is the probability that the coin is heads up? This is an example where equal levels of uncertainty for the two sides lead to the subjective answer of 1/2.

The *symmetry* of the experiment led to the classical definition of probability. An ideal die is symmetric. All sides are "equiprobable." When rolling a fair die, the probability of outcome ⚃ is a ratio of the number of *favorable* outcomes (in our example only one outcome is favorable) to the number of all possible outcomes, 1/6.[1]

Among several possible ways to define probability, three are outlined below.

Frequentist. An event's *probability* is the proportion of times that we would expect the event to occur if the experiment were repeated a large number of times.

Subjectivist. A subjective *probability* is an individual's degree of belief in the occurrence of an event.

Classical. An event's *probability* is the ratio of the number of favorable outcomes to possible outcomes in a (symmetric) experiment.

A formal definition of probability is axiomatic (Kolmogorov, 1933) and is a special case of measure theory in mathematics.

The events that are assigned probabilities can be considered as sets of outcomes. Table 3.1 uses a rolling die experiment to introduce the set notation among events:

To understand the probabilities in Table 3.1, consider a simple MATLAB code that will simulate rolling a fair die. A random number from (0, 1) is generated and multiplied by 6. This becomes a random number between 0 and 6. When this number is rounded up to the closest integer, the outcomes ⚀, ⚁, …, ⚃ are simulated. They are all equally likely. For example, the outcome ⚂ comes from the original number, which is in the range (3, 4), and this interval is one-sixth part of (0, 6). Formal justification of this fact requires the concept of uniform distribution, which will be covered in Chapter 5.

---

[1] This definition is criticized by philosophers because of the fallacy called a vicious circle in definition (*circulus vitiosus in definiendo*). One defines the notion of probability in terms of *equiprobable outcomes*.

Table 3.1 Notation in a rolling die experiment.

Term	Description	Example
Experiment	A phenomenon, action, or procedure where the outcomes are uncertain	A single roll of a balanced six-sided die
Sample space	Set of all possible outcomes in an experiment	$S = \{\boxed{\cdot}, \boxed{\cdot\cdot}, \boxed{\cdot\cdot\cdot}, \boxed{\cdot\cdot}, \boxed{\cdot\cdot\cdot}, \boxed{\cdots}\}$
Event	A collection of outcomes; a subset of S	$A = \{\boxed{\cdot\cdot\cdot}\}$ (3 dots show), $B = \{\boxed{\cdot\cdot\cdot}, \boxed{\cdot\cdot}, \boxed{\cdot\cdot\cdot}, \boxed{\cdots}\}$ (at least three dots show), $C = \{\boxed{\cdot}, \boxed{\cdot\cdot}\}$
Probability	A number between 0 and 1 assigned to an event	$P(A) = \frac{1}{6}$, $P(B) = \frac{4}{6} = \frac{2}{3}$, $P(C) = \frac{2}{6} = \frac{1}{3}$

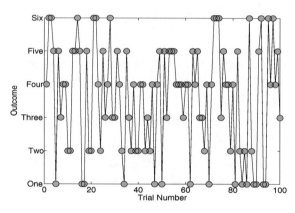

Fig. 3.1 MATLAB simulation of rolling a fair die. The first 100 outcomes $\{4,6,6,5,1,\ldots,4,5,3\}$ are shown.

The MATLAB code ◀ rollingdie1.m generates 50,000 outcomes and checks the proportion of those equal to 3, probA, those outcomes greater than or equal to 3, probB, and those smaller than 3, probC. The relative frequencies of these outcomes tend to their theoretical probabilities of 1/6, 2/3, and 1/3. Figure 3.1 shows the outcomes of the first 100 simulations described in the MATLAB code below.

```
% rollingdie1.m
outcomes = []; %keep outcomes here
M=50000 %# of rolls
for i= 1:M
  outcomes = [outcomes ceil( 6*rand )];
  % ceil(6*rand) rounds up (takes ceiling) of random
  % number from (0, 6), thus the outcomes 1, 2, 3, 4, 5, and 6
```

```
    % are equally likely
end
probA = sum((outcomes == 3))/M
    % probA = 0.1692
probB = sum((outcomes >= 3))/M
    % probB = 0.6693
probC = sum((outcomes < 3))/M
    % probC = 0.3307
```

Events in an experiment are sets containing the elementary outcomes, that is, distinctive outcomes of the experiment. Among all events in an experiment, two are special: a sure event and an impossible event. A *sure event* occurs *every time* an experiment is repeated and has a probability of 1. It consists of all outcomes and is equal to the sample space of the experiment, S. An *impossible event never* occurs when an experiment is performed and is usually denoted as $\emptyset$. It contains no elementary outcomes and its probability is 0.

For any event A, the probability that A will occur is a number between 0 and 1, inclusive:

$$0 \leq \mathbb{P}(A) \leq 1.$$

Also,

$$\mathbb{P}(\emptyset) = 0, \text{ and } \mathbb{P}(S) = 1.$$

The *intersection* $A \cap B$ of two events A and B occurs if both events A *and* B occur. The key word in the definition of the intersection is *and*. The intersection of two events $A \cap B$ is often written as a product AB. We will use both the $\cap$ and product notations.

The product of the events translates into the product of their probabilities only if the events are independent, meaning the outcome of one does not affect the outcome of the other. We will see later that relationship $\mathbb{P}(AB) = \mathbb{P}(A)\mathbb{P}(B)$ is the definition of the independence of events A and B.

Events are said to be *mutually exclusive* if they have no common elementary outcomes. In other words, it is impossible for both events to occur in a single trial of the experiment. For mutually exclusive events, $\mathbb{P}(A \cdot B) = \mathbb{P}(\emptyset) = 0$.

In the die-toss example, events $A = \{\boxed{\cdot\cdot}\}$ and $B = \{\boxed{\cdot\cdot}, \boxed{::}, \boxed{:\cdot:}, \boxed{::}\}$ are not mutually exclusive, since the elementary outcome $\{\boxed{\cdot\cdot}\}$ belongs to both of them. The events $A = \{\boxed{\cdot\cdot}\}$ and $C = \{\boxed{\cdot}, \boxed{\cdot\cdot}\}$ are mutually exclusive.

The *union* $A \cup B$ of two events A and B occurs if at least one of the events A or B occurs. The key word in the definition of the union is *or*.

For mutually exclusive events, the probability that at least one of them occurs is

$$\mathbb{P}(A \cup C) = \mathbb{P}(A) + \mathbb{P}(C).$$

For example, if the probability of event $A = \{\boxed{\cdot\cdot}\}$ is 1/6, and the probability of the event $C = \{\boxed{\cdot}, \boxed{\cdot\cdot}\}$ is 1/3, then the probability of A or C is

$$\mathbb{P}(A \cup C) = \mathbb{P}(A) + \mathbb{P}(C) = 1/6 + 1/3 = 1/2.$$

The *additivity* property is valid for any number of mutually exclusive events $A_1, A_2, A_3, \ldots$:

$$\mathbb{P}(A_1 \cup A_2 \cup A_3 \cup \ldots) = \mathbb{P}(A_1) + \mathbb{P}(A_2) + \mathbb{P}(A_3) + \ldots.$$

What is $\mathbb{P}(A \cup B)$ if events A and B are not mutually exclusive?

For any two events A and B, the probability that either A or B will occur is given by the *inclusion-exclusion* rule:

$$\mathbb{P}(A \cup B) = \mathbb{P}(A) + \mathbb{P}(B) - \mathbb{P}(A \cdot B). \tag{3.1}$$

If events A and B are exclusive, then $\mathbb{P}(A \cdot B) = 0$, and we get the familiar result $\mathbb{P}(A \cup B) = \mathbb{P}(A) + \mathbb{P}(B)$.

The inclusion-exclusion rule can be generalized to unions of an arbitrary number of events. For example, for three events $A, B,$ and C, the rule is

$$\begin{aligned}
\mathbb{P}(A \cup B \cup C) = {} & \mathbb{P}(A) + \mathbb{P}(B) + \mathbb{P}(C) \\
& - \mathbb{P}(A \cdot B) - \mathbb{P}(A \cdot C) - \mathbb{P}(B \cdot C) \\
& + \mathbb{P}(A \cdot B \cdot C).
\end{aligned} \tag{3.2}$$

For every event defined on a space of elementary outcomes, $\mathcal{S}$, we can define a counterpart event called its *complement*. The complement A^c of an event A consists of all outcomes that are in $\mathcal{S}$ but are not in A. The key word in the definition of a complement is *not*. In our example, A^c consists of the outcomes $\{\boxed{\cdot}, \boxed{\cdot\cdot}, \boxed{::}, \boxed{:\cdot:}, \boxed{::}\}$.

Events A and A^c are mutually exclusive by definition. Consequently,

$$\mathbb{P}(A \cup A^c) = \mathbb{P}(A) + \mathbb{P}(A^c).$$

Since we also know from its definition that A^c includes all outcomes in the sample space, $\mathcal{S}$, that are not in A, so that $\mathcal{S} = A \cup A^c$, it follows that

$$\mathbb{P}(A) + \mathbb{P}(A^c) = \mathbb{P}(\mathcal{S}) = 1.$$

For any pair of complementary events A and A^c,
$$\mathbb{P}(A) + \mathbb{P}(A^c) = 1, \quad \mathbb{P}(A) = 1 - \mathbb{P}(A^c), \text{ and } \quad \mathbb{P}(A^c) = 1 - \mathbb{P}(A).$$

These equations simplify the solutions of some probability problems. If $\mathbb{P}(A^c)$ is easier to calculate than $\mathbb{P}(A)$, then the equations above let us obtain $\mathbb{P}(A)$ indirectly.

Having defined the complement, we can prove (3.1). The argument is easy if event B is written as a union of two exclusive events, $B = (B \cap A^c) \cup (A \cap B)$. From this and the additivity property,

$$\mathbb{P}(B \cap A^c) = \mathbb{P}(B) - \mathbb{P}(A \cap B).$$

Since $A \cup B$ is equal to a union of exclusive events, $A \cup B = A \cup (B \cap A^c)$, by the additivity property of probability we obtain

$$\mathbb{P}(A \cup B) = \mathbb{P}(A) + \mathbb{P}(B \cap A^c) = \mathbb{P}(A) + \mathbb{P}(B) - \mathbb{P}(A \cap B).$$

This and some other probability properties are summarized in the following table.

Property	Notation
If event $\mathcal{S}$ will *always* occur, its probability is 1.	$\mathbb{P}(\mathcal{S}) = 1$
If event $\varnothing$ will *never* occur, its probability is 0.	$\mathbb{P}(\varnothing) = 0$
Probabilities are always between 0 and 1, inclusive.	$0 \leq \mathbb{P}(A) \leq 1$
If $A, B, C, \ldots$ are all mutually exclusive, then $\mathbb{P}(A \cup B \cup C \ldots)$ can be found by addition.	$\mathbb{P}(A \cup B \cup C \ldots) = \mathbb{P}(A) + \mathbb{P}(B) + \mathbb{P}(C) + \cdots$
The general *addition rule* for probabilities.	$\mathbb{P}(A \cup B) = \mathbb{P}(A) + \mathbb{P}(B) - \mathbb{P}(A \cdot B)$
Since A and A^c are mutually exclusive, and between them include all outcomes from $\mathcal{S}$, $\mathbb{P}(A \cup A^c)$ is 1.	$\mathbb{P}(A \cup A^c) = \mathbb{P}(A) + \mathbb{P}(A^c) = \mathbb{P}(\mathcal{S}) = 1$, and $\mathbb{P}(A^c) = 1 - \mathbb{P}(A)$

Of particular importance in assessing the probability of composite events are De Morgan's laws, which are simple algebraic relationships between events. The laws are named after Augustus De Morgan, a nineteenth-century British mathematician and logician.

For any set of n events $A_1, A_2, \ldots, A_n$,

$$(A_1 \cup A_2 \cup \cdots \cup A_n)^c = A_1^c \cap A_2^c \cap \cdots \cap A_n^c,$$
$$(A_1 \cap A_2 \cap \cdots \cap A_n)^c = A_1^c \cup A_2^c \cup \cdots \cup A_n^c.$$

De Morgan's laws can be readily demonstrated using Venn diagrams, as discussed in Section 3.4.

The following example shows how to apply De Morgan's laws.

Example 3.3. **Nanotubules and Cancer Cells.** One technique of killing cancer cells involves inserting microscopic synthetic rods, called carbon nanotubules, into the cell. When the rods are exposed to near-infrared light from a laser, they heat up, killing the cell, while cells without rods are left unscathed (Wong et al., 2005). Suppose that five nanotubules are inserted in a single cancer cell. Independently of each other, they become exposed to near-infrared light with probabilities 0.2, 0.4, 0.3, 0.6, and 0.5. What is the probability that the cell will be killed?

Let B be an event where a cell is killed and A_i an event where the ith nanotubule kills the cell. The cell is killed if $A_1 \cup A_2 \cup \cdots \cup A_5$ happens. In other words, the cell is killed if nanotubule 1 kills the cell, or nanotubule 2 kills the cell, etc. We consider the event where the cell is not killed and apply De Morgan's laws. De Morgan's laws state that $A_1^c \cup A_2^c \cup \cdots \cup A_n^c = (A_1 \cap A_2 \cap \cdots \cap A_n)^c$,

$$\mathbb{P}(B) = 1 - \mathbb{P}(B^c) = 1 - \mathbb{P}((A_1 \cup A_2 \cup \cdots \cup A_5)^c) = 1 - \mathbb{P}(A_1^c \cap A_2^c \cap \cdots \cap A_5^c)$$
$$= 1 - (1 - 0.2)(1 - 0.4)(1 - 0.3)(1 - 0.6)(1 - 0.5) = 0.9328.$$

Thus, the cancer cell will be killed with a probability of 0.9328.

Example 3.4. **Bonferroni Inequality.** As an example of the algebra of events and basic rules of probability, we derive the Bonferroni inequality. It will be revisited later in the text when calculating the significance level in simultaneous testing of multiple hypotheses (page 415).

The Bonferroni inequality states that for arbitrary events $A_1, A_2, \ldots, A_n$,

$$\mathbb{P}(A_1 \cap A_2 \cap \cdots \cap A_n) \geq \mathbb{P}(A_1) + \mathbb{P}(A_2) + \cdots + \mathbb{P}(A_n) - n + 1. \quad (3.3)$$

Start with n events A_i, $i = 1, \ldots, n$ and the event $A_1^c \cup A_2^c \cup \cdots \cup A_n^c$. The probability of any union of events is never larger than the sum of probabilities of individual events:

$$\mathbb{P}(A_1^c \cup A_2^c \cup \cdots \cup A_n^c) \leq \mathbb{P}(A_1^c) + \mathbb{P}(A_2^c) + \cdots + \mathbb{P}(A_n^c).$$

De Morgan's laws state that $A_1^c \cup A_2^c \cup \cdots \cup A_n^c = (A_1 \cap A_2 \cap \cdots \cap A_n)^c$ and

$$1 - \mathbb{P}((A_1 \cap A_2 \cap \cdots \cap A_n)^c) \leq (1 - \mathbb{P}(A_1)) + (1 - \mathbb{P}(A_2)) + \cdots + (1 - \mathbb{P}(A_n)),$$

leading to the inequality in (3.3).

Circuits. The application of basic probability rules involving unions, intersections, and complements of events can be quite useful. An example is the application in the reliability of a complex system consisting of many components that work independently. If a complex system can be expressed as a configuration of simple elements that are linked in a "serial" or "parallel" fashion, the reliability of such a system can be calculated by knowing the reliabilities of its constituents.

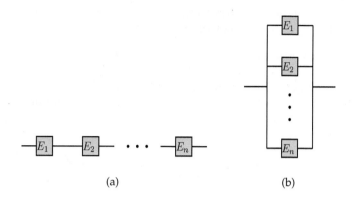

Fig. 3.2 (a) Serial connection modeled as $E_1 \cap E_2 \cap \cdots \cap E_n$. (b) Parallel connection modeled as $E_1 \cup E_2 \cup \cdots \cup E_n$.

Let a system S consist of n constituent elements $E_1, E_2, \ldots, E_n$ that can be interconnected in either a serial or a parallel fashion (Fig. 3.2). Suppose that elements E_i work in time interval T with probability p_i and fail with probability $q_i = 1 - p_i$, $i = 1, \ldots, n$. The following table gives the probabilities of working for elements in S:

Connection	Notation	Works with probability	Fails with probability
Serial	$E_1 \cap E_2 \cap \cdots \cap E_n$	$p_1 p_2 \ldots p_n$	$1 - p_1 p_2 \ldots p_n$
Parallel	$E_1 \cup E_2 \cup \cdots \cup E_n$	$1 - q_1 q_2 \ldots q_n$	$q_1 q_2 \ldots q_n$

If the system has both serial and parallel connections, then the probability of the system working can be found by the subsequent application of the probabilities for the union and intersection of events. Here is an example.

Example 3.5. **Circuit.** A complex system S is defined via

$$S = E_1 \cap [(E_2 \cap E_3) \cup (E_4 \cap (E_5 \cup E_6))] \cap E_7,$$

where the unreliable components E_i, $i = 1, \ldots, 7$ work and fail independently. The system is depicted in Figure 3.3. The components are operational in some fixed time interval $[0, T]$ with probabilities given in the following table.

Component	E_1	E_2	E_3	E_4	E_5	E_6	E_7
Probability of functioning well	0.9	0.5	0.3	0.1	0.4	0.5	0.8

We will find the probability that system S will work in $[0, T]$ first analytically, and then find an approximation by simulating the circuit in MATLAB and WinBUGS.

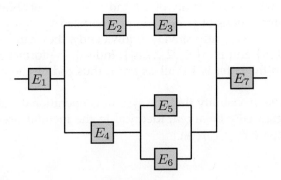

Fig. 3.3 Circuit $E_1 \cap [(E_2 \cap E_3) \cup (E_4 \cap (E_5 \cup E_6))] \cap E_7$.

To find the probability that system S works/fails, it is useful to create a table with probabilities $p_i = \mathbb{P}(\text{component } E_i \text{ works})$ and their complements $q_i = 1 - p_i$, $i = 1, \ldots, 7$:

Component	E_1 E_2 E_3 E_4 E_5 E_6 E_7
p_is	0.9 0.5 0.3 0.1 0.4 0.5 0.8
q_is	0.1 0.5 0.7 0.9 0.6 0.5 0.2

Next, we calculate step-by-step the probabilities of subsystems that ultimately add up to the final system. For example, we calculate the probability of working/failing for $S_1 = E_2 \cap E_3$, then $S_2 = E_5 \cup E_6$, then $S_3 = E_4 \cap S_2$, then $S_4 = S_1 \cup S_3$, and finally $S = E_1 \cap S_4 \cap E_7$.

Component	Probability of working	Probability of failing
$S_1 = E_2 \cap E_3$	$p_{s1} = 0.5 \cdot 0.3 = 0.15$	$q_{s1} = 1 - 0.15 = 0.85$
$S_2 = E_5 \cup E_6$	$p_{s2} = 1 - 0.3 = 0.7$	$q_{s2} = 0.6 \cdot 0.5 = 0.3$
$S_3 = E_4 \cap S_2$	$p_{s3} = 0.1 \cdot 0.7 = 0.07$	$q_{s3} = 1 - 0.07 = 0.93$
$S_4 = S_1 \cup S_3$	$p_{s4} = 1 - 0.7905 = 0.2095$	$q_{s4} = 0.85 \cdot 0.93 = 0.7905$
$S = E_1 \cap S_4 \cap E_7$	$p_S = 0.9 \cdot 0.2095 \cdot 0.8 = \mathbf{0.15084}$	$q_S = 1 - 0.15084 = 0.84916$

Thus the probability that the system will work in the time interval $[0, T]$ is 0.15084.

The MATLAB code that approximates this probability uses a random number generator to simulate the case where the simple elements "work" and binary operations to simulate intersections and unions. For example, the fact that e_1 is functioning well (working) with a probability of 0.9 is modeled by e1 = rand < 0.9. Note that the left-hand side of the equation e1 = rand < 0.9 is a logical expression that takes values TRUE (numerical value 1) and FALSE (numerical value 0). Given that the event $\{\text{rand} < 0.9\}$ is true 90% of the time, the value e1 represents the status of component E_1.

This will be 0 with a probability of 0.1 and 1 with a probability of 0.9. The unions and intersections of $e_1, e_2, \ldots, e_n$ are modeled as $(e_1 + e_2 + \cdots + e_n > 0)$ and $e_1 * e_2 * \cdots * e_n$, respectively. Equivalently, they can be modeled as $\max\{e1, e2, \ldots, e_n\}$ and $\min\{e1, e2, \ldots, e_n\}$. Indeed, the former is 1 if at least one e_i is 1, and the latter is 1 if all e_is are 1, thus coding the union and the intersection.

To assess the probability that the system is operational, subsystems are formed and gradually enlarged, identical to the method used to find the analytic solution (circuit.m).

```
% circuit.m
M=1000000;
s = 0;
for i = 1:M
e1 = rand < 0.9; e2 = rand < 0.5; e3 = rand < 0.3;
e4 = rand < 0.1; e5 = rand < 0.4; e6 = rand < 0.5;
e7 = rand < 0.8;
% ================
s1 = min(e2,e3);   %   or s1 = e2*e3;
s2 = max(e5,e6);   %   or s2= e5+e6>0;
s3 = min(e4,s2);   %   or s3 = e4*s2;
s4 = max(s1,s3);   %   or s4 = s1+s3 > 0;
st = min([e1;s4;e7]); %   or st=e1*s4*e7;
s = s + st;
end
works = s/M
fails = 1 - works

% works = 0.150944
% fails = 0.849056
```

Next, we repeat this simulation in WinBUGS. There are many differences between MATLAB and WinBUGS that go beyond the differences in the syntax. In MATLAB, we had an explicit loop to generate 10^6 runs; in WinBUGS this is done via the Model>Update tool and is not a part of the code. Also, the e_is in MATLAB are 0 and 1; in WinBUGS they are 1 and 2, since the outcomes are realizations of a categorical discrete random variable dcat, and this variable is coded by nonnegative integers: 1, 2, 3, ... For this reason we adjusted the probability of a system working as ps <- s - 1.

```
# circuit1.odc
model  {
for (i in 1:7){
e[i] ~ dcat(p[i,])
   }
s1 <- min(e[2],e[3])
s2 <- max(e[5],e[6])
s3 <- min(e[4],s2)
s4 <- max(s1,s3)
```

```
s <- min( min(e[1],s4) , e[7] )
ps <- s-1
}
DATA IN:
list(
 p = structure(.Data =
 c(0.1,0.9,    0.5,0.5,
   0.7,0.3,    0.9,0.1,
   0.6,0.4,    0.5,0.5,
   0.2,0.8) , .Dim = c(7,2) ) )

INITS NONE,  just 'gen inits'
```

The result of the simulations is close to the theoretical value.

	mean	sd	MC error	val2.5pc	median	val97.5pc	start	sample
ps	0.1508	0.3578	3.528E–4	0.0	0.0	1.0	10001	1000000

This is the first WinBUGS program in the text, and the reader is advised to consult Chapter 19, which discusses how communication with the WinBUGS program is structured and carried out. This comment has the mark "dangerous bend" because many students initially find the BUGS interface and programming intimidating.

3.3 Odds

Odds are alternative measures for the likelihood of events. If an event A has a probability $\mathbb{P}(A)$, then the odds of A are defined as

$$Odds(A) = \frac{\mathbb{P}(A)}{\mathbb{P}(A^c)}, \qquad \mathbb{P}(A) = \frac{Odds(A)}{Odds(A) + 1}.$$

From the classical definition of probability $\mathbb{P}(A) = \frac{\text{\# of favorable for A}}{\text{\# in the sample space}} = n_A/n$, the odds of A are defined as $Odds(A) = n_A/(n - n_A)$. For instance, the odds of event $A = \{\boxdot\}$ are $1/(6 - 1)$, one in five.

In economic decision theory, epidemiology, game theory, and some other areas, odds and odds ratios are preferred measures of quantifying and comparing events.

Example 3.6. **Odds for Circuit.** The odds that the circuit S in Example 3.5 is working are 17.76%, since $\mathbb{P}(S) = 0.15084$ and $Odds(S) = 0.15084/(1 - 0.15084) = 0.17763$.

3.4 Venn Diagrams*

Venn diagrams help in graphically presenting the algebra of events and in determining the probability of composite events involving unions, intersections, and complements. The diagrams are named after John Venn, the English logician who introduced the diagrams in his 1880 paper (Venn, 1880).

Venn diagrams connect sets and events in a graphical way – the events are represented as circles (squares, rectangles) and the notions of unions, intersections, complements, exclusiveness, implication, etc., among the events translate directly to the corresponding relations among the geometric areas. Exclusive events are represented by nonoverlapping circles, while the notion of causality among the events translates to the subset relation. The geometric areas representing the events are plotted in a large rectangle representing the sample space (sure event).

Panels (a) and (b) in Figure 3.4 show the union and intersection of events A and B, while panel (c) shows the complement of event A.

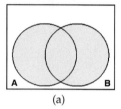 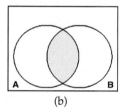 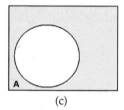

Fig. 3.4 (a) Union and (b) intersection of events A and B and (c) complement of event A.

It is possible to define more exotic operations with events. For example, the difference between events A and B, denoted as $A \backslash B$, is shown in Figure 3.5a. It is obvious from the diagram that $A \backslash B = A \cap B^c$. The symmetric difference (or exclusive union) of events A and B, denoted as $A \Delta B$, is an event in which either A or B happens, but not both (Fig. 3.5b). From the Venn diagram it is easy to see that $A \Delta B = (A \cap B^c) \cup (B \cap A^c) = (A \backslash B) \cup (B \backslash A)$.

Sometimes, the evidence for more complex algebraic relations between events can be established by Venn diagrams. Usually, a Venn diagram of the left-hand side in a relation is compared with the Venn diagram of the right-hand side, and if the resulting sets coincide, we have a "proof."

For example, one of De Morgan's laws for three events, $(A \cup B \cup C)^c = A^c \cap B^c \cap C^c$, can be demonstrated by Venn diagrams. Panel (a) in Figure 3.6 shows $A \cup B \cup C$, while panel (b) shows $A^c \cap B^c \cap C^c$. It is obvious that the sets in the two panels are complementary and De Morgan's law is "demonstrated."

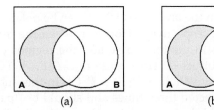

Fig. 3.5 Difference $A \backslash B$ and symmetric difference $A \Delta B$.

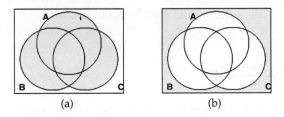

Fig. 3.6 De Morgan's Law: $(A \cup B \cup C)^c = A^c \cap B^c \cap C^c$.

Likewise, if we want to demonstrate the distributive law $A \cup (B \cap C) = (A \cup B) \cap (A \cup C)$, the Venn diagram argument is shown in Figure 3.7a–c. The set $A \cup (B \cap C)$ is shown in panel (a). Panels (b) and (c) show sets $A \cup B$ and $A \cup C$, respectively. Their intersection coincides with the set in panel (a).

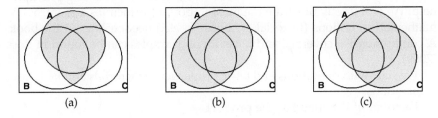

Fig. 3.7 Distributive law among events, $A \cup (B \cap C) = (A \cup B) \cap (A \cup C)$.

Proofs of this kind can be formalized with the help of mathematical logic and tautologies.

In addition to algebraic relations among events, Venn diagrams can help in finding the probability of the complex algebraic composition of events. The probability can be informally connected with the area of a set in a Venn diagram, and this connection is extremely useful. For example, for the result in (3.2), $\mathbb{P}(A \cup B \cup C) = \mathbb{P}(A) + \mathbb{P}(B) + \mathbb{P}(C) - \mathbb{P}(AB) - \mathbb{P}(AC) - \mathbb{P}(BC) + \mathbb{P}(ABC)$, an informal "proof" based on areas in a Venn diagram is simple and intuitive. The argument is as follows: if the probability is thought of as an area, then the area of $A \cup B \cup C$ can be obtained by adding the areas of A, B, and C, respectively. However, when the three areas are added, there is an excess in the total area, and the regions counted multiple times should be subtracted. Thus areas of $A \cap B$, $A \cap C$, and $B \cap C$ are subtracted from the sum $\mathbb{P}(A) + \mathbb{P}(B) + \mathbb{P}(C)$. In this subtraction, the area of $A \cap B \cap C$ is subtracted three times and should be "patched back." Alternatively, one could think about *painting* the set $A \cup B \cup C$ with a *single layer of paint*, and the total *amount of paint* used is the probability. Of course, the amount of paint needed to paint the universal event S is 1. Although very informal, such a discursion may be quite useful.

3.5 Counting Principles*

Many experiments can be modeled by a sample space with a finite number of equally likely outcomes. We discussed the experiment of rolling a die, in which the sample space had six equally likely outcomes. In finding the probability of an event defined on this sample space, we divided the number of outcomes favorable to A by 6. For example, the event $A = \{\boxdot, \boxdot, \boxdot\}$ (the number is even) has a probability of $3/6=1/2$. But what if 10 dice are simultaneously rolled and we were interested in the probability that the sum of numbers will be equal to 55? The problem here is to count how many of $6^{10} = 60,466,176$ possible equally likely outcomes produce the sum of 55, and a simple inspection of the sample space applicable for one or two dice is not feasible. In situations like this, combinatorial and counting principles help. We will briefly illustrate the most important principles and introduce mathematical notions (factorial, n-choose-k, etc.) needed later in the book. A comprehensive coverage and a wealth of examples can be found in Ross (2009).

We start with definitions and basic properties of factorials and n-choose-k operations.

Factorial $n!$ is defined as the product

$$n! = n(n-1)(n-2)\ldots 2 \cdot 1 = \prod_{i=1}^{n} i.$$

For example, $5! = 5 \cdot 4 \cdot 3 \cdot 2 \cdot 1 = 120$. By definition $0! = 1$.

An n-choose-k operation (or binomial coefficient) is defined as follows:

$$\binom{n}{k} = \frac{n(n-1)\ldots(n-k+1)}{k!} = \frac{n!}{(n-k)!k!}.$$

As the name indicates, n-choose-k is the number of possible subsets of size k from a set of n elements. For example, the number of different committees of size 3 formed from a group of 8 students is $\binom{8}{3} = \frac{8 \times 7 \times 6}{3 \times 2 \times 1} = 56$. In MATLAB the command for $\binom{n}{k}$ is `nchoosek(n,k)`. For example, `nchoosek(8,3)` results in 56.

The following properties follow directly from the definition of $\binom{n}{k}$:

$$\binom{n}{k} = \binom{n}{n-k},$$

$$\binom{n}{0} = 1 \quad \text{and} \quad \binom{n}{1} = n,$$

$$\binom{n}{k} + \binom{n}{k+1} = \binom{n+1}{k+1}.$$

Fundamental Counting Principle . If an experiment consists of k actions, and the ith action can be performed in n_i different ways, then the whole experiment can be performed in $n_1 \times n_2 \times \cdots \times n_k$ different ways. This is called the *multiplication counting rule* or *fundamental counting principle*.

Example 3.7. Item Inspection. Out of 15 items, 4 are defective. The items are inspected one by one. What is the probability that the ninth item was the last defective one?

Consider the arrangement of 11 conforming and 4 defective items. The number of all possible arrangements is $\binom{15}{4} = \binom{15}{11} = 1316$, as one chooses 4 places out of 15 to place defective items or, equivalently, 11 places out of 15 to place conforming items.

The number of favorable outcomes can be found by the multiplication rule. Favorable outcomes are defined as follows: among the first-selected eight items three are defective, the ninth position is occupied by a defective item, and none of the remaining six items is defective:

$$\binom{8}{3} \times 1 \times 1 = 56.$$

Note that the number of ways in which a defective item falls at the ninth position, and the number of ways where six fair items occupy positions 10 to 15, are 1 each. Thus, the required probability is $56/1365 = 0.041$.

There is also an *addition counting rule* that mimics the additive property of probability: If k events are exclusive and have $n_1, n_2, \ldots, n_k$ outcomes, then their union has $n_1 + n_2 + \cdots + n_k$ outcomes. If the events are not exclusive, this rule is known as the *inclusion–exclusion principle*. For instance, if two events are arbitrary, the inclusion–exclusion rule count for outcomes in their union is $n_1 + n_2 - n_{12}$, where n_{12} is the number of common outcomes. For three events the inclusion-exclusion rule is $n_1 + n_2 + n_3 - n_{12} - n_{13} - n_{23} + n_{123}$; see (3.2).

If the population has N subjects and a sample of size n is needed, then Table 3.2 summarizes the number of possible samples, given the sampling policy and importance of ordering.

We first introduce the necessary notation. When the order is important, the samples are called *variations* or *permutations*. One can think about variations as words in an alphabet, since for words the order of letters is important. By the fundamental counting principle, the number of variations with repetitions of N elements of length n is $\overline{V}_N^n = N^n$, since each of n places can be selected in N ways. The number of variations without repetition of N elements of length n is $V_N^n = N \times (N-1) \times \cdots \times (N-n+1) = N^{(n)}$, $n \leq N$. Note that $V_N^N = N!$ is the number of permutations of N distinct elements.

In combinations, the order in the sample is not important. If there is no repetition of elements, then $C_N^n = \binom{N}{n}$. If the repetition is possible, then $\overline{C}_N^n = \binom{N+n-1}{n}$.

Table 3.2 Number of variations/combinations when the selection of n from N elements is done with/without the repetition.

	Order important (variations or permutations)	Order not important (combinations)
Sampling w/ repetition	$\overline{V}_N^n = N^n$	$\overline{C}_N^n = \binom{N+n-1}{n}$
Sampling w/o repetition	$V_N^n = N(N-1)\ldots(N-n+1)$	$C_N^n = \binom{N}{n},\ n \leq N$

The number of permutations of N distinctive elements is $N!$, but if among N elements there are only k different elements, n_1 of type 1, n_2 of type 2, ..., n_k of type k, $(n_1 + n_2 + \cdots + n_k = N)$, then the number of different permutations is

$$\binom{N}{n_1, n_2, \ldots n_k} = \frac{N!}{n_1! n_2! \cdots n_k!}. \tag{3.4}$$

The number in (3.4) is also called the multinomial coefficient.

Example 3.8. **Probability by Counting.** What is the probability that in a six-digit license plate of a randomly selected car

(a) All digits will be different?
(b) Exactly two digits will be equal?
(c) At least three digits will be different?
(d) There will be exactly two pairs of equal digits?

We assume that any digit from 0 to 9 can be at any of the six positions in the six-digit plate number.

This example is solved by using the classical definition of probability. For each event in (a)–(d), the number of favorable outcomes will be divided by the number of possible outcomes. The number of all possible outcomes is common, $\overline{V}_{10}^6 = 10^6$.

(a) To find the number of favorable outcomes for the event where all digits are different, consider forming a six-digit number position by position. For the first position there are ten digits available, for the second nine (the digit used in the first position is eliminated as a choice for the second position), for the third eight, etc., for the last five. By the fundamental counting principle, the number of all favorable outcomes is the product $10 \times 9 \times 8 \times 7 \times 6 \times 5 = 10^{(6)} = V_{10}^6$, and the probability is

$$\frac{10^{(6)}}{10^6} = 0.1512.$$

(b) Out of ten digits choose one and place it on any two positions out of six available. This can be done in $10 \times \binom{6}{2} = 150$ ways. The remaining four positions could be chosen in $9 \times 8 \times 7 \times 6$ ways. Thus, the number of favorable outcomes is $150 \times 9 \times 8 \times 7 \times 6$, and the required probability is 0.4536.

(c) The opposite event for "at least three digits are different" is "all digits are the same" or "exactly two digits are the same," and we will find its probability first. The number of cases where all digits are the same is ten, while the number of cases where there are exactly two different digits is $\binom{10}{2} \times (2^6 - 2)$. Digits a and b can be selected in $\binom{10}{2}$ ways. Given the fixed selection, there are $2^6 - 2$ words in alphabet $\{a,b\}$ of length 6 where the words $aaaaaa$ and $bbbbbb$ are excluded. The probability of "at least three different digits" is

$$1 - \frac{10 + \binom{10}{2}(2^6 - 2)}{10^6} = 1 - 0.0028 = 0.9972.$$

(d) From ten digits first select two for the two pairs, and then an additional two digits for the remaining two places. There are four different digits: a,b for the two pairs and c,d for the remaining two places. The selection can be done in $\binom{10}{2} \times \binom{8}{2}$ ways. Once selected, the digits can be arranged in $\binom{6}{2,2,1,1} = \frac{6!}{2!2!1!1!}$ ways, by using permutations with repetitions as in (3.4). Thus, the probability is

$$\frac{\binom{10}{2} \times \binom{8}{2} \times \frac{6!}{2!\,2!\,1!\,1!}}{10^6} = 0.2268.$$

The following two important equations for probability calculation are a direct consequence of the combinatorial properties discussed in this section. They will be used later in the text when discussing the binomial and hypergeometric distributions and their generalizations.

Multinomial and Multihypergeometric Trials. Suppose that an experiment can result in m possible outcomes, $A_1, A_2, \ldots, A_m$, that have probabilities $\mathbb{P}(A_1) = p_1, \ldots, \mathbb{P}(A_m) = p_m$, $p_1 + \cdots + p_m = 1$. If the experiment is independently repeated n times, then the probability that event A_1 will appear exactly n_1 times, A_2 exactly n_2 times, $\ldots$, A_m exactly n_m times $(n_1 + \cdots + n_m = n)$ is

$$\binom{n}{n_1, n_2, \ldots n_m} p_1^{n_1} p_2^{n_2} \cdots p_m^{n_m}.$$

If a finite population of size m has k_1 subjects of type 1, $\ldots$, k_p subjects of type p, $(k_1 + \cdots + k_p = m)$, and n subjects are sampled at random, then the probability that x_1 will be of type 1, $\ldots$, x_p will be of type p $(x_1 + \cdots + x_p = n)$, is

$$\frac{\binom{k_1}{x_1}\binom{k_2}{x_2} \cdots \binom{k_l}{x_l}}{\binom{m}{n}}.$$

3.6 Conditional Probability and Independence of Events

Important contemporary applications of probability in bioengineering, medical diagnostics, systems biology, bioinformatics, etc., concern modeling and prediction of causal relationships in complex systems. The methodologies include influence diagrams, Bayesian networks, Granger causality, and related methods for which the notions of conditional probability, causality, and independence are fundamental. In this section, we discuss conditional probabilities and independence.

3.6.1 Conditioning and Product Rule

A *conditional probability* is the probability of one event if we have information that another event, typically from the same sample space, has occurred. In the die-toss example, the probability of event $A = \{\boxed{\because}\}$ is $\mathbb{P}(A) = \frac{1}{6}$. But what if we knew that event $B = \{\boxdot, \boxdot, \boxdot, \boxdot\}$ occurred? There are only four possible outcomes, only one of which is favorable for A. Thus, the probability of A given B is $\frac{1}{4}$. The conditional probability of A given B is denoted as $\mathbb{P}(A|B)$.

In general, the conditional probability of an event A given that B has occurred is equal to the probability of their intersection $\mathbb{P}(AB)$ divided by the probability of the event that we are conditioning upon, $\mathbb{P}(B)$. Of course, event B has to have a positive probability, $\mathbb{P}(B) > 0$, because conditioning upon an event of zero probability is equivalent to the indeterminacy $0/0$, as $0 \leq \mathbb{P}(AB) \leq \mathbb{P}(B)$:

$$\mathbb{P}(A|B) = \frac{\mathbb{P}(A \cdot B)}{\mathbb{P}(B)}, \text{ for } \mathbb{P}(B) > 0.$$

Figure 3.8 gives a graphical description of the conditional probability $\mathbb{P}(A|B)$. Once event A is conditioned by B, B "becomes the sample space" and B's Venn diagram expands by a factor of $\frac{1}{\mathbb{P}(B)}$. The intersection AB in the expanded B becomes event $A|B$.

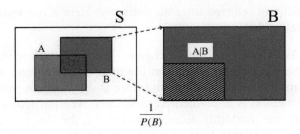

Fig. 3.8 Graphical illustration of conditional probability.

An event A is *independent* of B if the conditional probability of A given B is the same as the probability of A alone.

Events A and B are independent if

$$\mathbb{P}(A|B) = \mathbb{P}(A). \tag{3.5}$$

In the die-toss example, $\mathbb{P}(A) = \frac{1}{6}$ and $\mathbb{P}(A|B) = \frac{1}{4}$. Therefore, events A and B are not independent.

We saw that the probability of the union $A \cup B$ was $\mathbb{P}(A \cup B) = \mathbb{P}(A) + \mathbb{P}(B) - \mathbb{P}(AB)$. Now we are ready to introduce the general rule for the probability of an intersection.

The probability that events A and B will both occur is obtained by applying the *multiplication rule:*

$$\mathbb{P}(A \cdot B) = \mathbb{P}(A)\mathbb{P}(B|A) = \mathbb{P}(B)\mathbb{P}(A|B), \qquad (3.6)$$

where $\mathbb{P}(A|B)$ and $\mathbb{P}(B|A)$ are conditional probabilities of A given B and of B given A, respectively.

Only for independent events, equation (3.6) simplifies to

$$\mathbb{P}(A \cdot B) = \mathbb{P}(A)\mathbb{P}(B). \qquad (3.7)$$

Relationship (3.7) is also used to define independence, but (3.5) and (3.7) are equivalent.

With the repeated application of the multiplication rule, one can easily show

$$\mathbb{P}(A_1 A_2 \ldots A_n) = \mathbb{P}(A_1 | A_2 \ldots A_n) \; \mathbb{P}(A_2 | A_3 \ldots A_n) \; \ldots \mathbb{P}(A_{n-1} | A_n) \; P(A_n),$$

which is sometimes referred to as the *chain rule.* Here is one example of the use of the chain rule.

Example 3.9. **3+3 Dose Escalation Scheme.** In a dose-finding stage of clinical trials (Phase I) patients are given the drug at some dose, and if there is no dose-limiting toxicity (DLT), the dose is escalated. A version of the popular 3+3 method is implemented as follows: At a particular dose level, three patients are randomly selected and given the drug. If there is no DLT, the dose is escalated to the next higher one. If there are two or more DLTs, the escalation process is stopped. If there is exactly one DLT among the three patients, three new patients are selected at random and given the drug at the same dose. If there are no DLTs among these three new patients, the dose is escalated. If there is at least one DLT, the escalation process is stopped.

Assume that 30 patients are available for the trial at some fixed dose. If among them 4 will exhibit DLT at that dose, what is the probability that in the described step of the 3+3 procedure the dose will be escalated?

We assume that patients are selected and given the drug one by one. Denote by A_i the event that the ith patient will exhibit no DLT.

Then the dose will be escalated if the event

$$B = A_1 A_2 A_3 \; \cup \; A_1^c A_2 A_3 A_4 A_5 A_6 \; \cup \; A_1 A_2^c A_3 A_4 A_5 A_6 \; \cup \; A_1 A_2 A_3^c A_4 A_5 A_6$$

happens. Here, for example, the event $A_1 A_2^c A_3 A_4 A_5 A_6$ means that among the first three subjects the second experienced DLT, and that in the second group of three there was no DLT. Since the events $A_1 A_2 A_3$, $A_1^c A_2 A_3 A_4 A_5 A_6$, $A_1 A_2^c A_3 A_4 A_5 A_6$, and $A_1 A_2 A_3^c A_4 A_5 A_6$ are exclusive, the probability of their union is the sum of probabilities:

$$\mathbb{P}(B) = \mathbb{P}(A_1 A_2 A_3) + \mathbb{P}(A_1^c A_2 A_3 A_4 A_5 A_6)$$
$$+ \mathbb{P}(A_1 A_2^c A_3 A_4 A_5 A_6) + \mathbb{P}(A_1 A_2 A_3^c A_4 A_5 A_6).$$

For each of the probabilities the chain rule is needed. For example,

$$\mathbb{P}(A_1^c A_2 A_3 A_4 A_5 A_6) = 0.0739,$$

since

$$\mathbb{P}(A_1^c) \cdot \mathbb{P}(A_2 | A_1^c) \cdot \mathbb{P}(A_3 | A_1^c A_2) \cdot \mathbb{P}(A_4 | A_1^c A_2 A_3)$$
$$\cdot \mathbb{P}(A_5 | A_1^c A_2 A_3 A_4) \cdot \mathbb{P}(A_6 | A_1^c A_2 A_3 A_4 A_5)$$
$$= \frac{4}{30} \cdot \frac{26}{29} \cdot \frac{25}{28} \cdot \frac{24}{27} \cdot \frac{23}{26} \cdot \frac{22}{25}.$$

Thus,

$$\mathbb{P}(B) = \frac{26}{30} \cdot \frac{25}{29} \cdot \frac{24}{28} + \frac{4}{30} \cdot \frac{26}{29} \cdot \frac{25}{28} \cdot \frac{24}{27} \cdot \frac{23}{26} \cdot \frac{22}{25}$$
$$+ \frac{26}{30} \cdot \frac{4}{29} \cdot \frac{25}{28} \cdot \frac{24}{27} \cdot \frac{23}{26} \cdot \frac{22}{25} + \frac{26}{30} \cdot \frac{25}{29} \cdot \frac{4}{28} \cdot \frac{24}{27} \cdot \frac{23}{26} \cdot \frac{22}{25}$$
$$= 0.8620.$$

The dose will be escalated with probability 0.8620.

If counting rules are applied, the solution can be expressed as

$$\frac{\binom{26}{3}\binom{4}{0}}{\binom{30}{3}} + \frac{\binom{26}{2}\binom{4}{1}}{\binom{30}{3}} \times \frac{\binom{24}{3}\binom{3}{0}}{\binom{27}{3}}.$$

The conditional odds of A given that B occurred is

$$Odds(A|B) = \frac{\mathbb{P}(A|B)}{\mathbb{P}(A^c|B)} = \frac{\mathbb{P}(AB)}{\mathbb{P}(A^c B)}.$$

If events A and B are independent, then $Odds(A|B) = Odds(A)$.

For two events A and B the notions of exclusiveness $AB = \emptyset$ and independence $\mathbb{P}(AB) = \mathbb{P}(A)\mathbb{P}(B)$ are often considered equivalent by some students. Their argument can be summarized as follows: if two events do not share outcomes and their intersection is empty, then they must be independent. The contrary is true. If the events are exclusive and none are impossible, then they *must* be dependent. This can be demonstrated with the simple example of a coin-flipping experiment.

If A denotes tails up and B denotes heads up, then A and B are exclusive but dependent. If we have information that A happened, then we also have complete information that B did not happen.

If the sample spaces are different and the events are well separated in either time or space, their independence is intuitive. However, if the events share the sample space, it could be difficult to discern whether or not they are independent without resorting to the definition. The following example shows this:

Example 3.10. **Queen of Spades.** Let an experiment consist of drawing a card at random from a standard deck of 52 playing cards. Define events A and B as "the card is a ♠" and "the card is a queen." Are the events A and B independent? By definition, $P(A \cdot B) = P(Q♠) = \frac{1}{52}$. This is the product of $P(♠) = \frac{13}{52}$ and $P(Q) = \frac{4}{52}$, and events A and B in question are independent. In this situation, intuition provides no help. Now, pretend that the 2♡ is drawn and excluded from the deck prior to the experiment. Events A and B become dependent since

$$\mathbb{P}(A) \cdot \mathbb{P}(B) = \frac{13}{51} \cdot \frac{4}{51} \neq \frac{1}{51} = \mathbb{P}(A \cdot B).$$

The multiplication rule tells us how to find the probability for a composite event $(A \cdot B)$. The probability of $(A \cdot B)$ is used in the general *addition rule* for finding the probability of $(A \cup B)$.

Rule	Notation
Definitions The *conditional probability* of A given B is the probability of event A if event B occurred.	$\mathbb{P}(A|B)$
A is *independent* of B if the conditional probability of A given B is the same as the unconditional probability of A.	$\mathbb{P}(A|B) = \mathbb{P}(A)$
Multiplication rule The general *multiplication rule* for probabilities.	$P(A \cdot B) = \mathbb{P}(A)\mathbb{P}(B|A) = \mathbb{P}(B)\mathbb{P}(A|B)$
For *independent events* only, the multiplication rule is simplified.	$\mathbb{P}(A \cdot B) = \mathbb{P}(A)\mathbb{P}(B)$

3.6.2 Pairwise and Global Independence

If three events A, B, and C are such that any pair of them is exclusive, i.e., $AB = \emptyset$, $AC = \emptyset$, or $BC = \emptyset$, then the events are mutually exclusive, $ABC = \emptyset$. However, an analogous result does not hold for independence. Even if the events are *pairwise* independent for all three pairs A, B; A, C; and B, C, i.e., $\mathbb{P}(AB) = \mathbb{P}(A)\mathbb{P}(B)$, $\mathbb{P}(AC) = \mathbb{P}(A)\mathbb{P}(C)$, and $\mathbb{P}(BC) = \mathbb{P}(B)\mathbb{P}(C)$, they may not be independent in their totality. That is, it could happen that $\mathbb{P}(ABC) \neq \mathbb{P}(A)\mathbb{P}(B)\mathbb{P}(C)$.

Here is one example of such a triple:

Example 3.11. **Rolling a Tetrahedron.** The four sides of a tetrahedron (regular three-sided pyramid with four sides consisting of isosceles triangles) are denoted by 2, 3, 5, and 30, respectively. If the tetrahedron is "rolled," the number on the bottom side is the outcome of interest. The three events are defined as follows: A – the number on the bottom side is even, B – the number is divisible by 3, and C – the number is divisible by 5. The events are pairwise independent, but in totality, they are dependent.

The algebra is simple here, but what is the intuition? The "trick" is that events AB, AC, BC, and ABC all coincide. In other words, $\mathbb{P}(A|BC) = 1$ even though $\mathbb{P}(A|B) = \mathbb{P}(A|C) = \mathbb{P}(A)$.

The concept of independence/dependence is not transitive. At first glance, it may seem incorrect. One may argue, "If A depends on B, and

B depends on *C*, then *A* should depend on *C*, right?" We can demonstrate that this reasoning is not correct with a simple example.

Example 3.12. **Two $Q\diamondsuit$ and no $Q\clubsuit$.** Take a standard deck of 52 playing cards and replace the $Q\clubsuit$ with $Q\diamondsuit$. The deck still has 52 cards, two $Q\diamondsuit$ and no $Q\clubsuit$. From that deck draw a card at random and consider three events: *A* – the card is a queen, *B* – the card is red, and *C* – the card is a $\heartsuit$. It is easy to see that *A* and *B* are dependent, since $\mathbb{P}(AB) = 3/52$ does not equal $\mathbb{P}(A) \cdot \mathbb{P}(B) = 4/52 \cdot 27/52$. Events *B* and *C* are dependent as well, since event *C* is contained in *B*, and $\mathbb{P}(BC) = \mathbb{P}(C)$ does not equal $\mathbb{P}(B) \cdot \mathbb{P}(C)$. However, events *A* and *C* are independent, since $\mathbb{P}(AC) = \mathbb{P}(Q\heartsuit) = \frac{1}{52} = \mathbb{P}(A)\mathbb{P}(C) = \frac{13}{52} \cdot \frac{4}{52}$.
📝

3.7 Total Probability

The rule of total probability expresses the probability of an event *A* as the weighted average of its conditional probabilities. The events that *A* is conditioned upon need to be exclusive and should partition the sample space $\mathcal{S}$. Here are the definitions:

> Events $H_1, H_2, \ldots, H_n$ form a partition of the sample space $\mathcal{S}$ if
> (i) they are mutually exclusive ($H_i \cdot H_j = \emptyset$, $i \neq j$) and
> (ii) their union is the sample space $\mathcal{S}$, $\bigcup_{i=1}^{n} H_i = \mathcal{S}$.

The events $H_1, \ldots, H_n$ are usually called *hypotheses*. By this definition, it follows that $\mathbb{P}(H_1) + \cdots + \mathbb{P}(H_n) = 1 \; (= \mathbb{P}(\mathcal{S}))$.

Let the event of interest *A* happen under any of the hypotheses H_i with a known (conditional) probability $\mathbb{P}(A|H_i)$. Assume, in addition, that the probabilities of hypotheses $H_1, \ldots, H_n$ are known. $\mathbb{P}(A)$ can then be calculated using the *rule of total probability*.

> **Rule of Total Probability.**
>
> $$\mathbb{P}(A) = \mathbb{P}(A|H_1)\mathbb{P}(H_1) + \cdots + \mathbb{P}(A|H_n)\mathbb{P}(H_n) \qquad (3.8)$$

Thus, the probability of *A* is a weighted average of the conditional probabilities $\mathbb{P}(A|H_i)$ with weights given by $\mathbb{P}(H_i)$. Since H_is partition the sample space, the sum of the weights is 1.

The proof is simple. From $\mathcal{S} = H_1 \cup H_2 \cup \cdots \cup H_n$ it follows that

$$A = AS = A(H_1 \cup H_2 \cup \cdots \cup H_n) = AH_1 \cup AH_2 \cup \cdots \cup AH_n$$

(Fig. 3.9). Events AH_i are all exclusive, and after applying the additivity property, we obtain

$$\mathbb{P}(A) = \mathbb{P}(AH_1) + \mathbb{P}(AH_2) + \cdots + \mathbb{P}(AH_n).$$

Since each $\mathbb{P}(AH_i)$ is equal to $\mathbb{P}(A|H_i)\mathbb{P}(H_i)$ by the multiplication rule (3.6), the equality in (3.8) is true.

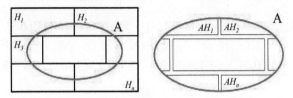

Fig. 3.9 $A = A(H_1 \cup H_2 \cup \cdots \cup H_n) = AH_1 \cup AH_2 \cup \cdots \cup AH_n$, and the events AH_i are exclusive.

Example 3.13. **Two-Headed Coin.** Out of 100 coins in a box, one has heads on both sides. The rest are standard fair coins. A coin is chosen at random from the box. Without inspecting whether it is fair or two-headed, the coin is flipped twice. What is the probability of getting two heads?

Let A be the event that both flips resulted in heads. Let H_1 denote the event (hypothesis) that a fair coin was chosen. Then, $H_2 = H_1^c$ denotes the hypothesis that the two-headed coin was chosen.

$$\begin{aligned}
\mathbb{P}(A) &= \mathbb{P}(A|H_1)\mathbb{P}(H_1) + \mathbb{P}(A|H_2)\mathbb{P}(H_2) \\
&= 1/4 \cdot 99/100 + 1 \cdot 1/100 = 103/400 = 0.2575.
\end{aligned}$$

The probability of the two flips resulting in tails is 0.2475 (check this!), which is slightly smaller than 0.2575. Is this an influence of the two-headed coin?

The next example is an interesting interplay between conditional and unconditional independence solved by the rule of total probability.

Example 3.14. **Accident Proneness.** Imagine a population with two types of individuals: N normal, and N^c accident prone. Suppose that 5/6 of these people are normal, so that, if we randomly select a person from this population, the probability that the chosen person will be normal is $\mathbb{P}(N) = 5/6$. Let A_i be the event that an individual has an accident in year i. For each individual, A_i is independent of A_j whenever $i \neq j$.

The accident probability is different for the two classes of individuals, $\mathbb{P}(A_i|N) = 0.01$ and $\mathbb{P}(A_i|N^c) = 0.1$. The chance of a randomly chosen individual having an accident in a given year is

$$\mathbb{P}(A_i) = \mathbb{P}(A_i|N)\mathbb{P}(N) + \mathbb{P}(A_i|N^c)\mathbb{P}(N^c)$$
$$= 0.01 \times 5/6 + 0.1 \times 1/6 = 0.025.$$

The probability that a randomly chosen individual has an accident in both the first and second year follows from the rule of total probability and the fact that A_1 and A_2 are independent for a given individual

$$\mathbb{P}(A_1 \cap A_2) = \mathbb{P}(A_1 \cap A_2|N)\mathbb{P}(N) + \mathbb{P}(A_1 \cap A_2|N^c)\mathbb{P}(N^c)$$
$$= \mathbb{P}(A_1|N)\mathbb{P}(A_2|N)\mathbb{P}(N) + \mathbb{P}(A_1|N^c)\mathbb{P}(A_2|N^c)\mathbb{P}(N^c)$$
$$= 0.01 \times 0.01 \times 5/6 + 0.1 \times 0.1 \times 1/6 = 0.00175.$$

Note that

$$\mathbb{P}(A_2|A_1) = \mathbb{P}(A_1 \cap A_2)/\mathbb{P}(A_1)$$
$$= 0.00175/0.025 = 0.07 \neq 0.025 = \mathbb{P}(A_2).$$

Therefore, A_1 and A_2 are not (unconditionally) independent!

3.8 Reassessing Probabilities: Bayes' Rule

The ideas in this section involve reassessing the probabilities of events when new evidence about related outcomes becomes available. Bayes' rule is named after Thomas Bayes, a nonconformist priest from the eighteenth century who was among the first to use conditional probabilities. He introduced "inverse" probabilities, which are the special case of what is now called *Bayes' rule* (Bayes, 1763). The general form was first used by Laplace (1774). Recall that the multiplication rule states

$$\mathbb{P}(AH) = \mathbb{P}(A)\mathbb{P}(H|A) = \mathbb{P}(H)\mathbb{P}(A|H).$$

This simple identity in association with the rule of total probability is the essence of Bayes' rule.

Bayes' Rule. Let the event of interest A happen under any of the hypotheses H_i with a known (conditional) probability $\mathbb{P}(A|H_i)$. Assume that the probabilities of hypotheses $H_1, \ldots, H_n$ are known (*prior* probabilities). Then the conditional (*posterior*) probability of the hypothesis H_i, $i = 1, 2, \ldots, n$, given that event A happened, is

$$\mathbb{P}(H_i|A) = \frac{\mathbb{P}(A|H_i)\mathbb{P}(H_i)}{\mathbb{P}(A)},$$

where

$$\mathbb{P}(A) = \mathbb{P}(A|H_1)\mathbb{P}(H_1) + \cdots + \mathbb{P}(A|H_n)\mathbb{P}(H_n).$$

The proof is simple:

$$\mathbb{P}(H_i|A) = \frac{\mathbb{P}(AH_i)}{\mathbb{P}(A)} = \frac{\mathbb{P}(A|H_i)\mathbb{P}(H_i)}{\mathbb{P}(A)},$$

where $\mathbb{P}(A)$ is given by the rule of total probability.

Although Bayes' rule is a simple formula for finding conditional probabilities, it is a precursor for a coherent "statistical learning" that will be discussed in the following chapters. It concerns the transition from prior probabilities of hypotheses to the posterior probabilities once new information about the sample space is obtained.

$$\mathbb{P}(H) \quad \overset{\text{BAYES' RULE}}{\longrightarrow} \quad \mathbb{P}(H|A)$$

Example 3.15. **Many Flips of a Possibly Two-Headed Coin.** Assume that out of N coins in a box, one has heads on both sides, and the remaining $N-1$ are fair. Assume that a coin is selected at random from the box and, without anyone inspecting what kind of coin it was, flipped k times. Every time the coin landed heads up. What is the probability that the two-headed coin was selected?

Let A_k denote the event where a randomly selected coin lands heads up k times. The hypotheses are H_1 – the coin is two-headed, and H_2 – the coin is fair. It is easy to see that $\mathbb{P}(H_1) = 1/N$ and $\mathbb{P}(H_2) = (N-1)/N$. The conditional probabilities are $\mathbb{P}(A_k|H_1) = 1$ for any k, and $\mathbb{P}(A_k|H_2) = 1/2^k$.

By the total probability rule,

$$\mathbb{P}(A_k) = \frac{2^k + N - 1}{2^k N},$$

and by Bayes' rule,

$$\mathbb{P}(H_1|A_k) = \frac{2^k}{2^k + N - 1}.$$

For $N = 1,000,000$ and $k = 1, 2, \ldots, 40$ the graph of posterior probabilities is given in Figure 3.10.

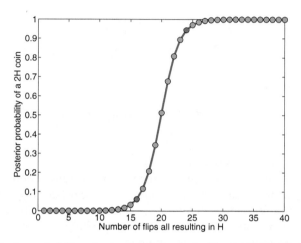

Fig. 3.10 Posterior probability of a two-headed coin for $N = 1,000,000$ if in k flips k heads appeared. The *red dots* show the posterior probabilities for $k = 16$ and $k = 24$, equal to 0.0615 and 0.9437, respectively.

Note that our prior probability $\mathbb{P}(H_1) = 0.000001$ jumps to a posterior probability of 0.9991 after observing 30 heads in a row. The code twoheaded.m calculates the probabilities and plots the graph in Figure 3.10. It is curious to observe how nonlinear the change of posterior probability is. This probability is quite stable for k up to 15 and after 25. The most rapid change is in the range $16 \leq k \leq 24$, where it increases from 0.0615 to 0.9437. This illustrates the "learning" ability of Bayes' rule.

Example 3.16. **Prosecutor's Fallacy.** The prosecutor's fallacy is a fallacy commonly occurring in criminal trials but also in other various arguments involving rare events. It consists of a subtle exchange of $\mathbb{P}(A|B)$ for $\mathbb{P}(B|A)$. We will explain it in the context of Example 3.15. Assume that out of $N = 1,000,000$ coins in a box, one is two-headed and "guilty." Assume that a coin is selected at random from the box and, without inspection, flipped $k = 15$ times. All $k = 15$ times the coin lands heads up. Based on this evidence, the "prosecutor" claims the selected coin is guilty because if it were "innocent," the outcome of $k = 15$ heads in a row would be extremely unlikely, with a probability of $\left(\frac{1}{2}\right)^{15} \approx 0.00003$. But in reality, the probability that the two-headed coin was selected and flipped is $\frac{1}{1+999999/2^{15}} \approx 0.03$, and the prosecutor is accusing an "innocent" coin with a probability of approximately 0.97.

✏️

Bayes' rule is even more revealing if expressed in terms of odds.

> The posterior odds of the hypothesis H_i are equal to the product of its prior odds and Bayes' factor (likelihood ratio),
>
> $$Odds(H_i|A) \; = \; BF \times Odds(H_i),$$
>
> where $BF = \mathbb{P}(A|H_i)/\mathbb{P}(A|H_i^c)$.

Thus, the "updater" is Bayes' factor BF, which represents the ratio of probabilities of the evidence (event A) under H_i and H_i^c. All available information from the experiment is contained in Bayes' factor, and Bayesian "learning" incorporates this information in a coherent way by transforming the prior odds to the posterior odds.

It is interesting to look at the log-odds equation:

$$\log Odds(H_i|A) = \log BF + \log Odds(H_i).$$

If the prior log-odds $\log Odds(H_i)$ increase/decrease for a constant C, then the posterior log-odds increase/decrease for the same constant, no matter what the Bayes' factor is. Likewise, if the log-Bayes factor increases/decreases for a constant C, then the log posterior odds increase/decrease for the same constant, no matter what the log prior odds are. This additivity property was used for constructing nomograms for fast approximate calculation of odds, mainly in a medical context.

The log BF was also termed *weight of evidence* by Alan Turing, who used similar techniques during the Second World War when breaking German "Enigma Machine" codes.

Example 3.17. **A Bridge Connection.** Figure 3.11 shows a circuit S that consists of components e_i, $i = 1,\ldots,7$, which work (and fail) independently of each other. Note that the connection of component e_7 is neither parallel nor serial. The components are operational in some time interval T with probabilities given in the following table:

Component	e_1	e_2	e_3	e_4	e_5	e_6	e_7
Probability of component working	0.3	0.8	0.2	0.2	0.5	0.6	0.4

We will calculate the posterior odds of e_7 working, given the information that circuit S is operational.

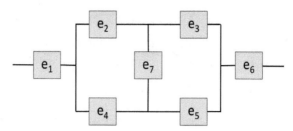

Fig. 3.11 "Bridge" connection of e_7.

Assume two hypotheses, H_1 – the component e_7 is operational, and $H_2 = H_1^c$ – e_7 is not operational. Under hypothesis H_1, circuit S can be expressed as

$$S|H_1 = e_1 \cap (e_2 \cup e_4) \cap (e_3 \cup e_5) \cap e_6,$$

while under H_1^c the expression is

$$S|H_1^c = e_1 \cap ((e_2 \cap e_3) \cup (e_4 \cap e_5)) \cap e_6.$$

By calculations similar to that in Example 3.5, we find that $\mathbb{P}(S|H_1) = 0.09072$ and $\mathbb{P}(S|H_1^c) = 0.04392$.

Note that $\mathbb{P}(H_1) = \mathbb{P}(e_7 \text{ works}) = 0.4$ and $\mathbb{P}(H_2) = \mathbb{P}(H_1^c) = 0.6$, so by the total probability rule, $\mathbb{P}(S) = 0.06264$. The prior odds of H_1 are $Odds(H_1) = 0.4/0.6 = 2/3$, and Bayes' factor is BF $= \mathbb{P}(S|H_1)/\mathbb{P}(S|H_1^c) = 2.06557$. The posterior odds of H_1 are $Odds(H_1|S) = \text{BF} \times Odds(H_1) = 2.06557 \times 2/3 = 1.37705$.

Thus, the odds of e_7 working increased from 0.66667 to 1.30705, after learning that the circuit is operational. See also ◀ bridge.m.

Example 3.18. **Subsequent Transfers.** In each of n boxes there are a white and b black balls. A ball is selected at random from the first box and placed into the second box. Then, from the second box a ball is selected at random and transferred to the third box, and so on. Finally, from the $(n-1)$th box a ball is selected at random and transferred to the nth box.

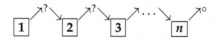

(a) After this series of consecutive transfers, a ball is selected from the nth box. What is the probability that this ball will be white?

(b) If the ball drawn from the fourth box was white, what is the probability that the first ball transferred had been white as well?

Let A_i denote the event that in the ith transfer the white ball was selected. Then

$$
\begin{aligned}
\mathbb{P}(A_n) &= \mathbb{P}(A_n|A_{n-1})\mathbb{P}(A_{n-1}) + \mathbb{P}(A_n|A_{n-1}^c)\mathbb{P}(A_{n-1}^c) \\
&= \frac{a+1}{a+b+1}\mathbb{P}(A_{n-1}) + \frac{a}{a+b+1}\mathbb{P}(A_{n-1}^c) \\
&= \frac{a+\mathbb{P}(A_{n-1})}{a+b+1}.
\end{aligned}
$$

Since $\mathbb{P}(A_1) = \frac{a}{a+b}$, we find that

$$
\mathbb{P}(A_2) = \mathbb{P}(A_3) = \cdots = \mathbb{P}(A_n) = \frac{a}{a+b}.
$$

(b) By Bayes' rule, $\mathbb{P}(A_1|A_4) = \frac{\mathbb{P}(A_4|A_1)\mathbb{P}(A_1)}{\mathbb{P}(A_4)} = \mathbb{P}(A_4|A_1)$.
Then,

$$
\begin{aligned}
\mathbb{P}(A_4|A_1) &= \mathbb{P}(A_2 A_3 A_4|A_1) + \mathbb{P}(A_2^c A_3 A_4|A_1) + \mathbb{P}(A_2 A_3^c A_4|A_1) + \mathbb{P}(A_2^c A_3^c A_4|A_1) \\
&= \frac{a+1}{a+b+1} \times \frac{a+1}{a+b+1} \times \frac{a+1}{a+b+1} + \frac{b}{a+b+1} \times \frac{a}{a+b+1} \times \frac{a+1}{a+b+1} \\
&\quad + \frac{a+1}{a+b+1} \times \frac{b}{a+b+1} \times \frac{a}{a+b+1} + \frac{b}{a+b+1} \times \frac{b}{a+b+1} \times \frac{a}{a+b+1} \\
&= \frac{(a+1)^3 + 2ab(a+1) + ab^2}{(a+b+1)^3}.
\end{aligned}
$$

3.9 Bayesian Networks*

We will discuss simple Bayesian networks in which the nodes are events. Many events linked in a causal network form a Bayesian net. Graphically, Bayesian networks are directed acyclic graphs (DAGs) where the nodes represent events and where the directed edges capture their hierarchy and dependence. Consider a simple graph in Figure 3.12.

Fig. 3.12 $A \longrightarrow B$ graph. A causes B or B is a consequence of A.

We would say that node A is a parent of B, B is a child of A, that A influences, or causes B, and B depends on A. This is captured by a directed edge (arrow) that leads from A to B. The term *acyclic* in DAG relates to the fact that a closed loop of dependencies is not allowed. In other words, there does not exist a path consisting of nodes $A_1, \ldots, A_n$ such that

$$\boxed{A} \longrightarrow \boxed{A_1} \longrightarrow \cdots \longrightarrow \boxed{A_n} \longrightarrow \boxed{A}.$$

The independence of two nodes in a DAG depends on their relative position in the graph as well as on the knowledge of other nodes (conditioning) in the graph.

Hard evidence for a node A is evidence that the outcome of A is known. Hard evidence about nodes is the information that we bring to the network, and it affects the probabilities of other nodes.

Bayesian networks possess a so-called Markovian property. The conditional distribution of any node depends only on its parental nodes. For instance, in the network

$$\boxed{A} \longrightarrow \boxed{B} \longrightarrow \boxed{C} \longrightarrow \boxed{D}$$

$\mathbb{P}(D|A,B,C) = \mathbb{P}(D|C)$, since C is a parental node of D.

The following simple example illustrates the influence of conditioning on independence.

Example 3.19. **Flips of Two Fair Coins.** Let A and B be outcomes of flips of two fair coins and C be an event that the two outcomes coincide. Thus, $\mathbb{P}(A = H) = \mathbb{P}(A = T) = 0.5$ and $\mathbb{P}(B = H) = \mathbb{P}(B = T) = 0.5$.

$$\boxed{A} \longrightarrow \boxed{C} \longleftarrow \boxed{B}$$

Nodes A and B are marginally independent (when we do not have evidence about C) but become dependent if the outcome of C is known:

$$\frac{1}{2} = \mathbb{P}(A = T, B = T|C) \neq \mathbb{P}(A = T|C)\mathbb{P}(B = T|C) = \frac{1}{2} \cdot \frac{1}{2}.$$

Example 3.20. **Alarm.** Your house has a security alarm system. The house is located in a seismically active area and the alarm system can be occasionally set off by an earthquake. You have two neighbors, Mary and John, who do not know each other. If they hear the alarm, they call you, but this is not guaranteed. They also call you from time to time just to chat.

Denote by $E, B, A, J,$ and M the events earthquake, burglary, alarm, John's call, and Mary's call took place, and by $E^c, B^c, A^c, J^c,$ and M^c the opposite events. The DAG of the network is shown in Figure 3.13.

The known (or elicited) conditional probabilities are as follows:

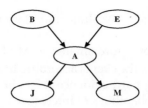

Fig. 3.13 Alarm Bayesian network.

A^c	A	Condition
0.999	0.001	$B^c\ E^c$
0.71	0.29	$B^c\ E$
0.06	0.94	$B\ E^c$
0.05	0.95	$B\ E$

B^c	B
0.999	0.001

E^c	E
0.998	0.002

J^c	J	condition
0.95	0.05	A^c
0.10	0.90	A

M^c	M	condition
0.99	0.01	A^c
0.30	0.70	A

We are interested in $\mathbb{P}(J,M|B)$, which is the probability that both John and Mary call, given the burglary happened.

We will first calculate this probability exactly and then find an approximation using WinBUGS. If E^* is either E or E^c and if A^* is either A or A^c, we have

$$\mathbb{P}(J,M|B) = \frac{1}{\mathbb{P}(B)} \times \mathbb{P}(B,J,M) = \frac{1}{\mathbb{P}(B)} \sum_{E^*,A^*} \mathbb{P}(B,E^*,A^*,J,M)$$

$$= \frac{1}{\mathbb{P}(B)}\{\mathbb{P}(B,E)\mathbb{P}(A|B,E)\mathbb{P}(J,M|A)$$
$$+\mathbb{P}(B,E^c)\mathbb{P}(A|B,E^c)\mathbb{P}(J,M|A)$$
$$+\mathbb{P}(B,E^c)\mathbb{P}(A^c|B,E^c)\mathbb{P}(J,M|A^c)$$
$$+\mathbb{P}(B,E)\mathbb{P}(A^c|B,E)\mathbb{P}(J,M|A^c)\}.$$

Given A^* (either A^c or A), $\mathbb{P}(J,M|A^*) = \mathbb{P}(J|A^*) \times \mathbb{P}(M|A^*)$, since John and Mary do not know each other, their calls can be considered independent. After substituting the probabilities with their numerical values from the tables above, we obtain

$$\mathbb{P}(J,M|B) = \frac{1}{0.001}\,(0.001 \cdot 0.002 \cdot 0.95 \cdot 0.90 \cdot 0.70 + 0.001 \cdot 0.998 \cdot 0.94 \cdot 0.90 \cdot 0.70$$
$$+\ 0.001 \cdot 0.998 \cdot 0.06 \cdot 0.05 \cdot 0.01 + 0.001 \cdot 0.002 \cdot 0.05 \cdot 0.05 \cdot 0.01)$$
$$=\ 0.5922.$$

Thus, in the case of burglary, both John and Mary will call with probability of 0.5922.

This probability can be approximated in MATLAB by simulation. The script ◢ alarm.m simulates the events according to known conditional probabilities and by changing hard evidence it can be applied to approximate unknown conditional probabilities of interest.

```
% alarm.m
% random numbers to default start
s = RandStream('mt19937ar','Seed',0);
RandStream.setGlobalStream(s);
%                    .
B=1000000; %number of simulations
bs=[]; es=[]; as=[]; js=[]; ms=[]; %save history
%set hard evidence
bh=1; %bulglary true
start simulation
for  i=1:B
  b=rand<=0.001;   %bulglary
  e=rand<=0.002;   %earthquake
  if(b)  if(e)  a=rand<=0.95; else a=rand<0.94; end;
         else  if(e) a=rand<=0.29; else a=rand<0.001; end;
  end
  if(a) j=rand<=0.9; else j=rand<=0.05; end; %john calls
  if(a) m=rand<=0.7; else m=rand<=0.01; end; %mary calls
  %hard evidence filter
          if(b == bh)
          bs=[bs b]; es=[es e]; as =[as a]; js=[js j]; ms=[ms m];
          end;
end
jm=sum(ms & js)        %608
tbs=sum(bs)            %1026
pjm=jm/tbs             %0.5926
```

Note that in 1,000,000 simulation runs, the burglary happened 1026 times. In these 1026 cases, both John and Mary called 608 times. This gives an estimate of $\mathbb{P}(J,M|B)$ as 0.5926.

This probability can also be approximated in WinBUGS. Note that $\mathbb{P}(J,M|B) = \mathbb{P}(J|M,B)\mathbb{P}(M|B)$ by the chain rule. The probabilities $\mathbb{P}(J|M,B)$ and $\mathbb{P}(M|B)$ will be approximated separately.

First, we approximate $\mathbb{P}(M|B)$ by fixing hard evidence for a burglary. WinBUGS will use the code burglary = 1 and burglary = 2 in the Data part to set the evidence that the burglary did not take place or that it took place, respectively. This is the only "hard evidence" here; all other nodes remain stochastic. The use of values 1, 2 instead of the expected 0, 1 is dictated by the categorical distribution dcat that takes only positive integers as realizations.

The WinBUGS code (alarm.odc) is as follows:

```
model alarm
    {
    burglary ~ dcat(p.burglary[]);
    earthquake ~ dcat(p.earthquake[]);
    alarm ~ dcat(p.alarm[burglary, earthquake, ])
    john ~ dcat(p.john[alarm,]);
    mary ~ dcat(p.mary[alarm,]);
    }

DATA
list(
    p.earthquake=c(0.998, 0.002),
    p.alarm = structure(.Data = c(0.999, 0.001,
                                  0.71,0.29,
                                  0.06,0.94,
                                  0.05,0.95),
                                 .Dim = c(2,2,2)),
p.john = structure(.Data = c(0.95,0.05,0.10,0.90),
                             .Dim = c(2,2)),
p.mary = structure(.Data = c(0.99,0.01,0.30,0.70),
                             .Dim = c(2,2)),
burglary = 2
)

INITS
list( earthquake = 1, alarm = 1, john = 1, mary = 1)
```

After 10,000 iterations, we obtain the mean value of M as $\mathbb{E}M = 2 \cdot p_M + 1 \cdot (1 - p_M) = 1.661$, that is, $\mathbb{P}(M|B) = p_M = 1.661 - 1 = 0.661$. Any of the 1s in the initial values earthquake = 1, alarm = 1, john = 1, mary = 1 can be replaced by 2, as this would not influence the final approximation.

Next, to estimate $\mathbb{P}(J|M,B)$, we change WinBUGS' data by setting $B = M = 2$. This is hard evidence that a burglary occurred and Mary called.

```
DATA
list(
    p.earthquake=c(0.998, 0.002),
    p.alarm = structure(.Data = c(0.999, 0.001,
                                  0.71,0.29,
                                  0.06,0.94,
                                  0.05,0.95), .Dim = c(2,2,2)),
p.mary = structure(.Data = c(0.99,0.01,0.30,0.70), .Dim = c(2,2)),
burglary = 2,
mary=2
)
```

and change the initial values to

```
list(earthquake = 1, alarm = 1, john = 1)
```

After 10,000 iterations, the mean value of J is obtained as $\mathbb{E}J = 2 \cdot p_J + 1 \cdot (1 - p_J) = 1.899$. Then, $\mathbb{P}(J|M,B) = p_J = 0.899$. Thus, the final result is

$$\mathbb{P}(J,M|B) = \mathbb{P}(J|M,B)\mathbb{P}(M|B) = p_J \cdot p_M = 0.899 \cdot 0.661 = 0.5942$$

(which approximates 0.5922, from the exact probability calculations).
✐

Bayesian networks can be useful in medical diagnostics if the conditional probabilities of the nodes are known. Here is the celebrated "Asia"
example.

Example 3.21. **Asia.** Lauritzen and Spiegelhalter (1988) discuss a fictitious
expert system for diagnosing a patient admitted to a chest clinic, who just
returned from a trip to Asia and is experiencing dyspnoea.[2] A graphical
model for the underlying process is shown in Figure 3.14, where each variable is binary. The WinBUGS code is shown below with the conditional
probabilities given as in Lauritzen and Spiegelhalter (1988).

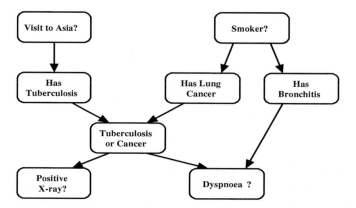

Fig. 3.14 Lauritzen and Spiegelhalter's (1988) Asia Bayes net: a fictitious *expert system*
representing the diagnosis of a patient having just returned from a trip to Asia and
showing *dyspnoea.*

```
model Asia;
    asia          ~ dcat(p.asia);
    smoking       ~ dcat(p.smoking[]);
    tuberculosis  ~ dcat(p.tuberculosis[asia,]);
    lung.cancer   ~ dcat(p.lung.cancer[smoking,]);
    bronchitis    ~ dcat(p.bronchitis[smoking,]);
    either        <- max(tuberculosis,lung.cancer);
    xray          ~ dcat(p.xray[either,]);
    dyspnoea      ~ dcat(p.dyspnoea[either,bronchitis,])
```

DATA

---

[2] Difficulty in breathing, often associated with lung or heart disease, and evincing in
shortness of breath. This is also called air hunger.

```
list(asia = 2, dyspnoea = 2,
     p.asia = c(0.99,0.01),    p.smoking = c(0.50,0.50),
     p.tuberculosis = structure(.Data = c(0.99,0.01,0.95,0.05),
                               .Dim = c(2,2)),
     p.bronchitis = structure(.Data = c(0.70,0.30,0.40,0.60),
                               .Dim = c(2,2)),
     p.lung.cancer = structure(.Data = c(0.99,0.01,0.90,0.10),
                               .Dim = c(2,2)),
     p.xray = structure(.Data = c(0.95,0.05,0.02,0.98),
                          .Dim = c(2,2)),
     p.dyspnoea = structure(.Data = c(0.9,0.1,
                          0.2,0.8,
                          0.3,0.7,
                          0.1,0.9), .Dim = c(2,2,2)))

INITS
list(smoking = 1, tuberculosis = 1,
         lung.cancer = 1, bronchitis = 1, xray = 1)
```

	mean	sd MC error	val2.5pc	median	val97.5pc	start	sample
bronchitis	1.812 0.3904 0.003988		1.0	2.0	2.0	2001	10000
lung.cancer	1.099 0.2985 0.003345		1.0	1.0	2.0	2001	10000
smoking	1.618 0.4859 0.004976		1.0	2.0	2.0	2001	10000
tuberculosis	1.095 0.2928 0.002706		1.0	1.0	2.0	2001	10000
xray	1.224 0.4171 0.004132		1.0	1.0	2.0	2001	10000

The results should be interpreted as follows: if the patient who visited Asia experiences dyspnoea (hard evidence in DATA: asia=2, dyspnoea = 2), then the probabilities of bronchitis, lung cancer, being a smoker, tuberculosis, and a positive X-ray, are 0.812, 0.099, 0.618, 0.095, and 0.224, respectively.

For practice, do the simulation of Asia example in MATLAB.

3.10 Exercises

3.1. **Event Differences.** Recall that the difference between events A and B was defined as $A \backslash B = A \cap B^c$. Using Venn diagrams, demonstrate that
(a) $A \backslash (A \backslash B) = A \cap B$ and $A \backslash (B \backslash A) = A$,
(b) $A \backslash (B \backslash C) = (A \cap C) \cup (A \cap B^c)$.

3.2. **Inclusion–Exclusion Principle in MATLAB.** From the set $\{1,2,3,\ldots,315\}$ a number is selected at random.
(a) Using MATLAB to count favorable outcomes, find the probability that the selected number is divisible by at least one of 3, 5, or 7.

(b) Compare this probability with a naïve solution $41/3 + 1/5 + 1/7 - 1/15 - 1/21 - 1/35 + 1/105 = 0.542857$, and show that the naïve solution is correct!

(c) Is the naïve solution correct for $\{1, 2, \ldots, N\}$ if $N = 316$?

(d) Is the naïve solution correct for any other N from $\{289, 290, \ldots, 340\}$? Plot this probability for $289 \leq N \leq 340$. Is there any symmetry in the plot?

3.3. **Gambling Fallacy.** An event that happened on August 18, 1913, in Le Grand Casino de Monte Carlo made headlines. The ball of a roulette wheel landed on "black" 26 times in a row. Out of 37 slots denoted by 0-36 (French roulettes have no a 00-slot), 18 slots (2, 4, 6, 8, 10, 11, 13, 15, 17, 20, 22, 24, 26, 28, 29, 31, 33, 35) are black, so the probability of a ball landing in black is 18/37.

(a) What is the probability that in the next 26 spins of a similar roulette wheel the ball lands on "black" every single time.

(b) After the ball landed in black slot 15 times in a row, the players in Le Grand Casino frantically started to bet on red, and that evening the Casino amassed a profit in millions of Francs. If one started to bet on black with \$1, what capital he/she will have after 26th consecutive black, if the Casino doubles the bet placed on winning color.

3.4. **Hexi.** There is a 10% chance that pure breed German shepherd Hexi is a carrier of canine hemophilia A. If she is a carrier, there is a 50–50 chance that she will pass the hemophiliac gene to a puppy.

Hexi has two male puppies and they have tested free of hemophilia. What is the probability that Hexi is a carrier, given this information about her puppies?

Hint: Passing the hemophiliac gene is independent between the puppies. If the puppies are male, then the only way they will get the hemophilia is from the mother carrier because hemophilia is X-chromosome-bound disorder.

3.5. **A Complex Circuit.** Figure 3.15 shows a circuit S that consists of identical components e_i, $i = 1, \ldots, 13$ that work (and fail) independently of each other. Any component is operational in some fixed time interval with probability 0.8.

(a) Calculate the probability that circuit S is operational.

(b) Write a MATLAB program that approximates the probability in (a) by simulation.

(c) Approximate the probability in (a) by WinBUGS simulations.

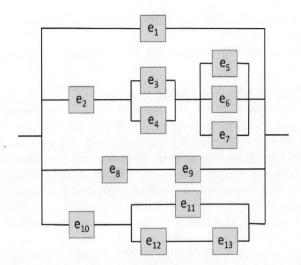

Fig. 3.15 Each of 13 independent components in the circuit is operational with probability 0.8.

3.6. **De Mere Paradoxes.** In 1654 the Chevalier de Mere asked Blaise Pascal (1623–1662) the following two questions:

(a) Why would it be advantageous in a game of dice to bet on the occurrence of a 6 in 4 trials but not advantageous in a game involving two dice to bet on the occurrence of a double 6 in 24 trials?

(b) In playing a game with three dice, why is a sum of 11 more advantageous than a sum of 12 when both sums are the result of six configurations:
11: (1, 4, 6), (1, 5, 5), (2, 3, 6), (2, 4, 5), (3, 3, 5), (3, 4, 4);
12: (1, 5, 6), (2, 4, 6), (2, 5, 5), (3, 3, 6), (3, 4, 5), (4, 4, 4)?
How would you respond to the Chevalier?

3.7. **Probabilities of Some Composite Events.** Show that for arbitrary events A, B,
(a) $\mathbb{P}(A \Delta B) = \mathbb{P}(A \cup B) - \mathbb{P}(AB) = \mathbb{P}(A) + \mathbb{P}(B) - 2\mathbb{P}(AB)$;
(b) $\mathbb{P}(A \Delta B) \geq |\mathbb{P}(A) - \mathbb{P}(B)|$.
(c) For arbitrary event C, $\mathbb{P}(AC \Delta BC) \leq \mathbb{P}(A \Delta B)$ and
(d) $(\mathbb{P}(A) + \mathbb{P}(B)) \frac{1}{1 + 2\mathbb{P}(AB)/(\mathbb{P}(A) + \mathbb{P}(B))} \leq \mathbb{P}(A \cup B) \leq \mathbb{P}(A) + \mathbb{P}(B)$.

3.8. **Deighton's Novel.** In his World War II historical novel *Bomber*, Len Dieghton argues that a pilot is "mathematically certain" to be shot down in 50 missions if the probability of being shot down on each mission is 0.02.

(a) Assuming independence of outcomes in each mission, is Deighton's reasoning correct?

(b) Find the probability of surviving all 50 missions without being shot down.

3.9. **Reliable System from Unreliable Components.** NASA is asking you to design a system that reliably performs a task on a space shuttle in the next 3 years with probability of $0.999999 = 1 - 10^{-6}$. In other words, the probability of failing during the next three years should not exceed one in a million. However, at your disposal you have components that in the next 3 years will fail with a probability of 0.2. Luckily, the weight and price of the components are not an issue, and you can combine/link them to increase the system's reliability.

(a) Should you link the components in a serial or parallel fashion to increase the probability of reliable performance?

(b) What minimal number of components should be linked as in (a) to satisfy NASA's requirement of 0.999999 probability of reliable performance?

3.10. **k-out-of-n Systems.** Suppose that n independent components constitute an engineering system. The system is called a k-out-of-n system if it works only when k or more components are operational. Parallel systems, for example, are 1-out-of-n systems, and series systems are n-out-of-n systems.

A particular 2-out-of-4 system has four independent components that are operational with probabilities $0.1, 0.8, 0.5$, and 0.4. What is the probability that it works?

3.11. **Székely's Reliability Problem.** This exercise is adapted from Székely (1986). Two systems, S_1 and S_2, are made of components A and B, which are operational with probabilities 0.9 and 0.1, respectively. The systems are serial,

$$S_1 : \boxed{A} \longrightarrow \boxed{B} \longrightarrow \boxed{A} \quad \text{and} \quad S_2 : \boxed{B} \longrightarrow \boxed{A} \longrightarrow \boxed{B},$$

and the components are independent. The systems are operational if any two neighboring components are operational, either $A - B$ or $B - A$, or both.

(a) Show that system S_2 is more reliable than S_1.

(b) At first glance the result in (a) is counterintuitive, since S_2 contains two components of low reliability. Can you provide an informal explanation?

3.12. **Dominos.** How many different dominos are in a set if the number of pips (dots) on the dominos ranges between (a) 0 and 6, (b) 0 and 8, and (c) 0 and 16, inclusive? For the three sets, find the probability of a domino selected at random having a "blank" side, i.e., having a side with no pips.

3.13. **"2+4" Dose Escalation Protocol.** When introducing a new drug in oncology and setting a dosage, a balance between efficacy and toxicity has to be found. The Phase I in a clinical trial is a dose finding. Usually, there are preset dosages, and based on the subjects' toxicity response, the dosage is either escalated or the escalation is stopped. Then, the current dosage adopted for the Phase II, which evaluates the drug's efficacy. Assume a "2+4" setup. Two subjects are selected and given the drug at the current (working) dosage. The dosage is escalated if neither subject exhibits the *dose-limiting toxicity* (DLT). If both exhibit the DLT, the escalation is stopped. If one subject exhibits the DLT, four new subjects are selected and given the drug at the current dosage. If no one subsequently exhibits the DLT, the dose is escalated. If at least one out of four exhibits DLT, the escalation is stopped.

In the group of 33 subjects available for the trial, five are sensitive and will exhibit the DLT if selected.

(a) Find the probability that a single application of the "2+4" scheme will result in a drug dosage escalation.

Hint: $E = A_1 A_2 \cup A_1^c A_2 A_3 A_4 A_5 A_6 \cup A_1 A_2^c A_3 A_4 A_5 A_6$, where E is escalation and A_i denote the event that ith selected person does not exhibit the DLT. Events A_i are independent, the three events in the union are exclusive.

(b) Simulate this probability in MATLAB. Is your simulation result close to the exact calculation in (a)?

As a hint, a skeleton of MATLAB code is given below.

```
% Set N -- number of simulation large, say N=10000;
escals=[]; %keep history
for i = 1:N
      %33 subjects, 5 sensitive will experience DLT.
      %WLOG, subjects 1-5 sensitive, 6-33 not sensitive.
      %Randomly permute 1:33, the permuted sequence is a.
   a = randperm(33);
   two = a(1:2); %selected first two
   s2=sum(two <= 5);
      %(two <= 5) is 0-1 vector of size 2;
      %s2 - the sum can be 0, 1, or 2, representing the
      % number of sensitive subjects among the selected
      switch s2
          case 0
                  escal=1;
          case 1
                  four=s(3:6);
                  s4=sum(four <= 5)
                  if s4=0
                      escal=1;
```

```
                    else
                        escal=0;
                    end
                otherwise
                        escal=0;
            end
        escals= [escals escal];
        end
        % How many 1's are in escals, what is the mean of escals?
```

3.14. **Counting Protocols.** Adel et al. (1993) applied various orders in drug combination sequence studies in search of a cure for human endometrial carcinoma. Four drugs A–D were evaluated for sequence-dependent inhibition of human tumor colony formation in soft agar.
(a) How many protocols are needed to evaluate all possible sequences of the four drugs?
(b) How many protocols are possible when only two drugs out of four are to be administered if the order of their administration is (i) important or (ii) not important?
(c) A fifth drug, E, is introduced. If drugs A and E cannot be given subsequent to each other because of cumulative toxicity concerns, how many protocols are possible if the order of drug administration is to be evaluated?

3.15. **Correlation between Events.** The correlation between events A and B is defined as

$$\mathrm{Corr}(A,B) = \frac{\mathbb{P}(A \cap B) - \mathbb{P}(A)\mathbb{P}(B)}{\sqrt{\mathbb{P}(A)(1 - \mathbb{P}(A))}\ \sqrt{\mathbb{P}(B)(1 - \mathbb{P}(B))}}.$$

Show that $\mathrm{Corr}(A,A) = 1$ and $\mathrm{Corr}(A,B) = \mathrm{Corr}(A^c,B^c)$.

3.16. **A Fair Gamble with a Possibly Loaded Coin.** Suppose that you have a coin for which you do not know the probability of its landing heads up. You suspect that the coin is loaded and that the probability of heads differs from $1/2$.
(a) Can you emulate a fair coin by flipping the possibly biased one?
(b) Can you emulate the rolling of a fair die by flipping the possibly biased coin?
Hint: You may need to flip the coin more than once.

3.17. **Easy Genetics.** Suppose that a certain organism contains four types of genes denoted with first four letters of the English alphabet and that each gene has dominant form (denoted with a capital letter), or recessive form (denoted by a lower case letter). Thus, xX and XX have different genotypes but the same phenotype because allele X in xX is dominant. In a mating, each organism contributes, at random, one of the genes (alleles) from each pair. The four contributions are mutually independent from pair to pair.

If parents have genotypes

$$\overline{\begin{array}{l} \text{AA bB cC DD} \\ \text{AA bB cc dD} \end{array}}$$

find the probability that the offspring will resemble (1) phenotypically and (2) genotypically:

(a) first parent
(b) second parent
(c) either parent
(d) neither parent.

Answer (a)–(d) [no calculation needed!] in the case

$$\overline{\begin{array}{l} \text{AA bB cc DD} \\ \text{aa bB CC dD} \end{array}}$$

Hint: Calculate probabilities of possible outcomes for each pair of alleles, and multiply (because of independence) along paths favorable for events in (a)–(b). Answers for (c)–(d) can be deduced from (a)–(b).

3.18. Neural Signal. A neuron will fire at random at any moment in $[0, T]$, with a probability of p. If up to time $t < T$ the neuron does not fire, what is the probability that it will fire in the remaining time, $(t, T]$?

3.19. Guessing. Subjects in an experiment are told that either a red or a green light will flash. Each subject is to guess which light will flash. The subject is told that the probability of a red light is 0.7, independently of guesses. Assume that the subject is a probability matcher, that is, guesses red with a probability of 0.7 and green with a probability of 0.3.
(a) What is the probability that the subject will guess correctly?
(b) Given that a subject guesses correctly, what is the probability that the light flashed red?

3.20. Propagation of Genes. The following example shows how the ideas of independence and conditional probability can be employed in studying genetic evolution. Consider a single gene that has two forms, *recessive (R)* and *dominant (D)*. Each individual in the population has two genes in his/her chromosomes and thus can be classified into the genotypes DD, RD, and RR. If an individual is drawn at random from the nth generation, then the probabilities of the three genotypes will be denoted by $p_n, 2r_n$, and q_n, respectively. (Clearly, $p_n + q_n + 2r_n = 1$.)

The problem is expressing the probabilities p_n, q_n, and r_n in terms of initial probabilities p_0, q_0, and r_0 and the method of reproduction. In *random Mendelian mating*, a single gene from each parent is selected at random and the selected pair determines the genotype of the offspring.

These selections are carried independently of each other from generation to generation.

Let M_n be the event that R is chosen from the male and F_n be the event that R is chosen from the counterpart female. Events M_n and F_n are independent and have the same probability. Thus,

$$\mathbb{P}(M_n) = \mathbb{P}(RR) \times \mathbb{P}(M_n|RR) + \mathbb{P}(RD) \times \mathbb{P}(M_n|RD) + \mathbb{P}(DD) \times \mathbb{P}(M_n|DD)$$
$$= \mathbb{P}(RR) \times 1 + \mathbb{P}(RD) \times 1/2 + \mathbb{P}(DD) \times 0$$
$$= q_n + 2r_n/2$$
$$= q_n + r_n$$

by the rule of total probability.

By the independence of M_n and F_n,

$$q_{n+1} = \mathbb{P}(M_n \cap F_n) = \mathbb{P}(M_n) \cdot \mathbb{P}(F_n) = (q_n + r_n)^2.$$

Similarly,

$$p_{n+1} = (p_n + r_n)^2$$

and

$$2r_{n+1} = 1 - p_{n+1} - q_{n+1}.$$

The last three equations govern the propagation of genotypes in this population.

Start with any initial probabilities p_0, q_0, and r_0. (Say, 0.3, 0.3, and 0.2; remember to check: $0.3 + 0.3 + 2 \cdot 0.2 = 1$.) Find iteratively (p_1, q_1, r_1), (p_2, q_2, r_2), and (p_3, q_3, r_3), and demonstrate that $p_1 = p_2 = p_3, q_1 = q_2 = q_3$, and $r_1 = r_2 = r_3$. The fact that the probabilities remain the same is known as the *Hardy–Weinberg law*. It does not hold if other factors (mutation, selection, dependence) are introduced into the model.

3.21. **Easy Conditioning.** Assume $\mathbb{P}(\text{rain today}) = 40\%$, $\mathbb{P}(\text{rain tomorrow}) = 50\%$, and $\mathbb{P}(\text{rain today and tomorrow}) = 30\%$. Given that it is raining today, what is the chance that it will rain tomorrow?

3.22. **Eye Color.** The eye color of a child is determined by a pair of genes, one from each parent. If {b} and {B} denote blue- and brown-eyed genes, then a child can inherit the following pairs: {bb}, {bB}, {Bb}, and {BB}. The {B} gene is dominant, that is, the child will have brown eyes when the pairs are {Bb}, {bB}, or {BB} and blue eyes only for the {bb} combination. A parent passes to a child either gene from his/her pair with equal probability.[3]

---

[3] This description is simplified, and in fact there are several genes affecting eye color and the amount of yellow and black pigments in the iris, leading to shades of colors including green and hazel.

Megan's parents are both brown-eyed, but Megan has blue eyes. Megan's brown-eyed sister is pregnant and her husband has blue eyes. What is the probability that the baby will have blue eyes?

3.23. **Dice.** In rolling ten fair dice we have information that at least one ⚁ appeared. What is the probability that there were at least two ⚁?

3.24. **Inflation and Unemployment.** Businesses commonly project revenues under alternative economic scenarios. For a stylized example, inflation could be high or low and unemployment could be high or low. There are four possible scenarios, with the following assumed probabilities:

Scenario	Inflation	Unemployment	Probability
1	High	High	0.16
2	High	Low	0.24
3	Low	High	0.36
4	Low	Low	0.24

(a) What is the probability of high inflation?
(b) What is the probability of high inflation if unemployment is high?
(c) Are inflation and unemployment independent?

3.25. **Multiple Choice.** A student answers a multiple choice examination question that has four possible answers. Suppose that the probability that the student knows the answer to a question is 0.80 and the probability that the student guesses is 0.20. If the student guesses, the probability of guessing the correct answer is 0.25.
(a) What is the probability that the fixed question will be answered correctly?
(b) If it is answered correctly, what is the probability that the student really knew the correct answer?

3.26. **Manufacturing Bayes.** A factory has three types of machines producing an item. The probabilities that the item is conforming if it is produced on the ith machine are given in the following table:

Type of machine	Probability of item conforming
1	0.94
2	0.95
3	0.97

The total production is distributed among the machines as follows: 30% is done on type 1, 50% on type 2, and 20% on type 3 machines. One item is selected at random from the production.
(a) What is the probability that it is conforming?
(b) If it is conforming, what is the probability that it was produced on a type 1 machine?

3.27. **Stanley.** Stanley takes an oral exam in statistics with several other students. He needs to answer the questions from an examination card drawn at random from the set of 20 cards. There are exactly 8 favorable cards among the 20 to which Stanley knows the answers. Stanley will get a grade of A if he knows the answers, that is, if he draws a favorable card. What is the probability that Stanley will get an A if he draws the card standing in line (a) first, (b) second, and (c) third?

3.28. **Kokomo, Indiana.** In Kokomo, IN, 65% of the people are conservative, 20% are liberal, and 15% are independent. Records show that in a particular election, 82% of conservatives voted, 65% of liberals voted, and 50% of independents voted. If a person from the city is selected at random and it is learned that she did not vote, what is the probability that the person is liberal?

3.29. **Mysterious Transfer.** Of two bags, one contains four white balls and three black balls and the other contains three white balls and five black balls. One ball is randomly selected from the first bag and placed unseen in the second bag.
(a) What is the probability that a ball now drawn from the second bag will be black?
(b) If the second ball is black, what is the probability that a black ball was transferred?

3.30. **Two Masked Robbers.** (Durrett, 2009) Two masked robbers try to rob a crowded bank during the lunch hour, but the teller presses a button that sets off an alarm and locks the front door. The robbers, realizing they are trapped, throw away their masks and disappear into the chaotic crowd. Confronted with 40 people claiming they are innocent, the police give everyone a lie detector test. Suppose that guilty people are detected with a probability of 0.85 and innocent people appear to be guilty with a probability of 0.08. What is the probability that Mr. Smith was one of the robbers given that the lie detector says he is a robber?

3.31. **Information Channel.** One of the three words AAAA, BBBB, and CCCC is transmitted via an information channel. The probabilities of these words being transmitted are 0.3, 0.5, and 0.2, respectively. Each letter is transmitted and received correctly with a probability of 0.6, independently of other letters. Since the channel is not perfect, the transmitted letter can change to one of the other two letters with an equal probability of 0.2. What is the probability that the word AAAA was submitted if the word ABCA is received?

3.32. **Quality Control.** An automatic machine in a small factory produces metal parts. Most of the time (90% according to long-term records), it produces 95% good parts, while the remaining parts have to be scrapped.

Other times, the machine slips into a less productive mode and only produces 70% good parts. The foreman observes the quality of parts that are produced by the machine and wants to stop and adjust the machine when she believes that the machine is not working well. Suppose that the first dozen parts produced are given by the sequence

s	u	s	s	s	s	s	s	s	u	s	u

where **s** is satisfactory and **u** is unsatisfactory. After observing this sequence, what is the probability that the machine is in its productive state? If the foreman wishes to stop the machine when the probability of "good state" is under 0.7, when should she stop it?

3.33. **Let's Make a Deal.** In Monty Hall's game show *Let's Make a Deal* there are three closed doors. Behind one of these doors is a car; behind the other two are goats. The contestant does not know where the car is, but Monty Hall does.
After the contestant picks a door Monty opens one of the remaining doors, one he knows does not hide the car. If the contestant has already chosen the door with the car behind, Monty is equally likely to open either of the two remaining doors.
After Monty has shown a goat behind the door that he opens, the contestant is given the option to switch doors.

(a) What is the probability of winning the car under the switching and non-switching strategies?
(b) Show analytically, using the formula of total probability that switching strategy gives the winning probability of $(n-1)/(n(n-2))$ in the case of n doors, a single car and $(n-1)$ goats. The non switching strategy trivially results in the winning probability of $1/n$.
Hint: Condition on the result of your first choice, H_1 : selected a door with the car behind, H_2: selected a door with a goat behind.
(c) Simulate Monty Hall game $B = 10,000$ times. How many cars would you win by adhering to the switching strategy? Non-switching strategy? Here is MATLAB skeleton script as a hint.

```
%set B large...
 cars=0;    carn=0; %set 0 at the beginning
for i = 1:B
a = randperm(3); %Monty places two goats and the car at random
%a(1) -goat, a(2) -goat, a(3) - car
i= randsample(1:3,1); %you select the door!

% SWITCH STRATEGY
if(i == a(1))   cars=cars+1; %a(2)-opened, switch to a(3), car!
elseif (i == a(2)) cars = cars + 1 ;%a(1) opened, switch to a(3), car!
else cars = cars + 0; %a(1)/a(2) opened, switch to a(2)/a(1), no car!
end
```

```
% NOT TO SWITCH STRATEGY
if(i == a(1))   ...
elseif (i==a(2)) ...
else ...
end
end

cars   %# of cars with switching
carn   %# of cars w/o switching
```

(d) Consider the following modification to the Monty Hall game. Monty realizes that most of the people understand that switching is beneficial and modifies the rules. If the contestant picks the door with the car, Monty would open a door with a goat and offer a switch. If the contestant selects a door with a goat, Monty goes backstage and flips a coin. If the coin comes up heads, Monty would just open the door that was selected and the contestant leaves with a goat. If the coin comes up tails, Monty would open another door with a goat and offer a switch. What are chances of getting the car in this modified version under switching and non-switching strategies? *Hint:* There are four hypotheses, defined by contestant's first pick and the coin's face.]

3.34. ***Marmota marmota.*** Cohas et al. (2009) explored the effect of heterozygosity on survival within a population of alpine marmots, *Marmota marmota*. In order to test this effect, individual marmots were genotyped at 16 loci and the alleles of those loci were studied. A locus is a location of a gene on a chromosome. An allele is an alternative form of a gene found on a locus. Alpine marmots were captured, genotyped, and released back in the wild. Sample size and alleles varied for each locus. Due to the large amount of data, only data of 7 loci from genotyping are provided in the table below.

	SS-Bibl1		SS-Bibl18		SS-Bibl20		SS-Bibl31		SS-Bibl4	
	Alleles	Freq	Alleles	Freq	Alleles	Freq	Alleles	Freq	Alleles	Freq
	95	0.16	133	<0.01	205	<0.01	157	0.50	175	0.12
	97	0.21	127	0.01	208	0.18	159	0.28	188	0.15
	101	0.45	143	0.34	216	0.39	161	0.17	190	0.70
	103	<0.01	145	0.14	218	0.33	163	0.05	192	0.03
	107	0.14	147	0.41	220	0.08				
	109	0.04	149	0.10	222	<0.01				
Number of Animals	692		698		684		692		690	

(a) Allele probability is estimated by the relative frequency of a specific mutated gene found within a locus. On locus SS-Bibl4, estimate the probability that allele 190 will not be present?

(b) In locus SS-Bibl1, allele 95 and 101 are mutually exclusive. Estimate the probability that a marmot will possess allele 95 or allele 101?
(c) Alleles 143 (locus SS-Bibl18), 208 (locus SS-Bibl20), and 159 (locus SS-Bibl31) are independent. Estimate the probability that a marmot will possess at least one of these alleles?
(d) Cohas et al. found that the death probability of a marmot with allele 190 is 0.01 and the death probability of a marmot without allele 190 is 0.1. What is the chance of a random marmot dying in this situation?
(e) Using Bayes' rule, estimate the probability of a dead marmot having allele 190 found on locus SS-Bibl4.

3.35. **Twins.**   Dizygotic (fraternal) twins have the same probability of each gender as in overall births, which is approximately 51% male, 49% female. Monozygotic (identical) twins must be of the same gender. Among all twin pregnancies, about 1/3 are monozygotic.
Find the probability of two girls in
(a) monozygotic pregnancy,
(b) dizygotic pregnancy, and
(c) dizygotic pregnancy given that we know that the gender of the babies is the same.
If Mary is expecting twins, but no information about the type of pregnancy is available, what is the probability that the babies are
(d) two girls,
(e) of the same gender.
(f) Find the probability that Mary's pregnancy is dizygotic if it is known that the babies are two girls.
Retain four decimal places in your calculations.

Hint: (b) genders are independent; (c) since A is subset of B, $A \cap B = A$ and $P(A|B) = P(A)/P(B)$; (d, e) total probability; (f) Bayes' rule.

3.36. **Left-handedness.**   Left-handedness can be an inconvenience for a portion of the human population and is a study of interest to researchers throughout the world. About 6–10% of the human population exhibits left-handedness. Studies suggest that left-handedness is due to genetics rather than environmental causes. It is found to be more common in males (about 8–12%) than females (about 5–9%), and also that the probability of left-handedness increases when a mother or father has it.
The table (McManus, 1991; Table 4, page 262) provides empirical probabilities of a child being left-handed depending on gender and maternal/paternal handedness.

Handedness		Left-handed Right-handed			
Father	Mother	Son		Daughter	
Right-handed	Right-handed	0.1049	0.8951	0.0863	0.9137
Right-handed	Left-handed	0.2150	0.7850	0.1955	0.8045
Left-handed	Right-handed	0.1849	0.8151	0.1744	0.8256
Left-handed	Left-handed	0.2940	0.7060	0.2839	0.7161

Using these probabilities, answer parts (a–c)

(a) Assume that parents are both right-handed. What is the probability that their child is right-handed? Probabilities of a son or daughter are assumed equal.

(b) Ann is left-handed and expecting a son. No information about handedness of baby's father Ben is available, but it is known that Ben's parents are right-handed. What is the probability that the baby is going be left-handed.

(c) Adam's parents are both right-handed. Adam's daughter Bethany and wife Claudia are both left-handed. What is the probability that Adam is right-handed. Does Claudia's handedness matter?

3.37. **Redundant Wiring.** In a circuit shown in Figure 3.16 the electricity is to move from point A to point B. The four independently working elements in the circuit are operational (and the current goes through) with probabilities given in the table

Element	e_1	e_2	e_3	e_4
Operational with prob	0.5	0.2	0.3	0.8

If an element fails, the current is not going through.

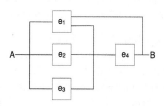

Fig. 3.16 Circuit for Exercise 3.37.

(a) Is it possible to save on the wire that connects the elements without affecting functionality of the network? Explain which part of wiring can be removed.

(b) Find the probability that the electricity will flow from A to B.

Hint: Although this configuration can be analyzed directly, it is simpler to to condition on the element e_1 and apply the Formula of Total Probability, as done in Example 3.17.

3.38. **Cross-linked System 1.** Each of the five components in a cross-linked system (shown in Fig. 3.17) is operational in a time interval $[0, T]$ with the probability of 0.6. The components are independent. Let E_i denote the event that ith component is operational at time T and E_i^c that it is not. Denote by A the event that the system is operational at time T.

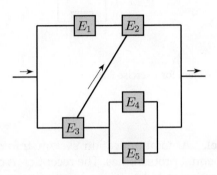

Fig. 3.17 Cross-linked system for Exercise 3.38.

(a) Find the probabilities of events $H_1 = E_2^c E_3^c$, $H_2 = E_2^c E_3$, $H_3 = E_2 E_3^c$, and $H_4 = E_2 E_3$. Do these four probabilities sum up to 1?
(b) What is the probability of the system being operational if H_1 is true; that is, what is $\mathbb{P}(A|H_1)$? Find also $\mathbb{P}(A|H_2)$, $\mathbb{P}(A|H_3)$ and $\mathbb{P}(A|H_4)$.
(c) Using results in (a) and (b), find $\mathbb{P}(A)$.

3.39. **Cross-linked System 2.** The five components in a cross-linked system (shown in Figure) are operational in a time interval Δ_T with probabilities

Component	E_1	E_2	E_3	E_4	E_5
Probability of working	0.8	0.2	0.6	0.7	0.5

The components are independent.
(a) Find the probability that the system is operational during the time period Δ_T.
(b) If the system was found operational during time interval Δ_T, what is the probability that component E_1 have been operational?

Hint: The problem combines circuit calculations and Total Probability/Bayes' formula. Discuss how the circuit simplifies when component E_1 works (hypothesis H_1) and, alternatively, when it fails (hypothesis H_2).

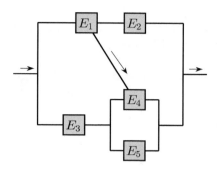

Fig. 3.18 Cross-linked system for Exercise 3.39.

3.40. **Ternary Channel.** A communication system transmits three signals, s_1, s_2, or s_3, with equal probabilities. The reception is corrupted by noise, causing the transmission to be changed according to the following table of conditional probabilities:

		Received		
		s_1	s_2	s_3
	s_1	0.75	0.1	0.15
Sent	s_2	0.098	0.9	0.002
	s_3	0.02	0.08	0.9

The entries in this table list the probability that s_j is received, given that s_i is sent, for $i, j = 1, 2, 3$. For example, if s_1 is sent, the conditional probability of receiving s_3 is 0.15.

(a) Compute the probabilities that s_1, s_2, and s_3 are received.

(b) Compute the probabilities $\mathbb{P}(s_i \text{ sent} \mid s_j \text{ received})$ for $i, j = 1, 2, 3$. (Complete the table.)

		Sent		
		s_1	s_2	s_3
	s_1	0.8641		
Received	s_2			0.0741
	s_3			

3.41. **Sprinkler Bayes Net.** Suppose that a sprinkler (S) or rain (R) can make the grass in your yard wet (W). The probability that the sprinkler was on depends on whether the day was cloudy (C). The probability of rain also depends on whether the day was cloudy. The DAG for events C, S, R, and W is shown in Figure 3.19.

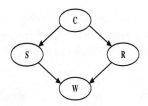

Fig. 3.19 Sprinkler Bayes net.

The conditional probabilities of the nodes are given in the following tables.

C^c	C
0.5	0.5

S^c	S	Condition
0.50	0.50	C^c
0.90	0.10	C

R^c	R	Condition
0.80	0.20	C^c
0.20	0.80	C

W^c	W	Condition
1	0	$S^c R^c$
0.10	0.90	$S^c R$
0.10	0.90	$S R^c$
0.01	0.99	$S R$

(a) Using WinBUGS, approximate the probabilities $\mathbb{P}(C|W)$, (b) $\mathbb{P}(S|W^c)$, and (c) $\mathbb{P}(C|R,W^c)$.

(b) Approximate the probabilities in (a) using MATLAB simulations.

3.42. **Diabetes in Pima Indians.** The Pima Indians have the world's highest reported incidence of diabetes. Since 1965, this population has participated in a longitudinal epidemiological study of diabetes and its complications. The examinations have included a medical history for diabetes and other major health problems. A population of women who were at least 21 years old, of Pima Indian heritage, and living near Phoenix, AZ, was tested for diabetes according to World Health Organization criteria. The following conditions ("events"), constructed from the database, can be related to a randomly selected subject from this population.

Event	Description
P	Three or more pregnancies
A	Older than the database median age
O	Heavier than the database median weight
D	Diagnosis of diabetes
G	High plasma glucose concentration in an oral glucose tolerance test
I	High 2-h serum insulin ($\mu U/ml$)
B	High blood pressure

The DAG in Figure 3.20 simplifies the proposal of Tom Bylander from the University of Texas at San Antonio, who used Bayesian networks and the Pima Indians Diabetes Database in a machine learning example.[4]

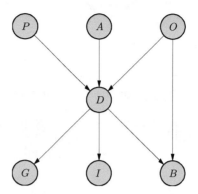

Fig. 3.20 Pima Indians diabetes Bayes net.

From 768 complete records, relative frequencies are used to approximate the conditional probabilities of the nodes. The probabilities are given in the following tables:

P^c	P
0.45	0.55

A^c	A
0.5	0.5

O^c	O
0.5	0.5

D^c	D	Condition
0.95	0.05	$P^c\ A^c\ O^c$
0.67	0.33	$P^c\ A^c\ O$
0.59	0.41	$P^c\ A\ O^c$
0.40	0.60	$P^c\ A\ O$

D^c	D	Condition
0.73	0.27	$P\ A^c O^c$
0.66	0.34	$P\ A^c\ O$
0.63	0.37	$P\ A\ O^c$
0.41	0.59	$P\ A\ O$

G^c	G	Condition
0.64	0.36	D^c
0.21	0.79	D

I^c	I	Condition
0.49	0.51	D^c
0.52	0.48	D

B^c	B	Condition
0.55	0.45	$O^c\ D^c$
0.58	0.42	$O^c\ D$
0.40	0.60	$O\ D^c$
0.49	0.51	$O\ D$

Suppose a female subject, older than 21, of Pima Indian heritage living near Phoenix, AZ, was selected at random. Using WinBUGS, approximate the probabilities (a) $\mathbb{P}(O|I)$, $\mathbb{P}(B|O^c, G)$, and $\mathbb{P}(G|B, A^c)$.

3.43. **A Simplified Probabilistic Model of Visual Pathway.** Our nervous system consists of specialized cells called neurons that are connected to one another in highly organized and specific ways. A neuron changes its membrane voltage in response to an external stimulus or a signal from an upstream neuron. For example, in the visual pathway, light excites photoreceptors in our eyes that are connected to more downstream

---

[4] http://www.cs.utsa.edu/~bylander/cs6243/bayes-example.pdf

neurons, and the activities of this neuronal network eventually elicit visual perceptions. The following diagram illustrates a simplified version of this connection (P = photoreceptor cell, B = bipolar cell, R = retinal ganglion cell, L = lateral geniculate nucleus ganglion cell, V = primary visual cortex simple cell):

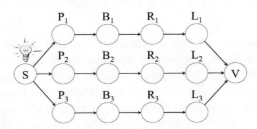

Fig. 3.21 Simplified probabilistic model of visual pathway.

Let events S and S^c denote, respectively, the presence and absence of a light stimulus in a small time interval Δt. Let events $P, B, R,$ and L mean that the corresponding cells produce a response (fire) in the time interval Δt. Suppose that the conditional probabilities of firing for three parallel branches $i = 1, 2,$ and 3 are the same:

$$
\begin{array}{ll}
\mathbb{P}(S) = 0.6 & \\
\mathbb{P}(P_i|S) = 0.95 & \mathbb{P}(P_i|S^c) = 0.1 \\
\mathbb{P}(B_i|P_i) = 0.95 & \mathbb{P}(B_i|P_i^c) = 0.05 \\
\mathbb{P}(R_i|B_i) = 0.9 & \mathbb{P}(R_i|B_i^c) = 0.02 \\
\mathbb{P}(L_i|R_i) = 0.8 & \mathbb{P}(L_i|R_i^c) = 0.08 \\
& i = 1, 2, 3
\end{array}
$$

In the schematic diagram in Figure 3.21, three L ganglion cells, L_1, L_2, and L_3, connect to one simple cell in the primary visual cortex, V. Assume that input from at least one of the L cells is needed for the V simple cell to respond with certainty. The V simple cell will not respond if the input there from L cells is absent.
The following questions may need the support of WinBUGS:
(a) What is the probability of the V cell responding given that event S (light stimulus present) has occurred?
(b) Assume that L_1 and R_3 have fired. What is the probability of S?
(c) Assume that V responded. What is the probability of S?

3.44. **Three Neurons 1.**   A neuron N is connected with three other neurons $N_1, N_2,$ and N_3 capable of producing a stimulus by firing, $\{N_1, N_2, N_3\} \longrightarrow N$. The probabilities of firing for neurons $N_1, N_2,$ and N_3 are 0.2 each and the actions of these three neurons are simultaneous and mutually independent. Neuron N receives a stimulus if at least one of $N_1, N_2,$ and N_3 fires.

(a) What is the probability that N receives a stimulus? Find this probability analytically.

(b) Approximate the probability in (a) using MATLAB simulation.

(c) Neuron N will fire with probability 0.9 if a stimulus is present and with probability 0.05 if a stimulus is not present. In the context of this problem what is the probability that N fires?

(d) If neuron N did not fire what is the probability that stimulus was not present.

3.45. **Three Neurons 2.**   Three neurons $N_1, N_2,$ and N_3 are connected as $N_1 \longrightarrow N_2 \longrightarrow N_3$. If a stimulus is present, the neuron will fire with probability 0.9. When a stimulus is not present, the neuron may still fire but with a small probability of 0.05. Firing of a neuron serves as a stimulus for the subsequent neuron. N_1 is given a stimulus.

(a) What is the probability that N_3 will fire? Find this probability exactly (analytically).

(b) Write a MATLAB code that approximates the probability in (a) by simulation.

(c) If N_3 did not fire, what is the probability that N_2 received stimulus.

MATLAB AND WINBUGS FILES USED IN THIS CHAPTER

http://statbook.gatech.edu/Ch3.Prob/

alarm.m, birthday.m, bridge.m, circ0.m, circ1.m, circ2.m, circuit.m, ComplexCircuit.m, demere.m, die.m, inclusionexclusion.m, mysterious.m, mystery.m, pima.m, rollingdie1.m, rollingdie2.m, sheriff.m, ternarychannel.m, twoheaded.m, venn0.m, venn1.m, venn2.m

alarm.odc, asia.odc, circuit1.odc, DeMere.odc, manufacturingbayes.odc, markcar.odc, misterioustransfers.odc, pima.odc, sprinkler.odc

---

CHAPTER REFERENCES

Adel, A. L., Dorr, R. T., and Liddil, J. D. (1993). The effect of anticancer drug sequence in experimental combination chemotherapy. *Cancer Invest.*, **11**, 1, 15–24.

Barbeau, E. (1993). The problem of the car and goats. *College Math. J.*, **24**, 2, 149–154.

Bayes, T. (1763). An essay towards solving a problem in the doctrine of chances. *Philos. Trans. R. Soc. Lond.*, *53*, 370–418.

Bonferroni, C. E. (1937). Teoria statistica delle classi e calcolo delle probabilita. In *Volume in Onore di Ricarrdo dalla Volta*, Universita di Firenza, 1–62.

Casscells, W., Schoenberger, A., and Grayboys, T. (1978). Interpretation by physicians of clinical laboratory results. *New Engl. J. Med.*, **299**, 999–1000.

Cohas, A., Bonenfant, C., Kempenaers, B., and Allainé, D. (2009). Age-specific effect of heterozygosity on survival in alpine marmots, *Marmota marmota*. *Mol. Ecol.*, **18**, 7, 1491–1503. doi:10.1111/j.1365-294X.2009.04116.x

Durrett, R. (2009). *Elementary Probability with Applications*. Cambridge University Press, New York.

Gillman, L. (1992). The car and the goats, *Am. Math. Mon.*, **99**, 1, 3–7.

Kolmogorov, A. N. (1933). *Grundbegriffe der Wahrscheinlichkeitsrechnung*, Springer, Berlin.

Laplace P. S. (1774). Mémoire sur la probabilité des causes par les évenements. *Mém. de l'Ac. R. des Sciences de Paris*, **6**, 621–656.

Lauritzen, S. L. and Spiegelhalter, D. L. (1988). Local computations with probabilities on graphical structures and their application to expert systems. *J. R. Stat. Soc. B*, **50**, 157–194.

McManus, I. (1991). The Inheritance of Left-Handedness. Ciba Foundation Symposia [serial online], **162**, 251–281. Available from: Science Citation Index, Ipswich, MA. Accessed October 22, 2012.

Piaget, J. and Inhelder B. (1976). *The Origin of the Idea of Chance in Children*. Norton, New York, 276 pp.

Ross, S. (2009). *A First Course in Probability*, 8th Edition. Prentice Hall.

Selvin, S. (1975). A Problem in Probability. *Am. Stat.*, **29**, 1, 67.

Székely, G. (1986). *Paradoxes in Probability Theory and Mathematical Statistics*, Académiai Kiadó, Budapest, Hungary.

Venn, J. (1880). On the employment of geometrical diagrams for the sensible representation of logical propositions. *Proc. Cambridge Philos. Soc.*, **4**, 47–59.

Wong, S. K. N., O'Connell, M., Wisdom, J. A., and Dai, H. (2005). Carbon nanotubes as multifunctional biological transporters and near-infrared agents for selective cancer cell destruction. *Proc. Natl. Acad. Sci.*, **102**, 11600–11605.

Chapter 4
Sensitivity, Specificity, and Relatives

Poetry teaches us music, metaphor, condensation and specificity.

– Walter Mosley

4.1 Introduction

This chapter introduces several notions fundamental for disease or device testing. The sensitivity, specificity, and positive and negative predictive values of a test are measures of the performance of a diagnostic test and are intimately connected with probability calculations (estimations) and Bayes' rule.

Although concepts such as "false positives" and "true negatives" are quite intuitive, many students and even health professionals have difficulties in assessing the associated probabilities. The following problem was posed by Casscells et al. (1978) to 60 students and staff at an elite medical school: *If a test to detect a disease whose prevalence is 1/1000 has a false positive rate of 5%, what is the chance that a person found to have a positive result actually has the disease, assuming you know nothing about the person's symptoms or signs?*

Assuming that the probability of a positive result given the disease is 1, the answer to this problem is approximately 2%. Casscells et al. found that only 18% of participants gave this answer. The most frequent response was 95%, presumably on the supposition that, because the error rate of the test is 5%, it must get 95% of results correct.

Examples of misconceptions of test precision measures, especially involving the sensitivity and positive predictive value, are abundant.

When a multiplicity of tests are possible and a researcher is to select the "best" test, the receiver operating characteristic (ROC) curve methodology is used. This methodology is especially useful in setting a threshold that separates positive and negative outcomes of a test.

4.2 Notation

Usually, n subjects are selected randomly from a given population. The population may not be very general; it could be a specific segment of subjects (patients in a hospital who were checked in 10+ days ago, subjects with a history of heart attack, etc.). Now suppose that the true disease status (disease present/absent) in all subjects is determined via a gold standard assessment and that we are interested in evaluating a particular test for the disease assuming that the gold standard results are always correct. For example, in testing for breast cancer (BC), a mammogram is used as a test while the battery of numerous patient symptoms, medical history, and biopsy results are used as a gold standard.

A positive test would not necessarily mean the disease is present but rather would mean that the test says the disease is present. For instance, a patient's mammogram appears to show breast cancer. A true-positive test result not only means that the test says the disease is present, but that the disease *really is present*. In this case a positive mammogram of a patient for which the gold standard indicates BC would be a true positive. In the same context, false positives, true negatives, and false negatives are defined in a corresponding manner.

By classifying the patients with respect to the test results and the true disease status, the following table can be constructed:

	Disease (D)	No disease (C)	Total
Test positive (P)	TP	FP	nP = TP + FP
Test negative (N)	FN	TN	nN = FN + TN
Total	nD = TP + FN	nC = FP + TN	n = nD + nC = nP + nN

where

TP	True positive (test positive, disease present)
FP	False positive (test positive, disease absent)
FN	False negative (test negative, disease present)
TN	True negative (test negative, disease absent)
nP	Total number of positives (TP + FP)
nN	Total number of negatives (TN + FN)
nD	Total number with disease present (TP + FN)
nC	Total number without disease present (TN + FP)
n	Total sample size (TP + FP + FN + TN)

The test's effectiveness is measured by the number of true positives and true negatives relative to the total number of cases. It turns out that these two numbers are used to define two complementary measures of test performance: *sensitivity* and *specificity*.

We will illustrate the defined numbers with the following example:

Example 4.1. **BreastScreen Victoria.** This 1994 study involved women who participated in the BreastScreen Victoria initiative in Victoria, Australia, where free biennial screenings for BC are provided to women aged 40 and older. The data provided by Kavanagh et al. (2000) show that among 96,420 asymptomatic women, 5,401 had positive and 91,019 negative mammogram results. The mammograms were read independently by two radiologists. In the case of disagreement over whether to recall, a consensus was reached or a third reader made the decision. The women were then recommended for routine rescreen or referred for assessment. Assessment might include clinical examination, further radiographs, ultrasound, or biopsy. After assessment, women might have a cancer diagnosed, be recommended for routine rescreening, or be recommended for further assessment (early review). Out of 96,420 women, 665 were diagnosed with BC. Of those 665 diagnosed, the mammogram was positive for 495 and negative for 160 women. The table summarizing the data is below.

	BC diagnosed	BC not diagnosed	Total
Mammogram positive	495	4906	5401
Mammogram negative	160	90859	91019
Total	665	95765	96420

Sensitivity is the ratio of the number of true positives and the number of subjects with a disease, while specificity is the ratio of the number of true negatives and the number of subjects without the disease. In the BC example sensitivity is $495/665 = 75.57\%$, and specificity is $90859/95765 = 94.88\%$.

Both measures have to be reported, since reporting only sensitivity or only specificity reveals little information about the test. There are two extreme cases. Imagine a test that classifies *all* subjects as positive – trivially the sensitivity is 100%. Since there are no negatives, the specificity is zero. Likewise, a test that classifies all subjects as negative has a specificity of 100% and zero sensitivity.

The following table summarizes the key notions:

Sensitivity (Se)	$Se = TP/(TP + FN) = TP/nD$
Specificity (Sp)	$Sp = TN/(FP + TN) = TN/nC$
Prevalence (Pre)	$(TP + FN)/(TP + FP + FN + TN) = nD/n$
Positive predictive value (PPV)	$PPV = TP/(TP + FP) = TP/nP$
Negative predictive value (NPV)	$NPV = TN/(TN + FN) = TN/nN$
Likelihood ratio positive (LRP)	$LRP = Se/(1-Sp)$
Likelihood ratio negative (LRN)	$LRN = (1-Se)/Sp$
Apparent prevalence (APre)	$APre = nP/n$
Accuracy (Ac)	$Ac = (TP + TN)/n$

The population prevalence of a disease is defined as the probability that a randomly selected person from this population will have the disease. As the table shows, the prevalence is estimated by $(TP + FN)/(TP + FP + FN + TN) = nD/n$. For the Victoria BC data the prevalence is $665/96420 = 0.0069$. This is a valid estimator only if the table is a summary of a representative sample of the population under analysis. In other words, the sample should have been taken at random and the tabulation made subsequently. This is not the case in many studies. The prevalence of some diseases in a general population is often so small that insisting on a random sample would require huge sample sizes in order to obtain a nonzero TP or FN table entries. When the table is made from available cases and controls (convenience samples), the prevalence for the population cannot be estimated from it.

A related quantity is the incidence of a disease in a population. It is defined as the probability that a randomly selected person from the subset of people not affected by the disease will develop the disease in a fixed time window (week, month, year). While the prevalence relates to the magnitude, the incidence provides information about the progression and dynamics of the disease.

Positive/Negative Predictive Values. One of the most important measures is the positive predictive value (PPV). Based on the table it can be estimated as the proportion of true positives among all positives, TP/nP. This is correct only if the population prevalence is well estimated by nD/n, that is, if the table is representative of its population. This is approximately the case for the Victoria BC data; the PPV is well estimated by $495/5401 = 0.0916$.

If the table is constructed from a convenience sample, the prevalence (Pre) would have to be provided as external information. Then the PPV is calculated as

$$PPV = \frac{Se \times Pre}{Se \times Pre + (1 - Sp) \times (1 - Pre)}.$$

This is simply the Bayes rule and will be discussed more in the next section.

Why is the PPV so important? Imagine an almost perfect test for a particular disease, with a sensitivity of 100% and specificity of 99%. If the prevalence of the disease in the population is 10%, then among ten positives there would be approximately one false positive. However, if the population prevalence is 1/10000, then for each true positive there would be approximately 100 false positives.

The negative predicted value (NPV) is the probability that a person with negative test result will not have the disease. For a person from the study, the NPV is estimated as $TN/(TN + FN)$. However, if the person is randomly selected from a population with the prevalence of disease Pre, then, by Bayes' rule:

$$NPV = \frac{Sp \times (1 - Pre)}{Sp \times (1 - Pre) + (1 - Se) \times Pre}.$$

Likelihood Ratio Positive/Negative. Diagnostic evidence provided by the test is sometimes judged by Likelihood Ratio Positive (LRP) and likelihood ratio negative (LRN). The LRP is sometimes called *disease rule-in*. It represents the extent by which a positive test result would increase the likelihood of the disease.

The LRN is sometimes called *disease rule-out*. It represents the extent by which a negative test result would decrease the likelihood of disease, or equivalently, increase the likelihood of no-disease. For example, if prior to the test the odds of disease are established (by symptoms, prevalence, etc.), the post-test odds are calculated as

Disease odds | Test positive $=$ LRP $\times$ Pretest disease odds;

Disease odds | Test negative $=$ LRN $\times$ Pretest disease odds.

In the context of BreastScreen Victoria, suppose that Ann and Betty are two participants who tested, respectively, positive and negative for BC. If prior to the test the odds that both Ann ad Betty had BC were 1 : 50, what are the odds after the test? The LRP for BreastScreen data is $\text{Se}/(1{-}\text{Sp}) = 0.7557/0.0512 = 14.7598$, while the LRN is $(1{-}\text{Se})/\text{Sp} = 0.2443/0.9488 = 0.2575$. For Ann the posttest odds of disease are $14.7598 \cdot 1/50 = 0.2952$. This is approximately 1 : 3.3876. In terms of the probabilities of disease, the pretest probability of $1/(50 + 1) = 0.0196$ increases to $1/(3.3876 + 1) = 0.2279$ after the positive result. For Betty the posttest odds of disease are $0.2575 \times 1/50$ which is 1 : 194.1748.

Diagnostic Odds Ratio. The ratio LRP/LRN is sometimes called diagnostic odds ratio (DOR) and represents a single index of test performance. The larger the DOR, the better the test. The DOR can be expressed as

$$\text{DOR} = \frac{\text{LRP}}{\text{LRN}} = \frac{\text{Se} \times \text{Sp}}{(1 - \text{Se})(1 - \text{Sp})} = \frac{\frac{\text{TP}}{\text{FP}}}{\frac{\text{FN}}{\text{TN}}} = \frac{\frac{\text{TP}}{\text{FN}}}{\frac{\text{FP}}{\text{TN}}} = \frac{\text{TP} \times \text{TN}}{\text{FN} \times \text{FP}}. \quad (4.1)$$

The DOR is interpreted as the odds of a person who tests positive to have disease divided by the odds of a person who tests negative to have the disease. It also can be interpreted, by regrouping TP, FP, FN, and TN in (4.1) in a different way, as the odds of a person with disease testing positive divided by the odds of a person without the disease testing positive.

Conditional Probability Notation. The definitions introduced in this section depend on the relative frequencies in the observed tables, and they are empirical. The theoretical counterparts are expressed in terms of probabilities. The analogy to this is the interplay of the probability of an event A, $\mathbb{P}(A)$, which is theoretical, and relative frequency of the event n_A/n, which is empirical.

Let T be the event that a subject tests positive and D, D^c the hypothesis that the subject does/does not have the disease. The sensitivity is the conditional probability $\mathbb{P}(T|D) = \mathbb{P}(T \cap D)/\mathbb{P}(D)$, which is estimated by $(\text{TP}/n)/(n\text{D}/n) = \text{TP}/n\text{D}$. Analogously, the specificity is $\mathbb{P}(T^c|D^c) = \mathbb{P}(T^c \cap D^c)/\mathbb{P}(D^c)$, which is estimated by $(\text{TN}/n)/(n\text{C}/n) = \text{TN}/n\text{C}$. We have argued that $\mathbb{P}(D)$, the population prevalence, cannot be estimated from the table unless the sample forming the table is representative of the population. In the case of "convenience" samples, however, the prevalence is evaluated separately or assumed known from other studies. If the sample is randomly obtained from the population, the prevalence can be estimated by $n\text{D}/n$. See also Exercise 4.10 for a related approach. The probability of a positive test is in fact given by the rule of total probability, $\mathbb{P}(T) = \mathbb{P}(T|D)\mathbb{P}(D) + \mathbb{P}(T|D^c)\mathbb{P}(D^c)$. Note that $\mathbb{P}(T)$ depends on the prevalence and is estimated by $n\text{P}/n$ for a table from a random sample.

Finally, the PPV and NPV are determined by Bayes' rule. For example, the PPV is

$$\mathbb{P}(D|T) = \frac{\mathbb{P}(T|D)\mathbb{P}(D)}{\mathbb{P}(T)} = \frac{\mathbb{P}(T|D)\mathbb{P}(D)}{\mathbb{P}(T|D)\mathbb{P}(D) + \mathbb{P}(T|D^c)\mathbb{P}(D^c)}.$$

The posttest disease odds ratio is LRP times the pretest odds ratio. Due to this property, the LRP is in fact Bayes' factor in the terminology of Chapters 3 and 8.

$$\frac{\mathbb{P}(D|T)}{\mathbb{P}(D^c|T)} = \frac{\mathbb{P}(T|D)\mathbb{P}(D)}{\mathbb{P}(T)} \bigg/ \frac{\mathbb{P}(T|D^c)\mathbb{P}(D^c)}{\mathbb{P}(T)}$$

$$= \frac{\mathbb{P}(T|D)\mathbb{P}(D)}{\mathbb{P}(T|D^c)\mathbb{P}(D^c)} = \frac{\mathbb{P}(T|D)}{\mathbb{P}(T|D^c)} \times \frac{\mathbb{P}(D)}{\mathbb{P}(D^c)}.$$

Thus, posterior disease odds = LRP × prior disease odds.

The definitions above are illustrated with an example where researchers test for acute pulmonary embolisms.

Example 4.2. **D-Dimer.** When a vein or artery is injured and begins to leak blood, a sequence of clotting steps and factors (called the coagulation cascade) are activated by the body to limit the bleeding and create a blood clot to plug the hole. During this process, threads of a protein called fibrin are produced. These threads are cross-linked (chemically glued together) to form a fibrin net that catches platelets and helps hold the forming blood clot together at the site of the injury. Once the area has had time to heal, the body uses a protein called plasmin to break the clot (thrombus) into small pieces so that it can be removed. The fragments of the disintegrating fibrin in the clot are called fibrin degradation products (FDPs). One of the FDPs produced is D-dimer, which consists of variously sized pieces of cross-linked fibrin. D-dimer is normally undetectable in the blood and is produced only after a clot has formed and is in the process of being broken down. Measurement of D-dimer can indicate problems in the body's clotting mechanisms. The data below consist of quantitative plasma D-dimer levels among patients undergoing pulmonary angiography for suspected pulmonary embolism (PE). The patients who exceed the threshold of 500 ng/mL are classified as positive for PE. The gold standard for PE is the pulmonary angiogram. Goldhaber et al. (1993), from Brigham and Women's Hospital at Harvard Medical School, considered a population of patients who are suspected of PE based on a battery of symptoms. The summarized data for 173 patients are provided in the table below.

	Acute PE	No PE present	Total
Test positive (D-dimer ≥ 500 ng/mL)	42	96	138
Test negative (D-dimer < 500 ng/mL)	3	32	35
Total	45	128	173

A simple MATLAB file ◀ sesp.m will calculate the sensitivity, specificity, prevalence, positive and negative predictive values, and degree of test accuracy.

```
function [se sp pre ppv npv ag] = sesp(tp, fp, fn, tn)
 %D-dimer as a test for acute PE (Goldhaber et al, 1993)
 % [s1, s2, p1, p2, p3, a, yi] = sesp(42,96,3,32)
 %
n = tp+tn+fn+fp; %total sample size
np = tp + fp; %total positive
nn = tn + fn; %total negative
nd = tp + fn; %total with disease
nc = tn + fp; %total control (without disease)
%--------------
se = tp/nd; %tp/(tp + fn):::sensitivity
sp = tn/nc; %tn/(tn + fp):::specificity
pre = nd/n; %(tp + fn)/(tp+tn+fn+fp):::prevalence
%only in for case when sample is random from the
%   population of interest. Otherwise, the prevalence
%   needed for calculating PPV and NPV is an input value
ppv = tp/np; %tp/(tp + fp):::positive predictive value
npv = tn/nn; %tn/(tn+fn):::negative predictive value
lrp = se/(1-sp); %:::likelihood ratio positive
lrn = (1-se)/sp; %:::likelihood ratio negative
ac = (tp+tn)/n;  %:::accuracy
yi = (se + sp - 1)/sqrt(2); %:::youden index
%--------------
disp(' Se Sp Pre PPV NPV LRP Ag Yi')
disp([se, sp, pre, ppv, npv, lrp, ag yi])
     %spacing in disp depends on the font size.
```

For the D-dimer data, the result is

```
[a b c d e f g] = sesp(42,96,3,32);
  Se     Sp    Pre    PPV    NPV    LRP    Ac     Yi
0.9333 0.2500 0.2601 0.3043 0.9143 1.2444 0.4277 0.1296
```

Goldhaber et al. (1993) conclude: "The results of our study indicate that quantitative plasma D-dimer levels can be useful in screening patients with suspected PE who require pulmonary angiography. Plasma D-dimer values less than 500 ng/mL may obviate the need for pulmonary angiography, particularly among medical patients for whom the clinical suspicion of PE is low. The plasma D-dimer value, assayed using a commercially available enzyme-linked immunosorbent assay kit, is a sensitive but nonspecific test for the presence of acute PE."

4.3 Combining Two or More Tests

Suppose that k independent tests for a particular condition are available and that their sensitivities and specificities are Se_1, Sp_1, Se_2, Sp_2, ..., Se_k, Sp_k. The independence here is conditional on the true disease status of the patient. If these tests could be combined, what would be the sensitivity/specificity of the combined test?

First, it is important to define how the tests are going to be combined. There are two main strategies: *parallel* and *serial*. When the tests are assumed conditionally independent, the calculations are similar to those in circuit problems from Chapter 3, page 82. Denote by Se and Sp the sensitivity and specificity of the combined test, respectively.

In the parallel strategy the combination is positive if at least one test is positive and negative if all tests are negative. Then, the sensitivity is calculated as the probability of a union and the specificity as the probability of an intersection:

Parallel combination (positive if at least 1 positive)

$Se = 1 - [(1 - Se_1) \times (1 - Se_2) \times \cdots \times (1 - Se_k)]$

$Sp = Sp_1 \times Sp_2 \times \cdots \times Sp_k$

It is easy to see that, in the parallel strategy, the sensitivity is larger than any individual sensitivity and the specificity smaller than any individual specificity.

In the serial strategy, the combination is positive if all tests are positive and negative if at least one test is negative. Then the sensitivity is calculated as the probability of an intersection and the specificity as the probability of a union:

Serial combination (positive if all positive)

$Se = Se_1 \times Se_2 \times \cdots \times Se_k$

$Sp = 1 - [(1 - Sp_1) \times (1 - Sp_2) \times \cdots \times (1 - Sp_k)]$

Note that the overall sensitivity is smaller than any individual sensitivity, while the specificity is larger than any individual specificity. There are other possible combinations as well as procedures that address bias and correlation among the individual tests.

Example 4.3. **Combining Two Tests for Sarcoidosis.** Parikh et al. (2008) provide an example of combining two tests for sarcoidosis. Sarcoidosis is an idiopathic multisystem granulomatous disease, where the diagnosis is made by a combination of clinical, radiological, and laboratory findings. The gold standard is a tissue biopsy showing noncaseating granuloma. Ocular sarcoidosis could be present as anterior, intermediate, posterior, or panuveitis, but none of these is pathognomonic. Therefore, one has to rely on ancillary testing to confirm the diagnosis.

An angiotensin-converting enzyme (ACE) test has a sensitivity of 73% and a specificity of 83% to diagnose sarcoidosis. An abnormal gallium scan has a sensitivity of 91% and a specificity of 84%. Though individually the specificity of either test is not impressive, for the serial combination the specificity becomes

$$Sp = 1 - (1 - 0.84) \times (1 - 0.83) = 1 - (0.16 \times 0.17) = 0.97.$$

The combination sensitivity becomes $0.73 \times 0.91 = 0.66$. Note that the overall specificity drastically improves, but at the expense of overall sensitivity.
🖉

The conditional independence of tests in the previous example is a limiting assumption. One could argue that two tests for the same disease are seldom independent by the very nature of the testing problem, even conditionally on the true disease status of the patient.

Combining Two Conditionally Dependent Tests.
Dependence between the sensitivities or specificities of pairs of tests affects the sensitivity and specificity of their combination. A positive dependence in test sensitivity reduces the sensitivity of parallel combination and a positive dependence in test specificity reduces the specificity of serial combination.

Assume that two tests that are not conditionally independent are to be combined. Let

$$c_{Se} = \frac{TP_{1 \cap 2}}{nD} - Se_1 \times Se_2, \quad \text{and} \quad c_{Sp} = \frac{TN_{1 \cap 2}}{nC} - Sp_1 \times Sp_2$$

be conditional covariances for sensitivity and specificity proportions. Here $TP_{1 \cap 2}$ is the number of instances when both tests simultaneously resulted positive among the diseased patients, and $TN_{1 \cap 2}$ is the number of instances when both tests resulted as negative among the controls. When the tests are conditionally independent, c_{Se} and c_{Sp} are 0.

Sensitivity and specificity for parallel and serial combinations of the tests are

Parallel combination
$Se = 1 - (1 - Se_1) \times (1 - Se_2) - c_{Se}$
$Sp = Sp_1 \times Sp_2 + c_{Sp}$
Serial combination
$Se = Se_1 \times Se_2 + c_{Se}$
$Sp = 1 - (1 - Sp_1) \times (1 - Sp_2) - c_{Sp}$

See also Exercise 4.17

In the following example, we show how to handle more complex batteries of tests in which the tests could be dependent. The example considers two tests and a parallel combination strategy, but it could be extended to any number of tests and to more general combination strategies.

The approach is based on a simulation because analytic solutions are typically computationally involved.

Example 4.4. **Simulation Approach.** Suppose that a testing procedure consists of two tests given in a sequence. Test A has a sensitivity of 0.9 and a specificity of 0.8. Test B has a sensitivity of 0.7 and a specificity of 0.9 for subjects who tested negative in test A and a sensitivity of 0.95 and a specificity of 0.6 for subjects who tested positive in test A.

Clearly, test A and test B are dependent. If a subject is declared positive when the result of at least one of the two tests was positive (parallel link), what is the overall sensitivity/specificity of the described testing procedure? The population prevalence is considered known and is used in the simulation of a patient's status, but it does not affect the overall sensitivity/specificity.

Note that, if a subject's status s is equal to 0/1 when the disease is absent/present, then the result of a test is s*(rand < se) + (1-s)*(rand > sp) for a known sensitivity and specificity, se, sp. The test outcome is binary, with 0/1 denoting a negative/positive test result.

The following MATLAB code (◢ simulatetesting2.m) considers 20,000 subjects from a population where the disease prevalence is 0.2. The estimated sensitivity/specificity was 0.97/0.72, but simulation results may vary slightly due to the random status of subjects.

```
nsubjects = 20000;
prevalence = 0.2;
se1  =0.9; sp1  = 0.8;  %se/sp of test1
se20 =0.7; sp20 = 0.9;  %se/sp of test2 if test1=0
se21 =0.95; sp21 = 0.6;  %se/sp of test2 if test1=1

tests = [];
ss=[]; tp=0; fp=0; fn=0; tn=0;
for i = 1:nsubjects
    %simulate a subject wp of disease equal to prevalence
    s = (rand < prevalence);
    %test the subject
    test1=s*(rand < se1) + (1-s)*(rand>sp1); %test is 0 or 1
    if (test1 == 0)
        test2=s*(rand < se20) + (1-s)*(rand>sp20);
    else
        test2=s*(rand < se21) + (1-s)*(rand>sp21);
    end
    %test = test1*test2;        %for serial
    test = (test1 + test2 > 0); %for parallel
    ss=[ss s];                  %save subject's status
    tests = [tests test];       %save subject's test
    %building the test table
    tp = tp + test*s;           %true positives
    fp = fp + test*(1-s);       %false positives
    fn = fn + (1-test)*s;       %false negatives
```

```
    tn = tn + (1-test)*(1-s);     %true negatives
end
% estimate overall Se/Sp from the table
sens = tp/(tp+fn)
spec = tn/(tn+fp)
```

Remark: In the previous discussion we assumed that a true disease status was known and that a perfect gold standard test was available. In many cases an error-free assessment does not exist, but a reference test, with known sensitivity Se_R and specificity Sp_R, can be used. By taking this reference test as a gold standard and by not accounting for its inaccuracy would lead to biases in evaluating a new test. Staquet et al. (1981) provide a solution based on Se_R, Sp_R, and concordance of results between the two tests.

Another approach to this problem is "discrepant resolution," in which the subjects for whom the reference and new test disagreed were subjected to a third "resolver" test. Although commonly used, the resolver method can be biased and can overestimate sensitivity and specificity of a new test significantly (Hawkins et al., 2001; Qu and Hadgu, 1998).

4.4 ROC Curves

The receiver operating characteristic (ROC) curve was first used during World War II for the analysis of radar signals. After WWII it was employed in signal detection theory and, subsequently, in a range of fields where testing is critical. The ROC curve is defined as a graphical plot of `sensitivity` vs. (1 - `specificity`) for a binary classifier system as its discrimination threshold (value that separates positives and negatives) varies.

Let us look at an ROC curve using the D-dimer example from the previous section. Mavromatis and Kessler (2001) report that in 18 publications (between 1988 and 1998) concerning D-dimer testing, the reported cut point for declaring the test positive ranged from 250 to 1000 ng/mL. What cut point should be recommended? To increase the apparently low specificity in the previous D-dimer analysis, suppose that the threshold for testing positive is increased from 500 to 650 ng/mL and that the data are distributed in the following way:

	Acute PE	No PE present	Total
Test positive (D-dimer $\geq$ 650 ng/mL)	31	33	64
Test negative (D-dimer $<$ 650 ng/mL)	14	95	109
Total	45	128	173

This new table results in the following `sesp` output:

```
[a b c d e f] = sesp(31,33,14,95);
  Se     Sp     Pre    PPV    NPV    LRP    Ag     Yi
0.6889 0.7422 0.2601 0.4844 0.8716 2.6721 0.7283 0.3048
```

Combining this with the output of the 500-ng/mL threshold, we get the vectors `1-sp = [0 1-0.7422 1-0.25 1]` and `se = [0 0.6889 0.9333 1]`.

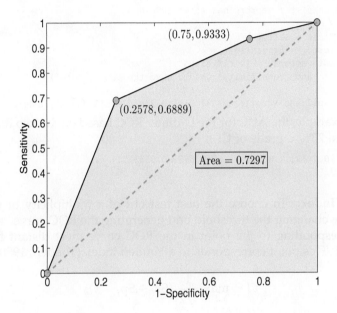

Fig. 4.1 Rudimentary ROC curve for D-dimer data based on two thresholds.

The code ⬛ `RocDdimer.m` plots this "rudimentary" ROC curve (Fig. 4.1). The curve is rudimentary because it is based on only two tests. Note that points (0,0) and (1,1) always belong to ROC curves. These two points correspond to the trivial tests in which all patients test negative or all patients test positive. The area under the ROC curve (AUC), is a well-accepted measure of test performance. The closer the area is to 1, the more unbalanced the ROC curve, implying that both sensitivity and specificity of the test are high. It is interesting that some researchers assign an academic scale to AUC as an informal measure of test performance.

AUC	Performance
0.9–1.0	A
0.8–0.9	B
0.7–0.8	C
0.6–0.7	D
0.0–0.6	F

The following MATLAB program calculates AUC when the vectors `csp` = 1 - specificity and `sensitivity` are supplied.

```matlab
function A = auc(csp, se)
%
% A = auc(csp,se) computes the area under the ROC curve
% where 'csp' and 'se' are vectors representing (1-specificity)
% and (sensitivity), used to plot the ROC curve
% The length of the vectors has to be the same

csp=csp(:); se = se(:);
if length(csp) ~= length(se)
error('Input vectors (1-specificity) ...
            and (sensitivity) should have the same length')
end
A = sum((csp(2:end)-csp(1:end-1)) .* (se(2:end)+se(1:end-1))/2 );
```

For example, the AUC for the D-dimer ROC based on the two thresholds is approx. 73%, a grade of C:

```matlab
auc([0, 1-0.7422, 1-0.25, 1],[0 0.6889 0.9333 1])
 ans = 0.7297
```

Youden Index. To choose the best test out of a multiplicity of tests obtained by changing the threshold and generating the ROC curve, select the test corresponding to the point in the ROC curve most distant from the diagonal. This point corresponds to a Youden index (Youden, 1950)

$$YI = \max_i \frac{1}{\sqrt{2}} (Se_i + Sp_i - 1),$$

where Se_i and Sp_i are, respectively, the sensitivity and specificity for the ith test. Thus, the Youden index is the distance of the most distant point $(1 - Sp_i, Se_i)$ on the ROC curve from the diagonal. This distance ranges between 0 and $\sqrt{2}/2$. In the expression for Youden index the constant $1/\sqrt{2}$ is taken because of geometric interpretation (distance from the diagonal), often the constant is omitted.

In the D-dimer example, the Youden index for the test with a 500-ng/mL threshold is 0.1296, compared to 0.3048 for the test with a 650-ng/mL threshold. Between the two tests, the test with the 650-ng/mL threshold is preferred.

*F***-measure.** Another objective criterion to choose from the multiplicity of thresholds in a test is *F*-measure or *F*-index. It is defined as a harmonic average of sensitivity and and positive predictive value,

$$F = \frac{2}{1/Se + 1/PPV}.$$

It is easy to see that *F* is the ratio of TP and an average of nD and nP,

$$F = \frac{TP}{(nD + nP)/2}.$$

The test that maximizes F-measure is favored.

In the D-dimer example, the F-measure for the test with a 500-ng/mL threshold is 0.4590, compared to 0.5688 for the test with a 650-ng/mL threshold. Between the two tests, the test with the 650-ng/mL threshold is preferred.

Example 4.5. **ADA.** *Adenosine deaminase* (ADA) is an enzyme involved in the breakdown of adenosine to uric acid. ADA levels were found to be elevated in the pleural fluid of patients with tuberculosis (TB) pleural effusion. Pleural effusion is a very common clinical problem. It may occur in patients of pulmonary TB, pneumonia, malignancy, congestive cardiac failure, cirrhosis of the liver, nephrotic syndrome, pulmonary infarction, and connective tissue disorders. TB is one of the primary causes of pleural effusion. Numerous studies have evaluated the usefulness of ADA estimation in the diagnosis of TB pleural effusion. However, the sensitivity and specificity of ADA estimation and the cutoff level used for distinguishing TB pleural effusion from non-TB pleural effusion have varied between studies. The data (given in ROCTBCA.XLS or ROC.mat) were collected by Dr. Mark Hopley of Chris-Hani Baragwanath Hospital (CHB, the largest hospital in the world), with the goal of critically evaluating the sensitivity and specificity of ADA estimation in the diagnosis of TB pleural effusion.

The data set consists of three columns:

Column 1 contains ADA levels.

Column 2 is an indicator of TB. The indicator is "1" if the patient had documented TB, and zero otherwise.

Column 3 is an indicator of documented carcinoma. Six patients who had both carcinoma and TB have been excluded from the analysis.

To create an empirical ROC curve, the following four steps are applied:

(i) The data are sorted according to the ADA level, with the largest values first.

(ii) A column is created where each entry gives the total number of TB patients with ADA levels greater than or equal to the ADA value for that entry.

(iii) A column equivalent to that from step 2 is created for patients with cancer.

(iv) Two new columns are created, containing the true positive frequency (TPF) and false positive frequency (FPF) for each entry. The TPF is calculated by taking the number of TB cases identified at or above the ADA level for the current entry and dividing by the total number of TB cases. The FPF is determined by taking the number of "false TB alarms" (cancer patients) at or above that level and dividing by the total number of such non-TB patients.

This description can be simply coded in MATLAB thanks to the cumulative summation (cumsum) command:

```
disp('ROC Curve Example')
set(0, 'DefaultAxesFontSize', 16);
fs = 15;
        % data file ADA.mat should be on path
load 'ADA.mat'
        % columns in ADA.mat are:
        % 1. ADA level (ordered decreasingly)
        % 2. indicator of case TB
        % 3. indicator of non-case CA
cumultruepos  = cumsum(ada(:,2));
cumulfalsepos = cumsum(ada(:,3));
        % these are true positives/false positives if the
        % cut-level is from the sequence ada(:,1).
tpf = cumultruepos/cumultruepos(end);    %sensitivity
fpf = cumulfalsepos/cumulfalsepos(end);  %1-specificity
plot(fpf,tpf) %ROC, sensitivity against (1-specificity)
xlabel('1 - specificity')
ylabel('sensitivity')
```

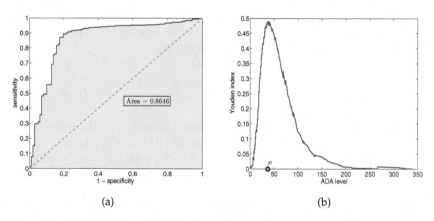

Fig. 4.2 (a) ROC curve for ADA data. (b) Youden index against ADA level.

Which ADA level should be recommended as a threshold? The Youden index for the ROC curve in Figure 4.2a is 0.4910, which corresponds to ADA level of 37, Figure 4.2b. For this particular threshold, the sensitivity and specificity are 0.8904 and 0.8039, respectively.

```
%youden index
yi = max((seth-cspth)/sqrt(2))        %0.4910
%ADA level corresponding to YI
ada((seth-cspth)/sqrt(2)== yi , 1)    %37
```

```
% sensitivity/specificity at YI
seth((seth-cspth)/sqrt(2)== yi)         %0.8904
1 - cspth((seth-cspth)/sqrt(2)== yi)    %0.8039
```

✎

4.5 Exercises

4.1. **Stacked Auditory Brainstem Response.** The failure of standard auditory brainstem response (ABR) measures to detect small (<1 cm) acoustic tumors has led to the use of enhanced magnetic resonance imaging (MRI) as the standard to screen for small tumors. The study by Don et al. (2005) investigated the suitability of the stacked ABR as a sensitive screening alternative to MRI for small acoustic tumors (SATs). The objective of the study was to determine the sensitivity and specificity of the stacked ABR technique for detecting SATs. A total of 54 patients were studied who had MRI-identified acoustic tumors that were either <1 cm in size or undetected by standard ABR methods, regardless of size. There were 78 nontumor normal-hearing subjects who tested as controls. The stacked ABR demonstrated 95% sensitivity and 88% specificity. Recover the testing table.

4.2. **Hypothyroidism.** Low values of a total thyroxine ($T4$) test can be indicative of *hypothyroidism* (Goldstein and Mushlin, 1987). Hypothyroidism is a condition in which the body lacks sufficient thyroid hormone. Since the main purpose of the thyroid hormone is to "run the body's metabolism," it is understandable that people with this condition will have symptoms associated with a slow metabolism. Over five million Americans have this common medical condition.

A total of 195 patients, among which 59 have confirmed hypothyroidism, have been tested for the level of $T4$. If the patients with a $T4$ level ≤ 5 are considered positive for hypothyroidism, the following table is obtained:

$T4$ value	Hypothyroid	Euthyroid	Total
Positive, $T4 \leq 5$	35	5	40
Negative, $T4 > 5$	24	131	155
Total	59	136	195

However, if the thresholds for $T4$ are 6, 7, 8, and 9, the following tables are obtained:

$T4$ value	Hypothyroid	Euthyroid	Total
Positive, $T4 \leq 6$	39	10	49
Negative, $T4 > 6$	20	126	146
Total	59	136	195

T4 value	Hypothyroid	Euthyroid	Total
Positive, $T4 \leq 7$	46	29	75
Negative, $T4 > 7$	13	107	120
Total	59	136	195

T4 value	Hypothyroid	Euthyroid	Total
Positive, $T4 \leq 8$	51	61	112
Negative, $T4 > 8$	8	75	83
Total	59	136	195

T4 value	Hypothyroid	Euthyroid	Total
Positive, $T4 \leq 9$	57	96	153
Negative, $T4 > 9$	2	40	42
Total	59	136	195

Notice that you can improve the sensitivity by moving the threshold to a higher $T4$ value; that is, you can make the criterion for a positive test less strict. You can improve the specificity by moving the threshold to a lower $T4$ value; that is, you can make the criterion for a positive test more strict. Thus, there is a trade-off between sensitivity and specificity.

(a) For the test that uses $T4 = 7$ as the threshold, find the sensitivity, specificity, positive and negative predictive values, likelihood ratio, and degree of agreement. You can use the code ◀ sesp.m.

(b) Using the given thresholds for the test to be positive, plot the ROC curve. What threshold would you recommend? Explain your choice.

(c) Find the area under the ROC curve. You can use the code ◀ auc.m.

4.3. **Alzheimer's.** A medical research team wished to evaluate a proposed screening test for Alzheimer's disease. The test was given to a random sample of 450 patients with Alzheimer's disease and to an independent sample of 500 subjects without symptoms of the disease.

The two samples were drawn from a population of subjects who are 65 years old or older. The results are as follows:

Test result	Diagnosed Alzheimer's	No Alzheimer's symptoms	Total
Positive test	436	5	441
Negative test	14	495	509
Total	450	500	950

(a) Using the numbers from the table, estimate $\mathbb{P}(T|D)$ and $\mathbb{P}(T^c|D^c)$. Interpret these probabilities in terms of the problem.

The probability of D (prevalence) is the rate of the disease in the relevant population (≥ 65 y.o.) and is estimated to be 11.3% (Evans 1990). Find $\mathbb{P}(D|T)$ (positive predicted value) using Bayes' rule. You cannot find $\mathbb{P}(D|T)$ using information from the table only – you need external info.

4.4. **Test for Being a Duchenne Muscular Dystrophy Carrier.** In Exercise 2.19, researchers used measures of pyruvate kinase and lactate dehydroginase to assess an individual's carrier status. The following table closely follows the authors' report:

	Woman carrier	Woman not carrier	Total
Test positive	56	6	62
Test negative	11	121	132
Total	67	127	194

(a) Find the sensitivity, specificity, and degree of agreement.
The sample is not representative of the general population for which the prevalence of carriers is 0.03%, or 3 in 10,000.
(b) With this information, find the PPV of the test, that is, the probability that a woman is a DMD carrier if she tested positive.
(c) What is the PPV if the table was constructed from a random sample of 194 subjects from a general population?
(d) Approximate the probability that among 15,000 women randomly selected from a general population, at least 2 are DMD carriers.

4.5. **Parkinson's Disease Statistical Excursions.** Parkinson's disease or, "shaking palsy," is a brain disorder that causes muscle tremor, stiffness, and weakness. Early symptoms of Parkinson's disease include muscular stiffness, a tendency to tire more easily than usual, and trembling that usually begins with a slight tremor in one hand, arm, or leg. This trembling is worse when the body is at rest, but will generally stop when the body is in use. For example, when the hand becomes occupied by "pill rolling," or when the thumb and forefinger are rubbed together as if rolling a pill (Fig. 4.3).

Fig. 4.3 "Pill rolling" stops muscle tremors in early Parkinson's disease.

In the later stages of Parkinson's disease, the affected person loses the ability to control his or her movements, making everyday activities hard to manage.

In a study by Schipper et al. (2008), 52 subjects, 20 with mild or moderate stages of Parkinson's disease and 32 age-matched controls, had whole blood samples analyzed using the near-infrared (NIR) spectroscopy and Raman spectroscopy methods. The data showed that the two independent biospectroscopy measurement techniques yielded similar and consistent results. In differentiating Parkinson's disease patients from the control group, Raman spectroscopy resulted in eight false positives and four false negatives. NIR spectroscopy resulted in four false positives and five false negatives.

(a) From the description above, construct tables for NIR spectroscopy and Raman spectroscopy containing TP, FP, FN, and TN.

(b) For both methods find the sensitivity and specificity. Assume that the prevalence of Parkinson's disease in the age group matching this group is 1/120 for the general population. For both methods, also find the PPV, that is, the probability that a person who tested positive and was randomly selected from the same age group in the general population has the disease if no other clinical information is available.

(c) Mr. Smith is one of the 52 subjects in the study, and he tested positive under a Raman spectroscopy test. What is the probability that Mr. Smith has the disease?

4.6. **Screening for Colorectal Cancer.** Immunochemical fecal occult blood screening has been used in Japan as an early screening practice for detecting colorectal cancer. The study described in Nakama (1997) was performed to assess the three testing methods of screening for colorectal cancer: one-day method, two-day method, and three-day method. The test involved 184 patients with biopsy confirmed colorectal cancer, and 368 healthy controls. The tests were conducted on three consecutive days, and the sensitivities and specificities of each day's test were evaluated.

Method	Sensitivity	Specificity
One-Day	67.9%	97.5%
Two-Day	88.0%	95.9%
Three-Day	90.1%	92.1%

(a) Fill out the three tables for TP, FP, TN, and FN based on the information available. Round table entries to the closest integer.

(b) If only one of the three tests is to be selected, which one would you recommend? Justify your choice.

(c) If a person who participated in this study is randomly chosen, what is the probability that this person has colorectal cancer if the "One-Day" test was positive?

(d) The prevalence of colorectal cancer among the adults in Japan is 1/200. If a person is randomly chosen from the adult population of Japan, what is the probability that this person is colorectal cancer free if the "Three-Day" test gave negative result and we do not have any additional information (symptoms, other tests, etc.)?

(e) Before the "Two-Day" test, Mr. Tanaka's odds of having colorectal cancer were 1/80, based on diet, lifestyle, and family history. If Mr. Tanaka tested positive, what are the odds after the test?

4.7. **Blood Tests in Diagnosis of Inflammatory Bowel Disease.** Cabrera-Abreu et al. (2004) explored the reliability of a panel of blood tests in screening for ulcerative colitis and Crohn's disease. The subjects were 153 children who were referred to a pediatric gastroenterology department with possible inflammatory bowel disease (IBD). Of these, 103 were found to have IBD (Crohn's disease 60, ulcerative colitis 37, indeterminate colitis 6). The 50 without IBD formed the controls. Blood tests evaluated several parameters, including hemoglobin, platelet count, ESR, CRP, and albumin. The optimal screening strategy used a combination of hemoglobin and platelet counts and "one of two abnormal" as the criterion for positivity. This was associated with a sensitivity of 90.3% and a specificity of 80.0%.

(a) Construct a table with TP, FP, FN, and TN rounded to the nearest integer.

(b) Find the prevalence and PPV if the prevalence can be assessed from the table (the table is obtained from a random sample from the population of interest).

4.8. **Carpal Tunnel Syndrome Tests.** Carpal tunnel syndrome is the most common entrapment neuropathy. The cause of this syndrome is hard to determine, but it can include trauma, repetitive maneuvers, certain diseases, and pregnancy.

Three commonly used tests for carpal tunnel syndrome are Tinel's sign, Phalen's test, and the nerve conduction velocity test. Tinel's sign and Phalen's test are both highly sensitive (0.97 and 0.92, respectively) and specific (0.91 and 0.88, respectively). The sensitivity and specificity of the nerve conduction velocity test are 0.93 and 0.87, respectively. Assume that the tests are conditionally independent.

Calculate the sensitivity and specificity of a combined test if combining is done

(a) in a serial manner;

(b) in a parallel manner.

(c) Find PPV for tests from (a) and (b) if prevalence of carpal tunnel syndrome is approximately 50 cases per 1000 subjects in the general population.

(d) Which of the tests from (a) and (b) would you recommend? Justify your answer.

4.9. **Hepatitic Scintigraphy.** A commonly used imaging procedure for detecting abnormalities in the liver is hepatitic scintigraphy. Drum and

Christacopoulos (1972) reported data on 344 patients who underwent scintigraphy and were later examined by autopsy, biopsy, or surgical inspection for a gold standard determination of the presence of liver pathology (parenchymal, focal, or infiltrative disease). The table summarizes the experimental results. Assume that this table is representative of the population of interest for this study.

	Liver disease (D)	No liver disease (C)	Total
Abnormal liver scan (P)	231	32	263
Normal liver scan (N)	27	54	81
Total	258	86	344

Find the sensitivity, specificity, prevalence, PPV, NPV, LRP, LRN, and concordance. Interpret the meaning of a LRP.

4.10. **Apparent Prevalence.** When the disease status in a sample is not known, the prevalence cannot be estimated directly. It is estimated using apparent prevalence. There is a distinction between the true prevalence (Pre – the proportion of a population with the disease) and apparent prevalence (APre – the proportion of the population that tests positive for the disease). If the estimators of sensitivity, specificity, and apparent prevalence are available, show that the estimator of prevalence is

$$Pre = \frac{APre + Sp - 1}{Se + Sp - 1}.$$

See also WinBUGS code ![icon]apre2pre.odc.

4.11. **HAAH Improves the Test for Prostate Cancer.** A new procedure based on a protein called human aspartyl (asparaginyl) beta-hydroxylase, or HAAH, adds to the accuracy of standard prostate-specific antigen (PSA) testing for prostate cancer. The findings were presented at the 2008 Genitourinary Cancers Symposium (Keith et al., 2008).

The research involved 233 men with prostate cancer and 43 healthy men, all over 50 years old. Results showed that the HAAH test had an overall sensitivity of 95% and specificity of 93%.

Compared to the sensitivity and specificity of PSA (about 40%), this test may prove particularly useful for men with both low and high PSA scores. In men with high PSA scores (4 to 10), the addition of HAAH information could substantially decrease the number of unnecessary biopsies, according to the authors.

(a) From the reported percentages, construct a table with true positives, false positives, true negatives, and false negatives. You will need to round to the nearest integer since the specificity and sensitivity were reported as integer percents.

(b) Suppose that for men aged 50+ in the United States, the prevalence of prostate cancer is 7%. Suppose that Jim Smith is randomly selected from this group and tested positive on the HAAH test. What is the probability that Jim has prostate cancer?

(c) Suppose that Bill Schneider is a randomly selected person from the sample of $n = 276$ $(= 233 + 43)$ subjects involved in the HAAH study. What is the probability that Bill has prostate cancer if he tests positive and no other information is available? What do you call this probability? What is different here from (b)?

4.12. **miRNA Identifies NSCLS.** Tremendous efforts have been made to develop cancer biomarkers by detecting circulating extracellular miRNAs directly released from tumors. Yet, until recently none of the cell-free biomarkers has been accepted to be used for early detection of non-small cell lung cancer (NSCLC).

Ma et al. (2015) investigated whether analysis of miRNA expressions of peripheral blood mononuclear cells (PBMC) has diagnostic value for NSCLC. They identified several PBMC miRNAs with a significantly altered expression level in subjects with NSCLC. In a training set of 84 patients with confirmed NSCLC and 69 cancer-free smokers, a panel of two miRNAs (miRs-19b-3p and -29b-3p) produced 72.62% sensitivity and 82.61% specificity in identifying NSCLC.

(a) From the data supplied, recover the table with TP, FP, TN, and FN.

(b) If Mr. Molina is one of the participants in the study and tested positive for NSCLC, what is the probability that he has the disease?

(c) Find F-measure and Youden Index for the test.

(d) The prevalence of undiagnosed NSCLC in the population of smokers aged over 60 is 0.3%. Mr. Ramussen and Ms. Antonetti are randomly selected from that population and tested for NSCLC. What is the probability that Mr. Ramussen has disease if he tested positive? What is the probability that Ms. Antonetti does not have the disease if she tested negative.

(e) Based on pretest symptoms there was 1:3 odds that Mr. Bhwana has disease. He tested positive. What are his posttest odds?

4.13. **Creatinine Kinase and Acute Myocardial Infraction.** In a study of 773 patients, Radack et al. (1986) used an elevated serum creatinine kinase concentration as a diagnostic test for acute myocardial infraction. The following thresholds of a diagnostic test have been suggested: 481, 361, 241, and 121 IU/l; if the creatine kinase concentration exceeds the selected threshold, the test for myocardial infraction is considered positive. The gold standard is dichotomized: myocardial infraction present (MIP) and myocardial infraction not present (MINP). Assume that the sample of 773 subjects is randomly selected from the population, so that the prevalence of the disease is estimated as 51/773.

	MIP	MINP	Total
≥ 481 IU/l	9	14	23
< 481 IU/l	42	708	750
≥ 361 IU/l	15	26	41
< 361 IU/l	36	696	732
≥ 241 IU/l	22	50	72
< 241 IU/l	29	672	701
≥ 121 IU/l	28	251	279
< 121 IU/l	23	471	494
Total	51	722	773

(a) For the test that uses 361 IU/l as a threshold, find the sensitivity, specificity, PPV, NPV, LRP, and degree of agreement.

(b) Using given thresholds, plot the ROC curve. What threshold would you suggest?

(c) Find the area under the ROC curve.

4.14. **FNAC and FNA-Tg Testing for CLN Metastases.** A key component for guiding clinical treatment and selecting surgical methods to combat thyroid cancer is the early detection of cervical lymph node (CLN) metastases. Diagnostic methods currently include US-guided fine-needle aspiration cytology (FNAC) and detection of thyroglobulin on FNA (FNA-Tg). Shi et al. (2015) provide data evaluating the specificity and sensitivity of these tests alone and combined.

(a) Using the numbers from the following table, determine the sensitivity, specificity, PPV, and NPV for the FNAC method. Give an interpretation of each proportion.

FNAC test result	CLN metastasis	No CLN metastasis	Total
Positive test	64	0	64
Negative test	30	54	84
Total	94	54	148

(b) The same patients from part (a) were also tested using FNA-Tg/serum Tg. This method was found to have a sensitivity of 91.5% and specificity of 88.9%. Fill out the following table for TP, FP, TN, and FN. (Round entries to the nearest integer)

FNATg/ Tg Test result	CLN metastasis	No CLN metastasis	Total
Positive test			
Negative test			
Total			148

(c) Find the sensitivity and specificity of a parallel combination of FNAC and FNA-Tg/serum Tg tests.

(d) Further results from the same group of patients finds a combination of tests with TP = 90 and TN = 52. A physician wants to use whichever

test (or combination of tests) gives the best accuracy. Which do you recommend?

4.15. **Asthma.** A medical research team wished to evaluate a proposed screening test for asthma. The test was given to a random sample of 100 patients with asthma and to an independent sample of 200 subjects without
symptoms of the disease.

The two samples were drawn from a population of subjects who were 50
years old or older. The results are as follows:

Test result	Asthma, D	No asthma, D^c	Total
Positive test, T	92	13	105
Negative test, T^c	8	187	195
Total	100	200	300

(a) Using the numbers from the table, estimate the sensitivity and specificity. Interpret these proportions in terms of the problem, one sentence
each.

(b) The probability of D (prevalence) as the rate of the disease in the
relevant population (≥ 50 y.o.) is estimated to be 6.3%. Find the PPV
using Bayes' rule.

4.16. *H. pylori* **and ELISA.** Marshal et al. (1999) evaluated an enzyme-linked
immunosorbent assay (ELISA) for the detection of anti-*Helicobacter pylori*
specific IgG antibodies in specimens of oral fluid. Antral biopsy specimens, serum and oral fluid samples were collected from 81 patients
attending for upper gastrointestinal endoscopy. The presence or absence
of current *H. pylori* infection was determined by culture, histology, and
urease detection that served as gold standard. Anti-*H. pylori* specific IgG
was detected in oral fluid by an ELISA developed for this study. In all,
34 (42%) of 81 patients were positive for *H. pylori* by the battery of gold
standard tests. The diagnosis was established by optical density (OD)
readings measured at a wavelength of 490 nm, taken on specially prepared plates with specimens.

Depending on the OD cut-point, the following list of sensitivities and
specificities was reported:

OD cut-point	Sensitivity	Specificity
0.1	1.00	0.43
0.2	1.00	0.70
0.3	0.94	0.85
0.4	0.91	0.85
0.5	0.82	0.87
0.6	0.76	0.91
0.7	0.74	0.94
0.8	0.71	0.96
0.9	0.62	0.96
1.0	0.62	0.96

In the same experiment, Marshall et al. also compared the described oral fluid ELISA performance to the standard serum ELISA. The following results were reported:

		H. pylori	No H. pylori
Oral fluid ELISA (OD=0.3)	+	32	7
	−	2	40
Serum ELISA	+	31	4
	−	3	43

(a) Using MATLAB, find the sequence of PPV for all OD thresholds. Using PPV and Se sequences, find the sequence of F-measures. What value of OD maximizes the F-measure?

(b) Marshal et al. state that OD = 0.3 should be used as a threshold in this test. Can this be confirmed by the Youden index calculation?

(c) Find the area under the ROC curve, AUC.

4.17. **Swine Toxoplasmosis.** Ingestion of undercooked infected pork is considered an important source of human toxoplasmosis. Serologic screening could have future use for surveillance of animal populations and certification of freedom from *Toxoplasma gondii*. Results on three serologic tests for the diagnosis of toxoplasmosis by modified agglutination test (MAT), enzyme-linked immunosorbent assay (ELISA), and Sabin–Feldman dye test (DT), are provided. The conditional covariances for sensitivity and specificity proportions, c_{Se} and c_{Sp}, are defined on page 142.

The data are a part of an extensive analysis by Gardner et al. (2000).

Test	Sensitivity	Specificity
ELISA	73%	86%
MAT	83%	90%
DT	54%	91%

Tests	c_{Se}	c_{Sp}
ELISA, MAT	0.08	0.04
MAT, DT	0.09	0.04
DT, ELISA	0.09	0.03

(a) Find sensitivity and specificity of a parallel combination of ELISA and MAT tests.

(b) Find sensitivity and specificity of a serial combination of MAT and DT tests.

(c) Which test would you favor: a parallel combination of ELISA and MAT or a serial combination of MAT and DT?

| MATLAB AND WINBUGS FILES AND DATA SETS USED IN THIS CHAPTER |

http://statbook.gatech.edu/Ch4.ROC/

auc.m, cancerslope.m, hypothyroidism.m, Kinaseandmi.m, rocada.m, RocDdimer.m, sesp.m, simulatetesting.m, simulatetseting2.m

apre2pre.odc

ADA.mat, pasi.dat, roccreatine.vi, ROCTBCA.XLS, slopesmammo.dat

---

CHAPTER REFERENCES

Cabrera-Abreu, J., Davies, P., Matek, Z., and Murphy, M. (2004). Performance of blood tests in diagnosis of inflammatory bowel disease in a specialist clinic. *Arch. Dis. Child.*, **89**, 69–71.

Casscells, W., Schoenberger, A., and Grayboys, T. (1978). Interpretation by physicians of clinical laboratory results. *New Engl. J. Med.*, **299**, 999–1000.

Don, M., Kwong, B., Tanaka, C., Brackmann, D., and Nelson, R. (2005). The stacked ABR: A sensitive and specific screening tool for detecting small acoustic tumors. *Audiol. Neurotol.*, **10**, 274–290.

Drum, D. E. and Christacopoulos, S. (1972). Hepatic seintigraphy in clinical decision making. *J. Nucl. Med.*, **13**, 908–915.

Evans, D. A. (1990). Estimated prevalence of Alzheimer's disease in the United States. *Milbank Q.*, **68**, 267–289.

Gardner, I. A., Stryhn, H., Lind, P., and Collins, M. T. (2000). Conditional dependence between tests affects the diagnosis and surveillance of animal diseases. *Prev. Vet. Med.*, **45**, 107–122.

Goldhaber, S. Z., Simons, G. R., Elliott, C. G., Haire, W. D., Toltzis, R., Blacklow, S. C., Doolittle M. H., and Weinberg, D. S. (1993). Quantitative plasma D-dimer levels among patients undergoing pulmonary angiography for suspected pulmonary embolism. *J. Am. Med. Assoc.*, **270**, 23, 819–822.

Goldstein, B. J. and Mushlin, A. I. (1987). Use of a single thyroxine test to evaluate ambulatory medical patients for suspected hypothyroidism. *J. Gen. Intern. Med.*, **2**, 20–24.

Hawkins, D. M., Garrett, J. A., and Stephenson, B. (2001). Some issues in resolution of diagnostic tests using an imperfect gold standard. *Stat. Med.*, **20**, 1987–2001.

Keith, S. N., Repoli, A. F., Semenuk, M., Harris, P. J., Ghanbari, H. A., and Lebowitz, M. S. (2008). HAAH identifies prostate cancer, regardless of PSA level. Presentation at 2008 Genitourinary Cancers Symposium (GCS), February 14–16, 2008, San Francisco, CA.

Ma, J., Lin, Y., Zhan, M., Mann, D. L., Stass, S. A., and Jiang, F. (2015). Differential miRNA expressions in peripheral blood mononuclear cells for diagnosis of lung cancer. *Laboratory Investigation*, **95**, 1197–1206. `doi:10.1038/labinvest.2015.88`

Marshall, B., Howat, A. J., and Wright, P. A. (1999). Oral fluid antibody detection in the diagnosis of Helicobacter pylori infection. *J. Med. Microbiol.*, **48**, 1043–1046.

Mavromatis, B. H. and Kessler, C. M. (2001). D-Dimer testing: the role of the clinical laboratory in the diagnosis of pulmonary embolism. *J. Clin. Pathol.*, **54**, 664–668.

Nakama, H., Kamijo, N., Fujimori, K., Fattah, A., and Zhang, B. (1997). Relationship between fecal sampling times and sensitivity and specificity of immunochemical fecal occult blood tests for colorectal cancer. *Dis. Colon Rectum*, **40**, 7, 781–784.

Parikh, R., Mathai, A., Parikh, S., Sekhar, G. C., and Thomas, R. (2008). Understanding and using sensitivity, specificity and predictive values. *Indian J. Ophthalmol.*, **56**, 1, 45–50.

Qu, Y. and Hadgu, A. (1998). A model for evaluating sensitivity and specificity for correlated diagnostic tests in efficacy studies with an imperfect reference test. *J. Am. Stat. Assoc.*, **93**, 920–928.

Radack, K. L., Rouan, G., Hedges, J. (1986). The likelihood ratio: An improved measure for reporting and evaluating diagnostic test results. *Arch. Pathol. Lab. Med.*, **110**, 689–693.

Schipper, H., Kwok, C.-S., Rosendahl, S. M., Bandilla, D., Maes, O., Melmed, C., Rabinovitch, D., and Burns, D. H. (2008). Spectroscopy of human plasma for diagnosis of idiopathic Parkinson's disease. *Biomark. Med.*, **2**, 3, 229–238.

Shi, J. H., Xu, Y. Y., Pan, Q. Z., Sui, G. Q., Zhou, J. P., and Wang, H. (2015). The value of combined application of ultrasound-guided fine needle aspiration cytology and thyroglobulin measurement for the diagnosis of cervical lymph node metastases from thyroid cancer. *Pak. J. Med. Sci.*, **31**, 5, 1152–1155. `doi:http://dx.doi.org/10.12669/pjms.315.6726`

Staquet, M., Rozencweig, M., Lee, Y. J., and Muggia, F. M. (1981). Methodology for assessment of new dichotomous diagnostic tests. *J. Chronic. Dis.*, **34**, 599–610.

Youden, W. J. (1950). Index for rating diagnostic tests. *Cancer*, **3**, 32–35.

Chapter 5
Random Variables

The generation of random numbers is too important to be left to chance.

— Robert R. Coveyou

WHAT IS COVERED IN THIS CHAPTER

- Definition of Random Variables and Their Basic Characteristics
- Discrete Random Variables: Bernoulli, Binomial, Poisson, Hypergeometric, Geometric, Negative Binomial, and Multinomial
- Continuous Random Variables: Uniform, Exponential, Gamma, Inverse Gamma, Beta, Double Exponential, Logistic, Weibull, Pareto, and Dirichlet
- Transformation of Random Variables
- Markov Chains

5.1 Introduction

Thus far we have been concerned with random experiments, events, and their probabilities. In this chapter we will discuss random variables and their probability distributions. The outcomes of an experiment can be associated with numerical values, and this association will help us arrive at the definition of a random variable.

A *random variable* is a variable whose numerical value is determined by the outcome of a random experiment.

Thus, a random variable is a mapping from the sample space of an experiment, S, to a set of real numbers. In this respect, the term *random variable* is a misnomer. The more appropriate term would be *random function* or *random mapping,* given that X maps a sample space S to real numbers. We generally denote random variables by capital letters $X, Y, Z, \ldots$.

Example 5.1. **Three Coin Tosses.** Suppose a fair coin is tossed three times. We can define several random variables connected with this experiment. For example, we can set X to be the number of heads, Y the difference between the number of heads and the number of tails, and Z an indicator that heads appeared, etc.

Random variables X, Y, and Z are fully described by their probability distributions, associated with the sample space on which they are defined.

For random variable X the possible realizations are 0 (no heads in three flips), 1 (exactly one head), 2 (exactly two heads), and 3 (all heads). Fully describing random variable X amounts to finding the probabilities of all possible realizations. For instance, the realization $\{X = 2\}$ corresponds to either outcome in the event $\{HHT, HTH, THH\}$. Thus, the probability of X taking value 2 is equal to the probability of the event $\{HHT, HTH, THH\}$, which is equal to 3/8. After finding the probabilities for other outcomes, we determine the distribution of random variable X:

X	0	1	2	3
Prob	1/8	3/8	3/8	1/8

The *probability distribution* of a random variable X is a table (assignment, rule, formula) that assigns probabilities to realizations of X, or sets of realizations.

Most random variables of interest to us will be the results of random sampling. There is a general classification of random variables that is based on the nature of realizations they can take. Random variables that take values from a finite or countable set are called *discrete random variables*. Random variable X from Example 5.1 is an example of a discrete random variable. Another type of random variable can take any value from an interval on a real line. These are called *continuous random variables*. The results of measurements are usually modeled by continuous random variables. Next, we will describe discrete and continuous random variables in a more structured manner.

5.2 Discrete Random Variables

Let random variable X take discrete values $x_1, x_2, \ldots, x_n, \ldots$ with probabilities $p_1, p_2, \ldots, p_n, \ldots$, $\sum_n p_n = 1$. The probability distribution function (PDF) is simply an assignment of probabilities to the realizations of X and is given by the following table.

$$\begin{array}{c|cccc} X & x_1 \; x_2 \; \cdots \; x_n \; \cdots \\ \hline \text{Prob} & p_1 \; p_2 \; \cdots \; p_n \; \cdots \end{array}.$$

The probabilities p_i sum up to 1: $\sum_i p_i = 1$. It is important to emphasize that discrete random variables can have an infinite number of realizations, as long as the infinite sum of the probabilities converges to 1. The PDF for discrete random variables is also called the probability mass function (PMF). The cumulative distribution function (CDF)

$$F(x) = P(X \leq x) = \sum_{n:x_n \leq x} p_n,$$

sums the probabilities of all realizations smaller than or equal to x. Figure 5.1a shows an example of a discrete random variable X with four values and a CDF as the sum of probabilities in the range $X \leq x$ shown in yellow.

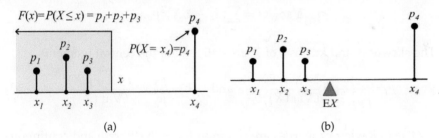

(a)   (b)

Fig. 5.1 (a) Example of a cumulative distribution function for discrete random variable X. The CDF is the sum of probabilities in the region $X \leq x$ (*yellow*). (b) Expectation as a point of balance for "masses" $p_1, \ldots, p_4$ located at the points $x_1, \ldots, x_4$.

The expectation of X is given by

$$\mathbb{E}X = x_1 p_1 + \cdots + x_n p_n + \cdots = \sum_n x_n p_n$$

and is a weighted average of all possible realizations with their probabilities as weights. Figure 5.1b illustrates the interpretation of the expectation as the point of balance for a system with weights $p_1, \ldots, p_4$ located at the locations $x_1, \ldots, x_4$.

The distribution and expectation of a function $g(X)$ are simple when X is discrete: one applies function g to realizations of X and retains the probabilities:

$$\frac{g(X)}{\text{Prob}}\,\bigg|\,\frac{g(x_1)\ g(x_2)\ \cdots\ g(x_n)\ \cdots}{p_1\qquad p_2\quad\cdots\quad p_n\quad\cdots}$$

and

$$\mathbb{E}g(X) = g(x_1)p_1 + \cdots + g(x_n)p_n + \cdots = \sum_n g(x_n)p_n.$$

The kth moment of a discrete random variable X is defined as

$$m_k = \mathbb{E}X^k = \sum_n x_n^k p_n,$$

and the kth central moment is

$$\mu_k = \mathbb{E}(X - \mathbb{E}X)^k = \sum_n (x_n - \mathbb{E}X)^k p_n.$$

The first moment is the expectation, $m_1 = \mathbb{E}X$, and the second central moment is the variance, $\mu_2 = \mathbb{V}\mathrm{ar}\,(X) = \mathbb{E}(X - \mathbb{E}X)^2$. Thus, the variance for a discrete random variable is

$$\mathbb{V}\mathrm{ar}\,(X) = \sum_n (x_n - \mathbb{E}X)^2 p_n.$$

The skewness and kurtosis of X are defined via the central moments as

$$\gamma = \frac{\mu_3}{\mu_2^{3/2}} = \frac{\mathbb{E}(X - \mathbb{E}X)^3}{(\mathbb{V}\mathrm{ar}\,(X))^{3/2}} \quad \text{and} \quad \kappa = \frac{\mu_4}{\mu_2^2} = \frac{\mathbb{E}(X - \mathbb{E}X)^4}{(\mathbb{V}\mathrm{ar}\,(X))^2}. \tag{5.1}$$

The following properties are common for both discrete and continuous random variables:

For any set of random variables $X_1, X_2, \ldots, X_n$

$$\mathbb{E}(X_1 + X_2 + \cdots + X_n) = \mathbb{E}X_1 + \mathbb{E}X_2 + \cdots + \mathbb{E}X_n. \tag{5.2}$$

For any constant c, $\mathbb{E}(c) = c$ and $\mathbb{E}(cX) = c\,\mathbb{E}X$.

The independence of two random variables is defined via the independence of events. Two random variables X and Y are independent if for arbitrary intervals A and B, the events $\{X \in A\}$ and $\{Y \in B\}$ are independent, that is, when

$$\mathbb{P}(X \in A,\ Y \in B) = \mathbb{P}(X \in A) \cdot \mathbb{P}(Y \in B),$$

holds.

If the random variables $X_1, X_2, \ldots, X_n$ are independent, then

$$\mathbb{E}(X_1 \cdot X_2 \cdot \ \ldots \ \cdot X_n) = \mathbb{E}X_1 \cdot \mathbb{E}X_2 \cdot \ \ldots \ \cdot \mathbb{E}X_n, \text{ and}$$
$$\text{Var}(X_1 + X_2 + \cdots + X_n) = \text{Var}\,X_1 + \text{Var}\,X_2 + \cdots + \text{Var}\,X_n. \qquad (5.3)$$

For a constant c, $\text{Var}(c) = 0$, and $\text{Var}(cX) = c^2\text{Var}\,X$.

If $X_1, X_2, \ldots, X_n, \ldots$ are independent and identically distributed random variables, we will refer to them as i.i.d. random variables.

The arguments behind these properties involve the linearity of the sums (for discrete variables) and integrals (for continuous variables). The independence of the X_is is critical for (5.3).

Moment-Generating Function. A particularly useful function for finding moments and for more advanced operations with random variables is the *moment-generating function*. For a random variable X, the moment-generating function is defined as

$$m_X(t) = \mathbb{E}e^{tX} = \sum_n p_n e^{tx_n}, \qquad (5.4)$$

which for discrete random variables has the form $m_X(t) = \sum_n p_n e^{tx_n}$. When the moment-generating function exists, it uniquely determines the distribution. If X has distribution F_X and Y has distribution F_Y, and if $m_X(t) = m_Y(t)$ for all t, then it follows that $F_X = F_Y$.

The name "moment-generating" is motivated by the fact that the kth derivative of $m_X(t)$ evaluated at $t = 0$ results in the kth moment of X, that is, $m_X^{(k)}(t) = \sum_n p_n x_n^k e^{tx_n}$, and $m_X^{(k)}(0) = \sum_n p_n x_n^k = \mathbb{E}X^k$. For example, if

X	0	1	3
Prob	0.2	0.3	0.5

then $m_X(t) = 0.2 + 0.3\,e^t + 0.5\,e^{3t}$. Since $m_X'(t) = 0.3\,e^t + 1.5\,e^{3t}$, the first moment is $\mathbb{E}X = m'(0) = 0.3 + 1.5 = 1.8$. The second derivative is $m_X''(t) = 0.3\,e^t + 4.5\,e^{3t}$, the second moment is $\mathbb{E}X^2 = m''(0) = 0.3 + 4.5 = 4.8$, and so on.

In addition to generating the moments, moment-generating functions satisfy

$$m_{X+Y}(t) = m_X(t)\,m_Y(t), \qquad (5.5)$$
$$m_{cX}(t) = m_X(ct),$$

which helps in identifying distributions of linear combinations of random variables whenever their moment-generating functions exist.

The properties in (5.5) follow from the properties of expectations. When X and Y are independent, e^{tX} and e^{tY} are independent as well, and by (5.3) $\mathbb{E}e^{t(X+Y)} = \mathbb{E}e^{tX}e^{tY} = \mathbb{E}e^{tX} \cdot \mathbb{E}e^{tY}$.

Example 5.2. **Apgar Score.** In the early 1950s, Dr. Virginia Apgar proposed a method to assess the health of a newborn child by assigning a grade referred to as the Apgar score (Apgar, 1953). It is given twice for each newborn, once at 1 min after birth and again at 5 min after birth.

Possible values for the Apgar score are 0, 1, 2, $\cdots$, 9, and 10. A child's score is determined by five factors: muscle tone, skin color, respiratory effort, strength of heartbeat, and reflex, with a high score indicating a healthy infant. Let the random variable X denote the Apgar score of a randomly selected newborn infant at a particular hospital. Suppose that X has a given probability distribution:

X	0	1	2	3	4	5	6	7	8	9	10
Prob	0.002	0.001	0.002	0.005	0.02	0.04	0.17	0.38	0.25	0.12	0.01

The following MATLAB program calculates (a) $\mathbb{E}X$, (b) $\mathbb{V}\mathrm{ar}\,(X)$, (c) $\mathbb{E}X^4$, (d) $F(x)$, (e) $\mathbb{P}(X < 4)$, and (f) $\mathbb{P}(2 < X \le 3)$:

```
X = 0:10;
p = [0.002  0.001  0.002  0.005  0.02  ...
       0.04  0.17 0.38  0.25  0.12  0.01];
EX = X * p'              %(a) EX = 7.1600
VarX = (X-EX).^2 * p'    %(b) VarX = 1.5684
EX4 = X.^4 * p'          %(c) EX4 = 3.0746e+003
ps = [0 cumsum(p)];
Fx = @(x)   ps( min(max( floor(x)+2, 1),12) );  %handle
Fx(3.45)                 %(d) ans = 0.0100
sum(p(X < 4))            %(e) ans = 0.0100
sum(p(X > 2 & X <= 3))   %(f) ans = 0.0050
```

Note that the CDF F is expressed as function handle `Fx` to a custom-made function.

Example 5.3. **Cells.** Randomly observed circular cells on a plate have a diameter D that is a random variable with the following PMF:

D	8	12	16
Prob	0.4	0.3	0.3

(a) Find the CDF for D.

(b) Find the PMF for the random variable $A = D^2\pi/4$ (the area of a cell). Show that $\mathbb{E}A \ne (\mathbb{E}D)^2\pi/4$. Explain.

(c) Find the variance $\text{Var}(A)$.

(d) Find the moment-generating functions $m_D(t)$ and $m_A(t)$. Find $\text{Var}(A)$ using its moment-generating function.

(e) It is known that a cell with $D > 8$ is observed. Find the probability of $D = 12$ taking into account this information.

Solution:

(a)

$$F_D(d) = \begin{cases} 0, & d < 8 \\ 0.4, & 8 \leq d < 12 \\ 0.7, & 12 \leq d < 16 \\ 1, & d \geq 16 \end{cases}$$

(b)

A	$8^2\pi/4$	$12^2\pi/4$	$16^2\pi/4$
Prob	0.4	0.3	0.3

A	16π	36π	64π
Prob	0.4	0.3	0.3

$\mathbb{E}A = 16\pi(\frac{4}{10}) + 36\pi(\frac{3}{10}) + 64\pi(\frac{3}{10}) = \frac{364\pi}{10} = 114.3540.$

$\mathbb{E}D = 8(\frac{4}{10}) + 12(\frac{3}{10}) + 16(\frac{3}{10}) = 116/10 = 11.6$

$\frac{(\mathbb{E}D)^2\pi}{4} = \frac{3364\pi}{100} \neq \frac{364\pi}{10}.$

The expectation is a linear operator, and such a "plug-in" operation would work only if the random variable A were a linear function of D, that is, if $A = \alpha D + \beta$, $\mathbb{E}A = \alpha\mathbb{E}D + \beta$. In our case, A is quadratic in D, and "passing" the expectation through the equation is not valid.

(c)

$$\text{Var}\,A = \mathbb{E}A^2 - (\mathbb{E}A)^2 = 1720\pi^2 - 1324.96\pi^2 = 395.04\pi^2,$$

since

A^2	$16^2\pi^2$	$36^2\pi^2$	$64^2\pi^2$
Prob	0.4	0.3	0.3

and $EA^2 = 1720\pi^2$.

(d) $m_D(t) = \mathbb{E}e^{tD} = 0.4e^{8t} + 0.3e^{12t} + 0.3e^{16t}$, and $m_A(t) = \mathbb{E}e^{tA} = 0.4e^{16\pi t} + 0.3e^{36\pi t} + 0.3e^{64\pi t}$.

From $m'_A(t) = 6.4e^{16\pi t} + 10.8e^{36\pi t} + 19.2e^{64\pi t}$, and $m''_A(t) = 6.4e^{16\pi t} + 10.8e^{36\pi t} + 19.2e^{64\pi t}$, we find $m'_A(0) = 36.4\pi$ and $m''_A(0) = 1720\pi$, leading to the result in (c).

(e) When $D > 8$ is true, only two values for D are possible, 12 and 16. These values are equally likely. Thus, the distribution for $D|\{D > 8\}$ is

$D|\{D > 8\}$	12	16
Prob	0.3/0.6	0.3/0.6

and $\mathbb{P}(D = 12|D > 8) = 1/2$. We divided 0.3 by 0.6 since $\mathbb{P}(D > 8) = 0.6$. From the definition of the conditional probability it follows that,

$\mathbb{P}(D=12|D>8) = \mathbb{P}(D=12,D>8)/\mathbb{P}(D>8) = \mathbb{P}(D=12)/\mathbb{P}(D>8) = 0.3/0.6 = 1/2.$

There are important properties of discrete distributions in which the realizations $x_1, x_2, \ldots, x_n$ are irrelevant and the focus is on the probabilities only, such as the measure of *entropy*. For a discrete random variable where the probabilities are $p = (p_1, p_2, \ldots, p_n)$ the (Shannon) entropy is defined as

$$\mathcal{H}(p) = -\sum_i p_i \log(p_i).$$

Entropy is a measure of the uncertainty of a random variable and for finite discrete distributions achieves its maximum when the probabilities of realizations are equal, $p = (1/n, 1/n, \ldots, 1/n)$.

For the distribution in Example 5.2, the entropy is 1.5812.

```
ps = [.002  .001  .002  .005  .02  .04   .17   .38  .25 .12 .01]
entropy = @(p)    -sum( p(p>0) .* log(p(p>0)))
entropy(ps) %1.5812
```

The maximum entropy for distributions with 11 possible realizations is 2.3979.

Jointly Distributed Discrete Random Variables. So far we have discussed probability distributions of a single random variable. As we delve deeper into this subject, a two-dimensional extension will be needed.

When two or more random variables constitute the coordinates of a random vector, their joint distribution is often of interest. For a random vector (X, Y) the joint distribution function is defined via the probability of the event $\{X \leq x, Y \leq y\}$,

$$F(x,y) = \mathbb{P}(X \leq x, Y \leq y).$$

The univariate case $\mathbb{P}(a \leq X \leq b) = F(b) - F(a)$ takes the bivariate form

$$\mathbb{P}(a_1 \leq X \leq a_2, b_1 \leq Y \leq b_2) = F(a_2, b_2) - F(a_1, b_2) - F(a_2, b_1) + F(a_1, b_1).$$

Marginal CDFs F_X and F_Y are defined as follows: for X, $F_X(x) = F(x, \infty)$ and for Y as $F_Y(y) = F(\infty, y)$.

For a discrete bivariate random variable, the PMF is

$$p(x,y) = \mathbb{P}(X = x, Y = y), \ \sum_{x,y} p(x,y) = 1,$$

while for marginal random variables X and Y the PMFs are

$$p_X(x) = \sum_y p(x,y), \quad p_Y(y) = \sum_x p(x,y).$$

The conditional distribution of X given $Y = y$ is defined as

$$p_{X|Y}(x|y) = p(x,y)/p_Y(y),$$

and, similarly, the conditional distribution for Y given $X = x$ is

$$p_{Y|X}(y|x) = p(x,y)/p_X(x).$$

When X and Y are independent, for any "cell" (x,y), $p(x,y) = \mathbb{P}(X = x, Y = y) = \mathbb{P}(X = x)\mathbb{P}(Y = y) = p_X(x)\,p_Y(y)$, that is, the joint probability of (x,y) is equal to the product of the marginal probabilities. If $p(x,y) = p_X(x)p_Y(y)$ holds for every (x,y), then X and Y are independent. The independence of two discrete random variables is fundamental for the inference in contingency tables (Chapter 12) and will be revisited later.

Example 5.4. PMF of a two-dimensional discrete random variable is given by the following table:

		Y		
		5	10	15
X	1	0.1	0.2	0.3
	2	0.25	0.1	0.05

The marginal distributions for X and Y are

X	1	2
Prob	0.6	0.4

and

Y	5	10	15
Prob	0.35	0.3	0.35

while the conditional distribution for X when $Y = 10$ and the conditional distribution for Y when $X = 2$ are

X|Y = 10	1	2
Prob	0.2 / 0.3	0.1 / 0.3

and

Y|X = 2	5	10	15
Prob	0.25 / 0.4	0.1 / 0.4	0.05 / 0.4

respectively. Here X and Y are not independent since

$$0.1 = \mathbb{P}(X = 1, Y = 5) \neq \mathbb{P}(X = 1)\mathbb{P}(Y = 5) = 0.6 \cdot 0.35 = 0.21.$$

For two independent random variables X and Y, $\mathbb{E}XY = \mathbb{E}X \cdot \mathbb{E}Y$; that is, the expectation of a product of random variables is equal to the product of their expectations.

The *covariance* of two random variables X and Y is defined as

$$\mathrm{Cov}(X,Y) = \mathbb{E}((X - \mathbb{E}X) \cdot (Y - \mathbb{E}Y)) = \mathbb{E}XY - \mathbb{E}X \cdot \mathbb{E}Y.$$

For a discrete random vector (X,Y), $\mathbb{E}XY = \sum_x \sum_y xy p(x,y)$, and the covariance is expressed as

$$\mathrm{Cov}(X,Y) = \sum_x \sum_y xy p(x,y) - \sum_x x p_X(x) \sum_y y p_Y(y).$$

It is easy to see that the covariance satisfies the following properties:

$$\mathrm{Cov}(X,X) = \mathbb{V}\mathrm{ar}\,(X),$$
$$\mathrm{Cov}(X,Y) = \mathrm{Cov}(Y,X), \text{ and}$$
$$\mathrm{Cov}(aX + bY, Z) = a\,\mathrm{Cov}(X,Z) + b\,\mathrm{Cov}(Y,Z).$$

For (X,Y) from Example 5.4 the covariance between X and Y is -1. The calculation is provided in the following MATLAB code. Note that the distribution of the product XY is found in order to calculate $\mathbb{E}XY$.

```
X = [1 2]; pX = [0.6 0.4]; EX = X * pX'              %EX = 1.4000
Y = [5 10 15]; pY = [0.35  0.3  0.35]; EY = Y*pY'    %EY =10
XY = [5 10 15 20 30];
pXY = [0.1   0.2+0.25   0.3  0.1  0.05];  EXY = XY * pXY'   %EXY = 13
CovXY = EXY - EX * EY                                %CovXY = -1
```

The *correlation* between random variables X and Y is the covariance normalized by the standard deviations:

$$\mathrm{Corr}(X,Y) = \frac{\mathrm{Cov}(X,Y)}{\sqrt{\mathbb{V}\mathrm{ar}\,X \cdot \mathbb{V}\mathrm{ar}\,Y}}.$$

In Example 5.4, the variances of X and Y are $\mathbb{V}\mathrm{ar}\,X = 0.24$ and $\mathbb{V}\mathrm{ar}\,Y = 17.5$. Using these values, we find that the correlation $\mathrm{Corr}(X,Y)$ is $-1/\sqrt{0.24 \cdot 17.5} = -0.488$. Thus, the random components in (X,Y) are negatively correlated.

5.3 Some Standard Discrete Distributions

5.3.1 *Discrete Uniform Distribution*

A random variable X that takes values from 1 to n with equal probabilities of $1/n$ is called a discrete uniform random variable. In MATLAB unidpdf and `unidcdf` are the PDF and CDF of X, while `unidinv` is its quantile. For example,

```
unidpdf(1:5, 5)
%ans =    0.2000    0.2000    0.2000    0.2000    0.2000

unidcdf(1:5, 5)
%ans =    0.2000    0.4000    0.6000    0.8000    1.0000
```

are the PDF and CDF of the discrete uniform distribution on $\{1,2,3,4,5\}$. From $\sum_{i=1}^{n} i = n(n+1)/2$, and $\sum_{i=1}^{n} i^2 = n(n+1)(2n+1)/6$, one can derive $\mathbb{E}X = (n+1)/2$ and $\mathbb{V}\text{ar}\,X = (n^2-1)/12$. One of the important uses of discrete uniform distribution is in nonparametric statistics (page 894).

Example 5.5. **Discrete Uniform: A Basis for Random Sampling.** Suppose that a population is finite and that we need a sample such that every subject in the population has an equal chance of being selected.

If the population size is N and a sample of size n is needed, then if replacement is allowed (each sampled object is recorded and then returned back to the population), there would be N^n possible equally likely samples. If replacement is not allowed or possible (all subjects in the selected sample are to be different, i.e., sampling is without replacement), then there would be $\binom{N}{n}$ different equally likely samples (see Section 3.5 for a definition of $\binom{N}{n}$).

The theoretical model for random sampling is the discrete uniform distribution. If replacement is allowed, each of $\{1,2,\ldots,N\}$ has a probability of $1/N$ of being selected. In the case of no replacement, possible subsets of n subjects can be indexed as $\{1,2,\ldots,\binom{N}{n}\}$ and each subset has a probability of $1/\binom{N}{n}$ of being selected.

In MATLAB, random sampling is achieved by the function `randsample`. If the population has n indexed subjects (from 1 to n), the indices in a random sample of size k are found as `indices=randsample(n,k)`.

If it is possible to code the entire population as a vector `population`, then taking a sample of size k is done by `y=randsample(population,k)`.

The default is set to sampling without replacement. For sampling with replacement, the flag for replacement should be `'true'`. If the sampling is done with replacement, it can be weighted with a nonnegative weight assigned to each subject in the population: `y=randsample(population,k,true,w)`. The size of weight vector `w` should be the same as that of `population`.

For instance,

```
randsample(['A' 'C' 'G' 'T'],50,true,[1 1.5 1.4  0.9])
%ans =   GCCTAGGGCATCCAAGTCGCGGCCGAGAATCAACGTTGCAGTGCTCAAAT
```

5.3.2 Bernoulli and Binomial Distributions

A simple Bernoulli random variable Y is dichotomous with $\mathbb{P}(Y = 1) = p$
and $\mathbb{P}(Y = 0) = 1 - p$ for some $0 \leq p \leq 1$ and is denoted as $Y \sim \mathcal{B}er(p)$. It is
named after Jakob Bernoulli (1654–1705), a prominent Swiss mathematician
and astronomer. Suppose that an experiment consists of n independent trials $(Y_1, \ldots, Y_n)$ in which two outcomes are possible (e.g., success or failure),
with $\mathbb{P}(\text{success}) = \mathbb{P}(Y = 1) = p$ for each trial. If $X = x$ is defined as the
number of successes (out of n), then $X = Y_1 + Y_2 + \cdots + Y_n$, and there are
$\binom{n}{x}$ arrangements of x successes and $n - x$ failures, each having the same
probability $p^x(1 - p)^{n-x}$. X is a *binomial* random variable with the PMF

$$p_X(x) = \binom{n}{x} p^x (1 - p)^{n-x}, \ x = 0, 1, \ldots, n.$$

This is denoted by $X \sim \mathcal{B}in(n, p)$. From the moment-generating function
$m_X(t) = (pe^t + (1 - p))^n$, we obtain $\mu = \mathbb{E}X = np$ and $\sigma^2 = \mathbb{V}ar\, X = np(1 - p)$.

   The cumulative distribution for a binomial random variable is not simplified beyond the sum, that is, $F(x) = \sum_{i \leq x} p_X(i)$. However, interval probabilities can be computed in MATLAB using `binocdf(x,n,p)`, which computes the CDF at value x. The PMF can also be computed in MATLAB
using `binopdf(x,n,p)`. In WinBUGS, the binomial distribution is denoted as
`dbin(p,n)`. Note the reversed order of parameters n and p.

Example 5.6. **Left-Handed Families.** About 10% of the world's population
is left-handed. Left-handedness is more prevalent in men (1/9) than in
women (1/13). Studies have shown that left-handedness is linked to the
gene LRRTM1, which affects the symmetry of the brain. In addition to
its genetic origins, left-handedness also has developmental origins. When
both parents are left-handed, a child has a probability of 0.26 of being left-
handed.

   Ten families in which both parents are left-handed and have a single
child are selected, and the ten children are inspected for left-handedness.
Let X be the number of left-handed children among the inspected. What is
the probability that X
   (a) Is equal to 3?
   (b) Falls anywhere between 3 and 6, inclusive?
   (c) Is at most 4?
   (d) Is not less than 4?
   (e) Would you be surprised if the number of left-handed children among
the ten inspected was eight or more? Why or why not?
   The solution is given by the following annotated MATLAB script:

```
% Solution
disp('(a) Bin(10, 0.26): P(X = 3)');
binopdf(3, 10, 0.26)
 % ans = 0.2563
disp('(b) Bin(10, 0.26): P(3 <= X <= 6)');
 %  using binopdf(x, n, p)
disp('(b)-using PDF');    binopdf(3, 10, 0.26) + ...
binopdf(4, 10, 0.26) + binopdf(5, 10, 0.26)+ binopdf(6, 10, 0.26)
 % using binocdf(x, n, p)
disp('(b)-using CDF'); binocdf(6, 10, 0.26) - binocdf(2, 10, 0.26)
 % ans = 0.4998
 %(c) at most four i.e., X <= 4
disp('(c) Bin(10, 0.26): P(X <= 4)'); binocdf(4, 10, 0.26)
 % ans = 0.9096
 %(d) not less than 4 is 4,5,...,10, or complement of <=3
disp('(d)  Bin(12, 0.7): P(X >= 4)'); 1-binocdf(3, 10, 0.26)
 % ans = 0.2479
disp('(e) Bin(10, 0.26): P(X >= 8)');
1-binocdf(7, 10, 0.26)
 % ans = 5.5618e-04
 % Yes, this would be a surprising outcome since
 % the probability of such an event is rather small
```

Panels (a) and (b) in Figure 5.2 show, respectively, the PMF and CDF for the binomial $Bin(10,0.26)$ distribution.

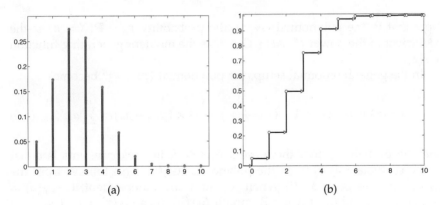

(a)

(b)

Fig. 5.2 Binomial $Bin(10,0.26)$: (a) PMF and (b) CDF.

How does one recognize that random variable X has a binomial distribution?

(a) It allows an interpretation as the sum of "successes" in n Bernoulli trials, for n fixed.

(b) The Bernoulli trials are independent.
(c) The Bernoulli probability p is constant for all n trials.

Next we discuss how to deal with a binomial-like framework in which condition (c) is violated.

Generalized Binomial Sampling*. Suppose that n independent experiments are performed and that an event A has a probability of p_i of appearing in the ith experiment.

We are interested in the probability that A appeared exactly k times in the n experiments. The binomial setup is not directly applicable since the probabilities of A differ from experiment to experiment. However, the binomial setup is useful as a hint on how to solve the general case. In the binomial setup the probability of k events A in n experiments is equal to the coefficient of z^k in the expansion of $G(z) = (pz + q)^n$. Indeed, $(pz + q)^n = p^n q^0 z^n + \cdots + \binom{n}{k} p^k q^{n-k} z^k + \cdots + npq^{n-1}z + p^0 q^n$.

The polynomial $G(z)$ is called the *probability-generating function*. If X is a discrete integer-valued random variable such that $p_n = \mathbb{P}(X = n)$, then its probability-generating function is defined as

$$G_X(z) = \mathbb{E}z^X = \sum_n p_n z^n.$$

Note that in the polynomial $G_X(z)$, the probability $p_n = \mathbb{P}(X = n)$ is the coefficient of the power z^n. Also, $G_X(e^z)$ is the moment-generating function $m_X(z)$.

In the general binomial setup, the polynomial $(pz + q)^n$ becomes

$$G_X(z) = (p_1 z + q_1) \times (p_2 z + q_2) \times \cdots \times (p_n z + q_n) = \sum_{i=0}^{n} a_i z^i, \quad (5.6)$$

and the probability that there are k events A in n experiments is equal to the coefficient a_k of z^k in the polynomial $G_X(z)$. This follows from the two properties of $G(z)$: (i) When X and Y are independent, $G_{X+Y}(z) = G_X(z)\,G_Y(z)$, and (ii) if X is a Bernoulli $Ber(p)$, then $G_X(z) = pz + q$.

Example 5.7. **System with Unreliable Components.** Let S be a system consisting of ten unreliable components that work and fail independently of each other. The components are operational in some fixed time interval $[0, T]$ with the probabilities

```
ps =[0.5 0.3 0.2 0.5 0.6 0.4 0.2 0.4 0.7 0.8];
```

Let a random variable X represent the number of components that remain operational after time T.

Find (a) the distribution for X and (b) $\mathbb{E}X$ and $\mathbb{V}\mathrm{ar}\,X$.

```
ps =[0.5 0.3 0.2 0.5 0.6 0.4 0.2 0.4 0.7 0.8];
qs = 1- ps;
all = [ps' qs'];
[m n]= size(all);
Gz = [1];  %initial
for i = 1:m
    Gz = conv(Gz, all(i,:) );
    % conv as polynomial multiplication
end
%at the end, Gz is the product of p_i x + q_i
%
sum(Gz) %the sum is 1
probs = Gz(end:-1:1);
k = 0:10
% probs=[0.0010  0.0117  0.0578  0.1547  0.2507 ...
% 0.2582  0.1716  0.0727  0.0188  0.0027  0.0002]
EX = k * probs' %expectation 4.6
EX2 = k.^2 * probs';
VX = EX2 - (EX)^2 %variance 2.12
```

Note that in the above script we used the convolution operation `conv` to multiply polynomials, as in

```
conv([2 -1],[1 3 2])
% ans =  2   5   1  -2,
```

which is interpreted as $(2z - 1) \cdot (z^2 + 3z + 2) = 2z^3 + 5z^2 + z - 2$.

From the MATLAB calculations we find that the probability-generating function $G(z)$ from (5.6) is

$$G(z) = 0.00016128z^{10} + 0.00268992z^9 + 0.01883264z^8 + 0.07273456z^7$$
$$+ 0.17155808z^6 + 0.25816544z^5 + 0.25070848z^4 + 0.15470576z^3$$
$$+ 0.05777184z^2 + 0.01170432z + 0.00096768,$$

and the random variable X, the number of operational items, has the following distribution (after rounding to four decimal places):

X	0	1	2	3	4	5	6	7	8	9	10
Prob	0.0010	0.0117	0.0578	0.1547	0.2507	0.2582	0.1716	0.0727	0.0188	0.0027	0.0002

The answers to (b) are $\mathbb{E}X = 4.6$ and $\mathbb{V}\mathrm{ar}\,X = 2.12$.

Note that a "solution" in which one finds the average of the component probabilities, `ps`, as $\bar{p} = \frac{1}{10}(0.5 + 0.3 + \cdots + 0.8) = 0.46$, and then applies the standard binomial calculation, will lead to the correct expectation, 4.6, because of linearity. However, the variance and probabilities for X would be different. For example, the probability $\mathbb{P}(X = 4)$ would be `binopdf(4,10,0.46)=0.2331`, while the correct value is 0.2507.

Example 5.8. **Surviving Pairs.** Daniel Bernoulli (1700–1782), a nephew of Jacob Bernoulli, posed and solved the following problem. If among N

married pairs there are m random deaths, what is the expected number of intact marriages?

Suppose that there are N pairs of balls denoted by $1,1, 2,2, \ldots, N,N$. If m balls are selected at random and removed, what is the expected number of intact pairs? Consider the pair i. Define a Bernoulli random variable Y_i equal to 1 if pair i remains intact after the removal of m balls, and 0 otherwise. Then the number of unaffected pairs N_m would be the sum of all Y_i, for $i = 1,\ldots,N$.

The probability that pair i is not affected by the removal of m balls is

$$\frac{\binom{2N-2}{m}}{\binom{2N}{m}} = \frac{\frac{(2N-2)(2N-3)...(2N-2-m+2)(2N-2-m+1)}{m!}}{\frac{2N(2N-1)...(2N-m+2)(2N-m+1)}{m!}} = \frac{(2N-m)(2N-m-1)}{2N(2N-1)},$$

and it is equal to $\mathbb{E}Y_i$. If N_m is the number of unaffected pairs, then

$$N_m = Y_1 + Y_2 + \cdots + Y_N$$
$$\mathbb{E}N_m = \mathbb{E}Y_1 + \mathbb{E}Y_2 + \cdots + \mathbb{E}Y_N = N\mathbb{E}Y_i = \frac{(2N-m)(2N-m-1)}{2(2N-1)}.$$

For example, if among $N = 1000$ couples there are 100 random deaths, then the expected number of unaffected couples is 902.4762. If among $N = 1000$ couples there are 1936 deaths, then a single couple is expected to remain intact.

Even though N_m is the sum of N Bernoulli random variables Y_i, each with the same probability $p = \frac{(2N-m)(2N-m-1)}{2N(2N-1)}$, it does not have a binomial distribution due to the dependence among Y_is.

5.3.3 Hypergeometric Distribution

Suppose a box contains m balls, k of which are white and $m - k$ of which are black. Suppose we randomly select and remove n balls from the box *without replacement*, so that when sampling is finished, there are only $m - n$ balls left in the box. If X is the number of white balls in n selected, then the probability that $X = x$ is

$$p_X(x) = \frac{\binom{k}{x}\binom{m-k}{n-x}}{\binom{m}{n}}, \quad x \in \{0,1,\ldots,\min\{n,k\}\}.$$

Random variable X is called hypergeometric and denoted by $X \sim \mathcal{HG}(m,k,n)$, where m,k, and n are integer parameters.

This PMF can be deduced by counting rules. There are $\binom{m}{n}$ different ways of selecting the n balls from a box with a total of m balls. From these (each equally likely), there are $\binom{k}{x}$ ways of selecting x white balls from the k white balls in the box and, similarly, $\binom{m-k}{n-x}$ ways of choosing the black balls. The probability $\mathbb{P}(X = x)$ is the ratio of these two numbers. The PDF and CDF of $\mathcal{HG}(40,15,10)$ are shown in Figure 5.3.

It can be shown that the mean and variance for the hypergeometric distribution are, respectively,

$$\mathbb{E}X = n\frac{k}{m} \quad \text{and} \quad \mathbb{V}\text{ar}\,X = n\frac{k}{m}\left(1 - \frac{k}{m}\right)\frac{m-n}{m-1}.$$

The MATLAB commands for hypergeometric CDF, PDF, quantile, and a random number are `hygecdf`, `hygepdf`, `hygeinv`, and `hygernd`. WinBUGS does not have a built-in command for a hypergeometric distribution.

Example 5.9. **CASES.** In a group of 40 people, 15 are "CASES" and 25 are "CONTROLS." A sample of 10 subjects is selected [(A) with replacement and (B) without replacement]. Find the probability $\mathbb{P}$(at least 2 subjects are CASES).

```
%Solution
%(A) - with replacement (binomial case);
%Let X be the number of CASES. The event
%X is at least 2 is the complement of X <= 1.
disp('(A) Bin(10, 15/40): P(X >= 2)');   1 - binocdf(1, 10, 15/40)
% ans = 0.9363
% or
 1 - binopdf(0, 10, 15/40) - binopdf(1, 10, 15/40)
% ans = 0.9363

%B - without replacement (hypergeometric case) hygecdf(x, m, k, n)
% where m size of population,
% k - number of cases among m,  and n sample size.
disp('(B) HyGe(40,15,10): P(X >=2)');  1 - hygecdf(1, 40, 15, 10)
% ans = 0.9600,   or
 1 - hygepdf(0, 40, 15, 10)- hygepdf(1, 40, 15, 10)
% ans = 0.9600
```

Example 5.10. **Capture–Recapture Models.** Suppose that an unknown number m of animals inhabit a particular region. To assess the population size, ecologists often apply the following capture–recapture scheme. They catch k animals, tag them, and release them back into the region. After some time, when the tagged animals are expected to be mixed well with the untagged, a second catch of size n is made. Suppose that x animals in the second sample are found to be tagged.

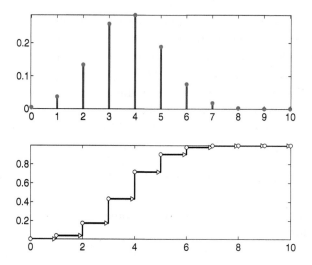

Fig. 5.3 The PDF and CDF for hypergeometric distribution with $m = 40, k = 15$, and $n = 10$.

If catching any animal is assumed equally likely, the number x of tagged animals in the second sample is hypergeometric $\mathcal{HG}(m,k,n)$. Ecologists use the observed ratio x/n as an approximation to k/m, from which m is estimated as

$$\hat{m} = \frac{k \times n}{x}.$$

A statistically better estimator of m (known as the Schnabel formula) is given as

$$\hat{m} = \frac{(k+1) \times (n+1)}{(x+1)} - 1.$$

In epidemiology and public health, capture–recapture methods use multiple, routinely collected, computerized data sources to estimate various population indexes.

For example, Gjini et al. (2004) investigated the number of matching records of pneumococcal meningitis among adults in England by comparing data from Hospital Episode Statistics (HES) and the Public Health Laboratory Services reconciled laboratory records (RLR). The time period covered was April 1996 to December 1999. The authors found 646 records in RLR and 737 in HES, and matching based on demographic information was possible in 296 cases.

By the capture–recapture method the estimated incidence is $\hat{m} = 646 \cdot 737/296 = 1608.5 \approx 1609$. If Schnabel's formula is used, then $\hat{m} \approx 1607$.

Thus, the total incidence of of pneumococcal meningitis in England between April 1996 to December 1999 is estimated to be 1607.

For large m, the hypergeometric distribution is close to binomial. More precisely, when $m \to \infty$ and $k/m \to p$, the hypergeometric distribution with parameters (m,k,n) approaches a binomial with parameters (n,p) for any value of x between 0 and n. It is also instructive to compare expressions for $\mathbb{E}X$ and $\mathbb{V}\text{ar}\,X$ for the two distributions.

```
format long
disp('(A)=(B) for large population');
1 - binocdf(1, 10, 150000/400000)    %ans = 0.936335370875895
1 - hygecdf(1, 400000, 150000, 10)   %ans = 0.936337703719839
```

We will use the hypergeometric distribution later in the book (i) in the Fisher exact test (page 602) and in Logrank test (page 818).

5.3.4 Poisson Distribution

This discrete distribution is named after Simeon Denis Poisson (1781–1840), French mathematician, geometer, and physicist.

The PMF for the Poisson distribution is

$$p_X(x) = \frac{\lambda^x}{x!}e^{-\lambda}, \quad x = 0,1,2,\ldots,$$

which is denoted by $X \sim \mathcal{P}oi(\lambda)$. From the moment-generating function $m_X(t) = \exp\{\lambda(e^t - 1)\}$ we have $\mathbb{E}X = \lambda$ and $\mathbb{V}\text{ar}\,X = \lambda$; the mean and the variance coincide.

The sum of a finite independent set of Poisson variables is also Poisson. Specifically, if $X_i \sim \mathcal{P}oi(\lambda_i)$, then $Y = X_1 + \cdots + X_k$ is distributed as $\mathcal{P}oi(\lambda_1 + \cdots + \lambda_k)$. If $X_1 \sim \mathcal{P}oi(\lambda_1)$ and $X_2 \sim \mathcal{P}oi(\lambda_2)$ are independent, then the distribution of X_1 given that $X_1 + X_2 = n$ is binomial $\mathcal{B}in\left(n, \frac{\lambda_1}{\lambda_1 + \lambda_2}\right)$ (Exercise 5.7).

Furthermore, the Poisson distribution is a limiting form for a binomial model, i.e.,

$$\lim_{n, np \to \infty, \lambda} \binom{n}{x} p^x (1-p)^{n-x} = \frac{1}{x!}\lambda^x e^{-\lambda}. \tag{5.7}$$

The MATLAB commands for Poisson CDF, PDF, quantile, and random number are poisscdf, poisspdf, poissinv, and poissrnd. In WinBUGS the Poisson distribution is denoted as dpois(lambda).

Example 5.11. **Poisson Model for EBs.** After 7 days of aggregation, the microscopy images of 2000 embryonic bodies (EBs) are used to assess their surface area size. The probability that the area of a randomly selected EB exceeds the critical size S_c is 0.001.

(a) Find the probability that the areas of exactly three EBs, among the 2000, exceed the critical size.

(b) Find the probability that the number of EBs exceeding the critical size is between three and eight, inclusively.

We use a Poisson approximation to the binomial probabilities since n is large, p is small, and product np is moderate.

```
%Solution
disp('Poisson(2): P(X=3)'); poisspdf(3, 2)
  %ans= 0.1804
disp('Poisson(2): P(3 <= X <= 8)'); poisscdf(8, 2)-poisscdf(2, 2)
  %ans= 0.3231
```

Figure 5.4 shows the PMF and CDF of the $Poi(2)$ distribution.

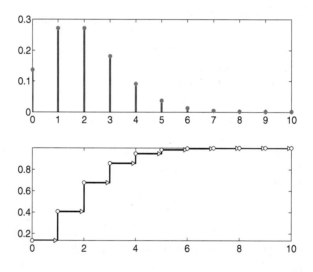

Fig. 5.4 (Top) Poisson probability mass function. (Bottom) Cumulative distribution function for $\lambda = 2$.

In the binomial sampling scheme, when $n \to \infty$ and $p \to 0$, so that $np \to \lambda$, binomial probabilities converge to Poisson probabilities.

The following MATLAB simulation demonstrates the convergence. In the binomial distribution, n is increasing as 2,000, 200,000, 20,000,000 and p is decreasing as 0.001, 0.00001, 0.0000001, so that np remains constant and equal to 2. Then, the binomial probabilities of $X = 3$ are compared to probability of $X = 3$ when X is distributed as Poisson with parameter $\lambda = 2$.

```
disp('P(X=3) for Bin(2000, 0.001), Bin(200000, 0.00001), ');
disp(' Bin(20000000, 0.0000001), and Poi(2) ');
format long
binopdf(3, 2000, 0.001)               % 0.180537328031786
binopdf(3, 200000, 0.00001)           % 0.180447946554779
binopdf(3, 20000000, 0.0000001)       % 0.180447058859339
poisspdf(3, 2)                        % 0.180447044315484
format short
```

Example 5.12. **Cold.** Suppose that the number of times during a year that an individual catches a cold can be modeled by a Poisson random variable with an expectation of 4. Further, suppose that a new drug based on vitamin C reduces this expectation to 3 (but the distribution still remains Poisson) for 90% of the population but has no effect on the remaining 10% of the population. We will calculate

(a) the probability that an individual taking the drug has two colds in a year if that individual is in part of the population that benefits from the drug;

(b) the probability that a randomly chosen individual has two colds in a year if that individual takes the drug; and

(c) the conditional probability that a randomly chosen individual is in the part of the population that benefits from the drug, given that the individual had two colds in the year during which he/she took the drug.

```
poisspdf(2,3)    %(Cold (a))
     %ans =  0.2240
poisspdf(2,3)*0.90 + poisspdf(2,4)*0.10   %(Cold (b))
     %ans = 0.2163
poisspdf(2,3)*0.90/(poisspdf(2,3)*0.90 + ...
                    poisspdf(2,4)*0.10) %(Cold (c))
     %ans = 0.9323
```

Example 5.13. **Imperfectly Observed Poisson.** Suppose that the number of particular experimental events in time interval $[0,T]$ has a Poisson distribution $\mathcal{P}oi(\lambda T)$. A student who is observing the experiment may fail to count some of the events. An event is counted with probability equal to p and missing one event is independent of missing or counting the others. What is the distribution of the number of events in $[0,T]$ that are counted?

By total probability formula,

$$\mathbb{P}(n \text{ events counted}) = \sum_{k=n}^{\infty} (\mathbb{P}(n \text{ events counted}|k \text{ events happened})$$

$$\times \mathbb{P}(k \text{ events happened})$$

$$= \sum_{k=n}^{\infty} \binom{k}{n} p^n (1-p)^{k-n} (\lambda T)^k \exp\{-\lambda T\}/k!$$

$$= \exp\{-\lambda T\}(p\lambda T)^n/n! \sum_{k=n}^{\infty} \frac{[(1-p)\lambda T]^{k-n}}{(k-n)!}$$

$$= (p\lambda T)^n \exp\{-p\lambda T\}/n!$$

after representing $\binom{k}{n}$ by factorials and observing that $\sum_{k=n}^{\infty} \frac{[(1-p)\lambda T]^{k-n}}{(k-n)!} = \sum_{v=0}^{\infty} \frac{[(1-p)\lambda T]^{v}}{v!} = \exp\{(1-p)\lambda T\}$. Thus, the number of counted events is again Poisson but with the rate $p\lambda T$.

5.3.5 Geometric Distribution

Suppose that independent trials are repeated and that in each trial the probability of a success is equal to $0 < p < 1$. We are interested in the number of failures X before the first success. The number of failures is a random variable with a geometric $\mathcal{G}e(p)$ distribution. Its PMF is given by

$$p_X(x) = p(1-p)^x, \quad x = 0, 1, 2, \ldots.$$

The expected value is $\mathbb{E}X = (1-p)/p$ and the variance is $\mathbb{V}\text{ar}\, X = (1-p)/p^2$. The moments can be found either directly or by the moment-generating function, which is

$$m_X(t) = \frac{p}{1 - (1-p)e^t}.$$

The geometric random variable possesses a "memoryless" property. That is, if we condition on the event $X \geq m$, for some nonnegative integer m, then for $n \geq m$, $\mathbb{P}(X \geq n | X \geq m) = \mathbb{P}(X \geq n - m)$ (Exercise 5.25). The MATLAB commands for geometric CDF, PDF, quantile, and random number are geocdf, geopdf, geoinv, and geornd. There are no special names for the geometric distribution in WinBUGS; the negative binomial can be used as dnegbin(p,1).

If instead of the number of failures before the first success (X) one is interested in the total number of experiments until the first success (Y), then the relationship is simple: $Y = X + 1$. In this formulation of the geometric distribution, $Y \sim \mathcal{G}eom(p)$, $\mathbb{E}Y = \mathbb{E}X + 1 = q/p + 1 = 1/p$, and $\mathbb{V}ar\,Y = \mathbb{V}ar\,X = (1 - p)/p^2$.

Example 5.14. **CASES I.** Let a subject constitute either a CASE or a CONTROL depending on whether the subject's level of LDL cholesterol is >160 mg/dL or ≤ 160 mg/dL, respectively. According to a recent National Health and Nutrition Examination Survey (NHANES III), the prevalence of CASES among white male Americans aged 20 and older (target population) is $p = 20\%$. Subjects are sampled (when the population is large, it is unimportant if the sampling is done with or without replacement) until the first CASE is found. The number of CONTROLS sampled before finding the first CASE is a geometric random variable with parameter $p = 0.2$ (Fig. 5.5).

(a) Find the probability that seven CONTROLS will be sampled before we come across the first CASE.

(b) Find the probability that the number of CONTROLS before the first CASE will fall between four and eight, inclusively.

```
disp('X ~ Geometric(0.2):P(X=7)');
geopdf(7, 0.2)
      %ans=0.0419
disp('X ~ Geometric(0.2):P(4 <= X <= 8)');
geocdf(8, 0.2) - geocdf(3,0.2)
      %ans=0.2754
```

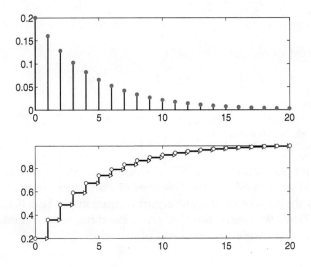

Fig. 5.5 Geometric (Top) PMF and (Bottom) CDF for $p = 0.2$.

Example 5.15. **Mingling Trees.** The degree to which the individual trees of two species are mingled together is an intrinsic property of a two-species population. Two species are said to be segregated if the individuals of each tend to have a member of their own species as nearest neighbor, rather than a member of the other species. To assess the segregation, Pielou (1961) developed a field experiment in which alternating uninterrupted runs of *Pseudotsuga menziesii* and *Pinus ponderosa* are measured along a narrow long belt.

The data in table give lengths of runs Y and their frequency.

Run length, Y	1	2	3	4	5	6	7	8	9	10	11	12
Frequency	21	20	21	4	6	6	2	3	3	1	0	1

(a) Assuming geometric $\mathcal{G}eom(1/3)$ for Y, or equivalently $\mathcal{G}e(1/3)$ distribution for $X = Y - 1$, find the mean $\mathbb{E}Y$, variance $\mathbb{V}ar\, Y$, and $\mathbb{P}(Y > 5 | Y > 2)$. Is this probability the same as $\mathbb{P}(Y > 5 - 2)$ (memoryless property)?

(b) What are sample counterparts of quantities from (a)?

```
%mingling.m
%(a)
[ex varx] = geostat(1/3)
ey = ex+1     %E(Y)=1/(1/3)=3
vary = varx  %Var(Y)=((1-1/3)/((1/3)^2)=6
%P(Y>5|Y>2)=P(Y>5)/P(Y>2)=P(X>=5)/P(X>=2)
(1-geocdf(4, 1/3))/(1-geocdf(1, 1/3))    %0.2963
% memoryles
%P(Y>5-2)=P(Y>3)=P(X>=3)
1-geocdf(2, 1/3)     %0.2963

%(b) empirical counterparts to (a)
Y=1:12;
freq   =[21  20  21   4   6   6   2   3   3   1   0   1];
n=sum(freq)       %88
ybar = sum(Y .* freq)/n  %3.3295
s2y = sum((Y -ybar).^2   .* freq)/(n-1)  %5.9936
sum(freq(Y >5))/sum(freq(Y>2))    %0.3404
sum(freq(Y>3))/n    %0.2955
```

Note that geometric $\mathcal{G}eom(1/3)$ distribution provides a good model for the data, as evidenced by the closeness of empirical moments and probabilities to their theoretical counterparts. Later in the text (Chapter 7 and Chapter 17) we will learn how to, given the data, set the model, estimate parameters, and assess the goodness of model fit.

5.3.6 Negative Binomial Distribution

The negative binomial distribution was formulated by Montmort (1714). Here we are dealing with independent trials again. This time we count the number of failures observed until a fixed number of successes ($r \geq 1$) occur. Let p be the probability of success in a single trial.

If we observe r consecutive successes at the start of the experiment, then the count of failures is $X = 0$ and $\mathbb{P}(X = 0) = p^r$. If $X = x$, then we have observed x failures and r successes in $x + r$ trials. There are $\binom{x+r}{x}$ different ways of arranging x failures in those $x + r$ trials, but we can only be concerned with those arrangements in which the last trial ended in a success. So there are really only $\binom{x+r-1}{x}$ equally likely arrangements. For any particular arrangement, the probability is $p^r(1-p)^x$. Therefore, the PMF is

$$p_X(x) = \binom{r+x-1}{x} p^r(1-p)^x, \quad x = 0, 1, 2, \ldots.$$

Sometimes this PMF is stated with $\binom{r+x-1}{r-1}$ in place of the equivalent $\binom{r+x-1}{x}$. This distribution is denoted as $X \sim \mathcal{NB}(r, p)$.

From its moment-generating function

$$m_X(t) = \left(\frac{p}{1 - (1-p)e^t} \right)^r,$$

the expectation of a negative binomial random variable is $\mathbb{E}X = r(1-p)/p$ and its variance is $\mathrm{Var}\, X = r(1-p)/p^2$.

Since the negative binomial $X \sim \mathcal{NB}(r, p)$ is a convolution (a sum) of r independent geometric random variables, $X = Y_1 + Y_2 + \cdots + Y_r$, $Y_i \sim \mathcal{G}(p)$, the mean and variance of X can be easily derived from the mean and variance of its geometric components Y_i, as in (5.2) and (5.3). Note also that $m_X(t) = (m_Y(t))^r$, where $m_Y(t) = \left(\frac{p}{1-(1-p)e^t} \right)$ is the moment-generating function of the component Y_i in the sum. This is a consequence of the fact that a moment-generating function for a sum of independent random variables is the product of the moment-generating functions of the components; see (5.5).

The distribution remains valid if r is not an integer, although an interpretation involving r successes is lost. For an arbitrary nonnegative r, the distribution is called a Pólya distribution or a generalized negative binomial distribution (although this second term can be ambiguous since several generalizations exist). The constant $\binom{r+x-1}{x} = \frac{(r+x-1)!}{x!(r-1)!}$ is replaced by $\frac{\Gamma(r+x)}{x!\Gamma(r)}$, keeping in mind that $\Gamma(n) = (n-1)!$ when n is an integer. The

Pólya distribution is used in ecology for inference about the abundance of species in nature.

The MATLAB commands for negative binomial CDF, PDF, quantile, and random number are `nbincdf`, `nbinpdf`, `nbininv`, and `nbinrnd`. In WinBUGS the negative binomial distribution is denoted as `dnegbin(p,r)`. Note the opposite order of parameters r and p compared to notation $\mathcal{NB}(r,p)$ and the order adopted by MATLAB.

Example 5.16. **CASES II.** Assume as in Example 5.14 that the prevalence of "CASES" in a large population is $p = 20\%$. Subjects are sampled, one by one, until seven CASES are found and then the sampling is stopped.

(a) What is the probability that the number of CONTROLS among all sampled subjects will be 18?

(b) What is the probability of observing more than the "expected number" of CONTROLS?

The number of CONTROLS X among all sampled subjects is a negative binomial, $X \sim \mathcal{NB}(7,0.2)$.

$$\mathbb{P}(X = 18) = \binom{25 + 7 - 1}{18} 0.2^7 (1 - 0.2)^1 8 = 0.0310.$$

Also, `nbinpdf(18,7,0.2)=0.0310`.

Thus, with a probability of 0.031 the number of CONTROLS sampled, before seven CASES are observed, is equal to 18.

(b) The expected number of CONTROLS is $\mathbb{E}X = 7\frac{0.8}{0.2} = 28$. The probability of $X > \mathbb{E}X$ is $\mathbb{P}(X > 28) = 1 - \mathbb{P}(X \le 28) = 1 - \sum_{x=0}^{28} \binom{7+x-1}{x} 0.8^x 0.2^7 = 0.4328$. In MATLAB $\mathbb{P}(X > 28)$ is calculated as `1-nbincdf(28,7,0.20)=0.4328`.

The tail probabilities of a negative binomial distribution can be expressed by binomial probabilities. If $X \sim \mathcal{NB}(r,p)$, then

$$\mathbb{P}(X > x) = \mathbb{P}(Y < r),$$

where $Y \sim \mathcal{B}in(x + r, p)$. In words, if we have not seen r successes after seeing x failures, then in $x + r$ experiments the number of successes will be less than r. In part (b) of the previous example, $r = 7, x = 28$, and $p = 0.20$, so

```
1 - nbincdf(28, 7, 0.20)   % 0.4328
binocdf(7-1, 28+7, 0.20)   % 0.4328
```

5.3.7 Multinomial Distribution

The binomial distribution was developed by counting the occurrences two complementary events, A and A^c, in n independent trials. Suppose, instead, that each trial results in one of $k > 2$ mutually exclusive events, $A_1, \ldots A_k$, so that $S = A_1 \cup \cdots \cup A_k$. One can define the vector of random variables $(X_1, \ldots, X_k)$ where a component X_i counts how many times A_i appeared in n trials. The defined random vector is called multinomial.

The probability mass function for $(X_1, \ldots, X_k)$ is

$$p_{X_1, \ldots, X_k}(x_1, \ldots, x_k) = \frac{n!}{x_1! \cdots x_k!} p_1{}^{x_1} \cdots p_k{}^{x_k},$$

where $p_1 + \cdots + p_k = 1$ and $x_1 + \cdots + x_k = n$. Since $p_k = 1 - p_1 - \cdots - p_{k-1}$, there are $k - 1$ free parameters to characterize the multinomial distribution, which is denoted by $X = (X_1, \ldots, X_k) \sim \mathcal{M}n(n, p_1, \ldots, p_k)$.

The mean and variance of the component X_i are the same as in the binomial case. It is easy to see that the marginal distribution for a component X_i is binomial since the events $A_1, \ldots, A_k$ can be grouped as A_i, A_i^c. Therefore, $\mathbb{E}(X_i) = np_i$, $\mathrm{Var}\,(X_i) = np_i(1 - p_i)$. The components X_i are dependent since they sum up to n. For $i \neq j$, the covariance between X_i and X_j is

$$\mathrm{Cov}(X_i, X_j) = \mathbb{E}X_i X_j - \mathbb{E}X_i \mathbb{E}X_j = -np_i p_j. \tag{5.8}$$

This is easy to verify if X_i and X_j are represented as the sums of Bernoullis $Y_{i1} + Y_{i2} + \cdots + Y_{ik}$ and $Y_{j1} + Y_{j2} + \cdots + Y_{jk}$, respectively. Since $Y_{im}Y_{jm} = 0$ (in a single trial A_i and A_j cannot occur simultaneously), it follows that

$$\mathbb{E}X_i X_j = (n^2 - n)p_i p_j.$$

Since $\mathbb{E}X_i \mathbb{E}X_j = n^2 p_i p_j$, the covariance in (5.8) follows.

If $X = (X_1, X_2, \ldots, X_k) \sim \mathcal{M}n(n, p_1, p_2, \ldots, p_k)$, then $X' = (X_1 + X_2, \ldots, X_k)$ $\sim \mathcal{M}n(n, p_1 + p_2, \ldots, p_k)$. This is called the *fusing* property of the multinomial distribution.

If $X_1 \sim \mathcal{P}oi(\lambda_1)$, $X_2 \sim \mathcal{P}oi(\lambda_2)$, $\ldots$, $X_n \sim \mathcal{P}oi(\lambda_n)$ are n independent Poisson random variables with parameters $\lambda_1, \ldots, \lambda_n$, then the conditional distribution of $X_1, X_2, \ldots, X_n$, given that $X_1 + X_2 + \cdots + X_n = n$, is $\mathcal{M}n(n, p_1, \ldots, p_k)$, where $p_i = \lambda_i / (\lambda_1 + \lambda_2 + \cdots + \lambda_n)$. This fact is used in modeling contingency tables with a fixed total and will be discussed in Chapter 12.

In MATLAB, the multinomial PMF is calculated by `mnpdf(x,p)`, where x is a $1 \times k$ vector of values, such that $\sum_{i=1}^{k} x_i = n$, and p is a $1 \times k$ vector of probabilities, such that $\sum_{i=1}^{k} p_i = 1$.

For example,

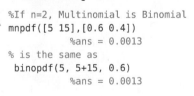

```
%If n=2, Multinomial is Binomial
mnpdf([5 15],[0.6 0.4])
          %ans = 0.0013
% is the same as
 binopdf(5, 5+15, 0.6)
          %ans = 0.0013
```

In WinBUGS, the multinomial distribution is coded as `dmulti(p[],n)`.

Example 5.17. **ABO Group Distribution.** Suppose that the probabilities of blood groups in a particular population are given as

O	A	B	AB
0.37	0.39	0.18	0.06

If eight subjects are selected at random from this population, what is the probability that
(a) $(O, A, B, AB) = (3, 4, 1, 0)$?
(b) $O = 3$?
In (a), the probability is

```
factorial(8) /(factorial(3) * ...
    factorial(4) * factorial(1) * factorial(0)) * ...
    0.37^3 * 0.39^4 * 0.18^1 * 0.06^0
%ans = 0.0591
%or
 mnpdf([3  4  1  0],[0.37 0.39 0.18 0.06])
%ans = 0.0591.
```

In (b), $O \sim Bin(8, 0.37)$ and $\mathbb{P}(O = 3) = 0.2815$.

5.3.8 *Quantiles*

Quantiles of random variables are defined as follows. A p-quantile (or $100 \times p$ percentile) of random variable X is the value x for which $F(x) = p$, if F is a monotone cumulative distribution function for X. For an arbitrary random variable, including discrete, this definition is not unique and modification is needed:

$$F(x) = \mathbb{P}(X \leq x) \geq p \quad \text{and} \quad \mathbb{P}(X \geq x) \geq 1 - p.$$

For example, the 0.05 quantile of a binomial distribution with parameters $n = 12$ and $p = 0.7$ is $x = 6$ since $\mathbb{P}(X \leq 6) = 0.1178 \geq 0.05$ and $\mathbb{P}(X \geq 6) = 1 - \mathbb{P}(X \leq 5) = 1 - 0.0386 = 0.9614 \geq 0.95$. Binomial $Bin(12, 0.7)$ and geometric $\mathcal{G}(0.2)$ quantiles are shown in Figure 5.6.

```
quab =[]; quag =[];
for p = 0.00:0.0001:1
    quab = [quab binoinv(p, 12, 0.7)];
    quag = [quag geoinv(p, 0.2)];
end
figure(1)
```

```
plot([0.00:0.0001:1],quab,'k-')
figure(2)
plot([0.00:0.0001:1],quag,'k-')
```

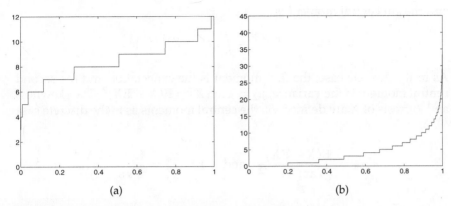

(a)   (b)

Fig. 5.6 (a) Binomial $\mathcal{B}in(12,0.7)$ and (b) geometric $\mathcal{G}(0.2)$ quantiles.

5.4 Continuous Random Variables

Continuous random variables take values within an interval (a,b) on a real line **R**. The probability density function (PDF) $f(x)$ fully specifies the variable. The PDF is nonnegative, $f(x) \geq 0$, and integrates to 1, $\int_R f(x)\,dx = 1$. The probability that X takes a value in an interval (a,b) (and for continuous random variables equivalently $[a,b),(a,b]$, or $[a,b]$) is $\mathbb{P}[X \in (a,b)] = \int_a^b f(x)dx$.

The CDF is

$$F(x) = \mathbb{P}(X \leq x) = \int_{-\infty}^{x} f(t)dt.$$

In terms of the CDF, $\mathbb{P}[X \in (a,b)] = F(b) - F(a)$.

The expectation of X is given by

$$\mathbb{E}X = \int_R xf(x)dx.$$

The expectation of a function of a random variable $g(X)$ is

$$\mathbb{E}g(X) = \int_R g(x)f(x)dx.$$

The kth moment of a continuous random variable X is defined as

$$m_k = \mathbb{E}X^k = \int_{\mathbf{R}} x^k f(x)dx,$$

and the kth central moment is

$$\mu_k = \mathbb{E}(X - \mathbb{E}X)^k = \int_{\mathbf{R}} (x - \mathbb{E}X)^k f(x)dx.$$

As in the discrete case, the first moment is the expectation and the second central moment is the variance, $\mu_2 = \mathbb{V}\mathrm{ar}\,(X) = \mathbb{E}(X - \mathbb{E}X)^2$. The skewness and kurtosis of X are defined via the central moments as in the discrete case (5.1),

$$\gamma = \frac{\mu_3}{\mu_2^{3/2}} = \frac{\mathbb{E}(X - \mathbb{E}X)^3}{(\mathbb{V}\mathrm{ar}\,(X))^{3/2}} \quad \text{and} \quad \kappa = \frac{\mu_4}{\mu_2^2} = \frac{\mathbb{E}(X - \mathbb{E}X)^4}{(\mathbb{V}\mathrm{ar}\,(X))^2}.$$

The moment-generating function of a continuous random variable X is

$$m(t) = \mathbb{E}e^{tX} = \int_{\mathbf{R}} e^{tx} f(x)dx.$$

Since $m^{(k)}(t) = \int_{\mathbf{R}} x^k e^{tx} f(x)dx$, $\mathbb{E}X^k = m^{(k)}(0)$. Moment-generating functions are related to Laplace transforms of densities. Since the bilateral Laplace transform of $f(x)$ is defined as

$$\mathcal{L}(f) = \int_{\mathbf{R}} e^{-tx} f(x)dx,$$

it holds that $m(-t) = \mathcal{L}(f)$.

The entropy of a continuous random variable with a density $f(x)$ is defined as

$$\mathcal{H}(X) = -\int_{\mathbb{R}} f(x) \log f(x)dx,$$

whenever this integral exists. Unlike the entropy for discrete random variables, $\mathcal{H}(X)$ can be negative and not necessarily invariant with respect to a transformation of X.

Example 5.18. **Markov's Inequality.** If X is a random variable that takes only nonnegative values, then for any positive constant a,

$$\mathbb{P}(X \geq a) \leq \frac{\mathbb{E}X}{a}. \tag{5.9}$$

Indeed,

$$\mathbb{E}X = \int_0^\infty x f(x) dx \geq \int_a^\infty x f(x) dx$$

$$\geq \int_a^\infty a f(x) dx$$

$$= a \int_a^\infty f(x) dx = a\mathbb{P}(X \geq a).$$

An average mass of a single cell of E. *coli* bacterium is 665 fg (femtogram, fg = 10^{-15}g). If a particular cell of E. *coli* is inspected, what can be said about the probability that its weight will exceed 1000 fg? According to Markov's inequality, this probability does not exceed $665/1000 = 0.665$.

Example 5.19. Durability of the Starr–Edwards Valve. The Starr–Edwards valve is one of the oldest cardiac valve prostheses in the world. The first aortic valve replacement (AVR) with a Starr–Edwards metal cage and silicone ball valve was performed in 1961. Follow-up studies have documented the excellent durability of the Starr–Edwards valve as an AVR. Suppose that the durability of the Starr–Edwards valve (in years) is a random variable X with density

$$f(x) = \begin{cases} ax^2/100, & 0 < x < 10, \\ a(x-30)^2/400, & 10 \leq x \leq 30, \\ 0, & \text{otherwise.} \end{cases}$$

(a) Find the constant a.
(b) Find the CDF $F(x)$ and sketch graphs of f and F.
(c) Find the mean and 60th percentile of X. Which is larger? Find the variance.
Solution: (a) Since $1 = \int_R f(x) dx$,

$$1 = \int_0^{10} ax^2/100 dx + \int_{10}^{30} a(x-30)^2/400 dx$$

$$= ax^3/300 \Big|_0^{10} + a(x-30)^3/1200 \Big|_{10}^{30}.$$

This gives $1000a/300 - 0 + 0 - (-20)^3 a/1200 = 10a/3 + 20a/3 = 10a = 1$, that is, $a = 1/10$. The density is

$$f(x) = \begin{cases} x^2/1000, & 0 < x < 10, \\ (x-30)^2/4000, & 10 \leq x \leq 30, \\ 0, & \text{otherwise.} \end{cases}$$

(b) The CDF is

$$F(x) = \begin{cases} 0, & x < 0, \\ x^3/3000, & 0 < x < 10, \\ 1 + (x - 30)^3/12000, & 10 \le x \le 30, \\ 1, & x \ge 30. \end{cases}$$

(c) The 60th percentile is a solution to the equation $1 + (x - 30)^3/12000 = 0.6$ and is $x = 13.131313\dots$. The mean is $\mathbb{E}X = 25/2$, and the 60th percentile exceeds the mean. $EX^2 = 180$; thus the variance is $\mathbb{V}\mathrm{ar}\, X = 180 - (25/2)^2 = 95/4 = 23.75$.

Example 5.20. **Soliton Waves and Sech Distribution.** Soliton waves were first described by John Scott Russell, a Scottish civil engineer. In August 1834 he was riding beside the Union Canal near Edinburgh, Scotland, and noticed a strange wave building up at the bow of a boat. After the boat stopped, the wave traveled on, "assuming the form of a large solitary elevation, a rounded, smooth and well-defined heap of water, which continued its course along the channel apparently without change of form or diminution of speed." Soliton waves appear within the ocean and the atmosphere, within magnets and super-cooled devices, within the ionized plasma of space, and in optical fibers, to list a few.

The envelope of a soliton wave (Fig. 5.7a), properly scaled, is a probability density as is described next. Let X be a continuous random variable with the density

$$f(x) = \frac{2}{e^{\pi x} + e^{-\pi x}}, \ x \in R. \tag{5.10}$$

This function is in fact hyperbolic secant of argument πx, motivating the name "sech,"

$$f(x) = \mathrm{sech}(\pi x), \ x \in R.$$

The density is shown in Figure 5.7b. The odd moments for this distribution are 0, and a few even moments are

$$EX^2 = 1/4, \quad EX^4 = 5/16, \quad EX^6 = 61/64, \quad EX^8 = 1385/256, \dots.$$

(a) What are the skewness and kurtosis of this distribution?

(b) Calculate the 0.25- and 0.75-quantiles of this distribution.

(c) Find the "width" of the sech envelope, defined as the length of the line segment at height 0.5 that falls inside the envelope, see Figure 5.7a.

(d) What is the probability of random variable X with sech distribution to fall within the "width" range?

Solution. Since this distribution is symmetric about 0, the central moments are equal to raw moments. The skewness $\gamma = 0$, and kurtosis is $\kappa = \frac{5/16}{(1/4)^2} = 5$.

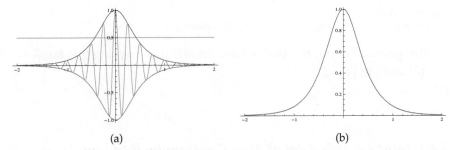

Fig. 5.7 (a) Soliton waves and (b) density of sech distribution.

(b) By representing (5.10) as

$$f(x) = \frac{2e^{\pi x}}{1 + (e^{\pi x})^2}$$

and taking the substitution $t = e^{\pi x}$ in the integral $F(x) = \int_{-\infty}^{x} f(t)dt$, we find the CDF,

$$F(x) = \frac{2}{\pi} \arctan(e^{\pi x}), \ x \in R.$$

Since $F(x)$ is monotone and one-to-one, its inverse is unique and represents a quantile function for this distribution. For $F(x) = p$, it is easy to find

$$x = \frac{1}{\pi} \log\left(\tan\left(\frac{\pi p}{2}\right) \right),$$

which for $p = 0.25$ gives $x_{0.25} = -0.2805$. Because of symmetry, $x_{0.75} = 0.2805$.

```
p=0.25;   x25=1/pi * log( tan(pi * p/2))    %-0.2805
```

(c) The solution of $f(x) = 1/2$ can be found in finite form, $x_{1/2} = \frac{1}{\pi}\log(2 \pm \sqrt{3}) = \pm 0.4192$. The length of segment inside the envelope is

$$x_2 - x_1 = \frac{1}{\pi}\log\frac{2 + \sqrt{3}}{2 - \sqrt{3}} = 0.8384.$$

In MATLAB,

```
fzero(@(x) sech(pi * x) - 1/2, 1)    % 0.4192
fzero(@(x) sech(pi * x) - 1/2,-1)    %-0.4192
```

(d) The required probability is 2/3. Numerically,

```
format long
sechcdf = @(x) 2/pi * atan( exp(pi * x));
```

```
x2=1/pi * log(2 + sqrt(3));
prob =sechcdf(x2)-sechcdf(-x2) %0.666666666666667
```

This probability can be obtained analytically by observing that $\tan\frac{\pi}{12} = 2 - \sqrt{3}$ and $\tan\frac{5\pi}{12} = 2 + \sqrt{3}$.

5.4.1 Joint Distribution of Two Continuous Random Variables

Two random variables X and Y are jointly continuous if there exists a nonnegative function $f(x,y)$ so that for any two-dimensional domain D,

$$\mathbb{P}((X,Y) \in D) = \int\int_D f(x,y)dxdy.$$

When such a two-dimensional density $f(x,y)$ exists, it is a repeated partial derivative of the cumulative distribution function $F(x,y) = \mathbb{P}(X \le x, Y \le y)$,

$$f(x,y) = \frac{\partial^2 F(x,y)}{\partial x\, \partial y}.$$

The marginal densities for X and Y are, respectively, $f_X(x) = \int_{-\infty}^{\infty} f(x,y)dy$ and $f_Y(y) = \int_{-\infty}^{\infty} f(x,y)dx$. The conditional distributions of X when $Y = y$ and of Y when $X = x$ are

$$f(x|y) = f(x,y)/f_Y(y) \text{ and } f(y|x) = f(x,y)/f_X(x).$$

The distributional analogy of the multiplication probability rule $\mathbb{P}(AB) = \mathbb{P}(A|B)\mathbb{P}(B) = \mathbb{P}(B|A)\mathbb{P}(A)$ is

$$f(x,y) = f(x|y)f_Y(y) = f(y|x)f_X(x). \qquad (5.11)$$

When X and Y are independent, the joint density is the product of marginal densities, $f(x,y) = f_X(x)f_Y(y)$. Conversely, if the joint density of (X,Y) can be represented as a product of marginal densities, X and Y are independent.

The definition of covariance and the correlation for X and Y coincides with the discrete case equivalents:

$$\text{Cov}(X,Y) = \mathbb{E}XY - \mathbb{E}X \cdot \mathbb{E}Y \quad \text{and} \quad \text{Corr}(X,Y) = \frac{\text{Cov}(X,Y)}{\sqrt{\mathbb{V}\text{ar}\,(X) \cdot \mathbb{V}\text{ar}\,(Y)}}.$$

Here, $\mathbb{E}XY = \int_{\mathbb{R}^2} xy f(x,y) dx dy$.

Example 5.21. **Probability, Marginals, and Conditional.** A two-dimensional random variable (X,Y) is defined by its density function, $f(x,y) = 2xe^{-x-2y}$, $x \geq 0, y \geq 0$.

(a) Find the probability that random variable (X,Y) falls in the rectangle $0 \leq X \leq 1, 1 \leq Y \leq 2$.

(b) Find the marginal distributions of X and Y.

(c) Find the conditional distribution of $X|\{Y = y\}$ Does it depend on y?

Solution: (a) The joint density separates variables x and y, therefore

$$\mathbb{P}(0 \leq X \leq 1,\ 1 \leq Y \leq 2) = \int_0^1 xe^{-x}dx \times \int_1^2 2e^{-2y}dy.$$

Since

$$\int_0^1 xe^{-x}dx = -xe^{-x}\Big|_0^1 + \int_0^1 e^{-x}dx = -e^{-1} - e^{-1} + 1 = 1 - 2/e,$$

and

$$\int_1^2 2e^{-2y}dy = -e^{-2y}\Big|_1^2 = -e^{-4} + e^{-2} = \frac{e^2 - 1}{e^4}.$$

then

$$\mathbb{P}(0 \leq X \leq 1,\ 1 \leq Y \leq 2) = \frac{e-2}{e} \times \frac{e^2 - 1}{e^4} \approx 0.0309.$$

(b) Since the joint density separates the variables, it is a product of marginal densities $f(x,y) = f_X(x) \times f_Y(y)$. This is an analytic way to state that components X and Y are independent. Therefore, $f_X(x) = xe^{-x}$, $x \geq 0$ and $f_Y(y) = 2e^{2y}$, $y \geq 0$.

(c) The conditional densities for $X|\{Y = y\}$ and $Y|\{X = x\}$ are defined as

$$f(x|y) = f(x,y)/f_Y(y) \quad \text{and} \quad f(y|x) = f(x,y)/f_X(x).$$

Because of independence of X and Y the conditional densities coincide with the marginal densities. Thus, the conditional density for $X|\{Y = y\}$ does not depend on y.

5.4.2 Conditional Expectation*

Conditional expectation of Y given $\{X = x\}$ is simply the expectation with respect to the conditional distribution,

$$\mathbb{E}(Y|X=x) = \int_{\mathbb{R}} y f(y|x) dy.$$

Since it depends on the value x taken by random variable X, conditional expectation is a function of x. When a particular realization of X is not specified, the conditional expectation of Y given X is denoted by $\mathbb{E}Y|X$ and represents a random variable.

The following properties of conditional expectation and variance are very important and useful in applications:

In general, $\mathbb{E}Y|X$ is a random variable for which

$$\mathbb{E}Y = \mathbb{E}(\mathbb{E}Y|X),$$ \hfill (5.12)
$$\mathbb{V}\mathrm{ar}\,Y = \mathbb{V}\mathrm{ar}\,(\mathbb{E}Y|X) + \mathbb{E}(\mathbb{V}\mathrm{ar}\,Y|X).$$

These two equations are sometimes called the Iterated Expectation Rule and Total Variance Rule.

Example 5.22. **Conditional Distributions, Expectations, and Variances.** Let a bivariate random variable (X,Y) have a uniform distribution on triangle $x \geq 0$, $y \geq 0$ and $x + y \leq 1$. The density is constant over the triangle, and the constant is a reciprocal of the triangle area,

$$f(x,y) = \begin{cases} 2, 0 \leq x, y, x+y \leq 1 \\ 0, \text{else} \end{cases}$$

The marginal density for X is obtained by integrating y form the joint density $f(x,y)$. Here variable y ranges from 0 to $1 - x$, and

$$f_X(x) = \int_0^{1-x} 2dy = 2y \Big|_0^{1-x} = 2(1-x),\ 0 \leq x \leq 1.$$

For $f_Y(y)$, the derivation is analogous, $f_Y(y) = 2(1-y)$, $0 \leq y \leq 1$. The means and variances of X (as well as Y) are

$$\mathbb{E}X = \int_0^1 2x(1-x)dx = \left(x^2 - \frac{2x^3}{3}\right)\bigg|_0^1 = 1 - \frac{2}{3} = \frac{1}{3},$$

$$\mathbb{V}ar\,X = \mathbb{E}X^2 - (\mathbb{E}X)^2 = \int_0^1 2x^2(1-x)dx - \frac{1}{9}$$

$$= \left(\frac{2x^3}{3} - \frac{x^4}{2}\right)\bigg|_0^1 - \frac{1}{9}$$

$$= \frac{2}{3} - \frac{1}{2} - \frac{1}{9} = \frac{1}{18}.$$

The conditional distribution of Y when $X = x$ is

$$f(y|x) = \begin{cases} \frac{1}{1-x}, & 0 \le y \le 1-x \\ 0, & \text{else} \end{cases}$$

The conditional expectation of Y given $\{X = x\}$ is

$$\mathbb{E}(Y|X=x) = \int_0^{1-x} \frac{y\,dy}{1-x} = \frac{y^2}{2(1-x)}\bigg|_0^{1-x} = \frac{1-x}{2}.$$

Since this is true for any x that X takes, the conditional expectation can be expressed in terms of X as

$$\mathbb{E}Y|X = \frac{1-X}{2},$$

and as such represents a random variable. It is straightforward to show (Exercise 5.23) that $\mathbb{V}ar\,(Y|X=x) = \frac{(1-x)^2}{12}$, that is,

$$\mathbb{V}ar\,Y|X = \frac{(1-X)^2}{12}.$$

We will check that the Iterated Expectation Rule and Total Variance Rule from (5.12) are satisfied. The iterate expectation is

$$\mathbb{E}(\mathbb{E}Y|X) = \mathbb{E}\frac{1-X}{2} = \frac{1-1/3}{2} = \frac{1}{3},$$

which coincides with $\mathbb{E}Y$. The total variance is

$$\mathbb{E}(\operatorname{Var}Y|X) + \operatorname{Var}(\mathbb{E}Y|X) = \mathbb{E}\frac{(1-X)^2}{12} + \operatorname{Var}\frac{1-X}{2}$$

$$= \frac{1}{12}(1 - 2\mathbb{E}X + \mathbb{E}X^2) + \frac{1}{4}\operatorname{Var}X$$

$$= \frac{1}{12}\left(1 - \frac{2}{3} + \frac{1}{6}\right) + \frac{1}{4} \cdot \frac{1}{18}$$

$$= \frac{6 - 4 + 1}{72} + \frac{1}{72} = \frac{1}{18},$$

which coincides with $\operatorname{Var}Y$.

✐

5.5 Some Standard Continuous Distributions

In this section we overview some popular, commonly used continuous distributions: uniform, exponential, gamma, inverse gamma, beta, double exponential, logistic, Weibull, Pareto, and Dirichlet. The normal (Gaussian) distribution will be just briefly mentioned here. Due to its importance, a separate chapter will cover the details of the normal distribution and its close relatives: χ^2, t, Cauchy, F, and lognormal distributions. Some other continuous distributions will be featured in the examples, exercises, and other chapters, such as Maxwell and Rayleigh distributions.

5.5.1 Uniform Distribution

A random variable X has a uniform $\mathcal{U}(a,b)$ distribution if its density is given by

$$f_X(x) = \begin{cases} \frac{1}{b-a}, & a \leq x \leq b, \\ 0, & \text{else}. \end{cases}$$

Sometimes, to simplify notation, the density can be written simply as

$$f_X(x) = \frac{1}{b-a}\mathbf{1}(a \leq x \leq b).$$

Here, $\mathbf{1}(A)$ is 1 if A is a true statement and 0 if A is false. Thus, for $x < a$ or $x > b$, $f_X(x) = 0$, since for those values of x the relation $a \leq x \leq b$ is false and $\mathbf{1}(a \leq x \leq b) = 0$. For $a = 0$ and $b = 1$, the distribution is called standard uniform.

The CDF of X is given by

$$F_X(x) = \begin{cases} 0, & x < a, \\ \frac{x-a}{b-a}, & a \leq x \leq b, \\ 1, & x > b. \end{cases}$$

The graphs of the PDF and CDF of a uniform $\mathcal{U}(-1,4)$ random variable are shown in Figure 5.8.

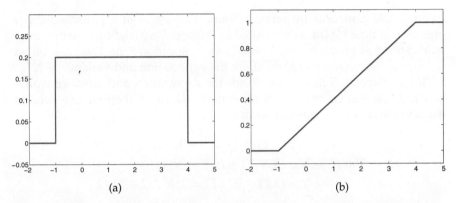

(a)   (b)

Fig. 5.8 (a) PDF and (b) CDF for uniform $\mathcal{U}(-1,4)$ distribution. The graphs are plotted as (a) `plot(-2:0.001:5, unifpdf(-2:0.001:5, -1, 4))` and (b) `plot(-2:0.001:5, unifcdf(-2:0.001:5, -1, 4))`.

The expectation of X is $\mathbb{E}X = \frac{a+b}{2}$ and the variance is $\mathbb{V}\text{ar}\,X = (b-a)^2/12$. The nth moment of X is given by $\mathbb{E}X^n = \frac{1}{n+1}\sum_{i=0}^{n} a^i b^{n-i}$. The moment-generating function for the uniform distribution is $m(t) = \frac{e^{tb}-e^{ta}}{t(b-a)}$.

If U is $\mathcal{U}(0,1)$, then $X = -\lambda\log(U)$ is an exponential random variable with scale parameter λ. The sum of two independent standard uniform random variables has triangular distribution,

$$f_X(x) = \begin{cases} x, & 0 \leq x \leq 1, \\ 2-x, & 1 \leq x \leq 2, \\ 0, & \text{else}. \end{cases}$$

This is sometimes called a "witch hat" distribution. The distribution of the sum of n independent standard uniforms random variables is known as the Irwing–Hall distribution.

The MATLAB commands for uniform CDF, PDF, quantile, and random number are `unifcdf`, `unifpdf`, `unifinv`, and `unifrnd`. In WinBUGS, the uniform distribution is coded as `dunif(a,b)`.

Example 5.23. **A Gauge That Rounds.** An absolute error E of a measurement read at a particular gauge has uniform $\mathcal{U}(0,1/2)$ distribution. This error is caused by gauge's rounding to the nearest integer. The mean and variance of E are $(0+1/2)/2 = 1/4$ and $(1/2-0)^2/12 = 1/48$. The probability

that in a single measurement the absolute error exceeds 0.3 is `1-unifcdf(0.3,`
`0, 1/2)` which is equal to 0.4. Since the density is 2 for values between 0 and
1/2, this probability can be easily visualized as an area of a rectangle with
basis $0.5 - 0.3 = 0.2$ and height 2.

Example 5.24. **Uniform Inspection Time.** Counts N at a particle counter
observed at time $t \geq 0$ are distributed as Poisson $\mathcal{P}oi(\lambda t)$. Suppose the count
is inspected at random time $T = t$. If the inspection time T is distributed
uniformly between 0 and b, what are the expectation and variance of N?

If the inspection time t was fixed, the expectation and variance would
be λt. When inspection time is random, $T \sim \mathcal{U}(0,b)$, then we use iterated
expectation and total variance as in (5.12),

$$\mathbb{E}N = \mathbb{E}(\mathbb{E}N|T) = \mathbb{E}(\lambda T) = \lambda b/2,$$
$$\mathbb{V}\text{ar}\, N = \mathbb{V}\text{ar}\,(\mathbb{E}N|T) + \mathbb{E}(\mathbb{V}\text{ar}\, N|T)$$
$$= \mathbb{V}\text{ar}\,(\lambda T) + \mathbb{E}(\lambda T) = \lambda^2 b^2/12 + \lambda b/2.$$

Note the overdispersion $\lambda^2 b^2/12$ due to randomness of the inspection
time.

5.5.2 Exponential Distribution

The probability density function for an exponential random variable is

$$f_X(x) = \begin{cases} \lambda e^{-\lambda x}, & x \geq 0, \\ 0, & \text{else}, \end{cases}$$

where $\lambda > 0$ is called the *rate* parameter. An exponentially distributed random variable X is denoted by $X \sim \mathcal{E}(\lambda)$. Its moment-generating function is
$m(t) = \lambda/(\lambda - t)$ for $t < \lambda$, and the mean and variance are $1/\lambda$ and $1/\lambda^2$,
respectively. The nth moment is $\mathbb{E}X^n = \frac{n!}{\lambda^n}$.

This distribution has several interesting features; for example, its *failure
rate*, defined as

$$\lambda_X(t) = \frac{f_X(t)}{1 - F_X(t)},$$

is constant and equal to λ.

The exponential distribution has an important connection to the Poisson
distribution. Suppose we observe i.i.d. exponential variates $(X_1, X_2, \dots)$ and
define $S_n = X_1 + \dots + X_n$. For any positive value t, it can be shown that

$\mathbb{P}(S_n < t < S_{n+1}) = p_Y(n)$, where $p_Y(n)$ is the probability mass function for a Poisson random variable Y with parameter λt.

Like a geometric random variable, an exponential random variable has the *memoryless property*, $\mathbb{P}(X \geq u + v | X \geq u) = \mathbb{P}(X \geq v)$ (Exercise 5.25).

The median value, representing a typical observation, is roughly 70% of the mean, showing how extreme values can affect the population mean. This is explicitly shown by the ease in computing the inverse CDF:

$$p = F(x) = 1 - e^{-\lambda x} \iff x = F^{-1}(p) = -\frac{1}{\lambda}\log(1-p).$$

The MATLAB commands for exponential CDF, PDF, quantile, and random number are `expcdf`, `exppdf`, `expinv`, and `exprnd`. MATLAB uses the alternative parametrization with $1/\lambda$ in place of λ. Thus, the CDF of random variable X with $\mathcal{E}(3)$ distribution evaluated at $x = 2$ is calculated in MATLAB as `expcdf(2,1/3)`. In WinBUGS, the exponential distribution is coded as `dexp(lambda)`.

Example 5.25. **Melanoma.** The 5-year cancer survival rate in the case of malignant melanoma of the skin at stage IIIA is 78%. Assume that the survival time T can be modeled by an exponential random variable with unknown rate λ. Given the 5-year survival rate, we will find the probability of a melanoma patient surviving more than 10 years.

Using the given survival rate of 0.78, we first determine the parameter of the exponential distribution – the rate λ. Since $\mathbb{P}(T > t) = \exp(-\lambda t)$, $\mathbb{P}(T > 5) = 0.78$ leads to $\exp\{-5\lambda\} = 0.78$, with solution $\lambda = -\frac{1}{5}\log(0.78)$, which can be rounded to $\lambda = 0.05$.

Next, we find the probability that the survival time exceeds 10 years, first directly using the CDF,

$$\mathbb{P}(T > 10) = 1 - F(10) = 1 - \left(1 - e^{-0.05 \cdot 10}\right) = \frac{1}{\sqrt{e}} = 0.6065,$$

and then by MATLAB. One should be careful when parameterizing the exponential distribution in MATLAB. MATLAB uses the scale parameter, a reciprocal of the rate λ.

```
1 - expcdf(10, 1/0.05)
 %ans =   0.6065
 %
 %Figures of PDF and CDF are produced by
time=0:0.001:30;
pdf = exppdf(time, 1/0.05);  plot(time, pdf, 'b-');
cdf = expcdf(time, 1/0.05);  plot(time, cdf, 'b-');
```

This is shown in Figure 5.9.

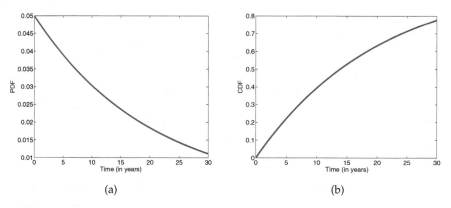

Fig. 5.9 Exponential (a) PDF and (b) CDF for rate $\lambda = 0.05$.

Example 5.26. **Minimum of n Exponential Lifetimes.** Let $n = 20$ independent components be connected in a serial system; that is, all components need to be operational for the system to work. The lifetime of each component is exponential $\mathcal{E}(\lambda)$ random variable, where $\lambda = 1/3$ is the rate parameter (in units of 1/year). What is the probability that the system remains operational for more than one month?

If T_i are lifetimes of system components, the system's lifetime is $T = \min\{T_1, T_2, \ldots, T_n\}$ because of the serial connection. When T_i are independent exponentials with rates λ_i, the system's lifetime T is also exponential with rate $\lambda = \sum_{i=1}^{n} \lambda_i$.

This is easy to see; for T to exceed t, each T_i has to exceed t,

$$\mathbb{P}(T > t) = \mathbb{P}(T_1 > t, T_2 > t, \ldots, T_n > t).$$

Due to the independence of T_i's, the probability above is

$$\prod_{i=1}^{n} \mathbb{P}(T_i > t) = \prod_{i=1}^{n} \exp\{-\lambda_i t\} = \exp\left\{-t \sum_{i=1}^{n} \lambda_i\right\}.$$

Thus, $T \sim \mathcal{E}\left(\sum_{i=1}^{n} \lambda_i\right)$.

In this example, all λ_i are equal and $T \sim \mathcal{E}(30 \cdot 1/3)$. We assume that 1 month is $1/12$ of a year, and

$$\mathbb{P}(T > 1/12) = \exp\{-10/12\} = 0.4346.$$

Even though each component will work for at least a month with probability of 97.26%, this probability for a serial system of 30 independent components scales down to 43.46%.

5.5.3 *Normal Distribution*

As we indicated at the start of this section, due to its importance, the normal distribution is covered in a separate chapter. Here we provide a definition and list a few important facts.

The probability density function for a normal (Gaussian) random variable X is given by

$$f_X(x) = \frac{1}{\sqrt{2\pi}\,\sigma}\,\exp\left\{-\frac{(x-\mu)^2}{2\sigma^2}\right\},$$

where μ is the mean and σ^2 is the variance of X. This will be denoted as $X \sim \mathcal{N}\left(\mu,\sigma^2\right)$. For $\mu = 0$ and $\sigma = 1$, the distribution is called the standard normal distribution. The CDF of a normal distribution cannot be expressed in terms of elementary functions and so defines a function of its own. For the standard normal distribution, the CDF is

$$\Phi(x) = \int_{-\infty}^{x} \frac{1}{\sqrt{2\pi}} \exp\left\{-\frac{t^2}{2}\right\} dt.$$

The standard normal PDF and CDF are shown in Figure 5.10a,b.

The moment-generating function is $m(t) = \exp\{\mu t + \sigma^2 t^2/2\}$. The odd central moments $\mathbb{E}(X - \mu)^{2k+1}$ are 0 because the normal distribution is symmetric about the mean. The even moments are

$$\mathbb{E}(X - \mu)^{2k} = \sigma^{2k}\,(2k - 1)!!,$$

where $(2k - 1)!! = (2k - 1) \cdot (2k - 3) \cdots 5 \cdot 3 \cdot 1.$

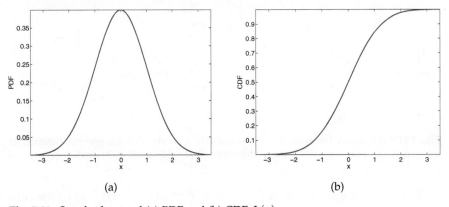

(a)   (b)

Fig. 5.10 Standard normal (a) PDF and (b) CDF $\Phi(x)$.

The MATLAB commands for normal CDF, PDF, quantile, and random number are `normcdf`, `normpdf`, `norminv`, and `normrnd`. In WinBUGS, the normal distribution is coded as `dnorm(mu,tau)`, where `tau` is a precision parameter, the reciprocal of variance.

5.5.4 Gamma Distribution

The gamma distribution is an extension of the exponential distribution. Prior to defining its density, we define the gamma function that is critical in normalizing the density. Function $\Gamma(x)$, defined via the integral $\int_0^\infty t^{x-1}e^{-t}dt$, $x > 0$, is called the gamma function (Fig. 5.11a). If n is a positive integer, then $\Gamma(n) = (n-1)!$. In MATLAB: `gamma(x)`.

Random variable X has a gamma $\mathcal{G}a(r,\lambda)$ distribution if its PDF is given by

$$f_X(x) = \begin{cases} \frac{\lambda^r}{\Gamma(r)}x^{r-1}e^{-\lambda x}, & x \geq 0, \\ 0, & \text{else.} \end{cases}$$

The parameter $r > 0$ is called the *shape* parameter, and $\lambda > 0$ is the *rate* parameter. Figure 5.11b shows gamma densities for $(r,\lambda) = (1,1/3), (2,2/3)$, and $(20,2)$.

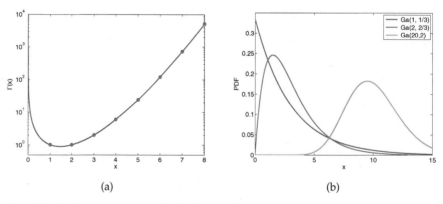

(a)   (b)

Fig. 5.11 (a) Gamma function, $\Gamma(x)$. The *red dots* are values of the gamma function at integers, $\Gamma(n) = (n-1)!$; (b) Gamma densities: $\mathcal{G}a(1,1/3)$, $\mathcal{G}a(2,2/3)$, and $\mathcal{G}a(20,2)$.

The moment-generating function is $m(t) = (\lambda/(\lambda - t))^r$, so in the case $r = 1$, the gamma distribution becomes the exponential distribution. From $m(t)$ we have $\mathbb{E}X = r/\lambda$ and $\mathbb{V}\text{ar}\,X = r/\lambda^2$.

If $X_1,\ldots,X_n$ are generated from an exponential distribution with (rate) parameter λ, it follows from $m(t)$ that $Y = X_1 + \cdots + X_n$ is distributed as gamma with parameters λ and n; that is, $Y \sim \mathcal{G}a(n,\lambda)$. A gamma distribution with an integer shape parameter is sometimes called Erlang's distribution. More generally, if $X_i \sim \mathcal{G}a(r_i,\lambda)$ are independent, then $Y = X_1 + \cdots + X_n$ is distributed as gamma with parameters λ and $r = r_1 + r_2 + \cdots + r_n$; that is, $Y \sim \mathcal{G}a(r,\lambda)$ (Exercise 5.24).

Often, the gamma distribution is parameterized with $1/\lambda$ in place of λ, and this alternative parametrization is used in MATLAB definitions. The CDF in MATLAB is gamcdf(x,r,1/lambda), and the PDF is gampdf(x,r,1/lambda). The function gaminv(p,r,1/lambda) computes the pth quantile of the $\mathcal{G}a(r,\lambda)$ random variable. In WinBUGS, $\mathcal{G}a(n,\lambda)$ is coded as dgamma(n,lambda).

Example 5.27. **Corneoretinal Potentials.** Emil du Bois-Reymond (1848) observed that the cornea of the eye is electrically positive relative to the back of the eye. This potential is not affected by the presence or absence of light, and its variability is critical in defining the electro-oculogram (EOG). Eye movements thus produce a moving (rotating) dipole source, and accordingly, signals that are indicative of the movement may be obtained.

Assume that corneoretinal potential is a random variable $X = Y + 0.35$ [mV], where Y is gamma distributed with shape parameter 3 and rate parameter 20 [1/mV] (or equivalently, scale parameter $1/20 = 0.05$ [mV]).

(a) What is the probability to observe corneoretinal potential X exceeding 0.5 [mV].

(b) If an observed corneoretinal potential exceeds x^*, it is recorded as significant. If, in the long run, we wish to label 1% largest potentials as significant, how should the threshold x^* be set?

```
%(a) P(X > 0.5)=P(Y+0.35 > 0.5)=P(Y > 0.15)
1-gamcdf(0.15, 3, 1/20)   %0.4232
%(b) 0.01=P(X > x*)=P(Y-0.35 > x*)=P(Y>x*-0.35).
%x*-35 is 0.99-quantile of gamma distribution with shape=3 and rate=20.
xstar = 0.35 + gaminv(0.99, 3, 1/20)     %0.7703
```

Thus, if modeled as gamma $\mathcal{G}a(3,20)$, the corneoretinal potential will exceed 0.5 with probability 0.4232, and will exceed $x^* = 0.7703$ with probability 0.01.

5.5.5 Inverse Gamma Distribution

Random variable X is said to have an inverse gamma $\mathcal{IG}(r,\lambda)$ distribution with parameters $r > 0$ and $\lambda > 0$ if its density is given by

$$f_X(x) = \begin{cases} \frac{\lambda^r}{\Gamma(r)x^{r+1}}e^{-\lambda/x}, & x \geq 0, \\ 0, & \text{else} . \end{cases}$$

The mean and variance of X are $\mathbb{E}X = \lambda/(r-1)$, $r > 1$, and $\mathrm{Var}\,X = \lambda^2/[(r-1)^2(r-2)]$, $r > 2$, respectively. If $X \sim \mathcal{G}a(r,\lambda)$, then its reciprocal X^{-1} is $\mathcal{IG}(r,\lambda)$ distributed. We will see that in the Bayesian context, the inverse gamma is a natural prior distribution for a scale parameter.

5.5.6 Beta Distribution

We first define two special functions: beta and incomplete beta. The beta function is defined as $B(a,b) = \int_0^1 t^{a-1}(1-t)^{b-1}dt = \Gamma(a)\Gamma(b)/\Gamma(a+b)$. In MATLAB, beta function is coded as beta(a,b). An incomplete beta is $B(x,a,b) = \int_0^x t^{a-1}(1-t)^{b-1}dt$, $0 \leq x \leq 1$. In MATLAB, betainc(x,a,b) represents the normalized incomplete beta, defined as $I_x(a,b) = B(x,a,b)/B(a,b)$. As we will see in a moment, $B(a,b)$ will be a normalizing constant in PDF, while $B(x,a,b)/B(a,b)$ coincides with CDF of beta distribution.

The density function for a beta random variable is

$$f_X(x) = \begin{cases} \frac{1}{B(a,b)}x^{a-1}(1-x)^{b-1}, & 0 \leq x \leq 1, \\ 0, & \text{else} , \end{cases}$$

where B is the beta function and $a,b \geq 0$. Because X is defined only in the interval $[0,1]$, the beta distribution is useful in modeling uncertainty or randomness in proportions or probabilities. A beta-distributed random variable is denoted by $X \sim \mathcal{B}e(a,b)$. The standard uniform distribution $\mathcal{U}(0,1)$ serves as a special case with $(a,b) = (1,1)$. The moments of beta distribution are

$$\mathbb{E}X^k = \frac{\Gamma(a+k)\Gamma(a+b)}{\Gamma(a)\Gamma(a+b+k)} = \frac{a(a+1)\ldots(a+k-1)}{(a+b)(a+b+1)\ldots(a+b+k-1)}$$

so that $\mathbb{E}(X) = a/(a+b)$ and $\mathrm{Var}\,X = ab/[(a+b)^2(a+b+1)]$.

In MATLAB, the CDF for a beta random variable (at $x \in (0,1)$) is computed as betacdf(x,a,b), and the PDF is computed as betapdf(x,a,b). The pth percentile is betainv(p,a,b). In WinBUGS, the beta distribution is coded as dbeta(a,b).

To emphasize the modeling diversity of beta distributions, we depict densities for a selection of (a,b), as in Figure 5.12.

If $U_1, U_2, \ldots, U_n$ is a sample from a uniform $\mathcal{U}(0,1)$ distribution, then the distribution of the kth component in the ordered sample is beta, $U_{(k)} \sim$

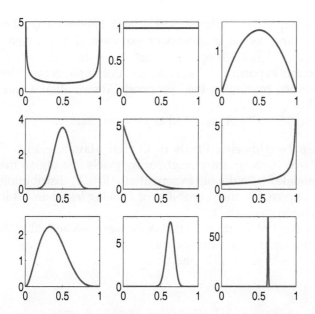

Fig. 5.12 Beta densities for (a,b) as $(1/2, 1,2)$, $(1,1)$, $(2,2)$, $(10,10)$, $(1,5)$, $(1, 0.4)$, $(3,5)$, $(50, 30)$, and $(5000, 3000)$.

$Be(k,n-k+1)$, for $1 \le k \le n$. Also, if $X \sim \mathcal{G}(m,\lambda)$ and $Y \sim \mathcal{G}(n,\lambda)$, then $X/(X+Y) \sim Be(m,n)$.

5.5.7 Double Exponential Distribution

A random variable X has double exponential $\mathcal{DE}(\mu,\lambda)$ distribution if its PDF and CDF are given by

$$f_X(x) = \frac{\lambda}{2} e^{-\lambda|x-\mu|},$$

$$F_X(x) = \begin{cases} \frac{1}{2}e^{\lambda(x-\mu)}, & x < \mu \\ 1 - \frac{1}{2}e^{-\lambda(x-\mu)}, & x \ge \mu \end{cases}, \quad -\infty < x < \infty, \lambda > 0$$

The expectation of X is $\mathbb{E}X = \mu$, and the variance is $\mathbb{V}\text{ar}\,X = 2/\lambda^2$. The moment-generating function for the double exponential distribution is

$$m(t) = \frac{\lambda^2 e^{\mu t}}{\lambda^2 - t^2}, \quad |t| < \lambda.$$

The double exponential distribution is also known as the *Laplace distribution*. If X_1 and X_2 are independent exponential $\mathcal{E}(\lambda)$, then $X_1 - X_2$ is distributed as $\mathcal{DE}(0,\lambda)$. Also, if $X \sim \mathcal{DE}(0,\lambda)$, then $|X| \sim \mathcal{E}(\lambda)$. In MATLAB the double exponential distribution is not implemented since it can be readily obtained by folding the exponential distribution about y-axis, see Figure 5.13a.

In WinBUGS, $\mathcal{DE}(\mu,\lambda)$ is coded as ddexp(mu,lambda).

Example 5.28. **Neighboring Pixels in Digital Mammograms.** The difference D between two arbitrary neighboring pixels in a digital mammogram image is modeled by a double exponential $\mathcal{DE}(0,\lambda)$ distribution.

(a) It is known that the probability of D being less than -4 is 0.3. Using this information calculate λ.

(b) Find the probability of D falling between -5 and 20.

(c) What are the mean and variance of D?

(d) Plot graphs of the PDF and CDF.

```
%mammopixels.m
dexppdf=@(x, mu, lambda) 1/2 * exppdf(abs(x-mu),1./lambda);
dexpcdf=@(x, mu, lambda) 1/2 + sign(x-mu)/2.*expcdf(abs(x-mu),1./lambda);
dexpinv=@(p,  mu, lambda) mu+sign(2*p-1).*expinv(abs(2*p-1),1./lambda);
dexprnd=@(mu,lambda,size) mu+exprnd(1./lambda,size)-exprnd(1./lambda,size);
dexpstat = @(mu, lambda) deal(mu, 2./lambda.^2);

% (a) 0.3=P(D<=-4)=0.5 * exp(- 4*lambda) -> lambda=-1/4*log(2*0.3)=0.1277
% To check:
dexpinv(0.3, 0, 0.1277)   %-4.0002
%(b) P( -5 < D < 20)
dexpcdf(20, 0, 0.1277) - dexpcdf(-5,0,0.1277)   %0.6971
%(c)
[m v]=dexpstat(0, 0.1277)   %m = 0,   v=122.6445
%(d)
mu=0;   lambda=0.1277
x = mu-5/lambda:0.001:mu+5/lambda;
figure;
plot(x, dexppdf(x, mu, lambda));
figure;
plot(x, dexpcdf(x,mu, lambda));
```

5.5.8 Logistic Distribution

The logistic distribution was first defined by Belgian mathematician Pierre Francois Verhulst (1804–1849) who, in 1838, used it in modeling population growth and coined the term *logistic*. Logistic distribution is used for models in pharmacokinetics, regression with binary responses, river discharge and

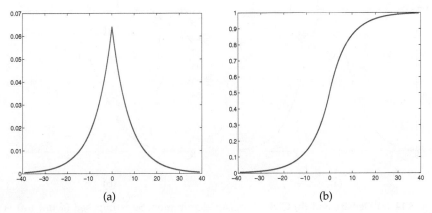

Fig. 5.13 (a) PDF and (b) CDF for $D \sim \mathcal{DE}(0, 0.1277)$.

rainfall in hydeology, neural networks, and machine learning, to list just a few modeling applications.

The logistic random variable can be introduced by a property of its CDF expressed by a differential equation. Let $F(x) = \mathbb{P}(X \leq x)$ be the CDF for which $F'(x) = F(x) \times (1 - F(x))$. One interpretation of this differential equation is as follows: For a Bernoulli random variable $\mathbf{1}(X \leq x) = \begin{cases} 1, X \leq x \\ 0, X > x \end{cases}$, the change in $\mathbb{E}\mathbf{1}(X \leq x)$ as a function of x, is equal to its variance. The solution in the class of CDFs is

$$F_X(x) = \frac{1}{1 + e^{-x}} = \frac{e^x}{1 + e^x},$$

which is called the logistic distribution. Its density is

$$f_X(x) = \frac{e^x}{(1 + e^x)^2} = \frac{e^{-x}}{(1 + e^{-x})^2}.$$

Graphs of $f_X(x)$ and $F_X(x)$ are shown in Figure 5.14. The mean of the distribution is 0 and the variance is $\pi^2/3$. For a more general logistic distribution given by the CDF

$$F_X(x) = \frac{1}{1 + e^{-(x-\mu)/\sigma}},$$

the mean is μ, variance $\pi^2\sigma^2/3$, skewness 0, and kurtosis 21/5. For the higher moments, one can use the moment-generating function

$$m(t) = \exp\{\mu t\} B(1 - \sigma t, 1 + \sigma t),$$

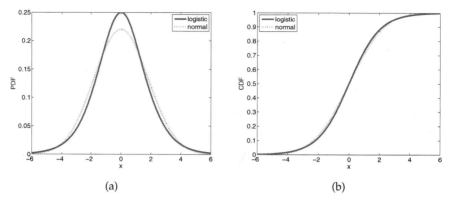

Fig. 5.14 (a) Density and (b) CDF of logistic distribution. Superimposed (*dotted red*) is the normal distribution with matching mean and variance, 0 and $\pi^2/3$, respectively.

where B is the beta function. In WinBUGS the logistic distribution is coded as `dlogis(mu,tau)`, where `tau` is the reciprocal of σ.

If X has a logistic distribution, then $\log(X)$ has a log-logistic distribution (also known as the Fisk distribution). The log-logistic distribution is used in economics (population wealth distribution) and reliability.

The logistic distribution will be revisited in Chapter 15, where we deal with logistic regression.

5.5.9 Weibull Distribution

The Weibull distribution is one of the most important distributions in survival theory and engineering reliability. It is named after Swedish engineer and scientist Waloddi Weibull after his publication in the early 1950s (Weibull, 1951).

The density of the two-parameter Weibull random variable $X \sim \mathcal{W}ei(r,\lambda)$ is given as

$$f_X(x) = \lambda r x^{r-1} e^{-\lambda x^r}, \; x > 0. \tag{5.13}$$

The CDF is given as $F_X(x) = 1 - e^{-\lambda x^r}$. Parameter r is the shape parameter, while λ is the rate parameter. Both parameters are strictly positive. In this form, Weibull $X \sim \mathcal{W}ei(r,\lambda)$ is a distribution of $X = Y^{1/r}$ for Y exponential $\mathcal{E}(\lambda)$.

In MATLAB, the Weibull distribution is parameterized by a and r, as in

$$f(x) = a^{-r} r x^{r-1} e^{-(x/a)^r}, \; x > 0. \tag{5.14}$$

Note that in this parametrization, a is the scale parameter and relates to λ as $\lambda = a^{-r}$. So when $a = \lambda^{-1/r}$, the CDF in MATLAB is `wblcdf(x,a,r)`, and the PDF is `wblpdf(x,a,r)`. The function `wblinv(p,a,r)` computes the pth quantile of the $\mathcal{Wei}(r,\lambda)$ random variable.

The (r,λ) parametrization of Weibull distribution is not as prevalent as the shape-scale parametrization from (5.14), but the likelihood in (5.13) is more convenient for Bayesian inference. In WinBUGS, $\mathcal{Wei}(r,\lambda)$ is coded as `dweib(r,lambda)`.

The Weibull distribution generalizes the exponential distribution ($r=1$) and Rayleigh distribution ($r=2$). Figure 5.15 shows the densities of the Weibull distribution for $r=2$ (blue), $r=1$ (red), and $r=1/2$ (black). In all three cases, $\lambda = 1/2$. The values for the scale parameter $a = \lambda^{-1/r}$ are $\sqrt{2}, 2$, and 4, respectively.

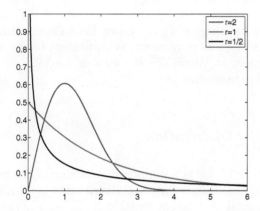

Fig. 5.15 Densities of Weibull distribution with $r=2$ (*blue*), $r=1$ (*red*), and $r=2$ (*black*). In all three cases, $\lambda = 1/2$.

The mean of a Weibull random variable X is $\mathbb{E}X = \frac{\Gamma(1+1/r)}{\lambda^{1/r}} = a\Gamma\left(1+\frac{1}{r}\right)$, and the variance is $\mathbb{Var}\,X = \frac{\Gamma(1+2/r)-\Gamma^2(1+1/r)}{\lambda^{2/r}} = a^2(\Gamma\left(1+\frac{2}{r}\right) - \Gamma^2\left(1+\frac{1}{r}\right))$. The kth moment is $\mathbb{E}X^k = \frac{\Gamma(1+k/r)}{\lambda^{k/r}} = a^k\Gamma\left(1+\frac{k}{r}\right)$.

5.5.10 Pareto Distribution

The Pareto distribution is named after the Italian economist Vilfredo Pareto (1848-1923). Some examples in which the Pareto distribution provides an exemplary model include wealth distribution in individuals, sizes of human settlements, visits to encyclopedia pages, and file size distribution of

Internet traffic that uses the TCP protocol. A random variable X has a Pareto $\mathcal{P}a(c,\alpha)$ distribution with parameters $0 < c < \infty$ and $\alpha > 0$ if its density is given by

$$f_X(x) = \begin{cases} \frac{\alpha}{c}\left(\frac{c}{x}\right)^{\alpha+1}, & x \geq c, \\ 0, & \text{else}. \end{cases}$$

The CDF is

$$F_X(x) = \begin{cases} 0, & x < c, \\ 1 - \left(\frac{c}{x}\right)^{\alpha}, & x \geq c. \end{cases}$$

The mean and variance of X are $\mathbb{E}X = \alpha c/(\alpha - 1)$, $\alpha > 1$, and $\operatorname{Var}X = \alpha c^2/[(\alpha - 1)^2(\alpha - 2)]$, $\alpha > 2$. The median is $m = c \cdot 2^{1/\alpha}$. If $X_1, \ldots, X_n$ are independent $\mathcal{P}a(c,\alpha)$, then $Y = 2c\sum_{i=1}^n \ln(X_i) \sim \chi^2$ with $2n$ degrees of freedom.

In MATLAB one can specify the generalized Pareto distribution, which for some selection of its parameters is equivalent to the aforementioned Pareto distribution. In WinBUGS, the code is `dpar(alpha,c)` (note the permuted order of parameters).

5.5.11 Dirichlet Distribution

The Dirichlet distribution is a multivariate version of the beta distribution in the same way that the multinomial distribution is a multivariate extension of the binomial. A random variable $X = (X_1, \ldots, X_k)$ with a Dirichlet distribution of $(X \sim \mathcal{D}ir(a_1, \ldots, a_k))$ has a PDF of

$$f(x_1, \ldots, x_k) = \frac{\Gamma(A)}{\prod_{i=1}^k \Gamma(a_i)} \prod_{i=1}^k x_i^{a_i-1},$$

where $A = \sum a_i$, and $x = (x_1, \ldots, x_k) \geq 0$ is defined on the simplex $x_1 + \cdots + x_k = 1$. Then

$$\mathbb{E}(X_i) = \frac{a_i}{A}, \quad \operatorname{Var}(X_i) = \frac{a_i(A - a_i)}{A^2(A+1)}, \quad \text{and} \quad \operatorname{Cov}(X_i, X_j) = -\frac{a_i a_j}{A^2(A+1)}.$$

The Dirichlet random variable can be generated from gamma random variables $Y_1, \ldots, Y_k \sim \mathcal{G}a(a,b)$ as $X_i = Y_i/S_Y$, $i = 1, \ldots, k$, where $S_Y = \sum_i Y_i$. The marginal distribution of a component X_i is $\mathcal{B}e(a_i, A - a_i)$. This is illustrated in the following MATLAB m-file that generates random Dirichlet vectors:

```
function drand = dirichletrnd(a,n)
```

```
% function drand = dirichletrnd(a,n)
% a - vector of parameters 1 x m
% n - number of random realizations
% drand - matrix m x n, each column one realization.
%---------------------------------------------------
a=a(:);
m=size(a,1);
a1=zeros(m,n);
for i = 1:m
    a1(i,:) = gamrnd(a(i,1),1,1,n);
end
for i=1:m
drand(i, 1:n )= a1(i, 1:n ) ./ sum(a1);
end
```

5.6 Random Numbers and Probability Tables

In older introductory statistics texts, many back-end pages have been devoted to various statistical tables. Several decades ago, many books of statistical tables were published. Also, the most respected of statistical journals occasionally published articles providing statistical tables.

In 1947 the RAND Corporation published the monograph *A Million Random Digits with 100,000 Normal Deviates*, which at the time was a state-of-the-art resource for simulation and Monte Carlo methods. The book can be found at http://www.rand.org/pubs/monograph_reports/MR1418.html.

These days, much larger tables of random numbers can be produced by a single line of code, resulting in a set of random numbers that can pass a battery of stringent randomness tests. With MATLAB and many other widely available software packages, statistical tables and tables of random numbers are now obsolete. For example, tables of binomial CDF and PDF for a specific n and p can be reproduced by

```
%n=12, p=0.7
disp('binocdf(0:12, 12, 0.7)');
binocdf(0:12, 12, 0.7)
disp('binopdf(0:12, 12, 0.7)');
binopdf(0:12, 12, 0.7)
```

We will show how to sample and simulate from a few distributions in MATLAB and compare empirical means and variances with their theoretical counterparts. The following annotated MATLAB code simulates from binomial, Poisson, and geometric distributions and compares theoretical and empirical means and variances:

```
%various_simulations.m
simu = binornd(12, 0.7, [1,100000]);
```

```
% simu is 10000 observations from Bin(12,0.7)
disp('simu = binornd(12, 0.7, [1,100000]); 12*0.7 - mean(simu)');

12*0.7 - mean(simu)  %0.001069
  %should be small since the theoretical mean is n*p
disp('simu = binornd(12, 0.7, [1,100000]); ...
                                  12 * 0.7 * 0.3 - var(simu)');

12 * 0.7 * 0.3 - var(simu)  %-0.008350
  %should be small since the theoretical variance is n*p*(1-p)

%% Simulations from Poisson(2)
poi = poissrnd(2, [1, 100000]);
disp('poi = poissrnd(2, [1, 100000]); mean(poi)');
mean(poi)   %1.9976
disp('poi = poissrnd(2, [1, 100000]); var(poi)');
var(poi)    %2.01501

%%% Simulations from Geometric(0.2)
geo = geornd(0.2, [1, 100000]);
disp('geo = geornd(0.2, [1, 100000]); mean(geo)');
mean(geo)  %4.00281
disp('geo = geornd(0.2, [1, 100000]); var(geo)');
var(geo)   %20.11996
```

5.7 Transformations of Random Variables*

When a random variable with known density is transformed, the result is a random variable as well. The question is how to find its distribution. The general theory for distributions of functions of random variables is beyond the scope of this text, and the reader can find comprehensive coverage in Ross (2010a, b).

We have already seen that, for a discrete random variable X, the PMF of a function $Y = g(X)$ is simply the table

$$\begin{array}{c|ccccc} g(X) & g(x_1) & g(x_2) & \cdots & g(x_n) & \cdots \\ \hline \text{Prob} & p_1 & p_2 & \cdots & p_n & \cdots \end{array}$$

in which only realizations of X are transformed while the probabilities are kept unchanged.

For continuous random variables the distribution of a function is more complex. In some cases, however, looking at the CDF is sufficient.

In this section we will discuss two topics: (i) how to find the distribution for a transformation of a single continuous random variable and (ii) how to approximate moments, in particular means and variances, of complex functions of many random variables.

Suppose that a continuous random variable has a density $f_X(x)$ and that a function g is monotone on the domain of f, with the inverse function h, $h = g^{-1}$. Then the random variable $Y = g(X)$ has a density

$$f_Y(y) = f(h(y))|h'(y)|. \tag{5.15}$$

If g is not one-to-one, but has k one-to-one inverse branches, $h_1, h_2, \ldots, h_k$, then

$$f_Y(y) = \sum_{i=1}^{k} f(h_i(y))|h_i'(y)|. \tag{5.16}$$

An example of a function which is not one-to-one is $g(x) = x^2$, for which inverse branches $h_1(y) = \sqrt{y}$ and $h_2(y) = -\sqrt{y}$ are one-to-one.

Example 5.29. Square Root of Exponential. Let X be a random variable with an exponential $\mathcal{E}(\lambda)$ distribution, where $\lambda > 0$ is the rate parameter. Find the distribution of the random variable $Y = \sqrt{X}$.

Here $g(x) = \sqrt{x}$ and $g^{-1}(y) = y^2$. The Jacobian is $|g^{-1}(y)'| = 2y$, $y \geq 0$. Thus,

$$f_Y(y) = \lambda e^{-\lambda y^2} \cdot 2y, \ y \geq 0, \lambda > 0,$$

which is known as the Rayleigh distribution.

An alternative approach to finding the distribution of Y is to consider the CDF:

$$F_Y(y) = \mathbb{P}(Y \leq y) = \mathbb{P}(\sqrt{X} \leq y) = \mathbb{P}(X \leq y^2) = 1 - e^{-\lambda y^2}$$

since X has the exponential distribution. The density is now obtained by taking the derivative of $F_Y(y)$,

$$f_Y(y) = (F_Y(y))' = 2\lambda y e^{-\lambda y^2}, \ y \geq 0, \lambda > 0.$$

The distribution of a function of one or many random variables is an ultimate summary. However, the result could be quite messy and sometimes the distribution lacks a closed form. Moreover, not all facets of the resulting distribution may be of interest to researchers; sometimes only the mean and variance are needed.

If X is a random variable with $\mathbb{E}X = \mu$ and $\mathbb{V}\mathrm{ar}\,X = \sigma^2$, then for a function $Y = g(X)$ the following approximation holds:

$$\mathbb{E}Y \approx g(\mu) + \frac{1}{2}g''(\mu)\sigma^2,$$

$$\mathbb{V}\mathrm{ar}\,Y \approx (g'(\mu))^2 \sigma^2. \tag{5.17}$$

If n independent random variables are transformed as $Y = g(X_1, X_2, \ldots, X_n)$, then

$$\mathbb{E}Y \approx g(\mu_1, \mu_2, \ldots, \mu_n) + \frac{1}{2}\sum_{i=1}^{n}\frac{\partial^2 g}{\partial x^2}(\mu_1, \mu_2, \ldots, \mu_n)\sigma_i^2,$$

$$\mathbb{V}\mathrm{ar}\,Y \approx \sum_{i=1}^{n}\left(\frac{\partial g}{\partial x_i}(\mu_1, \mu_2, \ldots, \mu_n)\right)^2 \sigma_i^2, \tag{5.18}$$

where $\mathbb{E}X_i = \mu_i$ and $\mathbb{V}\mathrm{ar}\,X_i = \sigma_i^2$.

The approximation for the mean $\mathbb{E}Y$ is obtained by the second-order Taylor expansion and is more precise than the approximation for the variance $\mathbb{V}\mathrm{ar}\,Y$, which is of the first order ("linearization"). The second-order approximation for $\mathbb{V}\mathrm{ar}\,Y$ is straightforward but involves third and fourth moments of Xs. Also, when the variables $X_1, \ldots, X_n$ are correlated, the factor $2\sum_{1 \le i < j \le n}\frac{\partial^2 g}{\partial x_i \partial x_j}(\mu_1, \ldots, \mu_n)\mathrm{Cov}(X_i, X_j)$ should be added to the expression for $\mathbb{V}\mathrm{ar}\,Y$ in (5.18).

If g is a complicated function, the mean $\mathbb{E}Y$ is often approximated by a first-order approximation, $\mathbb{E}Y \approx g(\mu_1, \mu_2, \ldots, \mu_n)$, that involves no derivatives.

Example 5.30. **String Vibrations.** In string vibration, the frequency of the fundamental harmonic is often of interest. The fundamental harmonic is produced by the vibration with nodes at the two ends of the string. In this case, the length of the string L is half of the wavelength of the fundamental harmonic. The frequency ω (in Hz) depends also on the tension of the string T, and the string mass M,

$$\omega = \frac{1}{2}\sqrt{\frac{T}{ML}}.$$

Quantities L, T, and M are measured imperfectly and are considered independent random variables. The means and variances are estimated as follows:

Variable (unit)	Mean	Variance
L (m)	0.5	0.0001
T (N)	70	0.16
M (kg/m)	0.001	10^{-8}

Approximate the mean μ_ω and variance σ_ω^2 of the resulting frequency ω.
The partial derivatives

$$\frac{\partial \omega}{\partial T} = \frac{1}{4}\sqrt{\frac{1}{TML}}, \quad \frac{\partial^2 \omega}{\partial T^2} = -\frac{1}{8}\sqrt{\frac{1}{T^3 ML}},$$

$$\frac{\partial \omega}{\partial M} = -\frac{1}{4}\sqrt{\frac{T}{M^3 L}}, \quad \frac{\partial^2 \omega}{\partial M^2} = \frac{3}{8}\sqrt{\frac{T}{M^5 L}},$$

$$\frac{\partial \omega}{\partial L} = -\frac{1}{4}\sqrt{\frac{T}{ML^3}}, \quad \frac{\partial^2 \omega}{\partial L^2} = \frac{3}{8}\sqrt{\frac{T}{ML^5}},$$

evaluated at the means $\mu_L = 0.5$, $\mu_T = 70$, and $\mu_M = 0.001$, are

$$\frac{\partial \omega}{\partial T}(\mu_L, \mu_T, \mu_M) = 1.3363, \quad \frac{\partial^2 \omega}{\partial T^2}(\mu_L, \mu_T, \mu_M) = -0.0095,$$

$$\frac{\partial \omega}{\partial M}(\mu_L, \mu_T, \mu_M) = -9.3541 \cdot 10^4, \quad \frac{\partial^2 \omega}{\partial M^2}(\mu_L, \mu_T, \mu_M) = 1.4031 \cdot 10^8,$$

$$\frac{\partial \omega}{\partial L}(\mu_L, \mu_T, \mu_M) = -187.0829, \quad \frac{\partial^2 \omega}{\partial L^2}(\mu_L, \mu_T, \mu_M) = 561.2486,$$

and the mean and variance of ω are

$$\boxed{\mu_\omega \approx 187.8117} \quad \text{and} \quad \boxed{\sigma_\omega^2 \approx 91.2857}.$$

The first-order approximation for μ_ω is $\frac{1}{2}\sqrt{\frac{\mu_T}{\mu_M \mu_L}} = 187.0829$.

5.8 Mixtures*

In modeling tasks it is sometimes necessary to combine two or more random variables in order to get a satisfactory model. There are two ways of combining random variables: by taking the linear combination $a_1 X_1 + a_2 X_2 + \ldots$ for which a density in the general case is often convoluted and difficult to express in a finite form, or by combining densities and PMFs directly.

For example, for two densities f_1 and f_2, the density $g(x) = \varepsilon f_1(x) + (1 - \varepsilon) f_2(x)$ is a mixture of f_1 and f_2 with weights ε and $1 - \varepsilon$. It is important for the weights to be nonnegative and add up to 1 so that $g(x)$ remains a density.

Very popular mixtures are point mass mixture distributions that combine a density function $f(x)$ with a point mass (Dirac) function δ_{x_0} at a value x_0. The Dirac functions belong to a class of special functions. Informally, one may think of δ_{x_0} as a limiting function for a sequence of functions

$$f_{n,x_0} = \begin{cases} n, & x_0 - \frac{1}{2n} < x < x_0 + \frac{1}{2n}, \\ 0, & \text{else}, \end{cases}$$

when $n \to \infty$. It is easy to see that for any finite n, f_{n,x_0} is a density since it integrates to 1; however, the function domain shrinks to a singleton x_0, while its value at x_0 goes to infinity.

For example, $f(x) = 0.3\delta_0 + 0.7 \times \frac{1}{\sqrt{2\pi}} \exp\{-\frac{x^2}{2}\}$ is a normal distribution contaminated by a point mass at zero with a weight 0.3.

5.9 Markov Chains*

You may have encountered statistical jargon containing the term "Markov chain." In Bayesian calculations the acronym MCMC stands for Markov chain Monte Carlo simulations, while in statistical models of genomes, hidden Markov chain models are popular. Here we give a basic definition and a few examples of Markov chains.

A sequence of random variables $X_0, X_1, \ldots, X_n, \ldots$, with values in the set of "states" $S = \{1, 2, \ldots\}$, constitutes a Markov chain if the probability of transition to a future state, $X_{n+1} = j$, depends only on the value at the current state, $X_n = i$, and not on any previous values $X_{n-1}, X_{n-2}, \ldots, X_0$. A popular way of putting this is to say that in Markov chains the future depends on the present and not on the past. Formally,

$$\mathbb{P}(X_{n+1} = j | X_0 = i_0, X_1 = i_1, \ldots, X_{n-1} = i_{n-1}, X_n = i) = \mathbb{P}(X_{n+1} = j | X_n = i) = p_{ij},$$

where $i_0, i_1, \ldots, i_{n-1}, i, j$ are the states from S. The probability p_{ij} is independent of n and represents the transition probability from state i to state j. In our brief coverage of Markov chains, we will consider chains with a finite number of states, N.

For states $S = \{1, 2, \ldots, N\}$, the transition probabilities form an $N \times N$ matrix $\mathbf{P} = (p_{ij})$. Each row of this matrix sums up to 1 since the probabilities of all possible moves from a particular state, including the probability of remaining in the same state, sum up to 1:

$$p_{i1} + p_{i2} + \cdots + p_{ii} + \cdots + p_{iN} = 1.$$

The matrix $\mathbf{P}$ describes the evolution and long-time behavior of the Markov chain it represents. In fact, if the distribution $\pi^{(0)}$ for the initial variable X_0 is specified, the pair $\pi^{(0)}, \mathbf{P}$ fully describes the Markov chain.

Matrix $\mathbf{P}^2$ gives the probabilities of transition in two steps. Its element $p_{ij}^{(2)}$ is $\mathbb{P}(X_{n+2} = j | X_n = i)$.

Likewise, the elements of matrix $\mathbf{P}^m$ are the probabilities of transition in m steps,

$$p_{ij}^{(m)} = \mathbb{P}(X_{n+m} = j | X_n = i),$$

for any $n \geq 0$ and any $i, j \in \mathcal{S}$.

If the distribution for X_0 is $\pi^{(0)} = \left(\pi_1^{(0)}, \pi_2^{(0)}, \ldots, \pi_N^{(0)} \right)$, then the distribution for X_n is

$$\pi^{(n)} = \pi^{(0)} \mathbf{P}^n. \tag{5.19}$$

Of course, if the state X_0 is known, $X_0 = i_0$, then $\pi^{(0)}$ is a vector of 0s except at position i_0, where the value is 1.

For n large, the probability $\pi^{(n)}$ "forgets" the initial distribution at state X_0 and converges to $\pi = \lim_{n \to \infty} \pi^{(n)}$. This distribution is called the stationary distribution of a chain and satisfies

$$\pi = \pi \mathbf{P}.$$

Operationally, to find stationary distribution, one solves the system

$$\begin{cases} (\mathbf{I} - \mathbf{P}) \pi' = 0 \\ \mathbf{1}' \pi = 1. \end{cases}$$

Result. If for a finite state Markov chain one can find an integer k so that all entries in $\mathbf{P}^k$ are strictly positive, then stationary distribution π exists.

Example 5.31. **Ehrenfest Model.** Ehrenfest model (Ehrenfest, 1907) illustrates the diffusion in gasses by considering random transition of molecules between two compartments.

Consider N balls numbered from 1 to N, distributed in two boxes, A and B. The system is in state i if i balls are in the box A (and $N - i$ balls in the box B). A number between 1 and N is randomly selected, and the ball with the selected number switches the boxes. The system constitutes a Markov chain, since the future state of the system depends on the present and not on the past states. We will analyze the case of $N = 4$.

Possible states of the system are $\{0, 1, 2, 3, 4\}$, so the MC has 5 states. The transition probabilities among the states are given as follows:

```
N=4;    %total number of particles
```

```
ns=N+1;   %number of MC states
%forming transition matrix
P=zeros(ns);
P(1,2)=1;      P(ns,ns-1)=1; %states 0 and N are ''reflective''
for j=2:ns-1
    i=j-1;                   %number of particles in box A
P(j, j-1)=i/N               %A -> B
P(j, j+1)=(N-i)/N           %B -> A
end
```

Therefore, the transition matrix is

$$\mathbf{P} = \begin{bmatrix} 0 & 1 & 0 & 0 & 0 \\ 1/4 & 0 & 3/4 & 0 & 0 \\ 0 & 1/2 & 0 & 1/2 & 0 \\ 0 & 0 & 3/4 & 0 & 1/4 \\ 0 & 0 & 0 & 1 & 0 \end{bmatrix}.$$

What is the most likely state of the system after $M = 11$ steps if all balls originally were in A?

```
pi0 =[0 0 0 0 1]  %probability of initial state i=4 is 1.
pi0 * P^11
%0     0.4995        0     0.5005         0
```

The most likely state is $i = 3$. For any even number of transitions, the most likely state is $i = 2$ with constant probability of $3/4$.

The stationary probabilities are found by solving the following system:

```
linsolve([[(eye(ns)-P)'; ones(1,ns)],[ zeros(ns,1);1]])
```

The stationary probabilities coincide with binomial $\mathcal{B}in(4+1,1/2)$ PDF.

```
st=[0.0625 0.25 0.375 0.25 0.0625]
st * P
%0.0625    0.2500    0.3750    0.2500    0.0625
```

MATLAB script ◀ ehrenfestsim.m simulates dynamic change of states in Ehrenfest model with $N = 20 \times 20 = 400$ particles, that are initially all in box A. Figure 5.16 summarizes the calculations. It shows the content of boxes A and B after 10,000 transitions, as well as the proportion of balls in each of the boxes.
✑

Example 5.32. **Point-Accepted Mutation.** Point-accepted mutation (PAM) implements a simple theoretical model for scoring the alignment of protein sequences. Specifically, at a fixed position, the rate of mutation at each moment is assumed to be independent of previous events. Then the evolution of this fixed position in time can be treated as a Markov chain, where the

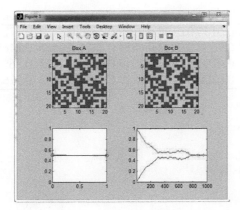

Fig. 5.16 Ehrenfest model simulation by `ehrenfestsim.m`. The top two panels show the contents of two boxes A and B after 10,000 transitions. The lower left panel shows the proportion of balls in the boxes (0 for A and 1 for B), and the lower right panel shows how the proportions changed over 10,000 transitions. The red curve is the proportion for box A.

PAM matrix represents its transition matrix. The original PAMs are 20×20 matrices describing the evolution of 20 standard amino acids (Dayhoff et al. 1978). As a simplified illustration, consider the case of a nucleotide sequence with only four states (A, T, G, and C). Assume that in a given time interval ΔT the probabilities that a given nucleotide mutates to each of the other three bases or remains unchanged can be represented by a 4×4 mutation matrix M:

$$M = \begin{pmatrix} & A & T & G & C \\ \hline A & 0.98 & 0.01 & 0.005 & 0.005 \\ T & 0.01 & 0.96 & 0.02 & 0.01 \\ G & 0.01 & 0.01 & 0.97 & 0.01 \\ C & 0.02 & 0.03 & 0.01 & 0.94 \end{pmatrix}$$

Consider the fixed position with the letter T at $t = 0$:

$$s_0 = (0 \ \ 1 \ \ 0 \ \ 0).$$

Then, at times Δ, 2Δ, 10Δ, 100Δ, 1000Δ, and 10000Δ, by (5.19), the probabilities of the nucleotides (A, T, G, C) are $s_1 = s_0 M$, $s_2 = s_0 M^2$, $s_{10} = s_0 M^{10}$, $s_{100} = s_0 M^{100}$, $s_{1000} = s_0 M^{1000}$, and $s_{10000} = s_0 M^{10000}$, as given in the following table:

Δ	2Δ	10Δ	100Δ	1000Δ	10000Δ
A 0.0100	0.0198	0.0909	0.3548	0.3721	0.3721
T 0.9600	0.9222	0.6854	0.2521	0.2465	0.2465
G 0.0200	0.0388	0.1517	0.2747	0.2651	0.2651
C 0.0100	0.0193	0.0719	0.1184	0.1163	0.1163

5.10 Exercises

5.1. **Phase I Clinical Trials and CTCAE Terminology.** In Phase I clinical trials, a safe dosage of a drug is assessed. In administering the drug, doctors are grading subjects' toxicity responses on a scale from 0 to 5. In CTCAE (Common Terminology Criteria for Adverse Events, National Institute of Health), Grade refers to the severity of adverse events. Generally, Grade 0 represents no measurable adverse events (sometimes omitted as a grade); Grade 1 events are mild; Grade 2 are moderate; Grade 3 are severe; Grade 4 are life-threatening or disabling; Grade 5 are fatal. This grading system inherently places a value on the importance of an event, although there is not necessarily "proportionality" among grades (a "2" is not necessarily twice as bad as a "1"). Some adverse events are difficult to "fit" into this point schema, but altering the general guidelines of severity scaling would render the system useless for comparing results between trials, which is an important purpose of the system.

Assume that based on a large number of trials (administrations to patients with renal cell carcinoma), the toxicity of the drug PNU (a murine Fab fragment of the monoclonal antibody 5T4 fused to a mutated superantigen staphylococcal enterotoxin A) at a particular fixed dosage is modeled by discrete random variable X,

X	0	1	2	3	4	5
Prob	0.620	0.190	0.098	0.067	0.024	0.001

Plot the PMF and CDF and find $\mathbb{E}X$ and $\mathbb{V}ar\,(X)$.

5.2. **Mendel and Dominance.** Suppose that a specific trait, such as eye color or left-handedness, in a person is dependent on a pair of genes, and suppose that D represents a dominant and d a recessive gene. Thus, a person having DD is pure dominant and dd is pure recessive while Dd is a hybrid. The pure dominants and hybrids are alike in outward appearance. A child receives one gene from each parent.
Suppose two hybrid parents have 4 children. What is the probability that 3 out of 4 children have outward appearance of the dominant gene.

5.3. **Chronic Kidney Disease.** Chronic kidney disease (CKD) is a serious condition associated with premature mortality, decreased quality of life, and increased healthcare expenditures. Untreated CKD can result in end-stage renal disease and necessitate dialysis or kidney transplantation. Risk factors for CKD include cardiovascular disease, diabetes, hypertension, and obesity. To estimate the prevalence of CKD in the United States (overall and by health risk factors and other characteristics), the CDC (CDC's MMWR Weekly, 2007; Coresh et al., 2003) analyzed the most recent data from the National Health and Nutrition Examination Survey (NHANES). The total crude (i.e., not age-standardized) CKD prevalence estimate for adults aged > 20 years in the United States was 17%. By age group, CKD was more prevalent among persons aged > 60 years (40%) than among persons aged 40–59 years (13%) or 20–39 years (8%).

(a) From the population of adults aged > 20 years, 10 subjects are selected at random. Find the probability that 3 of the selected subjects have CKD.

(b) From the population of adults aged > 60, 5 subjects are selected at random. Find the probability that at least one of the selected have CKD.

(c) From the population of adults aged > 60, 16 subjects are selected at random and it was found that 6 of them had CKD. From this sample of 16, subjects are selected at random, *one-by-one with replacement,* and inspected. Find the probability that among 5 inspected (i) exactly 3 had CKD; (ii) at least one of the selected have CKD.

(d) From the population of adults aged > 60 subjects are selected at random until a subject is found to have CKD. What is the probability that exactly 3 subjects are sampled.

(e) Suppose that persons aged > 60 constitute 23% of the population of adults older than 20. For the other two age groups, 20–39, and 40–59, the percentages are 42% and 35%. Ten people are selected at random. What is the probability that 5 are from the > 60 group, 3 from the 20–39 group, and 2 from the 40-59 group.

5.4. **Experimenting to See All Possible Outcomes.** In a chemical experiment two outcomes are possible, A and A^c, with probabilities p and $q = 1 - p$. A student is repeating the experiment until both A and A^c are observed.

(a) Find the distribution of random variable X, the number of experiments necessary to observe A and A^c.

(b) What is the expected number of experiments?

(c) If the expected number of experiments is 3, what can you say about p?

Hint: Use $P(X = k) = P(X > k - 1) - P(X > k)$, $k = 2,3,\ldots$. Argue that $P(X > k) = p^k + q^k$.

5.5. **Ternary Channel.** Refer to Exercise 3.40 in which a communication system was transmitting three signals, s_1, s_2, and s_3.
(a) If s_1 is sent $n = 1000$ times, find an approximation to the probability of the event that it was correctly received between 730 and 770 times, inclusive.
(b) If s_2 is sent $n = 1000$ times, find an approximation to the probability of the event that the channel did not switch to s_3 at all, that is, if 1,000 s_2 signals are sent and not a single s_3 was received. Can you use the same approximation as in (a)?

5.6. **Random Circular Sector with Cells.** On a circular plate, there are 400 randomly located cells. A part of the plate in the shape of a circular sector with central angle $\varphi = \frac{\pi}{100}$ (in radians) is selected at random.
Find an approximation to the probability that the number of cells in the selected sector is
(a) zero;
(b) 4 or more.
Hint: Argue that the number of cells in the selected area is Poisson with $\lambda = 2$.

5.7. **Conditioning a Poisson.** If $X_1 \sim Poi(\lambda_1)$ and $X_2 \sim Poi(\lambda_2)$ are independent, show that the distribution of X_1, given $X_1 + X_2 = n$, is binomial $Bin(n, \lambda_1/(\lambda_1 + \lambda_2))$.

5.8. **Rh+ Plates.** Assume that there are 6 plates with red blood cells, three are Rh+ and three are Rh−.
Two plates are selected (a) with, (b) without replacement. Find the probability that one plate out of the 2 selected/inspected is of Rh+ type.
Now, increase the number of plates keeping the proportion of Rh+ fixed to 1/2. For example, if the total number of plates is 10000, 5000 of each type, what are the probabilities from (a) and (b)?

5.9. **Your Teammate's Misconceptions about Density and CDF.** Your teammate thinks that if f is a probability density function for the continuous random variable X, then $f(10)$ is the probability that $X = 10$. (a) Explain to your teammate why his/her reasoning is false.
Your teammate is not satisfied with your explanation and challenges you by asking, "If $f(10)$ is not the probability that $X = 10$, then just what does $f(10)$ signify?" (b) How would you respond?
Your teammate now thinks that if F is a cumulative probability density function for the continuous random variable X, then $F(5)$ is the probability that $X = 5$. (c) Explain why your teammate is wrong.
Your teammate then asks you, "If $F(5)$ is not the probability of $X = 5$, then just what does $F(5)$ represent?" (d) How would you respond?

5.10. **Falls among Elderly.** Falls are the second leading cause of unintentional injury-related death for people of all ages and the leading cause for people 60 years and older in the United States. Falls are also the most costly injury among older persons in the United States.
One in three adults aged 65 years and older falls annually.
(a) Find the probability that 3 among 11 adults aged 65 years and older will fall in the following year.
(b) Find the probability that among 110,000 adults aged 65 years and older the number of falls will be between 36,100 and 36,700, inclusive. Find the exact probability by assuming a binomial distribution for the number of falls, and an approximation to this probability via de Moivre's theorem; see page 252.

5.11. **Cell Clusters in 3D Petri Dishes.** The number of cell clusters in a 3D Petri dish has a Poisson distribution with mean $\lambda = 5$. Find the percentage of Petri dishes that have (a) 0 clusters, (b) at least one cluster, (c) more than 8 clusters, and (d) between 4 and 6 clusters. Use MATLAB and poisspdf, poisscdf functions.

5.12. **Left-Handed Twins.** The identical twin of a left-handed person has a 76 % chance of being left-handed, implying that left-handedness has partly genetic and partly environmental causes. Ten identical twins of ten left-handed persons are inspected for left-handedness. Let X be the number of left-handed among the inspected. What is the probability that X
(a) falls anywhere between 5 and 8, inclusive;
(b) is at most 6;
(c) is not less than 6.
(d) Would you be surprised if the number of left-handed among the 10 inspected was 3? Why or why not?

5.13. **Pot Smoking Is Not Cool!** A nationwide survey of seniors by the University of Michigan reveals that almost 70% disapprove of daily pot smoking, according to a report in Parade, September 14, 1980. If 12 seniors are selected at random and asked their opinion, find the probability that the number who disapprove of smoking pot daily is
(a) anywhere from 7 to 9;
(b) at most 5;
(c) not less than 8.

5.14. **Power Supply.** A power supply is connected to 20 independent loads. Each load is ON 30% of the time and draws a current of 0.75 amps. Let X be a current in the power supply at a particular moment.
(a) If X exceeds 13 amps, the power supply is declared to be in a critical regime. What is the probability of this happening?
(b) Find the probability that X is below 5 amps.

(c) Find the expectation and variance of X.

5.15. **Emergency Help by Phone.** The emergency hotline in a hospital tries to answer questions to its patient support within 3 minutes. The probability is 0.9 that a given call is answered within 3 minutes and the calls are independent.

(a) What is the expected total number of calls that occur until the first call is answered late?

(b) What is the probability that exactly one of the next 10 calls is answered late?

5.16. **Min of Three.** Let X_1, X_2, and X_3 be three mutually independent random variables, with a discrete uniform distribution on $\{1,2,3\}$, given as $P(X_i = k) = 1/3$ for $k = 1,2$ and 3.

(a) Let $M = \min\{X_1, X_2, X_3\}$. What is the distribution (probability mass function) and cumulative distribution function of M?

(b) What is the distribution (probability mass function) and cumulative distribution function of random variable $R = \max\{X_1, X_2, X_3\} - \min\{X_1, X_2, X_3\}$.

5.17. **Cystic Fibrosis in Japan.** Some rare diseases, including those of genetic origin, are life-threatening or chronically debilitating diseases that are of such low prevalence that special combined efforts are needed to address them. An accepted definition of low prevalence is a prevalence of less than 5 in a population of 10,000. A rare disease has such a low prevalence in a population that a doctor in a busy general practice would not expect to see more than one case in a given year.

Assume that cystic fibrosis, which is a rare genetic disease in most parts of Asia, has a prevalence of 2 per 10,000 in Japan. What is the probability that in a Japanese city of 15,000 there are

(a) exactly 3 incidences,

(b) at least one incidence,

of cystic fibrosis.

5.18. **Random Variables as Models.** Tubert-Bitter et al. (1996) found that the number of serious gastrointestinal reactions reported to the British Committee on Safety of Medicines was 538 out of 9,160,000 prescriptions of the anti-inflammatory drug *Piroxicam*.

(a) What is the rate of gastrointestinal reactions per 10,000 prescriptions?

(b) Using the Poisson model with the rate λ as in (a), find the probability of exactly two gastrointestinal reactions per 10,000 prescriptions.

(c) Find the probability of finding at least two gastrointestinal reactions per 10,000 prescriptions.

5.19. **Jack and Jill, Poisson, and Bayes' Rule.** Jack and Jill are partners in a typing service. Jill handles 60% of the typing work in their partnership. She makes errors (uncorrected errors) at an average rate of one per 4 pages while Jack makes errors at a rate of one per page. Assume that for each typist these errors occur independently and at a constant rate throughout the paper. Assume, in addition, that for both typists the number of errors per page is well approximated by a Poisson distribution.

You submit a 5-page paper to the partnership for typing without knowing whether Jack or Jill will type it.
(a) It comes back error-free. What is the probability that Jack typed it?
(b) What is the probability that Jack typed the paper if 3 errors are found.

5.20. **Variance of Difference of Two Multinomial Components.** Let $(X_1, X_2, \ldots, X_k)$ be a discrete random vector with multinomial $\mathcal{M}n(n, p_1, \ldots, p_k)$ distribution. Show that the variance of $X_i - X_j$ is $n(p_i + p_j - (p_i - p_j)^2)$.

5.21. **A 2D PDF.** Let

$$f(x,y) = \begin{cases} \frac{3}{8}(x^2 + 2xy), & 0 \le x \le 1, \, 0 \le y \le 2 \\ 0, & \text{else} \end{cases}$$

be a bivariate PDF of a random vector (X, Y).
(a) Show that $f(x,y)$ is a density.
(b) Show that marginal distributions are $f_X(x) = \frac{3}{2}x + \frac{3}{4}x^2$, $0 \le x \le 1$, and $f_Y(y) = \frac{3+8y}{4+12y}$, $0 \le y \le 2$.
(c) Show $\mathbb{E}X = 11/16$ and $\mathbb{E}Y = 5/4$.
(d) Show that conditional distributions are

$$f(x|y) = \frac{3x(x+2y)}{1+3y}, \, 0 \le x \le 1, \quad \text{for any fixed } y \in [0,2],$$

$$f(y|x) = \frac{2y+x}{4+2x}, \, 0 \le y \le 2, \quad \text{for any fixed } x \in [0,1].$$

(e) Show that

$$\mathbb{E}X|Y = \frac{3+8Y}{4+12Y} \quad \text{and} \quad \mathbb{E}Y|X = \frac{8+3X}{6+3X}.$$

(f) Demonstrate that iterated expectation rule (5.12) is satisfied,

$$\mathbb{E}(\mathbb{E}X|Y) = 11/16 \quad \text{and} \quad \mathbb{E}(\mathbb{E}Y|X) = 5/4.$$

5.22. **2-D Density Tasks.** If

$$f(x,y) = \begin{cases} \frac{1}{4}xy(x+y)\exp\{-x-y\}, & 0 \le x < \infty, \, 0 \le y < \infty \\ 0, & \text{else} \end{cases}$$

Find
(a) marginal distribution $f_X(x)$,

(b) conditional distribution $f(x|y)$,

(c) expectation $\mathbb{E}X$, and

(d) conditional expectation $\mathbb{E}X|Y$.

(f) Are X and Y independent? Explain.

5.23. **Conditional Variance.** In the context of Example 5.22 show that

$$\mathbb{V}\mathrm{ar}\,(Y|X=x) = \frac{(1-x)^2}{12}.$$

5.24. **Additivity of Gammas.** If $X_i \sim \mathcal{G}a(r_i,\lambda)$ are independent, prove that $Y = X_1 + \cdots + X_n$ is distributed as gamma with parameters $r = r_1 + r_2 + \cdots + r_n$ and λ; that is, $Y \sim \mathcal{G}a(r,\lambda)$.

5.25. **Memoryless Property.** Prove that the geometric $\mathcal{G}e(p)$ distribution ($\mathbb{P}(X = x) = (1 - p)^x p, x = 0,1,2,\ldots$) and the exponential distribution ($\mathbb{P}(X \leq x) = 1 - e^{-\lambda x}$, $x \geq 0, \lambda \geq 0$) both possess the *Memoryless Property*; that is, they satisfy

$$\mathbb{P}(X \geq v|X \geq u) = \mathbb{P}(X \geq v - u), \ v \geq u.$$

5.26. **Rh System.** Rh antigens are transmembrane proteins with loops exposed at the surface of red blood cells. They appear to be used for the transport of carbon dioxide and/or ammonia across the plasma membrane. They are named for the rhesus monkey in which they were first discovered. There are a number of different Rh antigens. Red blood cells that are *Rh positive* express the antigen designated as D. About 15% of the population do not have RhD antigens and thus are *Rh negative*. The major importance of the Rh system for human health is to avoid the danger of RhD incompatibility between a mother and her fetus.

(a) From the general population 8 people are randomly selected and checked for their Rh factor. Let X be the number of Rh negative among the eight selected. Find $\mathbb{P}(X = 2)$.

(b) In a group of 16 patients, three members are Rh negative. Eight patients are selected at random. Let Y be the number of Rh negative among the eight selected. Find $\mathbb{P}(Y = 2)$.

(c) From the general population subjects are randomly selected and checked for their Rh factor. Let Z be the number of Rh positive subjects before the first Rh negative subject is selected. Find $\mathbb{P}(Z = 2)$.

(d) Identify the distributions of the random variables in (a), (b), and (c).

(e) What are the expectations and variances for the random variables in (a), (b), and (c)?

5.27. **Blood Types.** The prevalence of blood types in the US population is O+: 37.4%, A+: 35.7%, B+: 8.5%, AB+: 3.4%, O–: 6.6%, A–: 6.3%, B–: 1.5%, and AB–: 0.6%.

(a) A sample of 24 subjects is randomly selected from the US population. What is the probability that 8 subjects are O+? Random variable X
describes the number of O+ subjects among 24 selected. Find $\mathbb{E}X$ and
$\text{Var }X$.

(b) Among 16 subjects, eight are O+. From these 16 subjects, five are
selected at random as a group. What is the probability that among the
five selected at most two are O+?

(c) Use Poisson approximation to find the probability that among 500
randomly selected subjects the number of AB– subjects is at least 1.

(d) Random sampling from the population is performed until the first
subject with B+ blood type is found. What is the expected number of
subjects sampled?

5.28. **Variance of the Exponential.** Show that for an exponential random variable X with density $f(x) = \lambda e^{-\lambda x}$, $x \geq 0$, the variance is $1/\lambda^2$.
Hint: You can use the fact that $\mathbb{E}X = 1/\lambda$. To find $\mathbb{E}X^2$ you need to repeat
the integration-by-parts twice.

5.29. **Equipment Aging.** Suppose that the lifetime T of a particular piece of
laboratory equipment (in 1000 hour units) is an exponentially distributed
random variable such that $\mathbb{P}(T > 10) = 0.8$.
(a) Find the "rate" parameter, λ.
(b) What are the mean and standard deviation of the random variable
T?
(c) Find the median, the first and third quartiles, and the inter-quartile
range of the lifetime T. Recall that for an exponential distribution, you
can find any percentile exactly.

5.30. **A Simple Continuous Random Variable.** Assume that the measured
responses in an experiment can be modeled as a continuous random
variable with density

$$f(x) = \begin{cases} c - x, \, 0 \leq x \leq c \\ \quad 0, \quad \text{else} \end{cases}$$

(a) Find the constant c and sketch the graph of the density $f(x)$.
(b) Find the CDF $F(x) = \mathbb{P}(X \leq x)$, and sketch its graph.
(c) Find $\mathbb{E}(X)$ and $\text{Var}(X)$.
(d) What is $\mathbb{P}(X \leq 1/2)$?

5.31. **2D Continuous Random Variable Question.** A two-dimensional random variable (X,Y) is defined by its density function, $f(x,y) = Cxe^{-xy}$,
$0 \leq x \leq 1; 0 \leq y \leq 1$.
(a) Find the constant C.
(b) Find the marginal distributions of X and Y.

5.32. **Insulin Sensitivity.** The insulin sensitivity (SI), obtained in a glucose tolerance test is one of the patient responses used to diagnose type II diabetes. Leading a sedative lifestyle and being overweight are well-established risk factors for type II diabetes. Hence, body mass index (BMI) and hip to waist ratio (HWR = HIP/WAIST) may also predict an impaired insulin sensitivity. In an experiment, 106 males (coded 1) and 126 females (coded 2) had their SI measured and their BMI and HWR registered. Data (⊞ diabetes.xls) are available on the text web page. For this exercise you will need only the 8th column of the data set, which corresponds to the SI measurements.
(a) Find the sample mean and sample variance of SI.
(b) A gamma distribution with parameters α and β seems to be an appropriate model for SI. What α, β should be chosen so that the $\mathbb{E}X$ matches the sample mean of SI and $\mathbb{V}\mathrm{ar}\, X$ matches the sample variance of SI.
(c) With α and β selected as in (b), simulate a random sample from gamma distribution with a size equal to that of SI ($n = 232$). Use gamrnd. Compare two histograms, one with the simulated values from the gamma model and the second from the measurements of SI. Use 20 bins for the histograms. Comment on their similarities/differences.
(d) Produce a Q–Q plot to compare the measured SI values with the model. Suppose that you selected $\alpha = 3$ and $\beta = 3.3$, and that dia is your data set. Take $n = 232$ equally spaced points between [0,1] and find their gamma quantiles using gaminv(points,alpha,beta). If the model fits the data, these theoretical quantiles should match the ordered sample.
Hint: (i) Here MATLAB's parametrization of gamma density is used, $\alpha = r$ and $\beta = 1/\lambda$. In terms of α and β, $\mathbb{E}X = \alpha\beta$ and $\mathbb{V}\mathrm{ar}\, X = \alpha\beta^2$.
(ii) The plot of theoretical quantiles against the ordered sample is called a Q–Q plot. An example of producing a Q–Q plot in MATLAB is as follows:

```
xx = 0.5/232: 1/232: 1;
yy=gaminv(xx, 3, 3.3);
plot(yy, sort(dia(:,8)),'*')
hold on
plot(yy, yy,'r-')
```

5.33. **Correlation between a Uniform and Its Power.** Suppose that X has uniform $\mathcal{U}(-1,1)$ distribution and that $Y = X^k$.
(a) Show that for k even, $\mathbb{C}\mathrm{orr}(X,Y) = 0$.
(b) Show that for arbitrary k, $\mathbb{C}\mathrm{orr}(X,Y) \to 0$, when $k \to \infty$.

5.34. **Precision of Lab Measurements.** The error X in measuring the weight of a chemical sample is a random variable with PDF.

$$f(x) = \begin{cases} \frac{3x^2}{16}, & -2 < x < 2 \\ 0, & \text{otherwise} \end{cases}$$

(a) A measurement is considered to be *accurate* if $|X| < 0.5$. Find the probability that a randomly chosen measurement can be classified as accurate.

(b) Find and sketch the graph of the cumulative distribution function $F(x)$.

(c) The loss in dollars, which is caused by measurement error, is $Y = X^2$. Find the mean of Y (expected loss).

(d) Compute the probability that the loss is less than \$3.

(e) Find the median of Y.

5.35. **Lifetime of Cells.** Cells in the human body have a wide variety of life spans. One cell may last a day; another a lifetime. Red blood cells (RBC) have a lifespan of several months and cannot replicate, which is the price RBCs pay for being specialized cells. The lifetime of a RBC can be modeled by an exponential distribution with density $f(t) = \frac{1}{\beta}e^{-t/\beta}$, where $\beta = 4$ (in units of months). For simplicity, assume that when a particular RBC dies, it is instantly replaced by a newborn RBC of the same type. For example, a replacement RBC could be defined as any new cell born approximately at the time when the original cell died.

(a) Find the expected lifetime of a single RBC. Find the probability that the cell's life exceeds 150 days. *Hint:* Days have to be expressed in units of β.

(b) A single RBC and its replacements are monitored over the period of 1 year. How many deaths/replacements are observed on average? What is the probability that the number of deaths/replacements exceeds 5. *Hint:* Utilize a link between exponential and Poisson distributions. In simple terms, if lifetimes are exponential with parameter β, then the number of deaths/replacements in the time interval $[0, t]$ is Poisson with parameter $\lambda = t/\beta$. Time units for t and β have to be the same.

(c) Suppose that a single RBC and two of its replacements are monitored. What is the distribution of their total lifetime? Find the probability that their total lifetime exceeds 1 year. *Hint:* Consult the gamma distribution. If n random variables are exponential with parameter β, then their sum is gamma distributed with parameters $\alpha = n$ and β.

(d) A particular RBC is observed $t = 2.2$ months after its birth and is found to still be alive. What is the probability that the total lifetime of this cell will exceed 7.2 months?

5.36. **k-out-of-n and Weibull Lifetime.** Engineering systems of type k-out-of-n are described in Exercise 3.10. Suppose that a k-out-of-n system consists of n identical and independent elements for which the lifetime has Weibull distribution with parameters r and λ. More precisely, if T is a lifetime of a component,

$$P(T \geq t) = e^{-\lambda t^r}.$$

Time t is in units of months, and consequently, rate parameter λ is in units $(\text{month})^{-1}$. Parameter r is dimensionless.

Assume that $n = 20, k = 7, r = 3/2$ and $\lambda = 1/4$.

(a) Find the probability that the k-out-of-n system is working at time $t = 3$.

(b) Plot this probability as a function of time.

(c) At time $t = 3$ the system is found operational. What is the distribution of the number of failed components? What is the expected number of failed components?

Hint: For each component the probability of the system working at time t is $p = e^{-1/2 t^{3/2}}$. The probability that a k-out-of-n system is operational corresponds to the tail probability of binomial distribution: $\mathbb{P}(X \geq k)$, where X is the number of components working. Use binocdf and be careful about the discrete nature of the binomial distribution.

In part (c), first find the probability that a component fails in the time interval $[0,3]$. Denote this probability with f. Then, the number of failed components Y cannot exceed $n - k$, and given the independence of components, it is binomial. That is, $Y \sim \mathcal{Bin}(n - k, f)$.

5.37. **Silver-Coated Nylon Fiber.** Silver-coated nylon fiber is used in hospitals for its anti-static electricity properties, as well as for antibacterial and antimycotic effects. In the production of silver-coated nylon fibers, the extrusion process is interrupted from time to time by blockages occurring in the extrusion dyes. The time in hours between blockages, T, has an exponential $\mathcal{E}(1/10)$ distribution, where $1/10$ is the rate parameter.

Find the probabilities that

(a) a run continues for at least 10 hours,

(b) a run lasts less than 15 hours, and

(c) a run continues for at least 20 hours, given that it has lasted 10 hours. Use MATLAB and expcdf function. Be careful about the parametrization of exponentials in MATLAB.

5.38. **Xeroderma Pigmentosum.** Xeroderma pigmentosum (XP) was first described in 1874 by Hebra et al. XP is the condition characterized as dry, pigmented skin. It is a hereditary condition with an incidence of 1:250,000 live births (Robbin et al., 1974). In a city with a population of 1,000,000, find the distribution of the number of people with XP. What is the expected number? What is the probability that there are no XP-affected subjects?

5.39. **Failure Time.** Let X model the time to failure (in years) of a Beckman Coulter TJ-6 laboratory centrifuge. Suppose that the PDF of X is $f(x) = c/(3 + x)^3$ for $x \geq 0$.

(a) Find the value of c such that f is a legitimate PDF.

(b) Compute the mean and median time to failure of the centrifuge.

5.40. **Resistors.** If n resistors with resistances $R_1, R_2, \ldots, R_n$ are connected in-line, the total resistance R is

$$R = R_1 + R_2 + \cdots + R_n.$$

If the connection is parallel (resistors branch out from a single node, and join up again somewhere else in the circuit), then

$$1/R = 1/R_1 + 1/R_2 + \cdots + 1/R_n.$$

Suppose that resistances of two resistors are independent random variables with means $\mu_1 = 2\ \Omega$ and $\mu_2 = 3\ \Omega$ and variances $\sigma_1^2 = 0.02^2\ \Omega^2$ and $\sigma_2^2 = 0.01^2\ \Omega^2$.

Estimate the mean and variance of the total resistance if the resistors are connected

(a) in line;

(b) parallel.

(c) In the case of parallel connection, assume that $R_1 \sim \mathcal{IG}(r_1, \lambda)$ and $R_2 \sim \mathcal{IG}(r_2, \lambda)$, where $r_1, r_2 > 1$ and λ is the rate parameter. Thus, R is $\mathcal{IG}(r_1 + r_2, \lambda)$ and $ER = \lambda/(r_1 + r_2 - 1)$. How does this exact value compare to the first-order approximation

$$ER = g(ER_1, ER_2)$$

for $g(x_1, x_2) = 1/(1/x_1 + 1/x_2)$?

5.41. **Beta Fit.** Assume that the fraction of impurities in a certain chemical solution is modeled by a Beta $\mathcal{Be}(\alpha, \beta)$ distribution with known parameter $\alpha = 1$. The average fraction of impurities is 0.1.

(a) Find the parameter β.

(b) What is the standard deviation of the fraction of impurities?

(c) Find the probability that the fraction of impurities exceeds 0.25.

5.42. **Uncorrelated but Possibly Dependent.** Show that for any two random variables X and Y with equal second moments, the variables $Z = X + Y$ and $W = X - Y$ are uncorrelated. Note, that Z and W could be dependent.

5.43. **Nights of Mr. Jones.** If Mr. Jones had insomnia one night, the probability that he would sleep well the following night is 0.6; otherwise, he would have insomnia. If he slept well one night, the probabilities of sleeping well or having insomnia the following night would be 0.5 each.

On Monday night Mr. Jones had insomnia. What is the probability that he had insomnia on the following Friday night?

5.44. **Stationary Distribution of MC.** Consider a Markov chain with transition matrix

$$\mathbf{P} = \begin{pmatrix} 0 & 1/2 & 1/2 \\ 1/2 & 0 & 1/2 \\ 1/2 & 1/2 & 0 \end{pmatrix}.$$

(a) Show that all entries of $\mathbf{P}^2$ are strictly positive.

(b) Using MATLAB, find $\mathbf{P}^{100}$ and guess what the stationary distribution $\pi = (\pi_1, \pi_2, \pi_3)$ would be. Confirm your guess by solving the equation $\pi = \pi\mathbf{P}$, which gives the exact stationary distribution. *Hint:* The system $\pi = \pi\mathbf{P}$ needs a closure equation $\pi_1 + \pi_2 + \pi_3 = 1$.

5.45. **Influence of Two Previous Trials.** In a potentially infinite sequence of trials, the probability of success is 1/2, unless the previous two trials resulted in a success. In this case, the probability of success is 2/3. Code successes as 1 and failures as 0. Such binary sequence defines a MC where the states are 00, 01, 10, and 11; see Figure 5.17.

(a) Write down the transition matrix P.

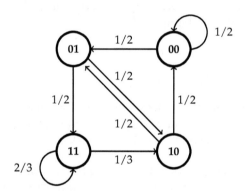

Fig. 5.17 Markov chain schematic graph

(b) Using MATLAB find P^{100} and argue that the stationary probabilities for 00, 01, 10, and 11 are 2/9, 2/9, 2/9, and 1/3, respectively. Confirm this numerical result by solving the system

$$\pi = \pi P,$$

where P is the transition matrix and $\pi = (\pi_1, \pi_2, \pi_3, \pi_4)$ is the row vector of stationary probabilities. Since P is not of full rank, the equation $\sum_i^4 \pi_i = 1$ completes the system.

HINT: `> linsolve([(eye(4)-P)'; ones(1,4)],[zeros(4,1); 1])`

(c) Argue that the proportion of successes in a long run is 5/9.

5.46. **Heat Production by a Resistor.** Joule's Law states that the amount of heat produced by a resistor is

$$Q = I^2 \, R \, T,$$

where
Q is heat energy (in Joules),
I is current (in Amperes),
R is resistance (in Ohms), and
T is duration of time (in seconds).
Suppose that in an experiment, I, R, and T are independent random variables with means $\mu_I = 10$ A, $\mu_R = 30$ Ω, and $\mu_T = 120$ s. Suppose that the variances are $\sigma_I^2 = 0.01$ A^2, $\sigma_R^2 = 0.02$ Ω^2, and $\sigma_T^2 = 0.001$ s^2. Estimate the mean μ_Q and the variance σ_Q^2 of the produced energy Q.

MATLAB AND WINBUGS FILES AND DATA SETS USED IN THIS CHAPTER

http://statbook.gatech.edu/Ch5.RanVar/

apgar.m, bookplots.m, circuitgenbin.m, corneoretinal.m, covcord2d.m, dexp.m, Discrete.m, empiricalcdf.m, histp.m, hyper.m, lefthanded.m, lifetimecells.m, mamopixels.m, markovchain.m, MCEhrenfest.m, melanoma.m, mingling.m, plotbino.m, plotsdistributions.m, plotuniformdist.m, randdirichlet.m, stringerror.m

hearttransplant1.odc, hearttransplant2.odc, lifetimecells.odc, simulationc.odc, simulationd.odc

diabetes.xls

CHAPTER REFERENCES

Apgar, V. (1953). A proposal for a new method of evaluation of the newborn infant. *Curr. Res. Anesth. Analg.*, **32**, (4), 260–267. PMID 13083014.

CDC (2007). *Morbidity and Mortality Weekly Report*. **56**, 8, 161–165.

Coresh, J., Astor, B. C., Greene, T., Eknoyan, G., and Levey, A. S. (2003). Prevalence of chronic kidney disease and decreased kidney function in the adult US population: 3rd national health and nutrition examination survey. *Am. J. Kidney Dis.*, **41**, 1–12.

Dayhoff, M. O., Schwartz, R., and Orcutt, B. C. (1978). A model of evolutionary change in proteins. Atlas of protein sequence and structure. *Nat. Biomed. Res. Found.*, **5**, Suppl. 3, 345–358.

du Bois-Reymond, E. H. (1848). *Untersuchungen Ueber Thierische Elektricität*, Vol. 1, G. Reimer, Berlin.

Ehrenfest, P. und T. (1907). Uber zwei bekannte Einwände gegen das Boltzmannsche H-Theorem. âĂŐPhysik. Z., **8**, 311–331.

Gjini, A., Stuart, J. M., George, R. C., Nichols, T., and Heyderman, R. S. (2004). Capture-recapture analysis and pneumococcal meningitis estimates in England. *Emerg. Infect. Dis.*, **10**, 1, 87–93.

Hebra F. and Kaposi M. (1874). On diseases of the skin including exanthemata. *New Sydenham Soc.*, **61**, 252–258.

Montmort, P. R. (1714). *Essai d'Analyse sur les Jeux de Hazards*, 2ed. Jombert, Paris.

Pielou, E.C. (1961). Segregation and symmetry in two-species populations as studied by nearest-neighbor relationships. *J. Ecol.*, **49**, 2, 255–269.

Robbin, J. H., Kraemer, K. H., Lutzner, M. A., Festoff, B. W., and Coon, H. P. (1974). Xeroderma pigmentosum: An inherited disease with sun sensitivity, multiple cutaneous neoplasms and abnormal DNA repair. *Ann. Intern. Med.*, **80**, 221–248.

Ross, M. S. (2010a). *A First Course in Probability*, Pearson Prentice-Hall.

Ross, M. S. (2010b) *Introduction to Probability Models*, 10th ed. Academic Press, Burlington.

Tubert-Bitter, P., Begaud, B., Moride, Y., Chaslerie, A., and Haramburu, F. (1996). Comparing the toxicity of two drugs in the framework of spontaneous reporting: A confidence interval approach. *J. Clin. Epidemiol.*, **49**, 121–123.

Weibull, W. (1951). A statistical distribution function of wide applicability. *J. Appl. Mech.*, **18**, 293–297.

Chapter 6
Normal Distribution

The adjuration to be normal seems shockingly repellent to me.

– Karl Menninger

WHAT IS COVERED IN THIS CHAPTER

- Definition of Normal Distribution, Bivariate Case
- Standardization, Quantiles of Normal Distribution, Sigma Rules
- Linear Combinations of Normal Random Variables
- Central Limit Theorem, de Moivre's Approximation
- Distributions Related to Normal: Chi-Square, Wishart, t, F, Lognormal, and Some Noncentral Distributions
- Transformations to Normality

6.1 Introduction

In Chapters 2 and 5 we occasionally referred to a normal distribution either informally (bell-shaped distributions/histograms) or formally, as in Section 5.5.3, where the normal density and its moments were briefly introduced. This chapter is devoted to the normal distribution due to its importance in statistics. What makes the normal distribution so important? The normal distribution is the proper statistical model for many natural and social phenomena. But even if some measurements cannot be modeled by the normal distribution (it could be skewed, discrete, multimodal, etc.), their sample means would closely follow the normal law, under very mild conditions. The central limit theorem covered in this chapter makes it possible to use probabilities associated with the normal curve to answer questions about the sums and averages in sufficiently large samples. This translates to the ubiquity of normality – many estimators, test statistics, and nonparametric tests covered in later chapters of this text are approximately normal, when sample sizes are not small (typically larger than 20 to 30), and this asymptotic normality is used in a substantial way. Several other important distributions can be defined through a normal distribution. Also, normality is a quite stable property – an arbitrary linear combination of normal random variables remains normal. The property of linear combinations of random variables preserving the distribution of their components is not shared by any other probability law and is a characterizing property of a normal distribution.

6.2 Normal Distribution

In 1738, Abraham de Moivre developed the normal distribution as an approximation to the binomial distribution, and it was subsequently used by Laplace in 1783 to study measurement errors and by Gauss in 1809 in the analysis of astronomical data. The name *normal* came from Quetelet, who demonstrated that many human characteristics distributed themselves in a bell-shaped manner (centered about the "average man," *l'homme moyen*), including such measurements as chest girths of 5,738 Scottish soldiers, the heights of 100,000 French conscripts, and the body weight and height of people he measured. From his initial research on height and weight has evolved the internationally recognized measure of obesity called the Quetelet index (QI), or body mass index (BMI), QI = (weight in kilograms)/(squared height in meters).

Table 6.1 provides frequencies of chest sizes of 5,738 Scottish soldiers as well as the relative frequencies. Using this now famous data set, Quetelet argued that many human measurements distribute as normal. Figure 6.1 gives a normalized histogram of Quetelet's data set with superimposed normal density in which the mean and the variance are taken as the sample mean (39.8318) and sample variance (2.0496^2).

The PDF for a normal random variable with mean μ and variance σ^2 is

$$f(x) = \frac{1}{\sqrt{2\pi\sigma^2}} e^{-\frac{1}{2\sigma^2}(x-\mu)^2}, \quad -\infty < x < \infty.$$

The distribution function is computed using integral approximation because no closed form exists for the antiderivative of $f(x)$; this is generally

Table 6.1 Chest sizes of 5738 Scottish soldiers, data compiled from the 13th edition of the *Edinburgh Medical Journal* (1817).

Size	Frequency	Relative frequency (in %)
33	3	0.05
34	18	0.31
35	81	1.41
36	185	3.22
37	420	7.32
38	749	13.05
39	1073	18.70
40	1079	18.80
41	934	16.28
42	658	11.47
43	370	6.45
44	92	1.60
45	50	0.87
46	21	0.37
47	4	0.07
48	1	0.02
Total	5738	99.99

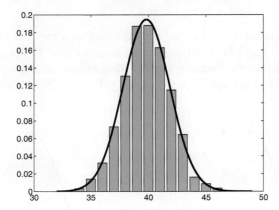

Fig. 6.1 Normalized bar plot of Quetelet's data set. Superimposed is the normal density with mean $\mu = 39.8318$ and variance $\sigma^2 = 2.0496^2$.

not a problem for practitioners because most software packages will compute interval probabilities numerically. In MATLAB, `normcdf(x, mu, sigma)` and `normpdf(x, mu, sigma)` calculate the CDF and PDF at x, and `norminv(p, mu, sigma)` computes the inverse CDF at given probability p, that is, the p-quantile. Equivalently, a normal CDF can be expressed in terms of a special function called the *error integral*:

$$\mathrm{erf}(x) = \frac{2}{\sqrt{\pi}} \int_0^x e^{-t^2} dt.$$

It holds that `normcdf(x)= 1/2+1/2*erf(x/sqrt(2))`. A random variable X with a normal distribution will be denoted $X \sim \mathcal{N}(\mu, \sigma^2)$.

In addition to software, CDF values are often given in tables. Such tables contain only quantiles and CDF values for the *standard* normal distribution, $Z \sim \mathcal{N}(0,1)$, for which $\mu = 0$ and $\sigma^2 = 1$. Such tables are sufficient since an arbitrary normal random variable X can be *standardized* to Z if its mean and variance are known:

$$X \sim \mathcal{N}(\mu, \sigma^2) \quad \longrightarrow \quad Z = \frac{X - \mu}{\sigma} \sim \mathcal{N}(0,1).$$

For a standard normal random variable, Z the PDF is denoted by ϕ, and CDF by Φ,

$$\Phi(x) = \int_{-\infty}^{x} \phi(t) \, dt = \int_{-\infty}^{x} \frac{1}{\sqrt{2\pi}} e^{-t^2/2} \, dt. \quad [\text{normcdf(x)}]$$

Suppose we are interested in the probability that a random variable X distributed as $\mathcal{N}(\mu, \sigma^2)$ falls between two bounds a and b, $\mathbb{P}(a < X < b)$. It is irrelevant whether the bounds are included or not since the normal distribution is continuous and $\mathbb{P}(a < X < b) = \mathbb{P}(a \leq X \leq b)$. Also, any of the bounds can be infinite.

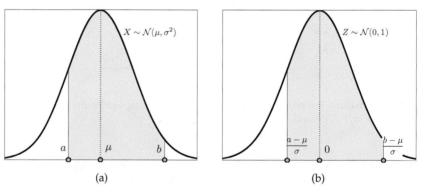

(a)   (b)

Fig. 6.2 Illustration of the relation $\mathbb{P}(a \leq X \leq b) = \mathbb{P}\left(\frac{a-\mu}{\sigma} \leq Z \leq \frac{b-\mu}{\sigma}\right)$.

In terms of Φ,

$$X \sim \mathcal{N}(\mu, \sigma^2):$$
$$\mathbb{P}(a \le X \le b) = \mathbb{P}\left(\frac{a-\mu}{\sigma} \le Z \le \frac{b-\mu}{\sigma}\right) = \Phi\left(\frac{b-\mu}{\sigma}\right) - \Phi\left(\frac{a-\mu}{\sigma}\right).$$

Figures 6.2 and 6.3 provide the illustration. In MATLAB:

```
normcdf((b-mu)/sigma) - normcdf((a - mu)/sigma)
        %or equivalently
normcdf(b, mu, sigma) - normcdf(a, mu, sigma)
```

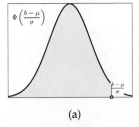 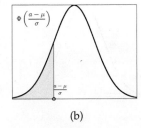 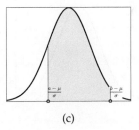

(a)   (b)   (c)

Fig. 6.3 Calculation of $\mathbb{P}(a \le X \le b)$ for $X \sim \mathcal{N}(\mu, \sigma^2)$. (a) $\mathbb{P}(X \le b) = \mathbb{P}(Z \le \frac{b-\mu}{\sigma}) = \Phi\left(\frac{b-\mu}{\sigma}\right)$; (b) $\mathbb{P}(X \le a) = \mathbb{P}(Z \le \frac{a-\mu}{\sigma}) = \Phi\left(\frac{a-\mu}{\sigma}\right)$; (c) $\mathbb{P}(a \le X \le b)$ as the difference of the two probabilities in (a) and (b).

Note that when the bounds are infinite, since Φ is a CDF,

$$\Phi(-\infty) = 0, \text{ and } \Phi(\infty) = 1.$$

Traditional statistics textbooks provide tables of cumulative probabilities for the standard normal distribution, $p = \Phi(x)$, for values of x typically between -3 and 3 with an increment of 0.01. The tables have been used in two ways: (i) directly, that is, for a given x the user finds $p = \Phi(x)$; and (ii) inversely, given p, one finds approximately what x gives $\Phi(x) = p$, which is of course a p-quantile of the standard normal. Given the limited precision of the tables, the results in direct and inverse uses have been approximate.

In MATLAB, the tables can be reproduced by a single line of code:

```
x=(-3:0.01:3)'; tables=[x normcdf(x)]
```

Similarly, the normal p-quantiles z_p defined as $p = \Phi(x_p)$ can be tabulated as

```
probs=(0.005:0.005:0.995)'; tables=[probs norminv(probs)]
```

There are several normal quantiles that are frequently used in the construction of confidence intervals and tests; these are the 0.9, 0.95, 0.975, 0.99, 0.995, and 0.9975 quantiles,

$$z_{0.9} = 1.28155 \approx 1.28 \quad z_{0.95} = 1.64485 \approx 1.64 \quad z_{0.975} = 1.95996 \approx 1.96$$
$$z_{0.99} = 2.32635 \approx 2.33 \quad z_{0.995} = 2.57583 \approx 2.58 \quad z_{0.9975} = 2.80703 \approx 2.81$$

For example, the 0.975 quantile of the normal is $z_{0.975} = 1.96$. This is equivalent to saying that 95% of the area below the standard normal density $\phi(x) = \frac{1}{\sqrt{2\pi}} \exp\{-x^2/2\}$ lies between –1.96 and 1.96. Note that the shortest interval containing $1 - \alpha$ probability is defined by quantiles $z_{\alpha/2}$ and $z_{1-\alpha/2}$ (see Figure 6.4 as an illustration for $\alpha = 0.05$). Since the standard normal density is symmetric about 0, $z_p = -z_{1-p}$.

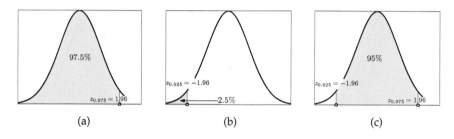

(a)   (b)   (c)

Fig. 6.4 (a) Normal quantiles (a) $z_{0.975} = 1.96$, (b) $z_{0.025} = -1.96$, and (c) 95% area between quantiles –1.96 and 1.96.

6.2.1 Sigma Rules

Sigma rules state that for any normal distribution, the probability that an observation will fall in the interval $\mu \pm k\sigma$ for $k = 1, 2$, and 3 is $68.27\%, 95.45\%$, and 99.73%, respectively. More precisely,

$$\mathbb{P}(\mu - \sigma < X < \mu + \sigma) = \mathbb{P}(-1 < Z < 1) = \Phi(1) - \Phi(-1) = 0.682689 \quad \approx 68.27\%$$
$$\mathbb{P}(\mu - 2\sigma < X < \mu + 2\sigma) = \mathbb{P}(-2 < Z < 2) = \Phi(2) - \Phi(-2) = 0.954500 \approx 95.45\%$$
$$\mathbb{P}(\mu - 3\sigma < X < \mu + 3\sigma) = \mathbb{P}(-3 < Z < 3) = \Phi(3) - \Phi(-3) = 0.997300 \approx 99.73\%$$

Have you ever wonder about the origin of the term *Six Sigma*? It does not involve $\mathbb{P}(\mu - 6\sigma < X < \mu + 6\sigma)$ as one may expect.

The Six Sigma doctrine is a standard according to which an item with measurement $X \sim \mathcal{N}(\mu, \sigma^2)$ should satisfy $X < 6\sigma$ to be conforming if μ is allowed to vary between -1.5σ and 1.5σ.

Thus, effectively, accounting for the variability in the mean, the Six Sigma constraint becomes

$$\mathbb{P}(X < \mu + 4.5\sigma) = P(Z < 4.5) = \Phi(4.5) = 0.99999660.$$

This means that only 3.4 items per million produced are allowed to exceed $\mu + 4.5\sigma$ (be defective). Such standard of quality was set by the Motorola Company in the 1980s, and it evolved into a doctrine for improving efficiency and quality in management.

6.2.2 Bivariate Normal Distribution*

When the components of a random vector have a normal distribution, we say that the vector has a multivariate normal distribution. For independent components, the density of a multivariate distribution is simply the product of the univariate densities. When components are correlated, the distribution involves the covariance matrix that describes the correlation. Next we discuss the bivariate normal distribution, which will be important later on, in the context of correlation and regression.

The pair (X,Y) is distributed as bivariate normal $\mathcal{N}_2(\mu_X, \mu_Y, \sigma_X^2, \sigma_Y^2, \rho)$ if the joint density is

$$f(x,y) = \frac{1}{2\pi\sigma_X\sigma_Y\sqrt{1-\rho^2}} \exp\left\{-\frac{1}{2(1-\rho^2)}\left[\frac{(x-\mu_x)^2}{\sigma_X^2}\right.\right.$$
$$\left.\left.-\frac{2\rho(x-\mu_x)(y-\mu_y)}{\sigma_X\sigma_Y} + \frac{(y-\mu_y)^2}{\sigma_Y^2}\right]\right\}. \tag{6.1}$$

The parameters $\mu_X, \mu_Y, \sigma_X^2, \sigma_Y^2$, and ρ are

$$\mu_X = \mathbb{E}(X), \ \mu_Y = \mathbb{E}(Y), \ \sigma_X^2 = \mathbb{V}\text{ar}(X), \ \sigma_Y^2 = \mathbb{V}\text{ar}(Y), \text{ and } \rho = \text{Corr}(X,Y).$$

One can define bivariate normal distribution with a density as in (6.1) by transforming two independent, standard normal random variables Z_1 and Z_2,

$$X = \mu_1 + \sigma_X Z_1,$$
$$Y = \mu_2 + \rho\sigma_Y Z_1 + \sqrt{1-\rho^2}\sigma_Y Z_2.$$

The marginal distributions in (6.1) are $X \sim \mathcal{N}(\mu_X, \sigma_X^2)$ and $Y \sim \mathcal{N}(\mu_Y, \sigma_Y^2)$. The bivariate normal vector (X,Y) has a covariance matrix

$$\Sigma = \begin{pmatrix} \sigma_X^2 & \sigma_X\sigma_Y\rho \\ \sigma_X\sigma_Y\rho & \sigma_Y^2 \end{pmatrix}. \tag{6.2}$$

The covariance matrix Σ is nonnegative definite. A sufficient condition for nonnegative definiteness in this case is $|\Sigma| \geq 0$ (see also Exercise 6.2).

Figure 6.5a shows the density of a bivariate normal distribution with mean

$$\mu = \begin{pmatrix} \mu_X \\ \mu_Y \end{pmatrix} = \begin{pmatrix} -1 \\ 2 \end{pmatrix}$$

and covariance matrix

$$\Sigma = \begin{pmatrix} 3 & -0.9 \\ -0.9 & 1 \end{pmatrix}.$$

Figure 6.5b shows contours of equal probability.

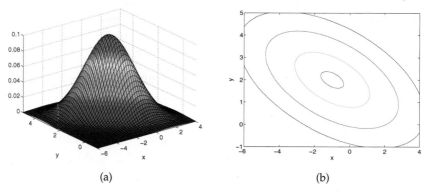

(a)   (b)

Fig. 6.5 (a) Density of bivariate normal distribution with mean `mu=[-1 2]` and covariance matrix `Sigma=[3 -0.9; -0.9 1]`. (b) Contour plots of a density at levels `[0.001 0.01 0.05 0.1]`

Several properties of bivariate normal are listed below:

(i) If (X,Y) is bivariate normal, then $aX + bY$ has a univariate normal distribution.

(ii) If (X,Y) is bivariate normal, then $(aX + bY, cX + dY)$ is also bivariate normal.

(iii) If the components in (X,Y) are such that $\text{Cov}(X,Y) = \sigma_X \sigma_Y \rho = 0$, then X and Y are independent.

(iv) Any bivariate normal pair (X,Y) can be transformed into a pair $(U,V) = (aX + bY, cX + dY)$ such that U and V are independent. If $\sigma_X^2 = \sigma_Y^2$, then one such transformation is $U = X + Y$, $V = X - Y$. For an arbitrary bivariate normal distribution, the rotation

$$U = X\cos\varphi - Y\sin\varphi$$
$$V = X\sin\varphi + Y\cos\varphi$$

makes components (U,V) independent if the rotation angle φ satisfies

$$\cot 2\varphi = \frac{\sigma_X^2 - \sigma_Y^2}{2\sigma_X \sigma_Y \rho}.$$

(v) If (X, Y) is bivariate normal, then the conditional distribution of Y when $X = x$ is normal with expectation and variance

$$\mu_X + \rho \frac{\sigma_Y}{\sigma_X}(x - \mu_X), \quad \text{and} \quad \sigma_Y^2(1 - \rho^2),$$

respectively. The linearity in x of the conditional expectation of Y will be the basis for linear regression, covered in Chapter 14. Also, the fact that $X = x$ is known decreases the variance of Y; indeed, $\sigma_Y^2(1 - \rho^2) \leq \sigma_Y^2$.

More generally, when the components of a p-dimensional random vector all have a normal distribution, we say that the vector has a multivariate normal distribution. For independent components, the density of a multivariate distribution is simply the product of the univariate normal densities. When the components are correlated, the distribution involves the covariance matrix that describes the correlation.

A random vector $\boldsymbol{X} = (X_1, \ldots, X_p)'$ has a multivariate normal distribution with parameters $\boldsymbol{\mu}$ and $\boldsymbol{\Sigma}$, denoted as $\boldsymbol{X} \sim \mathcal{MVN}_p(\boldsymbol{\mu}, \boldsymbol{\Sigma})$, if its density is

$$f(\boldsymbol{x}) = \frac{1}{(2\pi)^{p/2}|\boldsymbol{\Sigma}|^{1/2}} e^{-(1/2)(\boldsymbol{x}-\boldsymbol{\mu})'\boldsymbol{\Sigma}^{-1}(\boldsymbol{x}-\boldsymbol{\mu})},$$

where $\boldsymbol{x} \in \mathbb{R}^p$, and $\boldsymbol{\Sigma}$ is a non-negative definite $p \times p$ matrix. Here $|\boldsymbol{\Sigma}|$ is the determinant and $\boldsymbol{\Sigma}^{-1}$ the inverse of the covariance matrix $\boldsymbol{\Sigma}$.

6.3 Examples with a Normal Distribution

We provide two examples with typical calculations involving normal distributions, with solutions in MATLAB and WinBUGS.

Example 6.1. **IgE Concentration.** Total serum IgE (immunoglobulin E) concentration allergy tests allow for the measurement of the total IgE level in a serum sample. Elevated levels of IgE are associated with the presence of an allergy. An example of testing for total serum IgE is the PRIST (paper radioimmunosorbent test). This test involves serum samples reacting with IgE that has been tagged with radioactive iodine. The bound radioactive iodine, calculated upon completion of the test procedure, is proportional to the amount of total IgE in the serum sample. The determination of normal IgE levels in a population of healthy, nonallergic individuals varies by the fact that some individuals may have subclinical allergies and therefore have abnormal serum IgE levels. The log concentration of IgE (in IU/ml) in a cohort of healthy subjects is distributed as a normal $\mathcal{N}(9, (0.9)^2)$ random variable. What is the probability that in a randomly selected subject from the same cohort the log concentration will

(a) Exceed 10 IU/ml?

(b) Be between 8.1 and 9.9 IU/ml?

(c) Differ from the mean by no more than 1.8 IU/ml?

(d) Find the number x_0 such that the IgE log concentration in 90% of the subjects from the same cohort exceeds x_0.

(e) In what bounds (symmetric about the mean) does the IgE log concentration fall with a probability of 0.95?

(f) If the IgE log concentration is $\mathcal{N}(9, \sigma^2)$, find σ so that

$$\mathbb{P}(8 \leq X \leq 10) = 0.64.$$

Let X be the IgE log concentration in a randomly selected subject. Then $X \sim \mathcal{N}(9, 0.9^2)$. The solution is given by the following MATLAB code (ige.m):

```
%(a)
%P(X>10)= 1-P(X <= 10)
1-normcdf(10,9,0.9) %or    1-normcdf((10-9)/0.9)
%ans = 0.1333
%(b)
%P(8.1 <= X <= 9.9)
%P((8.1-9)/0.9 <= Z <= (9.9-9)/0.9)
%P(-1 <= Z <= 1) :::: Note the 1-sigma rule.
normcdf(9.9, 9, 0.9) - normcdf(8.1, 9, 0.9)
%or, normcdf((9.9-9)/0.9)-normcdf((8.1-9)/0.9)
%ans = 0.6827
%(c)
%P(9-1.8 <= X <= 9+1.8) = P(-2 <= Z <= 2)
%Note the 2-sigma rule.
normcdf(9+1.8, 9, 0.9) - normcdf(9-1.8, 9, 0.9)
% ans = 0.9545
%(d)
%0.90 = P(X > x0)=1-P(X <= x0)
%that is P(Z <= (x0-9)/0.9)=0.1
norminv(1-0.9,   9, 0.9)
%ans = 7.8466
%(e)
%P(9-delta <= X <= 9+delta)=0.95
[9-0.9*norminv(1-0.05/2),  9+0.9*norminv(1-0.05/2)]
%ans =  7.2360   10.7640
%(f)
%P(-1/sigma) <= Z <= 1/sigma)=0.64
%note that 0.36/2 + 0.64 + 0.36/2 = 1
1/norminv( 1 - 0.36/2 )
%ans = 1.0925
```

Example 6.2. **Aplysia Nerves.** In this example, easily solved analytically and using MATLAB, we will show how to use WinBUGS and obtain an approximate solution. The analysis is not Bayesian; WinBUGS will simply serve as a random number generator and the required probability and quantile will be found approximately by simulation.

Characteristics of Aplysia nerves in response to extension were examined by Koike (1987). Only the Aplysia nerve was easily elongated up to about five times its resting or relaxing length without impairing propagation of the action potential along the axon in the nerve. The conduction velocity along the elongated nerve increased linearly in proportion to the nerve length in a range from the relaxing length to about 1 to 1.5 times extension. For an expansion factor of 1.5, the conducting velocity factors are normally distributed with a mean of 1.4 and a standard deviation of 0.1. Using WinBUGS, we are interested in finding

(a) the proportion of Aplysia nerves elongated by a factor of 1.5 for which the conduction velocity factor exceeds 1.5;

(b) the proportion of Aplysia nerves elongated by a factor of 1.5 for which the conduction velocity factor falls in the interval $[1.35, 1.61]$; and

(c) the velocity factor x that is exceeded by 5% of Aplysia nerves elongated by a factor of 1.5.

```
#aplysia.odc
model{
mu <- 1.4
stdev <- 0.1
prec<- 1/(stdev * stdev)
y ~ dnorm(mu, prec)
#a
propexceed <- step(y - 1.5)
#b
propbetween <-  step(y-1.35)*step(1.61-y)
#c
#done in Sample Monitor Tool by
#selecting 95th percentile
}
```

There are no data to load; after the check model in Model>Specification go directly to compile, and then to gen inits. Update 10,000 iterations, and set in Sample Monitor Tool from Inference>Samples the nodes y, propexceed, and propbetween. For part (c) select the 95th percentile in Sample Monitor Tool under percentiles. Finally, run the Update Tool for 1,000,000 updates and check the results in Sample Monitor Tool by setting a star (*) in the node window and looking at stats.

	mean	sd	MC error	val2.5pc	median	val97.5pc	start	sample
propbetween	0.6729	0.4691	4.831E-4	0.0	1.0	1.0	10001	1000000
propexceed	0.1587	0.3654	3.575E-4	0.0	0.0	1.0	10001	1000000
y	1.4	0.1001	1.005E-4	1.204	1.4	1.565	10001	1000000

Here is the same computation in MATLAB.

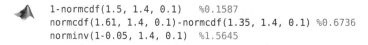

```
1-normcdf(1.5, 1.4, 0.1)    %0.1587
normcdf(1.61, 1.4, 0.1)-normcdf(1.35, 1.4, 0.1) %0.6736
norminv(1-0.05, 1.4, 0.1)  %1.5645
```

6.4 Combining Normal Random Variables

Any linear combination of independent normal random variables is also normally distributed. Thus, we need only keep track of the mean and variance of the variables involved in the linear combination, since these two parameters completely characterize the distribution. Let $X_1, X_2, \ldots, X_n$ be independent normal random variables such that $X_i \sim \mathcal{N}(\mu_i, \sigma_i^2)$; then for any selection of constants $a_1, a_2, \ldots, a_n$,

$$a_1 X_1 + a_2 X_2 + \cdots + a_n X_n = \sum_{i=1}^{n} a_i X_i \sim \mathcal{N}(\mu, \sigma^2),$$

where

$$\mu = a_1 \mu_1 + a_2 \mu_2 + \cdots + a_n \mu_n = \sum_{i=1}^{n} a_i \mu_i,$$

$$\sigma^2 = a_1^2 \sigma_1^2 + a_2^2 \sigma_2^2 + \ldots a_n^2 \sigma_n^2 = \sum_{i=1}^{n} a_i^2 \sigma_i^2.$$

Two special cases are important: (i) $a_1 = 1, a_2 = -1$ and (ii) $a_1 = \cdots = a_n = 1/n$. In case (i) we have a difference of two normals; its mean is the difference of the corresponding means and variance is a *sum* of two variances. Case (ii) corresponds to the arithmetic mean of normals, $\overline{X}$. For example, if $X_1, \ldots, X_n$ are i.i.d. $\mathcal{N}(\mu, \sigma^2)$, then the sample mean $\overline{X} = (X_1 + \cdots + X_n)/n$ has a normal $\mathcal{N}(\mu, \sigma^2/n)$ distribution. Thus, variances for X_is and $\overline{X}$ are related as

$$\sigma_{\overline{X}}^2 = \frac{\sigma^2}{n}$$

or, equivalently, for standard deviations

$$\sigma_{\overline{X}} = \frac{\sigma}{\sqrt{n}}.$$

Example 6.3. **The Piston Production Error.** The profile of a piston comprises a ring in which inner and outer radii X and Y are normal random variables, $\mathcal{N}(88,0.01^2)$ and $\mathcal{N}(90,0.02^2)$, respectively. The thickness $D = Y - X$ is the random variable of interest.

(a) Find the distribution of D.

(b) For a randomly selected piston, what is the probability that D will exceed 2.04?

(c) If D is averaged over a batch of $n = 64$ pistons, what is the probability that $\overline{D}$ will exceed 2.04? Exceed 2.004?

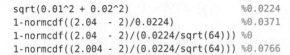

```
sqrt(0.01^2 + 0.02^2)                            %0.0224
1-normcdf((2.04  - 2)/0.0224)                    %0.0371
1-normcdf((2.04  - 2)/(0.0224/sqrt(64)))  %0
1-normcdf((2.004 - 2)/(0.0224/sqrt(64)))  %0.0766
```

Compare the probabilities of events $\{D > 2.04\}$ and $\{\overline{D} > 2.04\}$. Why is the probability of $\{\overline{D} > 2.04\}$ essentially 0, when the analogous probability for an individual measure D is 3.71%?

Example 6.4. **Diluting Acid.** In a laboratory, students are told to mix 100 ml of distilled water with 50 ml of sulfuric acid and 30 ml of C_2H_5OH. Of course, the measurements are not exact. The water is measured with a mean of 100 ml and a standard deviation of 4 ml, the acid with a mean of 50 ml and a standard deviation of 2 ml, and C_2H_5OH with a mean of 30 ml and a standard deviation of 3 ml. The three measurements are normally distributed and independent.

(a) What is the probability of a given student measuring out at least 103 ml of water?

(b) What is the probability of a given student measuring out between 148 and 157 ml of water plus acid?

(c) What is the probability of a given student measuring out a total of between 175 and 180 ml of liquid?

```
1 - normcdf(103, 100, 4)                         %0.2266
normcdf(157, 150, sqrt(4^2 + 2^2)) ...
   - normcdf(148, 150, sqrt(4^2 + 2^2))          %0.6139
normcdf(180, 180, sqrt(4^2 + 2^2 + 3^2 )) ...
   - normcdf(175, 180, sqrt(4^2 + 2^2 + 3^2))  %0.3234
```

Example 6.5. **Two Plate Assembly Simulation.** The following example is adapted from Banks et al. (1984). In assembly of two square 4×4 steel plates, comprising a part of a medical device, each plate has a hole drilled in its center. The plates are to be joined by a pin. Assembling machine adjusts the plates with respect to the lower left corner denoted as (0,0) in coordinate system xOy.

The coordinates of hole centers X_i and Y_i for ith plate ($i = 1,2$) are independent normally distributed random variables with mean 2 and standard deviation 0.001.

The hole diameters D_1 and D_2 are normally distributed with mean of 0.2 and standard deviation 0.0012, for both plates. The pin diameter, R, is also normally distributed with mean 0.195 and standard deviation of 0.0005.

(a) What proportion of pins will go through assembled plates? We will approximate this proportion by 1,000,000 simulated assembles using MATLAB. The clearance

$$\min\{D_1, D_2\} - \sqrt{(X_1 - X_2)^2 + (Y_1 - Y_2)^2} - R,$$

has to be positive for a successful assembly. Why is the second term needed?

(b) In an assembled pair of plates the pin will wobble if it is too loose. This wobbling will occur if

$$\min\{D_1, D_2\} - R \geq 0.006.$$

What fraction of assembled plates would not wobble? This is conditional probability, since we restrict attention on the assembled plates only. Thus in simulating this proportion we ignore the cases when the assembly was not possible.

The following MATLAB script estimates the desired proportions:

```
%Normal Probabilities by Simulation
rng(10,'twister')
M=1000000 ; %number of simulations
  clear = 0;
  clearnowobb=0;
for i = 1:M
    X1 = 2 + 0.001  * randn;
    Y1 = 2 + 0.001  * randn;
    X2 = 2 + 0.001  * randn;
    Y2 = 2 + 0.001  * randn;
    C=sqrt((X1-X2)^2 + (Y1-Y2)^2);
    D1=0.2+0.0012*randn;
    D2=0.2+0.0012*randn;
    D = min(D1, D2);
    R = 0.195 + 0.0005*randn;
    clear = clear + (D-C-R > 0);
    clearnowobb = clearnowobb + (D-C-R > 0)*(D-R<0.006);
end
  p1=clear/M               %(a) 0.9553
  p2 = clearnowobb/clear %(b) 0.9346
```

Thus, by simulation, we estimated that 95.53% of assembles are possible and that among the assembled plates 93.46% would not wobble.

✐

6.5 Central Limit Theorem

The central limit theorem (CLT) elevates the status of the normal distribution above other distributions. We have already seen that a linear combination of independent normals is a normal random variable itself. That is, if $X_1, \ldots, X_n \overset{\text{iid}}{\sim} \mathcal{N}(\mu, \sigma^2)$, then

$$\sum_{i=1}^{n} X_i \sim \mathcal{N}(n\mu, n\sigma^2), \quad \text{and} \quad \overline{X} = \frac{1}{n}\sum_{i=1}^{n} X_i \sim \mathcal{N}\left(\mu, \frac{\sigma^2}{n}\right).$$

The CLT states that $X_1, \ldots, X_n$ need not be normal in order for $\sum_{i=1}^{n} X_i$ or, equivalently, for $\overline{X}$ to be *approximately* normal. This approximation is quite good for n as low as 30. As we said, variables $X_1, X_2, \ldots, X_n$ need not be normal but must satisfy some conditions. For CLT to hold, it is sufficient for X_is to be independent, equally distributed, and have finite variances and, consequently, means. Other than that, the X_is can be arbitrary – skewed, discrete, etc. The conditions of i.i.d. and finiteness of variances are sufficient – more precise formulations of the CLT are beyond the scope of this text. Dasgupta (2008) provides comprehensive coverage.

CLT. Let $X_1, X_2, \ldots, X_n$ be i.i.d. random variables with a mean μ and finite variance σ^2. Then,

$$\sum_{i=1}^{n} X_i \overset{\text{approx}}{\sim} \mathcal{N}(n\mu, n\sigma^2) \quad \text{and} \quad \overline{X} = \frac{1}{n}\sum_{i=1}^{n} X_i \overset{\text{approx}}{\sim} \mathcal{N}\left(\mu, \frac{\sigma^2}{n}\right).$$

A special case of CLT involving Bernoulli random variables results in a normal approximation to binomials because the sum of many i.i.d. Bernoullis is at the same time exactly binomial and approximately normal. This approximation is handy when n is very large.

de Moivre (1738). Let $X_1, X_2, \ldots, X_n$ be independent Bernoulli $\mathcal{B}er(p)$ random variables with parameter p.
   Then,

$$Y = \sum_{i=1}^{n} X_i \overset{\text{approx}}{\sim} \mathcal{N}(np, npq)$$

and

$$\mathbb{P}(k_1 \leq Y \leq k_2) = \Phi\left(\frac{k_2 + 1/2 - np}{\sqrt{npq}}\right) - \Phi\left(\frac{k_1 - 1/2 - np}{\sqrt{npq}}\right),$$

where Φ is the CDF of standard normal random variable.

De Moivre's approximation is good if both np and nq exceed 10 and n exceeds 30. If that is not the case, a Poisson approximation to binomial (page 179) could be better.

The factors $1/2$ in de Moivre's formula are continuity corrections. For example, Y, which is discrete, is approximated with a continuous distribution. $\mathbb{P}(Y \leq k_2 + 1)$ and $\mathbb{P}(Y < k_2 + 1)$ are the same for a normal but not for a binomial distribution for which $\mathbb{P}(Y < k_2 + 1) = \mathbb{P}(Y \leq k_2)$. Likewise, $\mathbb{P}(Y \geq k_1 - 1)$ and $\mathbb{P}(Y > k_1 - 1)$ are the same for a normal but not for a binomial distribution for which $\mathbb{P}(Y > k_1 - 1) = \mathbb{P}(Y \geq k_1)$. Thus, $\mathbb{P}(k_1 \leq Y \leq k_2)$ for a binomial distribution is better approximated by $\mathbb{P}(k_1 - 1/2 \leq Y \leq k_2 + 1/2)$.

All approximations used to be much more important in the era before modern computing power was available. MATLAB is capable of calculating exact binomial probabilities for huge values of n, and for practical reasons de Moivre's approximation is obsolete. For example,

```
format long
binocdf(1999988765, 4000000000, 1/2)
%ans = 0.361195130797824
format short
```

However, the theoretical value of de Moivre's approximation is significant since many estimators and tests based on a binomial distribution can use well-developed normal distribution machinery for an analysis beyond the computation.

The following MATLAB program exemplifies the CLT by averages of simulated uniform random variables:

```
% Central Limit Theorem Demo
figure;
subplot(3,2,1)
hist(rand(1, 10000),40)          %histogram of 10000 uniforms
subplot(3,2,2)
hist(mean(rand(2, 10000)),40)    %histogtam of 10000
                                 %averages of 2 uniforms
subplot(3,2,3)
```

```
hist(mean(rand(3, 10000))),40)   %histogtam of 10000
                                 %averages of 3 uniforms
subplot(3,2,4)
hist(mean(rand(5, 10000))),40)   %histogtam of 10000
                                 %averages of 5 uniforms
subplot(3,2,5)
hist(mean(rand(10, 10000))),40)  %histogtam of 10000
                                 %averages of 10 uniforms
subplot(3,2,6)
hist(mean(rand(100, 10000))),40) %histogtam of 10000
                                 %averages of 100 uniforms
```

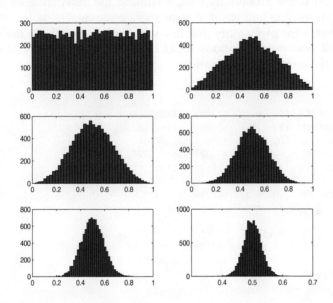

Fig. 6.6 Convergence to normal distribution shown via averages of 1, 2, 3, 5, 10, and 100 independent uniform (0,1) random variables.

Figure 6.6 shows the histograms of 10,000 simulations of averages of $k = 1, 2, 3, 5, 10$, and 100 uniform random variables. It is interesting to see the metamorphosis of a flat single uniform ($k = 1$), via a "witch hat distribution" ($k = 2$), into bell-shaped distributions close to the normal. For additional simulation experiments, see the script ◀ cltdemo.m.

Example 6.6. **Is Grandpa's Genetic Theory Valid?** The domestic cat's wild appearance is increasingly overshadowed by color mutations, such as black, white spotting, maltesing (diluting), red and tortoiseshell, shading, and Siamese pointing. By favoring the odd or unusually colored and marked cats over the "plain" tabby, people have consciously and unconsciously enhanced these color mutations over the course of domestication. Today, "colored" cats outnumber the wild looking tabby cats, and pure tabbies are becoming rare. Some may not be quite as taken by the coat of our domestic feline friends as Jason's grandpa is. He has a genetic theory that asserts that three-fourths of cats with more than three colors in their fur are female. A total of $n = 300$ three-color cats (TCCs) are observed and 86 are found to be male. If Jason's grandpa's genetic theory is true, then the number of male TCCs is binomial $B(300, 0.25)$, with an expectation of 75 and variance of $56.25 = 7.5^2$.

(a) What is the probability that, assuming Jason's grandpa's theory, one will observe 86 or more male cats? How does this finding support the theory?

(b) What is the probability that, assuming the independence of a cat's fur and gender, one will observe 86 or more male cats?

(c) What is the probability that one will observe exactly 75 male TCCs?

We will find exact solutions using binomial distribution and compare results with normal approximations.

```
format long  %for precise comparisons
%(a)
1 - binocdf(85, 300, 0.25) %0.08221654140000, exact
1 - normcdf(85, 75, 7.5)   %0.09121121972587
1 - normcdf(86, 75, 7.5)   %0.07123337741399
1 - normcdf(85.5, 75, 7.5) %0.08075665923377, approx
   %85.5 is taken as continuity-corrected argument
%(b)
1 - binocdf(85, 300, 0.5)  %0.99999999999998
 %virtually a sure event
%(c)
binopdf(75, 300, 0.25)     %0.05312831515720, exact
normcdf(75.5, 75, 7.5)-normcdf(74.5, 75, 7.5)
                           %0.05315292860073, approx
```

Example 6.7. **Avio Company.** The Avio Company sells 410 plane tickets for a 400-seater flight. Find the probability that the company overbooked the flight if a person who bought a ticket shows up at the gate with a probability of 0.96.

Each sold ticket can be thought of as an "experiment" where "success" means showing up at the gate for the flight. The number of people that show up X is binomial $Bin(410, 0.96)$. The following MATLAB script calculates the normal approximation:

```
410*0.96                         %393.6000
sqrt(410*0.96*0.04)              %3.9679
1-normcdf((400.5-393.6)/3.9679)  %0.0410
```

Notice that in this case the normal approximation is not very good since the exact binomial probability is 0.0329:

```
1-binocdf(400, 410, 0.96)          %0.0329
```

The reason is that the normal approximation works well when the probabilities are not close to 0 or 1, and here 0.96 is quite close to 1 for a given sample size of 410.

The Poisson approximation to the binomial performs better. The probability of missing the flight is $1 - 0.96 = 0.04$, and overbooking will happen if 9 or fewer passengers miss the flight:

```
%prob that 9 or less fail to show
poisscdf(9, 0.04*410)              %0.0355
```

6.6 Distributions Related to Normal

Four distributions – chi-square χ^2, t, F, and lognormal – are specially related to the normal distribution. This relationship is described in terms of functions of independent standard normal variables. Let $Z_1, Z_2, \ldots, Z_n$ be n independent standard normal (mean 0, variance 1) random variables. Then:

• The sum of squares $Z_1^2 + \cdots + Z_n^2$ is chi-square distributed with n degrees of freedom, χ_n^2:

$$\chi_n^2 \sim Z_1^2 + Z_2^2 + \cdots + Z_n^2.$$

• The ratio of a standard normal Z and the square root of an independent chi-square χ^2 random variable normalized by its number of degrees of freedom, has a t-distribution with n degrees of freedom, t_n:

$$t_n \sim \frac{Z}{\sqrt{\frac{\chi_n^2}{n}}}.$$

• The ratio of two independent chi-squares normalized by their respective numbers of degrees of freedom is distributed as an F:

$$F_{m,n} \sim \frac{\chi_m^2/m}{\chi_n^2/n}.$$

The degrees of freedom for F are m – *numerator df* and n – *denominator df*.

- As the name indicates, the lognormal ("log-is-normal") distribution is connected to a normal distribution via a logarithm function. If X has a lognormal disrtibution, then the distribution of $Y = \log X$ is normal.

A more detailed description of these four distributions follows next.

6.6.1 Chi-square Distribution

The probability density function for a chi-square random variable with parameter k, called the *degrees of freedom*, is

$$f_X(x) = \frac{(1/2)^{k/2} x^{k/2-1}}{\Gamma(k/2)} e^{-x/2}, \quad 0 \le x < \infty.$$

The chi-square distribution (χ^2) is a special case of the gamma distribution with parameters $r = k/2$ and $\lambda = 1/2$. Its mean and variance are $\mu = k$ and $\sigma^2 = 2k$, respectively.

If $Z \sim \mathcal{N}(0,1)$, then $Z^2 \sim \chi_1^2$, that is, a chi-square random variable with one degree of freedom. Furthermore, if $U \sim \chi_m^2$ and $V \sim \chi_n^2$ are independent, then $U + V \sim \chi_{m+n}^2$.

From these results it can be shown that if $X_1, \ldots, X_n \sim \mathcal{N}(\mu, \sigma^2)$ and $\overline{X}$ is the sample mean, then the *sample variance* $s^2 = \sum_i (X_i - \overline{X})^2 / (n-1)$ is proportional to a chi-square random variable with $n - 1$ degrees of freedom:

$$\frac{(n-1)s^2}{\sigma^2} \sim \chi_{n-1}^2. \tag{6.3}$$

This result was proved first by German geodesist Helmert (1876). The χ^2-distribution was previously defined by Abbe and Bienaymé in the mid-1800s.

The formal proof of (6.3) is beyond the scope of this text, but an intuition can be obtained by inspecting

$$\frac{(n-1)s^2}{\sigma^2} = \left(\frac{X_1 - \overline{X}}{\sigma}\right)^2 + \left(\frac{X_2 - \overline{X}}{\sigma}\right)^2 + \cdots + \left(\frac{X_n - \overline{X}}{\sigma}\right)^2$$

$$= (Y_1 - \overline{Y})^2 + (Y_2 - \overline{Y})^2 + \cdots + (Y_n - \overline{Y})^2,$$

where Y_i are independent normal $\mathcal{N}(\mu/\sigma, 1)$.

$$(Y_1 - \overline{Y})^2 + (Y_2 - \overline{Y})^2 = \left(\frac{Y_1 - Y_2}{\sqrt{2}}\right)^2 = Z_1^2, \quad \text{for } \overline{Y} = \frac{Y_1 + Y_2}{2},$$

$$(Y_1 - \overline{Y})^2 + (Y_2 - \overline{Y})^2 + (Y_3 - \overline{Y})^2 = \left(\frac{Y_1 - Y_2}{\sqrt{2}}\right)^2 + \left(\frac{Y_1 + Y_2 - 2Y_3}{\sqrt{6}}\right)^2 = Z_1^2 + Z_2^2,$$

$$\text{for } \overline{Y} = \frac{Y_1 + Y_2 + Y_3}{3},$$

etc.

Note that the right-hand sides are sums of squares of uncorrelated standard normal variables.

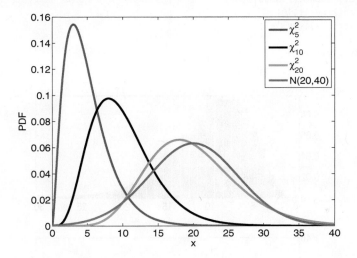

Fig. 6.7 χ^2-distribution with 5, 10, and 20 degrees of freedom. A normal $\mathcal{N}(20,40)$ distribution is superimposed to illustrate a good approximation to χ_n^2 by $\mathcal{N}(n,2n)$ for n large

In MATLAB, the CDF and PDF for a χ_k^2 are `chi2cdf(x,k)` and `chi2pdf(x,k)`, respectively. The pth quantile of the χ_k^2 distribution is `chi2inv(p,k)`.

Example 6.8. χ_{10}^2 **as a Sum of Squares of Ten Standard Normals.** In this example we demonstrate by simulation that the sum of squares of standard normal random variates follows the χ^2-distribution. In particular we compare $Z_1^2 + Z_2^2 + \cdots + Z_{10}^2$ with χ_{10}^2.

Figure 6.8, produced by the code in ◄ `nor2chi2.m`, shows a normalized histogram of the sums of squares of ten standard normals with a superimposed χ_{10}^2 density (above) and a Q–Q plot comparing the sorted generated sample with χ_{10}^2 quantiles (below). As expected, the simulated empirical distribution is very close to the theoretical chi-square distribution.

```
figure;
subplot(2,1,1)
  %form a matrix of standard normals 10 x 10000
  %square the entries, sum up columnwise, to
  % get a vector of 10000 chi2 with 10 df.
  histn(sum(normrnd(0,1,[10, 10000]).^2),0, 1,30)
    hold on
  plot((0.1:0.1:30), chi2pdf((0.1:0.1:30),10),'r-','LineWidth',2)
  axis tight
subplot(2,1,2)
  %check the Q-Q plot
  xx = sum(normrnd(0,1,[10, 10000]).^2);
  tt = 0.5/10000:1/10000:1;
  yy = chi2inv(tt,10);
  plot(sort(xx), yy,'*')
    hold on
  plot(yy, yy,'r-')
```

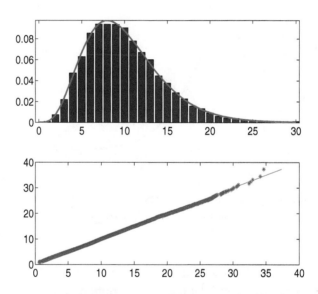

Fig. 6.8 Sum of 10 squared standard normals compared to χ^2_{10} distribution. Above: Normalized histogram with superimposed χ^2_{10} density (*red*); Below: Q–Q-plot of sorted sums against χ^2_{10} quantiles.

Example 6.9. **Targeting Meristem Cells.** A gene transfer system for meristem cells can be developed on the basis of a ballistic approach (Sautter, 1993). Instead of a macroprojectile, microtargeting uses the law of Bernoulli for acceleration of highly uniform-sized gold particles. The particle is aimed at an area as small as 150 μm in diameter, which corresponds to the size of a meristem. Suppose that a particle is fired at a meristem at the origin of a plane coordinate system, with units in microns. The particle lands at (X, Y), where X and Y are independent and each has a normal distribution with mean $\mu = 0$ and variance $\sigma^2 = 10^2$. The particle is successfully delivered if it lands within $\sqrt{738}\,\mu$m of the target (origin). What is the probability of this event? The particle is successfully delivered if $X^2 + Y^2 \leq 738$, or $(X/10)^2 + (Y/10)^2 \leq 7.38$. Since both $X/10$ and $Y/10$ have a standard normal distribution, random variable $(X/10)^2 + (Y/10)^2$ is χ^2_2-distributed. Since `chi2cdf(7.38,2)=0.975`, we conclude that the particle is successfully delivered with a probability of 0.975.

📎

A square root of chi-square random variable χ^2_k with k degrees of freedom is called chi (χ_k) random variable. The density of χ_k random variable X is

$$f_X(x) = \frac{2^{1-k/2} x^{k-1} e^{-x/2}}{\Gamma\left(\frac{k}{2}\right)}, \quad 0 \leq x < \infty.$$

The mean and variance of X are

$$\mathbb{E}X = \frac{\sqrt{2}\,\Gamma\left(\frac{k+1}{2}\right)}{\Gamma\left(\frac{k}{2}\right)} \quad \text{and} \quad \text{Var}\,X = k - (\mathbb{E}X)^2.$$

Special cases of χ-distribution are Rayleigh and Maxwell distributions. In their standard form (scale/rate = 1), these two distributions are χ_2 and χ_3 respectively. Absolute value of a standard normal random variable is χ_1 distributed.

A multivariate version of the χ^2-distribution is called a Wishart distribution. It is a distribution of random matrices that are symmetric and positive definite. As such, it is a proper model for normal covariance matrices, and we will see later its use in Bayesian inference involving bivariate normal distributions.

A $p \times p$ random matrix X has a Wishart distribution if its density is given by

$$f(X) = \frac{|X|^{(n-p-1)/2} \exp\{-\frac{1}{2} tr(\Sigma^{-1} X)\}}{2^{np/2} \pi^{p(p-1)/4} |\Sigma|^{n/2} \prod_{i=1}^{p} \Gamma\left(\frac{n+1-i}{2}\right)},$$

where Σ is the scale matrix and n is the number of degrees of freedom. Operator tr is the trace of a matrix, that is, the sum of its diagonal elements, and $|\Sigma|$ and $|X|$ are determinants of Σ and X, respectively.

For $p = 1$ and $\Sigma = 1$, the Wishart distribution is χ_n^2. In MATLAB, it is possible to simulate from the Wishart distribution as `wishrnd(Sigma,n)`. In WinBUGS, the Wishart distribution is coded as `dwish(R[,],n)`, where the precision matrix R is defined as Σ^{-1}.

6.6.2 *t-Distribution*

Random variable X has *t*-distribution with k degrees of freedom, $X \sim t_k$, if its PDF is

$$f_X(x) = \frac{\Gamma\left(\frac{k+1}{2}\right)}{\sqrt{k\pi}\,\Gamma(k/2)} \left(1 + \frac{x^2}{k}\right)^{-\frac{k+1}{2}}, \quad -\infty < x < \infty.$$

The *t*-distribution is similar in shape to the standard normal distribution except for having fatter tails. If $X \sim t_k$, then $\mathbb{E}X = 0$, $k > 1$ and $\mathbb{V}\mathrm{ar}\,X = k/(k-2)$, $k > 2$. For $k = 1$, the *t*-distribution coincides with the Cauchy distribution.

The *t*-distribution has an important role to play in statistical inference. With a set of i.i.d. $X_1, \ldots, X_n \sim \mathcal{N}(\mu, \sigma^2)$, we can standardize the sample mean using the simple transformation of $Z = (\overline{X} - \mu)/\sigma_{\overline{X}} = \sqrt{n}(\overline{X} - \mu)/\sigma$. However, if the variance is unknown, by using the same transformation, except for substituting the sample standard deviation s for σ, we arrive at a *t*-distribution with $n - 1$ degrees of freedom:

$$t = \frac{\overline{X} - \mu}{s/\sqrt{n}} \sim t_{n-1}.$$

More technically, if $Z \sim \mathcal{N}(0,1)$ and $Y \sim \chi_k^2$ are independent, then $t = Z/\sqrt{Y/k} \sim t_k$. In MATLAB, the CDF at x for a *t*-distribution with k degrees of freedom is calculated as `tcdf(x,k)`, and the PDF is computed as `tpdf(x,k)`. The pth percentile is computed with `tinv(p,k)`. In WinBUGS, the *t*-distribution is coded as `dt(mu,tau,k)`, where `tau` is a precision parameter and `k` is the number of degrees of freedom.

The *t*-distribution was originally found by German mathematician and astronomer Jacob Lüroth in 1876 (Lüroth, 1876). William Sealy Gosset rediscovered the *t*-distribution in 1908 and published the results under the pen name "Student."

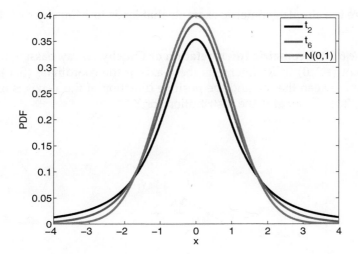

Fig. 6.9 t-distribution with 2 and 6 degrees of freedom. A standard normal distribution is superimposed as the solid red line.

6.6.3 Cauchy Distribution

The Cauchy distribution is a special case of the t-distribution; it is symmetric and bell-shaped like the normal distribution, but with much fatter tails. In fact, it is a popular distribution to use in nonparametric robust procedures and simulations because the distribution is so spread out; it has no mean and variance (none of the Cauchy moments exist). Physicists know this distribution as the *Lorentz distribution*. If $X \sim Ca(a,b)$, then X has a density

$$f_X(x) = \frac{1}{\pi} \frac{b}{b^2 + (x-a)^2}, \quad -\infty < x < \infty.$$

The standard Cauchy $Ca(0,1)$ distribution coincides with the t-distribution with 1 degree of freedom.

The Cauchy distribution is also related to the normal distribution. If Z_1 and Z_2 are two independent $\mathcal{N}(0,1)$ random variables, then their ratio $C = Z_1/Z_2$ is Cauchy, $Ca(0,1)$. Finally, if $C_i \sim Ca(a_i,b_i)$ for $i = 1, \ldots, n$, then $S_n = C_1 + \cdots + C_n$ is Cauchy distributed with parameters $a_S = \sum_i a_i$ and $b_S = \sum_i b_i$. The consequence of this additivity is interesting. If one observes n Cauchy $Ca(0,1)$ random variables $X_i, i = 1, \ldots, n$, and takes the average $\overline{X}$, the average is also Cauchy $Ca(0,1)$. This means that for Cauchy CLT does not hold; a single measurement is as precise as the average of any finite number of measurements.

Here is a simple geometric example that leads to a Cauchy distribution:

Example 6.10. **Geometric Interpretation of Cauchy.** A ray passing through the point $(-1,0)$ in R^2 intersects the y-axis at the coordinate $(0,Y)$. If the angle α between the ray and the positive direction of the x-axis is uniform $\mathcal{U}(-\pi/2, \pi/2)$, what is the distribution for Y?

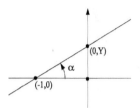

Fig. 6.10 If the angle α between the ray and x-axis is uniform $\mathcal{U}(-\pi/2, \pi/2)$, Y is Cauchy $\mathcal{C}a(0,1)$.

Here $Y = \tan\alpha$, $\alpha = h(Y) = \arctan(Y)$ and $h'(y) = \frac{1}{1+y^2}$. The density for uniform $\mathcal{U}(-\pi/2, \pi/2)$ is constant $1/\pi$ if $\alpha \in (-\pi/2, \pi/2)$, and 0 else. From (5.15),

$$f_Y(y) = \frac{1}{\pi}|h'(y)| = \frac{1}{\pi}\frac{1}{1+y^2},$$

which is the density of the Cauchy $\mathcal{C}a(0,1)$ distribution.
✐

6.6.4 F-Distribution

Random variable X has an F-distribution with m and n degrees of freedom, denoted as $F_{m,n}$, if its density is given by

$$f_X(x) = \frac{m^{m/2}n^{n/2}}{B(m/2,n/2)}\, x^{m/2-1}(n+mx)^{-(m+n)/2}, \quad x > 0.$$

The CDF of an F-distribution is not of closed form, but it can be expressed in terms of an incomplete beta function (page 206) as

$$F(x) = 1 - I_v(n/2, m/2), \quad v = n/(n+mx), \quad x > 0.$$

The mean is given by $\mathbb{E}X = n/(n-2), n > 2$, and the variance by $\mathbb{V}\text{ar }X = \frac{2n^2(m+n-2)}{m(n-2)^2(n-4)}, n > 4$.

If $X \sim \chi_m^2$ and $Y \sim \chi_n^2$ are independent, then $(X/m)/(Y/n) \sim F_{m,n}$. Because of this representation, m and n are often called, respectively, the *numerator* and *denominator* degrees of freedom. F and beta distributions are related. If $X \sim \mathcal{B}e(a,b)$, then $bX/[a(1-X)] \sim F_{2a,2b}$. Also, if $X \sim F_{m,n}$, then $mX/(n+mX) \sim \mathcal{B}e(m/2,n/2)$.

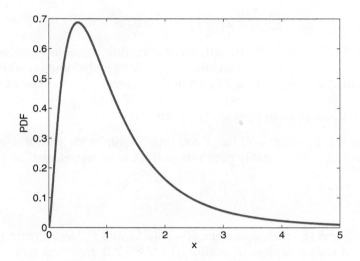

Fig. 6.11 $F_{5,10}$ PDF. `t2 = 0:0.005:5; plot(t2, fpdf(t2, 5, 10))`

The F-distribution is one of the most important distributions for statistical inference; in introductory statistical courses, the test for equality of variances, ANOVA, and multivariate regression are based on the F-distribution. For example, if s_1^2 and s_2^2 are sample variances of two independent normal samples with variances σ_1^2 and σ_2^2 and sizes m and n respectively, the ratio $\frac{s_1^2/\sigma_1^2}{s_2^2/\sigma_2^2}$ is distributed as $F_{m-1,n-1}$. The F-distribution is named after Sir Ronald Fisher, who in fact tabulated not F but $z = \frac{1}{2}\log F$. The F-distribution in its current form was first tabulated and used by George W. Snedecor, and the distribution is sometimes called Snedecor's F, or the Fisher–Snedecor F.

In MATLAB, the CDF at x for an F-distribution with m, n degrees of freedom is calculated as `fcdf(x,m,n)`, and the PDF is computed as `fpdf(x,m,n)`. The pth percentile is computed with `finv(p,m,n)`. Figure 6.11 provides a plot of a $F_{5,10}$ PDF.

6.6.5 *Noncentral χ^2, t, and F Distributions*

Noncentral χ^2, t, and F distributions are generalizations of standard χ^2, t, and F distributions. They are used mainly in power analysis of tests and sample size designs. For example, we will use noncentral t for power analysis of one-sample and two-sample t tests later in the text.

Random variable $\chi_n^2(\delta)$ has a *noncentral χ^2-distribution* with n degrees of freedom and parameter of noncentrality δ if it can be represented as

$$\chi_n^2(\delta) = Z_1 + Z_2 + \cdots + Z_{n-1} + X_n,$$

where $Z_1, Z_2, \ldots Z_{n-1}, X_n$ are independent random variables. Random variables $Z_1, \ldots, Z_{n-1}$ have a standard normal $\mathcal{N}(0,1)$ distribution while X_n is distributed as $\mathcal{N}(\delta,1)$. In MATLAB the noncentral χ^2 is denoted as ncx2pdf, ncx2cdf, ncx2inv, ncx2stat, and ncx2rnd for PDF, CDF, quantile, descriptive statistics, and random number generator.

Random variable $t_n(\delta)$ has a *noncentral t-distribution* with n degrees of freedom and noncentrality parameter δ if it can be represented as

$$t_n(\delta) = \frac{X}{\sqrt{\chi_n^2/n}},$$

where X and χ_n^2 are independent, $X \sim \mathcal{N}(\delta,1)$, and χ_n^2 has a (central) χ^2 distribution with n degrees of freedom. In MATLAB, functions nctpdf, nctcdf, nctinv, nctstat, and nctrnd, stand for PDF, CDF, quantile, descriptive statistics, and random number generator of the noncentral t.

Figure 6.12 plots the densities of noncentral t for values of the noncentrality parameter $-1, 0$, and 2. Noncentral t for $\delta = 0$ is a standard t-distribution.

Random variable $F_{m,n}(\delta)$ has a *noncentral F-distribution* with m, n degrees of freedom and parameter of noncentrality δ if it can be represented as

$$F_{m,n}(\delta) = \frac{\chi_m^2(\delta)/m}{\chi_n^2/n},$$

where $\chi_m^2(\delta)$ and χ_n^2 are independent, with noncentral (δ) and standard χ^2 distributions with m and n degrees of freedom, respectively. In MATLAB, functions ncfpdf, ncfcdf, ncfinv, ncfstat, and ncfrnd, stand for the PDF, CDF, quantile, descriptive statistics, and random number generator of the noncentral F.

The noncentral F will be used in Chapter 11 for power calculations in several ANOVA designs.

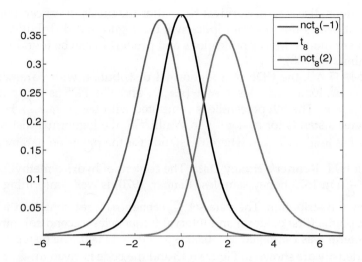

Fig. 6.12 Densities of noncentral $t_8(\delta)$ distribution for $\delta = -1, 0$, and 2.

6.6.6 Lognormal Distribution

A random variable X has a lognormal distribution with parameters μ and σ^2, $X \sim \mathcal{LN}(\mu, \sigma^2)$, if its density function is given by

$$f(x) = \frac{1}{x\sqrt{2\pi}\sigma} \exp\left\{-\frac{(\log x - \mu)^2}{2\sigma^2}\right\}, \quad x > 0.$$

If Y has a normal distribution, then $X = e^Y$ is lognormal.

Parameter μ is the mean and σ is the standard deviation of the distribution for the normal random variable $Y = \log X$, not the lognormal random variable X, and this can sometimes be confusing.

The moments of the lognormal distribution can be computed from the moment-generating function of the normal distribution. The nth moment is $\mathbb{E}(X^n) = \exp\{n\mu + n^2\sigma^2/2\}$, from which the mean and variance of X are

$$\mathbb{E}(X) = \exp\{\mu + \sigma^2/2\}, \quad \text{and} \quad \mathbb{V}\text{ar}(X) = \exp\{2(\mu + \sigma^2)\} - \exp\{2\mu + \sigma^2\}.$$

The median is $\exp\{\mu\}$ and the mode is $\exp\{\mu - \sigma^2\}$.

The lognormality is preserved under multiplication and division, i.e., the products and quotients of lognormal random variables remain lognormally distributed. If $X_i \sim \mathcal{LN}(\mu_i, \sigma_i^2)$, then $\prod_{i=1}^{n} X_i \sim \mathcal{LN}(\sum_{i=1}^{n} \mu_i, \sum_{i=1}^{n} \sigma_i^2)$.

Several biomedical phenomena are well modeled by a lognormal distribution, such as the age at onset of Alzheimer's disease, latent periods of infectious diseases, or survival time after diagnosis of cancer. For measurement errors that are multiplicative, the lognormal distribution is the convenient model. More applications and properties can be found in Crow and Shimizu (1988).

In MATLAB, the CDF of a lognormal distribution with parameters m and s is evaluated at x as `logncdf(x,m,s)`, and the PDF is computed as `lognpdf(x,m,s)`. The pth percentile is computed with `logninv(p,m,s)`. Here the parameter s stands for σ, not σ^2. In WinBUGS, the lognormal distribution is coded as `dlnorm(mu,tau)`, where `tau` stands for the precision parameter $\frac{1}{\sigma^2}$.

Example 6.11. **Renner's Honey Data.** The content of hydroxymethylfurfurol (HMF, $\frac{mg}{kg}$) in 1573 honey samples (Renner, 1970) is well conforming to the lognormal distribution. The data set 🖳 `renner.mat|dat` contains the interval midpoints (first column) and interval frequencies (second column). The parameter μ was estimated as -0.6084 and σ as 1.0040. The histogram and fitting density are shown in Figure 6.13 and the code is given in ◢ `renner.m`.

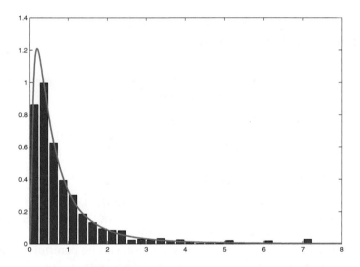

Fig. 6.13 Normalized histogram of Renner's honey data and lognormal distribution with parameters $\mu = -0.6083$ and $\sigma^2 = 1.0040^2$ that fits data well.

The goodness of such fitting procedures will be discussed in Chapter 17 more formally. Note that μ and σ are the mean and standard deviation of the logarithms of observations, not the observations themselves.

◢
```
load 'renner.dat'
% mid-intervals, int. length = 0.25
rennerx = renner(:,1);
% frequencies in the interval
```

```
rennerf = renner(:,2);
n = sum(renner(:,2)); % sample size (n=1573)
bar(rennerx,    rennerf./(0.25 * n))
hold on
    m = sum(log(rennerx) .* rennerf)/n    %m =-0.6083
    s = sqrt(   sum( rennerf .*(log(rennerx) - m).^2 )/n    )
    %s=1.0040
xx = 0:0.01:8;
yy = lognpdf(xx, m, s);
plot(xx, yy, 'r-','linewidth',2)
```

6.7 Delta Method and Variance-Stabilizing Transformations*

The CLT states that for independent identically distributed random variables $X_1, \ldots, X_n$ with mean μ and finite variance σ^2,

$$\sqrt{n}(\overline{X} - \mu) \overset{approx}{\sim} \mathcal{N}(0, \sigma^2),$$

where the symbol $\overset{approx}{\sim}$ means *distributed approximately as*. Other than for a finite variance, there are no restrictions on the type, distribution, or any other feature of random variables X_i.

For a function g,

$$\sqrt{n}\left(g(\overline{X}) - g(\mu)\right) \overset{approx}{\sim} \mathcal{N}(0, g'(\mu)^2 \sigma^2).$$

The only restriction on g is that the derivative evaluated at μ must be finite and nonzero.

This result is called the *delta method* and the proof, which uses a simple Taylor expansion argument, will be omitted since it also uses facts concerning the convergence of random variables not covered in the text.

Example 6.12. Reciprocal and Square of Sample Mean. For n large

$$1/\overline{X} \overset{approx}{\sim} \mathcal{N}\left(\frac{1}{\mu}, \frac{\sigma^2}{\mu^4}\right),$$

$$(\overline{X})^2 \overset{approx}{\sim} \mathcal{N}\left(\mu^2, 4\mu^2\sigma^2\right).$$

The delta method is useful for many asymptotic arguments. Now we focus on the selection of the transformation g that stabilizes the variance.

Important statistical methodologies often assume that observations have variances that are constant for all possible values of the mean. Observations coming from a normal $\mathcal{N}(\mu,\sigma^2)$ distribution would satisfy this requirement since σ^2 does not depend on the mean μ. However, constancy of variances with respect to the mean is rather an exception than the rule. For example, if random variates from the exponential $\mathcal{E}(\lambda)$ distribution are generated, then the variance $\sigma^2 = 1/\lambda^2$ depends on the mean $\mu = 1/\lambda$, as $\sigma^2 = \mu^2$.

For some important distributions we will find a transformation that will make the variance constant and thus uninfluenced by the mean. This will prove beneficial for a range of inferential statistical procedures covered later in the text (confidence intervals, testing hypotheses).

Suppose that the variance $\mathbb{V}\text{ar}\ X = \sigma_X^2(\mu)$ can be expressed as a function of the mean $\mu = \mathbb{E}X$. For $Y = g(X)$, $\mathbb{V}\text{ar}\ Y \approx [g'(\mu)]^2\sigma_X^2(\mu)$, see (5.18). The condition that the variance of Y is constant leads to a simple differential equation

$$[g'(\mu)]^2\sigma_X^2(\mu) = c^2$$

with the following solution:

$$g(x) = c\int \frac{dx}{\sigma_X(x)}\,dx. \qquad (6.4)$$

This is the theoretical basis for many proposed variance-stabilizing transformations. Note that $\sigma_X(x)$ in (6.4) is a function expressing the variance as a function of the mean.

Example 6.13. **Stabilizing Variance.** Suppose data are sampled from (a) Poisson $\mathcal{P}oi(\lambda)$, (b) exponential $\mathcal{E}(\lambda)$, and (c) binomial $\mathcal{B}in(n,p)$ distributions.

In (a), the mean and variance are equal, $\sigma^2(\mu) = \mu\ (= \lambda)$, and (6.4) becomes

$$g(x) = c\int \frac{dx}{\sqrt{x}}\,dx = 2c\sqrt{x} + d$$

for some constants c and d. Thus, as the variance-stabilizing transformation for Poisson observations we can take $g(x) = \sqrt{x}$.

In (b) and (c), $\sigma^2(\mu) = \mu^2$ and $\sigma^2(\mu) = \mu - \mu^2/n$, and, after solving the integral in (6.4), we find that the transformations are $g(x) = \log(x)$ and $g(x) = \arcsin\sqrt{x/n}$ (Exercise 6.19).

Example 6.14. **Box–Cox Transformation.** Box and Cox (1964) introduced a family of transformations, indexed by a parameter λ, applicable to positive data $X_1, \ldots, X_n$:

$$Y_i = \begin{cases} \frac{X_i^{\lambda}-1}{\lambda}, & \lambda \neq 0 \\ \log X_i, & \lambda = 0. \end{cases} \tag{6.5}$$

This transformation is mostly applied to responses in linear models exhibiting nonnormality or heterogeneity of variances (heteroscedasticity). For a properly selected λ, transformed data $Y_1, \ldots, Y_n$ may look "more normal" and amenable to standard modeling techniques. The parameter λ is selected by maximizing,

$$(\lambda - 1) \sum_{i=1}^{n} \log X_i - \frac{n}{2} \log \left[\frac{1}{n} \sum_{i=1}^{n} (Y_i - \overline{Y})^2 \right], \tag{6.6}$$

where Y_i are as given in (6.5) and $\overline{Y} = \frac{1}{n} \sum_{i=1}^{n} Y_i$. As an illustration, we apply the Box–Cox transformation to apparently skewed data of pyruvate kinase concentrations.

Exercise 2.19 featured a multivariate data set ▥ dmd.dat in which the fourth column gives pyruvate kinase concentrations in 194 female relatives of boys with Duchenne muscular dystrophy (DMD). The distribution of this measurement is skewed to the right (Fig. 6.14a). We will find the Box–Cox transformation to symmetrize the data (make it approximately normal). Panel (b) gives the values of likelihood in (6.6) for different values of λ. Note that (6.6) is maximized for λ approximately equal to –0.15. Figure 6.14c gives the histogram for data transformed by the Box–Cox transformation with $\lambda = -0.15$. The histogram is notably symmetrized. For details see ◀ boxcox.m.

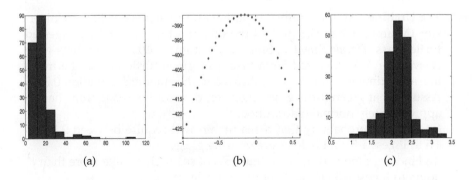

(a)   (b)   (c)

Fig. 6.14 (a) Histogram of row data of pyruvate kinase concentrations; (b) log-likelihood is maximized at $\lambda = -0.15$; and (c) histogram of Box–Cox-transformed data.

6.8 Exercises

6.1. **Standard Normal Calculations.** Random variable X has a standard normal distribution. What is larger, $\mathbb{P}(|X| \leq 0.7)$ or $\mathbb{P}(|X| \geq 0.7)$?

6.2. **Nonnegative Definiteness of Σ Constrains ρ.** A symmetric 2×2-matrix $A = [a\ b; b\ d]$ is nonnegative definite if $a \geq 0$ and $\det(A) = ad - b^2 \geq 0$. Show that condition $\det(\Sigma) \geq 0$ for Σ in (6.2), implies $-1 \leq \rho \leq 1$.

6.3. **Herrings.** The alewife (*Pomolobus pseudoharengus*, Wilson 1811) grows to maximum length of about 15 in., but adults average only about 10.5 in. long and about 8 oz. in weight; 16,400,000 fish taken in New England in 1898 weighed about 8,800,000 lbs.

Fig. 6.15 Alewife fish.

Assume that the length of an individual fish (Fig. 6.15) is normally distributed with mean 10.5 in. and standard deviation 1.6 in. and that the weight is distributed as χ^2 with 8 degrees of freedom.
(a) What percentage of fish are between 10.5 and 13 in. long?
(b) What percentage of fish weigh more than 10 oz.?
(c) Ten percent of fish are longer than x. Find x.

6.4. **Sea Urchins.** In a laboratory experiment, researchers at Barry University, (Miami Shores, FL) studied the rate at which sea urchins ingested turtle grass (*Florida Scientist*, Summer/Autumn 1991). The urchins were starved for 48 h, then fed 5-cm blades of green turtle grass. The mean ingestion time was found to be 2.83 h and the standard deviation 0.79 h. Assume that green turtle grass ingestion time for the sea urchins has an approximately normal distribution.
(a) Find the probability that a sea urchin will require between 2.3 and 4 h to ingest a 5-cm blade of green turtle grass.
(b) Find the time t^* (hours) so that 95% of sea urchins take more than t^* hours to ingest a 5-cm blade of green turtle grass.

6.5. **Pyruvate Kinase for Controls Is Normal.** Refer to Exercise 2.19. The histogram for PK response for controls, X, is fairly bell-shaped (as much as 142 observations show), so you decided to fit it with a normal distribution, $\mathcal{N}(12, 4^2)$.

(a) How would you defend the choice of a normal model that allows for negative values when the measured level is always positive?

(b) Find the probability that X falls between 4 and 20.

(c) Find the probability that X exceeds 20.

(d) Find the value x_0 so that 93% of all PK measurements exceed x_0.

6.6. **Leptin.** Leptin (from the Greek word *leptos*, meaning thin) is a 16-kDa hormone that plays a key role in regulating energy intake and energy expenditure, including the regulation (decrease) of appetite and (increase) of metabolism. Serum leptin concentrations can be measured in several ways. One approach is by using a radioimmunoassay in venous blood samples (Linco Research Inc., St Charles, MO). Several studies have consistently found women to have higher serum leptin concentrations than do men. For example, among US adults across a broad age range, the mean serum leptin concentration in women is approximately normal $\mathcal{N}(12.7 \ \mu g/L, (1.3 \ \mu g/L)^2)$ and in men approximately normal $\mathcal{N}(4.6 \ \mu g/L, (0.5 \ \mu g/L)^2)$.

(a) What is the probability that the concentration of leptin in a randomly selected US adult male exceeds 6 $\mu g/L$?

(b) What proportion of US women have concentration of leptin in the interval $12.7 \pm 2 \ \mu g/L$?

(c) What interval, symmetric about the mean 12.7 $\mu g/L$, contains leptin concentrations of 95% of adult US women?

6.7. **Pulse Rate.** The pulse rate of 1-month-old infants has a mean of 115 beats per minute and a standard deviation of 16 beats per minute.

(a) Explain why the average pulse rate in a sample of 64 1-month-old infants is approximately normally distributed.

(b) Find the mean and the variance of the normal distribution in (a).

(c) Find the probability that the average pulse rate of a sample of 64 will exceed 120.

6.8. **Side Effects.** One of the side effects of flooding a lake in northern boreal forest areas[1] (e.g., for a hydroelectric project) is that mercury is leached from the soil, enters the food chain, and eventually contaminates the fish. The concentration of mercury in fish will vary among individual fish because of differences in eating patterns, movements around the lake, etc. Suppose that the concentrations of mercury in individual fish follows an approximately normal distribution with a mean of 0.25 ppm and a standard deviation of 0.08 ppm. Fish are safe to eat if the mercury level is below 0.30 ppm. What proportion of fish are safe to eat?

---

[1] The northern boreal forest, sometimes also called the taiga or northern coniferous forest, stretches unbroken from eastern Canada westward throughout the majority of Canada to the central region of Alaska.

6.9. **Macrolepiota Procera.** The size of mushroom caps varies. While many species of *Marasmius* and *Collybia* are only 12 to 20 mm (1/2 to 3/4 in.) in diameter, some fungi are nearly 200 mm (8 in.) across. The cap diameter of parasol mushroom (*Macrolepiota procera*, Fig. 6.16) is a normal random variable with parameters $\mu = 230$ mm and $\sigma = 25$ mm.

Fig. 6.16 Parasol mushroom *Macrolepiota procéra.*

(a) What proportion of parasol caps has a diameter between 200 and 250 mm?

(b) Five percent of parasol caps are larger than x_0 in diameter. Find x_0.

6.10. **Duration of Gestation in Humans.** Altman (1980) quotes the following incident from the UK: "In 1949 a divorce case was heard in which the sole evidence of adultery was that a baby was born 349 days after the husband had gone abroad on military service. The appeal judges agreed that medical evidence was unlikely but scientifically possible." So the appeal failed. "Most people think that the husband was hard done by," Altman adds.

So let us judge the judges. The reported mean duration of an uncomplicated human gestation is between 266 and 288 days, depending on many factors but mainly on the method of calculation. Assume that population mean and standard deviations are $\mu = 280$ and $\sigma = 10$ days, respectively. In fact, smaller standard deviations have been reported, so 10 days is a conservative choice. The normal model fits the data reasonably well if the samples are large.

Under the normal $\mathcal{N}(\mu, \sigma^2)$ model, find the probability that a gestation period will be equal to or greater than 349 days.

6.11. **Tolerance Design.** Eggert (2005) provides the following engineering design question. A 5-in. diameter pin will be assembled into a 5.005-in. journal bearing. The pin manufacturing tolerance is specified to $t_{pin} = 0.003$ inch. A minimum clearance fit of 0.001 in. is needed.

Determine tolerance required of the hole, t_{hole}, such that 99.9% of the mates will exceed the minimum clearance. Assume that manufacturing variations are normally distributed. The tolerance is defined as 3 standard deviations.

6.12. **Ulnar Variance.** The lower arm is made up of two bones – the ulna and the radius. The length of these bones can lead to an ulnar variance, which can cause wrist pain, degenerative ailments, improper hand and wrist functioning.

This exercise uses data reported in Jung et al. (2001), who studied radiographs of the wrists of 120 healthy volunteers in order to determine the normal range of ulnar variance. The radiographs had been taken in various positions under both unloaded (static) and loaded (dynamic) conditions.

The ulnar variance in neutral rotation was modeled by normal distribution with a mean of $\mu = 0.74$ mm and standard deviation of $\sigma = 1.46$ mm. (a) What is the probability that a radiogram of a normal person will show negative ulnar variance in neutral rotation (ulnar variance, unlike the statistical variance, can be negative)?

The researchers modeled the maximum ulnar variance (UV_{max}) as normal $\mathcal{N}(1.52, 1.56^2)$ when gripping in pronation and minimum ulnar variance (UV_{min}) as normal $\mathcal{N}(0.19, 1.43^2)$ when relaxed in supination. (b) Find the probability that the mean dynamic range in ulnar variance, $C = UV_{max} - UV_{min}$, will exceed 1 mm.

6.13. **Independence of Sample Mean and Standard Deviation in Normal Samples.** Simulate 1000 samples from the standard normal distribution, each of size 100, and find their sample mean and standard deviation. (a) Plot a scatterplot of sample means vs. the corresponding sample standard deviations. Are there any trends? (b) Find the coefficient of correlation between sample means and standard deviations from (a) arranged as two vectors. Is the coefficient close to zero?

6.14. **Sonny and Multiple Choice Exam.** An instructor gives a 100-question multiple-choice final exam. Each question has 4 choices. In order to pass, a student has to have at least 35 correct answers. Sonny decides to guess at random on each question. What is the probability that Sonny will pass the exam?

6.15. **Amount of Liquid in a Bottle.** Suppose that the volume of liquid in a bottle of a certain chemical solution is normally distributed with a mean of 0.5 L and standard deviation of 0.01 L. (a) Find the probability that a bottle will contain at least 0.48 L of liquid. (b) Find the volume that corresponds to the 95th percentile.

6.16. **Marginals and Conditionals of a 2D Normal.** Find marginal and conditional densities $f_X(x)$, $f_Y(y)$, $f(x|y)$ and $f(y|x)$, if (X, Y) has density

$$f(x,y) = \frac{3\sqrt{3}}{\pi} \exp\left\{-4x^2 - 6xy - 9y^2\right\}.$$

6.17. **Meristem Cells in 3D.** Suppose that a particle is fired at a cell sitting at the origin of a spatial coordinate system, with units in microns. The particle lands at (X, Y, Z), where X, Y, and Z are independent, and each has a normal distribution with a mean of $\mu = 0$ and variance of $\sigma^2 = 250$. The particle is successfully delivered if it lands within $70\,\mu$m of the origin. Find the probability that the particle was not successfully delivered.

6.18. **Glossina morsitans.** *Glossina morsitans* (tsetse fly) is a large biting fly that inhabits most of midcontinental Africa. This fly is infamous as the primary biological vector (the meaning of vector here is epidemiological, not mathematical. A vector is any living carrier that transmits an infectious agent) of trypanosomes, which cause human sleeping sickness. The data in the table below are reported in Pearson (1914) and represent the frequencies of length in microns of trypanosomes found in *Glossina morsitans*.

Microns	15	16	17	18	19	20	21	22	23	24	25
Frequency	7	31	148	230	326	252	237	184	143	115	130

Microns	26	27	28	29	30	31	32	33	34	35	Total
Frequency	110	127	133	113	96	54	44	11	7	2	2500

The original data distinguished five different strains of trypanosomes, but it seems that the summary data set, as shown in the table, can be well approximated by a mixture of two normal distributions, $p_1 \mathcal{N}(\mu_1, \sigma_1^2) + p_2 \mathcal{N}(\mu_2, \sigma_2^2)$.

Using MATLAB's gmdistribution.fit identify the means of the two normal components, as well as their weights in the mixture, p_1 and p_2. Plot the normalized histogram and superimpose the density of the mixture.

Data can be found in ▣ glossina.mat.

6.19. **Stabilizing the Variance.** In Example 6.13 it was stated that the variance stabilizing transformations for exponential $\mathcal{E}(\lambda)$ and binomial $\mathcal{B}in(n, p)$ distributions are $g(x) = \log(x)$ and $g(x) = \arcsin\sqrt{\frac{x}{n}}$, respectively. Prove these statements.

6.20. **From Normal to Lognormal.** Derive the density of a lognormal distribution by transforming $X \sim \mathcal{N}(0,1)$ into $Y = \exp\{X\}$.

6.21. **Changing the Threshold for FPG.** Woolf and Rothmich (1998) report that a change of the diagnostic threshold for fasting plasma glucose (FPG) from 140 to 126 mg per dL, drastically increased the number of people diagnosed as diabetics:

Lowering the diagnostic threshold shifts the definition of diabetes into the central bulge of the distribution curve where the glucose level of most Americans falls. Among U.S. adults 40 to 74 years of age who have not been diagnosed with diabetes, 1.9 million have FPG levels of 126 to 140 mg per dL, which is almost as many as the number of people who have levels over 140 mg per dL. Under the new guidelines (ADA 1997), many Americans with FPG levels of 126 to 140 mg per dL, who previously would have been told that they had normal (or impaired) glucose tolerance, will now be informed that they harbor a disease.

Assume that the FPG of a randomly selected adult of age 40 to 74 from the US state of Georgia, can be modeled as lognormal $\mathcal{LN}(\mu, \sigma^2)$, where $\mu = 4.46$ and $\sigma^2 = 0.22^2$.
(a) Estimate how many people will fall in the range 126–140 if the population of adults of age 40 to 74 in Georgia is approximately 4 million.
(b) Find the FPG* level so that 95% of the population falls below FPG*.
(c) The lognormal model is not symmetric (lognormal distribution is positively skewed), so the mean is larger than the median. Find the median. In one sentence explain what this median represents in the terms of FPG.
Hint: In (a) you need first to estimate proportion of the population in 126–140 FPG range. MATLAB parametrizes lognormal distributions with μ and σ. Be careful about the mean and variance of FPG. They are **not** $\mu = 4.46$ and $\sigma^2 = 0.22^2$.

6.22. **The Square of a Standard Normal.** If $X \sim \mathcal{N}(0,1)$, show that $Y = X^2$ has a density of

$$f_Y(y) = \frac{1}{\sqrt{2}\,\Gamma\left(\frac{1}{2}\right)} y^{1/2 - 1} e^{-y/2}, \quad y \geq 0,$$

which is χ^2 with 1 degree of freedom.

MATLAB AND WINBUGS FILES AND DATA SETS USED IN THIS CHAPTER
http://statbook.gatech.edu/Ch6.Norm/

      acid.m, aviocompany.m, boxcox.m, ch2itf.m, cltdemo.m, glossina.m,
histn.m, ige.m, meanvarind.m, nor2chi2.m, piston.m, plot2dnormal.m,
plotnct.m, quetelet.m, renner.m, simulplates.m, tsetse.m

aplysia.odc

glossina.mat, renner.dat|mat

CHAPTER REFERENCES

Altman, D. G. (1980). Statistics and ethics in medical research: misuse of statistics is unethical. *Br. Med. J.*, **281**, 1182–1184.

Banks, J., Carson, J. S. II, and Nelson, B. (1984). *Discrete Event System Simulation*, 2nd ed., Prentice Hall, Upper Saddle River, NJ.

Casella, G. and Berger, R. (2002). *Statistical Inference*. Duxbury Press, Belmont, CA.

Crow E. L. and Shimizu K., eds. (1988). *Lognormal Distributions: Theory and Application*. Dekker, New York.

DasGupta, A. (2008). *Asymptotic Theory of Statistics and Probability*. Springer Texts in Statistics, Springer, New York.

Eggert, R. J. (2005). *Engineering Design*. Pearson Prentice Hall, Boston.

Helmert, F. R. (1876). Die Genauigkeit der Formel von Peters zur Berechnung des wahrscheinlichen Fehlers directer Beobachtungen gleicher Genauigkeit. *Astronom. Nachr.*, **88**, 113–132.

Jung, J. M., Baek, G. H., Kim, J. H., Lee, Y. H., and Chung, M. S. (2001). Changes in ulnar variance in relation to forearm rotation and grip. *J. Bone Joint Surg. Br.*, **83**, 7, 1029–1033. PubMed PMID: 11603517.

Koike, H. (1987). The extensibility of Aplysia nerve and the determination of true axon length. *J. Physiol.*, **390**, 469–487.

Lüroth, J. (1876). Vergleichung von zwei Werten des wahrscheinlichen Fehlers. *Astron. Nachr.*, **87**, 14, 209–220.

Pearson, K. (1914). On the probability that two independent distributions of frequency are really samples of the same population, with special reference to recent work on the identity of trypanosome strains. *Biometrika*, **10**, 1, 85–143.

Renner E. (1970). *Mathematisch-statistische Methoden in der praktischen Anwendung*. Parey, Hamburg.

Sautter, C. (1993). Development of a microtargeting device for particle bombardment of plant meristems. *Plant Cell Tiss. Org.*, **33**, 251–257.

Woolf, S. H. and Rothemich, S. F. (1998). New diabetes guidelines: A closer look at the evidence. *Am. Fam. Physician*, **58**, 6, 1287–1290.

Chapter 7
Point and Interval Estimators

A grade is an inadequate report of an inaccurate judgment by a biased and variable judge of the extent to which a student has attained an undefined level of mastery of an unknown proportion of an indefinite amount of material.

– Paul Dressel

WHAT IS COVERED IN THIS CHAPTER

- Moment-Matching and Maximum Likelihood Estimators
- Unbiased and Consistent Estimators
- Estimation of Mean and Variance
- Confidence Intervals
- Estimation of Population Proportions
- Sample Size Design by Length of Confidence Intervals
- Prediction and Tolerance Intervals
- Intervals for the Poisson Rate

7.1 Introduction

One of the primary objectives of inferential statistics is estimation of population characteristics, or descriptors, on the basis of limited information contained in a sample. The population descriptors are formalized by a statistical model, which can be postulated at various levels of specificity: a broad class of models, a parametric family, or a fully specific unique model.

Often, a functional or distributional form is fully specified but dependent on one or more parameters. Such a model is called parametric. When the model is parametric, the task of estimation is to find the best possible sample counterparts as estimators for the parameters and to assess the accuracy of the estimators.

The estimation procedure follows standard rules. Usually, a sample is taken and a *statistic*, as a function of observations, is calculated. The value of the statistic serves as a point estimator for the unknown population parameter. For example, responses in political pools observed as sample proportions are used to estimate the population proportion of voters in favor of a particular candidate. The associated model is binomial and the parameter of interest is the binomial proportion in the population.

The estimators for a parameter can be given as a single value – *point estimators* or as a range of values – *interval estimators*. For example, the sample mean is a point estimator of the population mean. Confidence intervals and credible sets in a Bayesian context are examples of interval estimators.

In this chapter, we first discuss general methods for finding estimators and then focus on estimation of specific population parameters: means, variances, proportions, rates, etc. Some estimators are universal; that is, they are not connected with any specific distribution. Universal estimators are a sample mean for the population mean and a sample variance for the population variance. However, for interval estimators and for Bayesian estimators, a knowledge of sampling distribution is critical.

In Chapter 2 we learned about many sample summaries that are good estimators for their population counterparts; these will be discussed further in this chapter. We have also seen some robust competitors based on order statistics and ranks; these will be discussed further in Chapter 18.

The methods for how to propose an estimator for a population parameter are discussed next. The methods will use knowledge of the form of population distribution or, equivalently, distribution of sample summaries treated as random variables.

7.2 Moment-Matching and Maximum Likelihood Estimators

We describe two approaches for devising point estimators: moment matching and maximum likelihood.

Matching Estimation. Matching theoretical descriptors, most often moments, with their empirical counterparts, is a natural way to propose an estimator. The theoretical moments of a random variable X with a density specified up to a parameter, $f(x|\theta)$, are functions of that parameter:

$$\mathbb{E}X^k = h(\theta).$$

For example, if the measurements have a Poisson distribution $\mathcal{P}oi(\lambda)$, the second moment $\mathbb{E}X^2$ is $\lambda + \lambda^2$, which is a function of λ. Here, $h(x) = x + x^2$.

Suppose we obtained a sample $X_1, X_2, \ldots, X_n$ from $f(x|\theta)$. The empirical counterparts for theoretical moments $\mathbb{E}X^k$ are sample moments

$$\overline{X^k} = \frac{1}{n} \sum_{i=1}^{n} X_i^k.$$

By matching the theoretical and empirical moments, an estimator $\hat{\theta}$ is found as a solution of the equation

$$\overline{X^k} = h(\theta).$$

For example, for the exponential distribution $\mathcal{E}(\lambda)$, the first theoretical moment is $\mathbb{E}X = 1/\lambda$. An estimator for rate parameter λ is obtained by solving the moment-matching equation $\overline{X} = 1/\lambda$, resulting in $\hat{\lambda}_{mm} = 1/\overline{X}$. Moment-matching estimators are not unique; different theoretical and sample moments can be matched. In the context of an exponential model, the second theoretical moment is $\mathbb{E}X^2 = 2/\lambda^2$, leading to an alternative matching equation,

$$\overline{X^2} = 2/\lambda^2,$$

with the solution

$$\hat{\lambda}_{mm} = \sqrt{\frac{2}{\overline{X^2}}} = \sqrt{\frac{2n}{\sum_{i=1}^{n} X_i^2}}.$$

The following simple MATLAB code simulates a sample of size 10^6 from an exponential distribution with rate parameter $\lambda = 3$, then calculates moment-matching estimators based on the first two moments.

```
Y = exprnd(1/3, 10e6, 1);
%parametrization in MATLAB is 1/lambda
1/mean(Y) %matching the first moment
  ans = 2.9981
sqrt(2/mean(Y.^2)) %matching the second moment
  ans = 2.9984
```

Example 7.1. Moment Matching for Gamma. Consider a sample from a gamma distribution with parameters r and λ. It is known that for $X \sim \mathcal{G}a(r,\lambda)$, $\mathbb{E}(X) = \frac{r}{\lambda}$, and $\mathbb{V}\text{ar}\,X = \mathbb{E}X^2 - (\mathbb{E}X)^2 = \frac{r}{\lambda^2}$. It is easy to see that

$$r = \frac{(\mathbb{E}X)^2}{\mathbb{E}X^2 - (\mathbb{E}X)^2} \quad \text{and} \quad \lambda = \frac{\mathbb{E}X}{\mathbb{E}X^2 - (\mathbb{E}X)^2}.$$

Thus, the moment-matching estimators are

$$\hat{r}_{mm} = \frac{(\overline{X})^2}{\overline{X^2} - (\overline{X})^2} \quad \text{and} \quad \hat{\lambda}_{mm} = \frac{\overline{X}}{\overline{X^2} - (\overline{X})^2}.$$

Matching estimation uses mostly moments, but any other statistic that is (i) easily calculated from a sample and (ii) whose population counterpart depends on parameter(s) of interest can be used in matching. For example, the sample/population quantiles can be used.

Example 7.2. **Melanoma Survival Rate.** In one study on cancer, the highest 5-year survival rate (90%) for women was for malignant melanoma of the skin. Assume that survival time T has an exponential distribution with an unknown rate parameter λ. Using quantiles, estimate λ.
    From

$$P(T > 5) = 0.90 \quad \Rightarrow \quad \exp\{-5 \cdot \lambda\} = 0.90$$

it follows that $\hat{\lambda} = 0.0211$.

Maximum Likelihood. An alternative method, which uses a functional form for distributions of measurements, is maximum likelihood estimation (MLE).
    The MLE was first proposed and used by R. A. Fisher in the 1920s and remains one of the most popular tools in estimation theory and broader statistical inference. The method can be formulated as an optimization problem involving the search for extrema when the model is considered as a function of parameters.
    Suppose that the sample $X_1,\ldots,X_n$ comes from a population with distribution $f(x|\theta)$ indexed by θ, which could be a scalar or a vector of parameters. Elements of the sample are independent, thus the joint distribution of $X_1,\ldots,X_n$ is a product of individual densities:

$$f(x_1,\ldots,x_n|\theta) = \prod_{i=1}^{n} f(x_i|\theta).$$

When the sample is observed, the joint distribution remains dependent upon the parameter,

$$L(\theta|X_1,\ldots,X_n) = \prod_{i=1}^{n} f(X_i|\theta), \qquad (7.1)$$

and, as a function of the parameter, L is called the *likelihood*. The value of the parameter θ that maximizes the likelihood $L(\theta|X_1,\ldots,X_n)$ is the MLE, $\hat{\theta}_{mle}$.

The problem of finding the maximum of L and the value $\hat{\theta}_{mle}$ at which L is maximized is an optimization problem. In some cases, the maximum can be found directly or with the help of the log transformation of L. Other times, the procedure must be iterative and the solution is an approximation. In some cases, depending on the model and sample size, the maximum is not unique or does not exist.

In the most common cases, maximizing the logarithm of likelihood, *log-likelihood*, is simpler than maximizing the likelihood directly. This is because the product in L becomes the sum when a logarithm is applied:

$$\ell(\theta|X_1,\ldots,X_n) = \log L(\theta|X_1,\ldots,X_n) = \sum_{i=1}^{n} \log f(X_i|\theta),$$

and finding an extremum of a sum is simpler. Since the logarithm is a monotonically increasing function, the maxima of L and ℓ are achieved at the same value $\hat{\theta}_{mle}$ (see Figure 7.1 for an illustration).

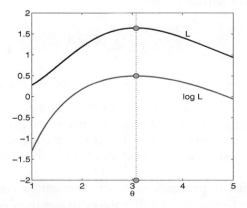

Fig. 7.1 Likelihood and log-likelihood of exponential distribution with rate parameter λ when the sample $X = [0.4, 0.3, 0.1, 0.5]$ is observed. The MLE is $1/\overline{X} = 3.077$.

Analytically,

$$\hat{\theta}_{mle} = \mathrm{argmax}_{\theta} \ell(\theta | X_1, \ldots, X_n),$$

and it can be found as a solution of

$$\frac{\partial \ell(\theta | X_1, \ldots, X_n)}{\partial \theta} = 0 \quad \text{subject to} \quad \frac{\partial^2 \ell(\theta | X_1, \ldots, X_n)}{\partial \theta^2} < 0.$$

In simple terms, the MLE makes the first derivative (with respect to θ) of the log-likelihood equal to 0 and the second derivative negative, which is a condition for a maximum.

As an illustration, consider the MLE of λ in the exponential model, $\mathcal{E}(\lambda)$. After $X_1, \ldots, X_n$ is observed, the likelihood becomes

$$L(\lambda | X_1, \ldots, X_n) = \prod_{i=1}^{n} \lambda e^{-\lambda X_i} = \lambda^n \exp \left\{ -\lambda \sum_{i=1}^{n} X_i \right\}.$$

The likelihood L is obtained as a product of densities $f(x_i | \lambda)$ where the arguments x_is are fixed observations X_i. The product is taken over all observations, as in (7.1). We can search for the maximum of L directly, but since it is a product of two terms involving λ, it is beneficial to look at the log-likelihood instead.

The log-likelihood is

$$\ell(\lambda | X_1, \ldots, X_n) = n \log \lambda - \lambda \sum_{i=1}^{n} X_i.$$

The equation to be solved is

$$\frac{\partial \ell}{\partial \lambda} = \frac{n}{\lambda} - \sum_{i=1}^{n} X_i = 0,$$

and the solution is $\hat{\lambda}_{mle} = \frac{n}{\sum_{i=1}^{n} X_i} = 1/\overline{X}$. The second derivative of the log-likelihood, $\frac{\partial^2 \ell}{\partial \lambda^2} = -\frac{n}{\lambda^2}$, is always negative; thus, the solution $\hat{\lambda}_{mle}$ maximizes ℓ, and consequently L. Figure 7.1 shows the likelihood and log-likelihood as functions of λ. For sample $X = [0.4, 0.3, 0.1, 0.5]$, the maximizing λ is $1/\overline{X} = 3.0769$. Note that both the likelihood and log-likelihood are maximized at the same value.

For the alternative parametrization of exponentials via a scale parameter, as in MATLAB, $f(x | \lambda) = \frac{1}{\lambda} e^{-x/\lambda}$, the estimator is, of course, $\hat{\lambda}_{mle} = \overline{X}$.

An important property of MLE is their *invariance property*.

Invariance Property of MLEs. Let $\hat{\theta}_{mle}$ be an MLE of θ and let $\eta = g(\theta)$, where g is an arbitrary function. Then $\hat{\eta}_{mle} = g(\hat{\theta}_{mle})$ is an MLE of η.

For example, if the MLE for λ in the exponential distribution was $1/\overline{X}$, then for a function of the parameter $\eta = \lambda^2 - \sin(\lambda)$ the MLE is $(1/\overline{X})^2 - \sin(1/\overline{X})$.

In MATLAB, the function mle finds the MLE when inputs are data and the name of a distribution with a list of options. The normal distribution is the default. For example, parhat = mle(data) calculates the MLE for μ and σ of a normal distribution, evaluated at vector data. One of the outputs is the confidence interval. For example, [parhat, parci] = mle(data) returns MLEs and 95% confidence intervals for the parameters. The confidence intervals, as interval estimators, will be discussed later in this chapter. The command [...] = mle(data,'distribution',dist) computes parameter estimations for the distribution specified by dist. Acceptable strings for dist are as follows:

```
'beta'              'bernoulli'                   'binomial'
'discrete uniform'  'exponential'                 'extreme value'
'gamma'             'generalized extreme value'   'generalized pareto'
'geometric'         'lognormal'                   'negative binomial'
'normal'            'poisson'                     'rayleigh'
'uniform'           'weibull'
```

Example 7.3. **MLE of Beta in MATLAB.** The following MATLAB commands show how to estimate parameters a and b in a beta distribution. We will simulate a sample of size 1,000 from a beta $\mathcal{B}e(2,3)$ distribution and then find MLEs of a and b from the sample.

```
a = betarnd( 2, 3,[1, 1000]);
thetahat = mle(a,'distribution', 'beta')
%thetahat = 1.9991    3.0267
```

It is possible to find the MLE using MATLAB's mle command for distributions that are not on the list. The code is given at the end of Example 7.4 in which moment-matching estimators and MLEs for parameters in a Maxwell distribution are compared.

Example 7.4. **Moment-Matching Estimators and MLEs in a Maxwell Distribution.** The Maxwell distribution models random speeds of molecules in thermal equilibrium as given by statistical mechanics. A random variable X with a Maxwell distribution is given by the probability density function

$$f(x|\theta) = \sqrt{\frac{2}{\pi}}\, \theta^{3/2}\, x^2\, e^{-\theta x^2/2}, \qquad \theta > 0, x > 0.$$

Assume that we observed velocities $X_1, \ldots, X_n$ and want to estimate the unknown parameter θ.

The following theoretical moments for the Maxwell distribution are available: the expectation $\mathbb{E}X = 2\sqrt{\frac{2}{\pi\theta}}$, the second moment $\mathbb{E}X^2 = 3/\theta$, and the fourth moment $\mathbb{E}X^4 = 15/\theta^2$. To find moment-matching estimators for θ, the theoretical moments are "matched" with their empirical counterparts $\overline{X}$, $\overline{X^2} = \frac{1}{n}\sum_{i=1}^{n} X_i^2$, and $\overline{X^4} = \frac{1}{n}\sum_{i=1}^{n} X_i^4$, and the resulting equations are solved with respect to θ:

$$\overline{X} = 2\sqrt{\frac{2}{\pi\theta}} \quad \Rightarrow \quad \hat{\theta}_1 = \frac{8}{\pi(\overline{X})^2},$$

$$\frac{1}{n}\sum_{i=1}^{n} X_i^2 = \frac{3}{\theta} \quad \Rightarrow \quad \hat{\theta}_2 = \frac{3n}{\sum_{i=1}^{n} X_i^2},$$

$$\frac{1}{n}\sum_{i=1}^{n} X_i^4 = \frac{15}{\theta^2} \quad \Rightarrow \quad \hat{\theta}_3 = \sqrt{\frac{15n}{\sum_{i=1}^{n} X_i^4}}.$$

To find the MLE of θ, we show that the log-likelihood has the form $\frac{3n}{2}\log\theta - \frac{\theta}{2}\sum_{i=1}^{n} X_i^2 +$ factor free of θ. The maximum of the log-likelihood is achieved at $\hat{\theta}_{\text{MLE}} = \frac{3n}{\sum_{i=1}^{n} X_i^2}$, which is the same as the moment-matching estimator $\hat{\theta}_2$.

Specifically, if $X_1 = 1.4$, $X_2 = 3.1$, and $X_3 = 2.5$ are observed, the MLE of θ is $\hat{\theta}_{\text{MLE}} = \frac{9}{17.82} = 0.5051$. The other two moment-matching estimators are $\hat{\theta}_1 = 0.4677$ and $\hat{\theta}_3 = 0.5768$.

In MATLAB, the Maxwell distribution can be custom-defined using a 'handle' to an anonymous function @:

```
maxwell = @(x,theta)  sqrt(2/pi)  *  ...
    theta.^(3/2) * x.^2   .* exp( - theta * x.^2/2);
mle([1.4 3.1 2.5], 'pdf', maxwell, 'start', rand)
    %ans = 0.5051
```

In most cases, taking the log of likelihood simplifies finding the MLE. Here is an example in which the maximization of likelihood was done without the use of derivatives.

Example 7.5. Suppose the observations $X_1 = 2$, $X_2 = 5$, $X_3 = 0.5$, and $X_4 = 3$ come from the uniform $\mathcal{U}(0,\theta)$ distribution. We are interested in estimating θ. The density for the single observation X is $f(x|\theta) = \frac{1}{\theta}\mathbf{1}(0 \le x \le \theta)$, and the likelihood, based on n observations $X_1, \ldots, X_n$, is

$$L(\theta|X_1, \ldots, X_n) = \frac{1}{\theta^n} \cdot \mathbf{1}(0 \le X_1 \le \theta) \cdot \mathbf{1}(0 \le X_2 \le \theta) \cdot \ \ldots \ \cdot \mathbf{1}(0 \le X_n \le \theta).$$

The product in the expression above can be simplified: if all Xs are less than or equal to θ, then their maximum $X_{(n)}$ is less than θ as well. Thus,

$$\mathbf{1}(0 \le X_1 \le \theta) \cdot \mathbf{1}(0 \le X_2 \le \theta) \cdot \ \dots \ \cdot \mathbf{1}(0 \le X_n \le \theta) \ = \ \mathbf{1}(X_{(n)} \le \theta).$$

Maximizing the likelihood now can be performed by inspection. In order to maximize $\frac{1}{\theta^n}$, subject to $X_{(n)} \le \theta$, we should take the smallest θ possible, and that θ is $X_{(n)} = \max X_i$. Therefore, $\hat{\theta}_{mle} = X_{(n)}$, and in this problem, the estimator is $X_{(4)} = X_2 = 5$.

An alternative estimator can be found by moment matching. It can be shown (the arguments are beyond the scope of this book) that in estimating θ in $\mathcal{U}(0, \theta)$, only $\max X_i$ should be used. What is the distribution of $\max X_i$?

We will find this distribution for general i.i.d. $X_i, i = 1, \dots, n$, with CDF $F(x)$ and PDF $f(x) = F'(x)$.

The CDF is, by definition,

$$G(x) = \mathbb{P}(\max X_i \le x) = \mathbb{P}(X_1 \le x, X_2 \le x, \dots, X_n \le x)$$

$$= \prod_{i=1}^{n} \mathbb{P}(X_i \le x) = (F(x))^n.$$

The reasoning in the equation above is as follows: If the maximum is $\le x$, then all X_i are $\le x$, and vice versa. The density for $\max X_i$ is $g(x) = G'(x) = n F^{n-1}(x) f(x)$, and the first moment is

$$\mathbb{E} \max X_i = \int_{\mathbb{R}} x g(x) \, dx = \int_{\mathbb{R}} x n F^{n-1}(x) f(x) \, dx.$$

For the uniform distribution $\mathcal{U}(0, \theta)$,

$$\mathbb{E} \max X_i = \int_0^\theta x \cdot n (x/\theta)^{n-1} \cdot 1/\theta \, dx = \frac{n}{\theta^n} \int_0^\theta x^n \, dx = \frac{n}{n+1} \theta.$$

The expectation of the maximum $\mathbb{E} \max X_i$ is matched with the largest order statistic in the sample, $X_{(n)}$. Thus, in solving the moment-matching equation, we obtain an alternative estimator for θ, $\hat{\theta}_{mm} = \frac{n+1}{n} X_{(n)}$. In this problem, $\hat{\theta}_{mm} = 25/4 = 6.25$. For a Bayesian estimator, see Example 8.6.
✐

7.3 Unbiasedness and Consistency of Estimators

Based on a sample $X_1, \dots, X_n$ from a population with distribution $f(x|\theta)$, let $\hat{\theta}_n = g(X_1, \dots, X_n)$ be a statistic that estimates the parameter θ. The statistic, or estimator, $\hat{\theta}_n$ as a function of the sample is a random variable. As a random variable, the estimator has an expectation of $\mathbb{E}\hat{\theta}_n$, a variance of $\mathbb{V}\text{ar}\,\hat{\theta}_n$, and its own distribution called a *sampling distribution*.

Example 7.6. **AB Blood-Group Proportion.** Suppose we are interested in finding the proportion of AB blood-group subjects in a particular geographic region. This proportion, θ, is to be estimated on the basis of the sample $Y_1, Y_2, \dots, Y_n$, each having a Bernoulli $Ber(\theta)$ distribution taking values 1 and 0 with probabilities θ and $1 - \theta$, respectively. The realization $Y_i = 1$ indicates the presence of the AB group in observation i. The sum $X = \sum_{i=1}^n Y_i$ is, by definition, binomial $Bin(n, \theta)$.

The estimator for θ is $\hat{\theta}_n = \overline{Y} = \frac{X}{n}$. It is easy to check that this estimator is both moment-matching ($\mathbb{E}Y_i = \theta$) and MLE (the likelihood is $\theta^{\sum Y_i}(1 - \theta)^{n - \sum Y_i}$). Thus, $\hat{\theta}_n$ has a binomial distribution with rescaled realizations $\{0, 1/n, 2/n, \dots, (n-1)/n, 1\}$, that is,

$$\mathbb{P}\left(\hat{\theta}_n = \frac{k}{n}\right) = \binom{n}{k}\theta^k(1 - \theta)^{n-k}, \quad k = 0, 1, \dots, n,$$

which is the estimator's sampling distribution.

It can be shown, by referring to a binomial distribution, that the expectation of $\hat{\theta}_n$ is the expectation of the binomial, $n\theta$, multiplied by $1/n$,

$$\mathbb{E}\hat{\theta}_n = \frac{1}{n} \times n\theta = \theta,$$

and that the variance is

$$\mathbb{V}\text{ar}\,\hat{\theta}_n = \left(\frac{1}{n}\right)^2 \times n\theta(1 - \theta) = \frac{\theta(1 - \theta)}{n}.$$

If $\mathbb{E}\hat{\theta}_n = \theta$, then the estimator $\hat{\theta}$ is called *unbiased*. The expectation is taken with respect to the sampling distribution. The quantity

$$b(\theta) = \mathbb{E}\hat{\theta}_n - \theta$$

is called the *bias* of $\hat{\theta}$.

The error in estimation can be assessed by various measures. The usual measure is the *mean squared error* (MSE).

The MSE is defined as

$$\text{MSE}(\hat{\theta}, \theta) = \mathbb{E}(\hat{\theta}_n - \theta)^2.$$

The MSE represents the expected squared deviation of the estimator from the parameter it estimates. This expectation is taken with respect to the sampling distribution of $\hat{\theta}_n$.

From the definition of MSE,

$$
\begin{aligned}
\mathbb{E}(\hat{\theta}_n - \theta)^2 &= \mathbb{E}(\hat{\theta}_n - \mathbb{E}\hat{\theta}_n + \mathbb{E}\hat{\theta}_n - \theta)^2 \\
&= \mathbb{E}(\hat{\theta}_n - \mathbb{E}\hat{\theta}_n)^2 - 2\mathbb{E}(\hat{\theta}_n - \mathbb{E}\hat{\theta}_n)(\mathbb{E}\hat{\theta}_n - \theta) + (\mathbb{E}\hat{\theta}_n - \theta)^2 \\
&= \mathbb{E}(\hat{\theta}_n - \mathbb{E}\hat{\theta}_n)^2 + (\mathbb{E}\hat{\theta}_n - \theta)^2.
\end{aligned}
$$

Consequently, the MSE can be represented as a sum of the variance of the estimator and its bias squared:

$$\text{MSE}(\hat{\theta}, \theta) = \text{Var}\,\hat{\theta} + b(\theta)^2.$$

The square root of the MSE is sometimes used; it is called the *root mean squared error* (RMSE). For example, in estimating the population proportion, the estimator $\hat{p} = X/n$, for the $X \sim \mathcal{Bin}(n,p)$ model, is unbiased, $\mathbb{E}(\hat{p}) = p$. In this case, the MSE is $\text{Var}\,(\hat{p}) = pq/n$, and the RMSE is $\sqrt{pq/n}$. Note that the RMSE is a function of the parameter. If parameter p is replaced by its estimator $\hat{p}$, then the RMSE becomes the *standard error, s.e.*, of the estimator. For binomial p, the standard error of $\hat{p}$ is $s.e.(\hat{p}) = \sqrt{\hat{p}\hat{q}/n}$.

Remark. The *standard error (s.e.)* of any estimator usually refers to a sample counterpart of its RMSE, which is a sample counterpart of standard deviation for unbiased estimators. For example, if $X_1, X_2, \dots, X_n$ are $\mathcal{N}(\mu, \sigma^2)$, then $s.e.(\overline{X}) = s/\sqrt{n}$.

Inspecting the variance of an unbiased estimator, when the sample size increases, allows for checking estimator's consistency. The consistency is a desirable property of estimators. Informally, it is defined as the convergence of an estimator, in a stochastic sense, to the parameter it estimates.

If, for an unbiased estimator $\hat{\theta}_n$, $\text{Var}\,\hat{\theta}_n \to 0$ when the sample size $n \to \infty$, the estimator is called *consistent*.

More advanced definitions of convergences of random variables, which are beyond the scope of this text, are required in order to deduce more precise definitions of asymptotic unbiasedness, weak and strong consistency. These definitions will not be discussed here.

Example 7.7. **Estimating Normal Variance.** Suppose that we are interested in estimating the parameter θ in a population with a distribution of $\mathcal{N}(0,\theta), \theta > 0$, and that the proposed estimator, when the sample $X_1, X_2, \ldots, X_n$ is observed, is $\hat{\theta} = \frac{1}{n} \sum_{i=1}^{n} X_i^2$.

It can be demonstrated that, when $X \sim \mathcal{N}(0,\theta)$, $\mathbb{E}X^2 = \theta$ and $\mathbb{E}X^4 = 3\theta^2$, by representing X as $\sqrt{\theta}Z$ for $Z \sim \mathcal{N}(0,1)$ and using the fact that $\mathbb{E}Z^2 = 1$ and $\mathbb{E}Z^4 = 3$.

The estimator $\hat{\theta} = \frac{1}{n} \sum_{i=1}^{n} X_i^2 = \overline{X^2}$ is unbiased and consistent. Since $\mathbb{E}\hat{\theta} = \frac{1}{n} \sum_{i=1}^{n} \mathbb{E}X_i^2 = \frac{1}{n} n\theta = \theta$, the estimator is unbiased. To show consistency, it is sufficient to demonstrate that the variance tends to 0 as the sample size increases. This is evident from

$$\mathbb{V}\mathrm{ar}\,\hat{\theta} = \frac{1}{n^2} \sum_{i=1}^{n} \mathbb{V}\mathrm{ar}\,X_i^2 = \frac{1}{n^2} 3n\theta^2 = \frac{3\theta^2}{n} \to 0, \text{ when } n \to \infty.$$

Alternatively, we can use the fact that $\frac{1}{\theta} \sum_{i=1}^{n} X_i^2$ has a χ_n^2-distribution, therefore the sampling distribution of $\hat{\theta}$ is a scaled χ_n^2, where the scaling factor is $\frac{1}{n\theta}$. The unbiasedness and consistency follow from $\mathbb{E}\chi_n^2 = n$ and $\mathbb{V}\mathrm{ar}\,\chi_n^2 = 2n$ by accounting for the scaling factor.
✐

Some important examples of unbiased and consistent estimators are provided next.

7.4 Estimation of a Mean, Variance, and Proportion

7.4.1 Point Estimation of Mean

For a sample $X_1, \ldots, X_n$ of size n we have already discussed the sample mean $\overline{X} = \frac{1}{n} \sum_{i=1}^{n} X_i$ as an estimator of location. A natural estimator of the population mean μ is the sample mean $\hat{\mu} = \overline{X}$. The estimator $\overline{X}$ is an "optimal" estimator of a mean in many different models/distributions and for many different definitions of optimality.

The estimator $\overline{X}$ varies from sample to sample. More precisely, $\overline{X}$ is a random variable with a fixed distribution depending on the common distribution of observations, X_i.

The following is true for *any* distribution in the population as long as $\mathbb{E}X_i = \mu$ and $\mathbb{V}\mathrm{ar}\,(X_i) = \sigma^2$ exist:

$$\mathbb{E}\overline{X} = \mu, \quad \mathbb{V}\mathrm{ar}\,(\overline{X}) = \frac{\sigma^2}{n}. \qquad (7.2)$$

The preceding equations are a direct consequence of independence in a sample and imply that $\overline{X}$ is an unbiased and consistent estimator of μ. If, in addition, we assume normality $X_i \sim \mathcal{N}(\mu, \sigma^2)$, then the sampling distribution of $\overline{X}$ is known exactly (page 248),

$$\overline{X} \sim \mathcal{N}\left(\mu, \frac{\sigma^2}{n}\right),$$

and the relations in (7.2) are apparent.

Chebyshev's Inequality and Strong Law of Large Numbers*. There are two general results in probability that theoretically justify the use of the sample mean $\overline{X}$ to estimate the population mean, μ. These are Chebyshev's inequality and strong law of large numbers (SLLN). We will briefly overview these results.

The Chebyshev inequality states that when $X_1, X_2, \ldots, X_n$ are i.i.d. random variables with mean μ and finite variance σ^2, the probability that $\overline{X}$ will deviate from μ is small,

$$\mathbb{P}(|\overline{X}_n - \mu| \geq \epsilon) \leq \frac{\sigma^2}{n\epsilon^2},$$

for any $\epsilon > 0$. The inequality is a direct consequence of (5.9) with $(\overline{X}_n - \mu)^2$ in place of X and ϵ^2 in place of a.

To translate this to specific numbers, we can choose ϵ small, say 0.000001. Assume that the X_is have a variance of 1. The Chebyshev inequality states that with n larger than the solution of $1/(n \times 0.0000001^2) = 0.9999$, the distance between $\overline{X}_n$ and μ will be smaller than 0.000001 with a probability of 99.99%. Admittedly, n here is an experimentally unfeasible number; however, for any small ϵ, finite σ^2, and "confidence" $1 - \frac{\sigma^2}{n\epsilon^2}$ close to 1, such n is finite.

The laws of large numbers state that, as a numerical sequence, $\overline{X}_n$ converges to μ. Care is nevertheless needed. The sequence $\overline{X}_n$ is not a sequence of numbers, but a sequence of random variables, which are functions defined on sample spaces $\mathcal{S}$. Thus, direct application of a calculus-type of convergence is not appropriate. However, for any fixed realization from the sample space $\mathcal{S}$, the sequence $\overline{X}_n$ becomes numerical and a traditional convergence can be stated. Thus, a correct statement for the so-called SLLN is

$$\mathbb{P}(\overline{X}_n \to \mu) = 1,$$

that is, viewed as an event, $\{\overline{X}_n \to \mu\}$ is a sure event – it happens with a probability of 1.

7.4.2 Point Estimation of Variance

To obtain some intuition, we start, once again, with a finite population: $y_1, \ldots, y_N$. The population variance is $\sigma^2 = \frac{1}{N} \sum_{i=1}^{N} (y_i - \mu)^2$, where $\mu = \frac{1}{N} \sum_{i=1}^{N} y_i$ is the population mean.

For a sample $X_1, X_2, \ldots, X_n$ that is observed, an estimator of variance σ^2 is

$$\hat{\sigma}^2 = \frac{1}{n} \sum_{i=1}^{n} (X_i - \mu)^2$$

for μ known, and

$$\hat{\sigma}^2 = s^2 = \frac{1}{n-1} \sum_{i=1}^{n} (X_i - \overline{X})^2$$

for μ not known, which is estimated by $\overline{X}$.

In the expression for s^2 we divide the sum by $n - 1$ instead of the "expected" n in order to ensure the unbiasedness of s^2, $\mathbb{E}s^2 = \sigma^2$. The proof of this fact is straightforward and does not require any distributional assumptions, except that the population variance σ^2 is finite.

Note that by the definition of variance, $\mathbb{E}(X_i - \mu)^2 = \sigma^2$ and $\mathbb{E}(\overline{X} - \mu)^2 = \sigma^2/n$.

$$(n-1)s^2 = \sum_{i=1}^{n} (X_i - \overline{X})^2$$

$$= \sum_{i=1}^{n} [(X_i - \mu) - (\overline{X} - \mu)]^2$$

$$= \sum_{i=1}^{n} (X_i - \mu)^2 - 2(\overline{X} - \mu) \sum_{i-1}^{n} (X_i - \mu) + n(\overline{X} - \mu)^2$$

$$= \sum_{i=1}^{n} (X_i - \mu)^2 - n(\overline{X} - \mu)^2, \quad \text{since} \sum_{i=1}^{n} (X_i - \mu) = n(\overline{X} - \mu).$$

Then,

$$
\begin{aligned}
\mathbb{E}(s^2) &= \frac{1}{n-1}\mathbb{E}(n-1)s^2 \\
&= \frac{1}{n-1}\mathbb{E}\Big[\sum_{i=1}^{n}(X_i - \mu)^2 - n(\overline{X} - \mu)^2\Big] \\
&= \frac{1}{n-1}\Big(n\sigma^2 - n\frac{\sigma^2}{n}\Big) \\
&= \frac{1}{n-1}(n-1)\sigma^2 = \sigma^2.
\end{aligned}
$$

When, in addition, the population is normal $\mathcal{N}(\mu,\sigma^2)$, then

$$
\frac{(n-1)s^2}{\sigma^2} \sim \chi^2_{n-1},
$$

meaning that the statistic $\frac{(n-1)s^2}{\sigma^2} = \sum_{i=1}^{n}\left(\frac{X_i - \overline{X}}{\sigma}\right)^2$ has a χ^2-distribution with $n-1$ degrees of freedom (see equation 6.3 and the related discussion).

For a sample from a normal distribution, the unbiasedness of s^2 is a consequence of the following two facts: $s^2 \sim \frac{\sigma^2}{n-1}\chi^2_{n-1}$ and $\mathbb{E}\chi^2_{n-1} = (n-1)$. The variance of s^2 is

$$
\mathbb{V}\mathrm{ar}\, s^2 = \left(\frac{\sigma^2}{n-1}\right)^2 \times \mathbb{V}\mathrm{ar}\, \chi^2_{n-1} = \frac{2\sigma^4}{n-1}, \tag{7.3}
$$

since $\mathbb{V}\mathrm{ar}\,\chi^2_{n-1} = 2(n-1)$. Unlike the unbiasedness result, $\mathbb{E}s^2 = \sigma^2$, which does not require a normality assumption, the result in (7.3) is valid only when observations come from a normal distribution. In the general case,

$$
\mathbb{V}\mathrm{ar}\, s^2 = \frac{\mu_4 - \mu_2^2}{n} - \frac{2(\mu_4 - 2\mu_2^2)}{n^2} + \frac{\mu_4 - 3\mu_2^2}{n^3}, \tag{7.4}
$$

where $\mu_k = \mathbb{E}(X - \mathbb{E}X)^k$ is kth central moment. It is easy to see how for a normal distribution, (7.4) becomes (7.3), since in this case $\mu_4 = 3\mu_2$ and $\mu_2 = \sigma^2$.

Although s^2 is an unbiased estimator for σ^2, s is not an unbiased estimator for σ, a fact that is often overlooked. If the population is normal, then $\sqrt{(n-1)/2}\frac{\Gamma((n-1)/2)}{\Gamma(n/2)}s$ is an unbiased estimator of σ. This bias correction for s is important when n is small; for n large the correction is negligible. For example, if $n = 50$, the unbiased estimator of σ is $1.0051\,s$.

As Figure 7.2 shows, the empirical distribution of normalized sample variances is close to a χ^2-distribution. We generated $M = 100{,}000$ samples of size $n = 8$ from a normal $\mathcal{N}(0, 5^2)$ distribution and found sample variances s^2 for each sample. The sample variances were multiplied by $n - 1 = 7$ and divided by $\sigma^2 = 25$. The histogram of these rescaled sample variances is plotted and the density of a χ^2-distribution with 7 degrees of freedom is superimposed in red. The code generating Figure 7.2 is given next.

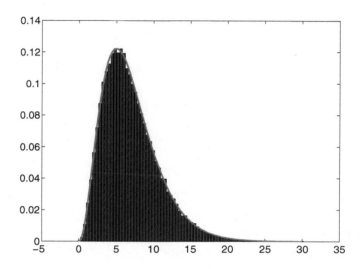

Fig. 7.2 Histogram of normalized sample variances $(n-1)s^2/\sigma^2$ obtained from $M = 100{,}000$ independent samples from $\mathcal{N}(0, 5^2)$, each of size $n = 8$. The density of a χ^2-distribution with 7 degrees of freedom is superimposed in *red*.

```
M=100000;  n = 8;
   X = 5 * randn([n, M]);
   ch2 = (n-1) * var(X)/25;
histn(ch2,0,0.4,30)
hold on
plot( (0:0.1:30), chi2pdf((0:0.1:30), n-1),'r-')
```

The code is efficient since a for-end loop is avoided. The simulated object X is an $n \times M$ matrix consisting of M columns (samples) of length n. The operator var(X) acts on columns of X producing M sample variances.

Several Robust Estimators of the Standard Deviation*. Suppose that a sample $X_1, \ldots, X_n$ is observed but its normality is not assumed. We discuss two estimators of the standard deviation that are calibrated by the normal distribution and are robust with respect to outliers and deviations from normality.

Gini's mean difference is defined as

$$G = \frac{2}{n(n-1)} \sum_{1 \le i < j \le n} |X_i - X_j|.$$

The statistic $G \frac{\sqrt{\pi}}{2}$ is an estimator of the standard deviation and is more robust to outliers than the standard statistic s.

A proposal by Croux and Rousseeuw (1992) involves absolute differences, as in Gini's mean difference estimator, but uses a kth-order statistic rather than the average. The estimator of σ is

$$Q = 2.2219 \, \{|X_i - X_j|, i < j\}_{(k)}, \quad \text{where } k = \binom{\lfloor n/2 \rfloor + 1}{2}.$$

The constant 2.2219 is used to calibrate the estimator, so that if the sample is a standard normal, then $Q = 1$. In calculating Q, all $\binom{n}{2}$ differences $|X_i - X_j|$ are ordered, and the kth in rank is selected and multiplied by 2.2219. This choice of k requires an additional multiplicative correction factor $n/(n + 1.4)$ for n odd, or $n/(n + 3.8)$ for n even.

MATLAB scripts ginimd.m and crouxrouss.m can be used to evaluate the estimators. The algorithm is naïve and uses a double loop to evaluate G and Q. The evaluation breaks down for sample sizes exceeding a few hundreds because of memory problems. A smarter algorithm that avoids looping is implemented in versions ginimd2.m and crouxrouss2.m. In these versions, the sample size can go up to 6,000.

In the next MATLAB session, we show how the robust estimators of the standard deviation perform. If 1,000 standard normal random variates are generated and one value is replaced with a clear outlier, say $X_{1000} = 20$, we will explore the influence of this outlier to both standard and robust estimators of the standard deviation. Note that s is quite sensitive, the outlier will inflate the estimator by almost 20%. The robust estimators are affected as well, but not as much as s.

```
x =randn(1, 1000);
x(1000)=20;
std(x)
% ans = 1.1999
s1 = ginimd2(x)
%s1 =1.0555
s2 = crouxrouss2(x)
%s2 =1.0287
iqr(x)/1.349
%ans = 1.0172
```

There are many other robust estimators of the variance/standard deviation. Good references containing extensive material on robust estimation are Wilcox (2005) and Staudte and Sheater (1990).

Estimation of Covariance. If $(X_1, Y_1), \ldots, (X_n, Y_n)$ are independent realizations of a bivariate random variable (X, Y), then an unbiased estimator of covariance

$$\sigma_{XY} = \mathbb{E}(X - \mathbb{E}X)(Y - \mathbb{E}Y)$$

is the sample covariance (page 32)

$$s_{XY} = \frac{1}{n-1} \sum_{i=1}^{n} (X_i - \overline{X})(Y_i - \overline{Y}).$$

In the case of normal distribution, the variance of this estimator is

$$\mathbb{V}ar(s_{XY}) = \frac{\sigma_X^2 \sigma_Y^2 + \sigma_{XY}^2}{n-1}.$$

7.4.3 Point Estimation of Population Proportion

It is natural to estimate the population proportion p by a sample proportion. The sample proportion is the MLE and moment-matching estimator for p.

For sample proportions a binomial distribution is used as the theoretical model. Let $X \sim \mathcal{B}in(n, p)$, where parameter p is unknown. The MLE of p based on a single observation X is obtained by maximizing the likelihood

$$\binom{n}{X} p^X (1-p)^{n-X}$$

or the log-likelihood

$$\text{factor free of } p + X \log(p) + (n - X) \log(1 - p).$$

The maximum is obtained by solving

$$(\text{factor free of } p + X \log(p) + (n - X) \log(1 - p))' = 0$$

$$\frac{X}{p} - \frac{n - X}{1 - p} = 0,$$

which after some algebra gives the solution $\hat{p}_{mle} = \frac{X}{n}$.

In Example 7.6, we argued that the exact distribution for X/n is a rescaled binomial and that the statistic is unbiased, with the variance converging to 0 when the sample size increases. These two properties define a consistent estimator.

7.5 Confidence Intervals

Whenever the sampling distribution of a point estimator $\hat{\theta}_n$ is continuous, then necessarily $\mathbb{P}(\hat{\theta}_n = \theta) = 0$. In other words, the probability that the estimator exactly matches the parameter it estimates is 0.

Instead of the point estimator, one may report two estimators, $L = L(X_1,\ldots, X_n)$ and $U = U(X_1,\ldots,X_n)$, so that the interval $[L,U]$ covers θ with a probability of $1 - \alpha$, for small α. In this case, the interval $[L,U]$ will be called a $(1 - \alpha)100\%$ confidence interval for θ.

For the construction of a confidence interval for a parameter, one needs to know the sampling distribution of the associated point estimator. The lower and upper interval bounds L and U depend on the quantiles of this distribution. We will derive the confidence interval for the normal mean, normal variance, population proportion, and Poisson rate. Many other confidence intervals, including differences, ratios, and some functions of statistics, are tightly connected to testing methodology and will be discussed in subsequent chapters.

Note that when the population is normal and $X_1,\ldots,X_n$ is observed, the exact sampling distributions of

$$Z = \frac{\overline{X} - \mu}{\sigma/\sqrt{n}} \quad \text{and}$$

$$t = \frac{\overline{X} - \mu}{s/\sqrt{n}} = \frac{\overline{X} - \mu}{\sigma/\sqrt{n}} \times \frac{1}{\sqrt{\frac{(n-1)s^2}{\sigma^2}/(n-1)}}$$

are standard normal and t_{n-1} distributions, respectively.

The expression for t is shown as a product to emphasize the construction of a t-distribution from a standard normal (in *blue*) and χ^2 (in *red*), as in page 255. When the population is not normal but n is large, both statistics Z and t have an approximate standard normal distribution, due to the CLT.

We saw that the point estimator for the population probability of a success is the sample proportion $\hat{p} = X/n$, where X is the number of successes in n trials. The statistic X/n is based on a binomial sampling scheme in which X has exactly a binomial $Bin(n,p)$ distribution. Using this exact distribution would lead to confidence intervals in which the bounds and confidence levels are discretized. The normal approximation to the binomial (CLT in the form of de Moivre's approximation) leads to

$$\hat{p} \overset{approx}{\sim} \mathcal{N}\left(p, \frac{p(1-p)}{n}\right), \tag{7.5}$$

and the confidence intervals for the population proportion p would be based on normal quantiles.

7.5.1 Confidence Intervals for the Normal Mean

Let $X_1, \ldots, X_n$ be a sample from a normal $\mathcal{N}(\mu, \sigma^2)$ distribution where the parameter μ is to be estimated and σ^2 is known.

Starting from the identity

$$\mathbb{P}(-z_{1-\alpha/2} \leq Z \leq z_{1-\alpha/2}) = 1 - \alpha$$

and the fact that $\overline{X}$ has a $\mathcal{N}(\mu, \frac{\sigma^2}{n})$ distribution, we can write

$$\mathbb{P}\left(-z_{1-\alpha/2}\frac{\sigma}{\sqrt{n}} + \mu \leq \overline{X} \leq z_{1-\alpha/2}\frac{\sigma}{\sqrt{n}} + \mu\right) = 1 - \alpha;$$

see Figure 7.3a for an illustration. Simple algebra gives

$$\overline{X} - z_{1-\alpha/2}\frac{\sigma}{\sqrt{n}} \leq \mu \leq \overline{X} + z_{1-\alpha/2}\frac{\sigma}{\sqrt{n}}, \tag{7.6}$$

which is a $(1 - \alpha)100\%$ confidence interval.

If σ^2 is not known, then a confidence interval with the sample standard deviation s in place of σ can be used. The z quantiles are valid for large n, but for small n ($n < 40$) we use t_{n-1} quantiles, since the sampling distribution for $\frac{\overline{X}-\mu}{s/\sqrt{n}}$ is t_{n-1}. Thus, for σ^2 unknown,

$$\overline{X} - t_{n-1,1-\alpha/2}\frac{s}{\sqrt{n}} \leq \mu \leq \overline{X} + t_{n-1,1-\alpha/2}\frac{s}{\sqrt{n}} \tag{7.7}$$

is the confidence interval for μ of level $1 - \alpha$.

Below is a summary of the above-stated intervals:

The $(1 - \alpha)\,100\%$ confidence interval for an unknown normal mean μ on the basis of a sample of size n is

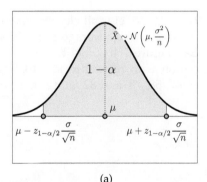 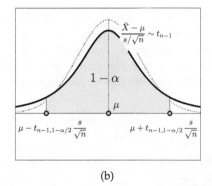

(a)   (b)

Fig. 7.3 (a) When σ^2 is known, $\overline{X}$ has a normal $\mathcal{N}(\mu, \sigma^2/n)$ distribution and $\mathbb{P}(\mu - z_{1-\alpha/2}\frac{\sigma}{\sqrt{n}} \leq \overline{X} \leq \mu + z_{1-\alpha/2}\frac{\sigma}{\sqrt{n}}) = 1 - \alpha$, leading to the confidence interval in (7.6). (b) If σ^2 is not known and s^2 is used instead, then $\frac{\overline{X}-\mu}{s/\sqrt{n}}$ is t_{n-1}, leading to the confidence interval in (7.7).

$$\left[\overline{X} - z_{1-\alpha/2}\frac{\sigma}{\sqrt{n}}, \ \overline{X} + z_{1-\alpha/2}\frac{\sigma}{\sqrt{n}} \right]$$

when the variance σ^2 is known, and

$$\left[\overline{X} - t_{n-1,1-\alpha/2}\frac{s}{\sqrt{n}}, \ \overline{X} + t_{n-1,1-\alpha/2}\frac{s}{\sqrt{n}} \right]$$

when the variance σ^2 is not known and s^2 is used instead.

Interpretation of Confidence Intervals. What does a "confidence of 95%" mean? A common misconception is that this means that the unknown mean falls in the calculated interval with a probability of 0.95. Such a probability statement is valid for credible sets in the Bayesian context, which will be discussed in Chapter 8.

The interpretation of the $(1 - \alpha)\, 100\%$ confidence interval is as follows. If a random sample from a normal population is selected a large number of times and the confidence interval for the population mean μ is calculated, the proportion of such intervals covering μ approaches $1 - \alpha$.

The following MATLAB code illustrates this. The code generates $M = 10{,}000$ random samples of size $n = 40$ from a normal population with a mean of $\mu = 10$ and a variance of $\sigma^2 = 4^2$; then it calculates a 95% confidence interval from each sample. It then counts how many of the intervals cover the mean μ, cover = 1, and finally finds their proportion, covers/M. The code was run consecutively several times and the following empirical confidences were obtained: 0.9461, 0.9484, 0.9469, 0.9487, 0.9502, 0.9482, 0.9502,

0.9482, 0.9530, 0.9517, 0.9503, 0.9514, 0.9496, 0.9515, etc., all clearly scattering around 0.95. Figure 7.4a plots the behavior of the coverage proportion when simulations range from 1 to 10,000. Figure 7.4b plots the first 100 intervals in the simulation and their position with respect to $\mu = 10$. The confidence intervals in simulations $17, 37, 47, 58, 78,$ and 82 fail to cover μ.

```matlab
M=10000;              %simulate M times
n = 40;               % sample size
alpha = 0.05;         %1-alpha = confidence
tquantile = tinv(1-alpha/2, n-1);
covers =[];
for i = 1:M
    X = 10 + 4*randn(1,n); %sample, mean=10, var =16
    xbar = mean(X); s = std(X);
    LB = xbar - tquantile * s/sqrt(n);
    UB = xbar + tquantile * s/sqrt(n);
    % cover=1 if the interval covers population mean 10
    if UB < 10 | LB > 10
            cover = 0;
    else
            cover = 1;
    end
        covers =[covers cover]; %saves cover history
end
sum(covers)/M %proportion of intervals covering the mean
```

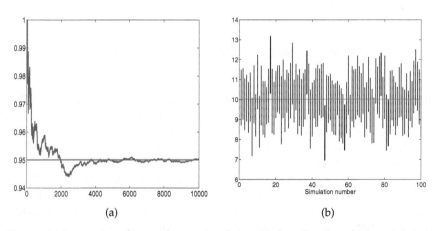

(a)   (b)

Fig. 7.4 (a) Proportion of intervals covering the mean plotted against the iteration number, as in plot(cumsum(covers)./(1:length(covers))). (b) First 100 simulated intervals. The intervals $17, 37, 47, 58, 78,$ and 82 fail to cover the true mean.

7.5.2 Confidence Interval for the Normal Variance

Earlier (page 256) we argued that the sampling distribution of $\frac{(n-1)s^2}{\sigma^2}$ was χ^2 with $n-1$ degrees of freedom. From the definition of χ^2_{n-1} quantiles,

$$1 - \alpha = \mathbb{P}(\chi^2_{n-1,\alpha/2} \leq \chi^2_{n-1} \leq \chi^2_{n-1,1-\alpha/2}),$$

as in Figure 7.5. Replacing χ^2_{n-1} with $\frac{(n-1)s^2}{\sigma^2}$, we get

$$1 - \alpha = \mathbb{P}\left(\chi^2_{n-1,\alpha/2} \leq \frac{(n-1)s^2}{\sigma^2} \leq \chi^2_{n-1,1-\alpha/2}\right).$$

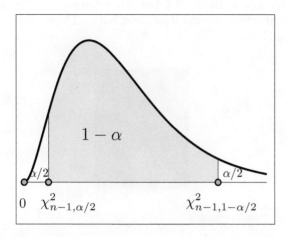

Fig. 7.5 Confidence interval for normal variance σ^2 is derived from $\mathbb{P}(\chi^2_{n-1,\alpha/2} \leq (n-1)s^2/\sigma^2 \leq \chi^2_{n-1,1-\alpha/2}) = 1 - \alpha$.

Simple algebra with the inequalities above (taking the reciprocal of all three parts, being careful about the direction of the inequalities, and multiplying everything by $(n-1)s^2$) gives

$$\frac{(n-1)s^2}{\chi^2_{n-1,1-\alpha/2}} \leq \sigma^2 \leq \frac{(n-1)s^2}{\chi^2_{n-1,\alpha/2}}.$$

The $(1-\alpha)\,100\%$ confidence interval for an unknown normal variance is

$$\left[\frac{(n-1)s^2}{\chi^2_{n-1,1-\alpha/2}}, \frac{(n-1)s^2}{\chi^2_{n-1,\alpha/2}} \right]. \tag{7.8}$$

Remark. If the population mean μ is known, then s^2 is calculated as $\frac{1}{n}\sum_{i=1}^{n}(X_i - \mu)^2$, and the χ^2 quantiles gain one degree of freedom (n instead of $n - 1$). This makes the confidence interval a bit tighter.

Example 7.8. Amanita muscaria. With its bright red, sometimes dinner-plate-sized caps, the fly agaric (*Amanita muscaria*) is one of the most striking of all mushrooms. The white warts that adorn the cap, the white gills, a well-developed ring, and the distinctive volva of concentric rings distinguish the fly agaric from all other red mushrooms. The spores of the mushroom print white, are elliptical, and have a larger axis in the range of 7 to 13 μm (Fig. 7.6).

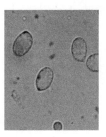

Fig. 7.6 Spores of *Amanita muscaria.*

Measurements of the diameter X of spores for $n = 51$ mushrooms are given in the following table:

10	11	12	9	10	11	13	12	10	11
11	13	9	10	9	10	8	12	10	11
9	10	7	11	8	9	11	11	10	12
10	8	7	11	12	10	9	10	11	10
8	10	10	8	9	10	13	9	12	9
9									

Assume that the measurements are normally distributed with mean μ and variance σ^2, but both parameters are unknown. The sample mean and variances are $\overline{X} = 10.098$, $s^2 = 2.1702$, and $s = 1.4732$. Also, the confidence interval would use an appropriate t-quantile, in this case `tinv(1-0.05/2, 51-1) = 2.0086`.

The 95% confidence interval for the population mean, μ, is

$$\left[10.098 - 2.0086 \times \frac{1.4732}{\sqrt{51}}, \; 10.098 + 2.0086 \times \frac{1.4732}{\sqrt{51}}\right] = [9.6836, 10.5124].$$

Thus, the unknown mean μ belongs to the interval $[9.6836, 10.5124]$ with confidence 95%. That means that if the sample is obtained many times and for each sample the confidence interval is calculated, 95% of the intervals would contain μ.

To find, say, the 90% confidence interval for the population variance, σ^2, we need χ^2 quantiles, `chi2inv(1-0.10/2, 51-1)` `= 67.5048`, and `chi2inv(0.10/2,` `51-1)` `= 34.7643`. According to (7.8), the interval is

$$[(51 - 1) \times 2.1702/67.5048, \; (51 - 1) \times 2.1702/34.7643] = [1.6074, 3.1213].$$

Thus, the interval $[1.6074, 3.1213]$ covers the population variance σ^2 with a confidence of 90%.

Example 7.9. A Confidence Interval for σ^2 by CLT. An alternative confidence interval for the normal variance is possible. Since by the CLT $s^2 \overset{approx}{\sim} \mathcal{N}\left(\sigma^2, \frac{2\sigma^4}{n-1}\right)$ (Can you explain why?), when n is not small, an approximate $(1 - \alpha)100\%$ confidence interval for σ^2 is

$$\left[s^2 - z_{1-\alpha/2} \cdot \frac{\sqrt{2}\, s^2}{\sqrt{n - 1}}, \; s^2 + z_{1-\alpha/2} \cdot \frac{\sqrt{2}\, s^2}{\sqrt{n - 1}}\right].$$

In Example 7.8, $s^2 = 2.1702$ and $n = 51$. A 90% confidence interval for the variance was $[1.6074, 3.1213]$. By normal approximation,

```
s2 = 2.1702; n=51; alpha = 0.1;
[s2 - norminv(1-alpha/2)*sqrt(2)* s2/sqrt(n-1), ...
 s2 + norminv(1-alpha/2)*sqrt(2)* s2/sqrt(n-1)]
%ans = 1.4563    2.8841
```

The interval $[1.4563, 2.8841]$ is shorter, compared to the standard confidence interval $[1.6074, 3.1213]$ obtained using χ^2 quantiles, as $1.4278 < 1.5139$. Insisting on equal-probability tails does not lead to the shortest interval since the χ^2-distribution is asymmetric. In addition, the approximate interval is centered at s^2. Why, then, is this interval not used? The coverage probability of a CLT-based interval is smaller than the nominal $1 - \alpha$, and unless n is large (>100, say), this discrepancy can be significant (Exercise 7.28).

7.5.3 Confidence Intervals for the Population Proportion

The sample proportion $\hat{p} = \frac{X}{n}$ has a range of optimality properties (unbiasedness, consistency); however, its realizations are discrete. For this reason, confidence intervals for p are obtained using the normal approximation, or connections of binomial with other continuous distributions, such as F.

Recall that for n large and np or nq not small (>10), the binomial X can be approximated by a $\mathcal{N}(np, npq)$ distribution. This approximation leads to $\frac{X}{n} \overset{approx}{\sim} \mathcal{N}\left(p, \frac{pq}{n}\right)$.

Note, however, that the standard deviation of $\hat{p}$, $\sqrt{\frac{pq}{n}}$, is not known, as it depends on p, and for the confidence interval we can use a plug-in estimator $\sqrt{\frac{\hat{p}\hat{q}}{n}}$ instead.

Let p be the population proportion and $\hat{p}$ the observed sample proportion. Assume that the smaller of $\frac{np}{q}$ and $\frac{nq}{p}$ is larger than 10. Then the $(1 - \alpha)100\%$ confidence interval for unknown p is

$$\left[\hat{p} - z_{1-\alpha/2}\sqrt{\frac{\hat{p}\,\hat{q}}{n}}, \ \hat{p} + z_{1-\alpha/2}\sqrt{\frac{\hat{p}\,\hat{q}}{n}}\right].$$

This interval is known as the Wald interval (Wald and Wolfowitz, 1939).

The Wald interval is used quite frequently, but its performance is suboptimal and even poor when p is close to 0 or 1. Figure 7.7a demonstrates the performance of Wald's 95% confidence interval for $n = 20$ and p ranging from 0.05 to 0.95 with a step of 0.01. The plot is obtained by simulation (waldsimulation.m). For each p, 100,000 binomial proportions are simulated, the Wald confidence intervals calculated, and the proportion of those intervals containing p is plotted. Notice that for nominal 95% confidence, the actual coverage probability may be much smaller, depending on p.

Unless the sample size n is very large, the Wald interval should not be used. The performance of Wald's interval can be improved by continuity corrections:

$$\left[\hat{p} - \frac{1}{2n} - z_{1-\alpha/2}\sqrt{\frac{\hat{p}\,\hat{q}}{n}}, \ \hat{p} + \frac{1}{2n} + z_{1-\alpha/2}\sqrt{\frac{\hat{p}\,\hat{q}}{n}}\right].$$

Figure 7.7b shows the coverage probability of Wald's corrected interval.

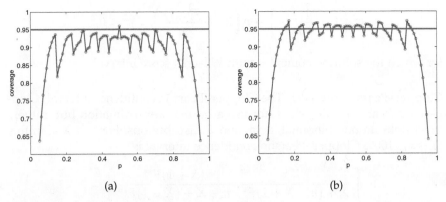

Fig. 7.7 (a) Simulated coverage probability for Wald's confidence interval for the true binomial proportion p ranging from 0.05 to 0.95, and $n = 20$. For each p, 100,000 binomial proportions are simulated, the Wald confidence intervals calculated, and the proportion of those containing p plotted. (b) The same as (a), but for the corrected Wald interval.

There is a range of intervals that have a performance superior to Wald's interval. An overview of several alternatives is provided next.

Adjusted Wald Interval. The adjusted Wald interval (Agresti and Coull, 1998) uses $p^* = \frac{X+2}{n+4}$ as an estimator of the proportion. Adding "two successes and two failures" was proposed by Wilson (1927):

$$\left[p^* - z_{1-\alpha/2} \sqrt{\frac{p^* q^*}{n+4}},\ p^* + z_{1-\alpha/2} \sqrt{\frac{p^* q^*}{n+4}} \right].$$

We will see in the next chapter that Wilson's proposal p^* has a Bayesian justification (page 343).

Wilson Score Interval. The Wilson score interval is another adjustment to the Wald interval based on the so-called Wilson-score test (Wilson, 1927; Hogg and Tanis, 2001):

$$\left[\frac{1}{1+z^2/n} \left(\hat{p} + \frac{z^2}{2n} - z \sqrt{\frac{\hat{p}\,\hat{q}}{n} + \frac{z^2}{4n^2}} \right),\ \frac{1}{1+z^2/n} \left(\hat{p} + \frac{z^2}{2n} + z \sqrt{\frac{\hat{p}\,\hat{q}}{n} + \frac{z^2}{4n^2}} \right) \right],$$

where z is $z_{1-\alpha/2}$. This interval can be obtained by solving the inequality

$$|\hat{p} - p| \leq z_{1-\alpha/2} \sqrt{\frac{p(1-p)}{n}}$$

with respect to p. After squaring the left- and right-hand sides and some algebra, we get the quadratic inequality

$$p^2 \left(1 + \frac{z^2_{1-\alpha/2}}{n} \right) - p \left(2\hat{p} + \frac{z^2_{1-\alpha/2}}{n} \right) + \hat{p}^2 \leq 0,$$

for which the solution coincides with Wilson's score interval.

Clopper–Pearson Interval. The Clopper–Pearson confidence interval (Clopper and Pearson, 1934) does not use a normal approximation but, rather, exact links among binomial, beta, and F distributions. For $0 < X < n$, the $(1 - \alpha) \cdot 100\%$ Clopper–Pearson confidence interval is

$$\left[\frac{X}{X + (n - X + 1)F^*}, \frac{(X + 1)F^{**}}{n - X + (X + 1)F^{**}} \right],$$

where F^* is the $(1 - \alpha/2)$-quantile of the F_{ν_1,ν_2}-distribution with $\nu_1 = 2(n - X + 1)$ and $\nu_2 = 2X$ and F^{**} is the $(1 - \alpha/2)$-quantile of the F_{ν_1,ν_2}-distribution with $\nu_1 = 2(X + 1)$ and $\nu_2 = 2(n - X)$. In terms of beta distribution, Clopper–Pearson interval takes a very simple form, its MATLAB code is `[betainv(alpha/2, X, n-X+1), betainv(1-alpha/2, X+1, n-X)]`.

When $X = 0$, the interval is $[0, 1 - (\alpha/2)^{1/n}]$ and for $X = n$, $[(\alpha/2)^{1/n}, 1]$.

Anscombe's ArcSin Interval. For $X \sim \mathcal{Bin}(n, p)$ Anscombe (1948) showed that if $p^* = \frac{X + 3/8}{n + 3/4}$, then the quantity

$$2\sqrt{n}(\arcsin \sqrt{p^*} - \arcsin \sqrt{p})$$

has an approximately standard normal distribution. From this result it follows that

$$\left[\sin^2 \left(\arcsin \sqrt{p^*} - \frac{z_{1-\alpha/2}}{2\sqrt{n}} \right), \sin^2 \left(\arcsin \sqrt{p^*} + \frac{z_{1-\alpha/2}}{2\sqrt{n}} \right) \right]$$

is the $(1 - \alpha)100\%$ confidence interval for p.

The next example shows the comparative performance of different confidence intervals for the population proportion.

Example 7.10. **Cyclosporine Reversal Study.** An interesting case study involved research on the therapeutic benefits of cyclosporine on patients with chronic inflammatory bowel disease (Crohn's disease). In a double-blind clinical trial, researchers reported (Brynskov et al., 1989) that out of 37 patients with Crohn's disease resistant to standard therapies, 22 improved after a 3-month period. This proportion was significantly higher than that for the placebo group (11/34). The study was published in the *New England Journal of Medicine.*

However, at the 6-month follow-up, no significant differences were found between the treatment group and the control. In the cyclosporine group, 30 patients *did not* improve, compared to 23 out of 34 in the placebo group (Brynskov et al., 1991). Thus, the proportion of patients who benefited in the cyclosporine group dropped from $\hat{p}_1 = 22/37 = 59.46\%$ at the 3-month to $\hat{p}_2 = 7/37 = 18.92\%$ at the 6-month follow-up. The researchers state: "We conclude that a short course of cyclosporin treatment does not result in long-term improvement in active chronic Crohn's disease."

To illustrate the performance of several introduced confidence intervals for the population proportion, we will find Wald's, Wilson's, Wilson score, Clopper–Pearson's, and Arcsin 95% confidence intervals for the proportion of patients who benefited in the cyclosporine group at the 3-month and 6-month follow-ups. Calculations are performed in MATLAB.

```
%Cyclosporine Clinical Trials
%
n = 37; %number of subjects in cyclosporine group
%   three months
X1 = 22;       p1hat = X1/n;    q1hat = 1-p1hat;
%   six months
X2 = 7;        p2hat = X2/n;    q2hat = 1- p2hat;
%==================================
%Wald Intervals
W3 = [p1hat   - norminv(0.975) * sqrt( p1hat * q1hat / n), ...
     p1hat   + norminv(0.975) * sqrt( p1hat * q1hat / n)]
W6 = [p2hat   - norminv(0.975) * sqrt( p2hat * q2hat / n), ...
     p2hat   + norminv(0.975) * sqrt( p2hat * q2hat / n)]
%W3 = 0.4364        0.75279
%W6 = 0.06299       0.31539
%==================================
% Wilson Intervals
    p1hats = (X1+2)/(n+4);    q1hats = 1-p1hats;
    p2hats = (X2+2)/(n+4);    q2hats = 1- p2hats;
Wi3 = [p1hats - norminv(0.975)*sqrt( p1hats * q1hats/(n+4)), ...
     p1hats + norminv(0.975) * sqrt( p1hats * q1hats/(n+4))];
Wi6 = [p2hats - norminv(0.975)*sqrt( p2hats * q2hats/(n+4)), ...
     p2hats + norminv(0.975) * sqrt( p2hats * q2hats/(n+4))];
% Wi3 =      0.43457         0.73617
% Wi6 =      0.092815        0.34621
%===========================
%Wilson Score Intervals
z=norminv(0.975);
Wis3 = [  1/(1 + z^2/n) *   (p1hat  + z^2/(2 * n)   - ...
    z * sqrt( p1hat * q1hat / n  + z^2/(4 * n^2))), ...
    1/(1 + z^2/n) * (p1hat  + z^2/(2 * n)  +  ...
    z * sqrt( p1hat * q1hat / n  + z^2/(4 * n^2)))];
Wis6 = [  1/(1 + z^2/n) *   (p2hat  + z^2/(2 * n)   - ...
    z * sqrt( p2hat * q2hat / n  + z^2/(4 * n^2))), ...
    1/(1 + z^2/n) * (p2hat  + z^2/(2 * n)  +  ...
    z * sqrt( p2hat * q2hat / n  + z^2/(4 * n^2)))];
%Wis3 =   0.43486       0.73653
%Wis6 =   0.0948        0.34205
%===========================
```

```
% Clopper-Pearson Intervals
       Fs = finv(0.975, 2*(n-X1 + 1), 2*X1);
       Fss = finv(0.975, 2*(X1+1), 2*(n-X1));
       CP3 = [ X1/(X1 + (n-X1+1).*Fs),  ...
           (X1+1).*Fss./(n - X1 + (X1+1).*Fss)];

       Fs = finv(0.975, 2*(n-X2 + 1), 2*X2);
       Fss = finv(0.975, 2*(X2+1), 2*(n-X2));
       CP6 = [ X2/(X2 + (n-X2+1).*Fs), ...
           (X2+1).*Fss./(n - X2 + (X2+1).*Fss)];
%CP3 = 0.421           0.75246
%CP6 = 0.079621        0.35155
%==========================================
% Anscombe ARCSIN intervals
%
p1h = (X1 + 3/8)/(n + 3/4);   p2h = (X2 + 3/8)/(n + 3/4);

AA3 = [(sin(asin(sqrt(p1h))-norminv(0.975)/(2*sqrt(n))))^2, ...
       (sin(asin(sqrt(p1h))+norminv(0.975)/(2*sqrt(n))))^2];
AA6 = [(sin(asin(sqrt(p2h))-norminv(0.975)/(2*sqrt(n))))^2, ...
       (sin(asin(sqrt(p2h))+norminv(0.975)/(2*sqrt(n))))^2];

%AA3 = 0.43235         0.74353
%AA6 = 0.085489        0.3366
```

Figure 7.8 shows the pairs of confidence intervals at the 3- and 6-month follow-ups. Wald's intervals are in black, Wilson's in red, the Wilson score in green, Clopper–Pearson's in magenta, and ArcSin in blue. Notice that for all methods, the confidence intervals at the 3- and 6-month follow-ups are well separated, suggesting a significant change in the proportions. There are differences among the intervals, in their centers and lengths, for a particular time of follow-up. However, as Figure 7.8 indicates, these differences are not large.

Next, we discuss the confidence interval for the probability of success when in n trials no successes have been observed.

7.5.4 Confidence Intervals for Proportions When $X = 0$

When the binomial probability is small, it is not unusual that out of n trials no successes are observed. How do we find a $(1 - \alpha)100\%$ confidence interval in such a case? The Clopper–Pearson interval is possible for $X = 0$, and it is given by $[0, 1 - (\alpha/2)^{1/n}]$.

Yet it is possible to establish an alternative interval based on the following consideration. First, we have $(1 - p)^n$ is as the probability of no success in n trials, and this probability is at least α:

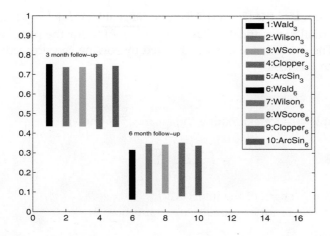

Fig. 7.8 Confidence intervals at 3- and 6-month follow-ups. Wald's intervals are in *black*, Wilson's in *red*, the Wilson Score in *green*, Clopper–Pearson's in *magenta*, and ArcSin in *blue*.

$$(1-p)^n \geq \alpha.$$

Since $n \log(1-p) \geq \log(\alpha)$ and $\log(1-p) \approx -p$, then

$$p \leq -\log(\alpha)/n.$$

This is a basis for the so-called $3/n$ rule: the 95% confidence interval for p is $[0, 3/n]$ if no successes have been observed since $-\log(0.05) = 2.9957 \approx 3$. By symmetry, the 95% confidence interval for p when n successes are observed in n experiments is $[1 - 3/n, 1]$. When n is small, this rule leads to intervals that are too wide to be useful. See Exercise 7.31 for a comparison of the Clopper–Pearson and $3/n$-rule intervals. We will argue in the next chapter that in the case where no successes are observed, one should approach the inference in a Bayesian manner.

7.5.5 Designing the Sample Size with Confidence Intervals

In all previous examples it was assumed that we had data in hand. Thus, we looked at the data after the sampling procedure had been completed. It is often the case that we have control over what sample size to adopt before the sampling. How large should the sample be? On one hand, a sample that is too small may affect the validity of our statistical conclusions. On the other hand, an unnecessarily large sample wastes money, time, and resources.

The length L of the $(1 - \alpha)100\%$ confidence interval is $L = 2z_{1-\alpha/2}\sigma/\sqrt{n}$ for the normal mean and $L = 2z_{1-\alpha/2}\sqrt{\hat{p}(1-\hat{p})/n}$ for the population proportion. The sample size n is determined by solving the preceeding equations when L is fixed.

(i) Sample size for estimating the mean: σ^2 is known:

$$n \geq \frac{4z_{1-\alpha/2}^2\sigma^2}{L^2},\qquad(7.9)$$

where L is the length of the interval.

(ii) Sample size for estimating the proportion:

$$n \geq \frac{4z_{1-\alpha/2}^2\hat{p}(1-\hat{p})}{L^2},\qquad(7.10)$$

where $\hat{p}$ is estimated or elicited.

Designing the sample size usually precedes the sampling. In the absence of data, $\hat{p}$ is elicited from experts or inferred from prior studies. In the absence of any information, the most conservative choice is $\hat{p} = 0.5$.

It is possible to express L^2 in the units of variance of observations, σ^2, for the normal and $p(1 - p)$ for the Bernoulli distribution. Therefore, it is sufficient to state that L/σ is $1/2$, for example, or that $L/\sqrt{p(1-p)}$ is $1/4$, and the required sample size can be calculated.

Example 7.11. **Cholesterol Level.** You are asked to design a cholesterol study experiment and you would like to estimate the mean cholesterol level of all students on a large metropolitan campus. You plan to take a random sample of n students and measure their cholesterol levels. Previous studies have shown that the standard deviation is 25, and you intend to use this value in planning your study. If a 99% confidence interval with a total length not exceeding 12 is desired, how many students should you include in your sample?

For a 99% confidence level, the normal 0.995 quantile is needed, $z_{0.995} = 2.58$. Then, $n \geq \frac{4\cdot2.5758^2\cdot25^2}{12^2} = 115.1892$, and desired sample size is 116 since 115.1892 should be rounded to the closest larger integer.

The *margin of error* is defined as half of the length of a 95% confidence interval for unknown proportion, location, scale, or some other population parameter of interest.

In popular use, the margin of error is usually connected with public opinion polls and represents the quantifiable sampling error built into well-designed sampling schemes. For estimating the true proportion of voters favoring a particular candidate, an approximate 95% confidence interval is

$$\left[\hat{p} - 1.96\sqrt{\hat{p}\,\hat{q}/n},\ \hat{p} + 1.96\sqrt{\hat{p}\,\hat{q}/n} \right],$$

where $\hat{p}$ is the sample proportion of voters favoring the candidate, $\hat{q} = 1 - \hat{p}$, 1.96 is the normal 97.5 percentile, and n is the sample size. Since $\hat{p}\,\hat{q} \le 1/4$, the margin of error, $1.96\sqrt{\hat{p}\,\hat{q}/n}$, is usually conservatively rounded to $\frac{1}{\sqrt{n}}$.

For example, if a survey of $n = 1600$ voters yields that 52% favor a particular candidate, then the margin of error can be estimated as $1/\sqrt{1600} = 1/40 = 0.025 = 2.5\%$ and is independent of the realized proportion of 52%.

However, if the true proportion is not close to $1/2$, the precision of the margin of error can be improved by selecting a less conservative upper bound on $\hat{p}\,\hat{q}$. For example, if a survey of $n = 1600$ citizens yields that 16% of them favor policy P, the margin of error can be estimated as $1.96 \cdot \sqrt{0.2 \cdot 0.8/1600} \approx 1/50 = 0.02 = 2\%$ provided that we are certain that the true proportion of citizens supporting policy P does not exceed 20%.

7.6 Prediction and Tolerance Intervals*

In addition to confidence intervals for parameters, a researcher may be interested in predicting future observations. This leads to prediction intervals.

We will focus on the prediction interval for predicting future observations from a normal population $\mathcal{N}(\mu, \sigma^2)$ once $X_1, \ldots, X_n$ have been observed. Any future observation will be denoted by X_{n+1}.

Consider $\overline{X}$ and X_{n+1}. These two random variables are independent and their difference has a normal distribution,

$$\overline{X} - X_{n+1} \sim \mathcal{N}(0, \sigma^2/n + \sigma^2),$$

thus, $Z = \frac{\overline{X} - X_{n+1}}{\sigma\sqrt{1+1/n}}$ has a standard normal distribution. This leads to $(1 - \alpha)100\%$ prediction intervals for X_{n+1}:

$$\left[\overline{X} - t_{n-1,1-\alpha/2}\, s\, \sqrt{1 + \frac{1}{n}},\ \overline{X} + t_{n-1,1-\alpha/2}\, s\, \sqrt{1 + \frac{1}{n}} \right].$$

When σ^2 is known, s is replaced by σ and $t_{n-1,1-\alpha/2}$ by $z_{1-\alpha/2}$.

Note that prediction intervals contain the factor $\sqrt{1 + \frac{1}{n}}$ in place of $\sqrt{\frac{1}{n}}$ in matching confidence intervals for the normal population mean. When n is large, the prediction interval can be substantially larger than the confidence interval. This is because the uncertainty about a future observation consists of (1) uncertainty about its mean and (2) uncertainty about the individual response.

Prediction intervals based on a random sample were used to predict the value of a future observation from the sampled population. In practice, the interest may be in the characteristics of a majority of the units in the population rather than a single unit or the overall mean. For example, a manufacturer of medical devices might want to learn the proportion of production for which a particular dimension falls within a given range.

Tolerance intervals (TI) are used for this purpose. A tolerance interval is constructed so that it would contain at least a specified proportion, say, $1 - \gamma$ of the population with a specified confidence, say, $1 - \alpha$. Such an interval is usually referred to as the $1 - \gamma$ content – $1 - \alpha$ coverage TI, or simply a $(1 - \gamma, 1 - \alpha)$ TI. The ends of a tolerance interval are called tolerance limits.

For normal populations, the two-sided interval is defined as

$$[\overline{X} - ks, \overline{X} + ks], \quad k = \sqrt{\frac{(n^2 - 1)\, z^2_{1-\gamma/2}}{n\, \chi^2_{n-1,\alpha}}} \qquad (7.11)$$

and interpreted as follows: With a confidence of $1 - \alpha$, the proportion $1 - \gamma$ of population measurements will fall between the lower and upper bounds in (7.11). The interval in (7.11) is called a $(1 - \gamma, 1 - \alpha)$-tolerance interval.

Example 7.12. **(0.95, 0.99)-Tolerance Interval.** For sample size $n = 20$, $\overline{X} = 12$, $s = 0.1$, confidence $1 - \alpha = 99\%$, and proportion $1 - \gamma = 95\%$, the tolerance factor k is calculated using the following MATLAB script:

```
n=20;
z = norminv(1-0.05/2)   %proportion of 1-0.05=0.95
  %z =  1.9600
xi = chi2inv(0.01, n-1)  %confidence 1-0.01=0.99
  %xi = 7.6327
k = sqrt( (n^2-1) * z^2/(n * xi) )
  %k = 3.1687
[12-k*0.1   12+k*0.1]
  %11.6831   12.3169
```

and the $(0.95, 0.99)$-tolerance interval is $[11.6831, 12.3169]$.

✎

For an example of a tolerance interval for a binomial X, see Exercise 7.37.

Example 7.13. **Distribution-Free Tolerance Intervals.** When the distribution of observations $X_1, \ldots, X_n$ is arbitrary, but continuous, and n is the smallest integer satisfying

$$(1 - \gamma/2)^n - \frac{1}{2}(1 - \gamma)^n \le \frac{\alpha}{2},$$

then the full range $(X_{(1)}, X_{(n)})$ is a $(1 - \gamma, 1 - \alpha)$ tolerance interval. For example, for a $(0.95, 0.95)$ tolerance interval in MATLAB, n can be found as

```
beta=0.05; alpha=0.05;
fzero(@(n) (1-beta/2)^n - 1/2*(1-beta)^n - alpha/2, 100)
%145.2464
```

This means that $(X_{(1)}, X_{(146)})$ is a $(0.95, 0.95)$ distribution-free tolerance interval.

✎

7.7 Confidence Intervals for Quantiles*

The confidence interval for a normal quantile is based on a noncentral t-distribution. Let $X_1, \ldots, X_n$ be a sample of size n with the sample mean $\overline{X}$ and sample standard deviation s.

we want to find a confidence interval on the population's pth quantile, $\mu + z_p \times \sigma$, with a confidence level of $1 - \alpha$.

The confidence interval is given by $[L, U]$, where

$$L = \overline{X} + s \cdot nct(\alpha/2, n - 1, \sqrt{n} \cdot z_p)/\sqrt{n},$$
$$U = \overline{X} + s \cdot nct(1 - \alpha/2, n - 1\sqrt{n} \cdot z_p)/\sqrt{n},$$

and $nct(q, df, nc)$ is the q-quantile of the noncentral t-distribution (page 264) with df degrees of freedom and noncentrality parameter nc.

The confidence intervals for quantiles can be based on order statistics when normality is not assumed. For example, instead of declaring the confidence interval for the mean, we should report the confidence interval on the median if the normality of the data is a concern. Let $X_{(1)}, X_{(2)}, \ldots, X_{(n)}$ be the order statistics of the sample. Then a $(1 - \alpha)100\%$ confidence interval for the median Me is

$$X_{(h)} \leq Me \leq X_{(n-h+1)}.$$

The value of h is usually tabulated. For large n ($n > 40$), a good approximation for h is an integer part of

$$\frac{n - z_{1-\alpha/2}\sqrt{n} - 1}{2}.$$

For example, if $n = 300$, the 95% confidence interval for the median is $[X_{(132)}, X_{(169)}]$ as demonstrated below:

```
n = 300;
h = floor( (n - 1.96 * sqrt(n) - 1)/2 )
% h = 132
n - h + 1
% ans =   169
```

7.8 Confidence Intervals for the Poisson Rate*

Recall that an observation X coming from $\mathcal{P}oi(\lambda)$ has both a mean and a variance equal to the rate parameter, $\mathbb{E}X = \mathbb{V}ar\, X = \lambda$. Also, Poisson random variables are additive in the rate parameter:

$$X_1, X_2, \ldots, X_n \sim \mathcal{P}oi(\lambda) \qquad \Rightarrow \qquad n\overline{X} = \sum_{i=1}^{n} X_i \sim \mathcal{P}oi(n\lambda). \quad (7.12)$$

The asymptotically shortest Wald-type $(1 - \alpha)100\%$ interval for λ is obtained using the fact that $Z = \sqrt{\frac{n}{\lambda}}\,(\overline{X} - \lambda)$ is approximately the standard normal. The inequality

$$\sqrt{\frac{n}{\lambda}}\,|\overline{X} - \lambda| \leq z_{1-\alpha/2}$$

leads to

$$\lambda^2 - \lambda\left(2\overline{X} + \frac{z_{1-\alpha/2}^2}{n}\right) + (\overline{X})^2 \leq 0,$$

which solves to

$$\left[\overline{X} + \frac{z_{1-\alpha/2}^2}{2n} - z_{1-\alpha/2}\sqrt{\frac{\overline{X}}{n} + \frac{z_{1-\alpha/2}^2}{4n^2}}, \right.$$

$$\left. \overline{X} + \frac{z_{1-\alpha/2}^2}{2n} + z_{1-\alpha/2}\sqrt{\frac{\overline{X}}{n} + \frac{z_{1-\alpha/2}^2}{4n^2}} \right] . \quad (7.13)$$

Other Wald-type intervals are derived from the fact that $\frac{\sqrt{\overline{X}} - \sqrt{\lambda}}{\sqrt{1/(4n)}}$ is approximately the standard normal. Then, the variance-stabilizing, modified variance-stabilizing, and recentered variance-stabilizing $(1 - \alpha)100\%$ confidence intervals are given as

$$\left[\overline{X} - z_{1-\alpha/2}\sqrt{\frac{\overline{X}}{n}}, \quad \overline{X} + z_{1-\alpha/2}\sqrt{\frac{\overline{X}}{n}} \right],$$

$$\left[\overline{X} + \frac{z_{1-\alpha/2}^2}{4n} - z_{1-\alpha/2}\sqrt{\frac{\overline{X}}{n}}, \quad \overline{X} + \frac{z_{1-\alpha/2}^2}{4n} + z_{1-\alpha/2}\sqrt{\frac{\overline{X}}{n}} \right], \quad \text{and}$$

$$\left[\overline{X} + \frac{z_{1-\alpha/2}^2}{4n} - z_{1-\alpha/2}\sqrt{\frac{\overline{X} + 3/8}{n}}, \quad \overline{X} + \frac{z_{1-\alpha/2}^2}{4n} + z_{1-\alpha/2}\sqrt{\frac{\overline{X} + 3/8}{n}} \right].$$

Details can be found in Barker (2002).

An alternative approach is based on the link between Poisson and chi-square distributions. Namely, if $X \sim \mathcal{P}oi(\lambda)$, then

$$\mathbb{P}(X > x) = \mathbb{P}(Y < 2\lambda), \text{ for } Y \sim \chi_{2x}^2$$

and the $(1 - \alpha)100\%$ confidence interval for λ when X is observed is

$$\left[\frac{1}{2} \chi_{2X,\alpha/2}^2, \frac{1}{2} \chi_{2(X+1),1-\alpha/2}^2 \right],$$

where $\chi_{2X,\alpha/2}^2$ and $\chi_{2(X+1),1-\alpha/2}^2$ are $\alpha/2$ and $1 - \alpha/2$ quantiles of the χ^2-distribution with $2X$ and $2(X + 1)$ degrees of freedom, respectively. By convention, $\chi_{0,\alpha}^2 = 0$. Due to the additivity property (7.12), the confidence interval changes slightly for the case of an observed sample of size n, $X_1, X_2, \ldots, X_n$. One finds $S = \sum_{i=1}^n X_i$, which is a Poisson with parameter $n\lambda$ and proceeds as in the single-observation case. Because the interval obtained is for $n\lambda$, the bounds should be divided by n to get the interval for λ:

$$\left[\frac{1}{2n} \, \chi^2_{2S,\alpha/2}, \, \frac{1}{2n} \, \chi^2_{2(S+1),1-\alpha/2} \right] . \qquad (7.14)$$

The interval in (7.14) is sometimes referred to as Garwood's interval (Garwood, 1936).

Example 7.14. **Counts of α-Particles.** Rutherford et al. (1930, pp. 171–172) provide descriptions and data on an experiment by Rutherford and Geiger (1910) on the collision of α-particles emitted from a small bar of polonium with a small screen placed at a short distance from the bar. The number of such collisions in each of 2,608 eight-minute intervals was recorded. The distance between the bar and screen was gradually decreased so as to compensate for the decay of radioactive substance.

X	0	1	2	3	4	5	6	7	8	9	10	11	≥ 12
Freq	57	203	383	525	532	408	273	139	45	27	10	4	2

It is postulated that because of the large number of atoms in the bar and the small probability of any of them emitting a particle, the observed frequencies should be well modeled by a Poisson distribution.

```
%Rutherford.m
X=[ 0  1  2  3  4  5  6  7  8  9  10  11  12 ];
fr=[ 57 203  383  525  532  408  273  139  45  27  10  4  2];
n = sum(fr); %number of experiments//time intervals
rfr = fr./n;      %relative frequencies %n=2608
xbar = X * rfr' ;  %lambdahat = xbar =  3.8704
tc = X * fr';     %total number of counts  tc = 10094
 %Recentered Variance Stabilizing
[xbar + (norminv(0.975))^2/(4*n) - ...
        norminv(0.975) * sqrt(( xbar + 3/8)/n )...
  xbar + (norminv(0.975))^2/(4*n)  + ...
        norminv(0.975) * sqrt( (xbar+ 3/8)/n )]
              %   3.7917     3.9498
 % Garwood's interval
[1/(2 *n) * chi2inv(0.025, 2 * tc)  ...
        1/(2 * n) * chi2inv(0.975, 2*(tc + 1))]
        %  3.7953     3.9467
```

The estimator for λ is $\hat{\lambda} = \overline{X} = 3.8704$, the Wald-type recentered variance stabilizing interval is $[3.7917, 3.9498]$, and the Garwood confidence interval is $[3.7953, 3.9467]$. The intervals are very close to each other and quite tight due to the large sample size.

7.9 Exercises

7.1. **Tricky Estimation.** A publisher gives the proofs of a new book to two different proofreaders, who read it separately and independently. The first proofreader found 60 misprints, the second proofreader found 70 misprints, and 50 misprints were found by both. Estimate how many misprints remain undetected in the book? *Hint:* Refer to Example 5.10.

7.2. **Laplace's Rule of Succession.** Laplace's Rule of Succession states that if an event appeared X times out of n trials, the probability that it will appear in a future trial is $\frac{X+1}{n+2}$.
(a) If $\frac{X+1}{n+2}$ is taken as an estimator for binomial p, compare the MSE of this estimator with the MSE of the traditional estimator, $\hat{p} = \frac{X}{n}$.
(b) Represent MSE from (a) as the sum of the estimator's variance and the bias squared.

7.3. **Neurons Fire in Potter's Lab.** The data set ⬚neuronfires.mat was compiled in Professor Steve Potter's lab at Georgia Tech. It consists of 989 firing times of a cell culture of neurons. The recorded firing times are time instances when a neuron sent a signal to another linked neuron (a spike). The cells from the cortex of an embryonic rat brain were cultured for 18 days on multielectrode arrays. The measurements were taken while the culture was stimulated at a rate of 1 Hz. It was postulated that firing times form a Poisson process; thus, the interspike intervals should have an exponential distribution.
(a) Calculate the interspike intervals T using MATLAB's diff command. Check the histogram for T and discuss its resemblance to the exponential density. By the moment-matching estimator, argue that exponential parameter λ is close to 1.
(b) According to (a), the model for interspike intervals is $T \sim \mathcal{E}(1)$. You are interested in the proportion of intervals that are shorter than 3, $T \le 3$. Find this proportion from the theoretical model $\mathcal{E}(1)$ and compare it to the estimate from the data. For the theoretical model, use expcdf and for empirical data use sum(T <= 3)/length(T).

7.4. **Moment Matching Uniform.** Let $X_1, X_2, \ldots, X_n$ be a sample from uniform $\mathcal{U}(\mu - \delta, \mu + \delta)$ distribution. Show that the moment-matching estimators of μ and δ are

$$\hat{\mu} = \overline{X} \quad \text{and} \quad \hat{\delta} = \sqrt{3 \left(\overline{X^2} - (\overline{X})^2 \right)},$$

where $\overline{X}$ and $\overline{X^2}$ are first and second sample moments.

7.5. **The MLE in a Discrete Case.** A sample $-1,1,1,0,-1,1,1,1,0,1,1,0,-1,1,1$ was observed from a population with a probability mass function of

X	−1	0	1
Prob	θ	2θ	$1-3\theta$

(a) What is the possible range for θ?
(b) What is the MLE for θ?
(c) How would the MLE look for a sample of size n?

7.6. **Two Thetas.** (a) A sample of size $n = 10$,

$$0,0,1,1,1,2,1,1,0, \text{ and } 2,$$

is obtained from a partially specified discrete distribution

X	0	1	2
Prob	θ	$1/2$	$1/2-\theta$

How would you estimate θ given the sample?

(b) A sample of size $n = 4$,

$$1.1,\ 0.7,\ 0.5,\ \text{and}\ 1.7,$$

is obtained from normal $\mathcal{N}(\theta, 0.6^2)$ distribution. As a confidence interval (CI) for θ, the interval $[0,2]$ is proposed. With what confidence does this interval contain θ?

7.7. **MLE for Two Continuous Distributions.** Find the MLE for parameter θ if the model for observations $X_1, X_2, \ldots, X_n$, is

$$\text{(a)} \quad f(x|\theta) = \frac{\theta}{x^2},\ 0 < \theta \leq x;$$

$$\text{(b)} \quad f(x|\theta) = \frac{\theta-1}{x^\theta},\ x \geq 1,\ \theta > 1.$$

7.8. **Match the Moment.** The geometric distribution (X is the number of failures before the first success) has a probability mass function

$$f(x|p) = (1-p)^x\, p,\ \ x = 0,1,2,\ldots.$$

Suppose $X_1, X_2, \ldots, X_n$ are observations from this distribution. It is known that $\mathbb{E}X_i = \frac{1-p}{p}$.
(a) What would you report as the moment-matching estimator if the sample $X_1 = 2, X_2 = 6, X_3 = 1$ were observed?
(b) What is the MLE for p?

7.9. **Weibull Distribution.** The two-parameter Weibull distribution is given by the density

$$f(x) = a\lambda^a x^{a-1} e^{-(\lambda x)^a}, \qquad a > 0, \lambda > 0, x \geq 0,$$

with mean and variance

$$\mathbb{E}X = \frac{\Gamma(1+1/a)}{\lambda}, \quad \text{and} \quad \mathbb{V}\text{ar}\, X = \frac{1}{\lambda^2}\left[\Gamma(1+2/a) - \Gamma(1+1/a)^2\right].$$

Assume that the "shape" parameter a is known and equal to $1/2$.
(a) Propose two moment-matching estimators for λ.
(b) If $X_1 = 1, X_2 = 3, X_3 = 2$, what are the values of the estimator?

Hint: Recall that $\Gamma(n) = (n-1)!$

7.10. **Rate Parameter of Gamma.** Let $X_1, \ldots, X_n$ be a sample from a gamma distribution given by the density

$$f(x) = \frac{\lambda^a x^{a-1}}{\Gamma(a)} e^{-\lambda x}, \qquad a > 0, \lambda > 0, x \geq 0,$$

where shape parameter a is known and rate parameter λ is unknown and of interest.
(a) Find the MLE of λ.
(b) Using the fact that $X_1 + X_2 + \cdots + X_n$ is also gamma distributed with parameters na and λ, find the expected value of the MLE from (a) and show that it is a biased estimator of λ.
(c) Modify the MLE so that it is unbiased. Compare MSEs for the MLE and the modified estimator.

7.11. **Estimating the Parameter of a Rayleigh Distribution.** If two random variables X and Y are independent of each other and normally distributed with variances equal to σ^2, then the variable $R = \sqrt{X^2 + Y^2}$ follows the Rayleigh distribution with scale parameter σ. An example of such a variable would be the distance of darts from the target in a dart-throwing game where the deviations in the two dimensions of the target plane are independent and normally distributed. The Rayleigh random variable R has a density

$$f(r) = \frac{r}{\sigma^2} \exp\left\{-\frac{r^2}{2\sigma^2}\right\}, \quad r \geq 0,$$

$$\mathbb{E}R = \sigma\sqrt{\frac{\pi}{2}} \qquad \mathbb{E}R^2 = 2\sigma^2.$$

(a) Find the two moment-matching estimators of σ.

(b) Find the MLE of σ.

(c) Assume that $R_1 = 3, R_2 = 4, R_3 = 2$, and $R_4 = 5$ are Rayleigh-distributed random observations representing the distance of a dart from the center. Estimate the variance of the horizontal error, which is theoretically a zero-mean normal.

(d) In Example 5.29, the distribution of a square root of an exponential random variable with a rate parameter λ was Rayleigh with the following density:

$$f(r) = 2\lambda r \exp\{-\lambda r^2\}.$$

To find the MLE for λ, can you use the MLE for σ from (b)?

7.12. **Monocytes among Blood Cells.** Eisenhart and Wilson (1943) report the number of monocytes in 100 blood cells of a cow in 113 successive weeks.

Monocytes	Frequency	Monocytes	Frequency
0	0	7	12
1	3	8	10
2	5	9	11
3	13	10	7
4	19	11	3
5	13	12	2
6	15	13+	0

(a) If the underlying model is Poisson, what is the estimator of λ?

(b) If the underlying model is binomial $Bin(100, p)$, what is the estimator of p?

(c) For the models specified in (a) and (b) find theoretical or "expected" frequencies.

Hint: Suppose the model predicts $\mathbb{P}(X = k) = p_k$, $k = 0, 1, \ldots, 13$. The expected frequency of $X = k$ is $113 \times p_k$. For a follow-up see Exercise 17.7.

7.13. **Estimation of θ in $\mathcal{U}(0, \theta)$.** Which of the two estimators in Example 7.5 is unbiased? Find the MSE of both estimators. Which one has a smaller MSE?

7.14. **Estimating the Rate Parameter in a Double Exponential Distribution.** Let $X_1, \ldots, X_n$ follow double exponential distribution with density

$$f(x|\theta) = \frac{\theta}{2} e^{-\theta|x|}, \quad -\infty < x < \infty, \ \theta > 0.$$

For this distribution, $\mathbb{E}X = 0$ and $\mathbb{V}ar(X) = \mathbb{E}X^2 = 2/\theta^2$. The double exponential distribution, also known as Laplace's distribution, is a model frequently encountered in statistics, see page 207.

(a) Find a moment-matching estimator for θ.

(b) Find the MLE of θ.

(c) Evaluate the two estimators from (a) and (b) for a sample $X_1 = -2, X_2 = 3, X_3 = 2$, and $X_4 = -1$.

7.15. **Reaction Times I.** A sample of 20 students is randomly selected and given a test to determine their reaction time in response to a given stimulus. Assume that individual reaction times are normally distributed. If the mean reaction time is determined to be $\overline{X} = 0.9$ (in seconds) and the standard deviation is $s = 0.12$, find the confidence intervals:

(a) 95% CI for the unknown population mean μ.

(b) 98.5% CI interval for the unknown population mean μ.

(c) 95% CI for the unknown population variance σ^2.

7.16. **Reaction Times II.** Under the conditions in the previous problem, assume that the population standard deviation was known to be $\sigma = 0.12$.

(a) Find the 98.5% CI for the unknown mean μ;

(b) Find the sample size necessary to produce a 95% CI for μ of length 0.07.

7.17. **Toxins.** An investigation on toxins produced by molds that infect corn crops was performed. A biochemist prepared extracts of the mold culture with organic solvents and then measured the amount of toxic substance per gram of solution. From 11 preparations of the mold culture, the following measurements of the toxic substance (in milligrams) were obtained: 3, 2, 5, 3, 2, 6, 5, 4.5, 3, 3, and 4.

Compute a 99% confidence interval for the mean weight of toxic substance per gram of mold culture. State the assumption you make about the population.

7.18. **Bias of s_*^2.** For a sample $X_1, \ldots, X_n$ from a $\mathcal{N}(\mu, \sigma^2)$ population, find the bias of $s_*^2 = \frac{1}{n} \sum_i (X_i - \overline{X})^2$ as an estimator of variance σ^2.

Using (7.3), show that the variance of s_*^2 is smaller than the variance of unbiased estimator s^2.

7.19. **COPD Patients.** Acute exacerbations of disease symptoms in patients with chronic obstructive pulmonary disease (COPD) often lead to hospitalizations and impose a great financial burdens on healthcare systems. A study by Ghanei et al. (2007) aimed to determine factors that may predict rehospitalization in COPD patients.

A total of 157 COPD patients were randomly selected from all COPD patients admitted to the chest clinic of Baqiyatallah Hospital during the year 2006. Subjects were followed for 12 months to observe the occurrence of any disease exacerbation that might lead to hospitalization. Over the 12-month period, 87 patients experienced disease exacerbation. The authors found significant associations between COPD exacerbation and monthly income, comorbidity score, and depression using logistic regression tools. We are not interested in these associations in this exercise, but we are interested in the population proportion of all COPD patients that experienced disease exacerbation over a 12-month period, p.

(a) Find an estimator of p based on the data available. What is an approximate distribution of this estimator?

(b) Find the 90% confidence interval for the unknown proportion p.

(c) How many patients should be sampled and monitored so that the 90% confidence interval as in (b) does not exceed 0.03 in length.

(d) The hospital challenges the claim by the local health system authorities that half of the COPD patients experience disease exacerbation in a 1-year period, claiming that the proportion is significantly higher. Can the hospital support their claim based on the data available? Use $\alpha = 0.05$. Would you reverse the decision if α were changed to 10%?

7.20. **Right to Die.** A Gallup Poll estimated the support among Americans for "right to die" laws. In the survey, 1528 adults were asked whether they favor voluntary withholding of life-support systems from the terminally ill. The results: 1238 said yes.

(a) Find the 99% confidence interval for the percentage of all adult Americans who are in favor of "right to die" laws.

(b) If the margin of error (half of the length of a 95% confidence interval, see page 310) is to be smaller than 0.01, what sample size is needed to achieve this requirement? Assume $\hat{p} = 0.8$.

7.21. **Exponentials Parameterized by the Scale.** A sample $X_1, \ldots, X_n$ was selected from a population that has an exponential $\mathcal{E}(\lambda)$ distribution with a density of $f(x|\lambda) = \frac{1}{\lambda} e^{-\frac{x}{\lambda}}$, $x \geq 0, \lambda > 0$. We are interested in estimating the parameter λ.

(a) What are the moment-matching and MLE estimators of λ based on $X_1, \ldots, X_n$?

(b) Two independent observations $Y_1 \sim \mathcal{E}(\lambda/2)$ and $Y_2 \sim \mathcal{E}(2\lambda)$ are available. Combine them (make a specific linear combination) to obtain an unbiased estimator of λ. What is the variance of the proposed estimator?

(c) Two independent observations $Z_1 \sim \mathcal{E}(1.1\lambda)$ and $Z_2 \sim \mathcal{E}(0.9\lambda)$ are available. An estimator of λ in the form $\hat{\lambda} = pZ_1 + (1-p)Z_2$, $0 \leq p \leq 1$ is proposed. What p minimizes the magnitude of bias of $\hat{\lambda}$? What p minimizes the variance of $\hat{\lambda}$?

7.22. **Bias in Estimator for Exponential λ Distribution.** If the exponential distribution is parameterized with λ as the scale parameter, $f(x|\lambda) = \frac{1}{\lambda} \exp\{-x/\lambda\}$, $x \geq 0, \lambda > 0$, (as in MATLAB), then $\hat{\lambda} = \overline{X}$ is an unbiased estimator of λ. However, if it is parameterized with λ as a rate parameter, $f(x|\lambda) = \lambda \exp\{-\lambda x\}$, $x \geq 0, \lambda > 0$, then $\hat{\lambda} = 1/\overline{X}$ is biased. Find the bias of this estimator. *Hint:* Argue that $1/\sum_{i=1}^{n} X_i$ has an inverse gamma distribution with parameters n and λ and take the expectation.

7.23. **Yucatan Miniature Pigs.** Ten adult male Yucatan miniature pigs were exposed to various durations of constant light ("Lighting"), then sacrificed after experimentally controlled time delay ("Survival"), as described in Dureau et al. (1996). Following the experimental protocol, entire eyes were fixed in Bouin's fixative for 3 days. The anterior segment (cornea, iris, lens, ciliary body) was then removed and the posterior segment divided into five regions: posterior pole (including optic nerve head and macula) ("P"), nasal ("N"), temporal ("T"), superior ("S"), and inferior ("I"). Specimens were washed for 2 days, embedded in paraffin, and subjected to microtomy perpendicular to the retinal surface. Every $200\,\mu$m, a 10-μm-thick section was selected, and 20 sections were kept for each retinal region. Sections were stained with hematoxylin. The outer nuclear layer (ONL) thickness was measured by an image-analyzing system (Biocom, Les Ulis, France), and three measures were performed for each section at regularly spaced intervals so that 60 measures were made for each retinal region. The experimental protocol for 11 animals was as follows (Lighting and Survival times are in weeks):

Animal	Lighting duration	Survival time
Control	0	0
1	1	12
2	2	10
3	4	0
4	4	4
5	4	6
6	8	0
7	8	4
8	8	8
9	12	0
10	12	4

The data set 🖳 pigs.mat contains the data structure pigs with

pigs.pc, pigs.p1,...,pigs.p10, representing the posterior pole measurements for the 11 animals. This data set and complete data yucatanpigs.dat can be found on the book's website page.

Observe the data pigs.pc and argue that it deviates from normality by using MATLAB's qqplot. Transform pigs.pc as x = (pigs.pc - 14)/(33 -14), to confine x between 0 and 1 and assume a beta $\mathcal{Be}(a,a)$ distribution. The MLE for a is complex (involves a numerical solution of equations with digamma functions), but the moment-matching estimator is straightforward.

Find a moment-matching estimator for a.

7.24. **Computer Games.** According to Hamilton (1990), certain computer games are thought to improve spatial skills. A mental rotations test, measuring spatial skills, was administered to a sample of school children after they had played one of two types of computer game. Construct 95% confidence intervals based on the following mean scores, assuming that the children were selected randomly and that the mental rotations test scores had a normal distribution in the population.
(a) After playing the "Factory" computer game: $\overline{X} = 22.47, s = 9.44, n = 19$.
(b) After playing the "Stellar" computer game: $\overline{X} = 22.68, s = 8.37, n = 19$.
(c) After playing no computer game (control group): $\overline{X} = 18.63, s = 11.13, n = 19$.

7.25. **Effectiveness in Treating Cerebral Vasospasm.** In a study on the effectiveness of hyperdynamic therapy in treating cerebral vasospasm, Pritz et al. (1996) reported on the therapy where success was defined as clinical improvement in terms of neurological deficits. The study reported 16 successes in 17 patients.
(a) Using the methods discussed in the text, find 95% confidence intervals for the success rate.
(b) Does any of the methods produce an upper bound larger than 1?
(c) How would you find the 95% confidence interval if the study reported 17 successes in 17 patients?

7.26. **Alcoholism and the Blyth–Still Confidence Interval.** Genetic markers were observed for a group of 50 Caucasian alcoholics in a study that aimed at determining whether alcoholism has (in part) a genetic basis. The antigen (marker) B15 was present in 5 alcoholics. Find the Blyth–Still 99% confidence interval for the proportion of Caucasian alcoholics having this antigen.

If either p or q is close to 0, then a precise $(1 - \alpha)100\%$ confidence interval for the unknown proportion p was proposed by Blyth and Still (1983). For $X \sim Bin(n, p)$,

$$
\left[\frac{(X - 0.5) + \frac{z^2_{1-\alpha/2}}{2} - z_{1-\alpha/2}\sqrt{(X - 0.5) - \frac{(X-0.5)^2}{n} + \frac{z^2_{1-\alpha/2}}{4}}}{n + z^2_{1-\alpha/2}}, \right.
$$

$$
\left. \frac{(X + 0.5) + \frac{z^2_{1-\alpha/2}}{2} + z_{1-\alpha/2}\sqrt{(X + 0.5) - \frac{(X+0.5)^2}{n} + \frac{z^2_{1-\alpha/2}}{4}}}{n + z^2_{1-\alpha/2}} \right]
$$

7.27. **Spores of *Amanita Phalloides*.** Exercise 2.4 provides measurements in μm of 28 spores of the mushroom *Amanita phalloides*.

Assuming normality of measurements, find the following:

(a) A point estimator for the unknown population variance σ^2. What is the sampling distribution of the point estimator?

(b) A 90% confidence interval for the population variance.

(c) (By MATLAB) the minimal sample size that ensures that the upper bound U of the 90% confidence interval for the variance is at most 30% larger than the lower bound L, that is, $U/L \leq 1.3$.

(d) Miller (1991) showed that the coefficient of variation in a normal sample of size n has an approximately normal distribution:

$$s/\overline{X} \overset{approx}{\sim} \mathcal{N}\left(\frac{\sigma}{\mu}, \frac{1}{n-1}\left(\frac{\sigma}{\mu}\right)^2\left[\frac{1}{2} + \left(\frac{\sigma}{\mu}\right)^2\right]\right).$$

Based on this asymptotic distribution, a $(1-\alpha)100\%$ confidence interval for the population coefficient of variation $\frac{\sigma}{\mu}$ is approximately

$$\left(\frac{s}{\overline{X}} - z_{1-\alpha/2}\frac{s}{\overline{X}}\sqrt{\frac{1}{n-1}\left[\frac{1}{2} + \left(\frac{s}{\overline{X}}\right)^2\right]}, \frac{s}{\overline{X}} + z_{1-\alpha/2}\frac{s}{\overline{X}}\sqrt{\frac{1}{n-1}\left[\frac{1}{2} + \left(\frac{s}{\overline{X}}\right)^2\right]}\right).$$

This approximation works well if n exceeds 10 and the coefficient of variation is less than 0.7. Find the 95% confidence interval for the population coefficient of variation σ/μ based on 28 spores measurements.

(e) Standardly used $(1-\alpha)100\%$ confidence interval for σ/μ is McKay's interval (McKay. 1932),

$$\left[\frac{s}{\overline{X}}\left[\left(\frac{u_1}{n} - 1\right)\left(\frac{s}{\overline{X}}\right)^2 + \frac{u_1}{n-1}\right]^{-1/2}, \frac{s}{\overline{X}}\left[\left(\frac{u_2}{n} - 1\right)\left(\frac{s}{\overline{X}}\right)^2 + \frac{u_2}{n-1}\right]^{-1/2}\right],$$

where $u_1 = \chi^2_{n-1,1-\alpha/2}$, and $u_2 = \chi^2_{n-1,\alpha/2}$.

Find the McKay's 95% confidence interval for σ/μ.

7.28. **CLT-Based Confidence Interval for Normal Variance.** Refer to Example 7.9. Using MATLAB, simulate a normal sample with mean 0 and variance 1 of size $n = 50$ and find if a 95% confidence interval for the population variance contains a 1 (the true population variance). Check this coverage for a standard confidence interval in (7.8) and for a CLT-based interval from Example 7.9. Repeat this simulation $M = 10000$ times, keeping track of the number of successful coverages. Show that the interval (7.8) achieves the nominal coverage, while the CLT-based interval has a smaller coverage of about 2%. Repeat the simulation for sample sizes of $n = 30$ and $n = 200$.

7.29. **Stent Quality Control.** A stent is a tube or mechanical scaffold used to counteract significant decreases in vessel or duct diameter by acutely propping open the conduit. Stents are often used to alleviate diminished blood flow to organs and extremities beyond an obstruction in order to maintain an adequate delivery of oxygenated blood.

In the production of stents, the quality control procedure aims to identify defects in composition and coating. Precision z-axis measurements (10 nm and greater) are obtained, along with surface roughness and topographic surface finish details, using a laser confocal imaging system (an example is the Olympus LEXT OLS3000). Samples of 50 stents from a production process are selected every hour. Typically, 1% of stents are nonconforming. Let X be the number of stents in the sample of 50 that are nonconforming. A production problem is suspected if X exceeds its mean by more than three standard deviations.

(a) Find the critical value for X that will implicate a production problem.

(b) Find an approximation for the probability that in the next-hour batch of 50 stents, the number X of nonconforming stents will be critical, i.e., will raise suspicion that the process has gone awry.

(c) Suppose now that the population proportion of nonconforming stents, p, is unknown. How would one estimate p by taking a 50-stent sample? Is the proposed estimator unbiased?

(d) Suppose now that a batch of 50 stents produced $X = 1$. Find the 95% confidence interval for p.

7.30. **Clopper–Pearson and $3/n$-Rule Confidence Intervals.** Using MATLAB, compare the performance of Clopper–Pearson and $3/n$-rule confidence intervals when $X = 0$. Use $\alpha = 0.001, 0.005, 0.01, 0.05, 0.1$ and $n = 10 : 10 : 200$. Which interval is superior and under what conditions?

7.31. **Fluid Overload in Hemodialysis.** The overload of fluid volume and hypertension are known to contribute to high cardiovascular morbidity and mortality seen in dialysis patients. The correct assessment of volume status is especially important as only a small increase in extracellular volume over prolonged periods of time can lead to a considerable cardiac strain and, as a consequence, to left ventricular hypertrophy. In clinical practice, volume overload is most often judged by a battery of clinical signs such as edema, dyspnea, hypertension, and coughing. A study by Ribitsch et al. (2012) compares volume overload in stable hemodialysis (HD) patients assessed by standard clinical judgment with data obtained from bioimpedance analysis.

Data set 💾 hemodialysis.dat|mat|xlsx provides measurements on 28 HD patients (17 males and 11 females) from the dialysis unit of the University Medical Center Graz. The variables are described in the following table:

Column	Variable	Unit	Description
1	M_0	(kg)	Pre-dialytic body mass
2	BMI	(kg/m^2)	Body mass index
3	P_0	(mmHg)	Pre-dialytic mean arterial pressure
4	P_1	(mmHg)	Post-dialytic mean arterial pressure
5	V_E	(L)	Extracelular volume
6	V_O	(L)	Volume overload
7	V_U	(L)	Delivered ultrafiltration volume
8	B_0	(pg/ml)	Pre-dialytic NT-pro-BNP
9	B_1	(pg/ml)	Post-dialytic NT-pro-BNP
10	S_W		Wizemann's clinical score

(a) Find the 95% CI for the population mean of the difference $D = P_1 - P_0$ in stable hemodialysis patients. Assume that this difference is normally distributed.

(b) Find the 90% CI for the population variance of V_0. Assume normality of V_0.

(c) Find the 99% CI for the population proportion of patients for which $B_1 > B_0$.

7.32. **Sensor Agreement.** A company producing an approved medical sensor A is applying to FDA for the approval of a new sensor B. Both sensors are prone to errors, and a gold standard is absent. The FDA is requesting that the new sensor is comparable to the one currently in use. Data are

		Sensor A	
		Result +	Result −
Sensor B	Result +	208	22
	Result −	11	5819

Find agreement rate $\hat{p}$, that is, the proportion of cases where the sensors agreed (both positive or both negative).

Calculate the 95% Clopper–Pearson CI for the population agreement rate p, and report the lower bound. To establish equivalence, the FDA requires for this lower bound to be at least 0.98. Is this the case?

7.33. **Seventeen Pairs of Rats, Carbon Tetrachloride, and Vitamin B.** In a widely cited experiment by Sampford and Taylor (1959), 17 pairs of rats were formed by selecting pairs from the same litter. All rats were given carbon tetrachloride, and one rat from each pair was treated with vitamin B_{12}, while the other served as a control. In 7 of 17 pairs, the treated rat outlived the control rat.

(a) Based on this experiment, estimate the population proportion p of pairs in which the treated rat would outlive the control rat.

(b) If the estimated proportion in (a) is the "true" population probability, what is the chance that in an independent replication of this experiment one will get exactly 7 pairs (out of 17) in which the treated rat outlives the control?

(c) Find the 95% confidence interval for the unknown p. Does the interval contain $1/2$? What does $p = 1/2$ mean in the context of this experiment, and what do you conclude from the confidence interval?

Would the conclusion be the same if in 140 out of 340 pairs the treated rat outlived the control?

(d) The length of the 95% confidence interval based on $n = 17$ in (c) may be too large. What sample size (number of rat pairs) is needed so that the 95% confidence interval has a length not exceeding $\ell = 0.2$?

7.34. **Hemocytometer Counts.** A set of 1,600 squares on a hemocytometer is inspected, and the number of cells is counted in each square. The number of squares with a particular count is given in the table below:

Count	0	1	2	3	4	5	6	7
# Squares	5	24	77	139	217	262	251	210
Count	8	9	10	11	12	13	14	15
# Squares	175	108	63	36	20	9	2	1

Assume that the count has a Poisson $\mathcal{P}oi(\lambda)$ distribution.

(a) Find an estimator of λ using method of moments.

(b) Find the 95% CI for λ. Compare solutions obtained by alternative intervals in (7.13) and (7.14).

7.35. **Predicting Alkaline Phosphatase.** Refer to BUPA liver disorder data, BUPA.dat|mat|xlsx. The second column gives measurements of alkaline phosphatase among 345 male individuals affected by liver disorder. If variable X represents the logarithm of this measurement, its distribution is symmetric and bell-shaped, so it can be assumed normal. From the data, $\overline{X} = 4.21$ and $s^2 = 0.0676$.

Suppose that a new patient with liver disorder just checked in. Find the 95% prediction interval for his log-level of alkaline phosphatase in the following cases:

(a) The population variance is known and equal to $1/15$.

(b) The population variance is not known.

(c) Compare the interval in (b) with a 95% confidence interval for the population mean. Why is the interval in (b) larger?

7.36. **CNFL for DSP.** Corneal nerve fiber length (CNFL), as measured using corneal confocal microscopy (CCM), can be used to reliably rule diabetic sensorimotor polyneuropathy (DSP) in or out, according to research published online on February 8, 2012, in *Diabetes Care*, doi:10.2337/dc11-1396. Part of the reported results can be summarized as follows:

	DSP	No DSP	Total
CNFL ≤ 140 (Positive)	28	20	48
CNFL > 140 (Negative)	5	100	105
Total	33	120	153

Find the 95% CI's for population sensitivity and specificity using:
(a) Wald's interval;
(b) Clopper–Pearson's interval; and
(c) Anscombe's ArcSin interval.
Which one is the shortest?
(d) It is desired to repeat the study and design sample sizes of DSP and control subjects that would lead to Wald-type 95% confidence intervals on sensitivity and specificity not exceeding 0.16 in length each.
Hint. Assume that data from the table can be used in assessing the sensitivity/specificity needed in the expression for sample size. Use the sample size formula in (7.10). Since the sensitivity gives sample size for cases and specificity gives sample size for controls, the total sample size for the new study should be the sum of the two sample sizes found.

7.37. **Tolerance Interval for Binomial** X. A $(1 - \gamma, 1 - \alpha)$ tolerance interval for binomial $X \sim Bin(n, p)$ is determined in two stages. In stage one, a $(1 - \alpha)100\%$ confidence interval on p is found, (p_L, p_U). In stage two, the tolerance bounds are determined via the quantiles of binomial distribution,

$$\left[F^{-1}\left(\frac{\gamma}{2}, n, p_L\right), F^{-1}\left(1 - \frac{\gamma}{2}, n, p_U\right) \right].$$

Here $F^{-1}(\alpha, n, p)$ is α-quantile of binomial $Bin(n, p)$ distribution, binoinv (alpha,n,p).

In a previous experiment, the number of "successes" was 46 out of 100 trials. What is the (0.95, 0.95) tolerance interval for number of successes in a future experiment with 100 trials?

MATLAB FILES AND DATA SETS USED IN THIS CHAPTER
http://statbook.gatech.edu/Ch7.Estim/

AmanitaCI.m, arcsinint.m, bickellehmann.m, clopperint.m, CLTvarCI.m, confintscatterpillar.m, crouxrouss.m, crouxrouss2.m, cyclosporine.m, dists2.m, estimweibull.m, ginimd.m, ginimd2.m, lfev.m, MaxwellMLE.m, MixtureModelExample.m, muscaria.m, plotlike.m, Rutherford.m, simuCI.m, tolerance.m, waldsimulation.m

amanita28.dat, hemodialysis.dat|xlsx, hypertension.dat, neuronfires.dat|mat|xlsx

---

CHAPTER REFERENCES

Agresti, A. and Coull, B. A. (1998). Approximate is better than "exact" for interval estimation of binomial proportions. *Am. Stat.*, **52**, 119–126.

Barker, L. (2002). A comparison of nine confidence intervals for a Poisson parameter when the expected number of events is ≤ 5. *Am. Stat.*, **56**, 85–89.

Blyth, C. and Still, H. (1983). Binomial confidence intervals. *J. Am. Stat. Assoc.*, **78**, 108–116.

Brynskov, J., Freund, L., Rasmussen, S. N., et al. (1989). A placebo-controlled, double-blind, randomized trial of cyclosporine therapy in active chronic Crohn's disease. *New Engl. J. Med.*, **321**, 13, 845–850.

Brynskov, J., Freund, L., Rasmussen, S. N., et al. (1991). Final report on a placebo-controlled, double-blind, randomized, multicentre trial of cyclosporin treatment in active chronic Crohn's disease. *Scand. J. Gastroenterol.*, **26**, 7, 689–695.

Clopper, C. J. and Pearson, E. S. (1934). The use of confidence or fiducial limits illustrated in the case of the binomial. *Biometrika*, **26**, 404–413.

Croux, C. and Rousseeuw, P. J. (1992). Time-efficient algorithms for two highly robust estimators of scale. *Comput. Stat.*, **1**, 411–428.

Dressel, P. L. (1957). Facts and fancy in assigning grades. *Basic College Quarterly*, **2**, 6–12.

Dureau, P., Jeanny, J.-C., Clerc, B., Dufier, J.-L., and Courtois, Y. (1996). Long term light-induced retinal degeneration in the miniature pig. *Mol. Vis.* **2**, 7.
http://www.emory.edu/molvis/v2/dureau.

Eisenhart, C. and Wilson, P. W. (1943). Statistical method and control in bacteriology. *Bact. Rev.*, **7**, 57–137,

Garwood, F. (1936). Fiducial limits for the Poisson distribution. *Biometrika*, **28**, 437–442.

Hamilton, L. C. (1990). *Modern Data Analysis: A First Course in Applied Statistics*. Brooks/Cole, Pacific Grove.

Hogg, R. V. and Tanis, E. A. (2001). *Probability and Statistical Inference*, 6th edn. Prentice-Hall, Upper Saddle River.

McKay, A. T. (1932). Distribution of the coefficient of variation and the extended *t*-distribution. *J. Roy. Statist. Soc.*, **95**, 695–698.

Miller, E. G. (1991). Asymptotic test statistics for coefficient of variation. *Comm. Stat. Theory Meth.*, **20**, 10, 3351–3363.

Pritz, M. B., Zhou, X. H., and Brizendine, E. J. (1996). Hyperdynamic therapy for cerebral vasospasm: a meta-analysis of 14 studies. *J. Neurovasc. Dis.*, **1**, 6–8.

Ribitsch, W., Stockinger, J., and Schneditz, D. (2012). Bioimpedance-based volume at clinical target weight is contracted in hemodialysis patients with a high body mass index. *Clinical Nephrology*, **77**, 5, 376–382.

Rutherford, E., Chadwick, J., and Ellis, C. D. (1930). *Radiations from Radioactive Substances*. Macmillan, London, pp. 171–172.

Rutherford, E. and Geiger, H. (1910). The probability variations in the distribution of α-particles (with a note by H. Bateman). *Philos. Mag.*, **6**, 20, 697–707.

Sampford, M. R. and Taylor, J. (1959). Censored observations in randomized block experiments. *J. Roy. Stat. Soc. Ser. B*, **21**, 214–237.

Staudte, R. G. and Sheater, S. J. (1990). *Robust Estimation and Testing*. Wiley, New York.

Wald, A. and Wolfowitz, J. (1939). Confidence limits for continuous distribution functions. *Ann. Math. Stat.*, **10**, 105–118.

Wilcox, R. R. (2005). *Introduction to Robust Estimation and Hypothesis Testing*, 2nd ed. Academic Press, San Diego.

Wilson, E. B. (1927). Probable inference, the law of succession, and statistical inference. *J. Am. Stat. Assoc.*, **22**, 209–212.

Chapter 8
Bayesian Approach to Inference

In 1954 I proved that the only sound methods were Bayesian; yet you continue to use non-Bayesian ideas without pointing out a flaw in either my premise or my proof, why?

– Leonard Jimmie Savage

WHAT IS COVERED IN THIS CHAPTER

- Bayesian Paradigm
- Likelihood, Prior, Marginal, Posterior, Predictive Distributions
- Conjugate Priors, Prior Elicitation
- Bayesian Computation
- Estimation, Credible Sets, Testing, Bayes Factor, Prediction

8.1 Introduction

Several paradigms provide a basis for statistical inference; the two most dominant are the *frequentist* (sometimes called classical, traditional, or Neyman–Pearsonian) and *Bayesian*. The term Bayesian refers to Reverend Thomas Bayes, a nonconformist minister interested in mathematics whose posthumously published essay (Bayes, 1763) is fundamental for this kind of inference. According to the Bayesian paradigm, the unobservable parameters in a statistical model are treated as random. Before data are collected, *prior distributions* are elicited to quantify our knowledge about the parameters. This knowledge comes from expert opinion, theoretical considerations, or previous similar experiments. When data are available, the prior distributions are updated to the *posterior distributions.* These are conditional distributions that incorporate the observed data. The transition from the prior to the posterior is possible via Bayes' theorem.

The Bayesian approach is relatively modern in statistics; it became influential with advances in Bayesian computational methods in the 1980s and 1990s.

Before launching into a formal exposition of Bayes' theorem, we revisit Bayes' rule for events (page 100). Prior to observing whether an event A has appeared or not, we set the probabilities of n hypotheses, $H_1, H_2, \ldots, H_n$, under which event A may appear. We called them *prior* probabilities of the hypotheses, $\mathbb{P}(H_1), \ldots, \mathbb{P}(H_n)$. Bayes' rule showed us how to update these prior probabilities to the posterior probabilities once we obtained information about event A. Recall that the posterior probability of the hypothesis H_i, given the evidence about A, was

$$\mathbb{P}(H_i|A) = \frac{\mathbb{P}(A|H_i)\mathbb{P}(H_i)}{\mathbb{P}(A)}.$$

Therefore, Bayes' rule gives a recipe for updating the prior probabilities of events to their posterior probabilities once additional information from the experiment becomes available. The focus of this chapter is on how to update prior knowledge about a model; however, this knowledge, or lack of it, is expressed in terms of probability distributions rather than by events.

Suppose that before the data are observed, a description of the population parameter θ is given by a probability density $\pi(\theta)$. The process of specifying the prior distribution is called *prior elicitation.* The data are modeled via the likelihood, which depends on θ and is denoted by $f(x|\theta)$. Bayes' theorem updates the prior $\pi(\theta)$ to the posterior $\pi(\theta|x)$ by incorporating observations x summarized via the likelihood:

$$\pi(\theta|x) = \frac{f(x|\theta)\pi(\theta)}{m(x)}. \tag{8.1}$$

Here, $m(x)$ normalizes the product $f(x|\theta)\pi(\theta)$ to be a density and is a constant once the prior is specified and the data are observed. Given the data x and the prior distribution, the posterior distribution $\pi(\theta|x)$ summarizes all available information about θ.

Although the equation in (8.1) is referred to as a theorem, there is nothing to prove there. Recall that the probability of intersection of two events A and B was calculated as $\mathbb{P}(AB) = \mathbb{P}(A|B)\mathbb{P}(B) = \mathbb{P}(B|A)\mathbb{P}(B)$ [multiplication rule in (3.6)]. By analogy, the joint distribution of X and θ, $h(x,\theta)$, would have two representations, as in (5.11), depending on the order of conditioning:

$$h(x,\theta) = f(x|\theta)\pi(\theta) = \pi(\theta|x)m(x),$$

and Bayes' theorem solves this equation with respect to the posterior $\pi(\theta|x)$.

To summarize, Bayes' rule updates the probabilities of events when new evidence becomes available, while Bayes' theorem provides the recipe for updating prior distributions of model's parameters once experimental observations become available.

$\mathbb{P}(\text{hypothesis})$	$\xrightarrow{\text{BAYES' RULE}}$	$\mathbb{P}(\text{hypothesis}|\text{evidence})$
$\pi(\theta)$	$\xrightarrow{\text{BAYES' THEOREM}}$	$\pi(\theta|\text{data})$

The Bayesian paradigm has many advantages, but the two most important are: (i) the uncertainty is expressed via the probability distribution and the statistical inference can be automated; thus, it follows a conceptually simple recipe embodied in Bayes' theorem, and (ii) available prior information is coherently incorporated into the statistical model describing the data.

The FDA guidelines document (FDA, 2010) recommends the use of a Bayesian methodology in the design and analysis of clinical trials for medical devices. This document eloquently outlines the reasons why a Bayesian methodology is recommended.

- Valuable prior information is often available for medical devices because of their mechanism of action and evolutionary development.
- The Bayesian approach, when correctly employed, may be less burdensome than a frequentist approach.
- In some instances, the use of prior information may alleviate the need for a larger sized trial. In some scenarios, when an adaptive Bayesian model is applicable, the size of a trial can be reduced by stopping the trial early when conditions warrant.
- The Bayesian approach can sometimes be used to obtain an exact analysis when the corresponding frequentist analysis is only approximate or is too difficult to implement.
- Bayesian approaches to multiplicity problems are different from frequentist ones and may be advantageous. Inferences on multiple endpoints and testing of multiple subgroups (e.g., race or sex) are examples of multiplicity.
- Bayesian methods allow for great flexibility in dealing with missing data.

In the context of clinical trials, an unlimited look at the accumulated data, when sampling is sequential in nature, will not affect the inference. In the frequentist approach, interim data analyses affect type I errors. The ability to stop a clinical trial early is important from the moral and economic viewpoints. Trials should be stopped early due to both futility, to save resources or stop an ineffective treatment, and superiority, to provide patients with the best possible treatments as fast as possible.

Bayesian models facilitate meta-analysis. Meta-analysis is a methodology for the fusion of results of related experiments performed by different researchers, labs, etc. An example of a rudimentary meta-analysis is discussed in Section 8.10.

8.2 Ingredients for Bayesian Inference

A density function for a typical observation X that depends on an unknown, possibly multivariate, parameter θ is called a model and denoted by $f(x|\theta)$. As a function of θ, $f(x|\theta) = L(\theta)$ is called the *likelihood*. If a sample $x = (x_1, x_2, \ldots, x_n)$ is observed, the likelihood takes a familiar form, $L(\theta|x_1, \ldots, x_n) = \prod_{i=1}^{n} f(x_i|\theta)$. This form was used in Chapter 7 to produce MLEs for θ.

Thus both terms model and likelihood are used to describe the distribution of observations. In the standard Bayesian inference the functional form of f is given in the same manner as in the classical parametric approach; the functional form is fully specified up to a parameter θ. According to the generally accepted *likelihood principle*, all information from the experimental data is summarized in the likelihood function, $f(x|\theta) = L(\theta|x_1, \ldots, x_n)$.

For example, if each datum $X|\theta$ were assumed to be exponential with the rate parameter θ and $X_1 = 2, X_2 = 3$, and $X_3 = 1$ were observed, then full information about the experiment would be given by the likelihood

$$\theta e^{-2\theta} \times \theta e^{-3\theta} \times \theta e^{-\theta} = \theta^3 e^{-6\theta}.$$

This model is $\theta^3 \exp\left\{-\theta \sum_{i=1}^{3} X_i\right\}$ if the data are kept unspecified, but in the likelihood function the expression $\sum_{i=1}^{3} X_i$ is treated as a constant term, as was done in the maximum likelihood estimation (page 283).

The parameter θ, with values in the parameter space Θ, is not directly observable and is considered a random variable. This is the key difference between Bayesian and classical approaches. Classical statistics consider the parameter to be a fixed number or vector of numbers, while Bayesians express the uncertainty about θ by considering it as a random variable. This random variable has a distribution $\pi(\theta)$ called the prior distribution. The prior distribution not only quantifies available knowledge, it also describes the uncertainty about a parameter before data are observed. If the prior distribution for θ is specified up to a parameter τ, $\pi(\theta|\tau)$, then τ is called a *hyperparameter*. Hyperparameters are parameters of a prior distribution, and they are either specified or may have their own priors. This may lead to a hierarchical structure of the model where the priors are arranged in a hierarchy.

The previous discussion can be summarized as follows:

> The goal in Bayesian inference is to start with prior information on the parameter of interest, θ, and update it using the observed data. This is achieved via Bayes' theorem, which gives a simple recipe for incorporating observations x in the distribution of θ, $\pi(\theta|x)$, called the *posterior* distribution. All information about θ coming from the prior distribution and the observations are contained in the posterior distribution. The posterior distribution is the ultimate summary of the parameter and serves as the basis for all Bayesian inferences.

According to Bayes' theorem, to find $\pi(\theta|x)$, we divide the *joint* distribution of X and θ $(h(x,\theta) = f(x|\theta)\pi(\theta))$ by the *marginal* distribution for X, $m(x)$, which is obtained by integrating out θ from the joint distribution $h(x,\theta)$:

$$m(x) = \int_{\Theta} h(x,\theta)d\theta = \int_{\Theta} f(x|\theta)\pi(\theta)d\theta.$$

The marginal distribution is also called the *prior predictive* distribution. Thus, in terms of the likelihood and the prior distribution only, the Bayes theorem can be restated as

$$\pi(\theta|x) = \frac{f(x|\theta)\pi(\theta)}{\int_{\Theta} f(x|\theta)\pi(\theta)d\theta}.$$

The integral in the denominator is a major hurdle in Bayesian computation, since for complex likelihoods and priors it could be intractable.

The following table summarizes the notation:

Likelihood, model	$f(x|\theta)$
Prior distribution	$\pi(\theta)$
Joint distribution	$h(x,\theta) = f(x|\theta)\pi(\theta)$
Marginal distribution	$m(x) = \int_{\Theta} f(x|\theta)\pi(\theta)d\theta$
Posterior distribution	$\pi(\theta|x) = f(x|\theta)\pi(\theta)/m(x)$

We illustrate these concepts by discussing a few examples in which the posterior distribution can be explicitly obtained. Note that the marginal distribution has the form of an integral, and in many cases these integrals cannot be found in a finite form. It is fair to say that the number of likelihood/prior combinations that lead to an explicit posterior is rather limited. However, in the general case, the posterior can be evaluated numerically or, as we will see later, a sample can be simulated from the posterior distribution. All of the, admittedly abstract, concepts listed above will be exemplified by several worked-out models. We start with the most important model in which both the likelihood and prior are normal.

Example 8.1. **Normal Likelihood with Normal Prior.**   The normal likelihood and normal prior combination is important because it is frequently used in practice. Assume that an observation X is normally distributed with mean θ and known variance σ^2. The parameter of interest, θ, is normally distributed as well, with its parameters μ and τ^2. Parameters μ and τ^2 are hyperparameters, and we will consider them given. Starting with our Bayesian model of $X|\theta \sim \mathcal{N}(\theta, \sigma^2)$ and $\theta \sim \mathcal{N}(\mu, \tau^2)$, we will find the marginal and posterior distributions. Before we start with a derivation of the posterior and marginal, we need a simple algebraic identity:

$$A(x-a)^2 + B(x-b)^2 = (A+B)(x-c)^2 + \frac{AB}{A+B}(a-b)^2, \qquad (8.2)$$

for $c = \frac{Aa+Bb}{A+B}$.

We start with the joint distribution of (X, θ), which is the product of two distributions:

$$h(x,\theta) = \frac{1}{\sqrt{2\pi\sigma^2}}\exp\left\{-\frac{1}{2\sigma^2}(x-\theta)^2\right\} \times \frac{1}{\sqrt{2\pi\tau^2}}\exp\left\{-\frac{1}{2\tau^2}(\theta-\mu)^2\right\}.$$

The exponent in the joint distribution $h(x,\theta)$ is

$$-\frac{1}{2\sigma^2}(x-\theta)^2 - \frac{1}{2\tau^2}(\theta-\mu)^2,$$

which, after applying the identity in (8.2), can be expressed as

$$-\frac{\sigma^2+\tau^2}{2\sigma^2\tau^2}\left(\theta - \left(\frac{\tau^2}{\sigma^2+\tau^2}x + \frac{\sigma^2}{\sigma^2+\tau^2}\mu\right)\right)^2 - \frac{1}{2(\sigma^2+\tau^2)}(x-\mu)^2. \quad (8.3)$$

Note that the exponent in (8.3) splits into two parts, one containing θ and the other θ-free. Accordingly the joint distribution $h(x,\theta)$ splits into the product of two densities. Since $h(x,\theta)$ can be represented in two ways, as $f(x|\theta)\pi(\theta)$ and as $\pi(\theta|x)m(x)$, and since we started with $f(x|\theta)\pi(\theta)$, the exponent in (8.3) corresponds to $\pi(\theta|x)m(x)$. Thus, the marginal distribution simply resolves to $X \sim \mathcal{N}(\mu, \sigma^2+\tau^2)$ and the posterior distribution of θ comes out to be

$$\theta | X \sim \mathcal{N}\left(\frac{\tau^2}{\sigma^2 + \tau^2}X + \frac{\sigma^2}{\sigma^2 + \tau^2}\mu, \frac{\sigma^2 \tau^2}{\sigma^2 + \tau^2}\right).$$

Below is a specific example of our first Bayesian inference.

Example 8.2. **Jeremy's IQ.** Jeremy, an enthusiastic bioengineering student, posed a statistical model for his scores on a standard IQ test. He thinks that, in general, his scores are normally distributed with unknown mean θ (true IQ) and a variance of $\sigma^2 = 80$. Prior (and expert) opinion is that the IQ of bioengineering students in Jeremy's school, θ, is a normal random variable, with mean $\mu = 110$ and variance $\tau^2 = 120$. Jeremy took the test and scored $X = 98$. The traditional estimator of θ would be $\hat{\theta} = X = 98$. The posterior is normal with a mean of $\frac{120}{80+120} \times 98 + \frac{80}{80+120} \times 110 = 102.8$ and a variance of $\frac{80 \times 120}{80+120} = 48$. We will see later that the mean of the posterior is Bayes' estimator of θ, and a Bayesian would estimate Jeremy's IQ as 102.8.

If n normal variates, $X_1, X_2, \ldots, X_n$, are observed instead of a single observation X, then the sample is summarized as $\overline{X}$ and the Bayesian model for θ is essentially the same as that for a single X, but with σ^2/n in place of σ^2. In this case, the likelihood and the prior are

$$\overline{X} | \theta \sim \mathcal{N}\left(\theta, \frac{\sigma^2}{n}\right) \text{ and } \theta \sim \mathcal{N}(\mu, \tau^2),$$

producing

$$\theta | \overline{X} \sim \mathcal{N}\left(\frac{\tau^2}{\frac{\sigma^2}{n} + \tau^2}\overline{X} + \frac{\frac{\sigma^2}{n}}{\frac{\sigma^2}{n} + \tau^2}\mu, \frac{\frac{\sigma^2}{n} \tau^2}{\frac{\sigma^2}{n} + \tau^2}\right).$$

Notice that the posterior mean

$$\frac{\tau^2}{\frac{\sigma^2}{n} + \tau^2}\overline{X} + \frac{\frac{\sigma^2}{n}}{\frac{\sigma^2}{n} + \tau^2}\mu$$

is a weighted average of the MLE $\overline{X}$ and the prior mean μ with weights $w = n\tau^2/(\sigma^2 + n\tau^2)$ and $1 - w = \sigma^2/(\sigma^2 + n\tau^2)$. When the sample size n increases, the contribution of the prior mean to the estimator diminishes as $w \to 1$. In contrast, when n is small and our prior opinion about μ is strong (i.e., τ^2 is small), the posterior mean remains close to the prior mean μ. Later, we will explore several more cases in which the posterior mean has a form of a weighted average of the MLE for the parameter and the prior mean.

Example 8.3. **Likelihood, Prior, and Posterior.** Suppose that $n = 10$ observations are coming from $\mathcal{N}(\theta, 10^2)$. Assume that the prior on θ is $\mathcal{N}(20, 20)$. For the observations
$\{2.944, -13.361, 7.143, 16.235, -6.917, 8.580, 12.540, -15.937, -14.409, 5.711\}$
the posterior is $\mathcal{N}(6.835, 6.667)$. The three densities: likelihood, prior, and posterior, are shown in Figure 8.1.

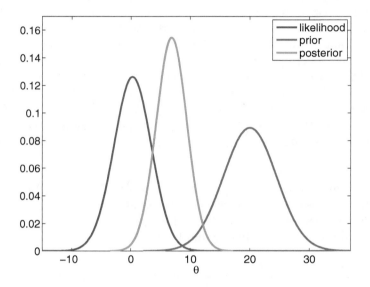

Fig. 8.1 The likelihood centered at MLE $\overline{X} = 0.2529$, $\mathcal{N}(0.2529, 10^2/10)$ (*blue*), $\mathcal{N}(20, 20)$ prior (*red*), and posterior for data $\{2.9441, -13.3618, \dots, 5.7115\}$ (*green*).

8.3 Conjugate Priors

A major technical difficulty in Bayesian analysis is finding an explicit posterior distribution, given the likelihood and prior. The posterior is proportional to the product of the likelihood and prior, but the normalizing constant, marginal $m(x)$, is often difficult to find since it involves integration.

In Examples 8.1 and 8.3, where the prior is normal, the posterior distribution remains normal. In such cases, the effect of likelihood is only to "update" the prior parameters and not to change the prior's functional form. We say that such priors are *conjugate* with the likelihood. Conjugacy is popular because of its mathematical convenience; once the conjugate likelihood/prior pair is identified, the posterior is found without integration.

The normalizing marginal $m(x)$ is selected such that $f(x|\theta)\pi(\theta)$ is a density from the same class to which the prior belongs. Operationally, one multiplies "kernels" of likelihood and priors, ignoring all multiplicative terms that do not involve the parameter. For example, a kernel of gamma $\mathcal{G}a(r,\lambda)$ density $f(\theta|r,\lambda) = \frac{\lambda^r\theta^{r-1}}{\Gamma(r)}e^{-\lambda\theta}$ would be $\theta^{r-1}e^{-\lambda\theta}$. We would write: $f(\theta|r,\lambda) \propto \theta^{r-1}e^{-\lambda\theta}$, where the symbol $\propto$ stands for "proportional to." Several examples in this chapter involve conjugate pairs (Examples 8.4 and 8.6).

In the pre-Markov chain Monte Carlo era, conjugate priors were extensively used (and overused and misused) precisely because of this computational convenience. Today, the general agreement is that simple conjugate analysis is of limited practical value since, given the likelihood, the conjugate prior has limited modeling capability.

There are quite a few instances of conjugacy. Table 8.1 gives several important cases. As a practice, you may want to derive the posteriors listed in the third column of the table. It is recommended that you consult Chapter 5 on functional forms of densities involved in the Bayesian model.

Table 8.1 Some conjugate pairs. Here $\mathbf{X}$ stands for a sample of size n, $X_1,\ldots,X_n$. For functional expressions of the densities and their moments refer to Chapter 5

Likelihood	Prior	Posterior
$X_i|\theta \sim \mathcal{N}(\theta,\sigma^2)$	$\theta \sim \mathcal{N}(\mu,\tau^2)$	$\theta|\mathbf{X} \sim \mathcal{N}\left(\frac{\tau^2}{\tau^2+\sigma^2/n}\overline{X} + \frac{\sigma^2/n}{\tau^2+\sigma^2/n}\mu, \frac{\tau^2\sigma^2/n}{\tau^2+\sigma^2/n}\right)$
$X_i|\theta \sim \mathcal{B}in(m,\theta)$	$\theta \sim \mathcal{B}e(\alpha,\beta)$	$\theta|\mathbf{X} \sim \mathcal{B}e(\alpha + \sum_{i=1}^n X_i, \beta + mn - \sum_{i=1}^n X_i)$
$X_i|\theta \sim \mathcal{P}oi(\theta)$	$\theta \sim \mathcal{G}a(\alpha,\beta)$	$\theta|\mathbf{X} \sim \mathcal{G}a(\alpha + \sum_{i=1}^n X_i, \beta + n)$
$X_i|\theta \sim \mathcal{N}B(m,\theta)$	$\theta \sim \mathcal{B}e(\alpha,\beta)$	$\theta|\mathbf{X} \sim \mathcal{B}e(\alpha + mn, \beta + \sum_{i=1}^n X_i)$
$X_i|\theta \sim \mathcal{G}a(1/2,1/(2\theta))$	$\theta \sim \mathcal{IG}(\alpha,\beta)$	$\theta|\mathbf{X} \sim \mathcal{IG}(\alpha + n/2, \beta + \frac{1}{2}\sum_{i=1}^n X_i)$
$X_i|\theta \sim \mathcal{U}(0,\theta)$	$\theta \sim \mathcal{P}a(\theta_0,\alpha)$	$\theta|\mathbf{X} \sim \mathcal{P}a(\max\{\theta_0,X_1,\ldots,X_n\},\alpha + n)$
$X_i|\theta \sim \mathcal{N}(\mu,\theta)$	$\theta \sim \mathcal{IG}(\alpha,\beta)$	$\theta|\mathbf{X} \sim \mathcal{IG}(\alpha + n/2, \beta + \frac{1}{2}\sum_{i=1}^n (X_i - \mu)^2)$
$X_i|\theta \sim \mathcal{G}a(v,\theta)$	$\theta \sim \mathcal{G}a(\alpha,\beta)$	$\theta|\mathbf{X} \sim \mathcal{G}a(\alpha + nv, \beta + \sum_{i=1}^n X_i)$
$X_i|\theta \sim \mathcal{P}a(c,\theta)$	$\theta \sim \mathcal{G}a(\alpha,\beta)$	$\theta|\mathbf{X} \sim \mathcal{G}a(\alpha + n, \beta + \sum_{i=1}^n \log(X_i/c))$

Example 8.4. **Binomial Likelihood with Beta Prior.** An easy, yet important, example of a conjugate structure is the binomial likelihood and beta prior. Suppose that we observed $X = x$ from a binomial $\mathcal{B}in(n,p)$ distribution,

$$f(x|\theta) = \binom{n}{x} p^x (1-p)^{n-x},$$

and that the population proportion p is the parameter of interest. If the prior on p is beta $\mathcal{Be}(\alpha,\beta)$ with hyperparameters α and β and density

$$\pi(p) = \frac{1}{B(\alpha,\beta)} p^{\alpha-1} (1-p)^{\beta-1},$$

the posterior is proportional to the product of the likelihood and the prior

$$\pi(p|x) = C \cdot p^x (1-p)^{n-x} \cdot p^{\alpha-1} (1-p)^{\beta-1} = C \cdot p^{x+\alpha-1} (1-p)^{n-x+\beta-1}$$

for some constant C. The normalizing constant C is free of p and is equal to $\frac{\binom{n}{x}}{m(x)B(\alpha,\beta)}$, where $m(x)$ is the marginal distribution.

By inspecting the expression $p^{x+\alpha-1}(1-p)^{n-x+\beta-1}$, it can be seen that the posterior density remains beta; it is $\mathcal{Be}(x+\alpha, n-x+\beta)$, and that the normalizing constant resolves to $C = 1/B(x+\alpha, n-x+\beta)$. From the equality of constants, it follows that

$$\frac{\binom{n}{x}}{m(x)B(\alpha,\beta)} = \frac{1}{B(x+\alpha, n-x+\beta)},$$

and one can express the marginal density as

$$m(x) = \frac{\binom{n}{x} B(x+\alpha, n-x+\beta)}{B(\alpha,\beta)},$$

which is known as a *beta-binomial distribution*.

8.4 Point Estimation

The posterior is the ultimate experimental summary for a Bayesian. The posterior location measures, especially the mean, are of great importance. The posterior mean is the most frequently used Bayes' estimator for a parameter. The posterior mode and median are alternative Bayes' estimators.

The posterior mode maximizes the posterior density in the same way that the MLE maximizes the likelihood. When the posterior mode is used as an estimator, it is called the maximum posterior (MAP) estimator. The MAP estimator is popular in some Bayesian analyses in part because it is computationally less demanding than the posterior mean or median. The reason for this is simple: to find a MAP, the posterior does not need to be fully specified because $\text{argmax}_\theta \pi(\theta|x) = \text{argmax}_\theta f(x|\theta)\pi(\theta)$, that is, the product of the likelihood and the prior as well as the posterior are maximized at the same point.

Example 8.5. **Binomial-Beta Conjugate Pair.** In Example 8.4 we argued that for the likelihood $X|\theta \sim Bin(n,\theta)$ and the prior $\theta \sim Be(\alpha,\beta)$, the posterior distribution is $Be(x+\alpha, n-x+\beta)$. The Bayes estimator of θ is the expected value of the posterior

$$\hat{\theta}_B = \frac{\alpha+x}{(\alpha+x)+(\beta+n-x)} = \frac{\alpha+x}{\alpha+\beta+n}.$$

This is actually a weighted average of the MLE, X/n, and the prior mean $\alpha/(\alpha+\beta)$,

$$\hat{\theta}_B = \frac{n}{\alpha+\beta+n} \cdot \frac{X}{n} + \frac{\alpha+\beta}{\alpha+\beta+n} \cdot \frac{\alpha}{\alpha+\beta}.$$

Notice that, as n becomes large, the posterior mean approaches the MLE because the weight $\frac{n}{n+\alpha+\beta}$ tends to 1. In contrast, when α or β or both are large compared to n, the posterior mean is close to the prior mean. Due to this interplay between n and prior parameters, the sum $\alpha+\beta$ is called the prior sample size, and it measures the influence of the prior as if additional experimentation was performed and $\alpha+\beta$ trials have been added. This is in the spirit of Wilson's proposal to "add two failures and two successes" to an estimator of proportion (page 305). Wilson's estimator can be seen as a Bayes estimator with a beta $Be(2,2)$ prior.

Large α indicates a small prior variance, since for fixed β, the variance of $Be(\alpha,\beta)$ is proportional to $1/\alpha^2$, and the prior is concentrated about its mean.

In general, the posterior mean will fall between the MLE and the prior mean. This was demonstrated in Example 8.1. As another example, suppose we flipped a coin four times and tails showed up on all four occasions. We are interested in estimating the probability of showing heads, θ, in a Bayesian fashion. If the prior is $\mathcal{U}(0,1)$, the posterior is proportional to $\theta^0(1-\theta)^4$, which is a beta $Be(1,5)$. The posterior mean *shifts* the MLE (0) toward the expected value of the prior (1/2) to get $\hat{\theta}_B = 1/(1+5) = 1/6$, which is a more reasonable estimator of θ than the MLE. Note that the $3/n$ rule produces a confidence interval for p of $[0,3/4]$, which is too wide to be useful (Section 7.5.4).

Example 8.6. **Uniform/Pareto Model.** In Example 7.5 we had the observations $X_1 = 2$, $X_2 = 5$, $X_3 = 0.5$, and $X_4 = 3$ from a uniform $\mathcal{U}(0,\theta)$ distribution. We are interested in estimating θ in Bayesian fashion. Let the prior on θ be Pareto $\mathcal{P}a(\theta_0,\alpha)$ for $\theta_0 = 6$ and $\alpha = 2$. Then the posterior is also Pareto $\mathcal{P}a(\theta^*, \alpha^*)$ with $\theta^* = \max\{\theta_0,\ X_{(n)}\} = \max\{6,5\} = 6$, and $\alpha^* = \alpha + n = 2 + 4 = 6$. The posterior mean is $\frac{\alpha^*\theta^*}{\alpha^*-1} = 36/5 = 7.2$, and the median is $\theta^* \cdot 2^{1/\alpha^*} = 6 \cdot 2^{1/6} = 6.7348$.

Figure 8.2 shows the prior (dashed red line) with the prior mean as a red dot. After observing $X_1, \ldots, X_4$, the posterior mode did not change since the elicited $\theta_0 = 6$ was larger than $\max X_i = 5$. However, the posterior has a smaller variance than the prior. The posterior mean is shown as a green dot, the posterior median as a black dot, and the posterior (and prior) mode as a blue dot.

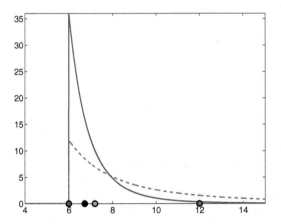

Fig. 8.2 Pareto $\mathcal{P}a(6,2)$ prior (*dashed red line*) and $\mathcal{P}a(6,6)$ posterior (*solid blue line*). The *red dot* is the prior mean, the *green dot* is the posterior mean, the *black dot* is the posterior median, and the *blue dot* is the posterior (and prior) mode.

Another widely used conjugate pair is Poisson–gamma pair.

Example 8.7. **Poisson–Gamma Conjugate Pair.** Let $X_1, \ldots, X_n$, given θ are Poisson $\mathcal{P}oi(\theta)$ with probability mass function

$$f(x_i|\theta) = \frac{\theta^{x_i}}{x_i!}e^{-\theta},$$

and $\theta \sim \mathcal{G}a(\alpha,\beta)$ is given by $\pi(\theta) \propto \theta^{\alpha-1}e^{-\beta\theta}$. Then

$$\pi(\theta|X_1, \ldots, X_n) = \pi(\theta|\sum X_i) \propto \theta^{\sum X_i + \alpha - 1}e^{-(n+\beta)\theta},$$

which is $\mathcal{G}a(\sum_i X_i + \alpha, n + \beta)$. The mean is $\mathbb{E}(\theta|X) = (\sum X_i + \alpha)/(n + \beta)$, and it can be represented as a weighted average of the MLE and the prior mean:

$$\mathbb{E}\theta|X = \frac{n}{n+\beta}\frac{\sum X_i}{n} + \frac{\beta}{n+\beta}\frac{\alpha}{\beta}.$$

Let us apply the this equation in a specific example. Let a rare disease have an incidence of X cases per 100,000 people, where X is modeled as Poisson, $X|\lambda \sim \mathcal{P}oi(\lambda)$, where λ is the rate parameter. Assume that for different cohorts of 100,000 subjects, the following incidences are observed: $X_1 = 2$, $X_2 = 0$, $X_3 = 0$, $X_4 = 4$, $X_5 = 0$, $X_6 = 1$, $X_7 = 3$, and $X_8 = 2$. The experts indicate that λ should be close to 2 and our prior is $\lambda \sim \mathcal{G}a(0.2, 0.1)$. We matched the mean, since for a gamma distribution the mean is $0.2/0.1 = 2$ but the variance $0.2/0.1^2 = 20$ is quite large, thereby expressing our uncertainty. By setting the hyperparameters to 0.02 and 0.01, for example, we would have variance of the gamma prior that is even larger. The MLE of λ is $\hat{\lambda}_{mle} = \overline{X} = 3/2$. The Bayes estimator is

$$\hat{\lambda}_B = \frac{8}{8+0.1}3/2 + \frac{0.1}{8+0.1}2 = 1.5062.$$

Note that since the prior was not informative, the Bayes estimator is quite close to the MLE.

✐

Normal-Inverse Gamma Conjugate Analysis. Let $y_1, y_2, \ldots, y_n$ be the observations from normal $\mathcal{N}(\mu, \sigma^2)$ distribution where both μ and σ^2 are of interest. For this problem there is a conjugate joint prior for (μ, σ^2), normal-inverse gamma $\mathcal{NIG}(\mu_0, c, a, b)$,

$$\pi(\mu, \sigma^2) = \pi(\mu|\sigma^2)\pi(\sigma^2) = \mathcal{N}(\mu_0, \sigma^2/c) \times \mathcal{IG}(a, b).$$

Note that apriori μ and σ^2 are not independent, their joint prior is not a product of densities that fully separates the variables.

Instead of variance σ^2, often the precision parameter $\tau = 1/\sigma^2$ is modeled and estimated. In many cases the estimation of τ is more stable than that of σ^2. From the definition of inverse-gamma it follows that if $\sigma^2 \sim \mathcal{IG}(a, b)$ then $\tau \sim \mathcal{G}a(a, b)$. Thus,

$$\pi(\mu, \tau) = \mathcal{N}\left(\mu_0, \frac{1}{c\tau}\right) \times \mathcal{G}a(a, b)$$

$$= \sqrt{\frac{c\tau}{2\pi}}\exp\left\{\frac{c\tau}{2}(\mu - \mu_0)^2\right\} \times \frac{b^a \tau^{a-1}}{\Gamma(a)}\exp\{-b\tau\}.$$

After observing $y = (y_1, \ldots, y_n)$, all inference depends on $\overline{y} = 1/n \sum_{i=1}^n y_i$ and $s^2 = \sum_{i=1}^n (y_i - \overline{y})^2/(n-1)$. Denote

$$SS = \sum_{i=1}^n (y_i - \overline{y})^2 + \frac{nc}{n+c}(\overline{y} - \mu_0)^2 = (n-1)s^2 + \frac{nc}{n+c}(\overline{y} - \mu_0)^2$$

When the likelihood is normal, the problem is conjugate and the posterior for (μ, σ^2) is $\mathcal{NIG}(\mu_0^*, c^*, a^*, b^*)$, or equivalently, $\mathcal{NG}(\mu_0^*, c^*, a^*, b^*)$ for (μ, τ).

The updated parameters (from prior to the posterior) are shown in the following table:

Prior	Posterior
μ_0	$\mu_0^* = \frac{c}{n+c}\mu_0 + \frac{n}{n+c}\overline{y}$
c	$c^* = c + n$
a	$a^* = a + n/2$
b	$b^* = b + SS/2$

Posterior expectations (Bayes' estimators) and variances for μ, τ, and σ^2 are:

$$\mathbb{E}(\mu|y) = \mu_0^* = \frac{c}{n+c}\mu_0 + \frac{n}{n+c}\overline{y},$$

$$\mathbb{V}\mathrm{ar}(\mu|y) = \frac{1}{n+c} \times \frac{SS + 2b}{n + 2a - 2}, \qquad n > 2 - 2a,$$

$$\mathbb{E}(\tau|y) = \frac{n + 2a}{SS + 2b},$$

$$\mathbb{V}\mathrm{ar}(\tau|y) = \frac{2n + 4a}{(SS + 2b)^2},$$

$$\mathbb{E}(\sigma^2|y) = \frac{SS + 2b}{n + 2a - 2}, \qquad n > 2 - 2a, \text{ and}$$

$$\mathbb{V}\mathrm{ar}(\sigma^2|y) = \frac{2(SS + 2b)^2}{(n + 2a - 2)^2(n + 2a - 4)}, \qquad n > 4 - 2a.$$

Example 8.8. **Jeremy and NIG Prior.** Suppose that Jeremy took the IQ test 6 times. His scores $(101, 98, 114, 105, 108, 111)$ are assumed to be a sample from a normal distribution with unknown mean μ and variance σ^2.

The prior on (μ, σ^2) is normal-inverse gamma with parameters $\mu_0 = 110$, $c = 1.5$, $a = 0.1$ and $b = 10$.

Using exact conjugate calculations, we find Bayes' estimators for μ and σ^2.

```
y = [ 101  98  114  105  108  111 ];
mu0 = 110; n=6; c=1.5; a=0.1;  b=10;
ybar = mean(y);
ss = (n-1) * var(y) + n*c/(n+c) * (ybar - mu0)^2;
%
muhat=  c/(n+c) * mu0  + n/(n+c) * ybar        %106.9333
varmuhat = 1/(n+c) * (ss + 2*b)/(n + 2*a -2)   %6.9989
stdmuhat = sqrt(varmuhat)                      %2.6456
tauhat = (n + 2 * a)/(ss + 2 * b)              %0.0281
vartauhat = 2 * (n  + 2 * a)/(ss  + 2 * b)^2   %2.5511e-04
```

```
stdtauhat = sqrt(vartauhat)                    %0.016
sigma2hat = (ss + 2 * b)/(n + 2*a - 2)         %52.4921
varsigma2hat = 2 * (ss + 2 * b)^2 /...
    ((n + 2*a - 2)^2 * (n + 2* a - 4))         %2.5049e+03
stdsigma2hat = sqrt(varsigma2hat)              %50.0492
```

Note that Bayes' estimator of μ is $\hat{\mu}_B = 106.9333$. The estimators of variance and precision are $\hat{\sigma}_B^2 = 52.4921$ and $\hat{\tau}_B = 0.0281$. In addition to estimators of these parameters, Bayesian model gives us the estimators of their variances varmuhat, varsigma2hat, and vartauhat and their standard deviations stdmuhat, stdsigma2hat, and stdtauhat.

8.5 Prior Elicitation

Prior distributions are carriers of prior information that is coherently incorporated via Bayes' theorem into an inference. At the same time, parameters are unobservable, and prior specification is subjective in nature. The subjectivity of specifying the prior is a fundamental criticism of the Bayesian approach. Being subjective does not mean that the approach is nonscientific, as critics of Bayesian statistics often insinuate. On the contrary, vast amounts of scientific information coming from theoretical and physical models, previous experiments, and expert reports guides the specification of priors and merges such information with the data for better inference.

In arguing about the importance of priors in Bayesian inference, Garthwhite and Dickey (1991) state that "expert personal opinion is of great potential value and can be used more efficiently, communicated more accurately, and judged more critically if it is expressed as a probability distribution."

In the last several decades Bayesian research has also focused on priors that were noninformative and robust; this was in response to criticism that results of Bayesian inference could be sensitive to the choice of a prior.

For instance, in Examples 8.4 and 8.5 we saw that beta distributions are an appropriate family of priors for parameters supported in the interval $[0,1]$, such as a population proportion. It turns out that the beta family can express a wide range of prior information. For example, if the mean μ and variance σ^2 for a beta prior are elicited by an expert, then the parameters (a,b) can be determined by solving $\mu = a/(a+b)$ and $\sigma^2 = ab/[(a+b)^2(a+b+1)]$ with respect to a and b:

$$a = \mu\left(\frac{\mu(1-\mu)}{\sigma^2} - 1\right), \quad \text{and} \quad b = (1-\mu)\left(\frac{\mu(1-\mu)}{\sigma^2} - 1\right). \quad (8.4)$$

If a and b are not too small, the shape of a beta prior resembles a normal distribution and the bounds $[\mu - 2\sigma, \mu + 2\sigma]$ can be used to describe the range of likely parameters. For example, an expert's claim that a proportion is unlikely to be higher than 90% can be expressed as $\mu + 2\sigma = 0.9$.

In the same context of estimating the proportion, Berry and Stangl (1996) suggest a somewhat different procedure:

(i) Elicit the probability of success in the first trial, p_1, and match it to the prior mean $\alpha/(\alpha + \beta)$.

(ii) Given that the first trial results in success, the posterior mean is $\frac{\alpha+1}{\alpha+\beta+1}$. Match this ratio with the elicited probability of success in a second trial, p_2, conditional upon the first trial's resulting in success. Thus, a system

$$p_1 = \frac{\alpha}{\alpha + \beta} \quad \text{and} \quad p_2 = \frac{\alpha + 1}{\alpha + \beta + 1}$$

is obtained that solves to

$$\alpha = \frac{p_1(1 - p_2)}{p_2 - p_1} \quad \text{and} \quad \beta = \frac{(1 - p_1)(1 - p_2)}{p_2 - p_1}. \tag{8.5}$$

See Exercise 8.15 for an application.

If one has no prior information, many noninformative choices are possible, such as invariant priors, Jeffreys' priors, default priors, reference priors, and intrinsic priors, among others. Informally speaking, a noninformative prior is one which is dominated by the likelihood, or that is "flat" relative to the likelihood.

Popular noninformative choices are the flat prior $\pi(\theta) = C$ for the location parameter (mean) and $\pi(\theta) = 1/\theta$ for the scale/rate parameter. A vague prior for the population proportion is proportional to $p^{-1}(1 - p)^{-1}$, $0 < p < 1$. This prior is sometimes called Zellner's prior and is equivalent of setting a flat prior on the $\mathrm{logit}(p) = \log \frac{p}{1-p}$. The listed priors are not proper probability distributions, that is, they are not bonafide densities because their integrals are not finite. However, Bayes' theorem usually leads to posterior distributions that are proper densities and on which Bayesian analysis can be carried out.

Jeffreys' priors (named after Sir Harold Jeffreys, English statistician, geophysicist, and astronomer) are obtained from a particular functional of a density (Fisher information), and they are also examples of vague and noninformative priors. For a binomial proportion, Jeffreys' prior is proportional to $p^{-1/2}(1 - p)^{-1/2}$, while for the rate of exponential distribution λ, Jeffreys' prior is proportional to $1/\lambda$. For a normal distribution, Jeffreys' prior on the mean is flat, while for the variance σ^2, it is proportional to $\frac{1}{\sigma^2}$.

Example 8.9. **Jeffreys' Prior on Exponential Rate Parameter.** If $X_1 = 1.7$, $X_2 = 0.6$, and $X_3 = 5.2$ come from an exponential distribution with a rate parameter λ, find the Bayes estimator if the prior on λ is $\frac{1}{\lambda}$.

The likelihood is $\lambda^3 e^{-\lambda \sum_{i=1}^{3} X_i}$ and the posterior is proportional to

$$\frac{1}{\lambda} \times \lambda^3 e^{-\lambda \sum_{i=1}^{3} X_i} = \lambda^{3-1} e^{-\lambda \sum X_i},$$

which is recognized as gamma $\mathcal{G}a\left(3, \sum_{i=1}^{3} X_i\right)$. The Bayes estimator, as a mean of this posterior, coincides with the MLE, $\hat{\lambda} = \frac{3}{\sum_{i=1}^{3} X_i} = \frac{1}{\overline{X}} = 1/2.5 = 0.4$.

Effective Sample Size in Prior Elicitation. In the previous discussion we used the notion *noninformative*, as a prior attribute in quite informal manner. For example, uniform, Jeffreys, and Zellner priors on binomial proportions have all been called noninformative.

It is possible to calibrate the amount of information a prior is carrying by assigning a sample size value to it. Informally, the information in a prior is "worth" the information contained in a sample of size m. We will call m the effective sample size (ESS).

The ESS is inferred mainly on conjugate pairs of distributions by comparing hyperparameters of the prior and posterior, or prior and posterior means.

(i) When the model is binomial, and the prior is beta $\mathcal{B}e(a,b)$, the prior mean is $a/(a+b)$ and the posterior mean is $(a+X)/(a+b+n)$, so ESS $= a + b$ is adopted.

(ii) Gamma $\mathcal{G}a(a,b)$ prior on Poisson rate λ is conjugate and the Bayes rule a/b without data goes to $(\sum_i X_i + a)/(b+n)$ with the data, so ESS $= b$.

(iii) In gamma $\mathcal{G}a(a,b)$ prior on normal precision $\tau = 1/\sigma^2$, the Bayes rules are a/b and $(a+n/2)/(b+1/2\sum_i(X_i - \mu)^2)$, so ESS $= 2a$.

(iv) For the normal mean with normal prior, ESS is σ^2/ξ^2, where σ^2 is variance of the likelihood, and ξ^2 is the variance of the prior.

Sometimes the historic data used to elicit priors and determine ESS are not of the same quality, rigor, or importance as the data in the experiment that is under analysis, and we may want to discount the ESS by a factor between 0 and 1, say k. That leads to replacing the priors above with $\mathcal{B}e(ka,kb)$, $\mathcal{G}a(ka,kb)$, or in the normal case, replacing ξ^2 by $k\xi^2$.

For an example of use of ESS in prior elicitation, see Example 10.3.

An applied approach to prior selection was taken by Spiegelhalter et al. (1994) in the context of clinical trials. They recommended a *community of priors* elicited from a large group of experts. A crude classification of community priors is as follows:

(i) Vague priors – noninformative priors, in many cases leading to posterior distributions proportional to the likelihood.

(ii) Skeptical priors – reflecting the opinion of a clinician unenthusiastic about the new therapy, drug, device, or procedure. This may be a prior of a regulatory agency.

(iii) Enthusiastic or clinical priors – reflecting the opinion of the proponents of the clinical trial, centered around the notion that a new therapy, drug, device, or procedure is superior. This may be the prior of the industry involved or of clinicians running the trial.

For example, the use of a skeptical prior when testing for the superiority of a new treatment would be a conservative approach. In equivalence tests, both skeptical and enthusiastic priors may be used. The superiority of a new treatment should be judged by a skeptical prior, while the superiority of the old treatment should be judged by an enthusiastic prior.

8.6 Bayesian Computation and Use of WinBUGS

If the selection of an adequate prior is the major conceptual and modeling challenge of Bayesian analysis, the major implementational challenge is computation. When the model deviates from the conjugate structure, finding the posterior distribution and the Bayes rule is all but simple. A closed-form solution is more the exception than the rule, and even for such exceptions, lucky mathematical coincidences, convenient mixtures, and other tricks are needed to uncover the explicit expression.

If classical statistics relies on optimization, Bayesian statistics relies on integration. The marginal needed to normalize the product $f(x|\theta)\pi(\theta)$ is an integral

$$m(x) = \int_{\Theta} f(x|\theta)\pi(\theta)d\theta,$$

while the Bayes estimator of $h(\theta)$ is a ratio of integrals,

$$\delta_\pi(x) = \int_{\Theta} h(\theta)\pi(\theta|x)d\theta = \frac{\int_{\Theta} h(\theta)f(x|\theta)\pi(\theta)d\theta}{\int_{\Theta} f(x|\theta)\pi(\theta)d\theta}.$$

The difficulties in calculating the above Bayes rule derive from the facts that (i) the posterior may not be representable in a finite form and (ii) the integral of $h(\theta)$ does not have a closed form even when the posterior distribution is explicit.

The last two decades of research in Bayesian statistics has contributed to broadening the scope of Bayesian models. Models that could not be handled before by a computer are now routinely solved. This is done by *Markov chain Monte Carlo* (MCMC) methods, and their introduction to the field of statistics revolutionized Bayesian statistics.

The MCMC methodology was first applied in statistical physics (Metropolis et al., 1953). Work by Gelfand and Smith (1990) focused on applications of MCMC to Bayesian models. The principle of MCMC is simple: one designs a Markov chain that samples from the target distribution. By simulating long runs of such a Markov chain, the target distribution can be well approximated. Various strategies for constructing appropriate Markov chains that simulate the desired distribution are possible: Metropolis–Hastings, Gibbs sampler, slice sampling, perfect sampling, and many specialized techniques. These are beyond the scope of this text, and the interested reader is directed to Robert (2001), Robert and Casella (2004), and Chen et al. (2000) for an overview and a comprehensive treatment.

In the examples that follow we will use WinBUGS for doing Bayesian inference when the models are not conjugate. Chapter 19 gives a brief introduction to the front end of WinBUGS. Three volumes of examples are a standard addition to the software; in the Examples menu of WinBUGS, see Spiegelhalter et al. (1996). It is recommended that you go over some of those examples in detail because they illustrate the functionality and modeling power of WinBUGS. A wealth of examples on Bayesian modeling strategies using WinBUGS can be found in the monographs of Congdon (2005, 2006, 2010, 2014), Lunn et al. (2013), and Ntzoufras (2009).

The following example is a WinBUGS solution of Example 8.2.

Example 8.10. **Jeremy's IQ in WinBUGS.** We will calculate a Bayes estimator for Jeremy's true IQ, θ, using simulations in WinBUGS. Recall that the model was $X \sim \mathcal{N}(\theta, 80)$ and $\theta \sim \mathcal{N}(100, 120)$. WinBUGS uses precision instead of variance to parameterize the normal distribution. Precision is simply the reciprocal of the variance, and in this example, the precisions are $1/120 = 0.00833$ for the prior and $1/80 = 0.0125$ for the likelihood. The WinBUGS code is as follows:

```
Jeremy in WinBUGS
model{
x ~ dnorm( theta, 0.0125)
theta ~ dnorm( 110, 0.008333333)
}
DATA
list(x=98)
INITS
list(theta=100)
```

Here is the summary of the MCMC output. The Bayes estimator for θ is rounded to 102.8. It is obtained as a mean of the simulated sample from the posterior.

	mean	sd	MC error	val2.5pc	median	val97.5pc	start	sample
theta	102.8	6.943	0.01991	89.18	102.8	116.4	1001	100000

Since this is a conjugate normal/normal model, the exact posterior distribution, $\mathcal{N}(102.8, 48)$, was easy to find, (Example 8.2). Note that in these simulations, the MCMC approximation, when rounded, coincides with the exact posterior mean. The MCMC variance of θ is $6.943^2 \approx 48.2$, which is close to the exact posterior variance of 48.

Example 8.11. **Uniform/Pareto Model in WinBUGS.** In Example 8.6, we found that a posterior distribution of θ, in a uniform $\mathcal{U}(0,\theta)$ model with a Pareto $\mathcal{P}a(6,2)$ prior, was Pareto $\mathcal{P}a(6,6)$. From the posterior, we found the mean, median, and mode to be 7.2, 6.7348, and 6, respectively. These are reasonable estimators of θ as location measures of the posterior.

```
Uniform with Pareto in WinBUGS
model{
for (i  in 1:n){
    x[i] ~ dunif(0, theta);
    }
theta ~ dpar(2,6)
}
DATA
list(n=4, x = c(2, 5, 0.5, 3) )
INITS
list(theta= 7)
```

Here is the summary of the WinBUGS output. The posterior mean was found to be 7.196 and the median 6.736. Apparently, the mode of the posterior was 6, as is evident from Figure 8.3. These approximations are close to the exact values found in Example 8.6.

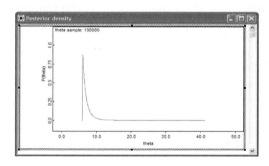

Fig. 8.3 Output from `Inference>Samples>density` shows MCMC approximation to the posterior distribution.

	mean	sd	MC error	val2.5pc	median	val97.5pc	start	sample
theta	7.196	1.454	0.004906	6.025	6.736	11.03	1001	100000

Example 8.12. **Jeremy, NIG Prior, and BUGS.** Using conjugate structure of the model in Example 8.8, we found the exact Bayes' estimator of μ as $\hat{\mu}_B = 106.9333$, and the estimators of variance and precision as $\hat{\sigma}_B^2 = 52.4921$ and $\hat{\tau}_B = 0.0281$. In addition to estimators of these parameters, Bayesian

model produced the estimators of their standard deviations: `stdmuhat=2.6456`, `stdsigma2hat=50.0492`, and `stdtauhat=0.016`. The following WinBUGS script calculates these estimators by MCMC simulation:

```
model{
for (i in 1:n){
    y[i] ~ dnorm(mu, tau)}
    tauc <-  c*tau
    mu ~ dnorm(mu0, tauc)
    tau ~ dgamma(a, b)
    sigma2 <- 1/tau
}
DATA
list( n=6, c=1.5,  mu0=110, a=0.1, b=10,
y=c(101, 98, 114, 105, 108, 111))
INITS
list( tau=0.01, mu=100)
```

	mean	sd	MC error	val2.5pc	median	val97.5pc	start	sample
mu	106.9	2.646	0.002655	101.6	106.9	112.2	1001	1000000
sigma2	52.48	49.61	0.06046	14.92	39.75	166.2	1001	1000000
tau	0.02813	0.01599	1.764E-5	0.00601	0.02516	0.06701	1001	1000000

Zero-Tricks in WinBUGS. Although the list of built-in distributions for specifying the likelihood or the prior in WinBUGS is rich (page 952), sometimes we encounter densities that are not on the list. How do we set the likelihood for a density that is not built into WinBUGS?

There are several ways, the most popular of which is the so-called zero-trick. Let f be an arbitrary model and $\ell_i = \log f(x_i|\theta)$ the log-likelihood for the ith observation. Then

$$\prod_{i=1}^{n} f(x_i|\theta) = \prod_{i=1}^{n} e^{\ell_i} = \prod_{i=1}^{n} \frac{(-\ell_i)^0 e^{-(-\ell_i)}}{0!} = \prod_{i=1}^{n} \mathbb{P}(Y_i = 0),$$

where Y_i are Poisson $\mathcal{P}oi(-\ell_i)$ random variables.

The WinBUGS code for a zero-trick can be written as follows:

```
for (i in 1:n){
zeros[i] <- 0
lambda[i] <- -llik[i] + 10000
   # Since lambda[i] needs to be positive as
   #    a Poisson rate, to ensure positivity
   #    an arbitrary constant C can be added.
   # Here we added C = 10000.
zeros[i] ~ dpois(lambda[i])
```

```
llik[i] <- ... write the log-likelihood function here
}
```

Example 8.13. **A Zero-Trick for Maxwell.** This example finds the Bayes estimator of parameter θ in a Maxwell distribution with a density of $f(x|\theta) = \sqrt{\frac{2}{\pi}} \, \theta^{3/2} \, x^2 \, e^{-\theta x^2/2}$, $x \geq 0, \theta > 0$. The moment-matching estimator and the MLE were discussed in Example 7.4. For a sample of size $n = 3$, $X_1 = 1.4$, $X_2 = 3.1$, and $X_3 = 2.5$ the MLE of θ was $\hat{\theta}_{\text{MLE}} = 0.5051$. The same estimator was found by moment-matching when the second moment was matched. The Maxwell density is not implemented in WinBUGS and we will use a zero-trick instead.

```
#Estimation of Maxwell's theta
#Using a zero-trick
model{
    for (i in 1:n){
    zeros[i] <- 0
    lambda[i] <- -llik[i] + 10000
    zeros[i] ~ dpois(lambda[i])
    llik[i] <- 1.5 * log(theta)-0.5 * theta * pow(x[i],2)
}
        theta ~ dgamma(0.1, 0.1) #non-informative choice
}
DATA
list(n=3, x=c(1.4, 3.1, 2.5))
INITS
list(theta=1)
```

	mean	sd	MC error	val2.5pc	median	val97.5pc	start	sample
theta	0.5115	0.2392	8.645E-4	0.1559	0.4748	1.079	1001	100000

Note that the Bayes estimator with respect to a vague prior dgamma(0.1, 0.1) is 0.5115.

Example 8.14. **Zero-Tricks for Priors.** The preceeding examples showed how to set a likelihood that is not supported in WinBUGS. Setting unsupported priors via a zero-trick is similar to setting likelihoods. Since there are no observations when setting the prior for parameter θ, we start with theta ~ dflat(). The rest is analogous to zero-trick construction for the likelihood.

We illustrate setting of the normal likelihood and normal prior using zero-tricks in Jeremy's IQ from Example 8.2.

```
#Jeremy with Zero-Tricks
model{
#normal likelihood
     z1   <- 0
     z1     ~ dpois(lambda1)
     #lambda1: -log(likelihood) + constant
     lambda1 <-  log(sigma) + 0.5*pow((y - theta)/sigma, 2) + 1000
     #setting normal prior
   theta          ~ dflat()
   z2 <- 0
   z2 ~ dpois(lambda2)
   #lambda2: -log(prior) + constant
   lambda2 <- log(tau) + 0.5*pow((theta-mu)/tau, 2) + 1000
}
DATA
list(y = 98, mu = 110,  sigma = 8.944272,  tau=10.954451)
INITS
list(theta=100)
```

	mean	sd	MC error	val2.5pc	median	val97.5pc	start	sample
theta	102.8	6.966	0.0436	89.19	102.7	116.5	1001	100000

Note that we added constant 1000 to both $-\log(\text{likelihood})$ and $-\log(\text{prior})$ to ensure that lambda1 and lambda2 are nonnegative as rates in zero-trick Poisson distributions. In this case it was not necessary to add any constants since log(sigma) and log(tau) were both positive, but care is needed if either tau or sigma is small.

8.7 Bayesian Interval Estimation: Credible Sets

The Bayesian term for an interval estimator of a parameter is *credible set*. Naturally, the measure used to assess the credibility of an interval estimator is the posterior distribution. Students learning concepts of classical confidence intervals often err by stating that "the probability that a particular confidence interval $[L, U]$ contains parameter θ is $1 - \alpha$." The correct statement seems more convoluted; one generates data from the underlying model many times and, for each generated data set, calculates the confidence interval. The proportion of confidence intervals covering the unknown parameter "tends to" $1 - \alpha$. The Bayesian interpretation of a credible set C is arguably more natural: the probability of a parameter belonging to set C is $1 - \alpha$. A formal definition follows.

Assume that set C is a subset of parameter space Θ. Then C is a *credible set* with credibility $(1 - \alpha)100\%$ if

$$\mathbb{P}(\theta \in C|X) = \mathbb{E}(I(\theta \in C)|X) = \int_C \pi(\theta|x)d\theta \geq 1 - \alpha.$$

If the posterior is discrete, then the integral is a sum, and

$$\mathbb{P}(\theta \in C | X) = \sum_{\theta_i \in C} \pi(\theta_i | x) \geq 1 - \alpha.$$

This is the definition of a $(1 - \alpha)100\%$ credible set. For a fixed posterior distribution and a $(1 - \alpha)100\%$ *credibility*, a credible set is not unique. We will consider two versions of credible sets: highest posterior density (HPD) and equal-tail credible sets.

HPD Credible Sets. For a given credibility level $(1 - \alpha)100\%$, the shortest credible set has obvious appeal. To minimize size, the sets should correspond to the highest posterior probability density areas.

Definition 8.1. The $(1 - \alpha)100\%$ HPD credible set for parameter θ is a set C, a subset of parameter space Θ of the form

$$C = \{\theta \in \Theta | \pi(\theta|x) \geq k(\alpha)\},$$

where $k(\alpha)$ is the largest constant for which

$$\mathbb{P}(\theta \in C | X) \geq 1 - \alpha.$$

Geometrically, if the posterior density is cut by a horizontal line at the height $k(\alpha)$, the credible set C is the projection on the θ-axis of the part of the line that lies below the density (Fig. 8.4).

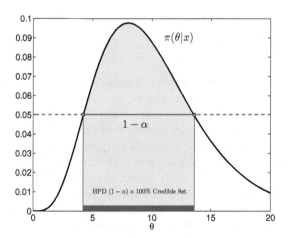

Fig. 8.4 Highest posterior density (HPD) $(1 - \alpha)100\%$ credible set (*blue*). The *area in yellow* is $1 - \alpha$.

Example 8.15. **Jeremy's IQ, Continued.** We are again back to Jeremy, the enthusiastic bioengineering student from Example 8.2 who used Bayesian inference in modeling his IQ test scores. For a score of X he was using a $\mathcal{N}(\theta, 80)$ likelihood, while the prior on θ was $\mathcal{N}(110, 120)$. After the score of $X = 98$ was recorded, the resulting posterior was normal $\mathcal{N}(102.8, 48)$.

Here, the MLE is $\hat{\theta} = 98$, and a 95% confidence interval is $[98 - 1.96\sqrt{80},\ 98 + 1.96\sqrt{80}] = [80.4692, 115.5308]$. The length of this interval is approximately 35. The Bayesian counterparts are $\hat{\theta} = 102.8$, and $[102.8 - 1.96\sqrt{48},\ 102.8 + 1.96\sqrt{48}] = [89.2207, 116.3793]$. The length of the 95% credible set is approx. 27. The Bayesian interval is shorter because the posterior variance is smaller than the likelihood variance; this is a consequence of the presence of prior information. Figure 8.5 shows the credible set (in blue) and the confidence interval (in red).

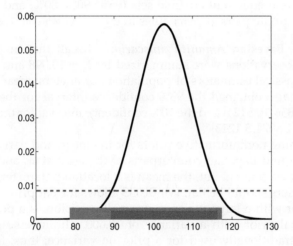

Fig. 8.5 HPD 95% credible set based on a density of $\mathcal{N}(102.8, 48)$ (*blue*). The interval in *red* is a 95% confidence interval based on the observation $X = 98$ and likelihood variance $\sigma^2 = 80$.

From the WinBUGS output table in Jeremy's IQ estimation example (page 351), the 95% credible set is $[89.18, 116.4]$.

	mean	sd	MC error	val2.5pc	median	val97.5pc	start	sample
theta	102.8	6.943	0.01991	89.18	102.8	116.4	1001	100000

Other posterior quantiles that lead to credible sets of different *credibility* levels can be specified in Sample Monitor Tool under Inference>Samples in Win-

BUGS. The credible sets from WinBUGS are HPD only if the posterior is symmetric and unimodal.

Equal-Tail Credible Sets. HPD credible sets may be difficult to find for asymmetric posterior distributions, such as gamma and Weibull, for example. Much simpler are *equal-tail credible sets* for which the tails have a probability of $\alpha/2$ each for a credibility of $1 - \alpha$. An equal-tail credible set may not be the shortest set, but to find it, we need only $\alpha/2$ and $1 - \alpha/2$ quantiles of the posterior. These two quantiles are the lower and upper bounds $[L, U]$:

$$\int_{-\infty}^{L} \pi(\theta|x)\,d\theta = \alpha/2, \qquad \int_{U}^{\infty} \pi(\theta|x)\,d\theta = 1 - \alpha/2.$$

Note that WinBUGS gives posterior quantiles from which one can directly establish several equal-tail credible sets (95%, 90%, 80%, and 50%) by selecting appropriate pairs of percentiles in the Sample Monitor Tool.

Example 8.16. **Bayesian *Amanita muscaria*.**   Recall that in Example 7.8 (page 302) observations were summarized by $\overline{X} = 10.098$ and $s^2 = 2.1702$, which are classical estimators of population parameters: mean μ and variance σ^2. We also obtained the 95% confidence interval for the population mean as $[9.6836, 10.5124]$ and the 90% confidence interval for the population variance as $[1.6074, 3.1213]$.

By assuming noninformative priors for the mean and variance, we use WinBUGS to find Bayesian counterparts of the estimators and confidence intervals. As we pointed out, the mean is a location parameter, and noninformative priors should be flat. WinBUGS allows for flat priors, mu~dflat(), but any prior with a large variance, or small precision, is a possibility. We take a normal prior with a variance of 10,000. The inverse gamma distribution is traditionally used for a prior on variance; thus, for precision as a reciprocal of variance, the gamma prior is appropriate. As we discussed earlier, gamma distributions with small parameters will have a large variance, thereby making the prior vague/noninformative. We selected prec~dgamma(0.001, 0.001) as a noninformative choice. This prior is noninformative because it is essentially flat; its variance is $0.001/(0.001)^2 = 1000$ (page 204). The WinBUGS program is simple:

```
model{
for ( i in 1:n ){
    amuscaria[i] ~ dnorm( mu, prec )
    }
    mu ~ dnorm(0, 0.00001)
    prec ~ dgamma(0.001, 0.001)
    sig2 <- 1/prec
}
DATA
```

```
list(n=51,amuscaria=c(10,11,12,9,10,11,13,12,10,11,11,13,9,10,
      9,10,8,12,10,11,9,10,7,11,8,9,11,11,10,12,10,8,7,11,12,
      10,9,10,11,10,8,10,10,8,9,10,13,9,12,9,9) )
INITS
list( mu =0, prec = 1 )
```

In WinBUGS' `Sample Monitor Tool` we asked for 2.5% and 97.5% posterior percentiles, which gives a 95% credible set and 5% and 95% posterior percentiles for the 90% credible set. The lower/upper bounds of the credible sets are given in boldface and the sets are [9.684,10.51] for the mean and [1.607,3.123] for the variance. The credible set for the mean is both HPD and equal-tail, but the credible set for the variance is only an equal-tail.

	mean	sd	MC error	val2.5pc	val5pc	val95pc	val97.5pc	start	sample
mu	10.1	0.2106	2.004E-4	**9.684**	9.752	10.44	**10.51**	1001	100000
prec	0.4608	0.09228	9.263E-5	0.2983	0.3202	0.6224	0.6588	1001	100000
sig2	2.261	0.472	4.716E-4	1.518	**1.607**	**3.123**	3.353	1001	100000

8.8 Learning by Bayes' Theorem

Bayesian statisticians often say: "Today's posterior is tomorrow's prior." This phrase captures the learning ability of Bayesian paradigm. As more data is acquired, Bayes' theorem updates our knowledge in a coherent manner.

We start with an example.

Example 8.17. **Leukemia Remission and 6-MP.** Freireich et al. (1963) conducted a remission maintenance therapy to compare 6-MP with placebo for prolonging the duration of remission in leukemia. From 42 patients affected with acute leukemia, but in a state of partial or complete remission, 21 pairs were formed. One randomly selected patient from each pair was assigned the maintenance treatment 6-MP, while the other patient received a placebo. Investigators monitored which patient stayed in remission longer. If that was a patient from the 6-MP treatment arm, this was recorded as a "success" (S); otherwise, it was a "failure" (F).

The results are given in the following table:

Pair	1	2	3	4	5	6	7	8	9	10
Outcome	S	F	S	S	S	F	S	S	S	S

11	12	13	14	15	16	17	18	19	20	21
S	S	S	F	S	S	S	S	S	S	S

The goal is to estimate p – the probability of success. Suppose we got information only on the first 10 subjects: 8 successes and 2 failures. When the prior on p is uniform, and the likelihood binomial, the posterior is proportional to $p^8(1-p)^2 \times 1$, which is a beta $\mathcal{Be}(9,3)$.

Suppose now that the remaining 11 observations became available (10 successes and 1 failure). If the posterior from the first stage serves as a prior in the second stage, the updated posterior is proportional to $p^{10}(1-p)^1 \times p^8(1-p)^2$ which is a beta $\mathcal{Be}(19,4)$.

By sequentially updating the prior we arrive to the same posterior as if all observations were available at the first place (18 successes and 3 failures). With a uniform prior, this would lead to the same beta $\mathcal{Be}(19,4)$ posterior. The final posterior would be the same even if the updating was done observation by observation. This exemplifies the *learning ability* of Bayes' theorem.

✐

Suppose that observations $x_1, \ldots, x_n$ from the model $f(x|\theta)$ are available and that prior on θ is $\pi(\theta)$. Then the posterior is

$$\pi(\theta|x) = \frac{f(x|\theta)\pi(\theta)}{\int f(x|\theta)\pi(\theta)d\theta},$$

where $x = (x_1, \ldots, x_n)$ and $f(x|\theta) = \prod_{i=1}^n f(x_i|\theta)$.

Suppose an that additional observation x_{n+1} is collected. Then

$$\pi(\theta|x, x_{n+1}) = \frac{f(x_{n+1}|\theta)\pi(\theta|x)}{\int f(x_{n+1}|\theta)\pi(\theta|x)d\theta}.$$

Bayes' theorem updates inference in a natural way: the posterior based on previous observations serves as a new prior.

8.9 Bayesian Prediction

Up to now, we have been concerned with Bayesian inference about population parameters. We are often faced with the problem of predicting a new observation X_{n+1} after $X_1, \ldots, X_n$ from the same population have been observed. Assume that the prior for parameter θ is elicited. The new observation would have a likelihood of $f(x_{n+1}|\theta)$, while the observed sample $X_1, \ldots, X_n$ will lead to a posterior of θ, $\pi(\theta|X_1, \ldots, X_n)$.

Then, the *posterior predictive distribution* for X_{n+1} can be obtained from the likelihood after integrating out parameter θ using the posterior distribution,

$$f(x_{n+1}|X_1,\ldots,X_n) = \int_{\Theta} f(x_{n+1}|\theta)\,\pi(\theta|X_1,\ldots,X_n)\,d\theta,$$

where Θ is the domain for θ. Note that the marginal distribution also integrates out the parameter, but using the prior instead of the posterior, $m(x) = \int_{\Theta} f(x|\theta)\pi(\theta)\,d\theta$. For this reason, the marginal distribution is sometimes called the *prior predictive* distribution.

The prediction for X_{n+1} is the expectation $\mathbb{E}X_{n+1}$, taken with respect to the predictive distribution,

$$\hat{X}_{n+1} = \int_{\mathbb{R}} x_{n+1} f(x_{n+1}|X_1,\ldots,X_n)\,dx_{n+1},$$

while the *predictive variance*,

$$\int_{\mathbb{R}} (x_{n+1} - \hat{X}_{n+1})^2 f(x_{n+1}|X_1,\ldots,X_n)\,dx_{n+1},$$

can be used to assess the precision of the prediction.

Example 8.18. **Exponential Survival Time.** Consider the exponential distribution $\mathcal{E}(\lambda)$ for a random variable X representing survival time of patients affected by a particular disease. The density for X is $f(x|\lambda) = \lambda\exp\{-\lambda x\}$, $x \geq 0$.

Suppose that the prior for λ is gamma $\mathcal{G}a(\alpha,\beta)$ with a density of $\pi(\lambda) = \frac{\beta^{\alpha}}{\Gamma(\alpha)}\lambda^{\alpha-1}\exp\{-\beta\lambda\}$, $\lambda \geq 0$.

The likelihood, after observing a sample $X_1,\ldots,X_n$ from $\mathcal{E}(\lambda)$ population, is

$$\lambda e^{-\lambda X_1}\cdot\ \ldots\ \cdot\lambda e^{-\lambda X_n} = \lambda^n \exp\left\{-\lambda\sum_{i=1}^{n} X_i\right\},$$

and the posterior is proportional to

$$\lambda^{n+\alpha-1}\exp\{-(\sum_{i=1}^{n} X_i + \beta)\lambda\},$$

which can be recognized as a gamma $\mathcal{G}a(\alpha + n, \beta + \sum_{i=1}^{n} X_i)$ distribution and completed as

$$\pi(\lambda|X_1,\ldots,X_n) = \frac{(\sum_{i=1}^{n} X_i + \beta)^{n+\alpha}}{\Gamma(n+\alpha)}\lambda^{n+\alpha-1}\exp\{-(\sum_{i=1}^{n} X_i + \beta)\lambda\},\ \lambda \geq 0.$$

The predictive distribution for a new X_{n+1} is

$$f(x_{n+1}|X_1,\ldots,X_n) = \int_0^\infty \lambda \exp\{-\lambda x_{n+1}\} \pi(\lambda|X_1,\ldots,X_n)\,d\lambda$$

$$= \frac{(n+\alpha)(\sum_{i=1}^n X_i + \beta)^{n+\alpha}}{(\sum_{i=1}^n X_i + \beta + x_{n+1})^{n+\alpha+1}}, \qquad x_{n+1} > 0.$$

Note that $X_{n+1} + \sum_{i=1}^n X_i + \beta$ is a Pareto $\mathcal{P}a(\sum_{i=1}^n X_i + \beta, n + \alpha)$, see page 212. The expected value for a new observation (a Bayesian prediction) is

$$\hat{X}_{n+1} = \int_0^\infty x_{n+1} f(x_{n+1}|X_1,\ldots,X_n)\,dx_{n+1} = \frac{\sum_{i=1}^n X_i + \beta}{n+\alpha-1}.$$

Also, the variance of the new observation is

$$\hat{\sigma}^2_{X_{n+1}} = \int_0^\infty (x_{n+1} - \hat{X}_{n+1})^2 f(x_{n+1}|X_1,\ldots,X_n)\,dx_{n+1}$$

$$= \frac{(\sum_{i=1}^n X_i + \beta)^2 (n+\alpha)}{(n+\alpha-1)^2(n+\alpha-2)}.$$

For example, if $X_1 = 2.1$, $X_2 = 5.5$, $X_3 = 6.4$, $X_4 = 8.7$, $X_5 = 4.9$, $X_6 = 5.1$, and $X_7 = 2.3$ are the observations, and $\alpha = 2$ and $\beta = 1$, then $\hat{X}_8 = 9/2$ and $\hat{\sigma}^2_{X_8} = 729/28 = 26.0357$. Figure 8.6 shows the posterior predictive distribution (solid blue line), observations (crosses), and prediction for the new observation (blue dot). The position of the mean of the data, $\overline{X} = 5$, is shown as a dotted red line.

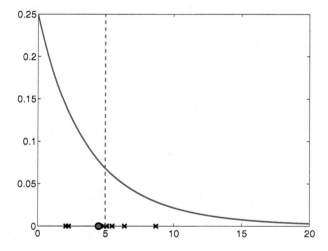

Fig. 8.6 Bayesian prediction (*blue dot*) based on the sample (*black crosses*) $X = [2.1, 5.5, 6.4, 8.7, 4.9, 5.1, 2.3]$ from the exponential distribution $\mathcal{E}(\lambda)$. The parameter λ is given a gamma $\mathcal{G}a(2,1)$ distribution and the resulting posterior predictive distribution is shown as a *solid blue line*. The position of the sample mean is plotted as a *dotted red line*.

The prediction $\hat{X}_{n+1}$ can be found in an alternative manner that avoids the need for explicit posterior predictive distribution. The following holds:

$$\hat{X}_{n+1} = \int_{\Theta} \mu(\theta)\pi(\theta|X_1,\ldots,X_n)d\theta, \qquad (8.6)$$

where $\mu(\theta) = \mathbb{E}^{X|\theta}X = \int xf(x|\theta)dx$ is the mean of X, as a function of the parameter.

When the parameter θ is in fact the the expectation, such as μ in $\mathcal{N}(\mu,\sigma^2)$ or λ in $\mathcal{P}oi(\lambda)$, the Bayes prediction for X_{n+1} is simply the posterior mean. To find $\hat{X}_{n+1}$ from Example 8.18 by (8.6), note that $\mu(\lambda) = 1/\lambda$ and the posterior is $\mathcal{G}a(\alpha + n, \beta + \sum_{i=1}^{n} X_i)$. Thus,

$$\hat{X}_{n+1} = \int_0^\infty \frac{1}{\lambda} \times \frac{\lambda^{n-\alpha-1}(\beta + \sum_{i=1}^n X_i)^{n-\alpha}}{\Gamma(\alpha+n)} \exp\{-(\beta + \sum_{i=1}^n X_i)\lambda\}d\lambda$$

$$= \frac{\beta + \sum_{i=1}^n X_i}{\alpha+n-1} \int_0^\infty \frac{\lambda^{(n-\alpha-1)-1}(\beta + \sum_{i=1}^n X_i)^{n-\alpha-1}}{\Gamma(\alpha+n-1)} \exp\{-(\beta + \sum_{i=1}^n X_i)\lambda\}d\lambda$$

$$= \frac{\beta + \sum_{i=1}^n X_i}{\alpha+n-1},$$

after using the identity $\Gamma(a) = (a-1)\Gamma(a-1)$. To find the Bayesian prediction in WinBUGS, one simply samples a new observation from a likelihood that has updated parameters.

Example 8.19. **Predicting the Exponential.** The WinBUGS program below implements Example 8.18; the observations are read within the `for` loop. However, if a new variable is simulated from the same likelihood, this is done for the current version of the parameter λ, and the mean of simulations approximates the posterior mean of the new observation.

```
model{
for (i in 1:7){
    X[i] ~ dexp(lambda)
}
lambda ~ dgamma(2,1)
Xnew ~ dexp(lambda)
}
DATA
list(X = c(2.1,  5.5,  6.4,  8.7,  4.9,  5.1,  2.3))
INITS
list(lambda=1, Xnew=1)
```

The output is

	mean	sd	MC error	val2.5pc	median	val97.5pc	start	sample
Xnew	4.499	5.09	0.005284	0.1015	2.877	18.19	1001	100000
lambda	0.25	0.08323	8.343E-5	0.1142	0.2409	0.4378	1001	100000

Note that the posterior mean for Xnew is well approximated, $4.499 \approx 4.5$, and that the standard deviation sd = 5.09 is close to $\sqrt{26.0357} = 5.1025$.

✎

8.10 Consensus Means*

Suppose that several labs are reporting measurements of the same quantity and that a consensus mean should be calculated. This problem appears in interlaboratory studies, as well as in multicenter clinical trials and various meta-analyses. In this section we provide a Bayesian solution to this problem and compare it with some classical proposals.

Let $Y_{ij}, j = 1,\ldots,n_i$; $i = 1,\ldots,k$ be measurements made at k laboratories, where n_i measurements come from lab i. Let $n = \sum_i n_i$ be the total sample size.

We are interested in estimating the mean that would properly incorporate information coming from all the labs, called the *consensus mean*. Why is the solution not trivial, and what is wrong with the average $\bar{Y} = 1/n \sum_i \sum_j Y_{ij}$?

There is nothing wrong, under the proper conditions: (a) variabilities within the labs must be equal and (b) there must be no variability between the labs.

When (a) is relaxed, proper pooling of the lab sample means is done via a Graybill–Deal estimator:

$$\bar{Y}_{gd} = \frac{\sum_{i=1}^{k} w_i \bar{Y}_i}{\sum_{i=1}^{k} w_i}, \quad w_i = \frac{n_i}{s_i^2}.$$

When both conditions (a) and (b) are relaxed, there are many competing classical estimators. For example, the Schiller–Eberhardt estimator is given by

$$\bar{Y}_{se} = \frac{\sum_{i=1}^{k} w_i \bar{Y}_i}{\sum_{i=1}^{k} w_i}, \quad w_i = \frac{1}{s_i^2/n_i + s_b^2},$$

where s_b^2 is an estimator of the variance between the labs, $s_b^2 = \frac{(\bar{y}_{max} - \bar{y}_{min})^2}{12}$. The Mandel–Paule is the same as the Schiller–Eberhardt estimator but with s_b^2 obtained iteratively.

The Bayesian approach is conceptually simple. Individual means as random variables are generated from a single distribution. The mean of this distribution is the consensus mean. In somewhat convoluted wording, the consensus mean is the mean of a hyperprior placed on the individual means.

Example 8.20. **Selenium in Powdered Milk.** The data on selenium in non-fat milk powder 📇 selenium.dat are adapted from Witkovsky (2001). Four independent measurement methods are applied. The Bayes estimator of the consensus mean is 108.8.

In the WinBUGS program below, the individual means theta[i] have a *t*-prior with location mu, precision tau, and 5 degrees of freedom. The choice of *t*-prior, instead of the usual normal, is motivated by robustness considerations.

```
model{
for (i in 1:n)
    {
    sel[i] ~ dnorm( theta[lab[i]], prec[lab[i]])
    }
for (i in 1:k)
    {
    theta[i] ~ dt(mu, tau,5) #individual means
    prec[i] ~ dgamma(0.0001, 0.0001)
    sigma2[i] <- 1/prec[i]
    }
mu ~ dt(0,0.0001,5) #consensus mean
tau ~ dgamma(0.0001,0.0001)
si2  <-1/tau
}


DATA
list(lab=c(1,1,1,1,1,1,1,1,     2,2,2,2,2,2,2,2,2,2,2,2,
3,3,3,3,3,3,3,3,3,3,3,3,3,           4,4,4,4,4,4,4,4),
sel = c(
115.7, 113.5, 103.3, 119.1, 114.2, 107.3,  91.2, 104.4,
108.6, 109.1, 107.2, 111.5, 100.6, 106.3, 105.9, 109.7,
                    111.1, 107.9, 107.9, 107.9,
107.6, 107.26,109.7, 109.7, 108.5, 106.5, 110.2, 108.3,
            110.5, 108.5, 108.8, 110.1, 109.4, 112.4,
118.7, 109.7, 114.7, 105.4, 113.9, 106.3, 104.8, 106.3),
                                      k=4, n=42)


INITS
list( mu=1, tau=1, prec=c(1,1,1,1), theta=c(1,1,1,1)  )
```

	mean	sd	MC error	val2.5pc	median	val97.5pc	start	sample
mu	108.8	0.6499	0.003674	107.6	108.9	110.0	5001	500000
si2	0.7252	9.456	0.02088	1.024E-4	0.01973	4.875	5001	500000
theta[1]	108.8	0.8593	0.003803	107.0	108.9	110.5	5001	500000
theta[2]	108.7	0.6184	0.004188	107.2	108.7	109.7	5001	500000
theta[3]	108.9	0.4046	0.00311	108.1	108.9	109.7	5001	500000
theta[4]	108.9	0.7505	0.003705	107.6	108.9	110.7	5001	500000

Next, we compare the Bayesian estimator with the classical Graybill–Deal and Schiller–Eberhardt estimators, 108.8892 and 108.7703, respectively. The Bayesian estimator falls between the two classical ones. A 95% credible set for the consensus mean is [107.6, 110].

```
lab1=[115.7, 113.5, 103.3, 119.1, 114.2, 107.3,  91.2, 104.4];
lab2=[108.6, 109.1, 107.2, 111.5, 100.6, 106.3, 105.9, 109.7,...
                           111.1, 107.9, 107.9, 107.9];
lab3=[107.6, 107.26,109.7, 109.7, 108.5, 106.5, 110.2, 108.3,...
                    110.5, 108.5, 108.8, 110.1, 109.4, 112.4];
lab4=[118.7, 109.7, 114.7, 105.4, 113.9, 106.3, 104.8, 106.3];

m = [mean(lab1)  mean(lab2)  mean(lab3)  mean(lab4)];
s = [std(lab1)   std(lab2)   std(lab3)   std(lab4) ];
ni=[8 12 14 8]; k=length(m);

%Graybill-Deal Estimator
wei = ni./s.^2; %weights
m_gd = sum(m .* wei)/sum(wei)   %108.8892

%Schiller-Eberhardt Estimator
z = sort(m);
sb2 = (z(k)-z(1))^2/12;
wei = 1./(s.^2./ni + sb2);%weights
m_se = sum(m .* wei)/sum(wei)   %108.7703
```

Borrowing Strength and Vague Priors. As popularly stated, the model in Example 8.20 allows for *borrowing strength* in the estimation of both the means θ_i and the variances σ_i^2. Even if some labs have extremely small sample sizes (as low as $n = 1$), the lab variances can be estimated through pooling via a hierarchical model structure. The prior distributions above are *vague*, which is appropriate when prior information in the form of expert opinion or historic data is not available.

Analyses conducted using vague priors can be considered objective and are generally accepted by classical statisticians. When prior information is available in the form of a mean and variance of μ, it can be included by simply changing the mean and variance of its prior, in our case the normal distribution. It is well known that Bayesian rules are sensitive with respect to changes in hyperparameters in light-tailed priors (e.g., normal priors).

If more robustness is required, a t-distribution with a small number of degrees of freedom can be substituted for the normal prior. Via MCMC sampling in WinBUGS we get a full posterior distribution of μ as the ultimate summary information.

8.11 Exercises

8.1. **Exponential Lifetimes.** A lifetime X (in years) of a particular device is modeled by an exponential distribution with unknown rate parameter θ. The lifetimes of $X_1 = 5$, $X_2 = 6$, and $X_3 = 4$ are observed. Assume that an expert familiar with this type of device suggests that θ has an exponential distribution with a mean of 3.
(a) Write down the MLE of θ for those observations.
(b) Elicit a prior according to the expert assumptions.
(c) For the prior in (b), find the posterior. Is the problem conjugate?
(d) Find the Bayes estimator $\hat{\theta}_{Bayes}$ and compare it with the MLE from (a). Discuss.
(e) Check if the following WinBUGS program gives an estimator of λ close to the Bayes estimator in (d):

```
model{
for (i in 1:n){
  X[i] ~ dexp(lambda)
  }
lambda ~ dexp(1/3)
  #note that dexp is parameterized
  #in WinBUGS by the rate parameter
}

DATA
list(n=3, X=c(5,6,4))

INITS
list(lambda=1)
```

8.2. **Fibrinogen.** Fibrinogen is a soluble plasma glycoprotein, synthesized by the liver, that is converted by thrombin into fibrin during blood coagulation. Marnie takes a blood test and finds that her level of fibrinogen is 217 mg/dL. The test results are accurate up to a random error, which is normal with mean 0 and standard deviation of 9 mg/dL.
The normal range of fibrinogen in plasma is 150–400 mg/dL, and Marnie puts a uniform prior over this range, dunif(150, 400).
(a) What is the Bayes estimator of the true level of fibrinogen given this uniform prior?

(b) Copy the `Inference>Samples>stats` output from WinBUGS. What is the 95% Credible Set for the parameter from (a)?

(c) What is the classical 95% CI for the parameter from (a)? (*Hint:* Sample Size = 1, σ known.) Compare the parameter estimates and 95% CI with Bayesian counterparts.

8.3. **Uniform/Pareto.** Suppose that $X = (X_1, \ldots, X_n)$ is a sample from $\mathcal{U}(0,\theta)$. Let θ have a Pareto $\mathcal{P}a(\theta_0,\alpha)$ prior. Show that the posterior distribution is $\mathcal{P}a(\max\{\theta_0, x_1, \ldots, x_n\}\, \alpha + n)$.

8.4. **Nylon Fibers.** Refer to Exercise 5.37, where times (in hours) between blockages of the extrusion process, T, had an exponential $\mathcal{E}(\lambda)$ distribution. Suppose that the rate parameter λ is unknown, but there are three measurements of interblockage times, $T_1 = 3$, $T_2 = 13$, and $T_3 = 8$.

(a) Estimate parameter λ using the moment-matching procedure. Write down the likelihood and find the MLE.

(b) What is the Bayes estimator of λ if the prior is $\pi(\lambda) = \frac{1}{\sqrt{\lambda}}$, $\lambda > 0$.

(c) Using WinBUGS find the Bayes estimator and 95% credible set if the prior is lognormal with parameters $\mu = 10$ and $\tau = \frac{1}{\sigma^2} = 0.0001$.

Hint: In (b) the prior is not a proper distribution, but the posterior is. Identify the posterior from the product of the likelihood from (a) and the prior.

8.5. **Gamma–Inverse Gamma.** Let $X \sim \mathcal{G}a\left(\frac{n}{2}, \frac{1}{2\theta}\right)$, so that X/θ is χ_n^2. Let $\theta \sim \mathcal{IG}(\alpha,\beta)$. Show that the posterior is $\mathcal{IG}(n/2 + \alpha, x/2 + \beta)$.

Hint: The likelihood is proportional to $\frac{x^{n/2-1}}{(2\theta)^{n/2}} e^{-x/(2\theta)}$ and the prior to $\frac{\beta^\alpha}{\theta^{\alpha+1}} e^{-\beta/\theta}$. Find their product and match the distribution for θ. There is no need to find the marginal distribution and apply Bayes' theorem since the problem is conjugate.

8.6. **Normal Precision–Gamma.** Suppose $X = -2$ was observed from a population distributed as $\mathcal{N}\left(0, \frac{1}{\theta}\right)$, and an analyst wishes to estimate the parameter θ. (Here θ is the reciprocal of the variance σ^2 and is called a *precision parameter*. Precision parameters are used in WinBUGS to parameterize the normal distribution). An MLE of θ does exist, but the analyst is tempted to estimate θ as $1/\hat{\sigma}^2$, which is troublesome since there is a single observation. Suppose the analyst believes that the prior on θ is $\mathcal{G}a(1/2,1)$.

(a) What is the MLE of θ?

(b) Find the posterior distribution and the Bayes estimator of θ. If the prior on θ is $\mathcal{G}a(r,\lambda)$, can you represent the Bayes estimator as the weighted average (sum of weights = 1) of the prior mean and the MLE?

(c) Find a 95% equal-tail credible set for θ. Use MATLAB to evaluate the quantiles of the posterior distribution.

(d) Using WinBUGS, numerically find the Bayes estimator from (b) and credible set from (c).

Hint: The likelihod is proportional to $\theta^{1/2}e^{-\theta x^2/2}$ while the prior is proportional to $\theta^{r-1}e^{-\lambda\theta}$.

8.7. **Jeremy and a Variance from a Single Observation.** Jeremy believes that his IQ test scores follow a normal distribution with mean 110 and unknown variance σ^2. He takes a test and scores $X = 98$.

(a) Show that inverse gamma prior $\mathcal{IG}(r,\lambda)$ is the conjugate for σ^2 if the observation X is normal $\mathcal{N}(\mu,\sigma^2)$ with μ known. What is the posterior?

(b) Find a Bayes estimator of σ^2 and its standard deviation in Jeremy's model if the prior on σ^2 is an inverse gamma $\mathcal{IG}(3,100)$.

(c) Use WinBUGS to solve this problem and compare the MCMC approximations with exact values from (b).

Hint: Express the likelihood terms of precision τ with gamma $\mathcal{G}a(r,\lambda)$ prior, but then calculate and monitor $\sigma^2 = \frac{1}{\tau}$. See also Exercise 8.6.

8.8. **Negative Binomial–Beta.** If $X = (X_1,\ldots,X_n)$ is a sample from $\mathcal{NB}(m,\theta)$ and $\theta \sim \mathcal{B}e(\alpha,\beta)$, show that the posterior for θ is a beta $\mathcal{B}e(\alpha + mn, \beta + \sum_{i=1}^n x_i)$ distribution.

8.9. **Poisson–Gamma Marginal.** In Example 8.7 on page 344, show that the marginal distribution for $\sum_{i=1}^n X_i$ is a generalized negative binomial, $\mathcal{NB}(\alpha,\beta/(n+\beta))$.

8.10. **Exponential–Improper.** Find Bayes' estimator for θ if a single observation X was obtained from a distribution with a density of $f(x|\theta) = \theta\exp\{-\theta x\}$, $x > 0, \theta > 0$. Assume priors (a) $\pi(\theta) = 1$ and (b) $\pi(\theta) = 1/\theta$.

8.11. **Bayes' Estimator in a Discrete Case.** Refer to the likelihood and data in Exercise 7.5.

(a) If the prior for θ is

θ	1/12	1/6	1/4
Prob	0.3	0.3	0.4

find the posterior and the Bayes estimator.

(b) What would the Bayes estimator look like for a sample of size n?

8.12. **Histocompatibility.** A patient who is waiting for an organ transplant needs a histocompatible donor who matches the patient's human leukocyte antigen (HLA) type. For a given patient, the number of matching donors per 1,000 National Blood Bank records is modeled as Poisson with an unknown rate λ. If a randomly selected group of 1,000 records showed exactly one match, estimate λ in a Bayesian fashion.

For λ assume the following:

(a) Gamma $\mathcal{G}a(2,1)$ prior.

(b) Flat prior $\lambda = 1$, for $\lambda > 0$.

(c) Invariance prior $\pi(\lambda) = \frac{1}{\lambda}$, for $\lambda > 0$.

(d) Jeffreys' prior $\pi(\lambda) = \frac{1}{\sqrt{\lambda}}$, for $\lambda > 0$.

Note that the priors in (b)–(d) are not proper densities (the integrals are not finite); nevertheless, the resulting posteriors are proper.

Hint: In all cases (a)–(d), the posterior is gamma. Write the product $\frac{\lambda^1}{1!} \exp\{-\lambda\} \times \pi(\lambda)$ and match the gamma parameters. The first part of the product is the likelihood when exactly one matching donor was observed.

8.13. **Hemocytometer Counts Revisited.** Refer to Exercise 7.36.

(a) Elicit gamma prior $\mathcal{G}a(\alpha, \beta)$ on λ for which the effective sample size (ESS) is 100 and expectation is 6. (*Hint:* ESS = β; $\mathbb{E}^\pi \lambda = \alpha/\beta$.)

(b) For the prior in (a), find an equal-tail credible set and compare it with confidence intervals from Exercise 7.36(b).

8.14. **Neurons Fire in Potter's Lab 2.** Data set ▣ neuronfires.mat consisting of 989 firing times in a cell culture of neurons was analyzed in Exercise 7.3. From this data set, the count of firings in consecutive 20-ms time intervals was recorded:

20	19	26	20	24	21	24	29	21	17
23	21	19	23	17	30	20	20	18	16
14	17	15	25	21	16	14	18	22	25
17	25	24	18	13	12	19	17	19	19
19	23	17	17	21	15	19	15	23	22

It is believed that the counts are Poisson distributed with unknown parameter λ. An expert believes that the number of counts in the 20-ms interval should be about 15.

(a) What is the likelihood function for these 50 observations?

(b) Using the information the expert provided, elicit an appropriate gamma prior. Is such a prior unique?

(c) For the prior suggested in (b), find the Bayes estimator of λ. How does this estimator compare to the MLE?

(d) Suppose now that the prior is lognormal with a mean of 15 (e.g., one possible choice is $\mu = \log(15) - 1/2 = 2.2081$ and $\sigma^2 = 1$). Using WinBUGS, find the Bayes estimator for λ. Recall that WinBUGS uses the precision parameter $\tau = 1/\sigma^2$ instead of σ^2.

8.15. **Eliciting a Beta Prior I.** This exercise is based on an example from Berry and Stangl (1996). An important prognostic factor in the early detection of breast cancer is the number of axillary lymph nodes. The surgeon will generally remove between 5 and 30 nodes during a traditional axillary dissection. We are interested in making an inference about the proportion of all nodes affected by cancer and consult the surgeon in order to elicit a prior.

The surgeon indicates that the probability of a selected node testing positive is 0.05. However, if the first node tested positive, the second will be found positive with an increased probability of 0.2.

(a) Using equations (8.5), elicit a beta prior that reflects the surgeon's opinion.

(b) If, in a particular case, two out of seven nodes tested positive, what is the Bayes estimator of the proportion of affected nodes when the prior in (a) is adopted?

8.16. **Eliciting a Beta Prior II.** A natural question for the practitioner in the elicitation of a beta prior is to specify a particular quantile. For example, we are interested in eliciting a beta prior with a mean of 0.8 such that the probability of exceeding 0.9 is 5%. Find hyperparameters a and b for such a prior. *Hint:* See file ◀ belicitor.m

8.17. **Eliciting a Weibull Prior.** Assume that the average recovery time for patients with a particular disease enters a statistical model as a parameter θ and that prior $\pi(\theta)$ needs to be elicited. Assume further that the functional form of the prior is Weibull $Wei(r, \lambda)$, so the elicitation amounts to specifying hyperparameters r and λ. A clinician states that the first and third quartiles for θ are $Q_1 = 10$ and $Q_3 = 20$ (in days). Elicit the prior. *Hint:* The CDF for the prior is $\Pi(\theta) = 1 - e^{-\lambda \theta^r}$, which with conditions on Q_1 and Q_3 leads to two equations $-e^{-\lambda \theta^r} = 0.75$ and $e^{-\lambda \theta^r} = 0.25$. Take the log twice to obtain a system of two equations with two unknowns r and $\log \lambda$.

8.18. **Bayesian Yucatan Pigs.** Refer to Example 7.23 (Yucatan Pigs). Using WinBUGS, find the Bayesian estimator of a and plot its posterior distribution.

8.19. **Eliciting a Normal Prior.** We elicit a normal prior $\mathcal{N}(\mu, \sigma^2)$ from an expert who can specify percentiles. If the 20th and 70th percentiles are specified as 2.7 and 4.8, respectively, how should μ and σ be elicited? *Hint:* If x_p is the pth quantile (100%pth percentile), then $x_p = \mu + z_p \sigma$. A system of two equations with two unknowns is formed with z_ps as norminv(0.20) = -0.8416 and norminv(0.70) = 0.5244.

8.20. **Is the Cloning of Humans Moral?** A recent Gallup poll estimates that about 88% of Americans oppose human cloning. Results are based on telephone interviews with a randomly selected national sample of $n = 1,000$ adults, aged 18 and older. In these 1,000 interviews, 882 adults opposed the cloning of humans.

(a) Write a WinBUGS program to estimate the proportion p of people opposed to human cloning. Use a noninformative prior for p.

(b) Pretend that the original poll had $n = 1,062$ adults, whereby results for 62 adults are missing. Estimate the number of people opposed to cloning among the 62 missing in the poll.

8.21. **Poisson Observations with Truncated Normal Rate.** A sample average of $n = 15$ counting observations was found to be $\overline{X} = 12.45$. Assume that each count comes from a Poisson $\mathcal{P}oi(\lambda)$ distribution. Using WinBUGS, find the Bayes estimator of λ if the prior on λ is a normal $\mathcal{N}(0, 10^2)$ constrained to $\lambda \geq 1$.
Hint: $n\overline{X} = \sum X_i$ is Poisson $\mathcal{P}oi(n\lambda)$.

8.22. **Counts of Alpha Particles.** In Example 7.14 we analyzed data from the experiment of Rutherford and Geiger on counting α-particles.
The counts, given in the table below, can be well modeled by a Poisson distribution.

X	0	1	2	3	4	5	6	7	8	9	10	11	≥ 12
Freq	57	203	383	525	532	408	273	139	45	27	10	4	2

(a) Find sample size n and sample mean $\overline{X}$. In calculations for $\overline{X}$, take ≥ 12 as 12.
(b) Elicit a gamma prior for λ with rate parameter $\beta = 5$ and shape parameter α selected in such a way that the prior mean is 7.
(c) Find the Bayes estimator of λ using the prior from (b). Is the problem conjugate? Use the fact that $\sum_{i=1}^{n} X_i \sim \mathcal{P}oi(n\lambda)$.
(d) Write a WinBUGS script that simulates the Bayes estimator for λ and compare its output with the analytic solution from (c).

8.23. **Credible Sets for Alpha Particles.** A Bayesian version of Garwood's interval in (7.14) is

$$\left[\frac{1}{2(n+b)} \chi^2_{2(S+a), \alpha/2}, \frac{1}{2(n+b)} \chi^2_{2(S+a+1), 1-\alpha/2} \right].$$

when the prior on λ is gamma $\mathcal{G}a(a, b)$.
(a) For gamma prior in Exercise 8.22 (b), find the Garwood interval that represents an equal-tail credible set.
(b) Compare the result in (a) with the credible set for λ from the Win-BUGS output in Exercise 8.22 (d).

8.24. **Hemocytometer Counts Revisited.** In Exercise 7.36 the Poisson rate of counts, λ, was estimated and 95% CIs were found.
(a) Elicit gamma $\mathcal{G}a(\alpha, \beta)$ prior on λ. Assume that the effective sample size EES is 20, and the prior mean is 6.
(b) Using WinBUGS and the prior from (a) find a 95% credible set for λ, and compare it to those from 7.36(b).
(c) Repeat calculations from (b) using normal $\mathcal{N}(0, 10^2)$ prior on λ, constrained to $\lambda \geq 3$. (*Hint:* `lambda ~ dnorm(0, 0.01) I(3,)`.)

8.25. **Rayleigh Estimation by Zero Trick.** Referring to Exercise 7.11, find the Bayes estimator of σ^2 in a Rayleigh distribution using WinBUGS.

Since the Rayleigh distribution is not on the list of WinBUGS distributions, you may use a Poisson zero trick with a negative log-likelihood as `negloglik[i] <- C + log(sig2) + pow(r[i],2)/(2 * sig2)`, where `sig2` is the parameter and `r[i]` are observations.

Since σ is a scale parameter, it is customary to put an inverse gamma on σ^2. This can be achieved by putting a gamma prior on $1/\sigma^2$, as in

```
sig2 <- 1/isig2
isig2~dgamma(0.1, 0.1)
```

where the choice of `dgamma(0.1, 0.1)` is noninformative.

8.26. **Jack and Jill, Poisson, and Bayes' Rule Revisited.** In Exercise 5.19 we assumed that Jack does *exactly* 40% of the work. This may be just an approximation. We could instead elicit a prior on this proportion that is beta with mean 0.4, say $p \sim \mathcal{B}e(4,6)$.

Write a WinBUGS script that will use this prior, and estimate the probabilities in Exercise 5.19 (a) and (b). Are the results close?
Hint:

```
model {
y ~ dpois(lambda)
lambda <- pages * rate[index]
index <- T + 1   #1 or 2, 1 for Jill, 2 for Jack
T ~ dbern(p)
p ~ dbeta(4,6)
rate[1] <- 1/4
rate[2] <- 1
}
```

where number of errors `y` and number of pages `pages` are inputs.

8.27. **Predictions in a Poisson/Gamma Model.** For a sample $X_1, \ldots, X_n$ from a Poisson $\mathcal{P}oi(\lambda)$ distribution and a gamma $\mathcal{G}a(\alpha, \beta)$ prior on λ,

(a) Prove that the marginal distribution is Pólya (a negative binomial with noninteger r, page 185), and identify its parameters.

(b) Show that the posterior predictive distribution for X_{n+1} is also a Pólya. Identify its parameters and find the prediction $\hat{X}_4$ for $X_1 = 4$, $X_2 = 5$, and $X_3 = 4.2$, $\alpha = 2$, and $\beta = 1$.

(c) Calculate the posterior mean for the data in (b). According to (8.6), this posterior mean is $\hat{X}_4$. Do the results from (b) and (c) agree?

(d) Support your findings in (b) and (c) with a WinBUGS simulation.

8.28. **Estimating Chemotherapy Response Rates.** An oncologist believes that 90% of cancer patients will respond to a new chemotherapy treatment and that it is unlikely that this proportion will be below 80%. Elicit a beta prior that models the oncologist's beliefs.
Hint: $\mu = 0.9$, $\mu - 2\sigma = 0.8$, and use equations (8.4).
During the trial, of the 30 patients treated, 22 responded. What are the likelihood and posterior distribution.

(a) Using MATLAB, plot the prior, likelihood, and posterior in a single figure.

(b) Using WinBUGS, find the Bayes estimator of the response rate and compare it to the posterior mean.

| MATLAB AND WINBUGS FILES AND DATA SETS USED IN THIS CHAPTER |

http://statbook.gatech.edu/Ch8.Bayes/

    BAint.m, belicitor.m, betaplots.m, HPDfigure.m, jeremy.m, nornorplot.m, ParetoUni.m, Predictive.m, selenium.m, [dir] matbugs

            coin.odc, copd.odc, ExeTransplant.odc, histocompatibility.odc, jeremy.odc|txt, jeremyminimal.odc, metalabs1.odc, metalabs2.odc, muscaria.odc, neurons.odc, pareto.odc, poistrunorm.odc, predictiveexample.odc, rayleigh.odc, rutherford.odc, selenium.odc, zerotrickjeremy.odc, ztNN.odc, ztNN1.odc, ztcoshprior.odc, ztmaxwell.odc

    selenium.dat

CHAPTER REFERENCES

Anscombe, F. J. (1962). Tests of goodness of fit. *J. Roy. Stat. Soc. B*, **25**, 81–94.

Bayes, T. (1763). An essay towards solving a problem in the doctrine of chances. *Philos. Trans. R. Soc. Lond.*, **53**, 370–418.

Berger, J. O. (1985). *Statistical Decision Theory and Bayesian Analysis*, 2nd ed. Springer, New York.

Berger, J. O. and Delampady, M. (1987). Testing precise hypothesis. *Stat. Sci.*, **2**, 317–352.

Berger, J. O. and Selke, T. (1987). Testing a point null hypothesis: the irreconcilability of *p*-values and evidence (with discussion). *J. Am. Stat. Assoc.*, **82**, 112–122.

Berry, D. A. and Stangl, D. K. (1996). Bayesian methods in health-related research. In: Berry, D. A. and Stangl, D. K. (eds.). *Bayesian Biostatistics*. Dekker, New York.

Chen, M.-H., Shao, Q.-M., and Ibrahim, J. (2000). *Monte Carlo Methods in Bayesian Computation*. Springer, New York.

Congdon, P. (2005). *Bayesian Models for Categorical Data*. Wiley, Hoboken, NJ.

Congdon, P. (2006). *Bayesian Statistical Modelling*, 2nd ed. Wiley, Hoboken, NJ.

Congdon, P. (2010). *Hierarchical Bayesian Modelling*. Chapman & Hall/CRC, Boca Raton, FL.

Congdon, P. (2014). *Applied Bayesian Modelling*, 2nd ed. Wiley, Hoboken, NJ.

FDA (2010). Guidance for the use of Bayesian statistics in medical device clinical trials. Center for Devices and Radiological Health Division of Biostatistics, Rockville, MD. http://www.fda.gov/downloads/MedicalDevices/ DeviceRegulationandGuidance/GuidanceDocuments/ucm071121.pdf

Finney, D. J. (1947). The estimation from individual records of the relationship between dose and quantal response. *Biometrika*, **34**, 320–334.

Freireich, E. J., Gehan, E., Frei, E., Schroeder, L. R., Wolman, I. J., Anbari, R., Burgert, E. O., Mills, S. D., Pinkel, D., Selawry, O. S., Moon, J. H., Gendel, B. R., Spurr, C. L., Storrs, R., Haurani, F., Hoogstraten, B., and Lee, S. (1963). The effect of 6-Mercaptopurine on the duration of steroid-induced remissions in acute leukemia: a model for evaluation of other potentially useful therapy. *Blood*, **21**, 699–716.

Garthwhite, P. H. and Dickey, J. M. (1991). An elicitation method for multiple linear regression models. *J. Behav. Decis. Mak.*, **4**, 17–31.

Gelfand, A. E. and Smith, A. F. M. (1990). Sampling-based approaches to calculating marginal densities. *J. Am. Stat. Assoc.*, **85**, 398–409.

Lindley, D. V. (1957). A statistical paradox. *Biometrika*, **44**, 187–192.

Lunn, D., Jackson, C., Best, N., Thomas, A., and Spiegelhalter, D. (2013). *The BUGS Book. A Practical Introduction to Bayesian Analysis.* CRC, Boca Raton.

Martz, H. and Waller, R. (1985). *Bayesian Reliability Analysis.* Wiley, New York.

Metropolis, N., Rosenbluth, A., Rosenbluth, M., Teller, A., and Teller, E. (1953). Equation of state calculations by fast computing machines. *J. Chem. Phys.*, **21**, 1087–1092.

Ntzoufras, I. (2009). *Bayesian Modeling Using WinBUGS.* Wiley, Hoboken, NJ.

Robert, C. (2001). *The Bayesian Choice: From Decision-Theoretic Motivations to Computational Implementation*, 2nd ed. Springer, New York.

Robert, C. and Casella, G. (2004). *Monte Carlo Statistical Methods*, 2nd ed. Springer, New York.

Sellke, T., Bayarri, M. J., and Berger, J. O. (2001). Calibration of p values for testing precise null hypotheses. *Am. Stat.*, **55**, 1, 62–71.

Spiegelhalter, D. J., Thomas, A., Best, N. G., and Gilks, W. R. (1996). *BUGS Examples Volume 1*, ver. 0.5. Medical Research Council Biostatistics Unit, Cambridge, UK (PDF document).

Chapter 9
Testing Statistical Hypotheses

If one in twenty does not seem high enough odds, we may, if we prefer it, draw the line at one in fifty (the 2 percent point), or one in a hundred (the 1 percent point). Personally, the writer prefers to set a low standard of significance at the 5 percent point, and ignore entirely all results which fail to reach this level. A scientific fact should be regarded as experimentally established only if a properly designed experiment rarely fails to give this level of significance.

– Ronald Aylmer Fisher

WHAT IS COVERED IN THIS CHAPTER

- Basic Concepts in Testing: Hypotheses, Errors of the First and Second Kind, Rejection Regions, Significance Level, *p*-Value, Power
- Bayesian Approach to Testing
- Testing the Mean in a Normal Population: *z* and *t* Tests
- Testing the Variance in a Normal Population
- Testing the Population Proportion
- Multiple Testing, Bonferroni Correction, and False Discovery Rate

9.1 Introduction

The two main tasks of inferential statistics are parameter estimation and testing statistical hypotheses. In this chapter we will focus on the latter.

Although the expositions on estimation and testing are separate, the two inference tasks are highly related, as it is possible to conduct testing by inspecting confidence intervals or credible sets. Both tasks can be unified via the so-called decision-theoretic approach in which both the estimator and the selection of a hypothesis represent an optimal action given the model, observations, and loss function.

Generally, any claim made about one or more populations of interest constitutes a *statistical hypothesis*. These hypotheses usually involve population parameters, the nature of the population, the relationsips between the populations, and so on. For example, we could hypothesize that:

- The mean of a population, μ, is 2, or
- Two populations have the same variance, or
- A population is normally distributed, or
- The means in four populations are the same, or
- Two populations are independent.

Procedures leading to either the acceptance[1] or rejection of statistical hypotheses are called statistical tests.

We will discuss two approaches: the frequentist (classical) approach, which is based on the Neyman–Pearson lemma, and the Bayesian approach, which assigns probabilities to hypotheses directly.

The Neyman–Pearson lemma is technical (details can be found in Casella and Berger, 2001), and the testing procedure based on it will be formulated as an algorithm or a testing "recipe." In fact, this recipe is a mix of Neyman–Pearson's and Fisher's approaches since it takes the best from both: a framework for power analysis from the Neyman–Pearsonian approach and better sensitivity to the observations from the Fisherian method.

In the Bayesian framework, one simply finds and reports the probability that a particular hypothesis is true given the observations. The competing hypotheses are assigned probabilities, and those with the larger probability are favored. Frequentist tests do not assign probabilities to hypotheses directly but rather to the statistic on which the test is based. This point will be emphasized later, since p-values are often mistaken for probabilities of hypotheses.

We start by discussing the terminology and algorithm of the frequentist testing framework.

---

[1] The use of jargon such as *accept a hypothesis* in the testing context should be avoided. The equivalent but conservative wording for *accept* would be: *there is not enough statistical evidence to reject*. We will use the terms "reject" and "do not reject" when appropriate, leaving the careful wording to practicing statisticians who could be liable for the undesirable consequences of their straightforward recommendations.

9.2 Classical Testing Problem

9.2.1 Choice of Null Hypothesis

The usual starting point in statistical testing is the formulation of statistical hypotheses. There will be at least (in most cases, exactly) two competing hypotheses. The hypothesis that reflects the current *state of nature*, adopted standard, or believed truth is denoted by H_0 and is termed the *null hypothesis*. The competing hypothesis, H_1, is called the *alternative* or *research hypothesis*. Sometimes, the alternative hypothesis is denoted by H_a.

In the classical testing approach it is important to carefully select which of the two hypotheses is assigned to be H_0, since the subsequent testing procedure depends on this assignment. The following "rule" describes the choice of H_0 and hints at the reason why it is termed the null hypothesis.

Rule: We want to establish an assertion about a population with substantive support obtained from the data. The negation of the assertion is taken to be the *null* hypothesis H_0, and the assertion itself is taken to be the research or alternative hypothesis H_1. In this context, the term *null* can be interpreted as a void research hypothesis.

The following example illustrates several hypothetical testing scenarios.

Example 9.1. **Hypothetical Testing Scenarios.** (a) A biomedical engineer wants to determine whether a new chemical agent provides a faster reaction than the agent currently in use. The new agent is more expensive, so the engineer would not recommend it unless its faster reaction is supported by experimental evidence. The reaction times are observed in several experiments prepared with the new agent. If the reaction time is denoted by the parameter θ, then the two hypotheses can be expressed in terms of that parameter. It is assumed that the reaction speed of the currently used agent is known, $\theta = \theta_0$. Null hypothesis H_0: The new agent is not faster ($\theta = \theta_0$). Alternative hypothesis H_1: The new agent is faster ($\theta > \theta_0$).

(b) A state labor department wants to determine if the current rate of unemployment varies significantly from the forecast of 8% made 2 months ago. Null hypothesis H_0: The current rate of unemployment is 8%. Alternative hypothesis H_1: The current rate of unemployment differs from 8%.

(c) A biomedical company claims that a new treatment is more effective than the standard treatment for prolonging the lives of terminal cancer patients. The standard treatment has been in use for a long time, and from reports in medical journals, the mean survival period is known to be 5.2 years. Null hypothesis H_0: The new treatment is as effective as the standard one, that is, the survival time θ is equal to 5.2 years. Alternative hypothesis H_1: The new treatment is more effective than the standard one, that is, $\theta > 5.2$.

(d) Katz et al. (1990) examined the performance of 28 students taking the SAT who answered multiple-choice questions without reading the referred passages. The mean score for the students was 46.6 (out of 100), with a standard deviation of 6.8. The expected score in random guessing is 20. Null hypothesis H_0: The mean score is 20. Alternative hypothesis H_1: The mean score is larger than 20.

(e) A pharmaceutical company claims that its best-selling painkiller has a mean effective period of at least 6 hours. Experimental data found that the average effective period was actually 5.3 hours. Null hypothesis H_0: The best-selling painkiller has a mean effective period of 6 hours. Alternative hypothesis H_1: The best-selling painkiller has a mean effective period of less than 6 hours.

(f) A pharmaceutical company claims that its generic drug has a mean AUC response equivalent to that of the innovative (brand name) drug. The regulatory agency considers two drugs bioequivalent if the population means in their AUC responses differ for no more than δ. Null hypothesis H_0: The difference in mean responses in AUC between the generic and innovative drugs is either smaller than $-\delta$ or larger than δ. Alternative hypothesis H_1: The absolute difference in the mean responses is smaller than δ; that is, the generic and innovative drugs are bioequivalent.

✐

When H_0 is stated as $H_0 : \theta = \theta_0$, the alternative hypothesis can be any of

$$\theta < \theta_0, \quad \theta \neq \theta_0, \quad \theta > \theta_0.$$

The first and third alternatives are one-sided, while the middle one is two-sided. Usually, the context of the problem indicates which one-sided alternative is appropriate. For example, if the pharmaceutical industry claims that the proportion of patients allergic to a particular drug is $p = 0.01$, then either $p \neq 0.01$ or $p > 0.01$ is a sensible alternative in this context, especially if the observed proportion $\hat{p}$ exceeds 0.01.

In the context of the bioequivalence trials, the research hypothesis H_1 states that the difference between the responses is tolerable, as in (f). There $H_0 : \mu_1 - \mu_2 < -\delta$ or $\mu_1 - \mu_2 > \delta$ and the alternative is $H_1 : -\delta \leq \mu_1 - \mu_2 \leq \delta$.

9.2.2 Test Statistic, Rejection Regions, Decisions, and Errors in Testing

Famous and controversial Cambridge astronomer Sir Fred Hoyle (1915–2001) once said: "I don't see the logic of rejecting data just because they seem incredible." The calibration of the credibility of data is done with respect to some theory or model; instead of rejecting data, the model should be questioned.

Suppose that a hypothesis H_0 and its alternative H_1 are specified, and a random sample from the population under research is obtained. As in the estimation context, an appropriate statistic is calculated from the random sample. Testing is carried out by evaluating the realization of this statistic. If the realization appears unlikely under the assumption stipulated by H_0, H_0 is rejected, since the experimental support for H_0 is lacking.

If a null hypothesis is rejected when it is actually true, then a *type I error*, or *error of the first kind*, is committed. If, however, an incorrect null hypothesis is not rejected, then a *type II error*, or *error of the second kind*, is committed. It is customary to denote the probability of a type I error as α and the probability of a type II error as β.

This is summarized in the table below:

	Decide H_0	Decide H_1
True H_0 probability	Correct action $1 - \alpha$	Type I error α
True H_1 probability	Type II error β	Correct action power $= 1 - \beta$

We will also use the notation $\alpha = P(H_1|H_0)$ to denote the probability that hypothesis H_1 is decided when in fact H_0 is true. Analogously, $\beta = P(H_0|H_1)$.

A good testing procedure minimizes the probabilities of errors of the first and second kind. However, minimizing both errors simultaneously, for a fixed sample size, is impossible. Controlling the errors is a trade-off; when α decreases, β increases, and vice versa. For this and other practical reasons, α is chosen from among several typical values: 0.01, 0.05, and 0.10.

Sometimes within testing problems there is no clear dichotomy: the *established truth* versus the *research hypothesis*, and both hypotheses may seem to be research hypotheses. For instance, the statements "The new drug is safe" and "The new drug is not safe" are both research hypotheses. In such cases H_0 is selected in such a way that the type I error is more severe than the type II error. If the hypothesis "The new drug is not safe" is chosen as H_0, then the type I error (rejection of a true H_0, "use unsafe drug") is more serious (at least for the patient) than the type II error (keeping a false H_0, "do not use a safe drug").

That is another reason why α is fixed as a small number; the probability of a more serious error should be controlled. The practical motivation for fixing a few values for α was originally the desire to keep the statistical tables needed to conduct a given test brief. This reason is now outdated since the "tables" are electronic and their brevity is not an issue.

9.2.3 Power of the Test

Recall that $\alpha = \mathbb{P}(\text{reject } H_0 | H_0 \text{ true})$ and $\beta = P(\text{reject } H_1 | H_1 \text{ true})$ are the probabilities of first- and second-type errors. For a specific alternative H_1, the probability $\mathbb{P}(\text{reject } H_0 | H_1 \text{ true})$ is the *power* of the test.

Power = $1 - \beta$ ($= \mathbb{P}(\text{reject } H_0 | H_1 \text{ true})$)

In plain terms, the power is measured by the probability that the test will reject a false H_0. To find the power, the alternative must be specific. For instance, in testing $H_0 : \theta = 0$, the alternative $H_1 : \theta = 2$ is specific but $H_1 : \theta > 0$ is not. A specific alternative is needed for the evaluation of the probability $\mathbb{P}(\text{reject } H_0 | H_1 \text{ true})$. The specific null and alternative hypotheses lead to the definition of *effect size*, a quantity that researchers want to set as a sensitivity threshold for a test.

Usually, the power analysis is prospective in nature. One plans the sample size and specifies the parameters in H_0 and H_1. This allows for the calculation of an error of the second kind β and the power as $1 - \beta$. This prospective power analysis is desirable and often required. In the real world of research and drug development, for example, no regulating agency will support a proposed clinical trial if the power analysis was not addressed.

Test protocols need sufficient sample sizes for the test to be sensitive enough to discrepancies from the null hypotheses. However, the sample sizes should not be unnecessarily excessive because of financial and ethical considerations (expensive sampling, experiments that involve laboratory animals). Also, overpowered tests may detect the effects of sizes irrelevant from a clinical or engineering standpoint.

The calculation of the power after data are observed and the test was conducted, known as *retrospective power*, is controversial (Hoenig and Heisey, 2001). After the sampling is done, more information is available. If H_0 was not rejected, the researcher may be interested in knowing if the sampling protocol had enough power to detect effect sizes of interest. Inclusion of this new information in the power calculation and the perception that the goal of retrospective analysis is to justify the failure of a test to reject the null hypothesis lead to the controversy referred to earlier. Some researchers argue that retrospective power analysis should be conducted in cases where H_0 was rejected "in order not to declare H_1 true if the test was underpowered." However, this argument only emphasizes the need for the power analysis to be done beforehand. Calculating effect sizes from the collected data may also lead to a low retrospective power of well-powered studies.

9.2.4 Fisherian Approach: p-Values

A lot of information is lost by reporting only that the null hypothesis should or should not be rejected at some significance level. Reporting a measure of support for H_0 is much more desirable. For this measure of support, the p-value is routinely reported despite controversy surrounding its meaning and use. The p-value approach was favored by Fisher, who criticized the Neyman–Pearsonian approach for reporting only a fixed probability of errors of the first kind, α, no matter how strong the evidence against H_0 was. Fisher also criticized the Neyman–Pearsonian paradigm for its need of an alternative hypothesis and for a power calculation that depends on unknown parameters.

A p-value is the probability of obtaining a value of the test statistic as extreme or more extreme (from the standpoint of the null hypothesis) than that actually obtained, given that the null hypothesis is true.

Equivalently, the p-value can be defined as the lowest significance level at which the observed statistic would be significant.

Advantage of Reporting p-Values. When a researcher reports a p-value as part of their research findings, users can judge the findings according to the significance level of their choice.

Decisions from a p-value:
- The p-value is less than α: reject H_0.
- The p-value is greater than α: do not reject H_0.

In the Fisherian approach, α is not connected to the error probability; it is a significance level against which the p-value is judged. The most frequently used value for α is 5%, though values of 1% or 10% are sometimes used as well. The recommendation of $\alpha = 0.05$ is attributed to Fisher (1926), whose "one-in-twenty" quote is provided at the beginning of this chapter. Although philosophically the p-values and error probabilities are quite different, there is a link. Since under H_0 the p-value is uniformly distributed on $[0,1]$, the probability of rejecting H_0 when $p < 0.05$ is equivalent to the statement that true H_0 was rejected with probability not exceeding 0.05.

A hypothesis may be rejected if the p-value is less than 0.05; however, a p-value of 0.049 is not the same evidence against H_0 as a p-value of 0.000001. Also, it would be incorrect to say that for any non-small p-value the null hypothesis is *accepted*. A large p-value indicates that the model stipulated under the null hypothesis is merely consistent with the observed data and that there could be many other such consistent models. Thus, the appropriate wording would be that the null hypothesis is *not rejected*. This point is further elaborated in Section 10.9.

Many researchers argue that the p-value is strongly biased against H_0 and that the evidence against H_0 derived from p-values not substantially smaller than 0.05 is rather weak. In Section 9.4 we discuss the calibration of p-values against Bayes factors and errors in testing.

The p-value is often confused with the probability of H_0, which it does not represent. As we stated, it is the probability that the test statistic will be more extreme than observed when H_0 is true. If the p-value is small, then an unlikely statistic has been observed that casts doubt on the validity of H_0.

9.3 Bayesian Approach to Testing

In frequentist tests, it was customary to formulate H_0 as $H_0 : \theta = 0$ versus $H_1 : \theta > 0$ instead of $H_0 : \theta \leq 0$ versus $H_1 : \theta > 0$, as one might expect. The reason was that we calculated the p-value under the assumption that H_0 is true, and this is why a precise null hypothesis was needed.

Bayesian testing is conceptually straightforward: The hypothesis with a higher posterior probability is favored. There is nothing special about the "null" hypothesis, and for a Bayesian, H_0 and H_1 are interchangeable.

Assume that Θ_0 and Θ_1 are two nonoverlapping sets for parameter θ. We assume that $\Theta_1 = \Theta_0^c$, although arbitrary nonintersecting sets Θ_0 and Θ_1 are easily handled. Let $\theta \in \Theta_0$ be the statement of the null hypothesis H_0 and let $\theta \in \Theta_1 = \Theta_0^c$ be the same for the alternative hypothesis H_1:

$$H_0 : \theta \in \Theta_0 \qquad H_1 : \theta \in \Theta_1.$$

Bayesian tests amount to a comparison of posterior probabilities of Θ_0 and Θ_1, the regions corresponding to the two competing hypotheses. If $\pi(\theta|x)$ is the posterior distribution, then the hypothesis corresponding to the smaller of

$$p_0 = \mathbb{P}(H_0|X) = \int_{\Theta_0} \pi(\theta|x)d\theta,$$

$$p_1 = \mathbb{P}(H_1|X) = \int_{\Theta_1} \pi(\theta|x)d\theta,$$

is rejected. Here $\mathbb{P}(H_i|X)$ is the notation for the posterior probability of hypothesis H_i, $i = 0,1$.

Conceptually, this approach differs from frequentist testing, where the p-value measures the agreement of data with the model postulated by H_0, but not the probability of H_0.

Example 9.2. **A Bayesian Test for Jeremy's IQ.** We return to Jeremy (Examples 8.2 and 8.10) and consider the posterior for the parameter θ, $\mathcal{N}(102.8, 48)$. Jeremy claims he had a bad day, and his true IQ is at least 105. The posterior probability of $\theta \geq 105$ is

$$p_0 = \mathbb{P}(\theta \geq 105|X) = \mathbb{P}\left(Z \geq \frac{105 - 102.8}{\sqrt{48}}\right) = 1 - \Phi(0.3175) = 0.3754,$$

less than $1/2$, so his claim is rejected in favor of $\theta < 105$.
🖉

We represent the prior and posterior odds in favor of the hypothesis H_0, respectively, as

$$\frac{\pi_0}{\pi_1} = \frac{\mathbb{P}(H_0)}{\mathbb{P}(H_1)} \quad \text{and} \quad \frac{p_0}{p_1} = \frac{\mathbb{P}(H_0|X)}{\mathbb{P}(H_1|X)}.$$

The *Bayes factor* in favor of H_0 is the ratio of the corresponding posterior to prior odds:

$$B_{01}^{\pi}(x) = \frac{\mathbb{P}(H_0|X)}{\mathbb{P}(H_1|X)} \bigg/ \frac{\mathbb{P}(H_0)}{\mathbb{P}(H_1)} = \frac{p_0/p_1}{\pi_0/\pi_1}. \tag{9.1}$$

In the context of Bayes' rule in Chapter 3 we discussed the Bayes factor (page 103). Its meaning here is analogous: the Bayes factor updates the prior odds of hypotheses to their posterior odds, after an experiment was conducted.

Example 9.3. **Jeremy Continued.** In the context of Example 9.2, the posterior odds in favor of H_0 are $\frac{0.3754}{1-0.3754} = 0.4652$, less than 1.
🖉

⌖ When the hypotheses are simple (i.e., $H_0 : \theta = \theta_0$ versus $H_1 : \theta = \theta_1$) and the prior is just the two-point distribution $\pi(\theta_0) = \pi_0$ and $\pi(\theta_1) = \pi_1 = 1 - \pi_0$, then the Bayes factor in favor of H_0 becomes the likelihood ratio:

$$B_{01}^{\pi}(x) = \frac{\mathbb{P}(H_0|X)}{\mathbb{P}(H_1|X)} \Big/ \frac{\mathbb{P}(H_0)}{\mathbb{P}(H_1)} = \frac{f(x|\theta_0)\pi_0}{f(x|\theta_1)\pi_1} \Big/ \frac{\pi_0}{\pi_1} = \frac{f(x|\theta_0)}{f(x|\theta_1)}.$$

If the prior is a mixture of two priors, ξ_0 under H_0 and ξ_1 under H_1, then the Bayes factor is the ratio of two marginal (prior-predictive) distributions generated by ξ_0 and ξ_1. Thus, if $\pi(\theta) = \pi_0\xi_0(\theta)\mathbf{1}(\theta \in \Theta_0) + \pi_1\xi_1(\theta)\mathbf{1}(\theta \in \Theta_1)$, then

$$B_{01}^{\pi}(x) = \frac{\dfrac{\int_{\Theta_0} f(x|\theta)\pi_0\xi_0(\theta)d\theta}{\int_{\Theta_1} f(x|\theta)\pi_1\xi_1(\theta)d\theta}}{\dfrac{\pi_0}{\pi_1}} = \frac{m_0(x)}{m_1(x)}.$$

As noted earlier, the Bayes factor measures the relative change in prior odds once the evidence is collected. Table 9.1 offers practical guidelines for Bayesian testing of hypotheses depending on the value of the log-Bayes factor (Jeffreys, 1961, Appendix B). One could use $B_{01}^{\pi}(x)$, but then $a < \log B_{10}(x) \le b$ becomes $-b \le \log B_{01}(x) < -a$. Negative values of the log-Bayes factor are handled by using symmetry and appropriately changed wording.

Table 9.1 Treatment of H_0 according to log-Bayes factor values: Jeffreys' scale (Jeffreys, 1961, page 432)

Value (log 10)	Evidence against H_0 is
$0 \le \log_{10} B_{10}(x) \le 0.5$	Poor
$0.5 < \log_{10} B_{10}(x) \le 1$	Substantial
$1 < \log_{10} B_{10}(x) \le 1.5$	Strong
$1.5 < \log_{10} B_{10}(x) \le 2$	Very strong
$\log_{10} B_{10}(x) > 2$	Decisive

Suppose $X|\theta \sim f(x|\theta)$ is observed and we are interested in testing

$$H_0 : \theta = \theta_0 \quad v.s. \quad H_1 : \theta <, \ne, > \theta_0.$$

⌖ If the priors on θ are continuous distributions, Bayesian testing of precise hypotheses in the manner we just discussed is impossible. With continuous priors, and subsequently continuous posteriors, the probability of a singleton $\theta = \theta_0$ is always 0, and the precise hypothesis is always rejected.

The Bayesian solution is to adopt a prior where singleton θ_0 has a probability of π_0 and the rest of the probability is spread on $\Theta \setminus \{\theta_0\}$ by a distribution $\xi(\theta)$ that is the prior under H_1. Thus, the prior on θ is a mixture of the point mass at θ_0 with a weight π_0 and a continuous density $\xi(\theta)$ on $\Theta \setminus \{\theta_0\}$, with a weight of $\pi_1 = 1 - \pi_0$. One can show that the marginal density for X is

$$m(x) = \pi_0 f(x|\theta_0) + \pi_1 m_1(x),$$

where

$$m_1(x) = \int_{\theta \in \Theta \setminus \{\theta_0\}} f(x|\theta)\xi(\theta)d\theta. \tag{9.2}$$

The posterior probability of the null hypothesis uses this marginal distribution and is equal to

$$
\begin{aligned}
\pi(\theta_0|x) &= \frac{f(x|\theta_0)\pi_0}{m(x)} = \frac{\pi_0 f(x|\theta_0)}{\pi_0 f(x|\theta_0) + \pi_1 m_1(x)} \\
&= \left(1 + \frac{\pi_1}{\pi_0} \cdot \frac{m_1(x)}{f(x|\theta_0)}\right)^{-1}.
\end{aligned} \tag{9.3}
$$

Example 9.4. **Improvement of Surgical Procedure.** In a disease in which the postoperative mortality is usually 10%, a surgeon devises a novel surgical technique. He implements the technique on 15 patients and has no fatalities.

A Bayesian wants to test a precise null hypothesis

$$H_0 : \theta = 0.1 \quad \text{versus} \quad H_1 : \theta < 0.1.$$

and adopts prior

$$\pi(\theta) = \pi_0 \cdot \mathbf{1}(\theta = 0.1) + \pi_1 \cdot 10 \cdot \mathbf{1}(0 \leq \theta < 0.1),$$

with equal prior probabilities of the hypotheses $\pi_0 = \pi_1 = 1/2$. What is the posterior probability of H_0? What is the Bayes factor B_{01}?

Here, the number of fatalities is binomial $\mathcal{B}in(15,\theta)$, the observed number of fatalities is $x = 0$, $\theta_0 = 0.1$, the likelihood is $f(x|\theta) = \binom{15}{x}\theta^x(1-\theta)^{15-x}$, and ξ from (9.2) is uniform on $[0,0.1)$. Note also that the parameter space is $\Theta = [0,1]$ and that $\Theta_0 = \{0.1\}$ and $\Theta_1 = [0,0.1)$. Then,

$$m_1(0) = \int_0^{0.1} \binom{15}{0}\theta^0(1-\theta)^{15} \cdot 10 \, d\theta = 10/16 \cdot (1 - 0.9^{16}) = 0.5092,$$

and by (9.3)

$$\pi(\theta_0|x) = \left[1 + \frac{0.509186}{0.1^0(1-0.1)^{15}}\right]^{-1} = 0.2879.$$

Since $\pi_0/\pi_1 = 1$, the Bayes factor $B_{01} = p_0/p_1 = 0.2879/(1-0.2879) = 0.4043$. The logarithm for basis 10 of B_{01} is approximately -0.39, that is, $\log_{10} B_{10} = 0.39$. Thus, the evidence against H_0 is poor (Table 9.1), or as Jeffreys (1961) phrases it: "not worth more than a bare mention."

The surgeon's claim is not substantiated by the evidence. Even if one finds the exact frequentist p-value, which in this case is $\mathbb{P}(X \le 0) = 0.9^{15} = 0.2059$ (see Exercise 9.23), the null hypothesis is not rejected at any reasonable significance level.

There is an alternate way of testing the precise null hypothesis in a Bayesian fashion. One could test the hypothesis $H_0 : \theta = \theta_0$ against the two-sided alternative by credible sets for θ. If θ_0 belongs to a 95% credible set for θ, then H_0 is not rejected. One-sided alternatives can be accommodated as well by one-sided credible sets. This approach is natural and mimics testing by confidence intervals; however, the posterior probabilities of hypotheses are not calculated.

Testing Using WinBUGS. WinBUGS generates samples from the posterior distribution. Testing hypotheses is equivalent to finding the relative frequencies of a posterior sample falling in competing regions Θ_0 and Θ_1. For example, if

$$H_0 : \theta \le 1 \quad \text{versus} \quad H_1 : \theta > 1$$

is tested in the WinBUGS program, the command `ph1<-step(theta-1)` will calculate the proportion of the simulated chain falling in Θ_1, that is, satisfying $\theta > 1$. The `step(x)` is equal to 1 if $x \ge 0$ and 0 if $x < 0$.

9.4 Criticism and Calibration of p-Values*

In a provocative article, Ioannidis (2005) states that *many published research findings are false* because statistical significance by a particular team of researchers is found. Ioannidis lists several reasons: "...a research finding is less likely to be true when the studies conducted in a field are smaller; when effect sizes are smaller; when there is a greater number and lesser pre-selection of tested relationships; where there is greater flexibility in designs, definitions, outcomes, and analytical modes; when there is greater financial and other interest and prejudice; and when more teams are involved in a scientific field in case of statistical significance." Certainly great responsibility for an easy acceptance of research (alternative) hypotheses can be attributed to the p-values. There are many objections to the use of raw p-values for testing purposes.

Since the p-value is the probability of obtaining the statistic as large or more extreme than observed, when H_0 is true, the p-values measure how consistent the data are with H_0, and they may not be a measure of support for a particular H_0.

Misinterpretation of the p-value as the error probability leads to a strong bias against H_0. What is the posterior probability of H_0 in a test for which the reported p-value is p? Berger and Sellke (1987) and Sellke et al. (2001) show that the minimum Bayes factor (in favor of H_0) for a null hypothesis having a p-value of p is $-e\,p\log p$. The Bayes factor transforms the prior odds π_0/π_1 into the posterior odds p_0/p_1, and if the prior odds are 1 (H_0 and H_1 equally likely *a priori*, $\pi_0 = \pi_1 = 1/2$), then the posterior odds of H_0 are not smaller than $-e\,p\log p$ for $p < 1/e \approx 0.368$:

$$\frac{p_0}{p_1} \geq -ep\log p, \quad p < 1/e, \; p_0 + p_1 = 1.$$

By solving this inequality with respect to p_0, we obtain a posterior probability of H_0 as

$$p_0 \geq \frac{1}{1 + (-e\,p\log p)^{-1}},$$

which also has a frequentist interpretation as a type I error, $\alpha(p)$. Now, the effect of bias against H_0, when judged by the p-value, is clearly visible. The type I error, α, always exceeds $(1 + (-e\,p\log p)^{-1})^{-1}$.

It is instructive to look at specific numbers. Assume that a particular test yielded a p-value of 0.01, which led to the rejection of H_0 with decisive evidence. However, if *a priori* we do not have a preference for either H_0 or H_1, the posterior odds of H_0 always exceed 12.53%. The frequentist type I error or, equivalently, the posterior probability of H_0 is never smaller than 11.13% – certainly not strong evidence against H_0.

Figure 9.1 (generated by ⬩ SBB.m) compares a p-value (dotted line) with a lower bound on the Bayes factor (red line) and a lower bound on the probability of a type I error α (blue line).

```
%SBB.m
sbb = @(p)  - exp(1) * p .* log(p);
alph = @(p) 1./(1 + 1./(-exp(1)*p.*log(p)) );
%
pp = 0.0001:0.001:0.15
plot(pp, pp, ':', 'linewidth',lw)
hold on
```

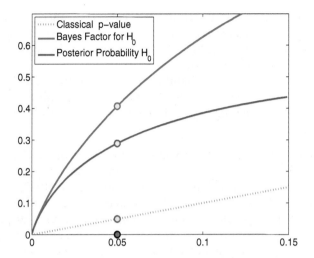

Fig. 9.1 Calibration of p-values. A p-value (*dotted line*) is compared with a lower bound on the Bayes factor (*red line*) and a lower bound on the frequentist type I error α (*blue line*). The bound on α is also the lower bound on the posterior probability of H_0 when the prior probabilities for H_0 and H_1 are equal. For the p-value of 0.05, the type I error is never smaller than 0.2893, while the Bayes factor in favor of H_0 is never smaller than 0.4072.

```
plot(pp, sbb(pp), 'r-','linewidth',lw)
plot(pp, alph(pp), '-','linewidth',lw)
```

The interested reader is directed to Berger and Sellke (1987), Schervish (1996), and Goodman (1999a,b, 2001), among many others, for a constructive criticism of p-values.

We start description of some important testing procedures by first discussing testing for the normal mean.

9.5 Testing the Normal Mean

Testing the normal mean is arguably the most important and fundamental statistical test. In this testing, we will distinguish between two cases depending on whether the population variance is known in advance (z-test) or not known (t-test). We will start with the case of known variance. Scenarios in which the population mean is unknown but the population variance would be known are not common, but not unrealistic. For example, a particular measuring equipment generating data has well-known precision characteristics specified by the factory but is not well calibrated.

9.5.1 z-Test

Let us assume that we are interested in testing the null hypothesis $H_0 : \mu = \mu_0$ on the basis of a sample $X_1, \ldots, X_n$ from a normal distribution $\mathcal{N}(\mu, \sigma^2)$, where the variance σ^2 is assumed known.

We know (page 248) that $\overline{X} \sim \mathcal{N}(\mu, \sigma^2/n)$ and that $Z = \frac{\overline{X} - \mu_0}{\sigma/\sqrt{n}}$ has the standard normal distribution if the null hypothesis is true, that is, if $\mu = \mu_0$. This statistic, Z, is used to test H_0, and the test is called a z-test. Statistic Z is compared to quantiles of the standard normal distribution.

The test can be performed using either (i) the rejection region or (ii) the p-value.

(i) The rejection region depends on the level α and the alternative hypothesis. For one-sided hypotheses, the tail of the rejection region follows the direction of H_1. For example, if $H_1 : \mu > 2$ and the level α is fixed, the rejection region is $[z_{1-\alpha}, \infty)$. For the two-sided alternative hypothesis $H_1 : \mu \neq \mu_0$ and significance level of α, the rejection region is two-sided, $(-\infty, z_{\alpha/2}] \cup [z_{1-\alpha/2}, \infty)$. Since the standard normal distribution is symmetric about 0 and $z_{\alpha/2} = -z_{1-\alpha/2}$, the two-sided rejection region is sometimes given as $(-\infty, -z_{1-\alpha/2}] \cup [z_{1-\alpha/2}, \infty)$.

The test is now straightforward. If statistic Z, calculated from the observations $X_1, \ldots, X_n$, falls within the rejection region, the null hypothesis is rejected. Otherwise, we say that hypothesis H_0 is not rejected.

(ii) As discussed earlier, the p-value gives a more refined analysis in testing than the "reject–do not reject" decision rule. The p-value is the probability of the rejection-region-like area cut by the observed Z (and, in the case of a two-sided alternative, by $-Z$ and Z) where the probability is calculated by the distribution specified by the null hypothesis.

The following table summarizes the z-test for $H_0 : \mu = \mu_0$ and $Z = \frac{\overline{X} - \mu_0}{\sigma/\sqrt{n}}$:

Alternative	α-level rejection region	p-value (MATLAB)
$H_1 : \mu > \mu_0$	$[z_{1-\alpha}, \infty)$	`1-normcdf(z)`
$H_1 : \mu \neq \mu_0$	$(-\infty, z_{\alpha/2}] \cup [z_{1-\alpha/2}, \infty)$	`2*normcdf(-abs(z))`
$H_1 : \mu < \mu_0$	$(-\infty, z_\alpha]$	`normcdf(z)`

9.5.2 Power Analysis of a z-Test

The power of a test is found against a specific alternative, $H_1 : \mu = \mu_1$. In a z-test, the variance σ^2 is known and μ_0 and μ_1 are specified by their respective H_0 and H_1.

The power is the probability that a z-test of level α will detect the effect of size e and, thus, reject H_0. The effect size is defined as $e = \frac{|\mu_0 - \mu_1|}{\sigma}$. Usually, μ_1 is selected such that effect e has a medical or engineering relevance.

Power of the z-test for $H_0 : \mu = \mu_0$ when μ_1 is the actual mean.
- One-sided test:

$$1 - \beta = \Phi\left(z_\alpha + \frac{|\mu_0 - \mu_1|}{\sigma/\sqrt{n}}\right) = \Phi\left(-z_{1-\alpha} + \frac{|\mu_0 - \mu_1|}{\sigma/\sqrt{n}}\right).$$

- Two-sided test:

$$1 - \beta = \Phi\left(-z_{1-\alpha/2} + \frac{(\mu_0 - \mu_1)}{\sigma/\sqrt{n}}\right) + \Phi\left(-z_{1-\alpha/2} + \frac{(\mu_1 - \mu_0)}{\sigma/\sqrt{n}}\right)$$

$$\approx \Phi\left(-z_{1-\alpha/2} + \frac{|\mu_0 - \mu_1|}{\sigma/\sqrt{n}}\right).$$

Typically the sample size is selected prior to the experiment. For example, it may be of interest to decide how many respondents to interview in a poll or how many tissue samples to process. We already selected sample sizes in the context of interval estimation to achieve a given interval size and confidence level.

In a testing setup, consider a problem of testing $H_0 : \mu = \mu_0$ using $\overline{X}$ from a sample of size n. Let the alternative have a specific value μ_1, i.e., $H_1 : \mu = \mu_1 (> \mu_0)$. Assume a significance level of $\alpha = 0.05$. How large should n be so that the power $1 - \beta$ is 0.90?

Recall that the power of a test is the probability that a false null will be rejected, $\mathbb{P}(\text{reject } H_0 | H_0 \text{ false})$. The null is rejected when $\overline{X} > \mu_0 + 1.645 \cdot \frac{\sigma}{\sqrt{n}}$. We want the power of 0.90 leading to $\mathbb{P}(\overline{X} > \mu_0 + 1.645 \cdot \frac{\sigma}{\sqrt{n}} | \mu = \mu_1) = 0.90$, that is,

$$\mathbb{P}\left(\frac{\overline{X} - \mu_1}{\sigma/\sqrt{n}} > \frac{\mu_0 - \mu_1}{\sigma/\sqrt{n}} + 1.645\right) = 0.9.$$

Since $\mathbb{P}(Z > -1.282) = 0.9$, it follows that $\frac{\mu_0 - \mu_1}{\sigma/\sqrt{n}} = 1.282 - 1.645 \Rightarrow n = \frac{8.567 \cdot \sigma^2}{(\mu_1 - \mu_0)^2}$.

In general terms, if we want to achieve the power $1 - \beta$ within the significance level of α for the alternative $\mu = \mu_1$, we need $n \geq \frac{(z_{1-\alpha} + z_{1-\beta})^2 \sigma^2}{(\mu_0 - \mu_1)^2}$ observations. For two-sided alternatives α is replaced by $\alpha/2$.

The sample size for fixed α, β, σ, μ_0, and μ_1 is

$$n = \frac{\sigma^2}{(\mu_0 - \mu_1)^2} (z_{1-\alpha} + z_{1-\beta})^2,$$

where σ is either known, estimated from a pilot experiment, or elicited from experts. If the alternative is two-sided, then $z_{1-\alpha}$ is replaced by $z_{1-\alpha/2}$. In this case, the sample size is approximate.

If σ is not known and no estimate exists, one can elicit the *effect size*, $e = |\mu_0 - \mu_1|/\sigma$, directly. This number is the distance between the competing means in units of σ. For example, for $e = 1/2$ we would like to find a sample size such that the difference between the true and postulated mean equal to $\sigma/2$ is detectable with a probability of $1 - \beta$.

9.5.3 Testing a Normal Mean When the Variance Is Not Known: t-Test

To test a normal mean when the population variance is unknown, we use the *t*-test. We are interested in testing the null hypothesis $H_0 : \mu = \mu_0$ against one of the alternatives $H_1 : \mu >, \neq, < \mu_0$ on the basis of a sample $X_1, \ldots, X_n$ from the normal distribution $\mathcal{N}(\mu, \sigma^2)$, where the variance σ^2 is unknown.

If $\overline{X}$ and s are the sample mean and standard deviation, then under H_0, which states that the true mean is μ_0, the statistic $t = \frac{\overline{X} - \mu_0}{s/\sqrt{n}}$ has a *t*-distribution with $n - 1$ degrees of freedom; see arguments on page 297.

The test can be performed either using (i) the rejection region or (ii) the *p*-value. The following table summarizes the test.

Alternative	α-level rejection region	*p*-value (MATLAB)
$H_1 : \mu > \mu_0$	$[t_{n-1,1-\alpha}, \infty)$	1-tcdf(t, n-1)
$H_1 : \mu \neq \mu_0$	$(-\infty, t_{n-1,\alpha/2}] \cup [t_{n-1,1-\alpha/2}, \infty)$	2*tcdf(-abs(t),n-1)
$H_1 : \mu < \mu_0$	$(-\infty, t_{n-1,\alpha}]$	tcdf(t, n-1)

It is sometimes argued that the *z*-test and the *t*-test are an unnecessary dichotomy and that only the *t*-test should be used. The population variance in the *z*-test is assumed "known," but this can be too strong an assumption. Most of the time when μ is not known, it is unlikely that the researcher would have definite knowledge about the population variance. Also, the *t*-test is more conservative and robust to deviations from normality than the z-test. However, the z-test has an educational value since the testing process and power analysis are easily formulated and explained. Moreover, when the sample size is large, say, larger than 100, the $z-$ and t-tests are practically indistinguishable, due to the Central Limit Theorem.

Example 9.5. **The Moon Illusion.** Kaufman and Rock (1962) stated that the commonly observed fact that the moon near the horizon appears larger than does the moon at its zenith (highest point overhead) could be explained on the basis of the greater *apparent* distance of the moon when at the horizon. The authors devised an apparatus that allowed them to present two artificial moons, one at the horizon and one at the zenith. Subjects were asked to adjust the variable horizon moon to match the size of the zenith moon, and vice versa. For each subject the ratio of the perceived size of the horizon moon to the perceived size of the zenith moon was recorded. A ratio of 1.00 would indicate no illusion, whereas a ratio other than 1.00 would represent an illusion. For example, a ratio of 1.50 would mean that the horizon moon appeared to have a diameter 1.50 times that of the zenith moon. Evidence in support of an illusion would require that we reject $H_0 : \mu = 1.00$ in favor of $H_1 : \mu > 1.00$.

Obtained ratio: 1.73 1.06 2.03 1.40 0.95 1.13 1.41 1.73 1.63 1.56

For these data,

```
x = [1.73, 1.06, 2.03, 1.40, 0.95, 1.13, 1.41, 1.73, 1.63, 1.56];
n = length(x)
t = (mean(x)-1)/(std(x)/sqrt(n))
  % t= 4.2976
crit = tinv(1-0.05, n-1)
  % crit=1.8331. RR = (1.8331, infinity)
pval = 1-tcdf(t, n-1)
  % pval = 9.9885e-004 < 0.05
```

As evident from the MATLAB output, the data do not support H_0, and H_0 is rejected.

A Bayesian solution implemented in WinBUGS is provided next. Each parameter in a Bayesian model should be assigned a prior distribution. Here we have two parameters, the mean μ, which is the population ratio, and σ^2, the unknown variance. The prior on μ is normal with mean 0 and variance $1/0.00001 = 100,000$. We also restricted the prior to be on the nonnegative domain (since negative ratios are not possible) by WinBUGS option `mu~dnorm(0,0.00001)I(0,)`. Such a large variance makes the normal prior essentially flat over $\mu \geq 0$. This means that our prior opinion on μ is vague, and the adopted prior is noninformative.

The prior on the precision, $1/\sigma^2$, is gamma with parameters 0.0001 and 0.0001. As we argued in Example 8.16, this selection of hyperparameters

makes the gamma prior essentially flat, and we are not injecting any prior
information about the variance.

```
model{
for (i in 1:n){
X[i] ~ dnorm(mu, prec)
}
  mu ~ dnorm(0, 0.00001) I(0, )
  prec ~ dgamma(0.0001, 0.0001)
  sigma <- 1/sqrt(prec)
  #TEST
  prH1 <- step(mu - 1)
  }
  DATA
  list(n=10, X=c(1.73, 1.06, 2.03, 1.40, 0.95,
             1.13, 1.41, 1.73, 1.63, 1.56) )

  INITS
  list(mu = 0, prec = 1)
```

	mean	sd	MC error	val2.5pc	median	val97.5pc	start	sample
mu	1.463	0.1219	1.26E-4	1.219	1.463	1.707	1001	100000
prH1	0.999	0.03115	3.188E-5	1.0	1.0	1.0	1001	100000
sigma	0.3727	0.101	1.14E-4	0.2344	0.354	0.6207	1001	100000

Note that the MCMC output in the previous example produced $\mathbb{P}(H_0) =$
0.001 and $\mathbb{P}(H_1) = 0.999$ and the Bayesian solution agrees with the classical.
Moreover, the posterior probability of hypothesis H_0 of 0.001 is quite close
to the p-value of 0.000998, which is often the case when the priors of the
Bayesian model are noninformative. Note also that posterior probability of the pro-
H_1 was estimated by the relative frequency of step(mu-1), that simulations.
portion of cases in which mu-1 resulted as positive in MC.

Example 9.6. **Hypersplenism and White Blood C** Hypersplenism
is a disorder that causes the spleen to rapid prematurely destroy
blood cells. In the general population the white blood cells per
mm^3 is normal with a mean of 7,200 deviation of $\sigma = 1,500$.

It is believed that hypersplen the leukocyte count. In a
sample of 16 persons affect nism, the mean white blood
cell count was found to be sample standard deviation was
$s = 1,682$.

Using WinBUC or probability of H_1 and estimate the
mean and vari population. The program in WinBUGS
will opera and the precision (reciprocal of the variable. T mes the precision of a single observation. In this
an since the original data are not avail-

case, knowledge of the population standard deviation σ will guide the setting of an informative prior on the precision. To keep the numbers manageable, we will express the counts in 1,000's, and $\overline{X}$ and s will be coded as 5.213 and 1.682, respectively. Since $s = 1.682, s^2 = 2.8291$, and $prec = 0.3535$, it is tempting to set the prior on the precision as precx~dgamma(0.3535,1) or precx~dgamma(3.535,10) since the mean of these priors will match the observed precision. However, this would be a "data-built" prior in the spirit of the empirical Bayes approach. We will use the fact that in the population σ was 1.5 and we will elicit the prior precx~dgamma(4.444,10) since $1/1.5^2 = 0.4444$.

```
model {
precxbar <- n * precx
xbar ~ dnorm(mu, precxbar)
mu ~   dnorm(0, 0.0001) I(0, )
   # sigma = 1.5, s^2 = 2.25, prec = 0.4444
   # X gamma(a,b) -> EX=a/b, Var X = a/b^2
precx ~ dgamma(4.444, 10 )
indh1 <- step(7.2 - mu)
sigx <- 1/sqrt(precx)
}


DATA
list(xbar = 5.213, n=16)

INITS
list(mu=1.000, precx=1.000)
```

	mean	sd	MC error	val2.5pc	median	val97.5pc	start	sample
indh1	0.9997	0.01643	3.727E-5	1.0	1.0	1.0	1001	200000
	5.212	0.4263	9.842E-4	4.367	5.212	6.064	1001	200000
	1.644	0.4486	0.001081	1.032	1.561	2.749	1001	200000

Note th

a clear winne posterior probability of H_1 is 0.9997 and this hypothesis is

9.5.4 Power Analysi

When an experiment is planne*t*
ance is unknown, as in the case of
the absolute difference $|\mu_0 - \mu_1|$ tha not available. Even if the vari-
be expressed in units of standard de d be elicited. Alternatively,
σ may not be necessary. Thus, at the d sider as significant can
analysis applicable to the z-test is also app plicit knowledge of
stage, the power
ective t-test.

Once the data are available and the test is performed, the sample mean and sample variance are available, and it becomes possible to assess the power retrospectively. We have already discussed controversies surrounding retrospective power analyses.

In a retrospective evaluation of the power, it is not recommended to replace $|\mu_0 - \mu_1|$ by $|\mu_0 - \overline{X}|$, as is sometimes done, but to simply update the elicited σ^2 with the observed variance. When σ is replaced by s, the expressions for calculating the power involve t and noncentral t-distributions. Here is an illustration.

Example 9.7. **Power in the t-Test.** Suppose that we are testing $H_0 : \mu = 10$ versus $H_1 : \mu > 10$, at a level $\alpha = 0.05$. A sample of size $n = 20$ gives $\overline{X} = 12$ and $s = 5$. We are interested in finding the power of the test against the alternative $H_1 : \mu = 13$.

The exact power is $\mathbb{P}(t \in RR | t \sim nct(df = n - 1, ncp = (\mu_1 - \mu_0)\sqrt{n}/\sigma))$, since under H_1, t has a noncentral t-distribution with $n - 1$ degrees of freedom and a noncentrality parameter $\frac{(\mu_1 - \mu_0)\sqrt{n}}{\sigma}$. "RR" denotes the rejection region.

```
n=20;  mu0 = 10; s=5; mu1= 13; alpha=0.05;
pow1 = nctcdf( -tinv(1-alpha, n-1), n-1,-abs(mu1-mu0)*sqrt(n)/s)
% or pow1=1-nctcdf(tinv(1-alpha, n-1),n-1,abs(mu1-mu0)*sqrt(n)/s)
% pow1 = 0.8266
%
pow = normcdf(-norminv(1-alpha) + abs(mu1-mu0)*sqrt(n)/s)
% or pow = 1-normcdf(norminv(1-alpha)-abs(mu1-mu0)*sqrt(n)/s)
% pow = 0.8505
```

For a large sample size, the power calculated as in the z-test approximates the exact power, but from the "optimistic" side, that is, by always overestimating it. In this MATLAB script we find a power of approx. 85%, which in an exact calculation (as above) drops to 82.66%.

For the two-sided alternative $H_1 : \mu \neq 10$, the exact power decreases,

```
pow2 = nctcdf(tinv(1-alpha/2, n-1), n-1,-abs(mu1-mu0)*sqrt(n)/s)  ...
           -nctcdf(tinv(1-alpha/2, n-1), n-1, abs(mu1-mu0)*sqrt(n)/s)
%pow2 =0.7210
```

When calculation of the noncentral t CDF is not available, a good approximation for the power is

$$
1 - \Phi\left(\frac{t_{n-1,\alpha} - |\mu_1 - \mu_0|\sqrt{n}/s}{\sqrt{1 + \frac{t^2_{n-1,1-\alpha}}{2(n-1)}}} \right).
$$

In our example,

```
1-normcdf((tinv(1-alpha,n-1)- ...
(mu1-mu0)/s * sqrt(n)/sqrt(1 + (tinv(1-alpha,n-1))^2/(2*n-2)))
%ans =     0.8209
```

The summary of retrospective power calculations for the t-test us listed below:

Power of the t-test for $H_0 : \mu = \mu_0$, when μ_1 is the actual mean.

- One-sided test:

$$1 - \beta = 1\text{-}nctcdf\left(t_{n-1,1-\alpha},\ n-1,\ \frac{|\mu_1 - \mu_0|}{s/\sqrt{n}}\right).$$

- Two-sided test:

$$1 - \beta = nctcdf\left(t_{n-1,1-\alpha/2},\ n-1,\ \frac{-|\mu_1 - \mu_0|}{s/\sqrt{n}}\right)$$

$$-nctcdf\left(t_{n-1,1-\alpha/2},\ n-1,\ \frac{|\mu_1 - \mu_0|}{s/\sqrt{n}}\right).$$

Here $nctcdf(x,df,\delta)$ is the CDF of a noncentral t-distribution, with df degrees of freedom and noncentrality parameter δ, evaluated at x. In MATLAB this function is `nctcdf(x,df,delta)`, see page 264

Example 9.8. **Sample Size in t-Test.** In Example 9.7 we were testing $H_0 : \mu = 10$ versus $H_1 : \mu > 10$, at a level $\alpha = 0.05$, where, for sample size $n = 20$ and $s = 5$, we found the power against the alternative $H_1 : \mu = 13$ to be 82.66%. What sample size is needed to increase this power to 95% in a future one-sided test with the same alternative, α and s?

```
mu0 = 10; mu1= 13; s=5; alpha=0.05; beta=0.05;
a = @(n) nctcdf( -tinv(1-alpha, n-1), n-1,-abs(mu1-mu0)*sqrt(n)/s)-(1-beta);
ssize=fzero(a, 20)   %31.4694
```

Thus, the sample of size 32 would ensure power of 95% in repeating the test from Example 9.7.

9.6 Testing the Multivariate Normal Mean*

Testing in the domain of multivariate data generalizes well-known univariate techniques. Conducting the univariate inference on the components of an observed data vector is not adequate since it ignores the covariance structure of observations. This naïve approach can lead to various biases. For example, the tests for individual component means $H_0 : \mu_1 = 3$ and $H_0' : \mu_2 = -1$ may not be significant, while the test $H_0'' : (\mu_1, \mu_2) = (3, -1)$ may turn out to be significant. This is because the evidence may accumulate across the components. On the other hand, in some situations a test on an individual component may turn significant, while the multivariate test involving that component may not be significant due to, again, the interplay with other components. In addition to this "borrowing of strength" from component to component, controlling the family-wise error of first kind is built in, whereas it could represent a problem when components are tested individually.

In this section we look at the multivariate extensions of a t-test, Hotelling's T-square test.

9.6.1 T-Square Test

Assume that a p-dimensional sample $X_1, \ldots, X_n$ is coming from multivariate normal distribution,

$$X_i \sim \mathcal{MVN}_p(\mu, \Sigma),$$

where μ is the parameter of interest, and the population covariance matrix Σ is unknown.

For some fixed μ_0, the testing $H_0 : \mu = \mu_0$ versus $H_1 : \mu \neq \mu_0$ is based on T^2 statistics,

$$T^2 = n(\overline{X} - \mu_0)' \, S^{-1} \, (\overline{X} - \mu_0),$$

where $\overline{X}$ and S are sample mean and sample covariance matrix. This statistic is sometimes called the Hotelling T-square in honor of Harold Hotelling, one of the pioneers in multivariate statistical inference. When H_0 is true, the scaled statistic $\frac{n-p}{p(n-1)} T^2$ follows an F-distribution with p and $n - p$ degrees of freedom.

The null hypothesis is rejected if $T^2 \geq \frac{p(n-1)}{n-p} F_{p,n-p,1-\alpha}$, where $F_{p,n-p,1-\alpha}$ is the $(1-\alpha)$ quantile of F-distribution with p and $n-p$ degrees of freedom.

A $100(1-\alpha)\%$ confidence region for μ consists of all such μ for which

$$(\overline{X} - \mu)' S^{-1} (\overline{X} - \mu) \leq \frac{p(n-1)}{n(n-p)} F_{p,n-p,1-\alpha}.$$

Remark. If $p=1$, we recover the standard t-statistic and CI. Indeed, note that for $t = \frac{\overline{X} - \mu_0}{s/\sqrt{n}}$,

$$t^2 = \left(\frac{\overline{X} - \mu_0}{s/\sqrt{n}} \right)^2 = n(\overline{X} - \mu_0)(s^2)^{-1}(\overline{X} - \mu_0),$$

which is the one-dimensional counterpart of T^2. The inference is also recovered since t and F distributions are connected, i.e., distributions t_n^2 and $F_{1,n}$ coincide. The confidence regions become standard t-confidence intervals as well, since for the quantiles, $(t_{n,1-\alpha/2})^2 = F_{1,n,1-\alpha}$.

A simultaneous $100(1-\alpha)\%$ confidence interval for all linear combinations $a'\mu = a_1\mu_1 + a_2\mu_2 + \cdots + a_p\mu_p$ is

$$\left[a'\overline{X} - \sqrt{\frac{p(n-1)}{n-p} F_{p,n-p,1-\alpha}} \sqrt{\frac{1}{n} a'Sa} \, , \right.$$

$$\left. a'\overline{X} + \sqrt{\frac{p(n-1)}{n-p} F_{p,n-p,1-\alpha}} \sqrt{\frac{1}{n} a'Sa} \, \right].$$

These simultaneous bounds are true for *any* number of arbitrary vectors *a*. By properly choosing vector *a*, various linear combinations of component means can be monitored.

Example 9.9. **Hook-Billed Kites.** Data set 🖳 bird.dat|mat|xlsx was analyzed by Johnson and Wichern (2002) and contains bivariate measurements on $n = 45$ female hook-billed kites. The data set contains three columns: bird number, tail length X_1, and wing length X_2. A bivariate normal distribution is assumed for $(X_1, X_2)'$. We are interested in testing $H_0 : \mu = (190, 275)'$ versus $H_1 : \mu \neq (190, 275)'$.

For this data set the sample mean is $\overline{X} = (193.6222, 279.7778)'$ and the sample covariance matrix is $S = \begin{bmatrix} 120.6949 & 122.3460 \\ 122.3460 & 208.5404 \end{bmatrix}$. MATLAB script

◀ bird.m performs the test and explores the relationship between individual and simultaneous testing.

```
%bird.m
load 'bird.mat'
x1 = bird(:,2); x2 = bird(:,3);
X=[x1 x2];
[n p]=size(X);
Xbar =transpose(mean(X))  %[193.6222;   279.7778]
S = cov(X)
    % 120.6949   122.3460
    % 122.3460   208.5404
mu0 = [190; 275];
T2 = n * (Xbar - mu0)'* inv(S) * (Xbar - mu0) %5.5431
F = (n-p)/(p*(n-1)) * T2  %2.7086
pval = 1-fcdf(F, p, n-p)  %0.078
```

We fail to reject H_0 at 5% significance level. However, if t-tests are performed on the individual components, the tests are significant.

```
%bird.m continued
t1 = (Xbar(1)-mu0(1))/sqrt(S(1,1)/n)  %2.2118
t2 = (Xbar(2)-mu0(2))/sqrt(S(2,2)/n)  %2.2194
p1 = 2*tcdf(-abs(t1), n-2)  %0.0323
p2 = 2*tcdf(-abs(t2), n-2)  %0.0318
```

If instead of mu0=[190; 275] we tested for mu0=[192; 283], the significance statements will be reversed. The T-square test will be significant, whereas the individual t-tests will not be significant. This situation was alluded to in the introduction of this section. The reasons for this discrepancy are illustrated in Figure 9.2.

Here, the 95% simultaneous confidence ellipse for population mean $\mu = (\mu_1, \mu_2)'$ is plotted together with individual 95% confidence intervals for μ_1 and μ_2.

The green dot corresponds to mu0=[190; 275] falls outside individual intervals and the corresponding componentwise t-tests are both significant. However, this point falls inside the confidence ellipse and the T-square test is not significant.

If the test is about mu0=[192; 283], then this point (red dot) falls outside the ellipse, but inside the individual confidence intervals.

```
%bird.m modified
mu0=[192; 283];
```

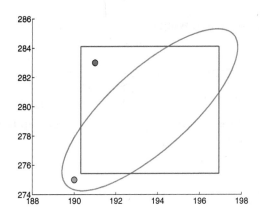

Fig. 9.2 Comparison of simultaneous and individual 95% confidence sets. The confidence ellipse contains mu0=[190; 275] (green dot); thus the individual tests are significant but not the multivariate. For mu0=[192; 283] (red dot), the significance results are reversed.

```
T2 = n * (Xbar - mu0)'* inv(S) * (Xbar - mu0) %13.5909
F = (n-p)/(p*(n-1)) * T2     %6.6410
pval = 1-fcdf(F, p, n-p)     %0.0031
%
t1 = (Xbar(1)-mu0(1))/sqrt(S(1,1)/n)   %0.9905
t2 = (Xbar(2)-mu0(2))/sqrt(S(2,2)/n)   %-1.4968
p1 = 2*tcdf(-abs(t1), n-2)   %0.3275
p2 = 2*tcdf(-abs(t2), n-2)   %0.1417
```

9.6.1.1 Power Analysis for T-Square Test

Suppose that we need to find the power of T-square test for testing $H_0 : \mu = \mu_0 = (0.3, 0.3)'$ against the alternative $H_1 : \mu = \mu_1 = (0.4, 0.4)'$ if the sample size of $n = 930$ is planned, and elicited covariance matrix is $\Sigma = \begin{bmatrix} 1 & 0.2 \\ 0.2 & 1 \end{bmatrix}$.

The effect size

$$D = \sqrt{(\mu_1 - \mu_0)'\Sigma^{-1}(\mu_1 - \mu_0)}.$$

is a multivariate analogue of Cohen's $d = |\mu_1 - \mu_0|/\sigma$, while the noncentrality parameter for F statistic (connected with T^2 via $F = (n - p)/p\ T^2/(n - 1), df = (p, n - p))$ is $\lambda = n \cdot D^2$.

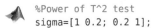

```
%Power of T^2 test
sigma=[1 0.2; 0.2 1];
```

```
n=930;
D2=[0.1; 0.1]' * inv(sigma) * [0.1; 0.1]
%D2 = 0.01667
%effect is D = sqrt(D2) = 0.1291
lambda = n * D2      %lambda = 15.5

power=1-ncfcdf( finv(1-0.05, 2, 930-2), 2, 930-2, lambda)
%power =  0.9501
```

Next, we will find the sample size so that effect $D = 0.2$ is found significant with the power of $1 - \beta = 0.90$ for $p = 2$ and $\alpha = 0.05$.

```
ssize=ceil(fzero(@(n) ncfcdf( finv(1-0.05, 2, n-2), ...
                 2, n-2, n*0.2^2)-(1-0.90), 1000))
%ssize = 320
```

The MATLAB script ◀ powerT2.m contains the calculations.

9.6.2 Test for Symmetry

In a multivariate context, tests for the equality of component means are called *tests of symmetry*. Let $\mu = (\mu_1, \mu_2, \ldots, \mu_p)'$ be the mean of $\mathcal{MVN}_p(\mu, \Sigma)$ from which a sample $X_1, X_2, \ldots, X_n$ is obtained. Assume that $p \geq 2$.

The hypothesis of symmetry

$$H_0 : \mu_1 = \mu_2 = \cdots = \mu_p,$$

can be expressed as

$$H_0 : C\mu = 0 \quad \text{versus} \quad H_1 : C\mu \neq 0$$

where C is *any* $(p - 1) \times p$ matrix, of rank $p - 1$ (rows are linearly independent), such that

$$C1 = 0, \quad \text{for } 1 = (1, 1, \ldots, 1)'.$$

Popular choices for C are,

$$
C = \begin{pmatrix}
1 & -1 & 0 & \ldots & 0 & 0 \\
0 & 1 & -1 & \ldots & 0 & 0 \\
0 & 0 & 1 & \ldots & 0 & 0 \\
 & \vdots & & & & \\
0 & 0 & 0 & \ldots & -1 & 0 \\
0 & 0 & 0 & \ldots & 1 & -1
\end{pmatrix}
\quad \text{or} \quad
C = \begin{pmatrix}
1 & -1 & 0 & \ldots & 0 & 0 \\
1 & 0 & -1 & \ldots & 0 & 0 \\
1 & 0 & 0 & \ldots & 0 & 0 \\
 & \vdots & & & & \\
1 & 0 & 0 & \ldots & -1 & 0 \\
1 & 0 & 0 & \ldots & 0 & -1
\end{pmatrix}.
$$

The test is based on

$$T^2 = n \; \overline{X}' \; C'(CSC')^{-1}C \; \overline{X}.$$

In this case,

$$\frac{n - (p - 1)}{(p - 1)\,(n - 1)}\, T^2 \sim F_{p-1,n-(p-1)},$$

which is used for the inference.

Example 9.10. **Cork Boring Data Revisited.** Consider data from Exercise 2.23 consisting of the weights of cork boring for 28 trees. We will test for the equality of component means (four directions: north, east, south, and west). In the MATLAB file below, we show that for two valid choices of C (rows sum up to 0) the value of the T^2 statistic remains the same.

```
%Rao's Cork Data
X =[   72 66 76 77;    60 53 66 63;    56 57 64 58;   41 29 36 38; ...
       32 32 35 36;    30 35 34 26;    39 39 31 27;   42 43 31 25; ...
       37 40 31 25;    33 29 27 36;    32 30 34 28;   63 45 74 63; ...
       54 46 60 52;    47 51 52 43;    91 79 100 75;  56 68 47 50; ...
       79 65 70 61;    81 80 68 58;    78 55 67 60;   46 38 37 38; ...
       39 35 34 37;    32 30 30 32;    60 50 67 54;   35 37 48 39; ...
       39 36 39 31;    50 34 37 40;    43 37 39 50;   48 54 57 43];
   [n p]=size(X);
   Xbar = mean(X)';  S=cov(X);
   %N E S W
   C=[ 1  -1  -1  1;      0  0  1  -1;      1  0  -1   0 ];
   T2 = n * Xbar' * C'  * inv(C * S * C')* C  * Xbar      %20.7420
   pval = 1-fcdf( (n-p+1)/((p-1)*(n-1)) * T2, p-1, n-p+1) %0.0023

   %invariance wrt C
   C1 =[1 -1 0 0;    1 0 -1 0;    1 0 0 -1];
   T2 = n * Xbar' * C1' * inv(C1 * S * C1')* C1 * Xbar  %20.7420
```

9.7 Testing the Normal Variances

When we discussed the estimation of the normal variance (Section 7.4.2), we argued that the statistic $(n - 1)s^2/\sigma^2$ had a χ^2-distribution with $n - 1$ degrees of freedom. The test for the normal variance is based on this statistic and its distribution.

Suppose we want to test $H_0 : \sigma^2 = \sigma_0^2$ versus $H_1 : \sigma^2 >, \neq, <, \sigma_0^2$. The test statistic is

$$\chi^2 = \frac{(n - 1)s^2}{\sigma_0^2}.$$

The testing procedure at the α level can be summarized by

Alternative	α-level rejection region	p-value (MATLAB)
$H_1 : \sigma > \sigma_0$	$[\chi^2_{n-1,1-\alpha}, \infty)$	1-chi2cdf(chi2,n-1)
$H_1 : \sigma \neq \sigma_0$	$[0, \chi^2_{n-1,\alpha/2}] \cup [\chi^2_{n-1,1-\alpha/2}, \infty)$	2*chi2cdf(min(chi2,1/chi2),n-1)
$H_1 : \sigma < \sigma_0$	$[0, \chi^2_{n-1,\alpha}]$	chi2cdf(chi2,n-1)

The power of the test against the specific alternative is the probability of the rejection region evaluated as if H_1 were a true hypothesis. For example, if $H_1 : \sigma^2 > \sigma_0^2$, and specifically if $H_1 : \sigma^2 = \sigma_1^2, \ \sigma_1^2 > \sigma_0^2$, then the power is

$$1 - \beta = \mathbb{P}\left(\frac{(n-1)s^2}{\sigma_0^2} \geq \chi^2_{1-\alpha,n-1}\Big|H_1\right) = \mathbb{P}\left(\frac{(n-1)s^2}{\sigma_1^2} \cdot \frac{\sigma_1^2}{\sigma_0^2} \geq \chi^2_{1-\alpha,n-1}\Big|H_1\right)$$

$$= \mathbb{P}\left(\chi^2 \geq \frac{\sigma_0^2}{\sigma_1^2}\chi^2_{1-\alpha,n-1}\right),$$

or in MATLAB:

```
power=1-chi2cdf(sigmasq0/sigmasq1*chi2inv(1-alpha,n-1),n-1).
```

For the one-sided alternative in the opposite direction and for the two-sided alternative, the procedure for finding the power is analogous. The sample size necessary to achieve a preassigned power can be found by trial and error or by using MATLAB's function fzero.

Example 9.11. **LDL-C Levels.** A new handheld device for assessing cholesterol levels in blood is presented for approval to the FDA. The variability of measurements obtained by the device for people with normal levels of LDL cholesterol is one of the measures of interest. A calibrated sample of size $n = 224$ of serum specimens with a fixed 130-level of LDL-C is measured by the device. The variability of measurements is assessed.

(a) If $s^2 = 2.47$ was found, test the hypothesis that the population variance is 2 (as achieved by a clinical computerized Hitachi 717 analyzer, with enzymatic, colorimetric detection schemes) against the one-sided alternative. Use $\alpha = 0.05$.

(b) Find the power of this test against the specific alternative, $H_1 : \sigma^2 = 2.5$.

(c) What sample size ensures the power of 90% in detecting the effect $\sigma_0^2/\sigma_1^2 = 0.8$ as significant.

```
n = 224; s2 = 2.47; sigmasq0 = 2; sigmasq1 = 2.5; alpha = 0.05;
%(a)
chisq = (n-1)*s2 /sigmasq0
  %test statistic chisq = 275.4050.
  %The alternative is H_1: sigma2 > 2
chi2crit = chi2inv( 1-alpha, n-1 )
  %one sided upper tail RR = [258.8365, infinity)
pvalue = 1 - chi2cdf(chisq, n-1) %pvalue = 0.0096
%(b)
power = 1-chi2cdf(sigmasq0/sigmasq1 * chi2inv(1-alpha, n-1), n-1 )
                        %power = 0.7708
%(c)
ratio = sigmasq0/sigmasq1  %0.8
pf = @(n) 1-chi2cdf( ratio * chi2inv(1-alpha, n-1), n-1 ) - 0.90;
ssize = fzero(pf, 300)      %342.5993 approx 343
```

9.8 Testing the Proportion

When discussing the CLT, and in particular the de Moivre theorem, we saw that the binomial distribution can be well approximated with the normal if n is large and $np(1 - p) > 5$.

Suppose that we observe n Bernoulli $\mathcal{B}er(p)$ random variables $Y_1, Y_2, \ldots, Y_n$, with p to be tested. The sum $X = Y_1 + \cdots + Y_n$ is $\mathcal{B}in(n, p)$ and sample proportion of Y's, $\hat{p} = \frac{X}{n}$ is the MLE of p. By the CLT, sample proportion $\hat{p}$ has an approximately normal distribution with mean p and variance $p(1 - p)/n$. This approximate normality will be used to construct the test.

Suppose that we are interested in testing $H_0 : p = p_0$ versus one of the three possible alternatives. When H_0 is true, the test statistic

$$Z = \frac{\hat{p} - p_0}{\sqrt{p_0(1 - p_0)/n}}$$

has approximately a standard normal distribution. The testing procedure is summarized in the following table:

Alternative	α-level rejection region	p-value (MATLAB)
$H_1 : p > p_0$	$[z_{1-\alpha}, \infty)$	`1-normcdf(z)`
$H_1 : p \neq p_0$	$-\infty, z_{\alpha/2}] \cup [z_{1-\alpha/2}, \infty)$	`2*normcdf(-abs(z))`
$H_1 : p < p_0$	$(-\infty, z_\alpha]$	`normcdf(z)`

Using the normal approximation one can derive that the power against the specific alternative $H_1 : p = p_1$ is

$$1 - \beta = \Phi \left[\frac{\sqrt{n}|p_1 - p_0| - z_{1-\alpha}\sqrt{p_0(1 - p_0)}}{\sqrt{p_1(1 - p_1)}} \right],$$

for the one-sided test. In the case of two-sided alternative, $z_{1-\alpha}$ is replaced by $z_{1-\alpha/2}$. The sample size needed to find the effect $|p_0 - p_1|$ significant $(1 - \beta)100\%$ of the time (i.e., the one-sided test would have a power of $1 - \beta$) is

$$n = \frac{\left(\sqrt{p_0(1 - p_0)}\, z_{1-\alpha} + \sqrt{p_1(1 - p_1)}\, z_{1-\beta} \right)^2}{(p_0 - p_1)^2}.$$

For the two sided alternative, $z_{1-\alpha}$ is replaced by $z_{1-\alpha/2}$. Note that specifying only $|p_1 - p_0|$ is not sufficient for sample size determination; both p_0 and p_1 need to be specified.

Example 9.12. **Proportion of Hemorrhagic-Type Strokes among American Indians.** The study described in the American Heart Association's news release of September 22, 2008, included 4,507 members of 13 American Indian tribes in Arizona, Oklahoma, and North and South Dakota. It found that American Indians have a stroke rate of 679 per 100,000, compared to 607 per 100,000 for African Americans and 306 per 100,000 for Caucasians. None of the participants, ages 45 to 74, had a history of stroke when they were recruited for the study from 1989 to 1992. Almost 60% of the volunteers were women.

During more than 13 years of follow-up, 306 participants suffered a first stroke, most of them in their mid-60s when it occurred. There were 263 strokes of the ischemic type, caused by a blockage that cuts off the blood supply to the brain, and 43 hemorrhagic (bleeding) strokes.

It is believed that in the general population one in five of all strokes is hemorrhagic.

(a) Test the hypothesis that the proportion of hemorrhagic strokes in the population of American Indians that suffered a stroke is lower than the national proportion of 0.2.

(b) What is the power of the test in (a) against the alternative $H_1 : p = 0.15$?

(c) What sample size ensures a power of 90% in detecting $p = 0.15$, if H_0 states $p = 0.2$?

Since $306 \times 0.2 > 10$, a normal approximation can be used.

```
z = (43/306 - 0.2)/sqrt(0.2 *(1- 0.2)/306)
% z = -2.6011

pval = normcdf(z)
% pval = 0.0046


%(b)
p0=0.2; p1=0.15; alpha=0.05; n=306;
power = normcdf((sqrt(n)*abs(p1-p0) - ...
        norminv(1-alpha)*sqrt(p0*(1-p0)))/sqrt(p1*(1-p1)) )
%0.7280

%(c)
beta = 0.1;
n=( sqrt(p0*(1-p0)) * norminv(1-alpha) + ...
    sqrt(p1*(1-p1)) * norminv(1-beta) )^2/(p1-p0)^2
%497.7779   approx 498
```

9.8.1 Exact Test for Population Proportions

In the previous section we used a normal approximation to the binomial distribution to test the population proportion via the familiar z-test. Since we assume a binomial model for the data, it is possible (and in the case of small $np(1 - p)$, e.g., < 5, necessary) to test for the proportion in an exact manner.

Here we operate not with $\hat{p} = X/n$ but with X that, under $H_0 : p = p_0$, has binomial $Bin(n, p_0)$ distribution. Thus, the statistic X takes a value k with probability

$$p_{0,n,k} = \binom{n}{k} p_0^k (1 - p_0)^{n-k}, \; k = 0, 1, \ldots, n.$$

For the one-sided alternative, say $H_1 : p < p_0$, we find k^* that is the maximum k for which $\mathbb{P}(X \le k) \le \alpha$. The hypothesis H_0 is rejected for X less than or equal to k^*, that is, the rejection region is $X \in \{0, 1, \ldots, k^*\}$. The level of this test is $\alpha^* = \mathbb{P}(X \le k^*)$. For the alternative $H_1 : p > p_0$ the critical region is $X \ge k^*$, where k^* is the minimum k for which $\mathbb{P}(X \ge k) \le \alpha$.

One of the difficulties in exact testing is that the significance level α^* can take only discrete values, since X is a discrete statistic, and none of these discrete values may match or even be close to the preassigned significance level α, say 0.05.

For the two-sided alternative, $H_0 : p \neq p_0$, the rejection region is $\{X \leq k_1^*\} \cup \{X \geq k_2^*\}$, where k_1^*, k_2^* are selected such that $\mathbb{P}(X \leq k_1^*) + \mathbb{P}(X \geq k_2^*) \leq \alpha$. The pair k_1^*, k_2^* is not unique, however, the choice where the probabilities of the two tails are similar (close to $\alpha/2$) is preferred.

It would be helpful to look at some numbers. For example, assume that in $n = 27$ trials we found X successes and are interested in testing $H_0 : p = 0.3$ at $\alpha = 0.05$. Under H_0, statistic $X \sim Bin(27, 0.3)$.

If the alternative is $H_1 : p > 0.3$, then the test with critical region $\{X \geq 14\}$ would have the level of `1-binocdf(13-1, 27, 0.3)=0.0359`. For the alternative $H_1 : p < 0.3$, the critical region $\{X \leq 3\}$ would have the level of `binocdf(3, 27, 0.3) = 0.0202`, while the test with critical region $\{X \leq 4\}$ would have the level of `binocdf(4, 27, 0.3) = 0.0591`, thus slightly exceeding 0.05. The exact $\alpha = 0.05$ level test is not possible here, so the test with $k^* = 3$ will be used since $0.0202 < 0.05$. We note that the exact tests could be randomized so that any α is achieved, but this theory is beyond the scope of this text.

If, for instance, $X = 5$ is observed, H_0 is not rejected since $X > k^* = 3$. The p-value is `binocdf(5, 27, 0.3) = 0.1358`.

For the two-sided alternative, $H_1 : p \neq 0.3$, and $\alpha = 0.05$, the values for k_1^* and k_2^* are 3 and 14, respectively, since `1-binocdf(14-1, 27, 0.3) = 0.0143`, and again X is not in rejection region. For this alternative, the achieved significance level is $0.0202 + 0.0143 = 0.0345 < 0.5$. The p-value is `2*min(binocdf(5, 27, 0.3), 1- binocdf(5-1, 27, 0.3)) = 0.2716`, so H_0 is not rejected.

These results are summarized in the table below where $p_{0,n,i} = \binom{n}{i} p_0^i (1 - p_0)^{n-i}$ are probabilities of $X = i$ under H_0.

Alternative	Critical region	p-value (MATLAB)
$H_1 : p < p_0$	$X \leq k^* = \max k : \sum_{i=0}^{k} p_{0,n,i} \leq \alpha$	`binocdf(X,n,p0)`
$H_1 : p \neq p_0$	$X \leq k_1^* = \max k : \sum_{i=0}^{k} p_{0,n,i} \leq \alpha/2$, or $X \geq k_2^* = \min k : \sum_{i=k}^{n} p_{0,n,i} \leq \alpha/2$	`2* min(binocdf(X,n,p0), 1-binocdf(X-1,n,p0))`
$H_1 : p > p_0$	$X \geq k^* = \min k : \sum_{i=k}^{n} p_{0,n,i} \leq \alpha$	`1-binocdf(X-1,n,p0)`

Example 9.13. **Proportion of Hemorrhagic Strokes: Exact Test.** In a follow-up study discussed in Example 9.12, out of 306 participants suffering a stroke, 43 of the strokes were of hemorrhagic type, and the rest of the is-chemic type. We tested hypotheses $H_0 : p = 0.2$ versus $H_1 : p < 0.2$ at $\alpha = 0.05$ level using the normal approximation, and found a p-value of 0.0046.

The results for the exact test are summarized in the annotated MATLAB code below:

```
pvalue = binocdf(43, 306, 0.20)      %0.0044
k=binoinv(0.05, 306, 0.2)            %k=50
```

```
kstar = k-1;                             %RRegion X <= k*;   k*=49
alphastar = binocdf(kstar, 306, 0.2)     %alpha*=0.0445<0.05
pow=binocdf(kstar, 306, 0.15)            %power against H1: p=0.15
                                         %pow = 0.7220
```

Note that the exact p-value (0.0044) is quite close to the p-value obtained by the normal approximation (0.0046). The achieved significance level α^* is $0.0445 < 0.05$. Note also that the power is 0.7220, which is slightly less than the power found using the normal approximation. In general, power analyses based on the normal approximation are more "optimistic."

Exact Sample Size in Testing the Proportion. Let $p_{1,n,k} = \binom{n}{k} p_1^k (1 - p_1)^{n-k}$ be the probabilities of $X = k$ under the precise alternative $H_1 : p = p_1$.

The power of an α-level test of $H_0 : p = p_0$ versus $H_1 : p = p_1$ for sample size n is

$$\sum_{k=0}^{n} \left[p_{1,n,k} \mathbf{1} \left(\sum_{i=k}^{n} p_{0,n,i} \leq \alpha \right) \right], \text{ when } H_1 : p = p_1 > p_0,$$

$$\sum_{k=0}^{n} \left[p_{1,n,k} \mathbf{1} \left(\sum_{i=0}^{k} p_{0,n,i} \leq \alpha \right) \right], \text{ when } H_1 : p = p_1 < p_0, \text{ and}$$

$$\sum_{k=0}^{n} \left[p_{1,n,k} \mathbf{1} \left(2 \cdot \min \left\{ \sum_{i=0}^{k} p_{0,n,i}, \sum_{i=k}^{n} p_{0,n,i} \right\} \leq \alpha \right) \right], \text{ when } H_1 : p = p_1 \neq p_0.$$

Here, $\mathbf{1}$ is an indicator, and $p_{0,n,i} = \binom{n}{i} p_0^i (1 - p_0)^{n-i}$ are binomial probabilities of $X = i$ under the null hypothesis. The sample size is now determined by increasing n until the power reaches the preassigned level of $1 - \beta$.

Example 9.14. **Proportion of Hemorrhagic Strokes: Exact Power and Sample Size.** In Example 9.13, we tested hypotheses $H_0 : p = 0.2$ versus $H_1 : p < 0.2$ at $\alpha = 0.05$ level using the exact binomial test. We also found the exact power, against the one-sided specific alternative $p = 0.15$, to be 0.7220.

Here, we repeat the power calculation in a more systematic fashion and also find the sample size necessary to achieve the power of 90% in a prospective test of the same hypotheses, at $\alpha = 0.05$ level.

```
n = 306; p0 = 0.2; p1 = 0.15; alpha = 0.05;
kargs = 0:n;
u =  binocdf(kargs, n, p0)  <= alpha;          %indicator
exactpower = sum( binopdf(kargs, n, p1).*u ) %0.7220
%sample size
beta = 0.1; %preset power of 90%
exactpower = 0; n = 10;
```

```
while exactpower < 1-beta
    n=n+1;
    kargs = 0:n;
    ind = binocdf(kargs, n, p0)   <= alpha;    %indicator
    exactpower = sum( binopdf(kargs, n, p1).* ind ) ;
end
 disp(['samplesize = ' num2str(n)])
%samplesize = 501
```

Thus, a sample of size 501 would be required to achieve the desired power. Note that in Example 9.12, a sample size of 498 was found using normal approximation. Show that, if the test is two-sided, for the same α, p_0, p_1, and n, the power would be 0.6078. Show also that, for the two-sided testing, with the same α, p_0, p_1 and $1 - \beta = 90\%$, the necessary sample size would be 619. For the two-sided test the indicator is `ind = 2*min(binocdf(kargs, n, p0),1-binocdf(kargs-1, n, p0)) <= alpha;`

9.8.2 Bayesian Test for Population Proportions

A Bayesian test for binomial proportion was already discussed in Example 9.4 on page 387.

In its simplest form, a Bayesian test requires a prior on population proportion p. In Example 9.4 the prior was uniform on $[0,0.1]$ with a point mass at $p = 0.1$.

In the context of Example 9.13, a beta prior with parameters 1 and 4 is elicited, so that the prior mean $\mathbb{E}^{\pi}p = 1/(1+4) = 0.2$ matches the mean under H_0. The following simple WinBUGS script conducts the test

$$H_0 : p \leq 0.2 \quad \text{versus} \quad H_1 : p > 0.2.$$

```
model{
X ~ dbin(p, n)
p ~ dbeta(1,4)
pH1 <- step(0.2-p)
    }
DATA
list(n=306, X=43)
#Generate Inits
```

The output variable pH1 gives the posterior probability of H_1.

	mean	sd	MC error	val2.5pc	median	val97.5pc	start	sample
p	0.1415	0.01974	1.9158E-5	0.1051	0.1407	0.1823	1001	1000000
pH1	0.9967	0.05694	5.675E-5	1.0	1.0	1.0	1001	1000000

Since population proportions are in $[0,1]$, typical prior on p is beta. Discussion on eliciting beta priors can be found on page 348. The following example uses Zellner's prior on p. Zellner's prior is in fact a flat prior on $\text{logit}(p)$ and it was also discussed on page 348.

Example 9.15. **eBay Story.** You decided to purchase a new Orbital Shaking Incubator for your research lab on eBay. A single seller is offering this item. The seller has positive feedback from 223 out of 230 responders.

(a) What is the 95% credible set for the population satisfaction rate with this seller, p?

(b) Test hypotheses (i) $H_0' : p \le 0.98$ vs. $H_1' : p > 0.98$ and (ii) $H_0'' : 0.96 \le p \le 0.99$ vs. $H_1'' = (H_0'')^c$.

```
model{
    Positives ~ dbin(p,n)
    # Zellner's 1/[p (1-p)] improper prior
    # set as flat prior on logit
    logit(p) <- eta
    eta ~ dflat()
    pH1prime <- step(0.98-p)
    pH1second <- 1-step(p-0.96)*step(0.99-p)
    }
DATA
list(n=230, Positives=223)
INITS
list(eta=0)
```

The output variables `pH1prime` and `pH1second` give the posterior probabilities of corresponding H_1's.

	mean	sd	MC error	val2.5pc	median	val97.5pc	start	sample
eta	3.533	0.3975	4.07E-4	2.823	3.508	4.379	1001	1000000
p	0.9696	0.01129	1.153E-5	0.9439	0.9709	0.9876	1001	1000000
pH1prime	0.8222	0.3823	3.77E-4	0.0	1.0	1.0	1001	1000000
pH1second	0.1956	0.3967	4.065E-4	0.0	0.0	1.0	1001	1000000

The 95% credible set for p is $[0.9439, 0.9876]$. The classical 95% Wald's confidence interval in this case is $[0.9474, 0.9918]$, which is slightly shifted right. The posterior for p is slightly skewed to the left, indicating that symmetry of likelihood assumed in normal approximation biases the interval; see Figure 9.3.

Note that H_1' and H_0'' have higher posterior probabilities, 0.8222 and $1 - 0.1956$, and should be favored.

The following example emphasizes the conditional nature of Bayesian inference and its conformity to the *likelihood principle*, which states that *all* information about the experimental results are summarized only in the likelihood.

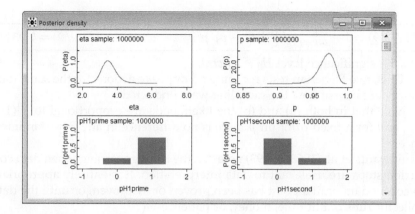

Fig. 9.3 Output from ebaystory0.odc. Posterior distribution for p appears slightly skewed to the left indicating that Wald type confidence intervals are biased. The bottom two bar-plots represent the posterior probabilities of hypotheses H_0', H_1' and H_0'', H_1'', respectively.

Example 9.16. **Savage's Disparity.** A Bayesian inference is based on data observed and not on data that could possibly be observed, or on the manner in which the sampling was conducted. This is the crux of the likelihood principle.

This is not the case in classical testing, and the argument first put forth by Jimmie Savage at the Purdue Symposium in 1962 emphasizes the difference.

Suppose a coin is flipped 12 times and 9 heads and 3 tails are obtained. Let p be the probability of heads. We are interested in testing whether the coin is fair against the alternative that it is more likely to come heads up, or

$$H_0 : p = 1/2 \quad \text{versus} \quad H_1 : p > 1/2.$$

The p-value for this test is the probability that one observes 9 or more heads if the coin is fair, that is, when H_0 is true.

Consider the following two scenarios:

(a) Suppose that the number of flips $n = 12$ was decided a priori. Then the number of heads X is binomial and under H_0 (fair coin) the p-value is

$$\mathbb{P}(X \geq 9) = 1 - \sum_{k=0}^{8} \binom{12}{k} p^k (1-p)^{12-k} = 1 - \mathtt{binocdf}(8, 12, 0.5) = 0.0730.$$

At a 5% significance level H_0 is *not rejected.*

(b) Suppose that the flipping is carried out until 3 tails have appeared. Let us call tails "success" and heads "failures." Then, under H_0, the number of failures (heads) Y is a negative binomial $\mathcal{NB}(3, 1/2)$ and the p-value is

$$\boxed{\mathbb{P}(Y \geq 9) = 1 - \sum_{k=0}^{8} \binom{3+k-1}{k}(1-p)^3 p^k = 1 - \texttt{nbincdf}(8,3,1/2) = 0.0327.}$$

At a 5% significance level H_0 *is rejected*.

Thus, two Fisherian tests recommend opposite actions for the same data simply because of how the sampling was conducted.

Note that in both (a) and (b) the likelihoods are proportional to $p^9(1 - p)^3$, and for a fixed prior on p there is no difference in any Bayesian inference.

Edwards et al. (1963, p. 193) note "... the rules governing when data collection stops are irrelevant to data interpretation. It is entirely appropriate to collect data until a point has been proven or disproven, or until the data collector runs out of time, money, or patience."

9.9 Multiplicity in Testing, Bonferroni Correction, and False Discovery Rate

Recall that when testing a single hypothesis H_0, a type I error is made if it is rejected, when it is actually true. The probability of making a type I error in a test is usually controlled to be smaller than a certain level of α, typically equal to 0.05.

When there are several null hypotheses, $H_{01}, H_{02}, \ldots, H_{0m}$, and all of them are tested simultaneously, one may want to control the type I error at some level α as well. In this scenario, a type I error is then made if at least one true hypothesis in the family of hypotheses being tested is rejected. Because it pertains to the family of hypotheses, this significance level is called the familywise error rate (FWER).

If the hypotheses in the family are independent, then

$$\text{FWER} = 1 - (1 - \alpha_i)^m,$$

where FWER and α_i are overall and individual significance levels, respectively.

For arbitrary, possibly dependent, hypotheses, the Bonferroni inequality (page 415) translates to

$$\text{FWER} \leq m\alpha_i.$$

Suppose $m = 15$ tests are conducted simultaneously. For an individual α_i of 0.05, the FWER is $1 - 0.95^{15} = 0.5367$. This means that the chance of claiming a significant result when there should not be one is larger than $1/2$. For possibly dependent hypotheses, the upper bound of FWER increases to 0.75.

> **Bonferroni Correction:** To control FWER $\leq \alpha$, one should reject all H_{0i} among $H_{01}, H_{02}, \ldots, H_{0m}$ for which the p-value is found smaller than α/m.

Thus, if for $n = 15$ arbitrary hypotheses we want an overall significance level of FWER ≤ 0.05, then the individual test levels should be set to $0.05/15 = 0.0033$.

Testing for significance with gene expression data from DNA microarray experiments involves simultaneous comparisons of hundreds or thousands of genes, and controlling the FWER by the Bonferroni method would require very small individual α_is. Yet, setting such small α levels decreases the power of individual tests and many false H_0 are not rejected. Therefore the Bonferroni correction is considered by many practitioners as overly conservative. Some call it a "panic approach."

Remark. If, in the context of interval estimation, k simultaneous interval estimates are desired with an overall confidence level $(1 - \alpha)100\%$, then each interval can be constructed with a confidence level $(1 - \alpha/k)100\%$, and the Bonferroni inequality would ensure that the overall confidence is at least $(1 - \alpha)100\%$.

Bonferroni–Holm Method. The Bonferroni–Holm method is an iterative procedure in which individual significance levels are adjusted to increase power and still control the FWER. One starts by ordering the p-values of all tests for $H_{01}, H_{02}, \ldots, H_{0m}$ and then compares the smallest p-value to α/m. If that p-value is smaller than α/m, then one should reject that hypothesis and compare the second ranked p-value to $\alpha/(m - 1)$. If this hypothesis is rejected, one should proceed to the third ranked p-value and compare it with $\alpha/(m - 2)$. This should be continued until the hypothesis with the smallest remaining p-value cannot be rejected. At this point the procedure stops and all hypotheses that have not been rejected at previous steps are retained.

Let $H_{(1)}, H_{(2)}, \ldots, H_{(m)}$ correspond to ordered p-values $p_{(1)}, p_{(2)}, \ldots, p_{(m)}$. For a given α, find minimum k such that

$$p_{(k)} > \frac{\alpha}{m + 1 - k}.$$

Reject hypotheses $H_{(1)}, \ldots, H_{(k-1)}$, and keep $H_{(k)}, \ldots, H_{(m)}$.

To better see this, let us assume that five hypotheses are to be tested with a FWER of 0.05. The five p-values are 0.09, 0.01, 0.04, 0.012, and 0.004. The smallest of these is 0.004. Since this is less than 0.05/5, hypothesis four is rejected. The next smallest p-value is 0.01, which is also smaller than 0.05/4. So this hypothesis is also rejected. The next smallest p-value is 0.012, which is smaller than 0.05/3, and this hypothesis is rejected. The next smallest p-value is 0.04, which is not smaller than 0.05/2. Therefore, the hypotheses with p-values of 0.004, 0.01, and 0.012 are rejected while those with p-values of 0.04 and 0.09 are not rejected.

False Discovery Rate. The false discovery rate paradigm (Benjamini and Hochberg, 1995) considers the proportion of falsely rejected null hypotheses (false discoveries) among the total number of rejections.

Controlling the expected value of this proportion, called the false discovery rate (FDR), provides a useful alternative that addresses low-power problems of the traditional FWER methods when the number of tested hypotheses is large. The test statistics in these multiple tests are assumed to be independent or positively correlated. Suppose that we are looking at the result of testing m hypotheses, among which m_0 are true. In the table that follows, V denotes the number of false rejections, and the FWER is $\mathbb{P}(V \geq 1)$:

	H_0 not rejected	H_0 rejected	Total
H_0 true	U	V	m_0
H_1 true	T	S	m_1
Total	W	R	m

If R denotes the number of rejections (declared significant genes, discoveries), then V/R, for $R > 0$, is the proportion of false rejected hypotheses. The FDR is

$$\mathbb{E}\left(\frac{V}{R}\middle| R > 0\right)\mathbb{P}(R > 0).$$

Let $p_{(1)} \leq p_{(2)} \leq \cdots \leq p_{(m)}$ be the ordered, observed p-values for the m hypotheses to be tested. Algorithmically, the FDR method finds k such that

$$k = \max\left\{i\middle| p_{(i)} \leq (i/m)\alpha\right\}. \tag{9.4}$$

The FDR is controlled at the α level if the hypotheses corresponding to $p_{(1)}, \ldots, p_{(k)}$ are rejected. If no such k exists, no hypothesis from the family is rejected. When the test statistics in the multiple tests are possibly negatively correlated as well, the FDR is modified by replacing α in (9.4) with $\alpha/(1 + 1/2 + \cdots + 1/m)$. The following MATLAB script (FDR.m) finds the critical p-value $p_{(k)}$. If $p_{(k)} = 0$, then no hypothesis is rejected.

 `function pk = FDR(p,alpha)`

```
%Critical p-value pk for FDR <= alpha.
%All hypotheses with p-value less than or equal
%to pk are rejected.
%if pk = 0 no hypothesis is to be rejected
m = length(p);      %number of hypotheses
po = sort(p(:));    %ordered p-values
i = (1:m)';         %index
pk = po(max(find(  po < i./m * alpha)));
%critical p-value
if ( isempty(pk)==1 )
    pk=0;
end
```

Suppose that we have 1,000 hypotheses and all hypotheses are true. Then their p-values represent a random sample from the uniform $\mathcal{U}(0,1)$ distribution. About 50 hypotheses would have a p-value of less than 0.05. However, for reasonable FDR levels (0.05–0.2) $p_{(k)} = 0$, as it should be since we do not want false discoveries.

```
p = rand(1000,1);
[FDR(p, 0.05), FDR(p, 0.2), FDR(p, 0.6), FDR(p, 0.92)]
%ans =      0      0    0.0022    0.0179
```

9.10 Exercises

9.1. **Public Health.** A manager of public health services in an area down-wind of a nuclear test site wants to test the hypothesis that the mean amount of radiation in the form of strontium-90 in the bone marrow (measured in picocuries) for citizens who live downwind of the site does not exceed that of citizens who live upwind from the site. It is known that "upwinders" have a mean level of strontium-90 of 1 picocurie. Measurements of strontium-90 radiation for a sample of $n = 16$ citizens who live downwind of the site were taken, giving $\overline{X} = 3$. The population standard deviation is $\sigma = 4$. Assume normality and use a significance level of $\alpha = 0.05$.
(a) State H_0 and H_1.
(b) Calculate the appropriate test statistic.
(c) Determine the critical region of the test.
(d) State your decision.
(e) What would constitute a type II error in this setup? Describe this in one sentence.

9.2. **Testing IQ.** We wish to test the hypothesis that the mean IQ of the students in a school system is 100. Using $\sigma = 15$, $\alpha = 0.05$, and a sample of 25 students the sample value $\overline{X}$ is computed. For a two-sided test find:
(a) The range of $\overline{X}$ for which we would not reject the hypothesis.

(b) If the true mean IQ of the students is 105, find the probability of falsely not rejecting $H_0 : \mu = 100$.

(c) What are the answers in (a) and (b) if the alternative is one-sided, $H_1 : \mu > 100$?

9.3. **Bricks.** A purchaser of bricks suspects that the quality of bricks is deteriorating. From past experience, the mean crushing strength of such bricks should be 400 pounds. A sample of $n = 100$ bricks yields a mean of 395 pounds and standard deviation of 20 pounds.

(a) Test the hypothesis that the mean quality has not changed against the alternative that it has deteriorated. Choose $\alpha = 0.05$.

(b) What is the p-value for the test in (a)?

(c) Suppose that the producer of the bricks contests your findings in (a) and (b). Their company suggests that you construct the 95% confidence interval for μ with a total length of no more than 4. What sample size is needed to construct such a confidence interval?

9.4. **Soybeans.** According to advertisements, a strain of soybeans planted on soil prepared with a specific fertilizer treatment has a mean yield of 500 bushels per acre. Fifty farmers planted the soybeans. Each used a 40-acre plot and reported the mean yield per acre. The mean and variance for the sample of 50 farms are $\bar{x} = 485$ and $s^2 = 10,045$. Use the p-value for this test to determine whether the data provide sufficient evidence to indicate that the mean yield for the soybeans is different from that advertised.

9.5. **Great White Shark.** One of the most feared predators in the ocean is the great white shark *Carcharodon carcharias*. Although it is known that the great white shark grows to a mean length of 14 ft. (record: 23 ft.), a marine biologist believes that the great white sharks off the Bermuda coast grow significantly longer due to unusual feeding habits. To test this claim, a number of full-grown great white sharks are captured off the Bermuda coast, measured, and then set free. However, because the capture of sharks is difficult, costly, and very dangerous, only five are sampled. Their lengths are 16, 18, 17, 13, and 20 ft.

(a) What assumptions must be made in order to carry out the test?

(b) Do the data provide sufficient evidence to support the marine biologist's claim? Formulate the hypotheses and test at a significance level of $\alpha = 0.05$. Provide solutions using both the rejection-region approach and the p-value approach.

(c) Find the power of the test against the specific alternative $H_1 : \mu = 17$.

(d) What sample size is needed to achieve the power of 0.90 in testing the preceding hypothesis if $\mu_1 - \mu_0 = 2$ and $\alpha = 0.05$. Pretend that the described experiment was a pilot study to assess the variability in data and adopt $\sigma = 2.5$.

(e) Provide a Bayesian solution using WinBUGS with noninformative priors on μ and $1/\sigma^2$ (precision). Compare with results from (b) and discuss.

9.6. **Serum Sodium Levels.** A data set compiled by Queen Elizabeth Hospital, Birmingham, and referenced in Andrews and Herzberg (1985), provides the results of analysis of 20 samples of serum measured for their sodium content. The average value for the method of analysis used is 140 ppm.

140	143	141	137	132	157	143	149	118	145
138	144	144	139	133	159	141	124	145	139

Is there evidence that the mean level of sodium in this serum is different from 140 ppm?

9.7. **Weight of Quarters.** The US Department of the Treasury claims that the procedure it uses to mint quarters yields a mean weight of 5.67 g with a standard deviation of 0.068 g. A random sample of 30 quarters yielded a mean of 5.643 g. Use an $\alpha = 0.05$ significance to test the claim that the mean weight is 5.67 g.
(a) What alternatives make sense in this setup? Choose one sensible alternative and perform the test.
(b) State your decision in terms of rejection region.
(c) Find the p-value and confirm your decision from (b).
(d) Would you change the decision if α were 0.01?

9.8. **Dwarf Plants.** A genetic model suggests that three-fourths of the plants grown from a cross between two given strains of seeds will be of the dwarf variety. After breeding 200 of these plants, 136 were of the dwarf variety.
(a) Does this observation strongly contradict the genetic model?
(b) Construct a 95% confidence interval for the true proportion of dwarf plants obtained from the given cross.
(c) Answer (a) and (b) using Bayesian arguments and WinBUGS.

9.9. **Eggs in a Nest.** The average number of eggs laid per nest each season by the Eastern Phoebe bird is a parameter of interest. A random sample of 70 nests was examined and the following results were obtained (Hamilton, 1990):

Number of eggs/nest	1	2	3	4	5	6
Frequency f	3	2	2	14	46	3

Test the hypothesis that the true average number of eggs laid per nest by the Eastern Phoebe bird is equal to five versus the two-sided alternative. Use $\alpha = 0.05$.

9.10. **Penguins.** A researcher is interested in testing whether the mean height of Emperor penguins (*Aptenodytes forsteri*) from a small island is less than $\mu = 45$ in., which is believed to be the average height for the whole Emperor penguin population. The heights were measured of 14 randomly selected adult birds from the island with the following results:

41	44	43	47	43	46	45	42	45	45	43	45	47	40

State the assumptions and hypotheses. Perform the test at the level $\alpha = 0.05$.

9.11. **Hypersplenism and White Blood Cell Count.** In Example 9.6, the belief was expressed that hypersplenism decreased the leukocyte count, so a Bayesian test was conducted. In a sample of 16 people affected by hypersplenism, the mean white blood cell count per mm^3 was found to be $\overline{X} = 5,213$. The sample standard deviation was $s = 1,682$.
(a) With this information, test $H_0 : \mu = 7,200$ versus the alternative $H_1 : \mu < 7,200$ using both the rejection region and the p-value. Compare the results with the WinBUGS output.
(b) Find the power of the test against the alternative $H_1 : \mu = 5,800$.
(c) What sample size is needed if, in a repeated study, a difference of $|\mu_1 - \mu_0| = 600$ is to be detected with a power of 80%? Use the estimate $s = 1,682$.

9.12. **Jigsaw.** An experiment with a sample of 18 nursery-school children involved the elapsed time required to put together a small jigsaw puzzle. The times in minutes were as follows:

3.1	3.2	3.4	3.6	3.7	4.2	4.3	4.5	4.7
5.2	5.6	6.0	6.1	6.6	7.3	8.2	10.8	13.6

(a) Calculate the 95% confidence interval for the population mean.
(b) Test the hypothesis $H_0 : \mu = 5$ against the two-sided alternative. Take $\alpha = 10\%$.

9.13. **Anxiety.** A psychologist has developed a questionnaire for assessing levels of anxiety. The scores on the questionnaire range from 0 to 100. People who obtain scores of 75 and greater are classified as *anxious*. The questionnaire has been given to a large sample of people who have been diagnosed with an anxiety disorder, and scores are well described by a normal model with a mean of 80 and a standard deviation of 5. When given to a large sample of people who do not suffer from an anxiety disorder, scores on the questionnaire can also be modeled as normal with a mean of 60 and a standard deviation of 10.
(a) What is the probability that the psychologist will misclassify a nonanxious person as anxious?

(b) What is the probability that the psychologist will erroneously label a truly anxious person as nonanxious?

9.14. **Aptitude Test.** An aptitude test should produce scores with a large amount of variation so that an administrator can distinguish between people with low aptitude and those with high aptitude. The standard test used by a certain university has been producing scores with a standard deviation of 5. A new test given to 20 prospective students produced a sample standard deviation of 8. Are the scores from the new test significantly more variable than scores from the standard? Use $\alpha = 0.05$.

9.15. **Rats and Mazes.** Eighty rats selected at random were taught to run through a new maze. All rats eventually succeeded in learning the maze, and the number of trials to perfect their performance was normally distributed with a sample mean of 15.4 and sample standard deviation of 2. Long experience with populations of rats trained to run a similar maze shows that the number of trials to attain success is normally distributed with a mean of 15.

(a) Is the new maze harder for rats to learn than the older one? Formulate the hypotheses and perform the test at $\alpha = 0.01$.

(b) Report the p-value. Would the decision in (a) be different if $\alpha = 0.05$?

(c) Find the power of this test for the alternative $H_1 : \mu = 15.6$.

(d) Assume that the experiment above was conducted to assess the standard deviation, and the result was 2. Design a sample size for a new experiment that will detect the difference $|\mu_0 - \mu_1| = 0.6$ with a power of 90%. Here $\alpha = 0.01$, and μ_0 and μ_1 are postulated means under H_0 and H_1, respectively.

9.16. **Hemopexin in DMD Cases I.** Refer to data set 📄 dmd.dat|mat|xls from Exercise 2.19. The measurements of hemopexin are assumed normal.

(a) Form a 95% confidence interval for the mean response of hemopexin h in a population of all female DMD carriers (carrier=1).

Although the level of pyruvate kinase seems to be the strongest single predictor of DMD, it is an expensive measure. Instead, we will explore the level of hemopexin, a protein that protects the body from oxidative damage. The level of hemopexin, in a general population of women of comparable age, is believed to be 85.

(b) Test the hypothesis that the mean level of hemopexin in the population of woman DMD carriers significantly exceeds 85. Use $\alpha = 5\%$. Report the p-value as well.

(c) What is the power of the test in (b) against the alternative $H_1 : \mu_1 = 89$.

(d) The data for this exercise come from a study conducted in Canada. If you wanted to replicate the test in the United States, what sample size would guarantee a power of 99% if H_0 were to be rejected whenever the difference from the true mean was 4, ($|\mu_0 - \mu_1| = 4$)? A small pilot study conducted to assess the variability of hemopexin level estimated the standard deviation as $s = 12$.

(e) Find the posterior probability of the hypothesis $H_1 : \mu > 85$ using WinBUGS. Use noninformative priors. Also, compare the 95% credible set for μ that you obtained with the confidence interval in (a).

Hint: The commands

```
%file dmd.mat should be on path
load 'dmd.mat';  hemo = dmd( dmd(:,6)==1, 3);
```

will distill the levels of hemopexin in carrier cases.

9.17. Haden's Data.

In the past, the blood counts were performed manually using the hemocytometers with microscopic grid
scoring. By properly diluting blood, counting all cells
in specified squares, and multiplying by the proper
conversion factor, the number of cells per cubic mil-
limeter can be approximated.

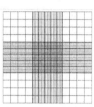

The Coulter principle[2] led to the availability of Coulter counters and
thereafter, the development of sophisticated automated blood-cell ana-
lyzers. The level of sophistication has been rising ever since.

The data set in MATLAB file haden.m comes from Haden (1923, Tables
1, 2, p. 770). It provides red blood cell count for 40 healthy men aged
18–50.

4.27	4.32	4.40	4.52	4.56	4.58	4.64	4.70	4.72	4.73
4.80	4.80	4.80	4.80	4.84	4.87	4.89	4.93	4.97	4.98
4.99	5.00	5.02	5.05	5.09	5.09	5.10	5.15	5.16	5.20
5.20	5.20	5.26	5.28	5.36	5.46	5.49	5.50	5.57	5.62

(a) Find 95% CI for the population mean.

(b) Test the hypothesis that the population mean from which Haden's
sample was taken is 5.1, versus the alternative that it is less than 5.1.

$$H_0 : \mu = 5.1 \quad \text{versus} \quad H_1 : \mu < 5.1$$

Find both the rejection region and the p-value.

(c) What is the power of this test against the alternative $H_1 : \mu = 4.9$?

(d) You are to determine the sample size for Haden's project so that a
0.05 level, two sided test rejects the null hypothesis with probability 0.95
whenever the true mean differs from 5.1 by more than 0.1. By assuming
that the population variance is $\sigma^2 = 0.16$, determine the sample size that
achieves the required power.

---

[2] The Coulter principle states that particles pulled through an orifice by an electric
current produce a change in electrical impedance that is proportional to the size of the
particle traversing the orifice. This is based on the principle that cells are relatively poor
conductors of electricity in relation to the diluent fluid.

9.18. **Retinol and a Copper-Deficient Diet.** The liver is the main storage site of vitamin A and copper. Inverse relationships between copper and vitamin A liver concentrations have been suggested. In Rachman et al. (1987) the consequences of a copper-deficient diet on liver and blood vitamin A storage in Wistar rats was investigated. Nine animals were fed a copper-deficient diet for 45 days from weaning. Concentrations of vitamin A were determined by isocratic high-performance liquid chromatography using UV detection. Rachman et al. (1987) observed in the liver of the rats fed a copper-deficient diet a mean level of retinol, in micrograms/g of liver, was $\overline{X} = 3.3$ and $s = 1.4$. It is known that the normal level of retinol in a rat liver is $\mu_0 = 1.6$.

(a) Find the 95% confidence interval for the mean level of liver retinol in the population of copper-deficient rats. Recall that the sample size was $n = 9$.

(b) Test the hypothesis that the mean level of retinol in the population of copper-deficient rats is $\mu_0 = 1.6$ versus a sensible alternative, either one-sided or two-sided, at the significance level $\alpha = 0.05$. Use both rejection region and p-value approaches.

(c) What is the power of the test in (b) against the alternative $H_1 : \mu = \mu_1 = 2.4$? Comment.

(d) Suppose that you are designing a new, larger study in which you are going to assume that the variance of observations is $\sigma^2 = 1.4^2$, as the limited nine-animal study indicated. Find the sample size so that the power of rejecting H_0 when an alternative $H_1 : \mu = 2.1$ is true is 0.80. Use $\alpha = 0.05$.

(e) Provide a Bayesian solution using WinBUGS.

9.19. **Rubidium.** Meltzer et al. (1973) demonstrated that there is a large variability in the amount of rubidium excreted each day, even when the amount of potassium ingested is controlled. However, when the rubidium excretion is computed as a ratio to potassium excretion, this variability is markedly diminished. Meltzer et al. concluded that the factors that normally control potassium flux operate at the same time to control rubidium flux.

The data consists of measurements on 17 hospitalized patients and represent the mean of naturally occurring rubidium-to-potassium ratio, in hundreds of mEq of Ru to mEq of K.

0.028	0.032	0.031	0.041	0.028
0.039	0.042	0.036	0.037	0.029
0.048	0.037	0.037	0.044	0.039
0.029	0.038			

Two published studies state that the ratio in healthy subjects is approx $\mu_0 = 0.036$.

(a) Assuming the normality of the ratio, test the hypothesis that population mean μ does not significantly differ from μ_0. Use $\alpha = 0.05$.

(b) How does your finding in (a) agree with the 95% CI for the population mean ratio? Is μ_0 in the confidence interval?

9.20. **Aniline.** Organic chemists often purify organic compounds by a method known as *fractional crystallization*. An experimenter wanted to prepare and purify 5 grams of aniline. It is postulated that 5 grams of aniline would yield 4 grams of acetanilide. Ten 5-gram quantities of aniline were individually prepared and purified.

(a) Test the hypothesis that the mean dry yield differs from 4 grams if the mean yield observed in a sample was $\overline{X} = 4.21$. The population is assumed normal with known variance $\sigma^2 = 0.08$. The significance level is set to $\alpha = 0.05$.

(b) Report the p-value.

(c) For what values of $\overline{X}$ will the null hypothesis be rejected at the level $\alpha = 0.05$?

(d) What is the power of the test for the alternative $H_1 : \mu = 3.6$ at $\alpha = 0.05$?

(e) If you are to design a similar experiment but would like to achieve a power of 90% versus the alternative $H_1 : \mu = 3.6$ at $\alpha = 0.05$, what sample size would you recommended?

9.21. **DNA Random Walks.** DNA random walks are numerical transcriptions of a sequence of nucleotides. The imaginary walker starts at 0 and goes one step up ($s = +1$) if a purine nucleotide (A, G) is encountered, and one step down ($s = -1$) if a pyramidine nucleotide (C, T) is encountered. Peng et al. (1992) proposed identifying coding/noncoding regions by measuring the irregularity of associated DNA random walks. A standard irregularity measure is the Hurst exponent H, an index that ranges from 0 to 1. Numerical sequences with H close to 0 are irregular, while the sequences with H close to 1 appear more smooth.

Figure 9.4 shows a DNA random walk in the DNA of a spider monkey (*Ateles geoffroyi*). The sequence is formed from a noncoding region and has a Hurst exponent of $H = 0.61$.

A researcher wishes to design an experiment in which n nonoverlapping DNA random walks of a fixed length will be constructed, with the goal of testing to see if the Hurst exponent for noncoding regions is 0.6.

The researcher would like to develop a test so that an effect $e = |\mu_0 - \mu_1|/\sigma$ will be detected with a probability of $1 - \beta = 0.9$. The test should be two-sided with a significance level of $\alpha = 0.05$. Previous analyses of noncoding regions in the DNA of various species suggest that exponent H is approximately normally distributed with a variance of approximately $\sigma^2 = 0.03^2$. The researcher believes that $|\mu_0 - \mu_1| = 0.02$ is a biologically meaningful difference. In statistical terms, a 5%-level test for $H_0 : \mu = 0.6$ versus the alternative $H_1 : \mu = 0.6 \pm 0.02$ should have a power

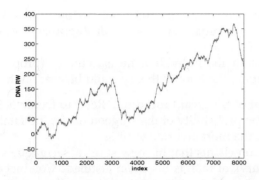

Fig. 9.4 A DNA random walk formed by a noncoding region from the DNA of a spider monkey. The Hurst exponent is 0.61.

of 90%. The preexperimentally assessed variance $\sigma^2 = 0.03^2$ leads to an effect size of $e = 2/3$.

(a) Argue that a sample size of $n = 24$ satisfies the power requirements. The experiment is conducted, and the following 24 values for the Hurst exponent are obtained:

```
H =[0.56  0.61   0.62   0.53   0.54  0.60   0.56   0.59 ...
     0.60  0.60   0.62   0.60   0.58  0.57   0.61   0.64 ...
     0.60  0.61   0.58   0.59   0.55  0.59   0.60   0.65 ];
%    [mean(H)  std(H)]  %%% 0.5917    0.0293
```

(b) Using the t-test, test H_0 against the two-sided alternative at the level $\alpha = 0.05$ using both the rejection-region approach and the p-value approach.

(c) What is the retrospective power of your test? Use the formula with a noncentral t-distribution and s found from the sample.

9.22. **Binding of Propofol.** Serum protein binding is a limiting factor in the access of drugs to the central nervous system. Disease-induced modifications of the degree of binding may influence the effect of anaesthetic drugs.

The protein binding of *propofol*, an intravenous anaesthetic agent that is highly bound to serum albumin, has been investigated in patients with chronic renal failure. Protein binding was determined by the ultrafiltration technique using an Amicon Micropartition System, MPS-1.

The mean proportion of unbound propofol in healthy individuals is 0.96, and it is assumed that individual proportions follow a beta distribution, $\mathcal{B}e(96,4)$. Based on a sample of size $n = 87$ patients with chronic renal failure, the average proportion of unbound propofol was found to be 0.93 with a sample standard deviation of 0.12.

(a) Test the hypothesis that the mean proportion of unbound propofol in a population of patients with chronic renal failure is 0.96 versus the one-sided alternative. Use $\alpha = 0.05$ and perform the test using both the rejection-region approach and the p-value approach. Would you change the decision if $\alpha = 0.01$?

(b) Even though the individual measurements (proportions) follow a beta distribution, the normal theory could be used in (a). Why?

9.23. **Improvement of Surgical Procedure.** Refer to Example 9.4.

(a) What is the probability of the surgeon having no fatalities in treating 15 patients if the mortality rate is 10%?

(b) The surgeon claims that his new surgical technique significantly improves the survival rate. Is his claim justified? Conduct the test and report the p-value. Note that np_0 here is small, so the z test based on normal approximation may not be accurate.

(c) What is the minimum number of patients the surgeon needs to treat without a single fatality in order to convince you that his procedure is a significant improvement over the old technique? Specify your criteria and justify your answer.

(d) Conduct the test in a Bayesian manner as in Example 9.4. Find the posterior probability of H_0 if the prior ξ on $[0,0.1)$ is $\xi(\theta) = 200\,\theta$.

9.24. **Cancer Therapy.** Researchers in cancer therapy often report only the number of patients who survive for a specified period of time after treatment rather than the patients' actual survival times. Suppose that 40% of the patients who undergo the standard treatment are known to survive 5 years. A new treatment is administered to 200 patients, and 92 of them are still alive after a period of 5 years.

(a) Formulate the hypotheses for testing the validity of the claim that the new treatment is more effective than the standard therapy.

(b) Test with $\alpha = 0.05$ and state your conclusion; use the rejection-region method.

(c) Perform the test by finding the p-value.

(d) What is the power of the test in (a) against the alternative $H_1 : p = 0.5$?

(e) What sample size is needed so that effect $p_1 - p_0 = 0.1$ is found significant in the $\alpha = 0.05$ level testing with the power of 90%? As before, $p_0 = 0.4$.

9.25. **Is the Cloning of Humans Moral?** The Gallup Poll estimates that 88% of Americans believe that cloning humans is morally unacceptable. Results are based on telephone interviews with a randomly selected national sample of $n = 1,000$ adults, aged 18 and older.

(a) Test the hypothesis that the true proportion is 0.9, versus the two-sided alternative, based on the Gallup data. Use $\alpha = 0.05$.

(b) Does 0.9 fall in the 95% confidence interval for the proportion?

(c) What is the power of this test against the alternative $H_1 : p = 0.85$?

9.26. **Smoking Illegal?** In a recent Gallup poll of Americans, fewer than a third of respondents thought smoking in public places should be made illegal, a significant decrease from the 39% who thought so in 2001.

The question used in the poll was: *Should smoking in all public places be made totally illegal?* In the poll, 497 people responded and 154 answered yes. Let p be the proportion of people in the US voting population supporting the idea that smoking in public places should be made illegal.

(a) Test the hypothesis $H_0 : p = 0.39$ versus the alternative $H_1 : p < 0.39$ at the level $\alpha = 0.05$.

(b) What is the 90% confidence interval for the unknown population proportion p?

9.27. **Spider Monkey DNA.** An 8,192-long nucleotide sequence segment taken from the DNA of a spider monkey (*Ateles geoffroyi*) is provided in the file ◀ dnatest.m.

(a) Find the relative frequency of adenine $\hat{p}_A$ as an estimator of the overall population proportion, p_A.

(b) Find a 99% confidence interval for p_A and test the hypothesis $H_0 : p_A = 0.2$ versus the alternative $H_1 : p_A > 0.2$. Use $\alpha = 0.05$.

MATLAB AND WINBUGS FILES AND DATA SETS USED IN THIS CHAPTER

http://statbook.gatech.edu/Ch9.Testing/

bayestestprecise.m, bird.m, ConfidenceEllipse.m, corkraotest.m, dnarw.m, dnatest.m, exactpowerprop.m, FDR.m, hemopexin1.m, hemoragic.m, hypersplenism.m, LDLCLevels.m, moon.m, powerT2.m, powers.m, SBB.m

bird.odc, hemopexin.odc, hemorrhagic.odc, hypersplenism.odc, moonillusion.odc, retinol.odc, shark.odc, spikes.odc, systolic.odc

bird.dat|mat|xlsx, dnadat.mat|txt, haden.mat, spid.dat

CHAPTER REFERENCES

Andrews, D. F. and Herzberg, A. M. (1985). *Data. A Collection of Problems from Many Fields for the Student and Research Worker*. Springer, New York.

Benjamini, Y. and Hochberg, Y. (1995) Controlling the false discovery rate: a practical and powerful approach to multiple testing. *J. R. Stat. Soc. B*, **57**, 289–300.

Berger, J. O. and Sellke, T. (1987). Testing a point null hypothesis: the irreconcilability of p-values and evidence (with discussion). *J. Am. Stat. Assoc.*, **82**, 112–122.

Casella, G. and Berger, R. (2001). *Statistical Inference*, 2nd ed. Duxbury Press, Belmont, CA.

Edwards, W., Lindman, H., and Savage, L. J. (1963). Bayesian statistical inference for psychological research. *Psychol. Rev.*, **70**, 193–242.

Fisher, R. A. (1925). *Statistical Methods for Research Workers*. Oliver and Boyd, Edinburgh.

Fisher, R. A. (1926). The arrangement of field experiments. *J. Ministry Agricult.*, **33**, 503–513.

Goodman, S. (1999a). Toward evidence-based medical statistics. 1: The p-value fallacy. *Ann. Intern. Med.*, **130**, 995–1004.

Goodman, S. (1999b). Toward evidence-based medical statistics. 2: The Bayes factor. *Ann. Intern. Med.*, **130**, 1005–1013.

Goodman, S. (2001). Of p-values and Bayes: a modest proposal. *Epidemiology*, **12**, 3, 295–297

Haden, R. L. (1923). Accurate criteria for differentiating anemias. *Arch. Intern. Med.*, **31**, 5, 766–780.

Hamilton, L. C. (1990). *Modern Data Analysis: A First Course in Applied Statistics*. Brooks/Cole, Pacific Grove, CA.

Hoenig, J. M. and Heisey, D. M. (2001). Abuse of power: the pervasive fallacy of power calculations for data analysis. *Am. Statist.*, **55**, 1, 19–24.

Ioannidis, J. P. (2005). Why most published research findings are false. PLoS Med 2(8): e124. doi:10.1371/journal.pmed.0020124.

Jeffreys, H. (1961). *Theory of Probability*, 3rd ed. Oxford University Press, Oxford, UK.

Johnson, R. A. and Wichern,D. W. (2002). *Applied Multivariate Statistical Analysis*, 5th ed. Prentice Hall, NY.

Kaufman, L. and Rock, I. (1962). The moon illusion, I. *Science*, **136**, 953–961.

Katz, S., Lautenschlager, G. J., Blackburn, A. B., and Harris, F. H. (1990). Answering reading comprehension items without passages on the SAT. *Psychol. Sci.*, **1**, 122–127.

Meltzer, H. L., Lieberman, K. W., Shelley, E. M., Stallone, F., and Fieve, R. R. (1973). Metabolism of naturally occurring *Rb* in the human: the constancy of urinary Rb-K. *Biochem Med.*, **7**, 2, 218–225. PubMed PMID: 4704456.

Peng, C. K., Buldyrev, S. V., Goldberger, A. L., Goldberg, Z. D., Havlin, S., Sciortino, E., Simons, M., and Stanley, H. E. (1992). Long-range correlations in nucleotide sequences. *Nature*, **356**, 168–170.

Rachman, F., Conjat, F., Carreau, J. P., Bleiberg-Daniel, F., and Amedee-Maneseme, O. (1987). Modification of vitamin A metabolism in rats fed a copper-deficient diet. *Int. J. Vitamin Nutr. Res.*, **57**, 247–252.

Schervish, M. (1996). *P*-values: what they are and what they are not. *Am. Stat.*, **50**, 203–206.

Sellke, T., Bayarri, M. J., and Berger, J. O. (2001). Calibration of p values for testing precise null hypotheses. *Am. Stat.*, **55**, 62–71.

Chapter 10
Two Samples

Given a choice between two theories, take the one which is funnier.

– Blore's Razor

10.1 Introduction

A two-sample inference is one of the most common statistical procedures used in practice. For example, a colloquial use of "*t*-test" usually refers to the comparison of means from two independent normal populations rather than a single-sample *t*-test. In this chapter we will test the equality of two normal means for independent and dependent (paired) populations as well as the equality of two variances and proportions. In the context of comparing proportions, we will discuss the risk and odds ratios. In testing the equality of means in independent normal populations, we will distinguish two cases: (i) when the underlying population variances are the same and (ii) when no assumption about the variances is made. In this second case the population variances may be different, or even equal, but simply no assumption about their equality enters the test. Each of the tests involves the difference or ratio of the parameters (means, proportions, variances), and for each difference/ratio we provide the $(1 - \alpha)100\%$ confidence interval. For selected tests we will include the power analysis. This chapter is intertwined with parallel Bayesian solutions whenever appropriate.

It is important to emphasize that the normality of populations and large samples for the proportions (for the CLT to hold) are critical for some tests. Later in the text, in Chapter 18, we discuss distribution-free counterpart tests that relax the assumption of normality (sign test, Wilcoxon signed-rank test, Wilcoxon–Mann–Whitney test) at the expense of efficiency if the normality holds.

10.2 Means and Variances in Two Independent Normal Populations

We start with an example that motivates the testing of two population means.

Example 10.1. **Lead Exposure.** It is hypothesized that blood levels of lead tend to be higher for children whose parents work in factories that use lead in manufacturing processes. Researchers examined lead levels in the blood of 12 children whose parents work in a battery manufacturing factory. The results for the exposed children $X_{11}, X_{12}, \ldots, X_{1,12}$ were compared to those of the control sample $X_{21}, X_{22}, \ldots, X_{2,15}$ consisting of 15 children selected randomly from families where the parents do not work in a factory that uses lead. It is assumed that the measurements are independent and come from normal populations. The resulting sample means and sample standard deviations are $\overline{X}_1 = 0.010$, $s_1 = 0.004$, $\overline{X}_2 = 0.006$, and $s_2 = 0.006$.

Obviously, the sample mean for the exposed children is higher than the sample mean in the control sample. But is this difference significant?

To state the problem in more general terms, we assume that two samples $X_{11}, X_{12}, \ldots, X_{1,n_1}$ and $X_{21}, X_{22}, \ldots, X_{2,n_2}$ are observed from populations with normal $\mathcal{N}(\mu_1, \sigma_1^2)$ and $\mathcal{N}(\mu_2, \sigma_2^2)$ distributions, respectively. We are interested in testing the hypothesis $H_0 : \mu_1 = \mu_2$ versus the alternative $H_1 : \mu_1 >, \neq, < \mu_2$ at a significance level α.

For the lead exposure example, the null hypothesis being tested is that the parents' workplace has no effect on their children's lead concentration; that is, the two population means will be the same:

$$H_0 : \mu_1 = \mu_2.$$

Here the populations are defined as all children that are exposed or not exposed. The alternative hypothesis H_1 may be either one- or two-sided. The two-sided alternative is simply $H_1 : \mu_1 \neq \mu_2$; the population means are not equal, and the difference can go either way. The choice of a one-sided hypothesis should be guided by the problem setup, and sometimes by the observations. In the context of this example, it would not make sense to take the one-sided alternative as $H_1 : \mu_1 < \mu_2$, stating that the concentration in the exposed group is smaller than that in the control. In addition, $\overline{X}_1 = 0.010$ and $\overline{X}_2 = 0.006$ are observed. Thus, the sensible one-sided hypothesis in this context is $H_1 : \mu_1 > \mu_2$.

There are two testing scenarios of population means that depend on an assumption about associated population variances. .

Scenario 1: Variances unknown but assumed equal. In this case, the joint σ^2 is estimated by both s_1^2 and s_2^2. The weighted average of s_1^2 and s_2^2 with weights w and $1 - w$ depending on group sample sizes n_1 and n_2,

$$s_p^2 = \frac{(n_1 - 1)s_1^2 + (n_2 - 1)s_2^2}{n_1 + n_2 - 2}$$

$$= \frac{n_1 - 1}{n_1 + n_2 - 2}s_1^2 + \frac{n_2 - 1}{n_1 + n_2 - 2}s_2^2 = ws_1^2 + (1 - w)s_2^2,$$

is called the *pooled sample variance*, and it better estimates the population variance than any individual s^2. The square root of s_p^2 is called the *pooled sample standard deviation* and is denoted by s_p. It can be shown that when H_0 is true, that is, when $\mu_1 = \mu_2$, the statistic

$$t = \frac{\overline{X}_1 - \overline{X}_2}{s_p \sqrt{1/n_1 + 1/n_2}} \qquad (10.1)$$

has a t-distribution with $df = n_1 + n_2 - 2$ degrees of freedom.

Scenario 2: No assumption about the variances (Behrens–Fisher Problem). In this case, when H_0 is true, that is, when $\mu_1 = \mu_2$, the statistic

$$t = \frac{\overline{X}_1 - \overline{X}_2}{\sqrt{s_1^2/n_1 + s_2^2/n_2}}$$

has an approximate t-distribution with

$$df = \frac{(s_1^2/n_1 + s_2^2/n_2)^2}{(s_1^2/n_1)^2/(n_1 - 1) + (s_2^2/n_2)^2/(n_2 - 1)} \qquad (10.2)$$

degrees of freedom. This is a special case of the so-called Welch–Satterthwaite formula, which approximates the degrees of freedom for a linear combination of chi-square random variables (Satterthwaite, 1946; Welch, 1948). When no assumptions about population variances are made, the two sample t-test is often refereed to as the Welch–Satterthwaite test.

For both scenarios:

Alternative	α-level rejection region	p-value
$H_1 : \mu_1 > \mu_2$	$[t_{df,1-\alpha}, \infty)$	`1-tcdf(t, df)`
$H_1 : \mu_1 \neq \mu_2$	$(-\infty, t_{df,\alpha/2}] \cup [t_{df,1-\alpha/2}, \infty)$	`2*tcdf(-abs(t), df)`
$H_1 : \mu_1 < \mu_2$	$(-\infty, t_{df,\alpha}]$	`tcdf(t, df)`

with $df = n_1 + n_2 - 2$ for Scenario 1, and df given by (10.2) for Scenario 2.

When the population variances are known, the proper statistic is Z:

$$Z = \frac{\overline{X}_1 - \overline{X}_2}{\sqrt{\sigma_1^2/n_1 + \sigma_2^2/n_2}},$$

with a normal $\mathcal{N}(0,1)$ distribution, and the proper statistical analysis involves normal quantiles as in the z-test. Given the fact that in realistic examples the variances are likely not known when the means are tested, the z-test is mainly used as an asymptotic test. When sample sizes n_1 and n_2 are large, the z-statistic can be used instead of t even if the variances are not known, due to the CLT. This approximation was more interesting in the past when computing was expensive, but these days there is no need for approximation; a t-test should be used for any sample size when population variances are not known.

In Example 10.1, the variances are not known. We may assume that they are either equal or possibly not equal based on the nature of the experiment, sampling, and some other nonexperimental factors. However, we may also formally test whether the population variances are equal prior to deciding on the testing scenario.

We briefly interrupt our discussion of testing the equality of means with an exposition on how to test the equality of variances in two normal populations.

Testing the Equality of Two Normal Variances. Selecting the "scenario" for testing the equality of normal means requires an assumption about the associated variances. This assumption can be guided by an additional test for the equality of two normal variances prior to testing the means. The variance-before-the-means testing is criticized mainly on the grounds that tests for variances are not as robust (with respect to deviations from normality) compared to the test for means, especially when the sample sizes are small or unbalanced. Note, also, that failing to reject the hypothesis of equality of variances does not imply that this hypothesis is confirmed, see the discussion at the beginning of Section 10.9. Although we agree with these criticisms, it should be noted that choosing the scenario seldom influences the resulting inference, except perhaps in borderline cases. See also the discussion on page 512. When in doubt, one should not make any assumptions about variances and should use the more conservative Scenario 2.

Suppose the samples $X_{11}, X_{12}, \ldots, X_{1,n_1}$ and $X_{21}, X_{22}, \ldots, X_{2,n_2}$ come from normal populations with distributions $\mathcal{N}(\mu_1, \sigma_1^2)$ and $\mathcal{N}(\mu_2, \sigma_2^2)$, respectively. To choose the strategy for testing the means, the following test of variances with the two-sided alternative is helpful:

$$H_0 : \sigma_1^2 = \sigma_2^2 \quad \text{versus} \quad H_1 : \sigma_1^2 \neq \sigma_2^2.$$

The testing statistic is the ratio of sample variances, $F = s_1^2 / s_2^2$, that has an F-distribution with $n_1 - 1$ and $n_2 - 1$ degrees of freedom when H_0 is true. The decision can be made based on a p-value that is equal to

```
p = 2 * min( fcdf(F, n1-1, n2-1), 1-fcdf(F, n1-1, n2-1) .)
```

Remark. The most popular method (recommended in many texts) for calculating the p-value in two-sided testing uses either the expression `2*fcdf(F,n1-1,n2-1)` or `2*(1-fcdf(F,n1-1,n2-1))`, depending on whether `F<1` or `F>1`. Although this approach leads to a correct p-value most of the time, it can lead to a p-value that exceeds 1 when the observed values of F are close to 1. This is clearly wrong since the p-value is a probability. Exercise 10.4 demonstrates such a case.

MATLAB has a built-in function, `vartest2`, for testing the equality of two normal variances.

Guided by the outcome of this test, either we assume that the population variances are the same, and for testing the equality of means, use a t-statistic with a pooled standard deviation and $df = n_1 + n_2 - 2$ degrees of freedom, or we use the t-test without making an assumption about the population variances and the degrees of freedom determined by the Welch–Satterthwaite formula in (10.2).

Next, we summarize the test for both the one- and two-sided alternatives. When $H_0 : \sigma_1^2 = \sigma_2^2$ and $F = s_1^2 / s_2^2$, the following table summarizes the test of the equality of normal variances against the one- or two-sided alternatives. Let $df_1 = n_1 - 1$ and $df_2 = n_2 - 1$.

Alternative	α-level rejection region	p-value
$H_1 : \sigma_1^2 > \sigma_2^2$	$[F_{df_1, df_2, 1-\alpha}, \infty)$	`1-fcdf(F,df1,df2)`
$H_1 : \sigma_1^2 \neq \sigma_2^2$	$[0, F_{df_1, df_2, \alpha/2}] \cup [F_{df_1, df_2, 1-\alpha/2}, \infty)$	`2*min( fcdf(F,df1,df2),` `(1-fcdf(F,df1,df2))`
$H_1 : \sigma_1^2 < \sigma_2^2$	$[0, F_{df_1, df_2, \alpha}]$	`fcdf(F,df1,df2)`

The F-test for testing the equality of variances assumes independent samples. Glass and Hopkins (1984, Sect. 13.9) give a test statistic for testing the equality of variances obtained from paired samples with a correlation coefficient r. The test statistic has a t-distribution with $n - 2$ degrees of freedom,

$$t = \frac{s_1^2 - s_2^2}{2 s_1 s_2 \sqrt{(1 - r^2)/(n - 2)}},$$

where s_1^2, s_2^2 are the two sample variances, n is the number of pairs of observations, and r is the correlation between the two samples. This test was first discussed in Pitman (1939).

Example 10.2. **Lead Exposure Continued.** In Example 10.1, the F-statistic for testing the equality of variances is 0.4444 and the hypothesis of equality of variances is not rejected at a significance level of $\alpha = 0.05$; the p-value is 0.1825.

```
n1 = 12; X1bar = 0.010; s1 = 0.004; % Exposed
n2 = 15; X2bar = 0.006; s2 = 0.006; % Nonexposed
% Testing equality of variances
Fstat = s1^2/s2^2                    % Fstat = 0.4444
% The p-value is
pval = 2*min(·fcdf(Fstat,n1-1,n2-1), 1-fcdf(Fstat,n1-1,n2-1))
                                     % pval =  0.1825
```

Back to Testing Two Normal Means. Guided by the previous test, we assume that the population variances are the same, and for the original problem of testing the means, we use the t-statistic normalized by the pooled standard deviation. The test statistic is

$$t = \frac{\overline{X}_1 - \overline{X}_2}{s_p \sqrt{1/n_1 + 1/n_2}}, \quad \text{where } s_p = \sqrt{\frac{(n_1 - 1)s_1^2 + (n_2 - 1)s_2^2}{n_1 + n_2 - 2}},$$

and it is t-distributed with $n_1 + n_2 - 2$ degrees of freedom.

```
%Lead Exposure Example Continued
sp = sqrt( ((n1-1)*s1^2 + (n2-1)*s2^2 )/(n1 + n2 - 2))    % sp =0.0052
df= n1 + n2 - 2                                            % df = 25
tstat = (X1bar - X2bar)/(sp * sqrt(1/n1 + 1/n2))          % tstat=1.9803
pvalue = 1 - tcdf(tstat, n1+n2-2)                         % pvalue = 0.0294
```

The null hypothesis of equality of the means is rejected at the 5% level since $p = 0.0294 < 0.05$.

Suppose that one wants to test H_0 using rejection regions. Since the alternative hypothesis is one-sided and right-tailed, as $\mu_1 - \mu_2 > 0$, the rejection region is $RR = [t_{n_1+n_2-2,1-\alpha}, \infty)$.

```
    tinv(1-0.05, df)   %ans =1.7081
```

By rejection-region arguments, the hypothesis H_0 is rejected since $t > t_{n_1+n_2-2,1-\alpha}$; that is, the observed value of statistic $t = 1.9803$ exceeds the critical value 1.7081.

Remark. Cochran and Cox (1957) proposed a method of testing two normal means with unequal variances by which the rejection region is based on a linear combination of t quantiles,

$$f_{1-\alpha} = \frac{(s_1^2/n_1)t_{n_1-1,1-\alpha} + (s_2^2/n_2)t_{n_2-1,1-\alpha}}{s_1^2/n_1 + s_2^2/n_2},$$

for one-sided alternatives. For the two-sided alternative, $1 - \alpha$ is replaced by $1 - \alpha/2$. This test is conservative, with the achieved level of significance smaller than the stated α.

10.2.1 Confidence Interval for the Difference of Means

Sometimes we might be interested in the $(1 - \alpha)100\%$ confidence interval for the difference of the population means. Such confidence intervals are easy to obtain, and they depend, as do the tests, on the assumption about the population variances. In general, the interval is

$$\left[\overline{X}_1 - \overline{X}_2 - t_{df,1-\alpha/2}\, s^*, \ \overline{X}_1 - \overline{X}_2 + t_{df,1-\alpha/2}\, s^*\right],$$

where for the equal variance case, $df = n_1 + n_2 - 2$ and $s^* = s_p\sqrt{1/n_1 + 1/n_2}$, and, for no assumption about the population variances case, df is the Welch–Satterthwaite value in (10.2) and $s^* = \sqrt{s_1^2/n_1 + s_2^2/n_2}$.

For the lead exposure example, the 95% confidence interval for $\mu_1 - \mu_2$ is $[-0.00016, 0.0082]$:

```
%Lead Exposure Example Continued
```

```
sp=sqrt(((n1-1)*s1^2 + (n2-1)*s2^2 )/(n1+n2-2))   % sp = 0.0052
df = n1 + n2 - 2                                  % df = 25
LB=X1bar-X2bar-tinv(0.975,df)*sp*sqrt(1/n1+1/n2)  % LB =-0.00016
UB=X1bar-X2bar+tinv(0.975,df)*sp*sqrt(1/n1+1/n2)  % UB = 0.0082
```

Note that this interval barely covers 0. A test for the equality of two means against the two-sided alternative can also be conducted by inspecting the confidence interval for their difference. For a two-sided test of level α, one finds the $(1 - \alpha)100\%$ confidence interval, and if this interval contains 0, the null hypothesis is not rejected. What may be concluded from the interval $[-0.00016, 0.0082]$ is the following: If instead of the one-sided alternative, that was found to be significant at the 5% level (the p-value was about 3%), we carry out the test against the two-sided alternative, the test of the same level would fail to reject H_0.

✎

MATLAB's toolbox "stats" has a built-in function, ttest2, that performs two sample t-tests.

10.2.2 Power Analysis for Testing Two Means

In testing $H_0 : \mu_1 = \mu_2$ against the two-sided alternative $H_1 : \mu_1 \neq \mu_2$, for the specific alternative $|\mu_1 - \mu_2| = \Delta$, an approximation of power is

$$1 - \beta = \Phi \left(z_{\alpha/2} + \frac{\Delta}{\sqrt{\frac{\sigma_1^2}{n_1} + \frac{\sigma_2^2}{n_2}}} \right) + 1 - \Phi \left(z_{1-\alpha/2} + \frac{\Delta}{\sqrt{\frac{\sigma_1^2}{n_1} + \frac{\sigma_2^2}{n_2}}} \right). \quad (10.3)$$

If the alternative is one-sided, say $H_1 : \mu_1 > \mu_2$, then $\Delta = \mu_1 - \mu_2$ and the power is

$$1 - \beta = 1 - \Phi \left(z_{1-\alpha} - \frac{\Delta}{\sqrt{\frac{\sigma_1^2}{n_1} + \frac{\sigma_2^2}{n_2}}} \right) = \Phi \left(z_\alpha + \frac{\Delta}{\sqrt{\frac{\sigma_1^2}{n_1} + \frac{\sigma_2^2}{n_2}}} \right). \quad (10.4)$$

The approximation is good if n_1 and n_2 are large, but it tends to overestimate the power for small to moderate values of n_1 and n_2.

Equations (10.3) and (10.4) are standardly used but are somewhat obsolete since the noncentral t-distribution (page 264) needed for an exact power is readily available.

We state the formulas in terms of MATLAB code:

```
1 - nctcdf(tinv(1-alpha/2,n1+n2-2), ...
                  n1+n2-2,Delta/(sp*sqrt(1/n1+1/n2)))...
  + nctcdf(tinv(alpha/2,n1+n2-2), ...
                  n1+n2-2,Delta/(sp*sqrt(1/n1+1/n2)))
```

For the one-sided alternative $H_1 : \mu_1 > \mu_2$ the code is

```
1 - nctcdf(tinv(1-alpha,n1+n2-2), ...
      n1+n2-2,Delta/(sp*sqrt(1/n1+1/n2)))
```

In testing the equality of means in two normal populations using independent samples, $H_0 : \mu_1 = \mu_2$, versus the one-sided alternative, the group sample size for fixed α, β is

$$n \geq \frac{2\sigma^2}{|\mu_1 - \mu_2|^2}(z_{1-\alpha} + z_{1-\beta})^2,$$

where σ^2 is the common population variance. If the alternative is two-sided, then $z_{1-\alpha}$ is replaced by $z_{1-\alpha/2}$. In that case, the sample size is approximate.

It is assumed that the group sample sizes are equal, namely, that the total sample size is $N = 2n$. If the variances are not the same, then

$$n \geq \frac{\sigma_1^2 + \sigma_2^2}{|\mu_1 - \mu_2|^2}(z_{1-\alpha} + z_{1-\beta})^2.$$

In the context of Example 10.1, let us find the power of the test against the alternative $H_1 : \mu_1 - \mu_2 = 0.005$.

The power for a one-sided α-level test against the alternative $H_1 : \mu_1 - \mu_2 = 0.005 (= \Delta)$ is given in (10.4). The normal approximation is used and s_1^2 and s_2^2 are plugged into the place of σ_1^2 and σ_2^2.

```
%Lead Exposure Example Continued
power = 1-normcdf(norminv(1-0.05)-0.005/sqrt(s1^2/n1+s2^2/n2) )
%power=  0.8271
power = normcdf(norminv(0.05)+0.005/sqrt(s1^2/n1+s2^2/n2) )
%power=  0.8271
```

Thus, the power is roughly 83%. This is an approximation that tends to overestimate the power. The exact power uses noncentral t calculations and is about 81%,

```
power=1-nctcdf(tinv(1-0.05,n1+n2-2),n1+n2-2,0.005/sqrt(s1^2/n1+s2^2/n2))
 %power = 0.8084
```

Suppose that we plan to design a future experiment to test the same phenomenon. When data are collected and analyzed, we would like for the $\alpha = 5\%$ test to achieve a power of $1 - \beta = 90\%$ against the specific alternative $H_1 : \mu_1 - \mu_2 = 0.005$. What sample size will be necessary? Since determining the sample size to meet a preassigned power and precision is prospective in nature, we assume that the previous data were obtained in a pilot study and that σ_1^2 and σ_2^2 are "known" and equal to the observed s_1^2 and s_2^2. The sample size formula is ,

$$n = \frac{(\sigma_1^2 + \sigma_2^2)(z_{1-\alpha} + z_{1-\beta})^2}{\Delta^2},$$

which in MATLAB gives

```
ssize = (s1^2 + s2^2)*(norminv(0.95)+norminv(0.9))^2/(0.005^2)
 % ssize = 17.8128 approx 18 each
```

The number of children is 18 per group if one wishes for the sample sizes to be the same, $n_1 = n_2$. In the following section we discuss the design with $n_2 = k \times n_1$, for some k. Such designs can be justified by different costs of sampling, where the meaning of "cost" may be more general than only the financial one.

Example 10.3. **Camera for Cataract.** The following problem is modified from Rosner (2010). A camera has been developed to detect the presence of a cataract more accurately. Using this camera, the gray level of each point (or pixel) in the lens of a human eye can be characterized into 256 graduations, where a gray level of 1 represents black and a gray level of 256 represents white. To test the camera, photographs were taken of 12 randomly selected normal eyes and 8 randomly selected cataractous eyes. The two groups consist of different people. The median gray level of each eye was computed over the 10,000+ pixels in the lens. The data are given below

Control	158 182 182 191 177 145 156 170 152 164 169 141
Cataract	161 140 136 171 116 149 143 152

Under the normality assumption we want to test if there is a significant difference in the gray levels between cataractous and normal eyes. Assume equal variances in the two populations.

(a) Can you find any statistical evidence showing that the two gray levels are significantly different? Report the p-value.

(b) Pretend that a new study is to be conducted for which the sample size is needed. From previous experiments you adopt $\sigma = 16$ for both groups. What sample size is needed to detect a difference of 10, in a 0.05-level testing against the two sided alternative, with power $1 - \beta = 80\%$.

```
%Camera for Cataract
Control  = [158, 182, 182, 191, 177, 145, 156, 170, 152, 164, 169, 141];
Cataract = [161, 140, 136, 171, 116, 149, 143, 152];
m1 = mean(Control)   %165.5833
m2 =mean(Cataract)   %146
n1  = 12; n2=8;
var1=var(Control)    %246.4470
var2=var(Cataract)   %277.1429
std1=std(Control)    %15.6986
std2=std(Cataract)   %16.6476
sp  = sqrt(    ((n1-1)*var1+ (n2-1)*var2)/(n1+n2-2)    )
  % sp=16.0743
t = (m1 - m2)/(sp * sqrt(1/n1 + 1/n2)) %2.6692
%(a)H1: mu_cataract ~= mu_control (two sided alternative)
pval = 2 *  tcdf(-abs(t), n1+n2-2)      %0.0156
%(b)Sample size
sigma=16; diff=10; alpha = 0.05; beta=0.2;
n = 2*sigma^2/diff^2 * (norminv(1-alpha/2)+norminv(1-beta))^2
%n=40.1863 approx 41.
```

How sure are we in the $n = 41$ recommendation?

We are sure that the wanted difference to be found is 10, but not sure about the common population variance σ^2. Thus, we elicit a prior on the precision $\tau = 1/\sigma^2$ as gamma $\mathcal{G}a(a,b)$. The number of observation used to set $\sigma^2 = 16^2$ was 20, so we set "effective sample size" as $2a = 20$, i.e., $a = 10$.

Elicited precision was $a/b = 1/256$, leading to $b = 2560$. The following WinBUGS file conducts the inference on the sample size

```
model{
   n   <- 2*pow((0.8416 + 1.96)*sigma/theta, 2) #n for 80% power
   tau  ~ dgamma(10, 2560)
   sigma <- 1/sqrt(tau)
   power <- phi(sqrt(41/2)*theta/sigma - 1.96) #power for n=41
   p80  <- step(power - 0.8)
}
DATA
list(theta=10)
INITS
list(tau=1)
```

	mean	sd	MC error	val2.5pc	median	val97.5pc	start	sample
n	44.68	15.74	0.04698	23.41	41.66	83.57	1001	100000
p80	0.48	0.4996	0.001512	0.0	0.0	1.0	1001	100000
power	0.7766	0.1216	3.652E-4	0.501	0.7937	0.9597	1001	100000
sigma	16.64	2.789	0.008333	12.21	16.29	23.07	1001	100000

Notice that classical power analysis for which a sample of size of $n = 41$ "ensured" the power of 80% is not adequate when the uncertainty about σ^2, or equivalently about precision $\tau = 1/\sigma^2$, is present. If $\sigma = 16$ is exactly specified, then the precision is $1/256$. Note that the uncertainty about precision in the Bayesian approach was introduced by a gamma $\mathcal{G}a(10,2560)$ prior that has a mean of $1/256$ and a small variance of $10/2560^2 = 1.5259e - 06$. Hyperparameter a was a surrogate for the effective sample size (ESS), which is to say, we value the prior information about τ as if 10 observations had been used. See page 349 for detailed discussion about ESS.

In the presence of uncertainty about σ, the sample size increased to 45. Moreover, the classical sample size of 41 will achieve the power of 80% with probability of only 0.48.

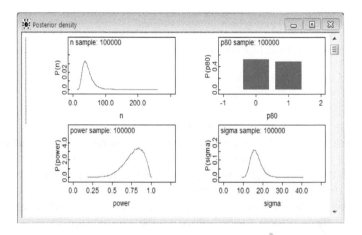

Fig. 10.1 Posterior distributions for sample size n, probability that power exceeds 80%, the power, and sigma are estimated in WinBUGS.

10.2.3 More Complex Two-Sample Designs

Suppose that we are interested in testing the equality of normal population means when the underlying variances in the two populations are σ_1^2 and σ_2^2, and not necessarily equal. Let us assume that the desired proportion of sample sizes to be determined is $k = n_2/n_1$, whereby, $n_2 = k \times n_1$. This proportion may be dictated by the cost of sampling or by the abundance of the populations. When equal group samples are desired, then $k = 1$,

$$n_1 = \frac{(\sigma_1^2 + \sigma_2^2/k)(z_{1-\alpha/2} + z_{1-\beta})^2}{|\mu_1 - \mu_2|^2}, \qquad n_2 = k \times n_1. \tag{10.5}$$

As before, μ_1, μ_2, σ_1^2, and σ_2^2 are unknown, and in the absence of any data, we can express $|\mu_1 - \mu_2|^2$ in units of $\sigma_1^2 + \sigma_2^2/k$ to elicit the effect size, d^2. However, if preliminary or historic samples are available, then μ_1, μ_2, σ_1^2, and σ_2^2 can be estimated by $\overline{X}_1$, $\overline{X}_2$, s_1^2, and s_2^2, respectively, and plugged into formula (10.6).

Example 10.4. **Two Amanitas.** Suppose that two independent samples of $m = 12$ and $n = 15$ spores of *A. pantherina* ("Panther") and *A. rubescens* ("Blusher"), respectively, are only a pilot study. It was found that the means are $\overline{X}_1 = 6.3$ and $\overline{X}_2 = 7.5$ with standard deviations of $s_1 = 2.12$ and $s_2 = 1.94$. All measures are in μm. Suppose that Blushers are twice as common as Panthers.

Determine the sample sizes for a future study that will find the difference obtained in the preliminary samples to be significant at the level $\alpha = 0.05$ with a power of $1 - \beta = 0.90$.

Here, based on the abundance of mushrooms, $2n_1 = n_2$ and $k = 2$. Substituting $\overline{X}_1$, $\overline{X}_2$, s_1^2, and s_2^2 into (10.6), we get

$$n_1 = \frac{(2.12^2 + 1.94^2/2)(z_{0.975} + z_{0.9})^2}{|6.3 - 7.5|^2} = 46.5260 \approx 47.$$

Thus, the effect size was $|6.3 - 7.5|/\sqrt{2.12^2 + 1.94^2/2} = 0.4752$, which corresponds to $d = 0.4752\sqrt{2}$.

The "plug-in" strategy applied above is in fact quite controversial. Proponents argue that in the absence of any information on μ_1, μ_2, σ_1^2, and σ_2^2, the "natural" approach is to use their MLEs, $\overline{X}_1$, $\overline{X}_2$, s_1^2, and s_2^2. Opponents say that one is looking for a sample size that will find the pilot difference to be significant at a level α with a preassigned power. They further argue that, due to routinely small sample sizes in pilot studies, the estimators for population means and variances can be unreliable. This unreliability is further compounded by the ratios and powers taken in calculating the sample size.

10.2.4 A Bayesian Test for Two Normal Means

Bayesian testing of two means simply analyzes the posterior distribution of the means difference, given the priors and the data. We will illustrate a Bayesian approach for a simple noninformative prior structure.

Let $X_{11}, X_{12}, \ldots, X_{1,n_1}$ and $X_{21}, X_{22}, \ldots, X_{2,n_2}$ be samples from normal $\mathcal{N}(\mu_1, \sigma_1^2)$ and $\mathcal{N}(\mu_2, \sigma_2^2)$ distributions, respectively. We are interested in the posterior distribution of $\theta = \mu_2 - \mu_1$ when $\sigma_1^2 = \sigma_2^2 = \sigma^2$. If the priors on

μ_1 and μ_2 are flat, $\pi(\mu_1) = \pi(\mu_2) = 1$, and the prior on the common σ^2 is noninformative, $\pi(\sigma^2) = 1/\sigma^2$, then the posterior of θ, after integrating out σ^2, is a t-distribution. That is to say, if $\overline{X}_1$ and $\overline{X}_2$ are the sample means and s_p is the pooled standard deviation, then

$$t = \frac{\theta - (\overline{X}_2 - \overline{X}_1)}{s_p \sqrt{1/n_1 + 1/n_2}} \tag{10.6}$$

has a t-distribution with $n_1 + n_2 - 2$ degrees of freedom (Box and Tiao, 1992, p. 103). Compare (10.6) with the distribution in (10.1). Although the two distributions coincide, they are conceptually different; (10.1) is the sampling distribution for the difference of sample means, whereas (10.6) gives the distribution for the difference of parameters.

In this case, the results of Bayesian inference coincide with frequentist results in the estimation of θ, in the confidence/credible intervals for θ, and in testing, as is usually the case when the priors are noninformative.

When σ_1^2 and σ_2^2 are not assumed equal and each has its own noninformative prior, finding the posterior in the previous model coincides with the Behrens–Fisher problem. The posterior is usually approximated (Patil's approximation method, as discussed by Box and Tiao, 1992, p. 107; Lee, 2004, p. 145).

When MCMC and WinBUGS are used in testing two normal means, we may entertain more flexible models. The testing becomes quite straightforward. We provide such an example next.

Example 10.5. **Microdamage in Bones.** Bone is a hierarchical composite material that provides our bodies with mechanical support and facilitates mobility, among other functions. Figure 10.2 shows the structure of bone tubecules. Damage in bone, in the form of microcracks, occurs naturally during daily physiological loading. The normal bone remodeling process repairs this microdamage, restoring, if not improving, biomechanical properties. Numerous studies have shown that microdamage accumulates as we age due to impaired bone remodeling. This accumulation contributes to a reduction in bone biomechanical properties such as strength and stiffness by disrupting the local tissue matrix.

In order to better understand the role of microdamage in bone tissue matrix properties as we age, a study was conducted in the lab of Dr. Robert Guldberg at the Georgia Institute of Technology. The interest was in the changes in microdamage progression in human bone between young and old female donors.

The data showing the score of normalized damage events are shown in the table below. There were $n_1 = 13$ donors classified as young (≤ 45 years old) and $n_2 = 17$ classified as old (>45 years old). To calculate the microdamage progression score, the counts of damage events (extensions, surface originations, widenings, and combinations) are normalized to the bone area and summed.

Fig. 10.2 Bone tubecules.

Young	Old
0.790 1.264	1.374 1.327
0.944 1.410	0.601 1.325
0.958 1.160	1.029 2.012
1.011 0.179	1.264 1.026
0.714	1.183 1.130
0.256	1.856 0.605
0.406	1.899 0.870
0.135	0.486 0.820
0.316	0.813

Assuming that the microdamage scores are normally distributed, test the hypothesis of equality of the population means, for young and old, against the one-sided alternative. The sensible one-sided alternative here is $H_1 : \mu_1 < \mu_2$, or equivalently, $H_1 : \mu_1 - \mu_2 < 0$. When posterior samples from μ_1 and μ_2 are obtained in WinBUGS, to assess the posterior probability of H_1, we will find the proportion of cases when $\theta = \mu_1 - \mu_2$ is negative. This can be done by WinBUGS function `step(-theta)`.

```
#microdamage.odc
model{
for (i in 1:n){
  score[i] ~ dnorm(mu[age[i]], prec[age[i]])
  }
mu[1] ~ dnorm(0, 0.00001)
mu[2] ~ dnorm(0, 0.00001)
prec[1] ~ dgamma(0.001, 0.001)
prec[2] ~ dgamma(0.001, 0.001)
theta <- mu[1] - mu[2]
r <- prec[1]/prec[2]
ph1 <- step(-theta)  #ph1=1 if theta < 0
ph0 <- 1-ph1
}


DATA
list(n=30,score=c(0.790, 0.944, 0.958, 1.011, 0.714, 0.256, 0.406,
     0.135, 0.316, 0.179, 1.264, 1.410, 1.160, 1.374,
     0.601, 1.029, 1.264, 1.183, 1.856, 1.899, 0.486,
     0.813, 0.820, 1.327, 1.325, 2.012, 1.026, 1.130,
     0.605, 0.870),
age = c(1,1,1,1,1,1,1,1,1,1,1,1,1,
     2,2,2,2,2,2,2,2,2,2,2,2,2,2,2))

INITS
list( mu = c(1,1), prec=c(1,1))
```

	mean	sd	MC error	val2.5pc	median	val97.5pc	start	sample
theta	−0.4194	0.1771	5.438E-4	−0.7688	−0.4192	−0.071	1001	100000
mu[1]	0.7339	0.1326	4.099E-4	0.4703	0.7338	0.9968	1001	100000
mu[2]	1.153	0.118	3.685E-4	0.9188	1.153	1.389	1001	100000
ph0	0.01011	0.1	3.112E-4	0.0	0.0	0.0	1001	100000
ph1	0.9899	0.1	3.112E-4	1.0	1.0	1.0	1001	100000
r	1.244	0.7517	0.002553	0.3456	1.071	3.155	1001	100000

The posterior probability of H_1 is 0.9899, so H_0 is rejected. Note that the credible interval for the difference θ is all negative, suggesting that the two-sided test would be significant (in Bayesian terms). The ratio of precisions (and variances) r has a credible set that contains 1; thus, the variances could be assumed equal. This assumption has no bearing on the Bayesian procedure, unlike the classical approach.

Remark. When the assumption $\sigma_1^2 = \sigma_2^2$ is not appropriate, the following independent priors can be assumed,

$$\pi(\mu_1) = 1, \ \pi(\mu_2) = 1, \ \pi(\sigma_1^2) = 1/\sigma_1^2, \ \text{and} \ \pi(\sigma_2^2) = 1/\sigma_2^2.$$

This leads to posterior distributions for μ_1 and μ_2 which are, in fact, independent t-distributions,

$$\frac{\mu_1 - \overline{X}_1}{s_1/\sqrt{n_1}} \sim t_{n_1-1}, \quad \text{and} \quad \frac{\mu_2 - \overline{X}_2}{s_2/\sqrt{n_2}} \sim t_{n_2-1}.$$

The inference about $\theta = \mu_1 - \mu_2$ can now be carried out by Bayesian simulation, as in Example 10.5.

10.3 Testing the Equality of Normal Means When Samples Are Paired

When comparing two treatments, it is desirable that the experimental units be as alike as possible so that the difference in responses can be attributed chiefly to the treatment. If a number of the relevant factors (age, gender, body mass index, presence of risk factors, etc.) vary in an uncontrolled manner, a large portion of variability in the response could be attributed to these factors rather than to the treatments. The concept of pairing, matching, or blocking is critical to eliminate nuisance variability and obtain better experimental designs.

Consider a sample consisting of paired elements, so that every element from population 1 has its match in population 2. A sample from population 1, $X_{11}, X_{12}, \ldots, X_{1,n}$, is thus paired with a sample from population 2, $X_{21}, X_{22}, \ldots, X_{2,n}$, so that a pair (X_{1i}, X_{2i}) represents the ith observation. Usually, observations in a pair are taken on the same subject, such as pretest–posttest or placebo–treatment, or on dependent subjects, such as brother–sister or two subjects with matching demographic characteristics. The examples are numerous, but most applications involve subjects with measurements taken at two different time points, during two different treatments, and so on. Sometimes, especially in industrial experimental design, this matching is called "blocking."

It is typically assumed that the samples come from normal populations with possibly different means μ_1 and μ_2 (subject to test) and with unknown variances σ_1^2 and σ_2^2. As linear combinations of normals, the differences $d_i = X_{1i} - X_{2i}$ are also normal,

$$d_i \sim \mathcal{N}(\mu_1 - \mu_2, \sigma_1^2 + \sigma_2^2 - 2 \cdot \sigma_{12}),$$

where $\sigma_{12} = \rho\sigma_1\sigma_2$ is the covariance, $\mathbb{E}\left[(X_1 - \mathbb{E}X_1)(X_2 - \mathbb{E}X_2)\right]$, as in (6.2). Let us define

$$t = \frac{\overline{d}}{s_d/\sqrt{n}}, \tag{10.7}$$

where $\overline{d}$ is an average of the differences d_i and s_d is the sample standard deviation of the differences. Here the sample size n relates to the number of pairs and not the total number of observations, which is $2n$.

Note that we can express s_d as $\sqrt{s_1^2 + s_2^2 - 2s_{12}}$, where s_{12} is the estimator of covariance between the samples:

$$s_{12} = \frac{1}{n-1} \sum_{i=1}^{n} (X_{1i} - \overline{X}_1)(X_{2i} - \overline{X}_2).$$

Also, s_{12} can be expressed as $s_1 s_2 r$, as on page 32.

For example, in MATLAB,

```
x1 = [2 4 5 6 5 7 8];
x2 = [6 8 6 5 3 4 2];
d = x1 - x2;
sd = std(d)    %ans =3.6904

co=cov(x1, x2)
  %co =    3.9048   -2.7857
  %       -2.7857    4.1429

sqrt(co(1,1) + co(2,2) - 2 * co(1,2))  %ans =3.6904
```

We remark that, for the paired samples problem, the assumption on the individual population variances is not necessary because we operate with the differences, and the population variance of the differences, $\sigma_d^2 = \sigma_1^2 + \sigma_2^2 - 2\sigma_{12}$, is unknown.

We are interested in testing the means, $H_0 : \mu_1 = \mu_2$, versus one of the three alternatives $H_1 : \mu_1 >, \neq, < \mu_2$. Under H_0 the test statistic t has a t-distribution with $n-1$ degrees of freedom. Thus, the test coincides with the one-sample t-test, where the sample consists of all differences and where H_0 is the hypothesis that the mean in the population of differences is equal to 0. The popular name for this test is the *paired t-test*, which can be summarized as follows:

Alternative	α-level rejection region	p-value
$H_1 : \mu_1 > \mu_2$	$[t_{n-1,1-\alpha}, \infty)$	`1-tcdf(t, n-1)`
$H_1 : \mu_1 \neq \mu_2$	$(-\infty, t_{n-1,\alpha/2}] \cup [t_{n-1,1-\alpha/2}, \infty)$	`2*tcdf(-abs(t), n-1)`
$H_1 : \mu_1 < \mu_2$	$(-\infty, t_{n-1,\alpha}]$	`tcdf(t, n-1)`

We can generalize this test to testing $H_0 : \mu_1 - \mu_2 = d_0$ versus the appropriate one- or two-sided alternative. The only modification needed is in the t-statistic (10.7), which now takes the form

$$t = \frac{\overline{d} - d_0}{s_d / \sqrt{n}},$$

and which under H_0 has a t-distribution with $n-1$ degrees of freedom.

If $\delta = \mu_1 - \mu_2$ is the difference between the population means, then the $(1 - \alpha)100\%$ confidence interval for δ is

$$\left[\bar{d} - t_{n-1,1-\alpha/2}\frac{s_d}{\sqrt{n}}, \ \bar{d} + t_{n-1,1-\alpha/2}\frac{s_d}{\sqrt{n}}\right].$$

Since matching (blocking) the observations eliminates the variability be-tween the subjects entering the inference, the paired t-test is preferred to a two-sample t-test whenever such a design is possible. For example, the case-control study design selects one observation or experimental unit as the "case" variable and obtains as "matched controls" one or more addi-tional observations or experimental units that are similar to the case, except for the variable(s) under study. When the treatments are assigned after sub-jects have been paired, this assignment should be random to avoid any potential systematic influences.

Example 10.6. **Psoriasis.** Woo and McKenna (2003) investigated the effect of broadband ultraviolet B (UVB) therapy and topical calcipotriol cream used together on areas of psoriasis. One of the outcome variables is the Psoriasis Area and Severity Index (PASI), where a lower score is better. The following table gives PASI scores for 20 subjects measured at baseline and after 8 treatments. Do these data provide sufficient evidence, at a 0.05 level of significance, to indicate that the combination therapy reduces PASI scores?

Subject	Baseline	After 8 treatments	Subject	Baseline	After 8 treatments
1	5.9	5.2	11	11.1	11.1
2	7.6	12.2	12	15.6	8.4
3	12.8	4.6	13	9.6	5.8
4	16.5	4.0	14	15.2	5.0
5	6.1	0.4	15	21.0	6.4
6	14.4	3.8	16	5.9	0.0
7	6.6	1.2	17	10.0	2.7
8	5.4	3.1	18	12.2	5.1
9	9.6	3.5	19	20.2	4.8
10	11.6	4.9	20	6.2	4.2

The data set is available as ▣ pasi.dat|xls|mat.

We will import the data into MATLAB and test the hypothesis that the PASI significantly decreased after treatment at a significance of $\alpha = 0.05$. We will also find a 95% confidence interval for the difference between the population means $\delta = \mu_1 - \mu_2$.

```
%psoriasis.m
baseline = [5.9 7.6 12.8 16.5 6.1 14.4 6.6 5.4 ...
```

```
    9.6 11.6 11.1 15.6 9.6 15.2 21 5.9 10 12.2 20.2 6.2];
after = [5.2 12.2 4.6 4 0.4 3.8 1.2 3.1 3.5 4.9 ...
    11.1 8.4 5.8 5 6.4 0 2.7 5.1 4.8 4.2];

d = baseline - after;
n = length(d);
dbar = mean(d)                    %dbar  = 6.3550
sdd= sqrt(var(d))                 %sdd   = 4.9309
tstat = dbar/(sdd/sqrt(n))  %tstat = 5.7637

   % Test using RR
critpt = tinv(0.95, n-1)     %critpt =1.7291
   % Rejection region (1.7291, infinity). Reject H_0 since
   % tstat=5.7637 falls in the rejection region.
   % Test using the p-value
p_value = 1-tcdf(tstat, n-1)  %p_value = 7.4398e-006
   % Reject H_0 at the level alpha=0.05
   % since the p_value = 0.00000744 < 0.05.
alpha = 0.05
LB =  dbar - tinv(1-alpha/2, n-1)*(sdd/sqrt(n))
   % LB = 4.0472
UB =  dbar + tinv(1-alpha/2, n-1)*(sdd/sqrt(n))
   % UB =  8.6628
   % 95% CI is  [4.0472, 8.6628]
```

Alternatively, we can use `d1=after-baseline`, but then we must be careful about choosing the "direction" of the H_1 and p-value calculations. In this case, the rejection region is $(-\infty, -1.7291)$ and the 95% confidence interval is $[-8.6628, -4.0472]$.

A Bayesian solution is given next:

```
model{
for(i in 1:n){
d[i] <- baseline[i] - after[i]
d[i] ~ dnorm(mu, prec)
}
mu ~ dnorm(0, 0.00001)
pH1 <- step(mu-0)
prec ~ dgamma(0.001, 0.001)
sigma2  <- 1/prec;
sigma <- 1/sqrt(prec)
}

DATA
list(n=20,
baseline = c(5.9, 7.6, 12.8, 16.5, 6.1, 14.4, 6.6,
  5.4,  9.6, 11.6 ,11.1, 15.6, 9.6, 15.2, 21, 5.9,
  10, 12.2, 20.2, 6.2),
after = c(5.2, 12.2, 4.6, 4, 0.4 , 3.8, 1.2, 3.1, 3.5,
  4.9, 11.1, 8.4, 5.8, 5, 6.4, 0, 2.7, 5.1, 4.8, 4.2))

INITS
list(mu=0, prec=1)
```

	mean	sd	MC error	val2.5pc	median	val97.5pc	start	sample
pH1	1.0	0.0	3.162E-13	1.0	1.0	1.0	1001	100000
mu	6.352	1.169	0.003657	4.043	6.351	8.666	1001	100000
prec	0.04108	0.01339	4.498E-5	0.01927	0.03959	0.07149	1001	100000
sigma	5.142	0.8912	0.003126	3.74	5.026	7.203	1001	100000
sigma2	27.23	10.0	0.03528	13.99	25.26	51.88	1001	100000

Let us compare the classical and Bayesian solutions. The estimator for the difference between population means is 6.3550 in the classical case and 6.352 in the Bayesian case. The standard deviations of the difference are close as well: the classical is $4.9309/\sqrt{20} = 1.1026$ and the Bayesian is 1.169.

The 95% confidence interval for the difference is $[4.0472, 8.6628]$, while the 95% credible set is $[4.043, 8.666]$. The posterior probability of H_1 is approx. 1, while the classical p-value (support for H_0) is 0.000007439. This closeness of classical and Bayesian results is expected given that the priors mu~dnorm(0,0.00001) and prec~dgamma(0.001,0.001) are fairly noninformative.

Next, we provide an example in which the measurements are taken on different subjects, but the subjects are matched with respect to some characteristics that can influence the response. Because of such matching, the sample is considered paired, and we say that the characteristics used for matching are "controlled." This is often done when the application of both treatments to a single subject is either impossible or leads to biased responses.

Example 10.7. **IQ-Test Pairing.** In a study concerning memorizing verbal sentences, children were first given an IQ test. One of the two children with the lowest scores was randomly assigned to a "noun-first" task, and the other to a "noun-last" task. The two next lowest IQ children were similarly assigned, one to a "noun-first" task, the other to a "noun-last" task, and so on, until all children were assigned. The data, recorded as scores on a word-recall task, are shown here, listed in order from lowest to highest IQ score:

Noun-first	12 21 12 16 20 39 26 29 30 35 38 34
Noun-last	10 12 23 14 16 8 16 22 32 13 32 35

Let μ_1 and μ_2 be the population means corresponding to "noun-first" and "noun-last" tasks. We will test the hypothesis $H_0 : \mu_1 - \mu_2 = 0$ against the two-sided alternative. The significance level is set to $\alpha = 5\%$.

Note that the two samples are not independent since the pairing is based on an ordered joint attribute, children's IQ scores. Thus, even though the subjects in the two groups are different, the paired t-test is appropriate.

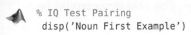

```
% IQ Test Pairing
  disp('Noun First Example')
```

```
nounfirst = [12  21  12  16  20  39  26  29  30  35  38  34];
nounlast  = [10  12  23  14  16   8  16  22  32  13  32  35];
d=nounfirst - nounlast;
dbar = mean(d)              %dbar = 6.5833
sd = std(d)                 %sd = 11.0409
n = length(d)               %n = 12
t = dbar/(sd/sqrt(n))       %t = 2.0655
pval = 2 * (1-tcdf(t, n-1)) %pval = 0.0633
```

The null hypothesis is not rejected against the two-sided alternative $H_1 : \mu_1 \neq \mu_2$ at the 5% level. However, for the one-sided alternative, in this case $H_1 : \mu_1 > \mu_2$, the p-value would be less than 5% and the null would be rejected. The testing is equivalent to a one-sample t-test against the alternative $\mu_1 - \mu_2 > 0$.

```
pval = 1-tcdf(t, n-1)       %pval = 0.0316
```

🖉

10.3.1 Sample Size in Paired t-Test

If in the paired t-test the variance of the differences, σ_d^2, is known, then under the alternative $H_1 : \mu_1 - \mu_2 = d^*$,

$$Z = \frac{\bar{d} - d^*}{\sigma_d / \sqrt{n}}$$

would have a standard normal distribution. Thus, to achieve a power of $1 - \beta$ by an α-level test that would reject H_0 if the effect size is $e = d^* / \sigma_d$, the one-sided test would require

$$n \geq \frac{(z_{1-\alpha} + z_{1-\beta})^2}{e^2} \tag{10.8}$$

observations.

As discussed on page 398, use of normal quantiles as in in (10.8) gives an approximate sample size, which in fact underestimates it. More precise calculations involve the noncentral t-distribution, as in the following example.

Example 10.8. **Sample Size for SBP Experiment.** Suppose that a study is to be designed for assessing the effect of a blood-pressure-lowering drug in middle-aged men. Each subject will have his systolic blood pressure (SBP) taken at the onset of the trial and after a 14-day regimen with the drug. In previous studies of related drugs, the variance of difference between the two measurements was found to be 300 (mmHg)^2. The new drug would be of interest if it reduced the SBP by 5 mmHg or more.

What sample size is needed so that a 5% level test detects a difference of 5 mmHg at least 90% of time?

Direct application of (10.8) with $\sigma_d^2 = 300$, $d^* = 5$, $\alpha = 0.05$, and $\beta = 0.1$ gives a sample size of 103 subjects. More precise calculations (using noncentral t from page 398) give a necessary sample size of 105.

```
n = (300 * (norminv(0.95) + norminv(0.9))^2 )/(5^2) %  102.7662
%using nctcdf
effect = 5/sqrt(300)    %0.2887
pf = @(n) 1-nctcdf( tinv(0.95, n-1), n-1, sqrt(n) * 0.2887) - 0.9
n = fzero(pf, 100)      %104.1163
```

10.3.2 Difference-in-Differences (DiD) Tests

Suppose that there are two treatment groups consisting of different and independent subjects. In each group two measurements are taken on each subject (e.g., at two time points, before an intervention and after the intervention). Let the first measurement serve as a baseline. We are interested in the change, that is in the difference between the second measurement and the baseline. For example, if the intervention is administration of a blood-pressure-lowering drug, then the difference between the post- and pre-intervention measurement can be attributed to the effect of the drug and is of interest. The two groups may be defined by a treatment level, demographic characteristics (gender, age, socioeconomic status), history of hypertension, etc.

Besides investigating the effect of the drug within each group, we might be interested in comparing the efficacy of the drug between the groups, which amounts to compare the differences as two sets of independent observations, thus the name difference-in-differences (DiD) test.

A DiD test is a two-sample t-test performed on the differences within the two groups.

Let the number of subjects in the first group be n_1 and in the second n_2. Next, let $(X_{11}, X_{21}), (X_{12}, X_{22}), \ldots, (X_{1,n_1}, X_{2,n_1})$ be the pairs of measurements in the group 1, and $(Y_{11}, Y_{21}), (Y_{12}, Y_{22}), \ldots, (Y_{1,n_2}, Y_{2,n_2})$ be the pairs for the group 2. We find the differences $d_{1i} = X_{2i} - X_{1i}, i = 1, \ldots, n_1$ for group 1, and $d_{2i} = Y_{2i} - Y_{1i}, i = 1, \ldots, n_2$, for group 2. The two sample t test is applied on $d_{11}, \ldots, d_{1,n_1}$ and $d_{21}, \ldots, d_{2,n_2}$. The scenarios could be either with the assumption of equal difference variances or without this assumption as in the two versions of t-test on page 431.

Let $\overline{d_1}$, $\overline{d_2}$, s_1, and s_2 be the sample means and standard deviations of the corresponding differences. Then, under the assumption of equal population difference variances,

$$t = \frac{\overline{d_1} - \overline{d_2}}{s_p\sqrt{1/n_1 + 1/n_2}},$$

has a t-distribution with $n_1 + n_2 - 2$ degrees of freedom. Here s_p is the pooled sample standard deviation,

$$s_p = \sqrt{\frac{(n_1 - 1)s_1^2 + (n_2 - 1)s_2^2}{n_1 + n_2 - 2}}.$$

The described DiD test is one of many possible tests involving the differences. There are experimental designs that lead to paired t-tests, more generally to ANOVA and repeated measure designs as we will see in Chapter 11 involving differences of measurements.

Example 10.9. **Blood Pressure and Calcium Supplementation.** Lyle et al. (1987) provide data on a randomized, double-blind, placebo-controlled trial that was conducted to examine the effect of calcium supplementation on blood pressure in normotensive African American ($n = 21$) and Caucasian ($n = 54$) men, aged 19 to 52 years. After a four-week baseline period of weekly blood pressure measurement, subjects were randomly assigned within racial groups to either a treatment (calcium, 1500 mg/day) or placebo group for a 12-week period. The authors state that calcium supplementation, in comparison with placebo, resulted in lower mean arterial pressure in normotensive men of both races during a 12-week period.

For the part of data consisting of $n = 21$ normotensive African American men, verify the authors' claim at significance level $\alpha = 0.05$. State your hypotheses, select the procedure, state assumptions, and conduct the test.

Calcium Group	Begin	107 110 123 129 112 111 107 112 136 102
$n_1 = 10$	End	100 114 105 112 115 116 106 102 125 104
Placebo Group	Begin	123 109 112 102 98 114 119 112 110 117 130
$n_2 = 11$	End	124 97 113 105 95 119 114 114 121 118 133

Assume that Begin and End data for both calcium and placebo groups are coming from normal distributions.

```
%lyle.m
CaBegin  = [107 110 123 129 112 111 107 112 136 102];
CaEnd    = [100 114 105 112 115 116 106 102 125 104];
%
PlaBegin = [123 109 112 102  98 114 119 112 110 117 130];
PlaEnd   = [124  97 113 105  95 119 114 114 121 118 133];
```

```
d1 = CaEnd-CaBegin;        n1=length(d1);
d2 = PlaEnd - PlaBegin;    n2=length(d2);
d1bar = mean(d1)  %-5
sd1=std(d1)        %8.7433
d2bar = mean(d2)   %0.6364
sd2=std(d2)        %5.8698
F=sd1^2/sd2^2      %2.2187
pval = 2*min(fcdf( F, n1-1, n2-1), 1-fcdf(F, n1-1, n2-1)) %0.2304
%
sdp = sqrt(   ((n1-1)*sd1^2 + (n2-1)*sd2^2)/(n1+n2-2))    %7.3719
t = (d1bar - d2bar)/( sdp * sqrt(1/n1 + 1/n2)) %-1.7499
pval= tcdf(t, n1+n2-2) %0.0481
```

10.4 Two Multivariate Normal Means*

Consider now the comparison of two multivariate normal means. We are interested in testing

$$H_0 : \mu_1 = \mu_2 \quad \text{versus} \quad H_1 : \mu_1 \neq \mu_2,$$

when samples of p-dimensional observations, $X_{11}, \ldots, X_{1n_1}$ and $X_{21}, \ldots, X_{2n_2}$, are obtained from two populations distributed as $\mathcal{MVN}(\mu_1, \Sigma_1)$ and $\mathcal{MVN}(\mu_2, \Sigma_2)$, respectively.

The covariance matrices Σ_1 and Σ_2 are assumed not known and are estimated from the samples via sample covariance matrices S_1 and S_2, as in page 36. However, if they are assumed equal, $\Sigma_1 = \Sigma_2$ $(= \Sigma)$, then the estimator of the common Σ is the pooled sample covariance matrix

$$S = \frac{(n_1 - 1)S_1 + (n_2 - 1)S_2}{n_1 + n_2 - 2}.$$

The statistic

$$T^2 = \frac{n_1 n_2}{n_1 + n_2}(\overline{X}_1 - \overline{X}_2)'S^{-1}(\overline{X}_1 - \overline{X}_2)$$

is formed. The properly scaled T^2 statistic follows F-distribution with p and $n_1 + n_2 - p - 1$ degrees of freedom, that is,

$$\frac{n_1 + n_2 - p - 1}{p(n_1 + n_2 - 2)} T^2 \sim F_{p, n_1+n_2-p-1}.$$

Thus,

$H_0 : \mu_1 = \mu_2$ is rejected at level α when

$$T^2 \geq \frac{p\,(n_1 + n_2 - 2)}{n_1 + n_2 - p - 1}\, F_{p,n_1+n_2-p-1,1-\alpha}.$$

where $F_{p,n_1+n_2-p-1,1-\alpha}$ is the $(1 - \alpha)$-quantile of F-distribution with p and $n_1 + n_2 - p - 1$ degrees of freedom.

When T^2 is observed, the p-value is given as

$$p\text{-value} = \mathbb{P}\left(F > \frac{n_1 + n_2 - p - 1}{p\,(n_1 + n_2 - 2)}\, T^2 \right),$$

for some F being F_{p,n_1+n_2-p-1} distributed. In MATLAB,

```
pval = 1-fcdf((n1+n2-p-1)/(p*(n1+n2-2))*T2, p, n1+n2-p-1).
```

Remark. It can be shown that for $p = 1$ the above procedures recover traditional univariate two sample inferences when the variances are unknown but assumed equal.

Example 10.10. **Leptoconops – Biting Flies.** In Exercise 2.22, a data set consisting of morphological characteristics of two species of biting flies *Leptoconops torrens* and *Leptoconops carteri* was analyzed. This data set contains 35 multivariate observations of each species given as a structure field `leptoconops.morpho` in MATLAB's data file ⬚ `leptoconops.mat`. The field `leptoconops.names` contains names of seven recorded morphological measures: `winglen`, `wingwid`, `papl3len`, `palp3wid`, `palp4len`, `ant12len`, and `ant14len`. The two species are identified in field `leptoconops.spec` where 0's correspond to *L.torrens* and 1's to *L.carteri*.

We are interested in testing the equality of multivariate means.

```
load leptoconops
taxom = leptoconops.morpho;
spec = leptoconops.spec;
X1 = taxom(spec==0,:);
X2 = taxom(spec==1,:);
varnames = 'winglen' ;   'wingwid' ;   'papl3len';   ...
    'palp3wid'; 'palp4len';   'ant12len' ;   'ant14len' ;
[n1 p1] = size(X1);  [n2 p2]=size(X2);
  if (p1 ~= p2)
     error('Dimensions different!')
  end
  p=p1
  %- - - - - - - - - - - - - - - - - - - - - - - - - - - - - - - -
  X1bar = mean(X1)';
  X2bar = mean(X2)';
 S1 = cov(X1); S2 = cov(X2);
```

```
 S = (  (n1-1)*S1 + (n2 - 1)*S2  )/(n1 + n2 - 2);
 %H0: mu1 = mu2     H1:  mu1 =/= mu2
 T2 = n1*n2/(n1+n2) * (X1bar-X2bar)' * inv(S) * (X1bar-X2bar)
 %106.1348
 pval = 1 - fcdf((n1+n2-p-1)/(p*(n1+n2-2))*T2, p, n1+n2-p-1)
 %1.2325 e-010
```

10.4.1 Confidence Intervals for Arbitrary Linear Combinations of Mean Differences

Suppose we are interested in confidence intervals for $a'(\mu_1 - \mu_2)$, where a is an arbitrary $p \times 1$ vector of constants. This may be of interest when Hotelling's T^2 test rejected H_0 and we are interested to find which components of the mean vectors are different. The bounds

$$
\left[a'(\overline{X}_1 - \overline{X}_2) - \sqrt{\frac{p\,(n_1 + n_2 - 2)}{n_1 + n_2 - p - 1}}\, F_{p,n_1+n_2-p-1,1-\alpha}\, \sqrt{\frac{n_1 + n_2}{n_1 n_2}}\, a'Sa \;, \right.
$$
$$
\left. a'(\overline{X}_1 - \overline{X}_2) + \sqrt{\frac{p\,(n_1 + n_2 - 2)}{n_1 + n_2 - p - 1}}\, F_{p,n_1+n_2-p-1,1-\alpha}\, \sqrt{\frac{n_1 + n_2}{n_1 n_2}}\, a'Sa \right]
$$

provide $(1 - \alpha)100\%$ simultaneous CIs for the components of $a'(\mu_1 - \mu_2)$. By properly choosing vector a, various differences can be monitored.

When Σ_1 and Σ_2 are not assumed equal, it is impossible to use T^2 statistic. However, if n_1 and n_2 are not small, and not very different from each other, approximate CIs are obtained as

$$
\left[a'(\overline{X}_1 - \overline{X}_2) - \sqrt{\chi^2_{p,1-\alpha}}\, \sqrt{a'\left(\frac{1}{n_1}S_1 + \frac{1}{n_2}S_2\right)a} \;, \right.
$$
$$
\left. a'(\overline{X}_1 - \overline{X}_2) + \sqrt{\chi^2_{p,1-\alpha}}\, \sqrt{a'\left(\frac{1}{n_1}S_1 + \frac{1}{n_2}S_2\right)a} \right],
$$

where $\chi^2_{p,1-\alpha}$ is the $(1 - \alpha)$ quantile of a chi-square distribution with p degrees of freedom.

Example 10.11. **Leptoconops, Continued.** Continuing Example 10.10, we find CIs for differences between the individual mean components for which 95% confidence level simultaneously holds. Figure 10.3 shows 95% CIs for individual component differences: $\mu_{1,1} - \mu_{2,1}, \ldots, \mu_{1,7} - \mu_{2,7}$.

```
figure(1)
 alpha = 0.05;
 for i= 1:7
     a=[0 0 0 0 0 0 0]';
     a(i) = 1;
 lb=a' * (X1bar - X2bar) - sqrt(  (n1 +n2)/(n1 * n2) * a' * S * a * ...
          p*(n1+n2-2)/(n1+n2-p-1) * finv(1-alpha, p, n1+n2-p-1) );
   ub=a' * (X1bar - X2bar) + sqrt( (n1 +n2)/(n1 * n2) * a' * S * a * ...
          p*(n1+n2-2)/(n1+n2-p-1) * finv(1-alpha, p, n1+n2-p-1) );
     [lb, ub]
     plot([i i],[lb ub],'r-','linewidth',lw)
     hold on
 end
 plot([0 8],[0 0],'k--')
 xlabel('Coordinate'); ylabel('95% CI for Differences')
```

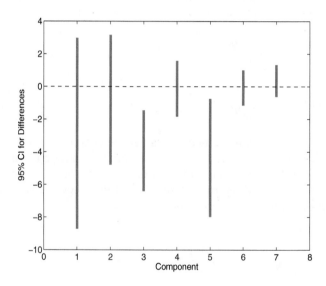

Fig. 10.3 CIs for individual component differences: $\mu_{1,1} - \mu_{2,1}, \ldots, \mu_{1,7} - \mu_{2,7}$.

10.4.2 Profile Analysis With Two Independent Groups*

When plotted componentwise so that component index falls on the x-axis and sample averages of the components on the y-axis, profiles of means are visualized. Suppose that it is of interest to compare profiles for two independent populations.

• (i) First, are the shapes of the profiles similar? That is, are the segments connecting adjacent means parallel?

To answer this question, consider $(p-1) \times p$ matrix C as discussed on page 403. In particular, focus on

$$C = \begin{pmatrix} 1 & -1 & 0 & \ldots & 0 & 0 \\ 0 & 1 & -1 & \ldots & 0 & 0 \\ 0 & 0 & 1 & \ldots & 0 & 0 \\ & \vdots & & & & \\ 0 & 0 & 0 & \ldots & -1 & 0 \\ 0 & 0 & 0 & \ldots & 1 & -1 \end{pmatrix},$$

where each row contains a pair $(1, -1)$ propagating with a shift. If $\mu = (\mu_1, \ldots, \mu_p)'$, then $C\mu = (\mu_1 - \mu_2, \mu_2 - \mu_3, \ldots, \mu_{p-1} - \mu_p)'$.

The profiles for the two populations can be tested as to whether they are parallel:

$$H_0 : C\mu_1 = C\mu_2 \quad \text{versus} \quad H_1 : C\mu_1 \neq C\mu_2.$$

The test is based on

$$T^2 = \frac{n_1 n_2}{n_1 + n_2} (\overline{X}_1 - \overline{X}_2)' \; C'(CSC')^{-1}C \; (\overline{X}_1 - \overline{X}_2),$$

where S is pooled covariance matrix. In this case,

$$\frac{n_1 + n_2 - p}{(p-1)(n_1 + n_2 - 2)} \; T^2 \sim F_{p-1, n_1+n_2-p}.$$

If the hypothesis of parallel profiles is not rejected, then we may be interested in:

• (ii) Whether the profiles are shifted?
• (iii) Whether the profiles are flat?

The hypothesis in (ii) is

$$H_0 : 1'\mu_1 = 1'\mu_2$$

for $1' = (1, 1, \ldots, 1)$, while the hypothesis in (iii) is

$$H_0 : C(\mu_1 + \mu_2) = 0.$$

10.4.3 Paired Multivariate Samples*

We will now discuss a multivariate extension of the paired t-test. This test is adequate in a range of situations involving pharmaceutical, educational, medical, or engineering practice. For example, one may need to compare multivariate results obtained in "pre" and "post" regime on the same subjects.

Assume that two p-dimensional samples, $X_{11}, \ldots, X_{1n}$ and $X_{21}, \ldots, X_{2n}$, are paired as $(X_{11}, X_{21}), \ldots, (X_{1n}, X_{2n})$, and that the differences

$$D_i = X_{1i} - X_{2i}, \; i = 1, \ldots, n$$

are formed. When the two samples are coming from multivariate normal distributions, the differences $D_1, \ldots, D_n$ are also multivariate normal,

$$D_i \sim \mathcal{MVN}_p(\delta, \Sigma_d),$$

where $\delta = \mu_1 - \mu_2$.

For some fixed δ_0, the test $H_0 : \mu_1 = \mu_2 + \delta_0$ versus $H_1 : \mu_1 \neq \mu_2 + \delta_0$ becomes a one-sample test $H_0 : \delta = \delta_0$ versus $H_0 : \delta \neq \delta_0$, so Hotelling's T^2 is applicable.

One forms

$$T^2 = n(\overline{D} - \delta_0)' S_d^{-1} (\overline{D} - \delta_0),$$

where $\overline{D}$ and S_d are the sample mean and sample covariance matrix of the differences. When H_0 is true, the scaled statistic $\frac{n-p}{p(n-1)} T^2$ follows an F-distribution with p and $n - p$ degrees of freedom.

The null hypothesis is rejected when $T^2 \geq \frac{p(n-1)}{n-p} F_{p,n-p,1-\alpha}$, where $F_{p,n-p,1-\alpha}$ is $(1 - \alpha)$ quantile of F-distribution with p and $n - p$ degrees of freedom. Of most interest is the case $\delta_0 = 0$, in which H_0 translates to the equality of population means.

A $100(1 - \alpha)\%$ confidence region for δ consists of all such δ for which

$$(\overline{D} - \delta)' S_d^{-1} (\overline{D} - \delta) \leq \frac{p(n-1)}{n(n-p)} F_{p,n-p,1-\alpha}.$$

Also, $100(1 - \alpha)\%$ simultaneous confidence intervals for individual $\delta_i = \mu_{1,i} - \mu_{2,i}$ are

$$\left[\overline{D}_i - \sqrt{\frac{p(n-1)}{n-p} F_{p,n-p,1-\alpha}} \sqrt{\frac{s_i^2}{n}}, \overline{D}_i + \sqrt{\frac{p(n-1)}{n-p} F_{p,n-p,1-\alpha}} \sqrt{\frac{s_i^2}{n}} \right],$$

where $\overline{D}_i$ is the ith component sample mean and $s_i^2 = s_{ii}$ is ith diagonal element of S_d.

Example 10.12. **Effluent Data.** Municipal wastewater treatment plants are required by law to monitor their discharges to rivers and streams on a regular basis. Concerns about reliability of data from one self-monitoring plant led to a study in which the samples of effluent were divided and sent to two labs for testing. One set of samples (X_1) was sent to a governmental lab and second (X_2) to a lab that the plant uses on a regular basis. Measurements of biochemical oxygen demand (BOD) and suspended solids (SS) were obtained from $n = 11$ independent sample splits. (Johnston & Wichern, Table 6.1).

Do the chemical analyses agree?

```
X1 = [6 27;   6 23;   18 64;    8 44;    11 30;   34 75;  ...
          28 26;   71 124;   43 54;   33 30;   20 14];
X2 = [25 15;   28 13;   36 22; 35 29;   15 31;   44 64;  ...
          42 30;   54 64;   34 56;   29 20;   39 21];
  D=X1-X2;
  [n p]=size(D)          %[11 2]
  Dbar = mean(D)'         %[-9.3636;    13.2727]
  Sd = cov(D)    %[199.2545    88.3091;    88.3091   418.6182]
  T2 = n * Dbar'* inv(Sd) * Dbar    %13.6393

  %Critical Region Reject H0: delta = 0 if T2 > CR
  CR =   p*(n-1)/(n-p)*finv(0.95, p, n-p)    %9.4589


  pval = 1-fcdf( (n-p)/(p*(n-1)) * T2 , p, n-p)    %0.0208

  [Dbar - sqrt(p*(n-1)/(n-p)*finv(0.95, p, n-p))*sqrt(diag(Sd)/n)  ...
          Dbar + sqrt(p*(n-1)/(n-p)*finv(0.95, p, n-p))*sqrt(diag(Sd)/n) ]
  %[-22.4533    3.7260];   [-5.7001    32.2456]
```

The analyses are not in agreement; the p-value found is about 2%. In contrast, note that componentwise 95% CIs of $[-22.4533, 3.7260]$ and $[-5.7001, 32.2456]$ both contain 0, making it impossible to state componentwise differences. This emphasizes the need to look at the problem as multivariate.

Next, a WinBUGS solution is provided. The file is available as effluent .odc. The T^2-like statistic is formed from the posterior counterparts, so it can be viewed as a posterior Mahalonobis distance between the populations.

```
        model{
        for (i in 1:n){
        D[i,1] <- X1[i,1]-X2[i,1];    D[i,2] <- X1[i,2]-X2[i,2];
                      D[i,1:p] ~ dmnorm(mu[], T[,]) }
        T[1:p , 1:p]   ~ dwish(R[ , ], p)
        S[1:p , 1:p]   <- inverse(T[ , ])
     mu[1:p] ~ dmnorm(mn[1:p],prec[,])
        # Distance counterpart of T2
for (k in 1:p) {a[k] <- inprod(T[k,],mu[])}
             T2 <- n*inprod(mu[],a[])
     # Simulating  random F0 ~ F(p, n-p)
     num<-p; denom<-n-p
```

```
   chisq1 ~ dchisqr(num);  chisq2 ~ dchisqr(denom)
    F0 <- (chisq1/num)/(chisq2/denom)
   # Comparing distance F with F0
    F <- (n-p)*T2/(p*(n-1))
    prob <- step(F0-F)
    #prop of times distance F < F0 = "expected" under H0
    }
DATA
list(p=2, n=11,
X1 = structure(.Data=c( 6, 27,    6, 23,   18, 64,   8, 44,
                       11, 30,   34, 75,   28, 26,  71,124,
                       43, 54,   33, 30,   20, 14), .Dim=c(11,2)),
X2 = structure(.Data=c(25, 15,   28, 13,    3, 22,  35, 29,
                       15, 31,   44, 64,   42, 30,  54, 64,
                       34, 56,   29, 20,   39, 21), .Dim=c(11,2)),
R = structure(.Data = c(1, 0,
0, 1), .Dim = c(2, 2)),
mn = c(0,0),
prec = structure(.Data = c(0.0001,0,
                           0,0.0001), .Dim = c(2,2)))
INITS
#Just generate Inits
```

The output from the WinBUGS is given below. Note that Bayes' estimators μ and S are close to the classical, however the closeness measures F and $T2$ differ. The Bayesian counterpart of classical p-value is prob. Its value of 0.03869 < 5% is in agreement with the significance results in the classical approach. Notice, however, that 95% credible set for mu[2] does not contain 0, unlike the classical componentwise 95% confidence interval.

	mean	sd	MC error	val2.5pc	median	val97.5pc	start	sample
F	8.826	5.051	0.01607	1.443	8.024	20.78	1001	100000
S[1,1]	267.8	141.2	0.476	110.2	233.9	626.7	1001	100000
S[1,2]	202.5	147.5	0.5175	17.54	171.0	576.3	1001	100000
S[2,1]	202.5	147.5	0.5175	17.54	171.0	576.3	1001	100000
S[2,2]	463.4	241.6	0.7971	191.9	404.0	1090.0	1001	100000
T2	19.61	11.22	0.03571	3.206	17.83	46.19	1001	100000
mu[1]	-6.384	4.937	0.01415	-16.27	-6.374	3.442	1001	100000
mu[2]	13.21	6.476	0.02033	0.2454	13.21	26.08	1001	100000
prob	0.03869	0.1929	5.927E-4	0.0	0.0	1.0	1001	100000

10.5 Two Normal Variances

We have already seen the test for the equality of variances from two normal populations when we discussed testing the equality of two independent normal means. We will not repeat the summary table (page 433) but will discuss how to find a confidence interval for the ratio of population variances, conduct power analyses, and provide some Bayesian considerations.

Let s_1^2 and s_2^2 be sample variances based on samples $X_{11}, X_{12}, \ldots, X_{1,n_1}$ and $X_{21}, X_{22}, \ldots, X_{2,n_2}$ from normal populations $\mathcal{N}(\mu, \sigma_1^2)$ and $\mathcal{N}(\mu_2, \sigma_2^2)$, respectively. The fact that the sampling distribution of $\frac{s_1^2}{\sigma_1^2} / \frac{s_2^2}{\sigma_2^2}$ is F, with $df_1 = n_1 - 1$ and $df_2 = n_2 - 1$ degrees of freedom was used in testing the equality of variances. The same statistic and its sampling distribution lead to a $(1 - \alpha)100\%$ confidence interval for σ_1^2/σ_2^2, as

$$\left[\frac{s_1^2/s_2^2}{F_{n_1-1,n_2-1,1-\alpha/2}}, \frac{s_1^2/s_2^2}{F_{n_1-1,n_2-1,\alpha/2}} \right].$$

This follows from

$$\mathbb{P}\left(F_{n_2-1,n_1-1,\alpha/2} \leq \frac{s_2^2}{\sigma_2^2} / \frac{s_1^2}{\sigma_1^2} \leq F_{n_2-1,n_1-1,1-\alpha/2} \right) = 1 - \alpha \qquad (10.9)$$

and the property of F quantiles,

$$F_{m,n,\alpha} = \frac{1}{F_{n,m,1-\alpha}}.$$

Power Analysis for the Test of Two Variances.[*]   For the case where $F = s_1^2/s_2^2$ is observed, and n_1 and n_2 are sample sizes, Desu and Raghavarao (1990) provide an approximation of the power of the test for the variance ratio,

$$1 - \beta \approx \Phi\left(\sqrt{\frac{2(n_1 - 1)(n_2 - 2)}{n_1 + n_2 - 2}} \, |\log(F)| - z_{1-\alpha} \right), \qquad (10.10)$$

where $z_{1-\alpha}$ is the $1 - \alpha$-quantile of the standard normal distribution. If the alternative is two-sided, the quantile $z_{1-\alpha/2}$ is used instead of $z_{1-\alpha}$.

The sample size necessary to achieve a power of $1 - \beta$ if the effect $\textit{eff} = \sigma_1^2/\sigma_2^2 \neq 1$ is to be detected by a test of level α is

$$n = \left(\frac{z_{1-\alpha} + z_{1-\beta}}{\log(\textit{eff})} \right)^2 + 2. \qquad (10.11)$$

This size is for each sample, so the total number of observations is $2n$. If unequal sample sizes are desired, the reader is referred to Zar (2010) and Desu and Raghavarao (1990).

Example 10.13. **Powers and Sample Sizes for Some Variance Tests.** In the context of Example 10.1, we next approximate the power of the two-sided, 5% level test of equality of variances against the specific alternative H_1 : $\sigma_1^2/\sigma_2^2 = 1.8$. Recall that the group sample sizes were $n_1 = 12$ and $n_2 = 15$. According to (10.10),

```
n1 = 12; n2=15; alpha = 0.05; F = 1.8;
normcdf(   sqrt(2 * (n1-1)*(n2 -2)/(n1 + n2 -2  )) ...
              * abs(log(F)) - norminv(1-alpha/2) )
 % 0.5112
```

and the power is about 51%.

Next we find the retrospective power of the test in Example 10.2.

```
n1 = 12; n2=15; alpha = 0.05;
s12 = 0.004^2;   s22 = 0.006^2;
F=s12^2/s22^2; %0.1975
normcdf(   sqrt(2 * (n1-1)*(n2 -2)/(n1 + n2 -2  )) ...
            * abs(log(F)) - norminv(1-alpha/2) )
 %  0.9998
```

The retrospective power of the test is quite high at 99.98%.

If the lead exposure trial from Example 10.1 were to be repeated, we would find the sample size that would guarantee that an effect of size 1.5 would be detected with a power of 90% in a two-sided 5% level test. Here the effect is defined as the ratio of population variances that is different than 1 and of interest to detect.

```
alpha = 0.05; beta = 0.1; eff = 1.5;
n = ( (norminv(1-alpha/2) + norminv(1-beta))/log(eff))^2 + 2
% n=66 (65.9130)
```

Therefore, each group will need 66 children.

A Noninformative Bayesian Solution. If the priors on the parameters are noninformative $\pi(\mu_1) = \pi(\mu_2) = 1, \pi(\sigma_1^2) = 1/\sigma_1^2$, and $\pi(\sigma_2^2) = 1/\sigma_2^2$, one can show that the posterior distribution of $(\sigma_1^2/\sigma_2^2)/(s_1^2/s_2^2)$ is F with $n_2 - 1$ and $n_1 - 1$ degrees of freedom. Since

$$\frac{\sigma_1^2/\sigma_2^2}{s_1^2/s_2^2} = \frac{s_2^2}{\sigma_2^2} \Big/ \frac{s_1^2}{\sigma_1^2},$$

the posterior distribution for $\frac{\sigma_1^2/\sigma_2^2}{s_1^2/s_2^2}$ (σ_1^2, σ_2^2 random variables and s_1^2, s_2^2 constants) and a sampling distribution of $\frac{s_2^2}{\sigma_2^2} \Big/ \frac{s_1^2}{\sigma_1^2}$ (s_1^2, s_2^2 random variables and

σ_1^2, σ_2^2 constants) coincide. Thus, a credible set based on this posterior coincides with a confidence interval by taking into account relation (10.9). If the priors on the parameters are more general, then the MCMC method can be used.

Example 10.14. **The Discovery of Argon.** Lord Rayleigh, following an observation by Henry Cavendish, performed a series of experiments measuring the density of nitrogen and recognized that atmospheric measurements give consistently higher results than chemical measurements (measurements from ammonia, oxides of nitrogen, etc.). This discrepancy of the order of 1/100 g was too large to be explained by the measurement error, which was of the order of approx. 2/10,000 g. Rayleigh postulated that atmospheric nitrogen contains a heavier constituent, and this led to the discovery of argon in 1895 (Ramsay and Rayleigh). Rayleigh's data, published in the *Proceedings of the Royal Society* in 1893 and 1894, are provided in the table below and shown as back-to-back histograms (Fig. 10.4).

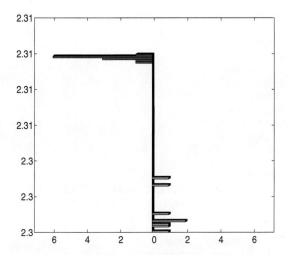

Fig. 10.4 Back-to-back histogram of Rayleigh's measurements. Measurements from the air are given on the *left-hand side*, while the measurements obtained from chemicals are on the *right*.

From air	2.31035	2.31026	2.31024	2.31012
	2.31027	2.31017	2.30986	2.31010
	2.31001	2.31024	2.31010	2.31028
From chemicals	2.30143	2.29890	2.29816	2.30182
	2.29869	2.29940	2.29849	2.29889

Assume that measurements are normal with means μ_1 and μ_2 and variances σ_1^2 and σ_2^2, respectively. Using WinBUGS and noninformative priors on the normal parameters, find 95% credible sets for

(a) $\theta = \mu_1 - \mu_2 - 0.01$ and

(b) $\rho = \frac{\sigma_2^2}{\sigma_1^2}$.

(c) Estimate the posterior probability of $\rho > 1$.

We are particularly interested in (b) and (c) since Figure 10.4 indicates that the variability in the measurements obtained from chemicals is much higher than in measurements obtained from the air.

The WinBUGS file argon.odc solves (a)–(c).

```
#Discovery of Argon
model{
  for(i in 1:n1) {
    fromair[i] ~ dnorm(mu1, prec1)
    }
  for (j in 1:n2){
    fromchem[j] ~ dnorm(mu2, prec2)
    }
  mu1 ~ dflat()
  mu2 ~ dflat()
  prec1 ~ dgamma(0.0001, 0.0001)
  prec2 ~ dgamma(0.0001, 0.0001)
  theta <- mu1 - mu2 - 0.01
  sig2air <- 1/prec1
  sig2chem <- 1/prec2
  rho <- sig2chem/sig2air
  ph1 <- step(rho - 1)
}


DATA
list(n1=12, n2 = 8,
      fromair = c(2.31035, 2.31026, 2.31024, 2.31012, 2.31027,
                  2.31017, 2.30986, 2.31010, 2.31001,
                  2.31024, 2.31010, 2.31028),
    fromchem = c(2.30143, 2.29890, 2.29816, 2.30182,
                 2.29869, 2.29940,
                 2.29849, 2.29889) )

INITS
list(mu1 = 0, mu2 = 0, prec1 = 10, prec2 = 10)
```

	mean	sd	MC error	val2.5pc	median	val97.5pc	start	sample
ph1	0.786	0.4101	4.473E-4	0.0	1.0	1.0	1001	1000000
rho	2.347	2.333	0.002684	0.4456	1.736	7.925	1001	1000000
theta	6.971E-4	0.002683	2.584E-6	−0.004626	6.983E-4	0.006017	1001	1000000

It is instructive to compare a frequentist test for the variance ratio with the WinBUGS output. The statistic $F = s_1^2/s_2^2 = 0.0099$ is strongly significant with a p-value of the order 10^{-9}. At the same time, the posterior probability of $\rho > 1$, a Bayesian equivalent to the F test, is only 0.786. The posterior probability of $\rho \leq 1$ is 0.214. Also, the 95% credible set for ρ contains 1. Even though a Bayesian would favor the hypothesis $\rho > 1$, the evidence against $\rho \leq 0$ is not as strong as in the classical approach.

10.6 Comparing Two Proportions

Comparing two population proportions is inarguably one of the most important tasks in statistical practice. For example, statistical support in clinical trials for new drugs, procedures, or medical devices almost always contains tests and confidence intervals involving two proportions: proportions of positive outcomes in control and treatment groups, proportions of readings within tolerance limits for proposed and currently approved medical devices, or proportions of cancer patients for which new and old treatment regimes manifested drug toxicity, to list just a few.

Sample proportions involve binomial distributions, and if sample sizes are not too small, the CLT implies their approximate normality. Let $X_1 \sim Bin(n_1, p_1)$ and $X_2 \sim Bin(n_2, p_2)$ be the observed numbers of "events" and $\hat{p}_1$ and $\hat{p}_2$ be the sample proportions. Then the difference $\hat{p}_1 - \hat{p}_2 = X_1/n_1 - X_2/n_2$ has an approximately normal distribution, when n_1 and n_2 are not too small, say, >20, with mean $p_1 - p_2$ and variance $p_1(1 - p_1)/n_1 + p_2(1 - p_2)/n_2$.

A Wald-type confidence interval can be constructed using this normal approximation. Specifically, the $(1 - \alpha)100\%$ confidence interval for the population proportion difference $p_1 - p_2$ is

$$\left[\hat{p}_1 - \hat{p}_2 - z_{1-\alpha/2} \sqrt{\frac{\hat{p}_1(1 - \hat{p}_1)}{n_1} + \frac{\hat{p}_2(1 - \hat{p}_2)}{n_2}}, \right.$$

$$\left. \hat{p}_1 - \hat{p}_2 + z_{1-\alpha/2} \sqrt{\frac{\hat{p}_1(1 - \hat{p}_1)}{n_1} + \frac{\hat{p}_2(1 - \hat{p}_2)}{n_2}} \right].$$

In testing $H_0 : p_1 = p_2$ against one of the alternatives, the test statistic is

$$Z = \frac{\hat{p}_1 - \hat{p}_2}{\sqrt{\frac{\hat{p}(1-\hat{p})}{n_1} + \frac{\hat{p}(1-\hat{p})}{n_2}}} = \frac{\hat{p}_1 - \hat{p}_2}{\sqrt{\hat{p}(1 - \hat{p})}\sqrt{\frac{1}{n_1} + \frac{1}{n_2}}},$$

where

$$\hat{p} = \frac{X_1 + X_2}{n_1 + n_2} = \frac{n_1}{n_1 + n_2}\hat{p}_1 + \frac{n_2}{n_1 + n_2}\hat{p}_2$$

is the pooled sample proportion. The pooled sample proportion is used since under H_0 the population proportions coincide and this common parameter should be estimated by all available data. By the CLT, the statistic Z has an approximately standard normal $\mathcal{N}(0,1)$ distribution.

Alternative	α-level rejection region	p-value
$H_1 : p_1 > p_2$	$[z_{1-\alpha}, \infty)$	`1-normcdf(z)`
$H_1 : p_1 \neq p_2$	$(-\infty, z_{\alpha/2}] \cup [z_{1-\alpha/2}, \infty)$	`2*normcdf(-abs(z))`
$H_1 : p_1 < p_2$	$(-\infty, z_\alpha]$	`normcdf(z)`

Example 10.15. **Vasectomies and Prostate Cancer.** Several studies have been conducted to analyze the relationship between vasectomy and prostate cancer. The study by Giovannucci et al. (1993) states that of 21,300 men who had not had a vasectomy, 69 were found to have prostate cancer, while of 22,000 men who had a vasectomy, 113 were found to have prostate cancer. Formulate hypotheses and perform a test at the 1% level.

```
x1=69; x2 = 113; n1 = 21300; n2 = 22000;
p1hat = x1/n1; p2hat  = x2/n2; phat = (x1 + x2)/(n1 + n2);
z=(p1hat - p2hat)/(sqrt(phat*(1-phat))*sqrt(1/n1 + 1/n2))
 % z = -3.0502
pval = normcdf(-3.0502)
 % pval = 0.0011
```

We tested $H_0 : p_1 = p_2$ versus $H_1 : p_1 < p_2$, where p_1 is the proportion of subjects with prostate cancer in the population of all subjects who had a vasectomy, while p_2 is the proportion of subjects with prostate cancer in the population of all subjects who did not have a vasectomy. Since the p-value was 0.0011, we concluded that vasectomy is a significant risk factor for prostate cancer.

The Sample Size for Testing the Two Proportions. The sample size required for a two-sided α-level test to detect the difference $\delta = |p_1 - p_2|$, with a power of $1 - \beta$, is

$$n \geq \frac{(z_{1-\alpha/2} + z_{1-\beta})^2 \times 2\bar{p}(1-\bar{p})}{\delta^2},$$

where $\bar{p} = (p_1 + p_2)/2$. The sample size n is for each group, so that the total number of observations is $2n$. If the alternative is one-sided, $z_{1-\alpha/2}$ is replaced by $z_{1-\alpha}$. This formula requires some preliminary knowledge about $\bar{p}$. In the absence of any information about the proportions, the most conservative choice for $\bar{p}$ is $1/2$.

Example 10.16. **Sample Size for Two Proportions.** An investigator believes that a control group would have an annual event rate of 30% and that the treatment would reduce this rate to 20%. She wants to design a study to be one-sided with a significance level of $\alpha = 0.05$ and a power of $1 - \beta = 0.85$. The necessary sample size per group is

```
n = 2 *(norminv(1-0.05)+norminv(1-0.15))^2 *0.25*(1-0.25)/0.1^2
        % n = 269.5988
```

This number is rounded to 270, so that the total sample size is $2 \times 270 = 540$. More precise sample sizes are

$$n' = \frac{\left(z_{1-\alpha/2}\sqrt{2\bar{p}(1-\bar{p})} + z_{1-\beta}\sqrt{p_1(1-p_1) + p_2(1-p_2)}\right)^2}{\delta^2},$$

with $z_{1-\alpha/2}$ replaced by $z_{1-\alpha}$ for the one-sided alternative.

Casagrande et al. (1978) proposed a correction to n' as

$$n'' = n'/4 \times \left(1 + \sqrt{1 + \frac{4}{n'\delta}}\right)^2,$$

while Fleiss et al. (1980) suggest $n''' = n' + \frac{2}{\delta}$.

In all three scenarios, however, preliminary knowledge about p_1 and p_2 is needed.

```
%Sample Size for Each Group:
n1 = (norminv(1-0.05) * sqrt(2 * 0.25 * 0.75)+ ...
      norminv(1-0.15) * sqrt(0.3*0.7+0.2*0.8))^2/0.1^2
%n1 = 268.2064
%Casagrande et al. 1978
n2 = n1/4 * (1 + sqrt(1 + 4/(n1 * 0.1)))^2   %n2 =287.8590
%Fleiss et al. 1980
n3 = n1 + 2/0.1     %n3 = 288.2064
```

The outputs n', n'', and n''' are rounded to 267, 288, and 289, so that the total sample sizes are 534, 576, and 578, respectively.

10.7 Risk Differences, Risk Ratios, and Odds Ratios

In epidemiological and population disease studies it is often the case that the findings are summarized as a table

	Disease (D)	No disease (C)	Total
Exposed (E)	a	b	$n_1 = a + b$
Unexposed (E^c)	c	d	$n_2 = c + d$
Total	$m_1 = a + c$	$m_2 = b + d$	$n = a + b + c + d$

In clinical trial studies, the risk factor status (E/E^c) can be replaced by a treatment/control or new treatment/old treatment, while the disease status (D/D^c) can be replaced by an improvement/nonimprovement.

Remark. In biostatistics, chiefly in the context of epidemiology, comparative studies leading to tabulated data can be *prospective* and *retrospective*. In a prospective study, a group of n disease-free individuals is identified and followed over a period of time. At the end of the study, the group, typically called the *cohort*, is assessed and tabulated with respect to disease development and exposure to the risk factor of interest.

In a retrospective study, groups of m_1 individuals with the disease (cases) and m_2 disease-free individuals (controls) are identified and their prior exposure histories are assessed. In this case, the table summarizes the numbers of exposure to the risk factor under consideration among the cases and controls. The retrospective studies are fast and less expensive, but sometimes the reliability of exposure history may be questionable.

10.7.1 Risk Differences

Let p_1 and p_2 be the population risks of a disease for exposed and unexposed (control) subjects. These are probabilities that the subjects will develop the disease during the fixed interval of time for the two groups, exposed and unexposed.

Let $\hat{p}_1 = a/n_1$ be an estimator of the risk of a disease for exposed subjects and $\hat{p}_2 = c/n_2$ be an estimator of the risk of that disease for control subjects. The $(1 - \alpha)100\%$ confidence interval for the risk difference coincides with the confidence interval for the difference of proportions from page 465:

$$\hat{p}_1 - \hat{p}_2 \ \pm \ z_{1-\alpha/2} \ \sqrt{\frac{\hat{p}_1(1-\hat{p}_1)}{n_1} + \frac{\hat{p}_2(1-\hat{p}_2)}{n_2}}.$$

Sometimes, better precision is achieved by a confidence interval with continuity corrections:

$$\left[\hat{p}_1 - \hat{p}_2 \pm (1/(2n_1) + 1/(2n_2)) - z_{1-\alpha/2} \sqrt{\frac{\hat{p}_1(1-\hat{p}_1)}{n_1} + \frac{\hat{p}_2(1-\hat{p}_2)}{n_2}}, \right.$$

$$\left. \hat{p}_1 - \hat{p}_2 \pm (1/(2n_1) + 1/(2n_2)) + z_{1-\alpha/2} \sqrt{\frac{\hat{p}_1(1-\hat{p}_1)}{n_1} + \frac{\hat{p}_2(1-\hat{p}_2)}{n_2}} \right],$$

where the sign of the correction factor $1/(2n_1) + 1/(2n_2)$ is taken as "+" if $\hat{p}_1 - \hat{p}_2 < 0$ and as "−" if $\hat{p}_1 - \hat{p}_2 > 0$. The recommended sample sizes for the validity of the interval should satisfy $\min\{n_1 p_1(1-p_1), n_2 p_2(1-p_2)\} \geq 10$. For large sample sizes, the difference between "continuity-corrected" and uncorrected intervals is negligible.

10.7.2 Risk Ratio

The risk ratio in a population is the quantity $R = p_1/p_2$. It is estimated by $r = \hat{p}_1/\hat{p}_2 = \frac{a/n_1}{c/n_2}$. The empirical distribution of r does not have a simple form, and, moreover, it is typically skewed (Fig. 10.5b). If the logarithm is taken, the risk ratio is "symmetrized," the log ratio is equivalent to the difference between logarithms, and, given the independence of populations, the CLT applies. It is evident in Figure 10.5c that the log risk ratios are approximated well by a normal distribution.

The following MATLAB code (simulrisks.m) simulates 10,000 pairs from $Bin(80, 0.21)$ and $Bin(60, 0.25)$ populations representing exposed and unexposed subjects. From each pair risks are assessed and histograms of risk differences, risk ratios, and log risk ratios are shown in Figure 10.5a–c.

```
disexposed = binornd(60, 0.25, [1 10000]);
disnonexposed = binornd(80, 0.21, [1 10000]);
p1s = disexposed/60; p2s =disnonexposed/80;
figure; hist(p1s - p2s, 25)
figure; hist(p1s./p2s, 25)
```

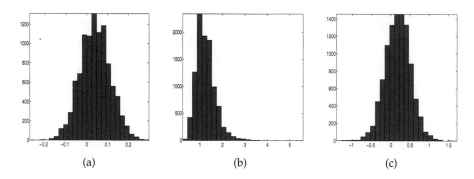

(a)   (b)   (c)

Fig. 10.5 Two samples of size 10,000 are generated from $\mathcal{B}in(80, 0.21)$ and $\mathcal{B}in(60, 0.25)$ populations and risks $\hat{p}_1$ and $\hat{p}_2$ are estimated for each pair. The panels show histograms of (a) risk differences, (b) risk ratios, and (c) log risk ratios.

```
figure; hist( log( p1s./p2s ), 25 )
```

Thus,

$$\log r \overset{\text{appr}}{\sim} \mathcal{N}(\log R, \sigma^2_{\log r}),$$

where $\sigma^2_{\log r}$ is estimated by $s^2_{\log r} = \frac{b}{an_1} + \frac{d}{cn_2}$. This estimator of variance is derived using the representation $\log r = \log \hat{p}_1 - \log \hat{p}_2$ and

$$\mathbb{V}\text{ar}\,(\log \hat{p}_1) = \left(\frac{1}{\hat{p}_1}\right)^2 \cdot \mathbb{V}\text{ar}\,\hat{p}_1 = \frac{b}{an_1},$$

by (5.17).

The $(1 - \alpha)100\%$ CI for the population log risk ratio $\log R$ is

$$\left[\log r - z_{1-\alpha/2}\sqrt{\frac{b}{an_1} + \frac{d}{cn_2}},\ \log r + z_{1-\alpha/2}\sqrt{\frac{b}{an_1} + \frac{d}{cn_2}}\right].$$

The confidence interval for the population risk ratio R is obtained by taking the exponents of the bounds:

$$\left[r\,\exp\left\{-z_{1-\alpha/2}\sqrt{\frac{b}{an_1} + \frac{d}{cn_2}}\right\},\ r\,\exp\left\{z_{1-\alpha/2}\sqrt{\frac{b}{an_1} + \frac{d}{cn_2}}\right\}\right].$$

Example 10.17. **Polio Vaccine Trial Revisited.** As an illustration, consider Example 1.4 on page 5. The risk ratio p_{vac}/p_{pla} was estimated to be $\frac{33/200745}{115/201229} = 0.2876$. This represents almost a fourfold reduction in risk of polio for vaccinated children in comparison with the placebo. The 95% confidence interval for population R is $[0.1953, 0.4236]$, which is well separated from 1. Recall, risk ratio of 1 represents no effect, and in this case the effect of vaccine is significant.

✐

10.7.3 Odds Ratios

For a particular proportion, p, the odds are defined as $\frac{p}{1-p}$, see page 85. For two proportions p_1 and p_2, the odds ratio is defined as $O = \frac{p_1/(1-p_1)}{p_2/(1-p_2)}$, and its sample counterpart is $o = \frac{\hat{p}_1/(1-\hat{p}_1)}{\hat{p}_2/(1-\hat{p}_2)} = \frac{ad}{bc}$

As evident in Figure 10.6, the odds ratio is symmetrized by the log transformation, and it is the log domain where the normal approximations are used. The sample variance for $\log o$ is $s_{\log o}^2 = \frac{1}{a} + \frac{1}{b} + \frac{1}{c} + \frac{1}{d}$. The derivation of $s_{\log o}^2$ is beyond the scope of this text, and it is due to Woolf (1955).

The $(1 - \alpha)100\%$ confidence interval for the log odds ratio is

$$\left[\log o - z_{1-\alpha/2}\sqrt{\frac{1}{a} + \frac{1}{b} + \frac{1}{c} + \frac{1}{d}},\ \log o + z_{1-\alpha/2}\sqrt{\frac{1}{a} + \frac{1}{b} + \frac{1}{c} + \frac{1}{d}}\right].$$

Of course, the confidence interval for the odds ratio is obtained by taking the exponents of the bounds:

$$\left[o \exp\left\{-z_{1-\alpha/2}\sqrt{\frac{1}{a} + \frac{1}{b} + \frac{1}{c} + \frac{1}{d}}\right\},\ o \exp\left\{z_{1-\alpha/2}\sqrt{\frac{1}{a} + \frac{1}{b} + \frac{1}{c} + \frac{1}{d}}\right\}\right].$$

For small sample sizes replacing counts $a, b, c,$ and d by $a + 1/2, b + 1/2, c + 1/2,$ and $d + 1/2$ leads to a more stable inference.

Some researchers argue that only odds ratios should be reported and used due to their superior properties over risk differences and risk ratios (Edwards, 1963; Mosteller, 1968). When risks $\hat{p}_1$ and $\hat{p}_2$ are small, the risk ratio and odds ratio are close, as in Example 1.4 where they almost coincide.

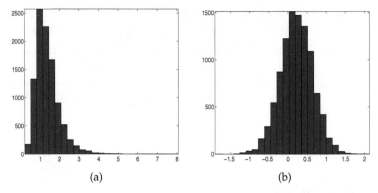

(a)   (b)

Fig. 10.6 For the data leading to Figure 10.5, the histograms of (a) odds ratios and (b) log odds ratios are shown.

However, if $\hat{p}_1$ and $\hat{p}_2$ are close to 1, risk differences and risk ratios can be very different. For example, if $\hat{p}_1 = 0.99$ and $\hat{p}_2 = 0.97$, $r = 1.0206$, but $o = 3.0619$.

The following table summarizes the estimators of three risk parameters:

	Risk difference	Risk ratio	Odds ratio
Parameter	$D = p_1 - p_2$	$R = p_1/p_2$	$O = \frac{p_1/(1-p_1)}{p_2/(1-p_2)}$
Estimator	$d = a/n_1 - c/n_2$	$r = (a/n_1)/(c/n_2)$	$o = (ad)/(bc)$
St. deviation	$s_d = \sqrt{\frac{ab}{n_1^3} + \frac{cd}{n_2^3}}$	$s_{\log r} = \sqrt{\frac{b}{an_1} + \frac{d}{cn_2}}$	$s_{\log o} = \sqrt{\frac{1}{a} + \frac{1}{b} + \frac{1}{c} + \frac{1}{d}}$

Interpretation of values for r or o are provided in the following table:

Value in	Effect of exposure
$[0,0.4)$	Strong benefit
$[0.4,0.6)$	Moderate benefit
$[0.6,0.9)$	Weak benefit
$[0.9,1.1]$	No effect
$(1.1,1.6]$	Weak hazard
$(1.6,2.5]$	Moderate hazard
> 2.5	Strong hazard

Example 10.18. **Framingham Data.** The table below gives the coronary heart disease status after 18 years, by level of systolic blood pressure (SBP). Levels of SBP ≥ 165 are considered as exposure to a risk factor.

SBP (mmHg)	Coronary disease	No coronary disease	Total
≥ 165	95	201	296
< 165	173	894	1067
Total	268	1095	1363

Find 95% confidence intervals for the risk difference, risk ratio, and odds ratio. The function ◢ risk.m calculates confidence intervals for risk differences, risk ratios, and odds ratios and will be used in this example.

```
function [stats] = risk(matr, alpha)
    %
    % Example of Use: Framingham Data
    % > [stats] = risk([95  201; 173  894])
    a=matr(1,1); b=matr(1,2); c=matr(2,1); d=matr(2,2);
    %
    if nargin < 5
        alpha=0.05;
    end
    %--------
    n1 = a + b;
    n2 = c + d;
    hatp1 = a/n1; hatp2 = c/n2;
    stats.p1 = hatp1;  stats.p2 = hatp2;
    %---------risk difference (d) and CI [rdl, rdu] ---------
    rd = hatp1 - hatp2;
    stdrd = sqrt(hatp1 * (1-hatp1)/n1 + hatp2 * (1- hatp2)/n2 );
    rdl = rd - norminv(1-alpha/2) * stdrd;
    rdu = rd + norminv(1-alpha/2) * stdrd;
    stats.rd=rd;
    stats.rdint = [rdl rdu];
    %---------risk ratio (r) and CI [rrl, rru] ---------
    rr = hatp1/hatp2;
    lrr = log(rr);
    stdlrr = sqrt(b/(a * n1) + d/(c*n2));
    lrrl = lrr - norminv(1-alpha/2)*stdlrr;
      rrl = exp(lrrl);
    lrru =  lrr + norminv(1-alpha/2)*stdlrr;
      rru = exp(lrru);
      stats.rr = rr;
      stats.rrint =[rrl rru];
    %---------odds ratio (o) and CI [orl, oru] -----------
    or = ( hatp1/(1-hatp1) )/(hatp2/(1-hatp2));
    lor = log(or);
    stdlor = sqrt(1/a + 1/b + 1/c + 1/d);
    lorl = lor - norminv(1-alpha/2)*stdlor;
      orl = exp(lorl);
    loru =  lor + norminv(1-alpha/2)*stdlor;
      oru = exp(loru);
      stats.or=or;
      stats.orint=[orl oru];
```

The solution is:

```
◢  [stats] = risk([95 201; 173 894])

%stats =
%       p1: 0.3209
%       p2: 0.1621
```

```
%        rd: 0.1588
%     rdint: [0.1012 0.2164]
%        rr: 1.9795
%     rrint: [1.5971 2.4534]
%        or: 2.4424
%     orint: [1.8215 3.2750]
```

Estimation of risk parameters in Bayesian fashion is conceptually straight-forward: posterior distributions for d, r and o are obtained and analyzed.

Example 10.19. **Retrospective Analysis of Smoking Habits.** This example is adopted from Johnson and Albert (1999), who use data collected in a study by Dorn (1954). A sample of 86 lung-cancer patients and a sample of 86 controls were questioned about their smoking habits. The two groups were chosen to represent random samples from a subpopulation of lung-cancer patients and an otherwise similar population of cancer-free individuals. Of the cancer patients, 83 out of 86 were smokers; among the control group, 72 out of 86 were smokers. The scientific question of interest was to assess the difference between the smoking habits in the two groups. Uniform priors on the population proportions were used as a noninformative choice.

```
model{
for(i in 1:2){
   r[i] ~ dbin(p[i],n[i])
   p[i] ~ dunif(0,1)
         }
RD <- p[1] - p[2]
RD.gt0 <- step(RD)

   RR <- p[1]/p[2]
   RR.gt1 <- step(RR - 1)

OR <-  (p[1]/(1-p[1]))/(p[2]/(1-p[2]))
OR.gt1 <- step(OR - 1)
}

DATA
list(r=c(83,72),n=c(86,86))
INITS
   #Generate Inits
```

	mean	sd	MC error	val2.5pc	median	val97.5pc	start	sample
OR	5.818	4.556	0.01398	1.556	4.613	17.29	1001	100000
OR.gt1	0.9978	0.04675	1.469E-4	1.0	1.0	1.0	1001	100000
RD	0.125	0.0455	1.478E-4	0.0385	0.1237	0.2179	1001	100000
RD.gt0	0.9978	0.04675	1.469E-4	1.0	1.0	1.0	1001	100000
RR	1.153	0.06276	2.038E-4	1.044	1.148	1.291	1001	100000
RR.gt1	0.9978	0.04675	1.469E-4	1.0	1.0	1.0	1001	100000
p[1]	0.9546	0.02209	7.06E-5	0.9022	0.958	0.9873	1001	100000
p[2]	0.8296	0.03991	1.26E-4	0.7444	0.8322	0.9002	1001	100000

Note that 95% credible sets for the risk ratio and odds ratio are above 1, and that the set for the risk difference does not contain 0. By all three measures the proportion of smokers among subjects with cancer is significantly larger than the proportion among the controls. In Bayesian testing the hypotheses $H_1' : p_1 > p_2$, $H_1'' : p_1/p_2 > 1$, and $H_1''' : \frac{p_1}{1-p_1} \big/ \frac{p_2}{1-p_2} > 1$ have posterior probabilities of 0.9978 each. Therefore, in this retrospective study, smoking status is indicated as a significant risk factor for lung cancer.

✐

10.7.4 Two Proportions from a Single Sample

The previous analysis for difference between population proportions assumes that the samples are independent. However, in some situations only one sample is available, and we are interested in inference about proportions of two attributes estimated form a single sample. For example, in a sample of 578 school age children, 466 are vaccinated against hepatitis B and 477 against varicella. We are interested in inference about the difference of the corresponding population proportions, $p_1 - p_2$. The standard deviation of $\hat{p}_1 - \hat{p}_2$ is equal to

$$\sigma_{\hat{p}_1 - \hat{p}_2} = \sigma_{\hat{p}_1' - \hat{p}_2'} = \sqrt{\frac{p_1' + p_2' - (p_1' - p_2')^2}{n}}, \tag{10.12}$$

where p_1' is the proportion of children who are vaccinated against hepatitis B but not against varicella, while p_2' is the population proportion of children vaccinated against varicella but not against hepatitis B. It is easy to verify $\hat{p}_1 - \hat{p}_2 = \hat{p}_1' - \hat{p}_2'$, The plug-in estimator of $\sigma_{\hat{p}_1 - \hat{p}_2}$ is

$$s_{\hat{p}_1 - \hat{p}_2} = \sqrt{\frac{\hat{p}_1' + \hat{p}_2' - (\hat{p}_1' - \hat{p}_2')^2}{n}}, \tag{10.13}$$

Thus, for a correct inference in the children's vaccination example, the marginal sums 466 and 477 are not sufficient. It is necessary to know the number of children who are vaccinated against both hepatitis B and varicella, since when this is known, the proportions $\hat{p}_1'$ and $\hat{p}_2'$ are unique and can be easily found. Suppose the number of children vaccinated against both hepatitis B and varicella is 458. Using this information we find $\hat{p}_1' = (466 - 458)/578 = 8/578$ and $\hat{p}_2' = (477 - 458)/578 = 19/578$. The standard deviation of $\hat{p}_1' - \hat{p}_2'$ is 0.0228. The 95% confidence interval for the difference is

$$[0.019 - 1.96 \cdot 0.0228, 0.019 + 1.96 \cdot 0.0228] = [0.0015, 0.0366].$$

Note that this interval does not contain 0, so we infer that the difference is significant.

If the proportions are treated as independent, the 95% CI is $[-0.0257, 0.0637]$ from which we would infer that the difference is not significant. This would be wrong.

This example is intimately related to McNemar's test discussed in Chapter 12. In the context of paired tables, each individual gives a pair of indices: indicators of hepatitis B and varicella vaccinations. This results in the following table:

		Varicella		
		Yes	No	Total
	Yes	458	8	**466**
Hepatitis B	No	19	93	112
	Total	**477**	101	578

Here the entry 458 is often called "pivotal cell." The "bolded" marginal sums are fixed.

If we are interested in testing $H_0 : p_1 = p_2$, a slight modification to $s_{\hat{p}_1 - \hat{p}_2}$ is needed. In this case the estimator for the standard deviation of proportion difference is

$$s_{\hat{p}_1 - \hat{p}_2} = \sqrt{\frac{\hat{p}_1' + \hat{p}_2'}{n}}, \tag{10.14}$$

since under H_0 the difference $p_1' - p_2' = p_1 - p_2$ is 0. For the two-sided alternative the p-value is 0.0343, which agrees with conclusions based on a CI.

The MATLAB file ◀ vaccination.m constructs a range of paired tables consistent with the marginal sums 466 and 477. Some of these tables result in significant differences among proportions and some do not, emphasizing the need to know the "pivotal cell."

Figure 10.7 shows the upper and lower bounds of a 95% CI for the problem for possible values of a "pivotal cell" that ranges between 365 and 466 under the fixed marginal sums. Note that when the pivotal cell exceeds 455 the corresponding CI does not contain 0. The intervals that contain 0 are presented with black end points while the intervals that do not contain 0 and lead to significant differences are plotted in red.

Example 10.20. **DNA from Spider Monkey.** An 8,196-long nucleotide sequence from spider monkey *Ateles geoffroyi* was obtained from the EMLB Nucleotide sequence alignment DNA database. It is of interest to test whether the proportion of A nucleotides, p_1, is the same as the proportion of T nucleotides, p_2, against the one-sided alternative $H_1 : p_1 < p_2$. Since the samples are not independent, equation (10.14) for finding the estimator

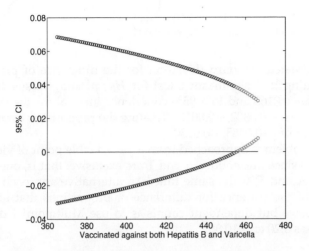

Fig. 10.7 Confidence interval for the difference in vaccination proportions. If the pivotal cell consistent with marginal sums is 456 or larger, the confidence interval does not contain 0 (plotted with red end-points).

of variance of the difference of proportions needs to be used. An annotated part of the script ◢ dnatest2.m contains the code for the test and finds a CI.

```
%excerpt from dnatest2.m
dna = ...
['gatctcttcttcgtggttagtattctttgtgttctgctaaagaactctttgcctaccctc'...
 'aggtttgcttaggtctacaatctactgtgagttgattttttttaatggacagtgagatgta'...
 'ggaattgtttctttcttatggatttccagtcattcagcaccatttattaaaaagagcccc'...
 ...
 'ctgtgagcccctggtcttggtgccttctgtgttatTaaatcaggtgttcctgtgtgtctat'...
 'agcgaagcccagatcccagcctgggagactctccagacaggccagcatcctccaccgggggg'...
 'gtgtgctacccccgccatggccatgctgacctccgcccctcggtggggactgccctgcct'...
 'gggacaggggccccagctggtttttccctgtg'];
% Point estimates
n = length(dna);                        %8192
pahat = sum( dna == 'a')/n;    %0.2144
pthat = sum( dna == 't')/n;    %0.2301

% H0: pa = pt    H1: pa < pt
% Var(pahat - pthat) = pa*qa/n + pt*qt/n + 2*pa*pt/n
%                    = [pa + pt - (pa - pt)^2]/n
%                    = (pa + pt)/n  [under (a) H0]
Zat = sqrt(n) * (pahat - pthat)/sqrt(pahat + pthat); %-2.1379
pat = normcdf(Zat); %0.0163
%--------------------------------
% For confidence intervals we use
% std(pahat - pthat)=sqrt [pa + pt - (pa - pt)^2]/n ,
% evaluated at pa=pahat and pt=pthat.
sat=sqrt((pahat + pthat - (pahat - pthat)^2)/n) %0.0074
```

```
[pahat - pthat - sat*norminv(0.975)     ...
     pahat - pthat + sat*norminv(0.975)]
  % -0.0302     -0.0013
```

The estimated standard deviation for the difference of proportions is 0.0074 leading to a significant z-test for $H_0 : p_1 = p_2$ versus $H_1 : p_1 < p_2$ with p-value 0.0163 and to a 95% confidence interval for the difference of proportions as $[-0.0302, -0.0013]$. Treating the proportions as independent leads to a CI of $[-0.0285, -0.0030]$.

If this problem is approached from a paired table point of view, the pivotal cell is 0 since nucleotides A and T are exclusive; that is, one nucleotide cannot be A and T at the same time. An alternative approach to find the estimator of the variance for difference of proportions that are obtained from exclusive but dependent counts is to use Multinomial distribution, see Exercises 5.20 and 10.30.

10.8 Two Poisson Rates*

There are several methods for devising confidence intervals on the differences or the ratios of two Poisson rates. We will focus on the method for the ratio that modifies well-known binomial confidence intervals.

Let $X_1 \sim \mathcal{P}oi(\lambda_1 t_1)$ and $X_2 \sim \mathcal{P}oi(\lambda_2 t_2)$ be two Poisson counts with rates λ_1 and λ_2 observed during time intervals of length t_1 and t_2. We are interested in the confidence interval for the ratio $\lambda = \lambda_1 / \lambda_2$.

Since X_1, given the sum $X_1 + X_2 = n$, is binomial $\mathcal{B}in(n, p)$ with $p = \frac{\lambda_1 t_1}{\lambda_1 t_1 + \lambda_2 t_2}$ (Exercise 5.7), the strategy is to find the confidence interval for p and, from its confidence bounds LB_p and UB_p, work out the bounds for the ratio λ :

$$LB_\lambda = \frac{LB_p}{1 - LB_p} \frac{t_2}{t_1}, \quad UB_\lambda = \frac{UB_p}{1 - UB_p} \frac{t_2}{t_1}.$$

For finding the LB_p and UB_p several methods were covered in Chapter 7. Note that in Chapter 7, $\hat{p} = X_1 / n$ and $n = X_1 + X_2$.

The design question can be addressed as well, but the sample size formulation needs to be expressed in terms of sampling durations t_1 and t_2. The sampling time frames t_1 and t_2, if assumed equal, can be determined on the basis of elicited precision for the confidence interval and preliminary estimates of the rates. Let $\bar{\lambda}_1$ and $\bar{\lambda}_2$ be pre-experimental assessments of the rates, and let the precision be elicited in the form of (a) the length of the interval $UB_\lambda - LB_\lambda = w$ or (b) the ratio of the bounds $UB_\lambda/LB_\lambda = w$.

Then, for achieving $(1-\alpha)100\%$ confidence with an interval of length w, the sampling time frame required is

(a)

$$t \quad (= t_1 = t_2) = \frac{z_{1-\alpha/2}^2 \left(1/\overline{\lambda}_1 + 1/\overline{\lambda}_2\right)}{\arcsin\left(\frac{\overline{\lambda}_2}{\overline{\lambda}_1} \times \frac{w}{2}\right)}$$

and

(b)

$$t \quad (= t_1 = t_2) = \frac{4z_{1-\alpha/2}^2 \left(1/\overline{\lambda}_1 + 1/\overline{\lambda}_2\right)}{\log^2(w)}.$$

Example 10.21. **Wire Failures.** Price and Bonett (2000) provide an example with data from Gardner and Ringlee (1968), who found that bare wire had $X_1 = 69$ failures in a sample of $t_1 = 1,079.6$ thousand foot-years, and a polyethylene-covered tree wire had $X_2 = 12$ failures in a sample of $t_2 = 467.9$ thousand foot-years. We are interested in a 95% confidence interval for the ratio of population failure rates.

The associated MATLAB file ◢ ratiopoissons.m calculates the 95% confidence interval for the ratio $\lambda = \lambda_1/\lambda_2$ using Wilson's proposal ("add two successes and two failures"). There, $\hat{p} = (X_1 + 2)/(n + 4)$, and the interval for p is $[0.7564, 0.9141]$. After transforming the bounds to the λ domain, the final interval is $[1.3461, 4.6147]$.

Suppose that we want to replicate this study using a new shipment of each type of wire. We want to estimate the failure rate ratio with 99% confidence and $UB_\lambda/LB_\lambda = 2$. Using $\overline{\lambda}_1 = 69/1079.6 = 0.0691$ and $\overline{\lambda}_2 = 12/467.9 = 0.0833$ as our planning estimates of λ_1 and λ_2, we would sample $t = \frac{4(2.5758)^2 \, (1/0.0691 + 1/0.0833)}{\log^2(2)} = 3018$ foot-years from each shipment. If we want to complete the study in k years, then we would sample $3018/k$ linear feet of wire from each shipment.

✐

◢
```
%CI for Ratio of Two Poissons
    X1=69; t1 = 1079.6;
    X2=12; t2=467.9;
    n=X1 + X2;
    phat = X1/n; %0.8519
    phat1 = (X1 +2)/(n + 4); %0.8353
    qhat1 = 1 - phat1;        %0.1647
% Agresti-Coull CI for prop was selected.
LBp=phat1-norminv(0.975)*sqrt(phat1*qhat1/(n+4)) %0.7564
```

```
UBp=phat1+norminv(0.975)*sqrt(phat1*qhat1/(n+4)) %0.9141
LBlam = LBp/(1 - LBp) * t2/t1; %back to lambda
UBlam = UBp/(1 - UBp) * t2/t1;
[LBlam, UBlam]        %[1.3461    4.6147]
%Frame size in Poisson Sampling
    lambar1 = 69/1079.6;  %0.0639
    lambar2 = 12/467.9;   %0.0256
    w = 2;
 td =4* norminv(0.995)^2 *(1/lambar1+1/lambar2)/...
     (asin(lambar2/lambar1 * w/2));        %3511.8
 tr =  4 * norminv(0.995)^2 *...
 ( 1/lambar1 + 1/lambar2 )/(log(w))^2;   %3018.1
```

Cox (1953) gives an approximate test and confidence interval for the ratio that uses an F-distribution. He shows that the statistic

$$F = \frac{t_1 \lambda_1}{t_2 \lambda_2} \frac{X_2 + 1/2}{X_1 + 1/2}$$

has an approximate F-distribution with $2X_1 + 1$ and $2X_2 + 1$ degrees of freedom. From this, an approximate $(1 - \alpha)100\%$ confidence interval for λ_1 / λ_2 is

$$\left[\frac{t_2}{t_1} \frac{X_1 + 1/2}{X_2 + 1/2} F_{2X_1+1,2X_2+1,\alpha/2}, \quad \frac{t_2}{t_1} \frac{X_1 + 1/2}{X_2 + 1/2} F_{2X_1+1,2X_2+1,1-\alpha/2} \right].$$

In the context of Example 10.21, the 95% confidence interval for the ratio λ_1 / λ_2 is $[1.3932, 4.7497]$.

```
%Cox
LBlamc= t2/t1*(X1+1/2)/(X2+1/2)*finv(0.025, 2*X1+1, 2*X2+1);
UBlamc= t2/t1*(X1+1/2)/(X2+1/2)*finv(0.975, 2*X1+1, 2*X2+1);
[LBlamc, UBlamc]   %1.3932    4.7497
```

Note that this interval does not contain 1, which is equivalent to a rejection of $H_0 : \lambda_1 = \lambda_2$ in a 5%-level test against the two-sided alternative.

The test of $H_0 : \lambda_1 = \lambda_2$ can be conducted using the statistic

$$F = \frac{t_1}{t_2} \frac{X_2 + 1/2}{X_1 + 1/2},$$

which under H_0 has an F-distribution with $df_1 = 2X_1 + 1$ and $df_2 = 2X_2 + 1$ degrees of freedom.

Alternative	α-level rejection region	p-value
$H_1 : \lambda_1 < \lambda_2$	$[F_{df_1,df_2,1-\alpha}, \infty)$	1-fcdf(F,df1,df2)
$H_1 : \lambda_1 \neq \lambda_2$	$[0, F_{df_1,df_2,\alpha/2}] \cup [F_{df_1,df_2,1-\alpha/2}, \infty)$	2*fcdf(min(F,1/F),df1,df2)
$H_1 : \lambda_1 > \lambda_2$	$[0, F_{df_1,df_2,\alpha}]$	fcdf(F,df1,df2)

In Example 10.21, the failure rate λ_1 for the bare wire is found to be significantly larger, with a p-value of 0.00066, than that of polyethylene-covered wire, λ_2.

```
%test against H_1: lambda1 > lambda2
pval =fcd(t1/t2*(X2+1/2)/(X1+1/2), 2*X1 + 1, 2*X2 + 1)
%6.6417e-004
```

10.9 Equivalence Tests*

In standard testing of two means, the goal is to show that one population mean is significantly smaller, larger, or different than the other. The null hypothesis is that there is no difference between the means. By not rejecting the null, the equality of means is not established – the test simply did not find enough statistical evidence for the alternative hypothesis. Absence of evidence is not evidence of absence.

United States Pharmacopeia (USP) <1033> Biological Assay Validation guidelines for compliance testing recommend equivalence testing, stating:

> "... a standard statistical approach used to demonstrate conformance to expectation and is called an equivalence test. It should not be confused with the practice of performing a significance test, such as a t-test, which seeks to establish a difference from some target value (e.g., 0% relative bias). A significance test associated with a p-value > 0.05 (equivalent to a confidence interval that includes the target value for the parameter) indicates that there is insufficient evidence to conclude that the parameter is different from the target value. This is not the same as concluding that the parameter conforms to its target value. The study design may have too few replicates, or the validation data may be too variable to discover a meaningful difference from target. Additionally, a significance test may detect a small deviation from target that is practically insignificant."

Thus, in many situations (drug and medical procedure testing, device performance, etc.), one wishes to test the equivalence hypothesis, which states that the population means or population proportions differ for no more than a small tolerance value preset by a regulatory agency. If, for example, manufacturers of a generic drug are able to demonstrate bioequivalence to the brand-name product, they do not need to conduct costly clinical trials in order to demonstrate the safety and efficacy of their generic product. More important, established bioequivalence protects the public from unsafe or ineffective drugs.

In this kind of inference it is desired that *no difference* constitutes the research hypothesis H_1 and that significance level α relates to the probability of falsely rejecting the hypothesis that there is a difference when in fact the means are equivalent. In other words, we want to control the type I error and design the power properly in this context.

In drug equivalence testing, typical measurements are the area under the concentration curve (AUC) or maximum concentration (C_{max}). The two drugs are bioequivalent if the population means of the AUC and C_{max} are sufficiently close.

Let η_T denote the population mean AUC for the generic (test) drug and let η_R denote the population mean for the brand-name (reference) drug. We are interested in testing

$$H_0 : \eta_T/\eta_R < \delta_L \quad \text{or} \quad \eta_T/\eta_R > \delta_U \qquad \text{versus} \qquad H_1 : \delta_L \le \eta_T/\eta_R \le \delta_U,$$

where δ_L and δ_U are the lower and upper tolerance limits, respectively. The FDA recommends $\delta_L = 4/5$ and $\delta_U = 5/4$ (FDA, 2001).

This hypothesis can be tested in the domain of original measurements (Berger and Hsu, 1996) or after taking the logarithm. This second approach is more common in practice since (i) AUC and C_{max} measurements are consistent with the lognormal distribution (the pharmacokinetic rationale based on multiplicative compartmental models) and (ii) normal theory can be applied to logarithms of observations. The FDA also recommends a log-transformation of data by providing three rationales: clinical, pharmacokinetic, and statistical (FDA, 2001, Appendix D).

Since for lognormal distributions the mean η is connected with the parameters of the associated normal distribution, μ and σ^2 (page 265), by assuming equal variances, we get $\eta_T = \exp\{\mu_T + \sigma^2/2\}$ and $\eta_R = \exp\{\mu_R + \sigma^2/2\}$. The equivalence hypotheses for the log-transformed data now take the form

$$H_0 : \mu_T - \mu_R \le \theta_L \quad \text{or} \quad \mu_T - \mu_R \ge \theta_U, \qquad \text{versus} \qquad H_1 : \theta_L < \mu_T - \mu_R < \theta_U,$$

where $\theta_L = \log(\delta_L)$ and $\theta_U = \log(\delta_U)$ are known constants. Note that if $\delta_U = 1/\delta_L$, then the bounds θ_L and θ_U are symmetric about zero, $\theta_L = -\theta_U$.

Equivalence testing is an active research area, and many classical and Bayesian solutions exist, as dictated by experimental designs in practice. The monograph by Wellek (2010) provides comprehensive coverage. We focus only on the case of testing the equivalence of two population means when the unknown population variances are the same.

TOST. Schuirmann (1981) proposed two one-sided tests (TOSTs) for testing bioequivalence. Two t-statistics are calculated:

$$t_L = \frac{\overline{X}_T - \overline{X}_R - \theta_L}{s_p \sqrt{1/n_1 + 1/n_2}} \quad \text{and} \quad t_U = \frac{\overline{X}_T - \overline{X}_R - \theta_U}{s_p \sqrt{1/n_1 + 1/n_2}},$$

where $\overline{X}_T$ and $\overline{X}_R$ are test and reference means, n_1 and n_2 are test and reference sample sizes, and s_p is the pooled sample standard deviation, as on page 431. Note that here, the test statistic involves the acceptable bounds

θ_L and θ_U in the numerator, unlike the standard two-sample t-test, where the numerator would be $\overline{X}_T - \overline{X}_R$.

The TOST is now carried out as follows.

(i) Using the statistic t_L, test $H_0' : \mu_T - \mu_R = \theta_L$ versus $H_1' : \mu_T - \mu_R > \theta_L$.

(ii) Using the statistic t_U, test $H_0'' : \mu_T - \mu_R = \theta_U$ versus $H_1'' : \mu_T - \mu_R < \theta_U$.

(iii) Reject H_0 at level α; that is, declare the drugs equivalent if *both* hypotheses H_0' and H_0'' are rejected at level α, which is, if

$$t_L > t_{n_1+n_2-2,1-\alpha} \quad \text{and} \quad t_U < t_{n_1+n_2-2,\alpha}.$$

Equivalently, if p_L and p_U are the p-values associated with statistics t_L and t_U, H_0 is rejected when $\max\{p_L, p_U\} < \alpha$.

Westlake's Confidence Interval. An equivalent methodology to test for equivalence is Westlake's confidence interval (Westlake, 1976). Bioequivalence is established at significance level α if a t-interval of confidence $(1 - 2\alpha)100\%$ is contained in the interval (θ_L, θ_U):

$$\left[\overline{X}_T - \overline{X}_R - t_{n_1+n_2-2,1-\alpha} \, s_p \, \sqrt{1/n_1 + 1/n_2}, \right.$$
$$\left. \overline{X}_T - \overline{X}_R + t_{n_1+n_2-2,1-\alpha} \, s_p \, \sqrt{1/n_1 + 1/n_2} \right] \in (\theta_L, \theta_U).$$

Here, the usual $t_{n_1+n_2-2,1-\alpha/2}$ is replaced by $t_{n_1+n_2-2,1-\alpha}$, and Westlake's interval coincides with the standard $(1 - 2\alpha)100\%$ confidence interval for a difference of normal means.

Example 10.22. **Equivalence of Generic and Brand-Name Drugs.** A manufacturer wishes to demonstrate that its generic drug for a particular metabolic disorder is equivalent to a brand-name drug. One indication of the disorder is an abnormally low concentration of levocarnitine, an amino acid derivative, in the plasma. Treatment with the brand-name drug substantially increases this concentration.

A small clinical trial is conducted with 43 patients, 18 in the brand-name drug arm and 25 in the generic drug arm. The following increases in the log-concentration of levocarnitine are reported:

Increase for brand-name drug	7 8 4 6 10 10 5 7 9 8
	6 7 8 4 6 10 8 9
Increase for generic drug	6 7 5 9 5 5 3 7 5 10
	2 5 8 4 4 8 6 11 7 5
	5 5 7 4 6

The FDA declares that bioequivalence among the two drugs can be established if the difference in response to the two drugs is within two units of the log-concentration. Assuming that the log-concentration measurements follow normal distributions with equal population variance, can these two drugs be declared bioequivalent within a tolerance of ± 2 units?

```
brandname = [7    8    4    6   10   10    5    7    9 ...
             8    6    7    8    4    6   10    8    9 ];
generic   = [6    7    5    9    5    5    3    7    5 ...
            10    8    5    8    4    4    8    6   11 ...
             7    5    5    5    7    4    6              ];

            xbar1 = mean(brandname)    %7.3333
            xbar2 = mean(generic)      %6.2000

            s1 = std(brandname)        %1.9097
            s2 = std(generic)          %1.9791

            n1 = length(brandname)     %18
            n2 = length(generic)       %25
%
sp = sqrt( ((n1-1)*s1^2 + (n2-1)*s2^2)/(n1 + n2 - 2))  % 1.9506
tL = (xbar1 - xbar2 -(-2))/(sp * sqrt( 1/n1 + 1/n2 ))  % 5.1965
tU = (xbar1 - xbar2 - 2 )/(sp * sqrt( 1/n1 + 1/n2 ))   %-1.4373
pL = 1-tcdf(tL, n1+ n2  - 2)    %2.9745e-006
pU = tcdf(tU, n1 + n2 - 2)      %0.0791
max(pL, pU)                     %0.0791  > 0.05, no equivalence

alpha = 0.05;
[xbar1-xbar2 - tinv(1-alpha, n1+n2-2)*sp*sqrt(1/n1+1/n2),...
    xbar1-xbar2 + tinv(1-alpha, n1+n2-2)*sp*sqrt(1/n1+1/n2)]
%  0.1186    2.1481
%  (0.1186, 2.1481) is not contained in (-2,2), no equivalence
```

Note that the equivalence of the two drugs was not established. The TOST did not simultaneously reject null hypotheses H_0' and H_0'', or, equivalently, Westlake's interval failed to be fully included in the preset tolerance interval $(-2,2)$.

Bayesian Solution. A Bayesian solution is conceptually straightforward. One finds the posterior distribution for the difference of the means, and evaluates the posterior probability of this difference falling in the interval $(-2,2)$. The posterior probability of $(-2,2)$ should be close to 1 (say, 0.95) in order to declare equivalence.

```
model{
for(i in 1:n) {
  increase[i] ~ dnorm(mu[type[i]],  prec)
  }
mu[1] ~ dnorm( 10, 0.00001)
mu[2] ~ dnorm( 10, 0.00001)
mudiff <- mu[1]-mu[2]
prec ~ dgamma(0.001, 0.001)
```

```
probint <- step( mudiff + 2) * step(2 - mudiff)
}

DATA
list( n=43, increase = c(7, 8, 4, 6, 10, 10, 5, 7, 9,
    8, 6, 7, 8, 4, 6, 10, 8, 9, 6, 7, 5, 9, 5, 5, 3, 7, 5,
    10, 8, 5, 8, 4, 4, 8, 6, 11, 7, 5, 5, 5, 7, 4, 6 ),
type = c(1, 1, 1, 1, 1, 1, 1, 1, 1, 1, 1, 1, 1, 1, 1, 1, 1, 1,
    2, 2, 2, 2, 2, 2, 2, 2, 2, 2, 2, 2, 2, 2, 2, 2, 2, 2,
    2, 2, 2, 2, 2, 2, 2))

INITS
list( mu = c(10, 10), prec = 1)
```

	mean	sd	MC error	val5.0pc	median	val95.0pc	start	sample
mudiff	1.133	0.6179	6.238E-4	0.117	1.133	2.147	10001	1000000
probint	0.9213	0.2693	2.766E-4	0.0	1.0	1.0	10001	1000000

The Bayesian analysis closely matches the findings by TOST and West-lake's interval. Note that the posterior probability of the tolerance interval $(-2,2)$ is 0.9213, short of 0.95. Also, the 90% credible set $(0.117, 2.147)$ is close to Westlake's interval $(0.1186, 2.1481)$. This closeness is expected because of the noninformative priors on the means and precision.

✐

10.10 Exercises

10.1. **Testing Piaget.** Two groups of elementary school students are taught mathematics by two different methods: traditional (group 1) and small group interactive teaching by discovery based on Piagetian theory (group 2). The results of a learning test are analyzed to test the difference in mean scores using the two methods. Group 1 had 16 students, while group 2 had 14 students. The scores are given below:

Groups	Scores
Traditional	80, 69, 85, 87, 74, 85, 95, 84, 87, 86, 82, 91, 79, 100, 83, 85
Piagetian	100, 89, 87, 76, 93, 68, 99, 100, 78, 99, 100, 74, 76, 97

Test the hypothesis that the methods have no influence on test scores against the alternative that the students in group 2 have significantly higher scores. Take $\alpha = 0.05$.

10.2. **Smoking and COPD.** It is well established that long-term cigarette smoking is associated with the activation of a cascade of inflammatory responses in the lungs that lead to tissue injury and dysfunction. This is manifested clinically as chronic obstructive pulmonary disease (COPD). It is believed that smoking causes approx. 80 to 90% of COPD cases.

Nine life-long nonsmoking, healthy volunteers (5 men and 4 women; mean age 22.0 ± 1.9 years $[\pm SD]$) and 11 healthy volunteers (5 men and 6 women; mean age 23.4 ± 0.9 years) with a 2.0 ± 1.2 pack-year cigarette smoking history were recruited from students attending the University of Illinois at Chicago. No participants had a history of chronic respiratory tract disorders, including asthma and COPD, and all denied symptoms of acute respiratory illness within 4 weeks preceding the study.

The study by Garey et al. (2004) found various summaries involving proteins, nitrite, and other inflammatory cascade signatures in exhaled breath condensate (EBC). Average nitrite concentration in EBC of non-smokers was found to be $\overline{X}_1 = 16,156$ (nmol/L) and the sample standard deviation was $s_1 = 7,029$ (nmol/L). For smokers, the mean nitrite concentration was $\overline{X}_2 = 24,672$ (nmol/L) and the sample standard deviation was $s_2 = 7,534$ (nmol/L).

Assuming that the population variances are the same, test the hypothesis that the nitrite concentration in EBC for smokers and nonsmokers are the same versus the one-sided alternative. Use $\alpha = 5\%$.

10.3. **Noradrenergic Activity.** Although loss of noradrenergic neurons in the locus ceruleus has been consistently demonstrated postmortem in Alzheimer's disease, several studies suggest that indexes of central noradrenergic activity increase with the severity of Alzheimer's disease in living patients. The research by Elrod et al. (1997) estimated the effect of Alzheimer's disease severity on central noradrenergic activity by comparing the CSF norepinephrine concentrations of subjects with Alzheimer's disease in early and advanced stages. Lumbar punctures were performed in 29 subjects with Alzheimer's disease of mild or moderate severity and 17 subjects with advanced Alzheimer's disease. Advanced Alzheimer's disease was defined prospectively by a mini-mental state score of less than 12. Norepinephrine was measured by radioenzymatic assay, and it is assumed that the measurements followed a normal distribution.

The CSF norepinephrine concentration for Alzheimer's-free subjects has a mean of 170 pg/ml. The patients with advanced Alzheimer's disease had a mean CSF norepinephrine concentration of 279 pg/ml, with a sample standard deviation of 122, while in those with mild to moderate severity the mean was 198 pg/ml with a standard deviation of 89. In both cases normality is assumed.

(a) Let μ_1 and μ_2 be the population means for CSF norepinephrine concentration for patients with advanced and mild-to-moderate severity, respectively. Test the hypotheses $H_0' : \mu_1 = 170$ and $H_0'' : \mu_2 = 170$ based on the available information.

(b) Test the hypothesis that the population variances are the same, $H_0 : \sigma_1^2 = \sigma_2^2$, versus the two-sided alternative.
(c) Test the hypothesis $H_0 : \mu_1 = \mu_2$. Choose the type of t-test based on the decision in (b).
In all cases use $\alpha = 0.05$.

10.4. Testing Variances.   Consider the following annotated MATLAB file.

```
%The two samples x and y are:
x = [0.34   0.52  -0.67  -0.98  -2.46   0.05   1.12 ...
     1.80  -0.51   0.88   0.29   0.29   0.66   0.06  -0.29];
y = [0.70  -0.61  -1.09   1.68  -0.62  -0.57   0.41];

%Test the equality of population variances
%against the two sided alternative.
%The built-in MATLAB function
[h p]=vartest2(x, y)
gives: %h=0 (choose H0) and p-value p=0.9727.
```

(a) Find the p-value using the MATLAB code in the table on page 433. Does it coincide with the vartest2 result?
(b) Show that $F = s_1^2/s_2^2 > 1$. In such a case the standard recommendation is to calculate the two-sided p-value as p = 2 * (1 - fcdf(F,n1-1, n2-1)), where n1=length(x) and n2=length(y). Show that for the samples from this exercise, this "p-value" exceeds 1. Can you explain what the problem is?

10.5. Mating Calls.   In a study of mating calls in the gray tree frogs *Hyla hrysoscelis* and *Hyla versicolor*, Gerhart (1994) reports that in a location in Lousiana the following data on the length of male advertisement calls have been collected:

	Sample size	Average duration	SD of duration	Duration range
Hyla chrysoscelis	43	0.65	0.18	0.36–1.27
Hyla versicolor	12	0.54	0.14	0.36–0.75

The two species cannot be distinguished by external morphology, but *H. chrysoscelis* are diploids while *H. versicolor* are tetraploids. The triploid crosses exhibit high mortality in larval stages, and if they attain sexual maturity, they are sterile. Females responding to the mating calls try to avoid mismatches.
Based on the data summaries provided, test whether the length of call is a discriminatory characteristic? Use $\alpha = 0.05$.

10.6. Fatigue.   According to the article "Practice and fatigue effects on the programming of a coincident timing response," published in the *Journal of Human Movement Studies* in 1976, practice under fatigued conditions distorts mechanisms that govern performance. An experiment was conducted using 15 college males who were trained to make a continuous horizontal right- to left-arm movement from a microswitch to a barrier, knocking over the barrier coincident with the arrival of a clock's second hand to the 6 o'clock position. The absolute value of the difference between the time, in milliseconds, that it took to knock over the barrier and the time for the second hand to reach the 6 o'clock position (500 ms) was recorded. Each participant performed the task five times under prefatigue and postfatigue conditions, and the sums of the absolute differences for the five performances were recorded as follows:

| Subject | Absolute time differences (ms) | |
	Prefatigue	Postfatigue
1	158	91
2	92	59
3	65	215
4	98	226
5	33	223
6	89	91
7	148	92
8	58	177
9	142	134
10	117	116
11	74	153
12	66	219
13	109	143
14	57	164
15	85	100

An increase in the mean absolute time differences when the task is performed under postfatigue conditions would support the claim that practice under fatigued conditions distorts mechanisms that govern performance. Assuming the populations to be normally distributed, test this claim at level $\alpha = 0.01$.

10.7. **Body Mass Index and Hirsutism.** Hirsutism is a condition that affects 5–10% of women and results in excessive terminal hair growth with a typical male pattern distribution. It is caused by an excessive production of androgen in fat tissue as well as increased insulin resistance. In a study by Fatermi et al. (2012) it was observed that hirsutism is more common in patients with a higher Body Mass Index. The study involved a large number of women (800), but here the summaries for 26 cases and 30 controls are provided. The sample sizes, sample means and standard deviations of BMI for case and control cases are:

Cases	Controls
$n_1 = 26, \overline{X}_1 = 24.57, s_1 = 3.04$	$n_2 = 30, \overline{X}_2 = 22.92, s_2 = 1.98$

(a) What hypotheses are of interest here and what test is appropriate? Write down the hypotheses to be tested.

(b) Are the population variances the same? Test the hypothesis of equality of variances against the two-sided alternative and decide on the testing scenario for the means. Use $\alpha = 0.05$.

(c) Test the equality of the means against the alternative that you believe is appropriate for this context. Report the p-value.

(d) You are interested in designing a follow-up test that will be sensitive to finding the difference in mean BMIs larger than 1.5. If population variances are $\sigma_1^2 = 3^2$ and $\sigma_2^2 = 2^2$, what sample size (per group) will achieve a power of 90% in one-sided testing at level $\alpha = 0.05$? *Hint:* Use the formula on page 438, with $\Delta = 1.5$.

10.8. **Mosaic Virus.** A single leaf is taken from each of 11 different tobacco plants. Each leaf is then divided in half and given one of two preparations of *mosaic* virus. Researchers wanted to examine if there is a difference in the mean number of lesions from the two preparations. Here are the raw data:

Plant	Prep 1	Prep 2	Plant	Prep 1	Prep 2
1	38	29	7	41	22
2	40	35	8	36	25
3	26	31	9	12	12
4	33	31	10	34	25
5	21	14	11	28	17
6	27	37			

(a) Is this experiment in accordance with a paired t-test setup?

(b) Test the hypothesis that the difference between the two population means, $\mu_1 - \mu_2$, is significantly positive. Assume normality and use $\alpha = 0.05$.

(c) What happens with the test in (b) if you assume that the two samples are independent, coming from normal populations with the same variance?

10.9. **Dopamine β-hydroxylase Activity.** Postmortem brain specimens from nine chronic schizophrenic patients and nine controls were assayed for activity of dopamine β-hydroxylase (DBH), the enzyme responsible for the conversion of dopamine to norepinephrine (Wyatt et al., 1975). The means and standard deviations of DBH activity in the hippocampus part of the brain are provided in the table. Assume that the data come from two normally distributed and independent populations.

	Schizophrenic subjects	Control subjects
Sample size	$n_1 = 9$	$n_2 = 9$
Sample mean	$\overline{X}_1 = 35.5$	$\overline{X}_2 = 39.8$
Sample standard deviation	$s_1 = 6.93$	$s_2 = 8.16$

(a) Test to determine if the mean activity is significantly lower for the schizophrenic subjects than for the control subjects. Use $\alpha = 0.05$.

(b) Construct a 99% confidence interval for the mean difference in enzyme activity between the two groups.

Solve the above in two ways: (i) by assuming that $\sigma_1 = \sigma_2$ and (ii) without such an assumption.

Wyatt et al. (1975) report that one of the control subjects with low DBH activity had unusually long death-to-morgue time (27 hours) and suggested excluding the subject from the study. The data for controls after exclusion were $n_2 = 8$, $\overline{X}_2 = 41.2$, and $s_2 = 7.52$. Repeat the test in (a) and (b) with these control data.

10.10. **5-HIAA Levels.** A rare and slow-growing form of cancer, carcinoid tumors, may develop anywhere in the body where neuroendocrine (hormone-producing) cells exist. Serotonin is one of the key body chemicals released by carcinoid tumors that are associated with carcinoid syndrome. The 5-hydroxyindoleacetic acid (5-HIAA) test is a 24-hour urine test that is specific to carcinoid tumors. Elevated levels of 5-HIAA, a by-product of serotonin decomposition, can be detected from a urine sample.

Ross and Roberts (1985) provide results of a case/control study of a morphologically specific type of carcinoid disorder that involves the mural and valvular endocardium on the right side of the heart, known as carcinoid heart disease. Out of a total of 36 subjects they investigate, urinary excretion of 5-hydroxyindoleacetic acid (5-HIAA) was measured on 28 subjects, 16 cases with carcinoid heart disease and 12 controls. The data recorded as level of 5-HIAA in milligrams per 24 hours, also discussed in Dawson–Saunders and Trapp (1994), are provided in the table below:

Patients	263	288	432	890
	450	1270	220	350
	283	274	580	285
	524	135	500	120
Controls	60	119	153	588
	124	196	14	23
	43	854	400	73

Assuming that the data come from respective normal distributions, compare the means of the two populations in both a classical and a Bayesian fashion. For the Bayes model use noninformative priors.

10.11. **Stress, Diet, and Acids.** In the study "Interrelationships between Stress, Dietary Intake, and Plasma Ascorbic Acid during Pregnancy," discussed by Walpole et al. (2007, p. 359), the plasma ascorbic acid levels of pregnant women were compared for smokers versus nonsmokers. Thirty-two healthy women, between 15 and 32 years old, in the last 3 months of pregnancy were selected for the study. Eight of the women were smokers. Prior to the lab tests, the participants were told to avoid food and vitamin supplements. From the blood samples, the following plasma ascorbic acid values of each subject were determined in milligrams per 100 ml:

Plasma ascorbic acid values		
Nonsmokers		Smokers
0.97	1.06	0.48
0.72	0.86	0.71
1.00	0.85	0.98
0.81	0.58	0.68
0.62	0.57	1.18
1.32	0.64	1.36
1.24	0.98	0.78
0.99	1.09	1.64
0.90	0.92	
0.74	0.78	
0.88	1.24	
0.94	1.18	

(a) Using WinBUGS, test the hypothesis of equality of levels of plasma ascorbic acid for the two populations. Use noninformative priors on population means and variances (precisions).

(b) Compare the results in (a) with a classical two-sample t-test with no assumption of equality of variances.

10.12. **A. pantherina and A. rubescens.** Making spore prints is an enormous help in identifying genera and species of mushrooms. To make a spore print, mushroom fans take a fresh, mature cap and lay it on a clean piece of glass. Left overnight or possibly longer, the cap should give you a good print. The Amanitas family is one that has the most poisonous (A. phalloides, A. verna, A. virosa, A. pantherina, etc.), as well as the most delicious (A. cesarea, A. rubescens) species. Two independent samples of $m = 12$ and $n = 15$ spores of A. pantherina ("Panther") and A. rubescens ("Blusher"), respectively, were analyzed. In both species of mushrooms the spores are smooth and elliptical, and the largest possible measurement was taken at the great axis of the ellipse. It was found that the means were $\overline{X}_1 = 6.3$ microns and $\overline{X}_2 = 7.5$ microns, with standard deviations of $s_1 = 2.12$ μm and $s_2 = 1.94$ μm.

(a) A researcher is interested in testing the hypothesis that the population mean sizes of spores for these two mushrooms, μ_1 and μ_2, are the same, versus the two-sided alternative. Use $\alpha = 0.05$.

(b) What sample sizes are needed so that the researcher is able to reject the null hypothesis of no effect (no difference between the population means) with the power of 0.90%, versus the "medium" standardized effect size of $d = 0.5$? The significance α is, as in (a), 0.05.

10.13. **Blood Volume in Infants.**   The total blood volume of normal new-born infants was estimated by Schücking (1879) who took into account the addition of placental blood to the circulation of the newborn infant when clamping of the umbilical cord is delayed. Demarsh et al. (1942) further studied the importance of early and late clamping. For 16 babies in whom the cord was clamped early the total blood, as a percentage of weight, on the third day is listed below:

13.8 8.0 8.4 8.8 9.6 9.8 8.2   8.0
10.3 8.5 11.5 8.2 8.9 9.4 10.3 12.6

For 16 babies in whom the cord was not clamped until the placenta began to descend, the corresponding figures are listed below:

10.4 13.1 11.4 9.0 11.9 16.2 14.0 8.2
13.0 8.8 14.9 12.2 11.2 13.9 13.4 11.9

(a) Do these two samples provide evidence of a significant difference between the blood volumes? Perform the test at $\alpha = 0.05$.
(b) Using WinBUGS, find the posterior probability of the hypothesis that there is no difference in blood volumes. Use noninformative priors.

10.14. **Biofeedback.**   In the past, many bodily functions were thought to be beyond conscious control. However, recent experimentation suggests that it may be possible for a person to control certain bodily functions if that person is trained in a program of *biofeedback* exercises. An experiment is conducted to show that blood pressure levels can be consciously reduced in people trained in this program. The blood pressure measurements (in millimeters of mercury) listed in the table represent readings before and after the biofeedback training of five subjects.

Subject	Before	After
1	137	130
2	201	180
3	167	150
4	150	153
5	173	162

(a) If we want to test whether the mean blood pressure decreases after the training, what are the appropriate null and alternative hypotheses?
(b) Perform the test in (a) with $\alpha = 0.05$.
(c) What assumptions are needed to assure the validity of the results?

10.15. Geriatric Assessment. Morrison (2004) reports the unpublished results of Perlin and Butler, two well-known geropsychologists.[1] Forty-nine elderly men participating in a study of aging were classified in two diagnostic categories: "senile factor present" and "no senile factor present" on the basis of comprehensive psychiatric examination. The Wechsler Adult Intelligence Scale had been administered to all subjects by an independent investigator, and certain subtests showed differences between the two groups.

	Group	
	No senile factor	Senile factor
	$n_1 = 37$	$n_2 = 12$
Information	12.57	8.75
Similarities	9.57	5.33
Arithmetic	11.49	8.50
Picture completion	7.97	4.75

The pooled sample covariance matrix S was found to be

$$S = \begin{bmatrix} 11.2553 & 9.4042 & 7.1489 & 3.3830 \\ & 13.5318 & 7.3830 & 2.5532 \\ & & 11.5744 & 2.6170 \\ & & & 5.8085 \end{bmatrix}.$$

after the assumption that variances/covariances in the two populations coincide.

(a) To test for the significance of the observed differences, it was proposed as the null hypothesis that the two groups arose from populations with a common mean vector. Assuming normality, test this hypothesis.

(b) Let $\mu_{3,1}$ and $\mu_{3,2}$ be the coordinates corresponding to the score in the arithmetic for the two groups. Show that 0 does not fall in individual 95% CI for $\mu_{3,1} - \mu_{3,2}$, but falls in the simultaneous CI for the same difference.

10.16. Hypertension. Dernellis and Panaretou (2002) examined a small number of subjects with hypertension and healthy control subjects. One of the variables of interest was the aortic stiffness index. Measures of this variable were calculated from the aortic diameter evaluated by M-mode echocardiography and blood pressure measured by a sphygmomanometer. Generally, physicians wish to reduce aortic stiffness. From $n_1 = 15$ patients with hypertension (group 1), the mean aortic stiffness index was $\overline{X}_1 = 16.16$ with a standard deviation of $s_1 = 4.29$. In the $n_2 = 16$ control subjects (group 2), the mean aortic stiffness index was $\overline{X}_2 = 10.53$ with a standard deviation of $s_2 = 3.33$.

---

[1] Dr Seymour Perlin MD, is well known for his research in suicide prevention; and Dr Robert N. Butler, MD is was founding director of NIA in 1976 and author of the bestseller book *Why Survive*. He coined the term *ageism*.

(a) Test the hypothesis that the population mean aortic stiffness indexes for the two groups differ using the two-sided alternative. Perform the test by both the rejection-region and p-value methods. Take $\alpha = 0.05$ and assume that the population variances are the same.

(b) What is the observed (retrospective) power of this test versus the specific alternative $\Delta = |\mu_2 - \mu_1| = 3$? Keep $\alpha = 0.05$.

(c) Pretend that this study was only a pilot study needed to design more elaborate clinical trials of the same type. What group sample sizes are needed to assure a 99% power versus the fixed alternative $\Delta = |\mu_2 - \mu_1| = 3$?

10.17. **Hemopexin in DMD Cases II.** Refer to Exercises 2.19 and 9.16 and data set ![disk] dmd.dat|mat|xls.

(a) Form a 95% confidence interval for the difference in mean responses of hemopexin h in two populations: population 1 consisting of all women who are not DMD (Duchenne muscular dystrophy) carriers (carrier=0) and population 2 consisting of all women who are DMD carriers (carrier=1).

(b) It is believed that the mean level of hemopexin in the population of women DMD carriers exceeds the mean level in women noncarriers by 10.

Test the one-sided alternative that the mean levels of hemopexin for the two populations differ by more than 10, that is, $H_0 : \mu_1 - \mu_2 = -10$ versus $H_1 : \mu_1 - \mu_2 < -10$. Use $\alpha = 5\%$. Report the p-value as well. Compare the population variances prior to deciding how to test the difference of means.

Hint: In testing the hypothesis $H_0 : \mu_1 - \mu_2 = C$, the associated t-statistic has $\overline{X}_1 - \overline{X}_2 - C$ as its numerator instead of $\overline{X}_1 - \overline{X}_2$.

10.18. **Risk of Stroke.** Abbott et al. (1986) evaluate the risk of stroke among smokers and nonsmokers in a 12-year prospective Honolulu Heart Program study. The data are given in the table below:

	Stroke yes	Stroke no	Total
Smoker	171	3264	3435
Nonsmoker	117	4320	4437
Total	288	7584	7872

Estimate

(a) the risk difference, and find the 95% confidence interval for the population risk difference;

(b) the risk ratio, and find the 95% confidence interval for the population risk ratio; and

(c) the odds ratio, and find the 95% confidence interval for the population odds ratio.

10.19. **CBG Test.**   In cardiovascular bypass grafting (CBG), patients' blood platelet reactivity was monitored prior to and 5 days after the operation. P-selectin expression was measured as the marker of platelet function in resting platelets, as well as in the cells agonized in vitro by thrombin. The goal of the research was to test if CBG operation affects blood platelet reactivity.

For $n = 20$ patients the P-selectin expressions before and 5-days after the surgery were recorded as

```
before = [44.2 73.8 70.5 60.8 58.7 52.1 45.7 67.2 61.3 71.4 ...
          61.2 64.0 57.4 55.7 54.7 52.5 63.7 67.8 53.6 48.5];
after  = [38.8 73.6 55.5 51.5 64.1 30.3 49.5 55.7 54.7 52.5 ...
          51.6 55.0 68.5 45.5 58.2 61.8 29.7 42.2 54.2 62.3];
```

Is the operation affecting platelet reactivity?

(a) Assume normal distributions for the two sets of measurements with means μ_b and μ_a. Test $H_0 : \mu_b = \mu_a$ against one-sided alternative $H_1 : \mu_b > \mu_a$.

(b) Find the 95% CI for the difference $\mu_b - \mu_a$.

(c) What sample size is needed to reject the null hypothesis in a one-sided $\alpha = 0.05$ level testing, with power of 90% whenever the true difference between the means exceeds 5? Assume that the standard deviations for Before and After measurements are 10 each and that correlation ρ between the measurements is 0.3. To express variance for the difference σ_d^2, use $\sigma_d^2 = \sigma_b^2 + \sigma_a^2 - 2\sigma_b\sigma_a\rho$.

10.20. **Cell Counts.**   Refer to Example 1.2 and assume that the data is approximately normal.

(a) Do automated and manual counts significantly differ for a fixed confluency level? Use a paired t-test with a two-sided alternative at significance level $\alpha = 0.05$.

(b) What are 95% confidence intervals for the population differences?

(c) If the difference between automated and manual counts constitutes an error, are the errors comparable for the two confluency levels? This is a DiD (difference-in-differences) test and is equivalent to a two-sample t-test when measurements are the differences. Use the two-sided alternative and $\alpha = 0.05$.

10.21. **Impulses from Crayfish.**   The crayfish is utilized in numerous neuroscience labs in order to explore the role of the central pattern generator (CPG) in locomotion. In an experiment with crayfish, you design glass electrodes to record from the abdominal ganglia nerves that contribute to the return and power strokes of the swimmerets. What this means is that the nerves you are recording from contain the motor output signals that tell the swimmerets to move water around their abdomen anterior or posterior. In the experimental setup, glass extracellular electrodes are used to measure the action potentials or voltage waveform. Due to this being a CPG, you normally expect bursting activity, but when you suction onto the nerve cord, you notice what seems to be regular spiking. A spike detector was applied on the voltage time trace, and the interspike intervals were recorded. The data were obtained under two treatments: (1) *Carbachol* (also known as *carbamylcholine*), a drug that binds and activates the acetylcholine receptor, was added to possibly induce faster spiking; and (2) control group. The data files ⬛ carbachol.dat and ⬛ control.dat containing the respective interspike times can be found at the book's Web site. Also, you can load both files by importing ⬛ frankdata.mat or ⬛ frankdata.xls. All interspike time measurements are given in seconds. (Thanks to Dr. Frank Lin for making the data available.)

(a) Find fundamental descriptive statistics for both samples carbachol and control. Is normality a reasonable assumption for the two populations?

(b) Find the 95% confidence interval for the population mean interspike time in the control case.

(c) It is believed that the spiking in the control population has a frequency of 5 Hz. Test this hypothesis by testing that the mean population interspike time is 0.2, versus the proper one-sided alternative. Use both RR and p-value approaches; $\alpha = 0.05$.

(d) Test the hypothesis that the mean interspike time in the control and carbachol populations are the same versus the one-sided alternative that states that carbachol increases the frequency of spikes (decreases the interspike times); $\alpha = 0.05$.

(e) Do (b) and (c) in a Bayesian fashion building on ⬛ spikes.odc, which contains control data and hints on how to set the priors. Compare and discuss the 95% confidence interval from (b) and the 95% credible set from BUGS. Compare tests, the p-value from (c), and the probability of H_0 from WinBUGS.

10.22. **Aerobic Capacity.**   The peak oxygen intake per unit of body weight, called the *aerobic capacity* of an individual performing a strenuous activity, is a measure of work capacity. For comparative study, measurements of aerobic capacities are recorded (Frisancho, 1975) for a group of 20 Peruvian highland natives and for a group of 10 Peruvian lowlanders acclimatized as adults to high altitudes. The measurements are taken on a bicycle ergometer at high altitude (ml kg^{-1} min^{-1}).

	Peruvian highland natives	Peruvian lowlanders acclimatized as adults
Sample mean	46.3	38.0
Sample st. deviation	5.0	5.2

(a) Test the hypothesis that the population mean aerobic capacities are the same versus the one-sided alternative. Assume an equality of population variances and take $\alpha = 0.05$.

(b) If you were to repeat this experiment, what sample size (per group) would give you a power of 90% to detect the difference between the means of magnitude 4, if you assumed that the common population variance was $\sigma^2 = 5^2 = 25$? The level of the test, α, is to be kept at 5%.

10.23. **Ibuprofen and Acute Sepsis.** Bernard et al. (1997) describe a randomized, double-blind, placebo-controlled trial of intravenous ibuprofen in 455 patients who had sepsis, defined as fever, tachycardia, tachypnea, and acute failure of at least one organ system.

The mortality rate was 92 out of 231 in the placebo group compared to 83 out of 224 in the ibuprofen group (at dose 10 mg per kilogram of body weight, given every six hours).

The authors conclude: "In patients with sepsis, treatment with ibuprofen reduces levels of prostacyclin and thromboxane and decreases fever, tachycardia, oxygen consumption, and lactic acidosis, but it does not prevent the development of shock or the acute respiratory distress syndrome and *does not improve survival.*"

Can you confirm that ibuprofen regimen did not significantly improve the survival rate, by analyzing the risk difference, and risk and odds ratios?

10.24. **Cataract and Diabetes.** Hiller and Kahn (1976) consider diabetes as a risk factor for cataracts and provide the results of a case/control study.

Diabetes	Cataract cases	No cataract	Total
Present	56	84	140
Absent	552	1927	2479
Total	608	2011	2619

Find 95% confidence intervals for the risk difference, risk ratio, and odds ratio.

10.25. **Beginnings of Antiseptic Surgeries.** Sir Joseph Lister (1827–1912), Professor of Surgery at Glasgow University, influenced by Pasteur's ideas, found that a wound wrapped in bandages treated by carbolic acid (phenol) would often not become infected. Here are Lister's data on treating open fractures and amputations:

Period	Carbolic acid	Results
1864–1866	No	Treated 34 patients, 15 died and 19 recovered
1867–1870	Yes	Treated 40 patients, 6 died and 34 recovered

(a) Find and interpret the risk difference, risk ratio, and odds ratio.
(b) Find the 95% CIs for population parameters estimated in (a).

10.26. **Reaction Times.** Researchers are interested in reactions times to different color light stimuli, specifically, green and red. A randomized design is proposed – each subject is given a number of trials, such as GGR-RGRRG, for example. As a measurement of reaction time for a subject we report an average speed of reaction to each color. The table below contains the measurements.

Subject	X (red)	Y (green)
1	18	22
2	16	20
3	23	29
4	30	35
5	32	27
6	30	29
7	31	33
8	25	29
9	27	31
10	21	24

Are the reaction times to red and green lights the same? Use a two-sided alternative and $\alpha = 0.05$.

10.27. **Kawasaki Disease Clinical Trials.** The team of researchers from RAISE Study Group (2012) report on a multicenter, prospective, randomized, open-label, blinded-endpoints trial in 74 hospitals in Japan between 2008 and 2010. Patients with severe Kawasaki disease were randomly assigned to receive either intravenous immunoglobulin (2 g/kg for 24 h and aspirin 30 mg/kg per day) or intravenous immunoglobulin plus prednisolone 2 mg/kg (instead of aspirin) per day. The outcome of interest was the development of coronary abnormalities (CA) over the study period. The following 2×2 table summarizes the results of the trial:

Treatment group	CA	No CA	Total
Immunoglobulin + Prednisolone	4	121	125
Immunoglobulin + Aspirin	28	95	123
Total	32	216	248

Report and interpret (in the context of this problem) the (a) risk difference, (b) risk ratio, and (c) odds ratio. Find the corresponding 95% confidence intervals.

10.28. **GREAT Trial.** Pocock and Spiegelhalter (1992a,b) examine the effect of anistreplase (a thrombolytic agent) on recovery from myocardial infraction. A total of 311 patients were randomized in two arms anistreplase and placebo (conventional treatment). The number of deaths (during 3 months following acute MI) in each arm is given in the table below

| | Event | | Total |
	Death	No death	
Treatment Anistreplase	14	149	163
Placebo	23	125	148
Total	37	274	311

(a) Find 95% confidence intervals for the risk difference, risk ratio, and odds ratio. You can use file risk.m from the text's website.
(b) Argue that the anistreplase treatment significantly lowers 3-month mortality. Base your arguments on confidence intervals from (a); no test is necessary.

10.29. **High/Low Protein Diet in Rats.**   Armitage and Berry (1994, p. 111) report data on the weight gain of 19 female rats between 28 and 84 days after birth. The rats were placed on diets with high (12 animals) and low (7 animals) protein content.

High protein	Low protein
134	70
146	118
104	101
119	85
124	107
161	132
107	94
83	
113	
129	
97	
123	

We want to test the hypothesis on dietary effect. Did a low protein diet result in significantly lower weight gain?
(a) What test should be used?
(b) Perform the test at an $\alpha = 0.05$ level.
(c) What sample size is needed (per diet) if $\sigma_1^2 = \sigma_2^2 = 450$ and we are interested in detecting the difference $\Delta = |\mu_1 - \mu_2| = 20$ as a significant deviation from H_0 95% of the time (power = 0.95). Use $\alpha = 0.05$.

10.30. **Spider Monkey DNA.** In Example 10.20 we tested for the difference of proportions of nucleotides. Using the fact that nucleotides are exclusive (e.g., one nucleotide cannot be at the same time A and T) and that the numbers of nucleotides follow multinomial distribution,
(a) Test that proportion p_A is significantly smaller than p_T.
(b) Demonstrate that proportions p_G and p_C are not significantly different.
Use $\alpha = 0.05$.

Hint: Since this is a multinomial model, the variance for difference $\hat{p}_A - \hat{p}_T$ needed for Z statistic is $\text{Var}\,(\hat{p}_A - \hat{p}_T) = p_A(1 - p_A)/n + p_T(1 - p_T)/n + 2p_A p_T/n$. Under H_0 from (a), both p_A and p_T are estimated by $(\hat{p}_A + \hat{p}_T)/2$, which leads to $Z = \sqrt{n}(\hat{p}_A - \hat{p}_T)/\sqrt{\hat{p}_A + \hat{p}_T}$.

10.31. **PBSC versus BM for Unrelated Donor Allogeneic Transplants.** We are interested in determining whether bone marrow (BM) is equivalent to peripheral blood stem cells (PBSC) in myeloablative unrelated donor transplantation, using the data from Eapen et al. (2007). The greater graft-versus-host disease burden of PBSC might make clinicians less likely to use PBSC in this context.

By using equivalence margins of $\pm 10\%$, and proportions of relapse within 6 months, test whether PBSC and BM are equivalent at level $\alpha = 0.05$.

Method	Number of patients	Relapsed after 6 months
BM	583	93 (16%)
PBSC	328	58 (18%)

Hint: By mimicking Example 10.22, devise a TOST using a normal approximation to the binomial:

$$z_L = \frac{\hat{p}_{BM} - \hat{p}_{PB} - (-\delta)}{\sqrt{\frac{\hat{p}_{BM}(1-\hat{p}_{BM})}{n_1} + \frac{\hat{p}_{PB}(1-\hat{p}_{PB})}{n_2}}}, \quad p_L = 1 - \Phi(z_L),$$

$$z_U = \frac{\hat{p}_{BM} - \hat{p}_{PB} - \delta}{\sqrt{\frac{\hat{p}_{BM}(1-\hat{p}_{BM})}{n_1} + \frac{\hat{p}_{PB}(1-\hat{p}_{PB})}{n_2}}}, \quad p_U = \Phi(z_U),$$

p-value $= \max\{p_L, p_U\}$.

10.32. **Risk Ratio Paradox.** A clinical trial is being conducted to study the benefits of fluoridation in reducing chances of caries in school children. In a control group, the number of children who developed caries over a 4-year period was $X_1 \sim Bin(n_1, p_1)$, while in the treatment group, the number of children who developed caries over the same 4-year period was $X_2 \sim Bin(n_2, p_2)$. The researchers believe that fluoridation is beneficial, but take flat priors for p_1, p_2 as independent beta $Be(1,1)$. In the clinical trial, it is obtained that $n_1 = 21, X_1 = 5$, and $n_2 = 14, X_2 = 3$; thus, data indicate that the fluoridation reduces the risk of caries by 2.38%, and the prior is neutral. Using WinBUGS find the posterior mean for the risk ratio p_2/p_1 and show that it is approximately 1.10. Thus, the Bayes estimate indicates that fluoridation increases chances of caries by 10%. Can you explain why this counterintuitive result is obtained?

10.33. **Hydrogels.** It has been demonstrated that poly(ethylene glycol) diacrylate (PEG-DA)-based hydrogels could serve as direct cell carriers for engineering soft orthopedic tissues. Recent studies have also shown that PEG-DA can be rendered degradable by introducing an enzyme-sensitive peptide sequence into the polymer chains.

The data from Dr. Temenoff's lab at Georgia Tech represent proportions of surviving cells after gel degradation by the enzyme, a bacterial collagenase, in durations of 30 and 60 minutes.

It is postulated that the proportions of live cells after 30 minutes and 60 minutes of exposure to the enzyme are equivalent. The data collected in eight independent experiments represent the number of cells alive among the total number of cells recovered, as in the table below.

30 minutes		60 minutes	
Alive	Total	Alive	Total
20,250	44,250	40,500	69,750
51,000	126,000	42,750	76,500
77,250	100,500	78,750	155,250
39,000	58,500	42,750	67,500

Form ratios `alive/total` for the two durations. The ratios could be assumed as approximately normal given the large number of cells recovered. Show that $H_0 : \mu_1 = \mu_2$ is not rejected against one- or two-sided alternatives. This, however, is not evidence of equivalence. Test the hypothesis of equivalence with equivalence margins $\theta_U = -\theta_L = 0.1$, that is, $H_0 : \mu_1 - \mu_2 < -0.1$ or $\mu_1 - \mu_2 > 0.1$ versus $H_1 : -0.1 < \mu_1 - \mu_2 < 0.1$. Use a Westlake interval and $\alpha = 0.05$.

10.34. **Bumpus' Sparrows Data.** After an unusually severe storm in February of 1898, a number of house sparrows, *Passer domesticus*, were brought to the Anatomical Laboratory of Brown University, Providence, Rhode Island. Seventy-two of these birds revived; sixty-four perished. This event is described by Hermon Carey Bumpus, the first PhD graduate of Clark University, whose paper (Bumpus, 1898) has served as an example of natural selection in action. The data set provided by Bumpus included several anatomic measurements on 136 birds (as data structure ⌨ bumpus.mat) and had been analyzed since by many diverse researches.

surv	1 if survived, 0 if perished
sex	11 = male young; 12 = male adult; 2 = female
lbt	Length (mm) from tip of the beak to the tip of the tail
ae	Alar extent (mm) from tip to tip of the extended wings
wei	Weight (g)
lbh	Length of beak and head (mm), from tip of the beak to the occiput
hum	Length of humerus [arm/wing bone] (in)
fem	Length of femur [thigh bone] (in)
tib	Length of tibiotarsus [leg bone linked to femur] (in)
wos	Width of skull (in), from the postorbital bone of one side to the postorbital bone of the other
kos	Length of keel of sternum [an extension of breastbone] (in)

By using 2-sample T^2 test find whether there exists a significant anatomic difference between the birds from the two groups defined by the survival status. Use the components sex, lbt, ae, wei, lbh, hum, fem, tib, wos, and kos. There is an agreement that lighter and shorter birds have a higher chance of survival. How this can be formally tested?

MATLAB AND WINBUGS FILES AND DATA SETS USED IN THIS CHAPTER

http://statbook.gatech.edu/Ch10.Two/

   argon.m, bihist.m, dnatest2.m, equivalence1.m, hemoc.m, hemopexin2.m, HIAA.m, leadexposure.m, miammt.m, microdamage.m, neanderthal.m, nounfirst.m, piaget.m, plazmaacid.m, plazmaacid2.m, ponvexam.m, psoriasis.m, ratiopoissons.m, risk.m, schizo.m, simulrisks.m, temenoff.m, vaccination.m

   argon.odc, braintissue.odc, cancerprop.odc, eBay.odc, equivalence.odc, microdamage.odc, plasma1.odc, plasma2.odc, psoriasis.odc, spikes.odc, stressacids.odc, twovars.odc

   carbachol.dat, control.dat, dmd.dat|mat|xls, frankdata.mat|xlsx, lice.xls, pasi.dat|mat|xls, PONV.mat|xlsx

---

CHAPTER REFERENCES

Abbott, R. D., Yin, Y., Reed, D. M., and Yano, K. (1986). Risk of stroke in male cigarette smokers. *N. Engl. J. Med.*, **315**, 717–720.

Armitage, P. and Berry, G. (1994). *Statistical Methods in Medical Research*, 3rd ed. Blackwell Science, Malden, MA.

Berger, R. L. and Hsu, J. C. (1996). Bioequivalence trials, intersection-union tests and equivalence confidence sets. *Stat. Sci.*, **11**, 4, 283–302.

Bernard, G. R., Wheeler, A. P., Russell, J. A., Schein, R., Summer, W. R., Steinberg, K. P., Fulkerson, W. J., Wright, P. E., Christman, B. W., Dupont, W. D., Higgins, S. B., and Swindell, B. B. (1997). The effects of ibuprofen on the physiology and survival of patients with sepsis. *N. Engl. J. Med.*, **336**, 912–918.

Bumpus, H. C. (1898). The elimination of the unfit as illustrated by the introduced sparrow, *Passer domesticus*. Biological Lectures at Woods Hole Marine Biological Laboratory, 11th Lecture, 209–225.

Casagrande, J. T., Pike M. C., and Smith, P. G. (1978). The power function of the exact test for comparing two binomial distributions. *Appl. Stat.*, **27**, 176–180.

Cochran, W. G. and Cox, G. M. (1957). *Experimental Designs*. Wiley, New York.

Cox, D. R. (1953). Some simple tests for Poisson variates. *Biometrika*, **40**, 3–4, 354–360.

Dawson-Saunders, B. and Trapp, R. G. (1994). *Basic and Clinical Biostatistics*, 2nd edn. Appleton and Lange, Norwalk, CT.

Demarsh, Q. B., Windle, W. F., and Alt, H. L. (1942). Blood volume of newborn infant in relation to early and late clamping of umbilical cord. *Am. J. Dis. Child.*, **63**, 6, 1123–1129.

Dernellis, J. and Panaretou, M. (2002). Effect of thyroid replacement therapy on arterial blood pressure in patients with hypertension and hyperthyroidism. *Am. Heart J.*, **143**, 718–724.

Desu, M. M. and Raghavarao, D. (1990). *Sample Size Methodology*. Academic, San Diego.

Dorn, H. F. (1954). The relationship of cancer of the lung and the use of tobacco. *Am. Stat.*, **8**, 7–13.

Eapen, M., Logan, B., Confer, D. L., Haagenson, M., Wagner, J. E., Weisdorf, D. J., Wingard, J. R., Rowley, S. D., Stroncek, D., Gee, A. P., Horowitz, M. M., and Anasetti, C. (2007). Peripheral blood grafts from unrelated donors are associated with increased acute and chronic graft-versus-host disease without improved survival. *Biol. Blood Marrow Transplant.*, **13**, 1461–1468.

Edwards, A. W. F. (1963). The measure of association in a 2x2 table. *J. Roy. Stat. Soc., A*, **126**, 109–114.

Elrod R., Peskind, E. R., DiGiacomo, L., Brodkin, K., Veith, R. C., and Raskind M. A. (1997). Effects of Alzheimer's disease severity on cerebrospinal fluid norepinephrine. *Am. J. Psychiatry*, **154**, 25–30.

Fatermi, N. F., Najafian, J., and Jazebi, N. (2012). Hirsutism and body mass index in a representative sample of Iranian people. *ARYA Artheroscl. J.*, **8**, 1, 43–45. PMID: 23056100 [PubMed] PMCID: PMC3448401

FDA. (2001). Guidance for Industry. Statistical Approaches to Establishing Bioequivalence. FDA-CDER. http://www.fda.gov/downloads/Drugs/GuidanceCompliance RegulatoryInformation/Guidances/ucm070244.pdf

Fleiss, J. L., Levin, B., and Paik, M. C. (2003). *Statistical Methods for Rates and Proportions*, 3rd edn. Wiley, New York.

Frisancho, A. R. (1975). Functional adaptation to high altitude hypoxia. *Science*, **187**, 313–319.

Gardner, E. S. and Ringlee, R. J. (1968). Line stormproofing receives critical evaluation. *Trans. Distrib.*, **20**, 59–61.

Garey, K. W., Neuhauser, M. M., Robbins, R. A., Danziger, L. H., and Rubinstein, I. (2004). Markers of inflammation in exhaled breath condensate of young healthy smokers. *Chest*, **125**, 1, 22–26. 10.1378/chest.125.1.22.

Gerhardt, H. C. (1994). Reproductive character displacement of female mate choice in the grey treefrog, *Hyla chrysoscelis*. *Anim. Behav.*, **47**, 959–969.

Giovannucci, E., Tosteson, T. D., Speizer, F. E., Ascherio, A., Vessey, M. P., and Colditz, G. A. (1993). A retrospective cohort study of vasectomy and prostate cancer in US men. *J. Am. Med. Assoc.*, **269**, 7, 878–882.

Glass, G. and Hopkins, K. (1984). *Statistical Methods in Education and Psychology*, 2nd ed. Allyn and Bacon, Boston, MA.

Hiller, R. and Kahn, H. A. (1976). Senile cataract extraction and diabetes. *Br. J. Ophthalmol.*, **60**, 4, 283–286.

Johnson, V. and Albert, J. (1999). *Ordinal Data Modeling*. Springer, New York.

Lyle, R. M., Melby, C. L., Hyner, G. C., Edmondson, J. W., Miller, J. Z., and Weinberger, M. H. (1987). Blood pressure and metabolic effects of calcium supplementation in normotensive white and black men. *JAMA*, **257**, 13, 1772–1776. doi:10.1001/jama.1987.03390130090035

Morrison, D. F. (2004). *Multivariate Statistical Methods*, 4th edn. Duxbury Press, Pacific Grove, CA

Mosteller, F. (1968). Association and estimation in contingency tables. *J. Am. Stat. Assoc.*, **63**, 1–28.

Pitman, E. (1939). A note on normal correlation. *Biometrika*, **31**, 9–12.

Pocock, S. J. and Spiegelhalter, D. J. (1992a). GREAT Group 1992 Feasibility, safety and efficacy of domiciliary thrombolysis by general practitioners: Grampian Region Early Anistreplase Trial. *Br. Med. J.*, **305**, 548–553.

Pocock, S. J. and Spiegelhalter, D. J. (1992b). Letter (untitled). *Br. Med. J.*, **305**, 1015.

Price, R. M. and Bonett, D. G. (2000). Estimating the ratio of two Poisson rates. *Comput. Stat. Data Anal.*, **34**, 3, 345–356.

RAISE Study Group (Kobayashi, T., et al.) (2012). Efficacy of immunoglobulin plus prednisolone for prevention of coronary artery abnormalities in severe Kawasaki disease (RAISE study): A randomized, open-label, blinded-endpoints trial. *Lancet*, **379**, 1613–1620.

Rosner, B. (2010). *Fundamentals of Biostatistics*, 7th ed. Duxbury Press, Pacific Grove, CA.

Ross, E. M. and Roberts, W. C. (1985). The carcinoid syndrome: comparison of 21 necropsy subjects with carcinoid heart disease to 15 necropsy subjects without carcinoid heart disease. *Am. J. Med.*, **79**, 3, 339–354.

Satterthwaite, F. E. (1946). An approximate distribution of estimates of variance components. *Biom. Bull.*, **2**, 110–114.

Schücking, A. (1879). Blutmenge des neugeborenen. *Berl. Klin. Wochenschr.*, **16**, 581–583.

Schuirmann, D. J. (1981). On hypothesis testing to determine if the mean of a normal distribution is contained in a known interval. *Biometrics*, **37**, 617.

United States Pharmacopeia (USP) <1033> Biological Assay Validation (2009). *Pharmacopeial Forum*, **35**, 2, 349–367.

Welch, B. L. (1947). The generalization of "Student's" problem when several different population variances are involved. *Biometrika*, **34**, 28–35.

Wellek, S. (2010). *Testing Statistical Hypotheses of Equivalence and Noninferiority*, 2nd edn. CRC Press, Boca Raton.

Walpole, R. E., Myers, R. H., Myers, S. L., and Ye, K. (2007). *Probability and Statistics for Engineers and Scientists*, 8th ed. Pearson-Prentice Hall, Englewood Cliffs.

Westlake, W. J. (1976). Symmetric confidence intervals for bioequivalence trials. *Biometrics*, **32**, 741–744.

Woo, W. K. and McKenna, K. E. (2003). Combination TL01 ultraviolet B phototherapy and topical calcipotriol for psoriasis: a prospective randomized placebo-controlled clinical trial. *Br. J. Dermatol.*, **149**, 146–150.

Woolf, B. (1955). On estimating the relationship between blood group and disease. *Ann. Human Genetics*, **19**, 251–253.

Wyatt, R. J., Schwarz, M. A., Erdelyi, E., and Barchas, J. D. (1975). Dopamine-hydroxylase activity in brains of chronic schizophrenic patients. *Science*, **187**, 368–370.

Zar, J. H. (2010). *Biostatistical Analysis*, 5th ed. Pearson-Prentice-Hall, Englewood Cliffs.

Chapter 11
ANOVA and Elements of Experimental Design

Design is art optimized to meet objectives.

– Shimon Shmueli

WHAT IS COVERED IN THIS CHAPTER

- ANOVA Model Formulation
- Contrasts and Multiple Comparisons
- Factorial Designs
- Randomized Block Designs and Repeated Measures Designs
- Nested Designs
- Sample Sizes in ANOVA
- Functional ANOVA
- Analysis of Means (ANOM)
- Gauge R&R ANOVA
- Testing Equality of Proportions and Poisson Means

11.1 Introduction

In Chapter 10 we discussed the test of equality of means from two populations, $H_0 : \mu_1 = \mu_2$. Under standard assumptions of normality and independence, the proper test statistic followed a t-distribution, and several tests (equal/different variances, paired/unpaired samples) shared the common name "two-sample t-tests."

Many experimental protocols involve more than two populations. For example, an experimenter may be interested in comparing the sizes of cells grown under several experimental conditions. At first glance, it seems that we can apply the t-test on all possible pairs of means. This "solution" would not be satisfactory since the probability of type I error for such a procedure is unduly large. For example, if the equality of four means is tested by testing the equality of $\binom{4}{2} = 6$ different pairs, each at the level of 5%, then the probability of finding a significant difference when in fact all means are equal is about 26.5%. We already discussed in Section 9.9 the problems of controlling type I error in multiple tests.

An appropriate procedure for testing hypotheses of equality of several means is the analysis of variance (ANOVA). ANOVA is probably one of the most frequently used statistical procedures, and the reasoning behind it is applicable to several other, seemingly different, problems.

11.2 One-Way ANOVA

ANOVA is an acronym for analysis of variance. Even though we are testing for differences among population means, the variability among the group means and the variability of the data within the treatment group are compared. This technique of splitting the total variability in data to *between* and *within* variabilities was first introduced by Fisher (1918, 1921).

Suppose we are interested in testing the equality of k means $\mu_1, \ldots, \mu_k$ characterizing k independent populations $\mathcal{P}_1, \ldots, \mathcal{P}_k$. From the ith population $\mathcal{P}_i$ a sample

$$y_{i1}, y_{i2}, \ldots, y_{in_i}$$

of size n_i is taken. Let $N = \sum_{i=1}^{k} n_i$ be the total sample size. The responses y_{ij} are modeled as

$$y_{ij} = \mu_i + \epsilon_{ij}, \ 1 \leq j \leq n_i; 1 \leq i \leq k, \qquad (11.1)$$

where ϵ_{ij} represents the "error" and quantifies the stochastic variability of difference between observation y_{ij} and the mean of the corresponding population, μ_i. Sometimes the μ_is are called *treatment means*, with index i representing one of the *treatments*.

The null hypothesis is that all population means μ_i are equal, and the procedure to test it is called one-way ANOVA. The assumptions underlying one-way ANOVA are as follows:

(i) All populations are normally distributed.

(ii) The variances in all populations are the same and constant, This is also known as the assumption of homoscedasticity.

(iii) The samples are mutually independent.

The assumptions (i)–(iii) can be expressed by the requirement that all y_{ij} in (11.1) must be i.i.d. normal $\mathcal{N}(\mu_i, \sigma^2)$ or, equivalently, all ϵ_{ij}s must be i.i.d. normal $\mathcal{N}(0, \sigma^2)$.

	Normal population			
	1	2 ...		k
Population mean	μ_1	$\mu_2 \ldots$		μ_k
Common variance	σ^2	$\sigma^2 \ldots$		σ^2

If the sample sizes are the same, i.e., $n_1 = n_2 = \cdots = n_k = N/k$, then the ANOVA is called *balanced*. It is often the case that many experiments are designed as a balanced ANOVA. During an experiment it may happen that a particular measurement is missing due to a variety of reasons, resulting in an *unbalanced* layout. Balanced designs are preferable because they lead to simpler computations and interpretations.

In terms of model (11.1), the null hypothesis to be tested is

$$H_0 : \mu_1 = \mu_2 = \cdots = \mu_k,$$

and the alternative is

$$H_1 : (H_0)^c \quad (\text{or} \quad \mu_i \neq \mu_j, \text{ for at least one pair } i, j).$$

Note that the alternative $H_1 \equiv (H_0)^c$ is any negation of H_0. Thus, for example, $\mu_1 > \mu_2 = \mu_3 = \cdots = \mu_k$ or $\mu_1 = \mu_2 \neq \mu_3 = \mu_4 = \cdots = \mu_k$ are valid alternatives. Later we will discuss how to assess the alternative if H_0 is rejected.

We can reparameterize the population mean μ_i as $\mu_i = \mu + \alpha_i$. Simply stated, the treatment mean is equal to the mean common for all treatments, called the *grand mean* μ, plus the effect of the population group, known as the *treatment effect* α_i. The hypotheses now can be restated in terms of treatment effects:

$$H_0 : \alpha_1 = \alpha_2 = \cdots = \alpha_k = 0.$$

The alternative is H_1: Not all α_is are equal to 0.

The representation $\mu_i = \mu + \alpha_i$ is not unique. We usually assume that $\sum_i \alpha_i = 0$. This is an identifiability assumption needed to ensure the uniqueness of the decomposition $\mu_i = \mu + \alpha_i$. Indeed, by adding and subtracting any number c, μ_i becomes $\mu + \alpha_i = (\mu + c) + (\alpha_i - c) = \mu' + \alpha'$, and uniqueness is assured by $\sum_i \alpha_i = 0$. This kind of constraint is sometimes called the *sum-to-zero* (STZ) constraint. There are other ways to ensure uniqueness; a sufficient assumption is, for example, to set $\alpha_1 = 0$. In this case $\mu = \mu_1$ becomes the reference or baseline mean. Prior to providing the procedure for testing H_0, we sum up the notation:

y_{ij}	jth observation from treatment i
$\overline{y}_i$	Sample mean from treatment i
$\overline{y}$	Average of all observations
μ_i	Population treatment mean
μ	Population grand mean
α_i	ith treatment effect
n_i	Size of ith sample
k	Number of treatments
N	Total sample size

11.2.1 ANOVA Table and Rationale for F-Test

The ANOVA table displays the data summaries needed for inference about the ANOVA hypothesis. It also provides an estimator of the variance in measurements and assesses the goodness of fit of an ANOVA model.

The variability in data follows the *fundamental ANOVA identity* in which the total sum of squares (SST) is represented as a sum of the treatment sum of squares ($SSTr$) and the sum of squares due to error (SSE):

$$SST = SSTr + SSE = SS_{\text{Between}} + SS_{\text{Within}},$$

$$\sum_{i=1}^{k}\sum_{j=1}^{n_i}(y_{ij} - \overline{y})^2 = \sum_{i=1}^{k} n_i(\overline{y}_i - \overline{y})^2 + \sum_{i=1}^{k}\sum_{j=1}^{n_i}(y_{ij} - \overline{y}_i)^2.$$

Here $\overline{y} = \frac{1}{N}\sum_{i=1}^{k}\sum_{j=1}^{n_i}y_{ij}$ is the mean of all observations, and $\overline{y}_i = \frac{1}{n_i}\sum_{j=1}^{n_i}y_{ij}$, $i = 1,\dots,k$, are the sample means.

The standard output from most statistical software includes degrees of freedom (DF), mean sum of squares, F ratio, and the corresponding p-value:

Source	DF	Sum of squares	Mean squares	F	p
Treatment	$k-1$	$SSTr$	$MSTr = SSTr/(k-1)$	$MSTr/MSE$	$\mathbb{P}(F_{k-1,N-k} > F)$
Error	$N-k$	SSE	$MSE = SSE/(N-k)$		
Total	$N-1$	SST			

The null hypothesis is rejected if the F-statistic is large compared to the $(1 - \alpha)$ quantile of an F-distribution with $k - 1$ and $N - k$ degrees of freedom. A decision can also be made by looking at the p-value.

Rationale. The mean square error, MSE, is an unbiased estimator of σ^2. Indeed,

$$MSE = \frac{1}{N-k}\sum_{i=1}^{k}\sum_{j=1}^{n_i}(y_{ij} - \bar{y}_i)^2 = \frac{1}{N-k}\sum_{i=1}^{k}\left[(n_i - 1)\frac{1}{n_i - 1}\sum_{j=1}^{n_i}(y_{ij} - \bar{y}_i)^2\right]$$

$$= \frac{1}{N-k}\sum_{i=1}^{k}(n_i - 1)s_i^2, \text{ and}$$

$$\mathbb{E}(MSE) = \frac{1}{N-k}\sum_{i=1}^{k}(n_i - 1)\mathbb{E}(s_i^2) = \frac{1}{N-k}\sum_{i=1}^{k}(n_i - 1)\sigma^2 = \sigma^2.$$

However, the mean square error due to treatments,

$$MSTr = \frac{SSTr}{k-1} = \frac{1}{k-1}\sum_{i=1}^{k}n_i(\bar{y}_i - \bar{y})^2,$$

is an unbiased estimator of σ^2 only when H_0 is true, that is, when all μ_i are the same. This follows from the fact that

$$\mathbb{E}(MSTr) = \sigma^2 + \frac{\sum_i n_i(\alpha_i)^2}{k-1},$$

where $\alpha_i = \mu_i - \bar{\mu}$ is the population effect of treatment i, and $\bar{\mu} = \frac{1}{N}\sum_{i=1}^{k}n_i\mu_i$. Since under H_0 $\alpha_1 = \alpha_2 = \cdots = \alpha_k = 0$, the $\mathbb{E}(MSTr)$ is equal to σ^2; thus, $MSTr$ is an unbiased estimator of variance. When H_0 is violated, not all α_i are 0, or, equivalently, $\sum \alpha_i^2 > 0$. Consequently, the ratio $MSTr/MSE$ quantifies the departure from H_0, and large values of this ratio are critical.

Example 11.1. **Coagulation Times.** To illustrate the one-way ANOVA, we work out an example involving coagulation times that is also considered by Box et al. (2005). Twenty-four animals are randomly allocated to four different diets, but the numbers of animals allocated to different diets are not the same. The blood coagulation time is measured for each animal. Does diet type significantly influence the coagulation time? The data and MATLAB solution are provided next:

```
times = [62, 60, 63, 59, 63, 67, 71, 64, 65, 66, 68, 66, ...
          71, 67, 68, 68, 56, 62, 60, 61, 63, 64, 63, 59];
diets = {'dietA','dietA','dietA','dietA','dietB','dietB',...
```

```
'dietB','dietB','dietB','dietB','dietC','dietC','dietC',...
'dietC','dietC','dietC','dietD','dietD','dietD',...
'dietD','dietD','dietD','dietD'};
[p,table,stats] = anova1(times, diets,'on')

% p = 4.6585e-005
% table =
%'Source'   'SS'     'df'    'MS'        'F'          'Prob>F'
%'Groups'   [228]    [ 3]    [   76]     [13.5714]    [4.6585e-005]
%'Error'    [112]    [20]    [5.6000]    []                    []
%'Total'    [340]    [23]    []          []                    []
%
%stats =
%     gnames: 4x1 cell
%          n: [4 6 6 8]
%     source: 'anova1'
%      means: [61 66 68 61]
%         df: 20
%          s: 2.3664
```

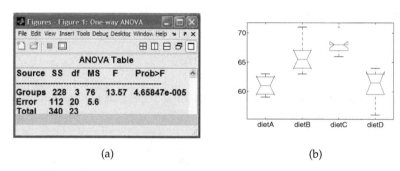

(a)   (b)

Fig. 11.1 (a) ANOVA table and (b) boxplots, as outputs of the anova1 procedure on the coagulation times data.

From the ANOVA table we conclude that the null hypothesis is not tenable, the diets significantly affect the coagulation time. The p-value is smaller than $0.5 \cdot 10^{-4}$, which indicates strong support for H_1.

The ANOVA table, featured in the output and Figure 11.1a, is a standard way of reporting the results of an ANOVA procedure. The SS column in the ANOVA table restates the fundamental ANOVA identity $SSTr + SSE = SST$ as $228 + 112 = 340$. The degrees of freedom for treatments, $k - 1$, and the error, $n - k$, are additive and their sum is total number of degrees of freedom, $n - 1$. Here, $3 + 20 = 23$. The column with mean square errors is obtained when the sums of squares are divided by their corresponding degrees of freedom. The ratio $F = MSTr/MSE$ is the test statistic distributed as F with $(3, 20)$ degrees of freedom. The observed $F = 13.5714$ exceeds the critical value finv(0.95, 3, 20)=3.0984, and H_0 is rejected. Recall that the rejection region is always right-tailed, in this case $[3.0984, \infty)$. The p-value is

`1-fcdf(13.5714, 3, 20)= 4.6585e-005`. Figure 11.1b shows boxplots of the co-agulation times by the diet type.

✎

Since the entries in ANOVA table are interrelated, it is possible to recover a table from only a few entries (e.g., Exercises 11.3, 11.7, and 11.14).

11.2.2 Testing the Assumption of Equal Population Variances

There are several procedures that test for the fulfillment of ANOVA's condition of *homoscedasticity*, that is, the condition that the variances are the same and constant over all treatments.

A reasonably sensitive and simple procedure is Cochran's test.

11.2.2.1 Cochran's Test.

Cochran's test (Cochran, 1941) rejects the hypothesis that k populations have the same variance if the statistic

$$C = \frac{s^2_{max}}{s^2_1 + \cdots + s^2_k}$$

is large. Here, $s^2_1, \ldots, s^2_k$ are sample variances in k samples, and s_{max} is the largest of $s^2_1, \ldots, s^2_k$. Cochran's test assumes equal sample sizes $n = n_1 = \cdots = n_k$. The hypothesis $H_0 : \sigma_1 = \sigma_2 = \cdots = \sigma_k$ is rejected at level α if $C > \frac{f}{f+k-1}$, where f is the (α/k)-quantile of F-distribution with $k - 1$ and $(k - 1)(n - 1)$ degrees of freedom, $F_{k-1,(k-1)(n-1),\alpha/k}$.

If the sample sizes are not equal, one can use the harmonic average of $n'_i s$, or a more precise 't Lam's generalization of Cochran's C ('t Lam, 2010).

11.2.2.2 Levene's Test.

Levene's test (Levene, 1960) hinges on the statistic

$$L = \frac{(N - k) \sum_{i=1}^{k} n_i (\overline{Z}_{i.} - \overline{Z}_{..})^2}{(k - 1) \sum_{i=1}^{k} \sum_{j=1}^{n_i} (Z_{ij} - \overline{Z}_{i.})^2},$$

where $Z_{ij} = |y_{ij} - \overline{y}_{i.}|$. To enhance the robustness of the procedure, the sample means could be replaced by the group medians, or trimmed means. The hypothesis $H_0 : \sigma_1 = \sigma_2 = \cdots = \sigma_k$ is rejected at level α if $L > F_{k-1,N-k,\alpha}$.

11.2.2.3 Bartlett's Test.

Another popular test for homoscedasticity is Bartlett's test (Bartlett, 1937). The statistic

$$
B = \frac{(N - k) \log s_p^2 - \sum_{i=1}^k (n_i - 1) \log s_i^2}{1 + \frac{1}{3(k-1)} \times \left(\sum_{i=1}^k \frac{1}{n_i - 1} - \frac{1}{N-k} \right)},
$$

where s_p^2 is the pooled sample variance $\frac{(n_1-1)s_1^2 + \cdots + (n_k-1)s_k^2}{N-k}$, has an approximately χ^2-distribution with $k - 1$ degrees of freedom. Large values of B are critical, i.e., H_0 is rejected at level α if $B > \chi^2_{k-1,\alpha}$.

In MATLAB, Bartlett's test is performed by the `vartestn(X)` command for samples formatted as columns of X, or as `vartestn(X, group)` for vector X, where group membership of Xs is determined by the vector `group`. Bartlett's test is the default. Levene's test is invoked by optional argument, `vartestn(...,'robust')`. In the context of Example 11.1, Bartlett's and Levene's tests are performed in MATLAB as

```
%  Coagulation Times: Testing Equality of Variances
[pval stats]=vartestn(times', diets','on')  %Bartlet
%   pval = 0.6441
%   chisqstat: 1.6680
%   df: 3

[pval stats]=vartestn(times', diets','on','robust')  %Levene
%   pval = 0.6237
%   fstat: 0.5980
%   df: [3 20]
```

According to these tests, the hypothesis of equal treatment variances is not rejected. Cochran's test agrees with Bartlett's and Levene's, giving a p-value of 0.6557.

Remark. As mentioned iwhen we compared the two means (page 433), the variances-before-means type procedures are controversial. A celebrated statistician George E. P. Box criticized checking assumptions of equal variances before testing the equality means, arguing that comparisons of means are quite robust procedures compared to a non-robust variance comparison (Box, 1953, p. 333). Aiming at Bartlett's test in particular, Box summarized his criticism as follows: "To make the preliminary test on variances is rather like putting to sea in a rowing boat to find out whether conditions are sufficiently calm for an ocean liner to leave port!"

11.2.3 *The Null Hypothesis Is Rejected. What Next?*

When H_0 is rejected, the form for the alternative is not obvious as in the case of two means, and thus we must further explore relationships between the individual means. We will discuss two post-test ANOVA analyses: (i) tests for contrasts and (ii) pairwise comparisons. They both make sense only if the null hypothesis is rejected; if H_0 is not rejected, then both tests for contrasts and pairwise comparisons are trivial.

11.2.3.1 Contrasts

A *contrast* is any linear combination of the population means,

$$C = c_1\mu_1 + c_2\mu_2 + \cdots + c_k\mu_k,$$

such that $\sum_{i=1}^{k} c_i = 0$.

For example, if $\mu_1, \ldots, \mu_5$ are means of $k = 5$ populations, the linear combinations $2\mu_1 - \mu_2 - \mu_4$, $\mu_3 - \mu_2$, and $\mu_1 + \mu_2 + \mu_3 - \mu_4 - 2\mu_5$, are all contrasts since $2 - 1 + 0 - 1 + 0 = 0$, $0 - 1 + 1 + 0 + 0 = 0$, and $1 + 1 + 1 - 1 - 2 = 0$.

In the ANOVA model, $y_{ij} = \mu_i + \epsilon_{ij}$, $i = 1, \ldots, k$; $j = 1, \ldots, n_i$, the sample treatment means $\bar{y}_i = \frac{1}{n_i} \sum_{j=1}^{n_i} y_{ij}$ are the estimators of the population treatment means μ_i. Let $N = n_1 + \cdots + n_k$ be the total sample size and $s^2 = MSE$ the estimator of variance.

The test for a contrast

$$H_0 : \sum_{i=1}^{k} c_i\mu_i = 0 \quad \text{versus} \quad H_1 : \sum_{i=1}^{k} c_i\mu_i <, \neq, > 0 \qquad (11.2)$$

is based on the test statistic that involves sample contrast $\sum_{i=1}^{k} c_i\bar{y}_i$

$$t = \frac{\sum_{i=1}^{k} c_i\bar{y}_i}{s\sqrt{\sum_{i=1}^{k} \frac{c_i^2}{n_i}}},$$

that has a t-distribution with $N - k$ degrees of freedom. Here $\hat{C} = \sum_{i=1}^{k} c_i\bar{y}_i$ is an estimator of contrast C and $s^2 \sum_{i=1}^{k} \frac{c_i^2}{n_i}$ is the sample variance of $\hat{C}$.

The $(1 - \alpha)100\%$ confidence interval for the contrast is

$$\left[\sum_{i=1}^{k} c_i \bar{y}_i - t_{N-k,1-\alpha/2} \cdot s \cdot \sqrt{\sum_{i=1}^{k} \frac{c_i^2}{n_i}}, \ \sum_{i=1}^{k} c_i \bar{y}_i + t_{N-k,1-\alpha/2} \cdot s \cdot \sqrt{\sum_{i=1}^{k} \frac{c_i^2}{n_i}} \right].$$

Sometimes contrast tests are called *single-degree F-tests* because of the link between t and F distributions. Recall that if random variable X has a t-distribution with n degrees of freedom, then X^2 has an F-distribution with 1 and n degrees of freedom. Thus, the test of contrast in (11.2) against the two-sided alternative can equivalently be based on the statistic

$$F = \frac{\left(\sum_{i=1}^{k} c_i \bar{y}_i \right)^2}{s^2 \sum_{i=1}^{k} \frac{c_i^2}{n_i}},$$

which has an F-distribution with 1 and $N - k$ degrees of freedom. This F-test is good only for two-sided alternatives since the direction of deviation from H_0 is lost by squaring the t-statistic.

Example 11.2. **Contrast for Coagulation Times.** As an illustration, let us test the hypothesis $H_0 : \mu_1 + \mu_2 = \mu_3 + \mu_4$ in the context of Example 11.1. The hypothesis above is a contrast since it can be written as $\sum_i c_i \mu_i = 0$ with $c = (1, 1, -1, -1)$. The following MATLAB code tests the contrast against the one-sided alternative and also finds the 95% confidence interval for $\sum_i c_i \mu_i$.

```
m = stats.means %[p,table,stats] = anova1(times, diets)
                %from Example Coagulation Times
c = [ 1  1 -1 -1 ];
L = c(1)*m(1) + c(2)*m(2)+c(3)*m(3) + c(4)*m(4) %L=-2
LL= m * c'  %LL=-2
stdL = stats.s * sqrt(c(1)^2/4+c(2)^2/6+c(3)^2/6+c(4)^2/8)
 %stdL = 1.9916
t = LL/stdL    %t =-1.0042

%test H_o: mu * c' = 0  H_1: mu * c' < 0
% p-value
tcdf(t, 20)  %0.1636

%or 95% confidence interval for population contrast
[LL -  tinv(0.975, 20)*stdL, LL +  tinv(0.975, 20)*stdL]
%   -6.1545    2.1545
```

The hypothesis $H_0 : \mu_1 + \mu_2 = \mu_3 + \mu_4$ is not rejected, and the p-value is 0.1636. Also, the 95% confidence interval for $\mu_1 + \mu_2 - \mu_3 - \mu_4$ is $[-6.1545, 2.1545]$. Note that the confidence interval contains 0.

Orthogonal Contrasts*. Two or more contrasts are called orthogonal if their sample counterpart contrasts are uncorrelated. Operationally, two contrasts $c_1\mu_1 + c_2\mu_2 + \cdots + c_k\mu_k$ and $d_1\mu_1 + d_2\mu_2 + \cdots + d_k\mu_k$ are orthogonal if, in addition to $\sum_i c_i = \sum_i d_i = 0$, the condition $c_1 d_1 + c_2 d_2 + \cdots + c_k d_k = \sum_i c_i d_i = 0$ holds. For unbalanced designs the condition is $\sum_i c_i d_i / n_i = 0$.

If there are k treatments, only $k - 1$ mutually orthogonal contrasts can be constructed. Any additional contrast can be expressed as a linear combination of the original $k - 1$ contrasts. For example, if

	Treatments			
Contrast	1	2	3	4
C_1	1	−1	−1	1
C_2	1	0	0	−1
C_3	0	1	−1	0

then the contrast $(1, -1, -3, 3)$ is $2C_1 - C_2 + C_3$. Any set of $k - 1$ orthogonal contrasts perfectly partitions the $SSTr$. If $SSC = \frac{(\sum c_i \bar{y}_i)^2}{\sum_i c_i^2 / n_i}$, then

$$SSTr = SSC_1 + SSC_2 + \cdots + SSC_{k-1}.$$

This gives a possibility of simultaneous testing of any subset $\leq k - 1$ of orthogonal contrasts. Of particular interest are orthogonal contrasts sensitive to polynomial trends among the ordered and equally spaced levels of a factor. For example, when the design is balanced, and $k = 4$:

	Equispaced levels			
Contrast	1	2	3	4
C_{Linear}	−3	−1	1	3
$C_{Quadratic}$	1	−1	−1	1
C_{Cubic}	−1	3	−1	3

11.2.3.2 Pairwise Comparisons

After rejecting H_0, an assessment of H_1 can be conducted by pairwise comparisons. As the name suggests, this is a series of tests for all pairs of means in k populations. Of course, there are $\binom{k}{2} = \frac{k(k-1)}{2}$ different tests.

A common error in doing pairwise comparisons is to perform $\frac{k(k-1)}{2}$ two-sample t-tests. The tests are dependent and the significance level α for simultaneous comparisons is difficult to control. This is equivalent in spirit to simultaneously testing multiple hypotheses and adjusting the significance level of each test to control overall significance level (page 415). For example, for $k = 5$, the Bonferroni procedure will require $\alpha = 0.005$ for individual comparisons in order to control the overall significance level at 0.05, which is clearly a conservative approach.

Tukey (1952, unpubl. IMS address; 1953, unpubl. mimeograph) proposed a test designed specifically for pairwise comparisons sometimes called the "honestly significant difference test." The Tukey method is based on the so-called studentized range distribution with quantiles q. The quantile used in a test or a confidence interval is $q_{v,k,1-\alpha}$, with α being the overall significance level (or $(1-\alpha)100\%$ overall confidence), k the number of treatments, and v the error degrees of freedom equal to $N-k$. The difference between two means μ_i and μ_j is significant if

$$|\overline{y}_i - \overline{y}_j| > q_{v,k,1-\alpha}\frac{s}{\sqrt{n}}, \tag{11.3}$$

where $s = \sqrt{MSE}$ and n is the treatment sample size for a balanced design. If the design is not balanced, then replace n in (11.3) by the harmonic mean of n_is, $n_h = \frac{k}{\sum_{i=1}^{k}1/n_i}$ (Tukey–Kramer procedure).

The function ◀ qtukey(v,k,p) (Trujillo-Ortiz and Hernandez-Walls, MATLAB Central #3469) approximates Tukey's quantiles for inputs $v = N - k$, k, and $p = 1 - \alpha$.

In biomedical experiments it is often the case that one treatment is considered a control and the only comparisons of interest are pairwise comparisons of all treatments with the control, forming a total of $k - 1$ comparisons. This is sometimes called the *many-to-one* procedure and it was developed by Dunnett (1955).

Let μ_1 be the control mean. Then for $i = 2, \ldots, k$ the mean μ_i is different than μ_1 if

$$|\overline{y}_i - \overline{y}_1| > d_{v,k-1,\alpha}\frac{s}{\sqrt{1/n_i + 1/n_1}},$$

where α is a joint significance test for $k - 1$ tests, and $v = N - k$.

It is recommended that the control treatment have more observations than other treatments. A discussion on the sample size necessary to perform Dunnett's comparisons can be found in Liu (1997). MATLAB function ◀ dunnett.m can be found on MATLAB Central (N. Pokala, File ID: #38157).

Example 11.3. **Multiple Comparisons for Coagulation Times.** In the context of Example 11.1, let us compare the means using the Tukey procedure. This is a default for MATLAB's command multcompare applied on the output stats in [p, table, stats] = anova1(times, diets). The multcompare command produces an interactive visual position for all means with their error bars and additionally gives an output with a confidence interval for the difference of trhe means of each pair. If the confidence interval contains 0, then the means are not statistically different, according to Tukey's procedure. For example, the 95% Tukey confidence interval for $\mu_2 - \mu_3$ is $[-5.8241, 1.8241]$, and the means μ_2 and μ_3 are "statistically the same." Conversely, Tukey's 95% interval for $\mu_1 - \mu_3$ is $[-11.2754, -2.7246]$, indicating that μ_1 is significantly smaller than μ_3.

```
multcompare(stats)  %[p,table,stats] = anova1(times, diets)
%Compares means: 1-2; 1-3; 1-4; 2-3; 2-4; 3-4
%ans =
%    1.0000     2.0000    -9.2754    -5.0000    -0.7246
%    1.0000     3.0000   -11.2754    -7.0000    -2.7246
%    1.0000     4.0000    -4.0560          0     4.0560
%    2.0000     3.0000    -5.8241    -2.0000     1.8241
%    2.0000     4.0000     1.4229     5.0000     8.5771
%    3.0000     4.0000     3.4229     7.0000    10.5771
```

We can also find the Tukey's confidence intervals by using qtukey.m. For example, the 95% confidence interval for $\mu_1 - \mu_2$ is

```
m=stats.means;
%1-2
[m(1)-m(2) - qtukey(20,4,0.95)*stats.s*sqrt(1/2 *(1/4+1/6)) ...
 m(1)-m(2) ...
 m(1)-m(2) + qtukey(20,4,0.95)*stats.s*sqrt(1/2*(1/4+1/6))]
%                   -9.3152    -5.0000    -0.6848
% Compare to: -9.2754    -5.0000    -0.7246 from multcompare
```

Although close to the output of multcompare, this interval differs due to a coarser approximation algorithm in qtukey.m.

In addition to Tukey and Dunnett, there is a range of other multiple comparison procedures. For example, Bonferroni is easy but too conservative. Since there are $\binom{k}{2}$ pairs among k means, replacing α by $\alpha^* = \alpha/\binom{k}{2}$ would control all the comparisons at level α. Scheffee's multiple comparison procedure provides a simultaneous $(1-\alpha)$-level confidence interval for all linear combinations of population means and as a special case all pairwise differences. Scheffee's $(1-\alpha)100\%$ confidence interval for $\mu_i - \mu_j$ is given by

$$|\bar{y}_i - \bar{y}_j| \pm s \sqrt{(k-1)F_{\alpha,k-1,N-k}\left(\frac{1}{n_i} + \frac{1}{n_j}\right)}.$$

Sidak's multiple comparison confidence intervals are

$$|\bar{y}_i - \bar{y}_j| \pm t_{N-k,\alpha^*/2} \, s \sqrt{\left(\frac{1}{n_i} + \frac{1}{n_j}\right)},$$

where $\alpha^* = 1 - (1 - \alpha)^{\frac{2}{k(k-1)}}$. Note that Sidak's comparisons are just slightly less conservative than Bonferroni's for which the $\alpha^* = \alpha / \binom{k}{2}$. Since $(1 - \alpha)^m = 1 - m\alpha + \frac{m(m-1)}{2}\alpha^2 - \dots$, Sidak's $1 - (1 - \alpha)^{\frac{2}{k(k-1)}}$ is approximately $\alpha / \binom{k}{2}$, which is the Bonferroni choice.

11.2.4 Bayesian Solution

Next we provide a Bayesian solution for the same problem. In the Win-BUGS code (anovacoagulation.odc) we stipulate that the data are normal with means equal to the grand mean, plus the effect of diet, mu[i]<-mu0 + alpha[diets[i]]. The priors on alpha[i] are noninformative and depend on the selection of identifiability constraint. Here the code uses *sum-to-zero*, a STZ constraint that fixes one of the αs, while the rest are given standard noninformative priors for the location. For example, α_1 is fixed as $-(\alpha_2 + \dots + \alpha_k)$, which explains the term sum-to-zero. Another type of constraint that ensures model identifiability is *corner* or CR constraint. In this case, "corner" value α_1 is set to 0. Then treatment 1 is considered as a baseline category.

The grand mean mu0 is given a noninformative prior as well. The parameter tau is a precision, that is, a reciprocal of variance. Traditionally, the noninformative prior on the precision is gamma with small parameters, in this case dgamma(0.001,0.001). From tau, the standard deviation is calculated as sigma<-sqrt(1/tau). Thus, the highlights of the code are (i) the indexing of alpha via diets[i], (ii) the identifiability constraints, and (iii) the choice of noninformative priors.

```
model{
for (i in 1:ntotal){
times[i] ~ dnorm( mu[i], tau )
mu[i] <-  mu0 + alpha[diets[i]]
}
#alpha[1] <- 0.0;       #CR Constraint
alpha[1] <- -sum( alpha[2:a] ); #STZ Constraint

mu0 ~ dnorm(0, 0.0001)
alpha[2] ~ dnorm(0, 0.0001)
alpha[3] ~ dnorm(0, 0.0001)
alpha[4] ~ dnorm(0, 0.0001)
tau ~ dgamma(0.001, 0.001)
sigma <- sqrt(1/tau)
}

DATA

list(ntotal = 24, a=4,
```

```
times =c(62, 60, 63, 59, 63, 67, 71, 64, 65, 66,
         68, 66, 71, 67, 68, 68, 56, 62, 60, 61, 63, 64, 63, 59),
diets = c(1,1,1,1, 2,2,2,2,2,2, 3,3,3,3,3,3,  4,4,4,4,4,4,4,4) )


INITS
list( mu0=0, alpha = c(NA,0,0,0), tau=1)
```

	mean	sd	MC error	val2.5pc	median	val97.5pc	start	sample
alpha[1]	–3.001	1.03	0.002663	–5.039	–3.002	–0.9566	1001	100000
alpha[2]	1.999	0.893	0.003573	0.2318	1.999	3.774	1001	100000
alpha[3]	4.001	0.8935	0.003453	2.232	4.002	5.779	1001	100000
alpha[4]	–2.999	0.8178	0.003239	–4.61	–2.999	–1.382	1001	100000
mu0	64.0	0.5248	0.00176	62.96	64.0	65.03	1001	100000
sigma	2.462	0.4121	0.001717	1.813	2.408	3.422	1001	100000
tau	0.1783	0.05631	2.312E-4	0.08539	0.1724	0.3043	1001	100000

Figure 11.2a summarizes the posteriors of treatment effects, $\alpha_1, \ldots, \alpha_4$, as boxplots. Once the simulation for ANOVA is completed in WinBUGS, this graphical output becomes available under Inference>Compare tab.

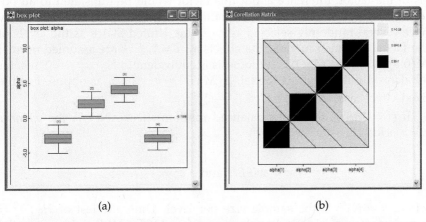

(a)   (b)

Fig. 11.2 (a) WinBUGS output from Inference>Compare. Boxplots of posterior realizations of treatment effects alpha. (b) Matrix of correlations among components of alpha.

WinBUGS can also estimate the correlations between the treatment effects. The correlation matrix below and its graphical representation (Fig. 11.2b) are outputs from Inference>Correlations.

	alpha[1]	alpha[2]	alpha[3]	alpha[4]
alpha[1]	1.0	–0.4095	–0.4056	–0.3694
alpha[2]	–0.4095	1.0	–0.3059	–0.2418
alpha[3]	–0.4056	–0.3059	1.0	–0.2476
alpha[4]	–0.3694	–0.2418	–0.2476	1.0

Note that off-diagonal correlations are negative, as expected because of the STZ constraint.

Expand the WinBUGS code ⚡anovacoagulation.odc to accommodate the six differences diff12 <- alpha[1]-alpha[2],..., diff34 <- alpha[3]-alpha[4]. Compare credible sets for the differences with MATLAB's multcompare output.

11.2.5 Fixed- and Random-Effect ANOVA

In Example 11.1 (Coagulation Times) the levels of the factor are fixed: $diet A$, ..., $dietD$. Such ANOVAs are called fixed-effect ANOVAs. Sometimes the number of factor levels is so large that a random subset is selected to serve as a set of levels. Then the inference is not concerned with these specific randomly selected levels but with the population of levels. For example, in measuring the response in animals to a particular chemical in food, a researcher may believe that the type of animal may be a factor. He/she would select a random but small number of different species as the levels of the factor. Inference about this factor would be translated to all potential species. In measuring the quality of healthcare, the researcher may have several randomly selected cities in the United States as the levels of the factor. For such models, the effects α_i, $i = 1,\ldots,k$ are assumed normal $\mathcal{N}(0,\sigma_\alpha^2)$, and the ANOVA hypothesis is equivalent to $H_0 : \sigma_\alpha^2 = 0$. Thus, for a balanced random-effect model, $\mathbb{E}(MSTr) = \sigma^2 + n\sigma_\alpha^2$, as opposed to the fixed-effect case, $\mathbb{E}(MSTr) = \sigma^2 + \frac{n}{k-1}\sum_{i=1}^{k}\alpha_i^2$.

If $s_1^2 = MSTr$, then in a balanced random-effect ANOVA the *variance components* are estimated as:

$$\hat{\sigma}^2 = s^2 = MSE \quad \text{and} \quad \hat{\sigma}_\alpha^2 = \frac{s_1^2 - s^2}{n},$$

where $n = N/k$ is the sample size per level. Thus, the test of $H_0 : \sigma_\alpha^2 = 0$ versus $H_1 : \sigma_\alpha^2 \neq 0$ is based on $s_1^2/s^2 = MSTr/MSE$, which has the F-distribution with $k-1, N-k$ degrees of freedom. Operationally, the random-effect and fixed-effect ANOVAs coincide, and the same ANOVA table can be used. The two differ mostly in the interpretation of the inference and the power analysis. Gauge R&R ANOVA in Sec. 11.11 is an example of a random-effect ANOVA.

11.3 Welch's ANOVA*

When population variances are different, the standard ANOVA may be misleading. The errors may go either way: either to reject true H_0 or fail to reject wrong H_0. Numerous approaches have been suggested to tackle the theoretical and practical aspects of heteroscedasticity. The Welch procedure (Welch, 1951) has proved to provide excellent Type I error control, superior power performance, and straightforward calculations. Unlike the traditional ANOVA, we will not assume equality of population variances, although the normality and independence is assumed.

Recall that one-way ANOVA was an extension of a two-sample t-test where population variances were assumed equal. Welch's ANOVA is an extension of Welch–Satterthwaite test, that is, the "Scenario 2" in a two-sample testing (page 431) where no explicit assumption on population variances was made.

Let, as in the traditional ANOVA setup, $y_{ij}, i = 1,\ldots,k; j = 1,\ldots,n_k$ be the observations from k populations with normal $\mathcal{N}(\mu_i, \sigma_i^2)$ distributions. Denote, as usual, $\bar{y}_i = \frac{1}{n_i}\sum_{j=1}^{n_i} y_{ij}$ and $s_i^2 = \frac{1}{n_i-1}\sum_{j=1}^{n_i}(y_{ij} - \bar{y}_i)^2$, the mean and variance of the ith sample.

Define

$$\omega_i = \frac{n_i}{s_i^2}, \ i = 1,\ldots,k$$

$$\hat{y} = \sum_{i=1}^{k} \omega_i \bar{y}_i \Big/ \sum_{i=1}^{k} \omega_i$$

$$Q = \sum_{i=1}^{k} \frac{1}{n_i - 1}\left(1 - \omega_i \Big/ \sum_{i=1}^{k}\omega_i\right)^2$$

$$v^2 = \frac{\sum_{i=1}^{k}\omega_i(\bar{y}_i - \hat{y})^2/(k-1)}{1 + 2(k-2)Q/(k^2-1)} \sim F_{k-1,v}, \ v = \frac{k^2-1}{3Q}$$

The statistic v^2 has an F-distribution with $k-1$ and $v = \frac{k^2-1}{3Q}$ degrees of freedom. Large values of F are critical for H_0.

Example 11.4. **Welch's Illustration.** As an illustration, we slightly modify an example discussed in Welch (1951, p. 335). The numerical characteristics of samples obtained from three normal populations are summarized in the table:

	Sample size	Mean	Variance
Sample 1	28	27.8	60.1
Sample 2	10	24.1	6.3
Sample 3	7	22.2	15.4

(The original example had sample sizes 20, 10, and 10). Using Welch's ANOVA, we will test the hypothesis that the population means are the same. The weights ω_i for the three samples are 0.4659, 1.5873, and 0.4545. This gives $\hat{y} = 24.4430$, $Q = 0.1512$ and the test statistic $v^2 = 3.7212$. This statistic has an F-distribution with $k - 1$ and $v = 17.6314$ degrees of freedom. The resulting p-value is 0.0449. Thus, at 5% level the test is barely significant. See ◢ welchexample.m. If the traditional ANOVA is applied (see Exercise 11.3 on how to conduct ANOVA when group means and variances are given), it results in a p-value of 0.0792, thus, which is not significant at 5% level.

Multiple Comparisons for Heterogeneous Variances. There are several procedures that we can use to compare pairs of means when the hypothesis of equality of population means is rejected. The simplest is to repeat a two-sample t-test with Welch–Satterthwaite degrees of freedom and a corrected significance level by Bonferroni or Šidák. For example, if there are k levels of a factor, then $C = \binom{k}{2}$ is the number of possible pairs of means. For the overall confidence of $(1 - \alpha)100\%$, the t-quantile in the repeated tests becomes $1 - \alpha/(2C)$ quantile for Bonferroni or $(1 - \alpha/2)^{1/C}$ quantile for Šidák's correction.

Simulation in Welch ANOVA. Using MATLAB, we simulate below an example that shows Welch's ANOVA performing superbly in a situation where the traditional ANOVA fails. The code is given in ◢ welchsimul.m. Consider balanced setup with 4 groups and group sample sizes of $n = 200$, obtained from normal distributions with mean 0 and variances $3^2, 1^2, 1^2$, and 1^2. In this case $H_0 : \alpha_1 = \alpha_2 = \alpha_3 = \alpha_4 = 0$ is correct, and in a repeated testing the p-values should be uniformly distributed on (0,1).

```
%Welch ANOVA Simulations: welchsimul.m
k=4; nk=200;
pvals=[]; polds=[];
M=50000;
for i = 1:M
    yy=randn(k, nk);
    y1=3*yy(1,:); y2= yy(2,:); ...
    y3 = yy(3,:); y4= yy(4,:);
yibar=[mean(y1) mean(y2) mean(y3) mean(y4)];
  ni=[nk nk nk nk];
  si2 = [var(y1) var(y2) var(y3) var(y4)];
%Welch's test
wei =ni./si2 ;
nwei=wei/sum(wei);
yhat= sum(nwei.*yibar);
Q=sum( (1-nwei).^2./(ni-1));
W=sum(wei.*(yibar - yhat).^2)/(k-1);
v2 = W/(1+2*(k-2)*Q/(k^2 -1));
nu = (k^2-1)/(3 * Q);
pval = 1-fcdf(v2, k-1, nu);
pvals=[pvals pval];
%ANOVA test
```

```
pold = anova1([y1 y2 y3 y4], ...
    [ones(1,nk)  2*ones(1,nk)  3*ones(1,nk)  4*ones(1,nk)],'off');
polds=[polds pold];
end
figure;
bihist(pvals,  polds, 50)
anovap=sum(polds<0.05)/M   % 0.0805
welchp=sum(pvals<0.05)/M   % 0.0501
```

In $M = 50{,}000$ simulations the p-values for both tests are saved and summarized via back-to-back histograms, as in Figure 11.3. Note the correct behavior of Welch's test: its p-values are uniformly distributed and the proportion of times H_0 was rejected in a nominal 5%-level testing was 0.0501. But for the same data, the regular ANOVA rejected correct H_0 more than 8% of the time. The non-uniformity of ANOVA-produced p-values is evident from Figure 11.3.

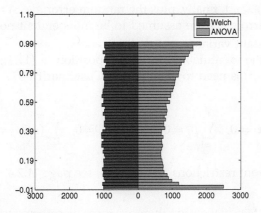

Fig. 11.3 Histograms of p-values in testing a correct H_0 by Welch's ANOVA (blue) and traditional ANOVA (green).

11.4 Two-Way ANOVA and Factorial Designs

Many experiments involve two or more factors. For each combination of factor levels an experiment is performed and the response recorded. We will discuss only factorial designs with two factors; the interested reader is directed to Kutner et al. (2004) for a comprehensive treatment of multifactor designs and incomplete factorial designs.

Denote the two factors by A and B and assume that factor A has a levels and B has b levels. Then for each $a \times b$ combination of levels we perform the experiment $n \geq 1$ times. Measurements at fixed levels of A and B are called replicates. Such a design will be called a *factorial design*. When factors are arranged in a factorial design, they are called *crossed*. If the number of replicates is the same for each cell (fixed levels for A and B), then the design is called balanced. We will be interested not only in how the factors influence the response, but also if the factors interact.

Suppose that the responses y_{ijk} are obtained under the ith level of factor A and the jth level of factor B. For each cell (i,j) one obtains n_{ij} replicates, and y_{ijk} is the kth replicate. The model for y_{ijk} is

$$y_{ijk} = \mu + \alpha_i + \beta_j + (\alpha\beta)_{ij} + \epsilon_{ijk},$$
$$i = 1,\ldots,a;\; j = 1,\ldots,b;\; k = 1,\ldots,n_{ij}. \qquad (11.4)$$

Thus, the observation y_{ijk} is modeled as the grand mean μ, plus the influence of factor A, α_i, plus the influence of factor B, β_j, plus the interaction term $(\alpha\beta)_{ij}$, and, finally, plus the random error ϵ_{ijk}. As in the one-way ANOVA, the errors ϵ_{ijk} are assumed to be independent normal with zero mean and constant variance σ^2 for all i, j, and k.

To ensure the identifiability of decomposition in (11.4), restrictions on α_is, β_js, and $(\alpha\beta)_{ij}$s need to be imposed. Customarily, STZ constraints are assumed

$$\sum_{i=1}^{a}\alpha_i = 0,\; \sum_{j=1}^{b}\beta_j = 0,\; \sum_{i=1}^{a}(\alpha\beta)_{i,j} = 0,\; \sum_{j=1}^{b}(\alpha\beta)_{i,j} = 0,$$

although different restrictions are possible, see page 11.2.4.

In a two-factor factorial design there are three hypotheses to be tested: effects of factor A,

$$H_0' : \alpha_1 = \alpha_2 = \cdots = \alpha_a = 0 \quad \text{versus} \quad H_1' = (H_0')^c,$$

effects of factor B,

$$H_0'' : \beta_1 = \beta_2 = \cdots = \beta_b = 0 \quad \text{versus} \quad H_1'' = (H_0'')^c,$$

and the interaction of A and B,

$$H_0''' : (\alpha\beta)_{11} = (\alpha\beta)_{12} = \cdots = (\alpha\beta)_{ab} = 0 \quad \text{versus} \quad H_1''' = (H_0''')^c.$$

The variability in observations follows the fundamental ANOVA identity in which the total sum of squares (SST) is represented as a sum of the A-treatment sum of squares (SSA), a sum of the B-treatment sum of squares (SSB), the interaction sum of squares ($SSAB$), and the sum of squares due to error (SSE). For a balanced design in which the number of replicates in all cells is n,

$$SST = SSA + SSB + SSAB + SSE$$

$$= \sum_{i=1}^{a}\sum_{j=1}^{b}\sum_{k=1}^{n}(y_{ijk} - \overline{y}_{...})^2$$

$$= bn\sum_{i=1}^{a}(\overline{y}_{i..} - \overline{y}_{...})^2 + an\sum_{j=1}^{b}(\overline{y}_{.j.} - \overline{y}_{...})^2 + n\sum_{i=1}^{a}\sum_{j=1}^{b}(\overline{y}_{ij.} - \overline{y}_{i..} - \overline{y}_{.j.} + \overline{y}_{...})^2$$

$$+ \sum_{i=1}^{a}\sum_{j=1}^{b}\sum_{k=1}^{n}(y_{ijk} - \overline{y}_{ij.})^2.$$

Here,

$$\overline{y}_{...} = \frac{1}{abn}\sum_{i=1}^{a}\sum_{j=1}^{b}\sum_{k=1}^{n}y_{ijk}, \quad \overline{y}_{i..} = \frac{1}{bn}\sum_{j=1}^{b}\sum_{k=1}^{n}y_{ijk},$$

$$\overline{y}_{.j.} = \frac{1}{an}\sum_{i=1}^{a}\sum_{k=1}^{n}y_{ijk}, \quad \text{and} \quad \overline{y}_{ij.} = \frac{1}{n}\sum_{k=1}^{n}y_{ijk}.$$

The point estimator for α_i effects is $\hat{\alpha}_i = \overline{y}_{i..} - \overline{y}_{...}$, for β_j effects it is $\hat{\beta}_j = \overline{y}_{.j.} - \overline{y}_{...}$, and for the interaction $(\alpha\beta)_{ij}$ it is $\widehat{(\alpha\beta)}_{ij} = \overline{y}_{ij.} - \overline{y}_{i..} - \overline{y}_{.j.} + \overline{y}_{...}$.

The degrees of freedom are partitioned according to the ANOVA identity as $abn - 1 = (a-1) + (b-1) + (a-1)(b-1) + ab(n-1)$. Outputs in standard statistical packages include degrees of freedom (DF), mean sum of squares, F-ratios, and their p-values.

Source	DF	SS	MS	F	p-value
Factor A	$a-1$	SSA	$MSA = \frac{SSA}{a-1}$	$F_A = \frac{MSA}{MSE}$	$\mathbb{P}(F_{a-1,ab(n-1)} > F_A)$
Factor B	$b-1$	SSB	$MSB = \frac{SSB}{b-1}$	$F_B = \frac{MSB}{MSE}$	$\mathbb{P}(F_{b-1,ab(n-1)} > F_B)$
A × B	$(a-1)(b-1)$	SSAB	$MSAB = \frac{SSAB}{(a-1)(b-1)}$	$F_{AB} = \frac{MSAB}{MSE}$	$\mathbb{P}(F_{(a-1)(b-1),ab(n-1)} > F_{AB})$
Error	$ab(n-1)$	SSE	$MSE = \frac{SSE}{ab(n-1)}$		
Total	$abn-1$	SST			

F_A, F_B, and F_{AB} are test statistics for H_0', H_0'', and H_0''', and their large values are critical. The rationale for these tests follows from the following expected values:

$$\mathbb{E}(MSE) = \sigma^2, \quad \mathbb{E}(MSA) = \sigma^2 + \frac{nb}{a-1}\sum_{i=1}^{a}\alpha_i^2, \quad \mathbb{E}(MSB) = \sigma^2 + \frac{na}{b-1}\sum_{j=1}^{b}\beta_j^2,$$

$$\text{and } \mathbb{E}(MSAB) = \sigma^2 + \frac{n}{(a-1)(b-1)}\sum_{i=1}^{a}\sum_{j=1}^{b}(\alpha\beta)_{ij}^2.$$

Example 11.5. **Insulin Therapy.** Insulin has anti-inflammatory effects, as evaluated by its ability to reduce plasma concentrations of cytokines. The cytokine content in several organs after endotoxin (lipopolysaccharide, LPS) exposure and the effect of hyperinsulinaemia was examined in a porcine model (Brix-Christensen et al., 2005). All animals (35 to 40 kg) were subject to general anaesthesia and ventilated for 570 minutes. There were two possible interventions:

LPS: Lipopolysaccharide infusion for 180 minutes.

HEC: Hyperinsulinemic euglycemic clamp in 570 minutes (from start). Insulin was infused at a constant rate and plasma glucose was clamped at a certain level by infusion of glucose.

LPS induces a systemic inflammation (makes the animals sick) and HEC acts as a treatment. There were four experimental cells: (1) only anaesthesia (no HEC, no LPS), (2) HEC, (3) LPS, and (4) HEC and LPS.

The responses are levels of interleukin-10 (IL-10, an anti-inflammatory cytokine) in the kidney after 330 minutes have elapsed. The table corresponds to a balanced design $n = 8$, although the original experiment was unbalanced with ten animals in group 1, nine in group 2, ten in group 3, and nine in group 4.

	No HEC	Yes HEC
	7.0607 4.7510	3.0693 2.1102
No LPS	2.6168 2.9530	1.6489 3.1004
	4.3489 3.6137	2.9160 4.1170
	3.6356 5.6969	2.9149 3.0229
	3.6911 4.5554	2.4159 1.8944
	4.3933 3.8447	3.1493 3.5133
Yes LPS	6.0513 1.3590	4.4462 4.6254
	4.2559 2.1449	2.8545 3.8967

```
%insulin.m
data2 = [...          %columns: IL10  LPS  HEC
7.0607  1 1;    2.6168  1 1;    4.3489  1 1;...
3.6356  1 1;    4.7510  1 1;    2.9530  1 1;...
3.6137  1 1;    5.6969  1 1;    3.0693  1 2;...
1.6489  1 2;    2.9160  1 2;    2.9149  1 2;...
2.1102  1 2;    3.1004  1 2;    4.1170  1 2;...
3.0229  1 2;    3.6911  2 1;    4.3933  2 1;...
6,0513  2 1;    4.2559  2 1;    4.5554  2 1;...
```

```
3.8447  2 1;   1.3590  2 1;   2.1449  2 1;...
2.4159  2 2;   3.1493  2 2;   4.4462  2 2;...
2.8545  2 2;   1.8944  2 2;   3.5133  2 2;...
4.6254  2 2;   3.8967  2 2];

IL10 = data2(:,1);   LPS=data2(:,2);   HEC=data2(:,3);
[p table stats terms] = anovan( IL10, {LPS,HEC}, ...
    'varnames',{'LPS','HEC'}, 'model','interaction')
```

The resulting ANOVA table provides the test for the two factors and their interaction.

Source	DF	SS	MS	F	p-value
LPS	1	0.0073	0.0073	0.0051	0.9436
HEC	1	7.2932	7.2932	5.0532	0.0326
LPS*HEC	1	2.1409	2.1409	1.4834	0.2334
Error	28	40.4124	1.4433		
Total	31	49.8539			

Note that factor LPS is insignificant. The associated F statistic is 0.0051 with p-value of 0.9436. The interaction (LPS*HEC) is insignificant as well (p-value of 0.2334), while HEC is significant (p-value of 0.0326).

Next, we generate the *interaction plots*.

```
%insulin.m continued
cell11 = mean(data2( 1: 8,1))    %L1 H1
cell12 = mean(data2( 9:16,1))    %L1 H2
cell21 = mean(data2(17:24,1))    %L2 H1
cell22 = mean(data2(25:32,1))    %L2 H2

figure;
plot([1 2],[cell11 cell12],'o','markersize',10, ...
    'MarkerEdgeColor','k','MarkerFaceColor','r')
hold on
plot([1 2],[cell11 cell12],'r-', 'linewidth',3)
plot([1 2],[cell21 cell22],'o','markersize',10, ...
    'MarkerEdgeColor','k','MarkerFaceColor','k')
plot([1 2],[cell21 cell22],'k-', 'linewidth',3)
title('Lines for LPS=1 (red) and LPS = 2 (black)')
xlabel('HEC')


figure;
plot([1 2],[cell11 cell21],'o','markersize',10, ...
    'MarkerEdgeColor','k','MarkerFaceColor','r')
hold on
plot([1 2],[cell11 cell21],'r-','linewidth',3)
plot([1 2],[cell12 cell22],'o','markersize',10, ...
    'MarkerEdgeColor','k','MarkerFaceColor','k')
plot([1 2],[cell12 cell22],'k-','linewidth',3)
title('Lines for HEC=1 (red) and HEC = 2 (black)')
xlabel('LPS')
```

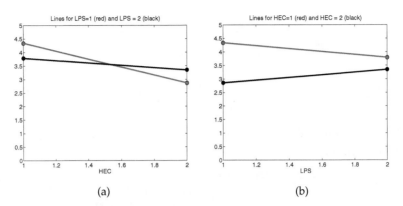

Fig. 11.4 Interaction plots of LPS against HEC (*left*) and HEC against LPS (*right*) to explore the additivity of the model.

Figure 11.4 presents treatment mean plots, also known as interaction plots. The x-axis contains the levels of the factors, in this case both factors have levels 1 and 2. The y-axis contains the means of response (IL-10). The circles in both plots correspond to the cell means.

For example, in Figure 11.4a x-axis has two levels of factor HEC. The means of IL-10 for LPS=1 are connected by the red line, while the means for LPS=2 are connected by the black line. When the lines on the plots are approximately parallel, the interaction between the factors is absent. Thus, the interaction plots serve as exploratory tools to check if the interaction term should be included in the ANOVA model. Note that some interaction between LPS and HEC is present (the lines are not perfectly parallel); however, this interaction was found not statistically significant (*p*-value of 0.2334).

Next, we visualize the ANOVA fundamental identity.

```
%insulin.m continued
SSA = table{2,2}; SSB = table{3,2}; SSAB=table{4,2};
SSE=table{5,2}; SST = table{6,2};

%Display the budget of Sums of Squares
H=figure;
set(H,'Position',[400 400 400 400]);
y=[0 0 1 1];
hold on
h1=fill([0 SST SST 0],y,'c');
y=y+1;
h2=fill([0 SSA SSA 0],y,'y');
h3=fill([0 SSB SSB 0]+SSA,y,'r');
h4=fill([0 SSAB SSAB 0]+SSA+SSB,y,'g');
h5=fill([0 SSE SSE 0]+SSA+SSB+SSAB,y,'b');
y=y+1;
h6=fill([0 SST SST 0],y,'w');
hold off
```

```
legend([h1 h2 h3 h4 h5],'SST','SSA','SSB','SSAB','SSE',...
    'Location','NorthWest')
title('Sums of Squares')
```

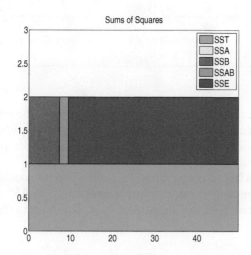

Fig. 11.5 Budget of sums of squares for Insulin example.

Figure 11.5 shows the budget of sums of squares in this design. It is a graphical representation of $SST = SSA + SSB + SSAB + SSE$. It shows the contributions to the total variability by the factors, their interaction and the error. Note that the sum of squares attributed to LPS (in yellow) is not visible in the plot. This is because its relative contribution to SST is very small, $0.0073/49.8539 < 0.00015$.

The ANOVA table and both Figs. 11.4 and 11.5 are generated by file ◀ insulin.m. For a Bayesian solution of this example consult ✳insulin.odc. ✏

11.4.1 Two-Way ANOVA: One Observation per Cell

When a two-way ANOVA has a single observation per cell, then there is no variability within the the cell and SSE is always 0. However, if two factors do not interact, the $MSAB$ has expectation σ^2 and can be used in place of MSE. This kind of design is not rare and could be dictated by experimental resources, both financial and material, and time constraints.

With the additivity assumption and single observation per cell the model from (11.4) becomes

$$y_{ij} = \mu + \alpha_i + \beta_j + \epsilon_{ij}, i = 1, \dots, a; \ j = 1, \dots, b.$$

The resulting ANOVA table is

Source	DF	SS	MS	F	p-value
Factor A	$a-1$	SSA	$MSA = \frac{SSA}{a-1}$	$F_A = \frac{MSA}{MSE}$	$\mathbb{P}(F_{a-1,(a-1)(b-1)} > F_A)$
Factor B	$b-1$	SSB	$MSB = \frac{SSB}{b-1}$	$F_B = \frac{MSB}{MSE}$	$\mathbb{P}(F_{b-1,(a-1)(b-1))} > F_B)$
Error	$(a-1)(b-1)$	SSE	$MSE = \frac{SSE}{(a-1)(b-1)}$		
Total	$ab-1$	SST			

Whether it is reasonable to assume additivity can be tested by Tukey's test. This procedure tests a particular deviation from additivity: the presence of a multiplicative interaction $(\alpha\beta)_{ij} = \delta \alpha_i \beta_j$. Given the form of interaction, the hypotheses are $H_0 : \delta = 0$ versus $H_1 : \delta \neq 0$.

Tukey's test is based on partitioning the SSE to two components, sum of squares corresponding to multiplicative interaction $SSAB^*$ and the remainder sum of squares, $SSRem$. The sum of squares assessing the multiplicative interaction is

$$SSAB^* = \frac{\left(\sum_{i=1}^{a} \sum_{j=1}^{b} y_{ij} \hat{\alpha}_i \hat{\beta}_j \right)^2}{\sum_{i=1}^{a} \hat{\alpha}_i^2 \times \sum_{j=1}^{b} \hat{\beta}_j^2}$$

where $\hat{\alpha}_i = \overline{y}_{i.} - \overline{y}_{..}$ and $\hat{\beta}_j = \overline{y}_{.j} - \overline{y}_{..}$ are estimators of α_i and β_j.

The Tukey's test for additivity is based on statistic

$$F = \frac{SSAB^*/1}{SSRem/(ab - a - b)}$$

which has an F-distribution with 1 and $ab - a - b$ degrees of freedom, when H_0 is true.

If the test is significant, then one may try to diminish interactions by transforming the observations (square root, log, or Box Cox transforms can be applied).

Example 11.6. **Yield of Turnips.** Quenouille (1953) gives an example of a two way design with a single observation per cell. The table consists of yields of turnips (cwt per 3 acre plots), with dependence on level of phosphate and level of liming that represent two factors affecting the growth.

Level of	Level of liming		
phosphate	0	1	2
0	706	998	589
1	1028	1025	998
2	928	1111	961
4	1185	1111	1014
8	1120	980	877
16	1050	1143	1053

Test for presence of multiplicative interaction using Tukey's test for additivity.

```
% turnip.m
a=6;    b=3;
Y=[706   1028   928   1185   1120   1050  ...
   998   1025  1111   1111    980   1143  ...
   589    998   961   1014    877   1053]; %Yields of turnip
pho = [ 0 1 2 4 8 16  0 1 2 4 8 16  0 1 2 4 8 16]; %Pho levels
lim = [ 0 0 0 0 0  0  1 1 1 1 1  1  2 2 2 2 2  2]; %Lim levels
[p tab stats]=anovan(Y, pho,lim);
Yi=reshape(Y,[a b]);   %Data as a matrix
alphahat = stats.coeffs(2:a+1);
betahat=stats.coeffs(a+2:a+b+1);
sse=tab{4,2} %9.3575e+04, SSE from ANOVA table output
d=0; s=0;
for i=1:a
    for j=1:b
        d=d+Yi(i,j)*alphahat(i)*betahat(j);
        s=s+alphahat(i)^2 * betahat(j)^2;
    end
end
% or avoiding the double loop by Kronecker prod
% d=sum(sum(Yi .* kron( alphahat, betahat' )))
% s =sum(sum(kron( alphahat.^2, betahat'.^2)))
ssab=d^2/s                      %3.2671e+04
ssrem = sse - ssab              %6.0904e+04
f = ssab/(ssrem/(a*b-a-b))      %4.8280
p = 1-fcdf(f, 1, a*b -a -b)     %0.0556
```

Since p-value is 0.0556, the null hypothesis of additivity is not rejected at 5% significance level. ✎

11.5 Blocking

In many cases the design can account for the variability due to subjects or to experimental runs and focus on the variability induced by the treatments that constitute the factor of interest.

Example 11.7. **Blocking by Rats.** A researcher wishes to determine whether or not four different testing procedures produce different responses to the concentration of a particular poison in the blood of experimental rats. To minimize the influence of rat-to-rat variability, the biologist selects four rats from the same litter. Each of the four rats is given the same dose of poison per gram of body weight and then samples of their blood are tested by the four testing methods administered in random order. There are four different litters each containing four rats for a total of 16 animals involved.

```
concentration = [9.3 9.4 9.2 9.7 9.4 9.3 9.4 9.6 ...
   9.6 9.8 9.5 10.0 10.0 9.9 9.7 10.2];
procedure = [1 2 3 4 1 2 3 4 1 2 3 4 1 2 3 4];
litter = [1 1 1 1 2 2 2 2 3 3 3 3 4 4 4 4];
[p,table,stats,terms]=anovan(concentration,{procedure,litter},...
'varnames',char('Procedure','Rat'))

%p = 1.0e-003 *
%
%      0.8713
%      0.0452
%
%table =
%   'Source'    'Sum Sq.'  'd.f.'  'Singular?' 'Mean Sq.'
%   'Procedure' [0.3850]   [ 3]    [   0]      [ 0.1283]
%   'Rat'       [0.8250]   [ 3]    [   0]      [ 0.2750]
%   'Error'     [0.0800]   [ 9]    [   0]      [ 0.0089]
%   'Total'     [1.2900]   [15]    [   0]            []
%
%   'F'                 'Prob>F'
%   [14.4375]           [8.7127e-004]
%   [30.9375]           [4.5233e-005]
%         []                  []
%         []                  []
```

In the above code, procedure is the factor of interest and litter is the blocking factor. Note that both the procedure and litter factors are highly significant at levels 0.0008713 and 0.0000452, respectively. We are interested in significant differences between the levels of procedure factor, but not between the litters or individual animals. However, it is desirable that the blocking factor turns out to be significant, since in that case we would have accounted for significant variability attributed to blocks and separated it from the variability attributed to the test procedures. This makes the test more accurate.

To emphasize the benefits of blocking, we provide a nonsolution by treating this problem as a one-way ANOVA layout. This time, we fail to find any significant difference between the testing procedures (p-value 0.2196). Clearly, this approach is incorrect on other grounds: the condition of independence among the treatments, required for ANOVA, is violated.

```
[p,table,stats] = anova1(concentration,procedure)

%p = 0.2196
%
%table =
%'Source'      'SS'       'df'      'MS'        'F'        'Prob>F'
%'Groups'     [0.3850]    [ 3]    [0.1283]    [1.7017]    [0.2196]
%'Error'      [0.9050]    [12]    [0.0754]       []          []
%'Total'      [1.2900]    [15]       []          []          []
```

In many cases the blocking is done by batches of material, animals from the same litter, matching the demographic characteristics of the patients, etc. The repeated measures design is a form of block design where the blocking is done by subjects, individual animals, etc.

11.6 Repeated Measures Design

Repeated measures designs represent a generalization of the paired t-test to designs with more than two groups/treatments. In repeated measures designs the blocks are usually subjects, motivating the name "within-subject ANOVA" that is sometimes used. Every subject responds at all levels of the factor of interest, i.e., treatments. For example, in clinical trials, the subjects' responses could be taken at several time instances.

Such designs are sometimes necessary and have many advantages. The most important advantage is that the design controls for the variability between the subjects, which is usually not of interest. In simple words, subjects serve as their own controls, and the variability between them does not "leak" into the variability between the treatments. Another advantage is operational. Compared with factorial designs, repeated measures need fewer participants.

In the repeated measures design the independence between treatments is violated. Naturally, the responses of a subject are dependent on each other across treatments.

11.6.1 ANOVA Table for Repeated Measures

Subject	Treatment 1	2	...	k	Subject totals
1	y_{11} y_{12}		$\cdots$	y_{1k}	$y_{1\cdot}$
2	y_{21} y_{22}		$\cdots$	y_{2k}	$y_{2\cdot}$
...					...
n	y_{n1} y_{n2}		$\cdots$	y_{nk}	$y_{n\cdot}$
Treatment totals	$y_{\cdot 1}$ $y_{\cdot 2}$		$\cdots$	$y_{\cdot k}$	$y_{\cdot\cdot}$

The repeated measures model is

$$y_{ij} = \mu + \beta_i + \alpha_j + \epsilon_{ij}, \; i = 1, \ldots, n; \; j = 1, \ldots, k,$$

in which the hypothesis $H_0 : \alpha_j = 0$, $j = 1, \ldots, k$, is of interest. Here, subject effects β_is are random and assumed to be normal $\mathcal{N}(0, \sigma_\beta^2)$. The treatment effects α_j are fixed and satisfy identifiability constraint $\sum_{j=1}^{k} \alpha_j = 0$. The errors ϵ_{ij} and effects β_i are mutually independent.

In the repeated measures design, total sum of squares $SST = \sum_{i=1}^{n} \sum_{j=1}^{k} (y_{ij} - \bar{y}_{..})^2$ partitions in the following way:

$$SST = SS_{BetweenSubjects} + SS_{WithinSubjects} = SSB + [SSA + SSE].$$

Equivalently,

$$\sum_{i=1}^{n} \sum_{j=1}^{k} (y_{ij} - \bar{y}_{..})^2 = k \sum_{i=1}^{n} (\bar{y}_{i.} - \bar{y}_{..})^2 + \sum_{i=1}^{n} \sum_{j=1}^{k} (y_{ij} - \bar{y}_{i.})^2$$

$$= k \sum_{i=1}^{n} (\bar{y}_{i.} - \bar{y}_{..})^2 + \left[n \sum_{j=1}^{k} (\bar{y}_{.j} - \bar{y}_{..})^2 + \sum_{i=1}^{n} \sum_{j=1}^{k} (y_{ij} - \bar{y}_{i.} - \bar{y}_{.j} + \bar{y}_{..})^2 \right].$$

The degrees of freedom are split as

$$kn - 1 = (n - 1) + n(k - 1) = (n - 1) + [(k - 1) + (n - 1) \cdot (k - 1)].$$

The ANOVA table is

Source	DF	SS	MS	F	p
Factor A	$k - 1$	SSA	$MSA = \frac{SSA}{k-1}$	$F_A = \frac{MSA}{MSE}$	$\mathbb{P}(F_{k-1, (k-1)(n-1)} > F_A)$
Subjects B	$n - 1$	SSB	$MSB = \frac{SSB}{n-1}$	$F_B = \frac{MSB}{MSE}$	$\mathbb{P}(F_{n-1, (k-1)(n-1)} > F_B)$
Error	$(k-1)(n-1)$	SSE	$MSE = \frac{SSE}{(k-1)(n-1)}$		
Total	$kn - 1$	SST			

The test statistic for Factor A ($H_0 : \alpha_j = 0, j = 1,\ldots,k$) is $F_A = MSA/MSE$ for which the p-value is $p = \mathbb{P}(F_{k-1,(k-1)(n-1)} > F_A)$. Usually we are not interested in $F_B = MSB/MSE$; however, its significance would mean that blocking by subjects was efficient in accounting for some variability, thus making the inference about Factor A more precise. If a formal test for the significance of subject effects is needed, the hypotheses are written as

$$H_0 : \sigma_\beta^2 = 0 \quad \text{vs.} \quad H_1 : \sigma_\beta^2 > 0,$$

since β's are random effects and $\beta_i \sim \mathcal{N}(0, \sigma_\beta^2)$ (see page 520).

Example 11.8. **Kidney Dialysis.** Eight patients each underwent three different methods of kidney dialysis (Daugridas and Ing, 1994). The following values were obtained for weight change in kilograms between dialysis sessions:

Patient	Treatment 1	Treatment 2	Treatment 3
1	2.90	2.97	2.67
2	2.56	2.45	2.62
3	2.88	2.76	1.84
4	1.73	1.20	1.33
5	2.50	2.16	1.27
6	3.18	2.89	2.39
7	2.83	2.87	2.39
8	1.92	2.01	1.66

Test the null hypothesis that there is no difference in mean weight change among treatments. Use $\alpha = 0.05$.

```
%dialysis.m
weich=[  2.90  2.97  2.67 ;...
         2.56  2.45  2.62 ;...
         2.88  2.76  1.84 ;...
         1.73  1.20  1.33 ;...
         2.50  2.16  1.27 ;...
         3.18  2.89  2.39 ;...
         2.83  2.87  2.39 ;...
         1.92  2.01  1.66 ];

subject=[1 2 3 4 5 6 7 8  ...
     1 2 3 4 5 6 7 8  1 2 3 4 5 6 7 8];
treatment = [1 1 1 1 1 1 1 1 ...
     2 2 2 2 2 2 2 2  3 3 3 3 3 3 3 3];
[p table stats terms] = ...
anovan(weich(:),{subject, treatment},'varnames',...
    {'Subject' 'Treatment'} )

%'Source'    'Sum Sq.' 'd.f.' 'Mean Sq.'     'F'    'Prob>F'
```

```
%'Subject'    [5.6530]   [ 7]   [0.8076] [11.9341][6.0748e-005]
%'Treatment' [1.2510]   [ 2]   [0.6255] [ 9.2436][      0.0028]
%'Error'     [0.9474]   [14]   [0.0677]         []            []
%'Total'     [7.8515]   [23]         []         []            []

SST = table{5,2}; SSE = table{4,2};
SSA = table{3,2}; SSB = table{2,2};
SSW = SST - SSB;
```

Since the hypothesis of equality of treatment means is rejected (p-val = 0.0028), one may look at the differences of treatment effects to find out which means are different. Command `multcompare(stats,'dimension',2)` will perform multiple comparisons along the second dimension, `Treatment`, and produces:

```
%1.0000    2.0000   -0.1917    0.1488   0.4892
%1.0000    3.0000    0.2008    0.5413   0.8817
%2.0000    3.0000    0.0521    0.3925   0.7329
```

This output is interpreted as $\alpha_1 - \alpha_2 \in [-0.1917, 0.4892]$, $\alpha_1 - \alpha_3 \in [0.2008, 0.8817]$ and $\alpha_2 - \alpha_3 \in [0.0521, 0.7329]$ with simultaneous confidence of 95%. Thus, by inspecting which interval contains 0, we conclude that treatment means 1 and 2 are not significantly different, while the mean for treatment 3 is significantly smaller from the means for treatments 1 and 2. For a Bayesian solution consult 👆dialysis.odc.

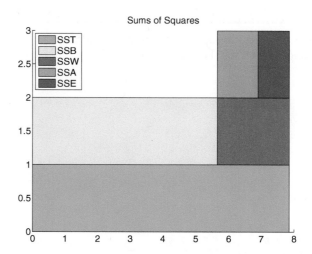

Fig. 11.6 Budget of sums of squares for Kidney Dialysis example.

Figure 11.6 shows the budget of sums of squares. Note that the *SSTr* and *SSE* comprise *SSW* (sum of squares within), while *SSBl* (variability due to subjects) is completely separated from *SSTr* and *SSE*. If the blocking by subjects is ignored and the problem is considered as a one-way ANOVA, the subject variability will be a part of SSE leading to wrong inference about the treatments (Exercise 11.16).

11.6.2 Sphericity Tests

Instead of the independence condition, as required in ANOVA, another condition is needed for repeated measures in order for an inference to be valid. This is the condition of *sphericity* or *circularity*. In simple terms, the sphericity assumption requires that all pairs of treatments be positively correlated in the same way. Another way to express sphericity is that variances of all pairwise differences between treatments are the same.

When an F-test in a repeated measures scenario is performed, many statistical packages (SAS, SPSS, etc.) automatically generate corrections for violations of sphericity. Examples are the Greenhouse–Geisser, the Huynh–Feldt, and lower-bound corrections. These packages correct for sphericity by altering the degrees of freedom, thereby altering the p-value for the observed F-ratio.

Opinions differ about which correction is the best, and this depends on what one wants to control in the analysis. An accepted universal choice is to use the Greenhouse–Geisser correction, ϵ_{GG}. Violation of sphericity is more serious when ϵ_{GG} is smaller. When $\epsilon_{GG} > 0.75$, the Huynh–Feldt correction ϵ_{HF} is recommended.

An application of Greenhouse–Geisser's and Huynh–Feldt's corrections, ϵ_{GG} and ϵ_{HF}, respectively, is provided in the script below (circularity.m) in the context of Example 11.8.

```
weich=[ 2.90  2.97  2.67 ;    2.56  2.45  2.62 ;...
        2.88  2.76  1.84 ;    1.73  1.20  1.33 ;...
        2.50  2.16  1.27 ;    3.18  2.89  2.39 ;...
        2.83  2.87  2.39 ;    1.92  2.01  1.66 ];
n = size(weich,1);      %number of subjects, n=8
k = size(weich,2);      %number of treatments, k=3
Sig = cov(weich);       %covariance matrix of weich
md = trace(Sig)/k;      %mean of diagonal entries of Sig
ma = mean(mean(Sig));   %mean of all components in Sig
mr = mean(Sig');        %row means of Sig
A = (k*(md-ma))^2;
B = (k-1)*(sum(sum(Sig.^2))-2*k*sum(mr.^2)+k^2*ma^2);
epsGG = A/B             %Greenhouse-Geisser epsilon 0.7038
epsHF = (n*(k-1)*epsGG-2)/((k-1)*((n-1)-(k-1)*epsGG))
                        %Huynh-Feldt epsilon 0.8281
%Corrections based on Trujillo-Ortiz et al. functions
%epsGG.m and epsHF.m available on MATLAB Central.
```

```
F = 9.2436; %F statistic for testing treatment differences
p = 1-fcdf(F,k-1,(n-1)*(k-1))  %original pvalue 0.0028
%
padjGG = 1-fcdf(F,epsGG*(k-1),epsGG*(n-1)*(k-1)) %0.0085
padjHF = 1-fcdf(F,epsHF*(k-1),epsHF*(n-1)*(k-1)) %0.0053
```

Note that both degrees of freedom in the F statistic for testing the treatments are multiplied by correction factors, which increased the original p-value. In this example the corrections for circularity did not change the original decision of rejection of hypothesis H_0 stating the equality of treatment effects.

Remark. To check whether the Repeated Measures design is an appropriate model one can

(i) inspect the QQ-plot of residuals, $e_{ij} = y_{ij} - \bar{y}_{i.} - \bar{y}_{.j} + \bar{y}_{..}$ plotted against normal quantiles;

(ii) overlay profiles $y_{i1}, y_{i2}, \ldots, y_{i,k}$ for all n subjects. Since the repeated measures model is additive by definition, the profiles should appear approximately parallel.

11.7 Nested Designs*

In the factorial design two-way ANOVA, two factors are *crossed*. This means that at each level of factor A we get measurements under all levels of factor B, that is, all cells in the design are nonempty. In Example 11.5, two factors, HEC and LPS, given at two levels each ("yes" and "no"), form a 2×2 table with four cells. The factors are crossed, meaning that for each combination of levels we obtained observations. Sometimes this is impossible to achieve due to the nature of the experiment.

Suppose, for example, that four diets are given to mice and that we are interested in the mean concentration of a particular chemical in the tissue. Twelve experimental animals are randomly divided into four groups of three and each group put on a particular diet. After 2 weeks the animals are sacrificed and from each animal the tissue is sampled at five different random locations. The factors "diet" and "animal" cannot be crossed. After a single dietary regime, taking measurements on an animal requires its sacrifice, thus repeated measures designs are impossible.

The design is nested, and the responses are

$$y_{ijk} = \mu + \alpha_i + \beta_{j(i)} + \epsilon_{ijk}, \quad i = 1, \ldots, 4; \ j = 1, \ldots, 3; \ k = 1, \ldots, 5,$$

where μ is the grand mean, α_i is the effect of the ith diet, and $\beta_{j(i)}$ is the effect of animal j, which is nested within treatment i.

For a general balanced two-factor nested design,

$$y_{ijk} = \mu + \alpha_i + \beta_{j(i)} + \epsilon_{ijk}, \quad i = 1, \ldots, a; \ j = 1, \ldots, b; \ k = 1, \ldots, n,$$

the identifiability constraints are $\sum_{i=1}^{a} \alpha_i = 0$, and for each i, $\sum_{j=1}^{b} \beta_{j(i)} = 0$. The ANOVA identity $SST = SSA + SSB(A) + SSE$ is

$$\sum_{i=1}^{a}\sum_{j=1}^{b}\sum_{k=1}^{n}(y_{ijk} - \bar{y}_{...})^2$$

$$= bn\sum_{i=1}^{a}(\bar{y}_{i..} - \bar{y}_{...})^2 + n\sum_{i=1}^{a}\sum_{j=1}^{b}(\bar{y}_{ij.} - \bar{y}_{...})^2 + \sum_{i=1}^{a}\sum_{j=1}^{b}\sum_{k=1}^{n}(y_{ijk} - \bar{y}_{ij.})^2.$$

The degrees of freedom are partitioned according to the ANOVA identity as $abn - 1 = (a-1) + a(b-1) + ab(n-1)$. The ANOVA table is shown below:

Source	DF	SS	MS
A	$a-1$	SSA	$MSA = \frac{SSA}{a-1}$
B(A)	$a(b-1)$	SSB	$MSB = \frac{SSB}{a(b-1)}$
Error	$ab(n-1)$	SSE	$MSE = \frac{SSE}{ab(n-1)}$
Total	$abn-1$	SST	

Notice that the table does not provide the F-statistics and p-values. This is because the inferences differ depending on whether the factors are fixed or random.

The test for the main effect, H_0 : all $\alpha_i = 0$, is based on $F = MSA/MSE$ if both factors A and B are fixed, and on $F = MSA/MSB(A)$ if at least one of the factors is random. The test for H_0 : all $\beta_{j(i)} = 0$ is based on $F = MSB(A)/MSE$ in both cases. In the mouse-diet example, factor B (animals) is random.

Example 11.9. **Nested Mice.** Suppose that the data for the mouse-diet study are given as

Diet	1			2			3			4		
Animal	1	2	3	1	2	3	1	2	3	1	2	3
$k=1$	65	68	56	74	69	73	65	67	72	81	76	77
$k=2$	71	70	55	76	70	77	74	59	63	75	72	69
$k=3$	63	64	65	79	80	77	70	61	64	77	79	74
$k=4$	69	71	68	81	79	79	69	66	69	75	82	79
$k=5$	73	75	70	72	68	68	73	71	70	80	78	66

There are 12 mice in total, and the diets are assigned 3 mice each. On each mouse 5 measurements are taken. As we pointed out, this does not constitute a design with crossed factors since, for example, animal 1 under diet 1 differs from animal 1 under diet 2. Rather, the factor animal is nested within the factor diet. The following script is given in ◀ nesta.m:

```matlab
yijk =[...
        65  68  56    74  69  73    65  67  72    81  76  77 ;...
        71  70  55    76  70  77    74  59  63    75  72  69 ;...
        63  64  65    79  80  77    70  61  64    77  79  74 ;...
        69  71  68    81  79  79    69  66  69    75  82  79 ;...
        73  75  70    72  68  68    73  71  70    80  78  66  ];

    a = 4;
    b = 3;
    n = 5;

    %matrices of means (y..., y_i.., y_ij.)
    yddd = mean(mean(yijk)) * ones(n, a*b)
    yijd = repmat(mean(yijk), n, 1) .
    %yidd--------------------
        m=mean(yijk);
        mm=reshape(m', b, a);
        c=mean(mm);
        d=repmat(c',1,b);
        e=d';
    yidd = repmat(e(:)',n,1)

SST = sum(sum((yijk - yddd).^2) )    %2.3166e+003
SSA = sum(sum((yidd - yddd).^2) )    %1.0227e+003
SSB_A = sum(sum((yijd - yidd).^2) )  %295.0667
SSE = sum(sum((yijk - yijd).^2) )    %998.8

MSA = SSA/(a-1)              %340.9111
MSB_A = SSB_A/(a * (b-1))    %36.8833
MSE = SSE/(a * b * (n-1))    %20.8083

%A fixed B(A) random //// 0r A random B(A) random
FArand = MSA/MSB_A    %9.2430
pa = 1- fcdf(FArand, a-1, a*(b-1))    %0.0056
%
FB_A = MSB_A/MSE    %1.7725
pb_a =  1- fcdf(FB_A,  a*(b-1), a*b*(n-1))    %0.1061
```

Figure 11.7 shows the budget of sums of squares for this example. From the analysis we conclude that the effects of factor A (diet) were significant ($p = 0.0056$), while the effects of factor B (animal) were not significant ($p = 0.1061$).

MATLAB's built-in anovan can handle nested designs. Here is a solution that uses anovan. For sintax details consult MATLAB's help.

```matlab
%%Need to recode the data and factors as row vectors
yijk =[...
        65  68  56    74  69  73    65  67  72    81  76  77 ...
        71  70  55    76  70  77    74  59  63    75  72  69 ...
        63  64  65    79  80  77    70  61  64    77  79  74 ...
```

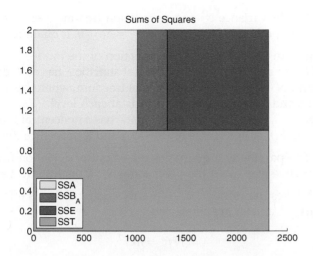

Fig. 11.7 The budget of sums of squares for the mouse-diet study.

```
     69   71   68     81   79   79     69   66   69     75   82   79 ...
     73   75   70     72   68   68     73   71   70     80   78   66  ];
  diet = repmat([1 1 1 2 2 2 3 3 3 4 4 4], 1, 5);
  animal = repmat([1 2 3],1, 20);

mynest=[0 0;1 0];      %2 nested in 1, matrix element (2,1) is 1.
%animals within diets are random, need to declare 'random',[2]
[p table stats]=...
anovan(yijk,{diet, animal},'nested',mynest,'random',[2],'varnames',{'diet',
'animal'}) table
% Analysis of Variance
%    Source        Sum Sq.   d.f.   Mean Sq.      F          Prob>F
%    diet          1022.7333    3    340.9111    9.243       0.00560
%    animal(diet)   295.0667    8     36.8833    1.7725      0.10609
%    Error           998.8     48     20.8083
%    Total          2316.6     59
```

Remarks. (1) Note that the nested model does not have the interaction term. This is because the animals are nested within the diets. (2) For designs that are not balanced, the calculations are substantially more complex and a regression model should be used whenever possible.

11.8 Power Analysis in ANOVA

To design a sample size for an ANOVA test, one needs to specify the significance level, desired power, and a precision or effect. Precision is defined in terms of ANOVA variance σ^2 and population treatment effects α_i coming from the null hypothesis $H_0 : \alpha_1 = \alpha_2 = \cdots = \alpha_k = 0$. It quantifies the extent of deviation from H_0 and usually is a function of the ratio $\frac{\sum n_i \alpha_i^2}{\sigma^2}$.

Under H_0, for a balanced design, the test statistic F has an F-distribution with $k - 1$ and $N - k = k(n - 1)$ degrees of freedom, where k is the number of treatments and n the number of subjects at each level.

However, if H_0 is not true, the test statistic has a noncentral F-distribution with $k - 1$ and $k(n - 1)$ degrees of freedom and a noncentrality parameter $\lambda = n \frac{\sum_i \alpha_i^2}{\sigma^2}$. The parameter λ quantifies the extent of deviation from H_0.

An alternative way to set the precision is via Cohen's effect size, which for ANOVA takes the form $f^2 = \frac{1/k \sum_i \alpha_i^2}{\sigma^2}$. Note that Cohen's effect size and noncentrality parameter are connected via

$$\lambda = N f^2 = n k f^2.$$

Since determining the sample size is a prospective task, information about σ^2 and α_i may not be available. In the context of ANOVA, Cohen (1988) recommends effect sizes of $f^2 = 0.1^2$ as small, $f^2 = 0.25^2$ as medium, and $f^2 = 0.4^2$ as large.

The power in ANOVA is, by definition,

$$1 - \beta = \mathbb{P}(F^{nc}_{k-1, N-k}(\lambda) > F_{k-1, k(n-1), 1-\alpha}), \qquad (11.5)$$

where $F^{nc}_{k-1, N-k}(\lambda)$ is a random variable with a noncentral F-distribution with $k - 1$ and $N - k$ degrees of freedom and noncentrality parameter λ. The quantity $F_{k-1, N-k, 1-\alpha}$ is the $1 - \alpha$ quantile of a standard F-distribution with $k - 1$ and $N - k$ degrees of freedom.

The interplay among the power, sample size, and effect size is illustrated by the following example:

Example 11.10. **Sample Size in One-Way ANOVA.** Suppose $k = 4$ treatment means are to be compared at a significance level of $\alpha = 0.05$. The experimenter is to decide how many replicates n to run at each level, so that the null hypothesis is rejected with a probability of at least 0.9 if $f^2 = 0.0625$ or if $\sum_i \alpha_i^2$ is equal to $\sigma^2/4$.

For $n = 60$, that is, $N = 4 \times 60 = 240$, the power is calculated in a one-line command:

```
1-ncfcdf(finv(1-0.05, 4-1, 4*(60-1)), 4-1, 4*(60-1), 15)   %0.9122
```

Here we used $\lambda = n k f^2 = 60 \times 4 \times 0.0625 = 15$.

We could try different values of n (group sample sizes) to achieve the desired power; this change affects only two arguments in Eq. (11.5): $k(n -$

1) and $\lambda = nkf^2$. Alternatively, we could use MATLAB's built-in function `fzero(fun,x0)` which tries to find a zero of `fun` near some initial value `x0`.

```
k=4;   alpha = 0.05;   f2 = 0.0625;
f = @(n) 1-ncfcdf( finv(1-alpha, k-1,k*n-k),k-1,k*n-k,  n*k*f2 ) - 0.90;
ssize = fzero(f, 100) %57.6731
%Sample size of n=58 (per treatment) ensures the power of 90% for
%the effect size f^2=0.0625.
```

Thus, sample size of $n = 58$ will ensure the power of 90% for the specified effect size. The function `fzero` is quite robust with respect to the specification of the initial value; in this case the initial value was $n = 100$.

If we wanted to plot the power for different sample sizes (Fig. 11.8), the following simple MATLAB script would do it. The inputs are the k number of treatments and the significance level α. The specific alternative H_1 is such that $\sum_i \alpha_i^2 = \sigma^2/4$, so that $\lambda = n/4$.

```
k=4;            %number of treatments
alpha = 0.05;   %significance level
y=[];           %set values of power for n
for n=2:100
  y =[y 1-ncfcdf(finv(1-alpha, k-1, k*(n-1)), ...
                        k-1, k*(n-1), n/4)];
end
plot(2:100, y,'b-','linewidth',3)
xlabel('Group sample size n'); ylabel('Power')
```

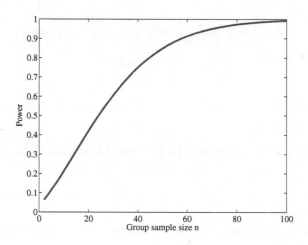

Fig. 11.8 Power for $n \leq 100$ (size per group) in a fixed-effect ANOVA with $k = 4$ treatments and $\alpha = 0.05$. The alternative H_1 is defined as $\sum_i \alpha_i^2 = \sigma^2/4$ so that the parameter of noncentrality λ is equal to $n/4$.

In the preceding analysis, the total sample size is $N = k \times n = 240$.

Sample Size in ANOVA in Terms of Maximal Difference between any Two Means. Let $\delta = \max_{i,j} |\mu_i - \mu_j|$ be the maximal difference between the population means that we would like the ANOVA test to detect and reject H_0. In terms of δ, the noncentrality parameter is expressed as

$$\lambda = \frac{n\delta^2}{2\sigma^2},$$

and the sample size determination proceeds as in (11.5) with this λ. Some researchers prefer eliciting the precision in terms of maximal detectable difference δ, rather than in terms of global $\sum_i \alpha_i^2$.

Power Analysis for Random-Effect ANOVA. In random-effect ANOVA the alternative hypothesis is $H_1 : \sigma_\alpha^2 \neq 0$. When H_1 is true, the ratio

$$F = \frac{MSTr}{MSE(1 + n\sigma_\alpha^2/\sigma^2)}$$

has a standard F-distribution with $k - 1$ and $k(n - 1)$ degrees of freedom. Here n is the level sample size and $N = nk$ is the total sample size. Consequently, the power calculation does not require non-central F-distribution. Simply,

$$1 - \beta = \mathbb{P}\left(\frac{MSTr}{MSE} > F_{k-1,k(n-1),1-\alpha} \middle| H_1 \text{ true}\right)$$
$$= \mathbb{P}\left(F > \frac{F_{k-1,k(n-1),1-\alpha}}{1 + n\sigma_\alpha^2/\sigma^2}\right).$$

Thus, the effect size should be elicited in units of ratios of variance components, σ_α^2/σ^2.

Example 11.11. **Power Analysis in Random-Effect ANOVA.** Assume that we are interested in power and sample size in an ANOVA test with $k = 5$ random levels, an effect of size $e = \sigma_\alpha^2/\sigma^2$, and $\alpha = 0.05$.

(i) With what probability will the effect $e = 1/2$ be detected if $n = 16$ observations per level are available?

(b) What sample size is needed so that the effect $e = 1/3$ is detected with power 85%?

```
%(a)
effect= 1/2; k=5; n=16; alpha=0.05;
pow = 1 - fcdf(finv(1-alpha, k-1, k*(n-1))/(1 + n*effect), k-1, k*(n-1))
```

```
%pow=0.9819
%(b)
k=5; effect=1/3; alpha=0.05; power=0.85;
f = @(n)  1 - fcdf(finv(1-alpha, k-1, k*(n-1))/(1 + n*effect), ...
                                    k-1, k*(n-1))-power;
ssize = fzero(f, 20)
%ssize = 18.8027 approx 19 per level, total ss is 5*19=95.
```

Power and Sample Size in Welch's ANOVA. Suppose that we want to find power in a one-way Welch ANOVA with $k = 4$ levels, where variances are 1,4, 9, and 16 and sample sizes are 12, 24, 48, and 96 respectively. The non-centrality parameter is $\lambda = \frac{\sum_i \alpha_i^2}{(\sum_i 1/\omega_i)/k}$. Since $(\sum_i 1/\omega_i)/k$ can be expressed as $(\sum_i s_i^2/n_i)/k$, the noncentrality parameter is the squared distance between H_1 and H_0 in units of an average of variances for the group means. When $\sum_i \alpha_i^2 = 2$, (e.g., if the group treatment effects are $(1,0,0,-1)$), then $\lambda = 13.2414$ and the power is $1 - \beta = 0.8542$. This is calculated by MATLAB script ◀ welchanovapower.m.

```
k=4;   alpha = 0.05;
ni=[12 24 48 96];
si2=[1 4 9 16];
alphai=[1  0 0 -1];
sai2=sum(alphai.^2); %effect = 2
wei=ni./si2;   recwei=1./wei;
nwei=wei/sum(wei) %0.4091   0.2045   0.1818   0.2045
Q=sum( (1-nwei).^2./(ni-1)) %0.0802
df = (k^2-1)/(3 * Q)   %62.3773
lambda = sai2/mean(recwei) %13.2414
%lambda is the effect in units of an average of
%variances for the group means
pow = 1-ncfcdf( finv(1-alpha, k-1, df), k-1, df, lambda)  %0.8542
```

If the power analysis is used to design sample size and n_i are desired to be all equal, $n_i = n$, then $f^2 = \frac{\frac{1}{k}\sum_i \alpha_i}{\frac{1}{k}\sum_i \sigma_i^2}$, which further simplifies to f^2 on page 542 when all σ_i^2 are equal to σ^2.

Sample Size for Multifactor ANOVA. A power analysis for multifactor ANOVA is usually done by selecting the most important factor and evaluating the power for that factor. Operationally, this is the same as the previously discussed power analysis for one-way ANOVA, but with modified error degrees of freedom to account for the presence of other factors.

Example 11.12. **Power Analysis for Multifactor ANOVA.** Assume a two-factor, fixed-effect ANOVA. The test for factor A at $a = 4$ levels is to be evaluated for its power. Factor B is present and has $b = 3$ levels. Assume a balanced design with 4×3 cells, with $n = 20$ subjects in each cell. The total number of subjects in the experiment is $N = 3 \times 4 \times 20 = 240$.

For $\alpha = 0.05$, a medium effect size $f^2 = 0.0625$, and $\lambda = Nf^2 = 20 \times 4 \times 3 \times 0.0625 = 15$, the power is 0.9121.

```
1-ncfcdf(finv(1-0.05, 4-1, 4*3*(20-1)), 4-1, 4*3*(20-1), 15)    %0.9121
```

Alternatively, if the cell sample size n is required for a specified power, we can use function `fzero`.

```
a=4; b=3; alpha = 0.05; f2=0.0625;
f = @(n) 1-ncfcdf( finv(1-alpha, a-1,a*b*(n-1)), ...
                a-1, a*b*(n-1), a*b*n*f2) - 0.90;
ssize = fzero(f, 100) %19.2363
%sample size of 20 (by rounding 19.2363 up) ensures 90% power
%given the effect size and alpha
```

Sample Size for Repeated Measures Design. In a *repeated measures* design, each of the k treatments is applied to every subject. Thus, the total sample size is equal to a treatment sample size, $N = n$. Suppose that ρ is the correlation between scores for any two levels of the factor and it is assumed constant. Then the power is calculated by Eq. (11.5), where the noncentrality parameter is modified as

$$\lambda = \frac{n \sum_i \alpha_i^2}{(1-\rho)\sigma^2} = \frac{nkf^2}{1-\rho}.$$

Example 11.13. **Power Analysis for Repeated Measures Design.** Suppose that $n = 25$ subjects go through $k = 3$ treatments and that a correlation between the treatments is $\rho = 0.6$. This correlation comes from the experimental design; the measures are repeated on the same subjects. Then, for the medium effect size ($f = 0.25$ or $f^2 = 0.0625$), the achieved power is 0.8526.

```
n=25;  k=3;  alpha=0.05; rho=0.6;
f = 0.25;  %medium effect size f^2=0.0625
lambda = n * k * f^2/(1-rho)      %11.7188
power = 1-ncfcdf( finv(1-alpha, k-1, (n-1)*(k-1)),...
     k-1, (n-1)*(k-1), lambda)      %0.8526
```

If the power is specified at 85% level, then the number of subjects is obtained as

```
k=3;  alpha=0.05; rho=0.6; f=0.25;
pf = @(n) 1-ncfcdf( finv(1-alpha, k-1, (n-1)*(k-1)),...
            k-1, (n-1)*(k-1), n*k*f^2/(1-rho) ) - 0.85;
ssize = fzero(pf, 100) %24.8342 (n=25 after rounding)
```

✐

Often the sphericity condition is not met. This happens in longitudinal studies (repeated measures taken over time) where the correlation between measurements on days 1 and 2 may differ from the correlation between days 1 and 4, for example. In such a case, average correlation $\bar{\rho}$ is elicited and used; however, all degrees of freedom in central and noncentral F, as well as λ, are multiplied by sphericity parameter ϵ.

Example 11.14. **Penalizing the Violation of Sphericity.** Suppose, as in Example 11.13, that $n = 25$ subjects go through $k = 3$ treatments, and that an average correlation between the treatments is $\bar{\rho} = 0.6$. If the sphericity is violated and parameter ϵ is estimated as $\epsilon = 0.7$, the power of 85.26% from Example 11.13 drops to 74.64%.

```
n=25;  k=3;  alpha=0.05; barrho=0.6; eps=0.7;
f = 0.25;
lambda =  n * k * f^2/(1-barrho)  %11.7188
power = 1-ncfcdf( finv(1-alpha, eps*(k-1), eps*(n-1)*(k-1)),...
    eps * (k-1), eps*(n-1)*(k-1), eps*lambda)    %0.7464
```

If the power is to remain at 85% level, then the number of subjects should increase from 25 to 32.

```
k=3;  alpha=0.05; barrho=0.6; eps=0.7;  f = 0.25;
pf = @(n) 1-ncfcdf( finv(1-alpha, eps*(k-1), eps*(n-1)*(k-1)),...
        eps*(k-1), eps*(n-1)*(k-1),  eps*n * k * f^2/(1-barrho) ) - 0.85;
ssize = fzero(pf, 100) %31.8147
```

✐

11.9 Functional ANOVA*

Functional linear models have become popular recently because many responses are functional in nature. For example, in an experiment in neuroscience, observations could be functional responses, and rather than applying the experimental design on some summary of these functions, one could use the densely sampled functions as data.

We provide a definition for the one-way case, which is a "functionalized" version of the standard one-way ANOVA.

Suppose that for any fixed $t \in T \subset \mathbb{R}$, the observations $\mathbf{y}$ are modeled by a fixed-effect ANOVA model:

$$y_{ij}(t) = \mu(t) + \alpha_i(t) + \epsilon_{ij}(t), \ i=1,\ldots,k, \ j=1,\ldots,n_i; \ \sum_{i=1}^{k} n_i = N, \ (11.6)$$

where $\epsilon_{ij}(t)$ are independent $\mathcal{N}(0,\sigma^2)$ errors. When i and j are fixed, we assume that functions $\mu(t)$ and $\alpha_i(t)$ are square-integrable functions. To ensure the identifiability of treatment functions α_i, one typically imposes

$$(\forall t) \sum_i \alpha_i(t) = 0. \tag{11.7}$$

In real life the measurements $\mathbf{y}$ are often taken at equidistant times t_m. The standard least square estimators for $\mu(t)$ and $\alpha_i(t)$

$$\hat{\mu}(t) = \overline{y}(t) = \frac{1}{n} \sum_{i,j} y_{ij}(t), \tag{11.8}$$

$$\hat{\alpha}_i(t) = \overline{y}_i(t) - \overline{y}(t), \tag{11.9}$$

where $\overline{y}_i(t) = \frac{1}{n_i} \sum_j y_{ij}(t)$, are obtained by minimizing the discrete version of LMSSE, (e.g., Ramsay and Silverman, 1997, p. 141):

$$LMSSE = \sum_t \sum_{i,j} [y_{ij}(t) - (\mu(t) + \alpha_i(t))]^2, \tag{11.10}$$

subject to the constraint $(\forall t) \sum_i n_i \alpha_i(t) = 0$.

The fundamental ANOVA identity becomes a functional identity,

$$SST(t) = SSTr(t) + SSE(t), \tag{11.11}$$

with $SST(t) = \sum_{i,l}[y_{il}(t) - \overline{y}(t)]^2$, $SSTr(t) = \sum_i n_i [y_i(t) - \overline{y}(t)]^2$, and $SSE(t) = \sum_{i,l}[y_{il}(t) - \overline{y}_i(t)]^2$.

For each t, the function

$$F(t) = \frac{SSTr(t)/(k-1)}{SSE(t)/(N-k)} \tag{11.12}$$

is distributed as noncentral $F_{k-1,N-k}\left(\frac{\sum_i n_i \alpha_i^2(t)}{\sigma^2}\right)$. Estimation of $\mu(t)$ and $\alpha_j(t)$ is straightforward, and estimators are given in Eqs. (11.8).

The testing of hypotheses involving functional components of the standard ANOVA method is hindered by dependence and dimensionality problems. Testing requires dimension reduction and this material is beyond the scope of this book. The interested reader can consult Ramsay and Silverman (1997) and Fan and Lin (1998).

Example 11.15. **FANOVA in Tumor Physiology.** Experiments carried out in vitro with tumor cell lines have demonstrated that tumor cells respond to radiation and anticancer drugs differently, depending on the environment. In particular, available oxygen is important. Efforts to increase the level of oxygen within tumor cells have included laboratory rats with implanted tumors breathing pure oxygen. Unfortunately, animals breathing pure oxygen may experience large drops in blood pressure, enough to make this intervention too risky for clinical use.

Mark Dewhirst, Department of Radiation Oncology at Duke University, sought to evaluate carbogen (95% pure oxygen and 5% carbon dioxide) as a breathing mixture that might improve tumor oxygenation without causing a drop in blood pressure. The protocol called for making measurements on each animal over 20 minutes of breathing room air, followed by 40 minutes of carbogen breathing. The experimenters took serial measurements of oxygen partial pressure (PO_2), tumor blood flow (LDF), mean arterial pressure (MAP), and heart rate. Microelectrodes, inserted into the tumors (one per animal), measured PO_2 at a particular location within the tumor throughout the study period. Two laser Doppler probes, inserted into each tumor, provided measurements of blood flow. An arterial line into the right femoral artery allowed measurement of MAP. Each animal wore a face mask for administration of breathing gases (room air or carbogen). [See Lanzen et al. (1998) for more information about these experiments.]

Nine rats had tumors transplanted within the quadriceps muscle, which we will denote by TM. For comparison, the studies also included eight rats with tumors transplanted subcutaneously (TS) and six rats without tumors (N) in which measurements were made in the quadriceps muscle. The data are provided in 📁 oxigen.dat.

Figure 11.9 show some of the data (PO_2). The plots show several features, including an obvious rise in PO_2 at the 20-minute mark among some of the animals. No physiologic model exists that would characterize the shapes of these profiles mathematically. The primary study question concerned evaluating the effect of carbogen breathing on PO_2. The analysis was complicated by the knowledge that there may be acute changes in PO_2 after carbogen breathing starts. The primary question of interest is whether the tumor tissue behaves differently than normal muscle tissue or whether a tumor implanted subcutaneously responds to carbogen breathing differently than tumor tissue implanted in muscle tissue in the presence of acute jumps in PO_2.

The analyses concern inference on changes in some physiologic measurements after an intervention. The problem for the data analysis is how best to define "change" in order to allow for the inference desired by the investigators.

From a statistical modeling point of view, the main issues concern building a flexible model for the multivariate time series y_{ij} of responses and providing for formal inferences on the occurrence of change at time t^* and the "equality" of the PO_2 profiles. From the figures it is clear that the main challenge arises from the highly irregular behavior of responses. Neither physiologic considerations nor any exploratory data analysis motivates any parsimonious parametric form. Different individuals seem to exhibit widely varying response patterns. Still, it is clear from inspection of

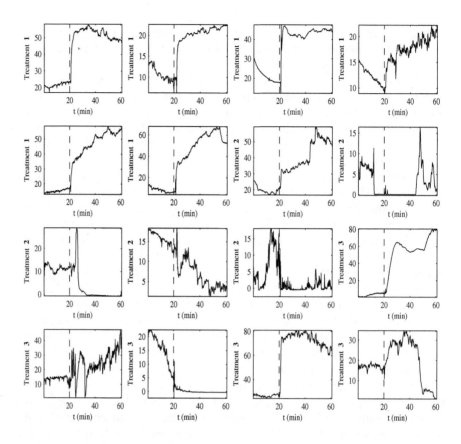

Fig. 11.9 PO$_2$ measurements. Notice that despite a variety of functional responses and a lack of a simple parametric model, at time $t^* = 20'$ the pattern generally changes.

the data that for some response series a definite change takes place at time t^*.

Figure 11.10 shows the estimators of components in the functional ANOVA model. As can be discern from the figure, adding $\mu(t)$ and $\alpha_2(t)$ will lead to a relatively horizontal expected profile for group i, each fitted curve canceling the other to some extent. Files ◀ miceP02.m and fanova.m support this example.

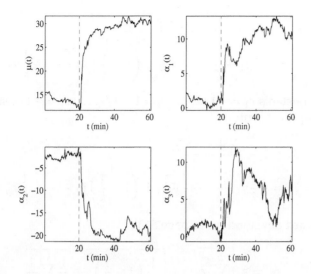

Fig. 11.10 Functional ANOVA estimators.

11.10 Analysis of Means (ANOM)*

Some statistical applications, notably in the area of quality improvement, involve a comparison of treatment means to determine which means are significantly different *from their overall average*. For example, a biomedical engineer might run an experiment to investigate which of six concentrations of an agent produces a different output, in the sense that the average measurement for each concentration differs from the overall average.

Questions of this type are answered by the analysis of means (ANOM), which is a method for making multiple comparisons, sometimes referred to as "multiple comparisons with the weighted mean." The ANOM answers different question than the ANOVA; however, it is related to ANOVA via multiple tests involving contrasts (Halperin et al., 1955; Ott, 1967).

The ANOM procedure consists of multiple testing of k hypotheses $H_{0i} : \alpha_i = 0$, $i = 1, \ldots, k$ versus the two-sided alternative. Since the testing is simultaneous, the Bonferroni adjustment to the type I error is used. Here α_i are, as in ANOVA, k population treatment effects $\mu_i - \mu$, $i = 1, \ldots, k$, which are estimated as

$$\hat{\alpha}_i = \bar{y}_i - \bar{y},$$

in the usual ANOVA notation. The population effect α_i can be represented via the treatment means as

$$\alpha_i = \mu_i - \mu = \mu_i - \frac{\mu_1 + \cdots + \mu_k}{k}$$

$$= -\frac{1}{k}\mu_1 - \cdots - \frac{1}{k}\mu_{i-1} + \left(1 - \frac{1}{k}\right)\mu_i - \cdots - \frac{1}{k}\mu_k .$$

Since the constants $c_1 = -1/k, \ldots, c_i = 1 - 1/k, \ldots, c_k = -1/k$ sum up to 0, $\hat{\alpha}_i$ is an empirical contrast,

$$\hat{\alpha}_i = \bar{y}_i - \frac{\bar{y}_1 + \cdots + \bar{y}_k}{k}$$

$$= -\frac{1}{k}\bar{y}_1 - \cdots - \frac{1}{k}\bar{y}_{i-1} + \left(1 - \frac{1}{k}\right)\bar{y}_i - \cdots - \frac{1}{k}\bar{y}_{k'}$$

with standard deviation (as in page 513),

$$s_{\hat{\alpha}_i} = s\sqrt{\sum_{j=1}^{k}\frac{c_j^2}{n_j}} = s\sqrt{\frac{1}{n_i}\left(1 - \frac{1}{k}\right)^2 + \frac{1}{k^2}\sum_{j \neq i}\frac{1}{n_j}}.$$

Here n_i are treatment sample sizes, $N = \sum_{i=1}^{k} n_i$ is the total sample size, and $s = \sqrt{MSE}$.

All effects $\hat{\alpha}_i$ falling outside the interval

$$\left[-s_{\hat{\alpha}_i} \times t_{N-k,1-\alpha/(2k)}, \; s_{\hat{\alpha}_i} \times t_{N-k,1-\alpha/(2k)}\right],$$

or equivalently, all treatment means $\bar{y}_i$ falling outside of the interval

$$\left[\bar{y} - s_{\hat{\alpha}_i} \times t_{N-k,1-\alpha/(2k)}, \; \bar{y} + s_{\hat{\alpha}_i} \times t_{N-k,1-\alpha/(2k)}\right],$$

correspond to rejection of $H_{0i}: \alpha_i = \mu_i - \mu = 0$. The Bonferroni correction $1 - \alpha/(2k)$ in the t-quantile $t_{N-k,1-\alpha/(2k)}$ controls the significance of the procedure at level α (page 415).

Example 11.16. **ANOM on Coagulation Times.**   We revisit Example 11.1 and perform an ANOM analysis. In this example $\bar{y}_1 = 61, \bar{y}_2 = 66, \bar{y}_3 = 68$, and $\bar{y}_4 = 61$. The grand mean is $\bar{y} = 64$. Standard deviations for $\hat{\alpha}$ are $s_{\hat{\alpha}_1} = 0.9736, s_{\hat{\alpha}_2} = 0.8453, s_{\hat{\alpha}_3} = 0.8453$, and $s_{\hat{\alpha}_4} = 0.7733$, and the t-quantile corresponding to $\alpha = 0.05$ is $t_{1-0.05/8,20} = 2.7444$. This leads to ANOM bounds, $[61.328, 66.672], [61.680, 66.320], [61.680, 66.320]$, and $[61.878, 66.122]$.

Only the second treatment mean $\bar{y}_2 = 66$ falls in its ANOM interval $[61.680, 66.320]$; see Figure 11.11 (generated by ◢ anomct.m). Consequently, the second population mean μ_2 is not significantly different from the grand mean μ.

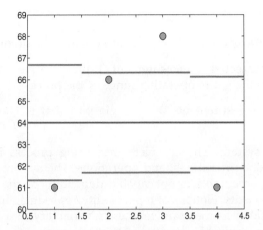

Fig. 11.11 ANOM analysis on Coagulation Times. The *blue* line is the overall mean, the *red* lines are the ANOM interval bounds, and the *green* dots are the treatment means. Note that only the second treatment mean falls in its interval.

11.11 The Capability of a Measurement System (Gauge R&R ANOVA)*

The gauge R&R methodology is concerned with the capability of measuring systems. Over time gauge R&R methods have evolved, and now two approaches have become widely accepted: (i) the average and range method, also known as the AIAG method, and (ii) the ANOVA method. We will focus on the ANOVA method and direct interested readers to Montgomery (2005) for comprehensive coverage of the topic.

In a typical gauge R&R study, several operators measure the same parts in random order. Most studies involve two to five operators and five to ten parts. There are usually several trial repetitions, which means that an operator performs multiple measurements on the same part.

The ANOVA method of analyzing measurement data provides not only the estimates for repeatability or equipment variation, reproducibility or appraiser variation, and part-to-part variation but also accounts for possible interaction components.

We first define several key terms in gauge R&R context:

Gauge or gage: Any device that is used to obtain measurements.

Part: An item subjected to measurement. Typically a part is selected at random from the entire operating range of the process.

Trial: A set of measurements on a single part that is taken by an operator.

Measurement system: The complete measuring process involving gauges, operators, procedures, and operations. The system is evaluated as capable, acceptable, or not capable, depending on the variabilities of its components. Notions of repeatability, reproducibility (R&R), and part variability as the main sources of variability in a measurement system, are critical for its capability assessment.

Repeatability: The variance in measurements obtained with a measuring instrument when used several times by an appraiser while measuring the identical characteristic on the same part.

Reproducibility: The variation in measurements made by different appraisers using the same instrument when measuring an identical characteristic on the same part.

The measurements in a gauge R&R experiment are described as an ANOVA model

$$y_{ijk} = \mu + P_i + O_j + PO_{ij} + \epsilon_{ijk}, \quad i = 1, \ldots, p, \ j = 1, \ldots, o, \ k = 1, \ldots, n,$$

where P_i, O_j, PO_{ij}, and ϵ_{ijk} are independent random variables that represent the contributions of parts, operators, part–operator interaction, and random error to the measurement. We assume that the design is balanced, that there are p parts and o operators, and that for each part/operator combination there are n trials.

This is an example of a random-effect, two-way ANOVA, since the factors parts and operators are both randomly selected from the population consisting of many parts and operators. Except for the grand mean μ, which is considered a constant, the random variables P_i, O_j, PO_{ij}, ϵ_{ijk} are considered independent zero-mean normal with variances σ_P^2, σ_O^2, σ_{PO}^2, and σ^2. Then the variability of measurement y_{ijk} splits into the sum of four variances:

$$\mathbb{V}\mathrm{ar}\,(y_{ijk}) = \sigma_P^2 + \sigma_O^2 + \sigma_{PO}^2 + \sigma^2.$$

As in Section 11.4, the mean squares in the ANOVA table are obtained as

$$MSP = \frac{SSP}{p-1}, \ MSO = \frac{SSO}{o-1}, \ MSPO = \frac{SSPO}{(p-1)(o-1)}, \ MSE = \frac{SSE}{po(n-1)},$$

where the sums of squares are standardly defined (page 525). Since this is a random-effect ANOVA, it holds that

$$\mathbb{E}(MSP) = \sigma^2 + n\sigma_{PO}^2 + on\sigma_P^2,$$
$$\mathbb{E}(MSO) = \sigma^2 + n\sigma_{PO}^2 + pn\sigma_O^2,$$
$$\mathbb{E}(MSPO) = \sigma^2 + n\sigma_{PO}^2,$$
$$\mathbb{E}(MSE) = \sigma^2.$$

The derivation of these expectations is beyond the scope of this text, but a thorough presentation of it can be found in Montgomery (1984) or Kutner et al. (2005). The expectations above, by moment matching, lead to the estimates

$$\hat{\sigma}_P^2 = \frac{MSP - MSPO}{on},$$
$$\hat{\sigma}_O^2 = \frac{MSO - MSPO}{pn},$$
$$\hat{\sigma}_{PO}^2 = \frac{MSPO - MSE}{n},$$
$$\hat{\sigma}^2 = MSE.$$

In the context of R&R analysis, the definitions of empirical variance components are as follows:

$$\hat{\sigma}_{Repeat}^2 = \hat{\sigma}^2$$
$$\hat{\sigma}_{Reprod}^2 = \hat{\sigma}_O^2 + \hat{\sigma}_{PO}^2$$
$$\hat{\sigma}_{Gauge}^2 = \hat{\sigma}_{Repeat}^2 + \hat{\sigma}_{Reprod}^2$$
$$= \hat{\sigma}^2 + \hat{\sigma}_O^2 + \hat{\sigma}_{PO}^2$$
$$\hat{\sigma}_{Total}^2 = \hat{\sigma}_{Gauge}^2 + \hat{\sigma}_{Part}^2$$

Next we provide several measures of the capability of a measurement system.

Number of Distinct Categories. The measure called signal-to-noise ratio, in the context of R&R, is defined as

$$SNR = \sqrt{\frac{2 \times \hat{\sigma}^2_{Part}}{\hat{\sigma}^2_{Gauge}}}.$$

The *SNR* rounded to the closest integer defines the number of distinct categories (*NDC*) measure, which is a resolution of the measurement system.

The *NDC* informally indicates how many "categories" the measurement system is able to differentiate. If $NDC = 0$ or 1, then the measurement system is useless. If $NDC = 2$, then the system can differentiate only between two categories ("small" and "large," in T-shirt terminology). If $NDC = 3$, then the system can distinguish "small," "medium," and "large," and so on. If $NDC \geq 5$, then the measurement system is capable. If $NDC \leq 1$, then the measurement system is not capable. Otherwise, the measurement system is evaluated as acceptable.

Percent of R&R Variability. The percent of R&R variability (*PRR*) measures the size of the R&R variation relative to the total data variation. It is defined as

$$PRR = \sqrt{\frac{\hat{\sigma}^2_{Gauge}}{\hat{\sigma}^2_{Total}}}.$$

If $PRR < 10\%$, then the measurement system is capable. If $PRR > 30\%$, the measurement system is not capable; the system is acceptable for $10\% \leq PRR \leq 30\%$. Also, the individual contribution of repeatability and reproducibility variances entering the summary measure *PRR* are of interest. For instance, a large repeatability variation, relative to the reproducibility variation, indicates a need to improve the gauge. A high reproducibility variance relative to repeatability indicates a need for better operator training.

MATLAB's function ◀ gagerr(y,part,operator) performs a gauge repeatability and reproducibility analysis on measurements in vector y collected by operator on part. As in the anovan command, the number of elements in part and operator should be the same as in y. There are many important options in gagerr, and we recommend that the user carefully consult the function help.

Example 11.17. **Measurements of Thermal Impedance.** This example of a gauge R&R study comes from Houf and Berman (1988) and Montgomery (2005). The data, in the table below, represent measurements on thermal impedance (in °C per Watt $\times$ 100) of a power module for an induction motor starter. There are ten parts, three operators, and three measurements per part.

| Part | Operator 1 | | | Operator 2 | | | Operator 3 | | |
number	Test 1	Test 2	Test 3	Test 1	Test 2	Test 3	Test 1	Test 2	Test 3
1	37	38	37	41	41	40	41	42	41
2	42	41	43	42	42	42	43	42	43
3	30	31	31	31	31	31	29	30	28
4	42	43	42	43	43	43	42	42	42
5	28	30	29	29	30	29	31	29	29
6	42	42	43	45	45	45	44	46	45
7	25	26	27	28	28	30	29	27	27
8	40	40	40	43	42	42	43	43	41
9	25	25	25	27	29	28	26	26	26
10	35	34	34	35	35	34	35	34	35

Using ANOVA R&R analysis, evaluate the capability of the measurement system by assuming that parts and operators possibly interact. See Exercise 11.33 for the case where parts and operators do not interact (additive ANOVA model).

We will first analyze the problem as a two-way ANOVA (◀ RandR2.m) and then, for comparison, provide MATLAB's output from the function gagerr.

```
%model with interaction
impedance = [...
     37 38 37 41 41 40 41 42 41 42 41 43 42 42 42 43 42 43  ...
     30 31 31 31 31 31 29 30 28 42 43 42 43 43 43 42 42 42  ...
     28 30 29 29 30 29 31 29 29 42 42 43 45 45 45 44 46 45  ...
     25 26 27 28 28 30 29 27 27 40 40 40 43 42 42 43 43 41  ...
     25 25 25 27 29 28 26 26 26 35 34 34 35 35 34 35 34 35]' ;

% forming part and operator vectors.
a = repmat([1:10],9,1);  part = a(:);
b = repmat([1:3], 3,1);  operator = repmat(b(:),10,1);

[p table stats terms] = anovan( impedance,{part, operator},...
 'model','interaction','varnames',{'part','operator'} )

MSE = table{5,5} %0.5111
MSPO =table{4,5} %2.6951
MSO = table{3,5} %19.6333
MSP = table{2,5} %437.3284
p = stats.nlevels(1); o = stats.nlevels(2);
n=length(impedance)/(p * o);%p=10, O=3, n=3

s2Part = (MSP - MSPO)/(o * n)   %48.2926
s2Oper = (MSO - MSPO)/(p * n)   %0.5646
s2PartOper = (MSPO - MSE)/n     %0.7280
s2 = MSE                        %0.5111

s2Repeat = s2                   %0.5111
s2Repro = s2Oper + s2PartOper   %1.2926
s2Gage = s2Repeat + s2Repro     %1.8037
s2Tot = s2Gage  + s2Part        %50.0963
```

```
%percent variation due to part
ps2Part = s2Part/s2Tot   %0.9640

% signal-to-noise ratio; > 5 measuring system is capable
snr = sqrt( 2 * ps2Part/(1-ps2Part))  %7.3177
% snr rounded is NDC (number of distinct categories).
ndc = round(snr)   %7


% percent of R&R variability
prr = sqrt(s2Gage/s2Tot)  %0.1897
```

Therefore, the measuring system is capable by $NDC = 7 \geq 5$ but falls in the "gray" zone $(10, 30)$ according to the PRR measure. MATLAB's built-in function gagerr produces a detailed output:

```
gagerr(impedance,{part, operator},'model','interaction')

% Source          Variance   %Variance   sigma 5.15*sigma 5.15*sigma
% ===========================================================================
% Gage R&R            1.80       3.60      1.34      6.92       18.97
%    Repeatability    0.51       1.02      0.71      3.68       10.10
%    Reproducibility  1.29       2.58      1.14      5.86       16.06
%      Operator       0.56       1.13      0.75      3.87       10.62
%      Part*Operator  0.73       1.45      0.85      4.39       12.05
% Part               48.29      96.40      6.95     35.79       98.18
% Total              50.10     100.00      7.08     36.45
%  --------------------------------------------------------------------
%          Number of distinct categories (NDC)      :   7
%          % of Gage R&R of total variations (PRR):   18.97
```

When operators and parts do not interact, an additive ANOVA model,

$$y_{ijk} = \mu + P_i + O_j + \epsilon_{ijk}, \quad i = 1,\ldots,p, \; j = 1,\ldots,o, \; k = 1,\ldots,n,$$

should be used. In this case, the estimators of variances are

$$\hat{\sigma}_P^2 = \frac{MSP - MSE}{on},$$

$$\hat{\sigma}_O^2 = \frac{MSO - MSE}{pn},$$

$$\hat{\sigma}^2 = MSE,$$

and $\hat{\sigma}_{Reprod}^2$ is simply reduced to $\hat{\sigma}_O^2$.

Exercise 11.33 solves the problem in Example 11.17 without the PO-interaction term and compares the analysis with the output from gagerr.

11.12 Testing Equality of Several Proportions

An ANOVA-type hypothesis can be considered in the context of several proportions. Consider k independent populations from which k samples of size $n_1, n_2, \ldots, n_k$ are taken. We record a binary attribute, say, 1 or 0. Let $X_1, X_2, \ldots, X_k$ be the observed number of 1s and $\hat{p}_1 = X_1/n_1, \ldots, \hat{p}_k = X_k/n_k$ the sample proportions.

To test $H_0 : p_1 = p_2 = \cdots = p_k$ against the general alternative $H_1 = H_0^c$, the statistic

$$\chi^2 = \sum_{i=1}^{k} \frac{(X_i - n_i \overline{p})^2}{n_i \overline{p}(1 - \overline{p})}$$

is formed. Here

$$\overline{p} = \frac{X_1 + X_2 + \cdots + X_k}{n_1 + n_2 + \cdots + n_k}$$

is a pooled sample proportion. Under H_0 all proportions are the same and equal to p, so $\overline{p}$ is the best estimator. The statistic χ^2 has an approximate χ^2-distribution with $k - 1$ degrees of freedom. This approximation is considered good if all $n_i \overline{p}$ exceed 0.5 and no more than 20% of n_is are smaller than 5.

If H_0 is rejected, a Marascuillo procedure (Marascuillo and Serlin, 1988) can be applied to compare individual pairs of population proportions. For k populations there are $k(k-1)/2$ tests, and the two proportions p_i and p_j are different if

$$|\hat{p}_i - \hat{p}_j| > \sqrt{\chi^2_{k-1,1-\alpha}} \times \sqrt{\hat{p}_i(1 - \hat{p}_i)/n_i + \hat{p}_j(1 - \hat{p}_j)/n_j}. \quad (11.13)$$

Example 11.18. **Gender and Hair Color.** Zar (2010) provides data on gender proportions in samples of subjects grouped by the color of their hair.

	Hair color			
Gender	Black	Brown	Blond	Red
Male	32	43	16	9
Female	55	65	64	16
Total	87	108	80	25

We will test the hypothesis that population proportions of males are the same for the four groups.

```
%Gender Proportions and Hair Color
Xi = [32  43   16   9];
ni = [87 108   80 25];
pi = Xi./ni     %0.3678    0.3981    0.2000    0.3600
```

```
pbar = sum(Xi)/sum(ni)        %0.3333
chi2 = sum(  (Xi - ni*pbar).^2./(ni * pbar * (1-pbar)) ) %8.9872
pval = 1-chi2cdf(chi2, 4-1)  %0.0295
```

Thus, with a p-value of about 3%, the hypothesis of homogeneity of population proportions is rejected. Using the condition in (11.13) show that the difference between $\hat{p}_2$ and $\hat{p}_3$ is significant and responsible for rejecting H_0.

Remark. We will see later (Chapter 12) that this test and the test for homogeneity in $2 \times c$ contingency tables are equivalent. Here we assumed that the sampling design involved fixed totals n_i. If the sampling was fully random (i.e., no totals for hair color nor males/females were prespecified), then H_0 would be the hypothesis of independence between gender and hair color.

For Tukey-type multiple comparisons and testing proportion trends, see Zar (2010) and Conover (1999).

11.13 Testing the Equality of Several Poisson Means*

When under in each of k treatments the observations are multiple counts, the standard ANOVA is often inappropriately applied to test for the equality of means. For counting observations the Poisson model is more adequate than the normal, and using a standard ANOVA methodology may be problematic. For example, when an ANOVA hypothesis of equality of means is rejected, but the observations are in fact Poisson, multiple comparisons may be invalid due to unequal treatment variances, which for Poisson observations are equal to the means. Next, we describe an approach that is appropriate for Poisson observations.

Suppose that in treatment i we observe n_i Poisson counts,

$$X_{ij} \sim \mathcal{P}oi(\lambda_i), \quad i = 1, \ldots, k, \, j = 1, \ldots, n_i.$$

Denote by O_i the sum of all counts in treatment i, $O_i = \sum_{j=1}^{n_i} X_{ij}$, and by O the sum of counts across all treatments, $O = \sum_{i=1}^{k} O_i$. The total number of observations is $N = \sum_{i=1}^{k} n_i$.

One can show that the distribution of the vector $(O_1, O_2, \ldots, O_k)$, given the total sum O, is multinomial (page 187),

$$(O_1, O_2, \ldots, O_k | O) \sim \mathcal{M}n(O, \boldsymbol{p}),$$

where $\boldsymbol{p} = (p_1, p_2, \ldots, p_k)$ is defined via

$$p_i = \frac{n_i \lambda_i}{\sum_{i=1}^{k} n_i \lambda_i}.$$

Thus, when the hypothesis $H_0 : \lambda_1 = \lambda_2 = \cdots = \lambda_k$ is true, then $p_i = n_i/N$. Suppose that counts O_i are observed. Define expected counts E_i as

$$E_i = p_i \times O = \frac{n_i}{N} \times O, \quad i = 1, \ldots, k.$$

Then

$$\chi^2 = \sum_{i=1}^{k} \frac{(O_i - E_i)^2}{E_i}$$

is Pearson's test statistic for testing H_0. When H_0 is true, it has an approximately χ^2-distribution with $k - 1$ degrees of freedom. Alternatively, one could use the likelihood statistic

$$G^2 = 2 \sum_{i=1}^{k} O_i \log \frac{O_i}{E_i},$$

which under H_0 also has the χ^2-distribution with $k - 1$ degrees of freedom. Large values of χ^2 or G^2 are critical for H_0.

Example 11.19. **Twenty Plates with Cells.** Assume that in an experiment three treatments are applied on 20 plates populated with a large number of cells, the first treatment on 7, the second on 5, and the third on 8 plates.

The following table gives the numbers of cells per plate that responded to the treatments:

Treatment 1	Treatment 2	Treatment 3
1 6 4 2 3	5 5 9 2 7	2 3 0 2 1
8 2		4 5 2

We assume a Poisson model and wish to test the equality of mean counts, $H_0 : \lambda_1 = \lambda_2 = \lambda_3$.

The solution is provided in ◀ poissonmeans.m. The hypothesis of the equality of means is rejected with a p-value of about 1.3%. Note that standard ANOVA fails to reject H_0, for the p-value is 6.15%. This example demonstrates the inadequacy of the standard ANOVA for this kind of data.

```
ncells =  [1 6 4 2 3 8 2  5 5 9 2 7  2 3 0 2 1 4 5 2]';
agent  =  [1 1 1 1 1 1 1  2 2 2 2 2  3 3 3 3 3 3 3 3]';
ncells1=ncells(agent==1)
ncells2=ncells(agent==2)
ncells3=ncells(agent==3)
k = 3; %treatments
n1 = length(ncells1); n2 = length(ncells2); n3= length(ncells3);
```

```
ni=[n1 n2 n3]    %7     5     8
N = sum(ni) %20
O1 = sum(ncells1); O2 = sum(ncells2); O3 = sum(ncells3);
Oi=[O1  O2  O3]  %26    28    19
O = sum(Oi) %73
%expected
Ei = O .* ni/N         %25.55    18.25    29.2
%Poisson chi2
chi2 = sum( (Oi - Ei).^2 ./ Ei )        %8.7798
%Likelihood G2
G2 = 2 * sum( Oi .* log(Oi./Ei ) )      %8.5484
pvalchi2 = 1 - chi2cdf(chi2, k-1)       %0.0124
pvalG2 = 1 - chi2cdf(G2, k-1)           %0.0139

% If the problem is treated as ANOVA
[panova table] = anova1(ncells, agent)

% panova = 0.0615
% table =
%      'Source'  'SS'         'df'   'MS'        'F'        'Prob>F'
%      'Groups'  [ 32.0464]   [ 2]   [16.0232]   [3.3016]   [0.0615]
%      'Error'   [ 82.5036]   [17]   [ 4.8532]   []         []
%      'Total'   [114.5500]   [19]   []          []         []
```

11.14 Exercises

11.1. **Nematodes.** Some varieties of nematodes, roundworms that live in the soil and are frequently so small that they are invisible to the naked eye, feed on the roots of lawn grasses and crops such as strawberries and tomatoes. This pest, which is particularly troublesome in warm climates, can be treated by the application of nematocides. However, because of the size of the worms, it is very difficult to measure the effectiveness of these pesticides directly. To compare four nematocides, the yields of equal-size plots of one variety of tomatoes were collected. The data, given as yields in pounds per plot, are shown in the table below:

Nematocide A	Nematocide B	Nematocide C	Nematocide D
18.6	18.7	19.4	19.0
18.4	19.0	18.9	18.8
18.4	18.9	19.5	18.6
18.5	18.5	19.1	18.7
17.9		18.5	

(a) Write a statistical model for ANOVA and state H_0 and H_1 in terms of your model.

(b) What is your decision if $\alpha = 0.05$?

(c) For what values of α will your decision be different than that in (b)?

11.2. **Cell Folate Levels in Cardiac Bypass Surgery.** Altman (1991, p. 208) provides data on 22 patients undergoing cardiac bypass surgery. The patients were randomized to one of three groups receiving the following treatments:

Treatment 1. Patients received a 50% nitrous oxide and 50% oxygen mixture continuously for 24 hours.

Treatment 2. Patients received a 50% nitrous oxide and 50% oxygen mixture only during the operation.

Treatment 3. Patients received no nitrous oxide but received 35% to 50% oxygen for 24 hours.

The measured responses, given as red cell folate levels (ng/ml) for the three groups after 24 hours of ventilation, are shown below:

Treat1	243 251 275 291 347 354 380 392
Treat2	206 210 226 249 255 273 285 295 309
Treat3	241 258 270 293 328

The question of interest is whether the three ventilation methods result in a different mean red cell folate level. If the hypothesis of equality of treatment means is rejected, which means are significantly different?

11.3. **MTHFR C677T Genotype and Levels of Homocysteine and Folate.** A study by Ozturk et al. (2005) considered the association of methylenetetrahydrofolate reductase (MTHFR) C677T polymorphisms with levels of homocysteine and folate. A total of $N = 815$ middle-aged and elderly subjects were stratified by MTHFR C677T genotype ($k = 3$) and their measurements summarized in the table below:

Characteristics	CC: $n_1 = 312$ $\overline{X}_1$ (s_1)	CT: $n_2 = 378$ $\overline{X}_2$ (s_2)	TT: $n_3 = 125$ $\overline{X}_3$ (s_3)
Homocysteine, nmol/l	14.1 (1.9)	14.2 (2.3)	15.3 (3.0)
Red blood cell folate, nmol/l	715 (258)	661 (236)	750 (348)
Serum folate, nmol/l	13.1 (4.7)	12.3 (4.2)	11.4 (4.4)

(a) Using a one-way ANOVA, test the hypothesis that the population homocysteine levels for the three genotype groups are the same. Use $\alpha = 0.05$.

Hint: Since the raw data are not given, calculate

$$MSTr = \frac{1}{k-1} \sum_{i=1}^{k} n_i (\overline{X}_i - \overline{X})^2 \text{ and } MSE = \frac{1}{N-k} \sum_{i=1}^{k} (n_i - 1) s_i^2,$$

for

$$N = n_1 + \cdots + n_k \text{ and } \overline{X} = \frac{n_1 \overline{X}_1 + \cdots + n_k \overline{X}_k}{N}.$$

Statistic F is the ratio $MSTr/MSE$. It has $k-1$ and $N-k$ degrees of freedom.

(b) For red blood cell folate measurements complete the following ANOVA table:

```
SS                      DF              MS                  F            p
=====================================================================================
SSTr=                   DF1=            MSTr=469904.065     F=           p=
SSE =                   DF2=            MSE = 69846.911
============================================
SST=                    DF=
```

11.4. **Computer Games.** In Exercise 7.24, mental rotations test scores were provided for three groups of children:

Group 1 ("Factory" computer game): $\overline{X}_1 = 22.47, s_1 = 9.44, n_1 = 19$.
Group 2 ("Stellar" computer game): $\overline{X}_2 = 22.68, s_2 = 8.37, n_2 = 19$.
Control (no computer game): $\overline{X}_3 = 18.63, s_3 = 11.13, n_3 = 19$.

Assuming a normal distribution of scores in the population and equal population variances, test the hypothesis that the population means are the same at a 5% significance level.

11.5. **Beetles.** The following data were extracted from a more extensive study by Sokal and Karten (1964). The data represent mean dry weights (in milligrams) of three genotypes of beetles, *Tribolium castaneum*, reared at a density of 20 beetles per gram of flour. The four independent measurements for each genotype are recorded below:

Genotypes		
++	+b	bb
0.958	0.986	0.925
0.971	1.051	0.952
0.927	0.891	0.829
0.971	1.010	0.955

Using a one-way ANOVA, test whether the genotypes differ in mean dry weight. Take $\alpha = 0.01$.

11.6. **ANOVA Table from Summary Statistics.** When accounts of one-way ANOVA designs are given in journal articles or technical reports, the data and the ANOVA table are often omitted. Instead, means and standard deviations for each treatment group are given, along with the F-statistic, its p-value, and the decision on whether to reject or not. One can build the ANOVA table from this summary information and thus verify the author's interpretation of the data.

The results below are from an experiment with $n_1 = n_2 = n_3 = n_4 = 10$ observations on each of $k = 4$ groups:

Treatment n_i	$\overline{X}_i$	s_i
1	10 100.40	11.68
2	10 103.00	11.58
3	10 107.10	10.05
4	10 114.80	10.61

(a) Write the model for this experiment and state the null hypothesis in terms of the parameters of the model.

(b) Use the information in the table above to show that $SSTr = 1186$ and $SSE = 4357$.

(c) Construct the ANOVA table for this experiment, with standard columns SS, df, MS, F, and p-value.

Hint: $SSTr = n_1(\overline{X}_1 - \overline{X})^2 + n_2(\overline{X}_2 - \overline{X})^2 + n_3(\overline{X}_3 - \overline{X})^2 + n_4(\overline{X}_4 - \overline{X})^2$, for $\overline{X} = \sum_{i=1}^4 n_i \overline{X}_i / \sum_{i=1}^4 n_i$ and $SSE = (n_1 - 1)s_1^2 + (n_2 - 1)s_2^2 + (n_3 - 1)s_3^2 + (n_4 - 1)s_4^2$; see also Exercise 11.3.

Part (c) shows that it is possible to use treatment means and standard deviations to reconstruct the ANOVA table, find its F-statistic, and test the null hypothesis. However, without knowing the individual data values, one cannot carry out some other important statistical analyses (e.g., residual analysis).

11.7. **Protein Content in Milk for Three Diets.** The data from Diggle et al. (1994) are used to compare the content of protein in milk of 79 cows randomly assigned to three different diets: barley, a barley-and-lupins mixture, and lupins alone.

The original data set is longitudinal in nature (protein monitored weekly for several months), but for the purpose of this exercise we took the protein concentration at 6 weeks following calving for each cow.

```
b=[3.7700    3.0000    2.9300    3.0500    3.6000    3.9200...
   3.6600    3.4700    3.2100    3.3400    3.5000    3.7300...
   3.4900    3.1600    3.4500    3.5200    3.1500    3.4200...
   3.6200    3.5700    3.6500    3.7100    3.5700    3.6300...
   3.6000];
m=[3.4000    3.8000    3.2900    3.7100    3.2800    3.3800...
   3.5700    2.9000    3.5500    3.5500    3.0400    3.4000...
   3.1500    3.1300    3.2500    3.1500    3.1000    3.1700...
   3.5000    3.5700    3.4700    3.4500    3.2600    3.2400...
   3.7000    3.0500    3.5400];
l=[3.0700    3.1200    2.8700    3.1100    3.0200    3.3800...
   3.0800    3.3000    3.0800    3.5200    3.2500    3.5700...
   3.3800    3.0000    3.0300    3.0600    3.9600    2.8300...
   2.7400    3.1300    3.0500    3.5500    3.6500    3.2700...
   3.2000    3.2700    3.7000];
resp = [b'; m'; l'];      %single column vector of all responses
class = [ones(25,1); 2*ones(27,1); 3*ones(27,1)]; %class vector
[pval, table, stats] = anova1(resp, class)
[cintsdiff,means] = multcompare(stats)    %default  comparisons
```

Partial output is given below.

(a) Fill in the empty spaces in the output.

```
%pval =
%    0.0053
%
%table
%'Source'    'SS'        'df'      'MS'      'F'      'Prob>F'
%'Groups'  [_____]  [_____]  [0.3735]  [_____]  [_____]
%'Error'   [_____]  [_____]  [_____]
%'Total'   [5.8056]    [_____]
%
%stats =
%    gnames: 3x1 cell
%         n: [25 27 27]
%    source: 'anova1'
%     means: [3.4688 3.3556 3.2293]
%        df: 76
%         s: 0.2580
%
% cintsdiff =
%    1.0000    2.0000   -0.0579    0.1132    0.2844
%    1.0000    3.0000    0.0684    0.2395    0.4107
%    2.0000    3.0000   -0.0416    0.1263    0.2941
%
% means =
%    3.4688    0.0516
%    3.3556    0.0497
%    3.2293    0.0497
```

(b) What is H_0 here and is it rejected? Use $\alpha = 5\%$.

(c) From the output of `cintsdiff`, discuss how the population means differ.

11.8. **Tasmanian Clouds.** The data 🖳 clouds.txt provided by OzDASL were collected in a cloud-seeding experiment in Tasmania between mid-1964 and January 1971. Analysis of these data is discussed in Miller et al. (1979).

The rainfalls are period rainfalls in inches. Variables TE and TW are the east and west target areas, respectively, while CN, CS, and CNW are the corresponding rainfalls in the north, south, and northwest control areas, respectively. S stands for seeded and U for unseeded. Variables C and T are averages of control and target rainfalls. Variable DIFF is the difference T-C.

(a) Use an additive two-way ANOVA to estimate and test the treatment effects of Season and Seeded.

(b) Repeat the analysis from (a) after adding the interaction term.

11.9. **Red Clover Varieties.** Successful production of red clover depends to a considerable extent on selecting the best varieties for a particular farm. For that reason, varieties are compared in trial plots on Minnesota Agricultural Experiment Station fields at Grand Rapids, Morris, and Rosemount. Varieties are grown in replicated plots at each location. These plots are handled so that the factors affecting yield and other characteristics are as nearly the same for all varieties at each location as possible. Minnesota Agricultural Experiment Station scientists seeded red clover at three locations in 1995. The trials were harvested at Grand Rapids, Morris, and Rosemount in 1996 and at Rosemount and Morris in 1997 and 1998. Severe winter injury destroyed the trial at Grand Rapids. The table below (Ehlke and Vellekson, 1999; Tab. 1, p. 3) provides data on yield of red clover, in tons of dry matter per acre, seeded at 3 locations in 1995. Data is also available in data structure ⌨ redclover.mat.

	Yield						
	Grand Rapids	Morris			Rosemount		
	1966	1966	1967	1968	1966	1967	1968
Variety							
Arlington	3.7	2.6	3.3	2.9	3.2	2.0	4.8
Astred	3.3	2.2	2.2	2.7	2.5	1.8	3.3
Cinnamon	4.0	3.1	3.7	3.0	3.4	2.1	5.3
Marathon	4.3	3.2	3.5	2.6	3.4	1.7	4.5
Randolph	3.8	3.7	3.5	2.8	3.8	2.0	4.7
Scarlett	3.0	3.0	3.5	2.8	3.7	1.8	4.8

(a) Test the hypothesis that the mean yields for the six clover varieties are the same regardless of year and location. Take $\alpha = 5\%$.
(b) Using a two-way ANOVA with factors Variety and Location, show that neither factor is significant. Is the interaction significant?
(c) Using a two-way ANOVA with factors Variety and Year, show that the effect of Year is significant. Which years significantly differ?

11.10. **Cochlear Implants.** Woodworth (2004) describes and analyzes data from an interesting experiment involving cochlear implants. A cochlear implant is a small, complex electronic device that can help to provide a sense of sound to a person who is profoundly deaf or severely hard of hearing. The implant consists of an external portion that sits behind the ear and a second portion that is surgically placed under the skin. Traditional hearing aids amplify sounds so that they may be detected by damaged ears. Cochlear implants bypass damaged portions of the ear and directly stimulate the auditory nerve. Signals generated by the implant are sent by way of the auditory nerve to the brain, which recognizes the signals as sound. Hearing through a cochlear implant is different from normal hearing and takes time to learn or relearn. However, it allows many people to recognize warning signals, understand other sounds in the environment, and enjoy a conversation in person or by telephone. Eighty-one profoundly deaf subjects in this experiment received one of three different brands of cochlear implant (A/B/G3). Brand G3 is a third-generation device, while brands A and B are second-generation devices.

The research question was to determine whether the three brands differed in levels of speech recognition.

The data file is given in ▣ cochlear.xlsx. The variables are as follows:

Name	Description
ID	Patient ID number
Age	Patient's age at the time of implantation
Device	Type of cochlear implant (A/B/G3)
YrDeaf	Years of profound deafness prior to implantation
CS	Consonant recognition score, sound only
CV	Consonant recognition score, visual only
CSV	Consonant recognition score, sound and vision
SNT	Sentence understanding score
VOW	Vowel recognition score
WRD	Word recognition score
PHN	Phoneme recognition score

Run three separate ANOVAs with three response variables, which are different tests of speech recognition:

(a) CSV, audiovisual consonant recognition (subjects hear triads like "ABA," "ATA," and "AFA" and have to pick the correct consonant on a touch screen);

(b) PHN, phoneme understanding (number of correct phonemes in random 5- to 7-word sentences like "The boy threw the ball."); and

(c) WRD, word recognition (number of words recognized in a list of random, unrelated monosyllabic words "ball," "dog," etc.).

11.11. **Does the Honeybee Change the Concentration of Nectar while en Route to the Hive?** Park (1932) investigated the means employed by the honeybee to eliminate excess water from nectar in the process commonly known as the ripening of honey.

Syrup of approx. 40% concentration was fed to the bees. The concentration in their honey sacs was determined upon their arrival at the hive (distance 0.5 mi). The decreases recorded in the table are classified to six batches according to time of collection. The question to be answered is this: were significant differences introduced by changes in the time of gathering the data, or can the six batches be considered random samples from a homogeneous population?

Batch 1	Batch 2	Batch 3	Batch 4	Batch 5	Batch 6
−1.06	−0.99	−0.64	1.60	−1.08	−2.50
−1.01	−0.59	−0.34	−0.75	−0.53	−0.65
−0.96	−1.04	0.06	−2.10	−2.18	−1.10
−1.11	−0.39	0.01	−1.10	−1.08	−0.65
−0.91	−0.39	−1.49	−0.65	−0.38	−1.85
−1.06	−0.94	−0.94	−0.55	1.97	−0.55
−0.64	−0.64	−0.34	−0.60	−1.43	−1.20
−0.74	−0.44	−0.19	−0.20		−1.15
	−1.14	−0.39	−0.75		−0.45
		−0.44	−0.60		−1.00

Data can be found in bees.mat, a data structure with two fields:
bees.difs and bees.batch.
(a) Conduct a one-way ANOVA test and state your conclusions.
(b) A test in MATLAB that will check equality of population variances is
vartestn, (see also p. 511). Conduct this test. Was the standard ANOVA
assumption of equal variances violated?
(c) A counterpart test that does not require equality of variances is
Welch's ANOVA, Section 11.3. Write a code in MATLAB that applies
Welch's ANOVA and apply it on Park's data. What do you conclude?

11.12. **SiRstv: NIST's Silicon Resistivity Data.** Measurements of bulk resistivity of silicon wafers were made at NIST with five probing instruments
(columns in the data matrix) on each of 5 days (rows in the data matrix). The wafers were doped with phosphorous by neutron transmu-
tation doping in order to have nominal resistibility of 200 ohm · cm.
Measurements were carried out with four-point DC probes according to
ASTM Standard F84-93 and described in Ehrstein and Croarkin (1984).

1	2	3	4	5
196.3052	196.3042	196.1303	196.2795	196.2119
196.1240	196.3825	196.2005	196.1748	196.1051
196.1890	196.1669	196.2889	196.1494	196.1850
196.2569	196.3257	196.0343	196.1485	196.0052
196.3403	196.0422	196.1811	195.9885	196.2090

(a) Test the hypothesis that the population means of measurements produced by these five instruments are the same at $\alpha = 5\%$. Use MATLAB
to produce an ANOVA table.
(b) Pretend now that some measurements for the first and fifth instruments are misrecorded (in italics):

1	2	3	4	5
196.3052	196.3042	196.1303	196.2795	*196.1119*
196.2240	196.3825	196.2005	196.1748	196.1051
196.2890	196.1669	196.2889	196.1494	196.1850
196.2569	196.3257	196.0343	196.1485	196.0052
196.3403	196.0422	196.1811	195.9885	*196.1090*

Test the same hypothesis as in (a). If H_0 is rejected, perform Tukey's multiple comparisons procedure.

11.13. **Dorsal Spines of *Gasterosteus aculeatus*.** Bell and Foster (1994) were interested in the effect of predators on dorsal spine length evolution in *Gasterosteus aculeatus* (threespine stickleback, Fig. 11.12). Dorsal spines are thought to act as an antipredator mechanism.

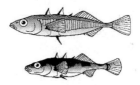

Fig. 11.12 *Gasterosteus aculeatus.*

To examine this issue, researchers sampled eight sticklebacks from each of Benka Lake (no predators), Garden Bay Lake (some predators), and Big Lake (lots of predators). Their observations on spine length (in millimeters) are provided in the following table.

Benka Lake	Garden Bay Lake	Big Lake
4.2	4.4	4.9
4.1	4.6	4.6
4.2	4.5	4.3
4.3	4.2	4.9
4.5	4.4	4.7
4.4	4.2	4.4
4.5	4.5	4.5
4.3	4.7	4.4

They would like to know if spine lengths differ among these three populations and apply ANOVA with lakes as "treatments."

(a) Test the hypothesis that the mean lengths in the three populations are the same at level $\alpha = 0.05$. State your conclusion in terms of the problem.

(b) Would you change the decision in (a) if $\alpha = 0.01$? Explain why or why not.

11.14. **Incomplete ANOVA Table.** In the context of a balanced two factor ANOVA, recover entries a–m.

Source of variation	Sum of squares	Degrees of freedom	Mean square	F	p-value
A	256.12	2	a	4.18	b
B	c	3	12.14	d	e
A×B	217.77	f	g	h	i
Error	j	k	l		
Total	m	119			

11.15. **Maternal Behavior in Rats.** To investigate the maternal behavior of laboratory rats, researchers separated rat pups from their mother and recorded the time required for the mother to retrieve the pups. This longitudinal study was run with six mothers when their litters were 5-, 20-, and 35-day-old. The pups were moved a fixed distance from the mother and the time of retrieval (in seconds) was recorded (Adapted from Montgomery, 1984):

5 days	15 40 25 15 20 18
20 days	30 55 20 25 23 20
35 days	30 85 50 24 45 40

State the inferential problem, pose the hypotheses, check for sphericity, and perform the test at $\alpha = 0.05$. State your conclusions.

11.16. **Comparing Dialysis Treatments.** In Example 11.8 pretend that the three columns of measurements are independent, that is, that 24 independent patients were randomly assigned to one of the three treatments, 8 patients to each treatment. Test the null hypothesis that there is no difference in mean weight change among the treatments. Use $\alpha = 0.05$. Compare results with those in Example 11.8 and comment.

11.17. **Materials Scientist Assessing Tensile Strength.** A materials scientist wishes to test the effect of four chemical agents on the strength of a particular type of cloth. There might be variability from one bolt to another, so the scientist decides to use a randomized block design, with the bolts of cloth considered as blocks. She selects five bolts and applies all four chemicals in random order to each bolt. The resulting tensile strengths follow:

	Chemical
Bolt	1 2 3 4
1	73 73 75 73
2	68 67 68 71
3	74 75 78 75
4	71 72 74 73
5	67 70 70 69

Analyze the data and draw appropriate conclusions.

11.18. **Magnesium Ammonium Phosphate and Chrysanthemums.** Walpole et al. (2007) provide data from a study on the effect of magnesium ammonium phosphate on the height of chrysanthemums, which was conducted at George Mason University in order to determine a possible optimum level of fertilization, based on the enhanced vertical growth response of the chrysanthemums. Forty chrysanthemum seedlings were assigned to 4 groups, each containing 10 plants. Each was planted in a similar pot containing a uniform growth medium. An increasing concentration of $MgNH_4PO_4$, measured in grams per bushel, was added to each plant. The 4 groups of plants were grown under uniform conditions in a greenhouse for a period of 4 weeks. The treatments and the respective changes in heights, measured in centimeters, are given in the following table:

Treatment			
50 g/bu	100 g/bu	200 g/bu	400 g/bu
13.2	16.0	7.8	21.0
12.4	12.6	14.4	14.8
12.8	14.8	20.0	19.1
17.2	13.0	15.8	15.8
13.0	14.0	17.0	18.0
14.0	23.6	27.0	26.0
14.2	14.0	19.6	21.1
21.6	17.0	18.0	22.0
15.0	22.2	20.2	25.0
20.0	24.4	23.2	18.2

(a) Do different concentrations of $MgNH_4PO_4$ affect the average attained height of chrysanthemums? Test the hypothesis at the level $\alpha = 0.10$.
(b) For $\alpha = 10\%$, perform multiple comparisons using multcompare.
(c) Find the 90% confidence interval for the contrast $\mu_1 - \mu_2 - \mu_3 + \mu_4$.

11.19. **Color Attraction for *Oulema melanopus*.** Some colors are more attractive to insects than others. Wilson and Shade (1967) conducted an experiment aimed at determining the best color for attracting cereal leaf beetles (*Oulema melanopus*). Six boards in each of four selected colors (lemon yellow, white, green, and blue) were placed in a field of oats in

July. The following table (modified from Wilson and Shade, 1967) gives data on the number of cereal leaf beetles trapped:

Board color	Insects trapped
Lemon yellow	45 59 48 46 38 47
White	21 12 14 17 13 17
Green	37 32 15 25 39 41
Blue	16 11 20 21 14 7

(a) Based on computer output, state your conclusions about the attractiveness of these colors to the beetles. See also Fig. 11.13a.
In MATLAB:

```
ntrap=[ 45,  59,  48,  46,  38,  47,  21,  12,  14,  17,...
        13,  17,  37,  32,  15,  25,  39,  41,  16,  11,...
        20,  21,  14,   7];
color={'ly','ly','ly','ly','ly','ly',...
       'wh','wh','wh','wh','wh','wh',...
       'gr','gr','gr','gr','gr','gr',...
       'bl','bl','bl','bl','bl','bl'};
[p, table, stats] = anova1(ntrap, color)
multcompare(stats)

%
%'Source'  'SS'             'df'          'MS'          'F'        'Prob>F'
%'Groups'  [4.2185e+003]    [ 3]   [1.4062e+003] [30.5519]   [1.1510e-007]
%'Error'   [   920.5000]    [20]   [   46.0250]       []               []
%'Total'   [5.1390e+003]    [23]             []       []               []
%
%Pairwise Comparisons
%    1.0000    2.0000     20.5370    31.5000    42.4630
%    1.0000    3.0000      4.7037    15.6667    26.6297
%    1.0000    4.0000     21.3703    32.3333    43.2963
%    2.0000    3.0000    -26.7963   -15.8333    -4.8703
%    2.0000    4.0000    -10.1297     0.8333    11.7963
%    3.0000    4.0000      5.7037    16.6667    27.6297
```

(b) The null hypothesis is rejected. Which means differ? See Fig. 11.13b.
(c) Perform an ANOM analysis. Which means are different from the (overall) grand mean?

11.20. **Raynaud's Phenomenon.** Raynaud's phenomenon is a condition resulting in a discoloration of the fingers or toes after exposure to temperature changes or emotional events. Skin discoloration is caused by an abnormal spasm of the blood vessels, diminishing blood supply to the local tissues. Kahan et al. (1987) investigated the efficacy of the calcium-channel blocker nicardipine in the treatment of Raynaud's phenomenon. This efficacy was assessed in a prospective, double-blind, randomized, crossover trial in 20 patients. Each patient received 20 mg nicardipine or

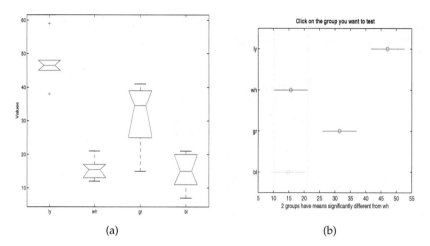

Fig. 11.13 (a) Barplots of ntrap for the four treatments. (b) A snapshot of the interactive plot for multiple comparisons.

placebo three times a day for 2 weeks and then was crossed over for 2 weeks. To suppress any carryover effect, there was a 1-week washout period between the two treatments. The researchers were interested in seeing if nicardipine significantly decreased the frequency and severity of Raynaud's phenomenon as compared with the placebo. To control for the order of drug administration effect, 10 randomly selected subjects received drug first, and the remaining 10 placebo first. The data consist of a number of attacks in 2 weeks.

| | Period 1 | Period 2 | | Period 1 | Period 2 |
Subject	Nicardipine	Placebo	Subject	Placebo	Nicardipine
1	16	12	11	18	12
2	26	19	12	12	4
3	8	20	13	46	37
4	37	44	14	51	58
5	9	25	15	28	2
6	41	36	16	29	18
7	52	36	17	51	44
8	10	11	18	46	14
9	11	20	19	18	30
10	30	27	20	44	4

Download the MATLAB format of these data (▲ raynaud.m) and perform an ANOVA test in MATLAB. Select an additive model, with the number of attacks y_{ijk} represented as

$$y_{ijk} = \mu + \alpha_i + \beta_j + \gamma_{k(j)} + \epsilon_{ijk}.$$

Here μ is the grand mean, α_i, $i = 1, \ldots, 20$ is the drug effect, β_j, $j = 1, 2$ is the order effect, $\gamma_{k(j)}$, $k = 1, \ldots, 10$ is the effect of subject nested in order j, and ϵ_{ijk} is the zero-mean normal random error.

Note that subjects are nested in order treatments, that is a subject cannot receive both drug and placebo first. We are interested in testing the null hypothesis about the drugs,

$$H_0 : \alpha_1 = \alpha_2 = 0,$$

as well as the effect of the order,

$$H_0' : \beta_1 = \beta_2 = 0.$$

The remaining hypothesis $H_0'' : \gamma_k(j) = 0$, $k = 1, \ldots, 10$, $j = 1, 2$, is lateral for this study.

(a) Conduct a nested ANOVA analysis using MATLAB's anovan with option 'nested'. Test hypotheses H_0 and H_0'. Describe your findings.

(b) What proportion of variability in the data y_{ijk} is explained by the ANOVA model given above?

11.21. **Leptograpsus Crabs.** Venables and Ripley (2002) analyzed the data from Campbell and Mahon (1974) on the morphology of rock crabs *Leptograpsus variegatus* collected at Fremantle, W. Australia ($32°S, 117°E$). Campbell and Mahon wanted to show that *Leptograpsus variegatus ver. Caeruleus* (blue form) and *Leptograpsus variegatus ver. Aurantius* (orange form) are in fact two distinct species based on the comparison of several morphologic measurements on the carapace.

The data on 50 specimens of each sex of each of two color forms were collected by Campbell and Mahon. File 📊 crabs.mat is MATLAB's data structure consisting of the following fields:

crabs.sp	Species, coded B (blue) or O (orange)
crabs.sex	Coded as M or F
crabs.FL	Frontal lip of carapace (mm)
crabs.RW	Rear width of carapace (mm)
crabs.CL	Length along the midline (mm)
crabs.CW	Maximum width of carapace (mm)
crabs.BD	Body depth (mm)
crabs.score	First principal component scores from FL, RW, CL, CW, and BD.

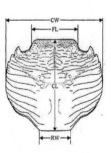

(a) Using a two-way ANOVA with factors sp and sex and response score, explore whether the interaction between the factors is significant. Report the p-value. Use MATLAB's function 🏄 anovan.

(b) If the interaction is found significant, do one-way ANOVAs separately for male and female crabs. Split the data accordingly and provide two ANOVA tables, one for each sex. What are the p-values associated with sp factor?

11.22. **Simvastatin.** In a Quantitative Physiology Lab at Georgia Tech, students were asked to find a therapeutic model to test on MC3T3-E1 cell line to enhance osteoblastic growth. The students found a drug called Simvastatin, a cholesterol lowering drug to test on these cells. Using a control and three different concentrations 10^{-9}M, 10^{-8}M, and 10^{-7}M, cells were treated with the drug. These cells were plated on four 24-well plates with each well plate having a different treatment. To test for osteoblastic differentiation, a pNPP assay was used to test for alkaline phosphatase activity. The higher the alkaline phosphatase activity, the better the cells are differentiating and become more bone like. This assay was performed 6 times total within 11 days. Each time the assay was performed, four wells from each plate were used. The data (⊞ simvastatin.dat) are provided in the following table:

| | Time | | | | | |
Concentration	Day 1	Day 3	Day 5	Day 7	Day 9	Day 11
	0.062	0.055	0.055	1.028	0.607	0.067
A	0.517	0.054	0.059	1.067	0.104	0.093
(Control)	0.261	0.056	0.062	1.128	0.163	0.165
	0.154	0.063	0.062	0.855	0.109	0.076
	0.071	0.055	0.067	0.075	0.068	0.347
B	0.472	0.060	1.234	0.076	0.143	0.106
$(10^{-7}M)$	0.903	0.057	1.086	0.090	0.108	0.170
	0.565	0.056	0.188	1.209	0.075	0.097
	0.068	0.059	0.092	0.091	0.098	0.115
C	0.474	0.070	0.096	0.218	0.122	0.085
$(10^{-8}M)$	0.063	0.090	0.123	0.618	0.837	0.076
	0.059	0.064	0.091	0.093	0.142	0.085
	0.066	0.447	0.086	0.248	0.108	0.290
D	0.670	0.091	0.076	0.094	0.105	0.090
$(10^{-9}M)$	0.076	0.079	0.082	0.215	0.093	0.518
	0.080	0.071	0.080	0.401	0.580	0.071

In this design there are two crossed factors: concentration and time.
(a) Using a two-way ANOVA, show that the interaction between factors time and concentration is significant.
(b) Since the presence of a significant interaction affects the tests for main effects, conduct a conditional test for the concentration factor if time is fixed at day 3. Also, conduct the conditional test for time factor if the concentration level is C.

11.23. **Differences in Differences.** Bolton (1984) provides data on 6 patients in a clinical trial involving two drugs and a placebo. In a randomized order the drugs are administered to all 6 patients, allowing for washout period between the administrations. For each treatment and each patient, pre-drug and post-drug measurements have been taken and are shown below:

	Placebo		Drug 1		Drug 2	
Patient	Pre	Post	Pre	Post	Pre	Post
1	180	176	170	161	172	165
2	140	142	143	140	140	141
3	175	174	180	176	182	175
4	120	128	115	120	122	122
5	165	165	176	170	171	166
6	190	183	200	195	192	185

This table (without the Patient column) is given in the file ⬛ bolton.dat|mat. It is of interest to compare the effectiveness of the treatments.

(a) Compare the treatments using only Post values.

(b) Test the effectiveness of the treatments by comparing the changes from the baseline, Post minus Pre. If significant, which treatments differ?

11.24. **Skull variations in** *Canis lupus L.* Data set ⬛ wolves.dat|mat provides skull morphometric measurements on wolves (*Canis lupus L.*) coming from two geographic locations: Rocky Mountains (rm) and the Arctic (ar). The original source of data is Jolicoeur (1959), but subsequently many authors used this data to illustrate various multivariate statistical procedures.

The goal of Jolicoeur's study was to determine how the location and gender affect the skull shape among the wolf populations. There were 9 response variables:

Variable	Measure
y_1	Palatal length
y_2	Postpalatal length
y_3	Zygomatic width
y_4	Palatal width outside the first upper molars
y_5	Palatal width inside the second upper molars
y_6	Width between the postglenoid foramina
y_7	Interorbital width
y_8	Least width of the braincase
y_9	Crown length of the first upper molar

These 9-dimensional measurements can be assigned to one of the four groups determined by a combination of location and gender: rmm, rmf, arm, and arf.

(a) Using ANOVA, test the equality of the group means for the response y_1.

(b) Find principal component scores for this data and plot the first principal component against the second. Represent the measurements from different groups by different markers/colors. Using ANOVA, test the equality of the group population means for the first principal component. Compare the means using the `multcompare` procedure.

11.25. **Isoflavones and ALP Activity.** A team of bioengineering students wanted to see the effect of Daidzein and Genistein, two isoflavones from soybeans, and time period cultured has on the Alkaline Phosphatase Activity (ALP) of osteoblastic cells. An increase in ALP activity would indicate these immature bone cells are differentiating and moving towards matrix maturation. The team cultures 3 identical well plates with wells of control, cells with Daidzein treatment, and cells with Genistein treatment in each. The ALP was measured on day 2 for well plate 1, day 6 for well plate 2, and day 10 for well plate 3. Data, in a data-structure format, is provided in 🖳 isoflavon.mat (Courtesy of Michelle Rost). The variables are

---
isoflavon.absorbance isoflavon.drug isoflavon.day isoflavon.ALP

The Absorbance reading was used to calculate ALP activity. Drug was coded as 1 for Control, 2 for Daidzein, and 3 for Genistein. ALP Activity was multiplied by a factor of 10^6.

(a) Was there a significant interaction between drug treatment and the time?

(b) Did the drug treatment or the time have an effect on the ALP activity? If so, which drug treatment group or time period had an inhibitory or stimulatory effect?

(c) Analyze the residuals of this model. Do they appear normal?

(d) Using WinBUGS, analyze the problem assuming STZ constraints in ANOVA model.

(d) Repeat the analysis in (a-c) using the logarithm of ALP activity as a response.

11.26. **Master of Light.** Albert Michelson devised a clever way to increase precision of measuring the speed of light (Michelson, 1927) by rotating mirrors. In his experiment the light was sent from Mount Wilson, reflected back from Mount San Antonio, and the time of flight was recorded. Back-engineered, when the speed of light is assumed known, crude calculations for an octagonal rotating mirror look as follows: The round trip distance between Mount Wilson and Mount San Antonio is $d = 70,800$ m. Dividing d by speed of light $c = 2.997 \times 10^{10}$ m/s results in time of flight $t = 2.36 \times 10^{-4}$ s. When an octagonal mirror rotates at 529 rev/s, the time between successive surfaces is $(1/529)/8 = 2.36 \times 10^{-4}$, the same as t. Thus, the return image of light should fall in the same place. Michelson was able to control the mirror rotation rate and align perfectly the locations of sent and received lights.

A diagram of his equipment and data can be found in Michelson (1927, p.4, http://adsabs.harvard.edu/full/1927ApJ....65....1M). The data in the table below represent the two last digits in his measurements and can be thought of as observations between 0 and 99. Data set michelson.dat contains the data in MATLAB format.

Material/Sides	Type of rotating mirror				
	Glass/8	Glass/12	Glass/16	Steel/12	Steel/8
	47 47	42 18	3 39 66 21	18 9	30 21
	38 62	36 45	27 66 27 33	12 30	33 18
	29 59	33 30	48 15 9 24	30 27	12 33
	92 44	0 27	3 7 6 39	30 39	24 23
	41 47	18 27	27 27 42 18	18 27	57 39
	44 41	57 66	42 37 12 63	48 24	44 33
		48 24	69 24 30 27	18	30 24
		15	63 15 42 42		24 30
			60		

(a) Test whether the means of measurements for different types of mirrors are statistically the same.

(b) Which treatment level(s) is(are) different. Find 95% confidence intervals for the differences of means, $\mu_i - \mu_j$, $1 \leq i < j \leq 5$.

(c) Consider now data for only four treatment levels Glass/8, Glass/12, Steel/8, and Steel/12. Recode the data to conduct a two-way ANOVA with factors Material (levels Glass and Steel) and Sides (levels 8 and 12). Pretend you are a data-analyst in Michelson's lab. If Michelson was interested whether the factors Material and Sides interact, how would you advise him?

11.27. **Multifactor Twisting Experiment.** Johnson (2000) provides data on a multifactor experiment studying the effects of ingot location (A), slab position (B), specimen preparation (C), and twisting temperature (D) on the number of turns required to break the steel specimen by twisting (y). For each of the $3 \times 2 \times 2 \times 3 = 36$ experimental conditions two measurements are recorded, so the total sample size is $n = 72$.

The data set twisting.dat contains 5 columns:

Column 1 Factor A at 3 levels: Top, Middle, Bottom coded by 1, 2, and 3
Column 2 Factor B at two levels coded by 1 and 2
Column 3 Factor C at two levels: Grind and Turn coded by 1 and 2
Column 4 Factor D at 3 levels: 2100°F, 2200°F, and 2300°F coded by 1, 2, and 3
Column 5 Response y: Numbers of turns

(a) Using anovan, fit the model for y with second-order interactions.

(b) Discuss the ANOVA table. Which main effects are significant at 5% level? Is any interaction significant?

(c) Show that with 2 observations per experimental condition (cell), the power for the Cohen's medium effect $f^2 = 0.0625$ in a 0.05-level testing for factor A (3 levels) is only 43.79%. What number of observations per cell is required to achieve the power of at least 85%?

Hint: Show first that with n observations per cell for the model with second-order interactions only, the error degrees of freedom (dfe) is $36n - 20$. In general, if in a 4-way balanced ANOVA with n observations per cell where factors A-D have a-d levels, the total sample size is $N = abcdn$ and the dfe is equal to dftotal $-$ dfa $-$ dfb $-$ dfc $-$ dfd $-$ dfab $-$ dfac $-$ dfad $-$ dfbc $-$ dfbd $-$ dfcd. The dftotal is $N - 1$, main effect A has dfa $= a - 1$ degrees of freedom, and second-order interaction A*B has dfab $= (a - 1)(b - 1)$ degrees of freedom. Degrees of freedom for other main effects and interactions are calculated similarly.

Following Example 11.9, the power for n observations per cell is

```
1-ncfcdf( finv(1-0.05,3-1,36*n- 20),3-1,36*n-20, 36*n*0.0625).
```

11.28. **The Honeybee Hierarchical.** In the context of Exercise 11.11 we are interested in the decrease of syrup concentration when the two different concentrations of 60% and 64% are explored. Under each concentration 3 batches of syrup are made. The batches are nested within each concentration and considered as random.

Concentration 60%			Concentration 64%		
Batch 1	Batch 2	Batch 3	Batch 1	Batch 2	Batch 3
−1.62	−0.94	−2.13	−0.41	−1.35	−1.72
−0.29	−1.04	−1.65	−0.46	−1.20	−2.77
−0.62	−1.21	−3.18	−1.71	−3.50	−2.22
−0.67	−1.04	−1.73	−0.86	−2.05	−1.45
−0.47	−0.89	−2.23	−1.33	−1.38	−1.27
−0.77	−0.99	−1.88	−1.06	−3.12	−0.85
−0.67	−0.39	−1.73	−1.29	−2.05	−2.02
−1.72	−0.59	−2.58	−1.33	−1.00	−3.42
−0.67	−0.89	−1.28	−0.96	−3.12	−1.42
−1.07	−1.24	−2.08	−4.26	−2.68	−1.89

Show that when basic ANOVA (in this case two-sample t-test) is applied, the concentrations show significant differences. When the nested ANOVA is conducted, the concentrations are no longer significantly different. Provide an explanation.

11.29. **Antitobacco Media Campaigns.** Since the early 1990s, the US population has been exposed to a growing number and variety of televised antitobacco advertisements. By 2002, 35 states had launched antitobacco media campaigns.

Four states (factor A) participated in an antitobacco awareness study. Each state independently devised an antitobacco program. Three cities (factor B) within each state were selected for participation, and 10 households within each city were randomly selected to evaluate effectiveness of the program. All members of the selected households were interviewed, and a composite index was formed for each household measuring the extent of antitobacco awareness. The data are given in the table (the larger the index, the greater the awareness):

State i	1			2			3			4		
City j	1	2	3	1	2	3	1	2	3	1	2	3
Household k 1	42	26	33	48	56	44	23	47	48	70	56	35
2	56	38	51	54	65	34	31	39	40	61	49	30
3	35	42	43	48	71	39	34	45	43	57	58	41
4	40	29	49	57	66	40	28	51	48	69	55	52
5	54	44	42	51	64	51	30	38	51	71	47	39
6	49	43	39	51	51	47	36	47	57	49	50	44
7	51	45	41	56	60	34	41	39	40	61	49	30
8	48	39	47	50	72	41	35	44	41	57	61	40
9	52	30	45	50	60	42	28	44	49	69	54	52
10	54	40	51	49	63	49	33	46	49	67	52	45

(a) Discuss and propose the analyzing methodology.

(b) Assume that a nested design with fixed factor effects is appropriate. Provide the ANOVA table.

(c) Test whether or not the mean awareness differ for the four states. Use $\alpha = 0.05$. State the alternatives and draw a conclusion.

11.30. **Orthosis.** The data 🖥️ Ortho.dat were acquired and computed by Dr. David Amarantini and Dr. Martin Luc (Laboratoire Sport et Performance Motrice, Grenoble University, France). The purpose of recording such data was an interest to better understand the processes underlying movement generation under various levels of an externally applied moment to the knee. In this experiment, stepping-in-place was a relevant task to investigate how muscle redundancy could be appropriately used to cope with an external perturbation while complying with the mechanical requirements related to balance control and/or minimum energy expenditure. For this purpose, 7 young male volunteers wore a spring-loaded orthosis of adjustable stiffness under 4 experimental conditions: a control condition (without orthosis), an orthosis condition (with the orthosis only), and two conditions (Spring1, Spring2) in which stepping in place was perturbed by fitting a spring-loaded orthosis onto the right knee joint.

For each stepping-in-place replication, the resultant moment was computed at 256 time points equally spaced and scaled so that a time interval corresponds to an individual gait cycle. A typical moment observation is therefore a one-dimensional function of normalized time t so that $t \in [0,1]$. The data set consists in 280 separate runs and involves $j = 7$

subjects over $i = 4$ described experimental conditions, replicated $k = 10$ times for each subject. Moment plots over gait cycles are shown in Figure 11.14. Model the data as arising from a fixed-effects FANOVA model

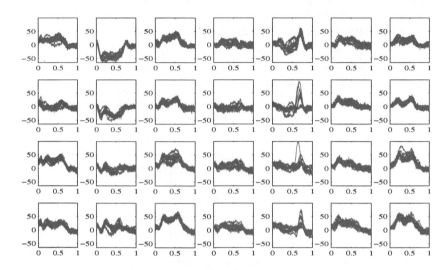

Fig. 11.14 Orthosis data set: The panels in rows correspond to *Treatments* while the panels in columns correspond to *Subjects*; there are 10 repeated measurements in each panel.

with 2 qualitative factors (*Subjects* and *Treatments*), 1 quantitative factor (*Time*) and 10 replications for each level combination.

If the functional ANOVA model is $\mathbb{E}y_{ijk}(t) = \mu(t) + \alpha_i(t) + \beta_j(t)$, find and plot functional estimators for the grand mean $\mu(t)$ and treatment effects $\alpha_1(t) - \alpha_4(t)$.

11.31. **Bone Screws.** Bone screws are the most commonly used type of orthopedic implant. Their precision is critical and several dimensions are rigorously checked for conformity (outside and root diameters, pitch, tread angle, etc.). The screws are produced in lots of 20,000. To confirm the homogeneity of production, 6 lots were selected and from each lot a sample of size 100 was obtained. The following numbers of nonconforming screws are found in the six samples: 20, 17, 23, 9, 15, and 14.
(a) Test the hypothesis of homogeneity of proportions of nonconforming screws in the six lots.
(b) If the null hypothesis in (a) is rejected, find which lots differ. Assume $\alpha = 0.05$.

11.32. **R&R Study.** Five parts are measured by two appraisers using the same measuring instrument. Each appraiser measured each part three times.

Find the relevant parameters of the measuring system and assess its capability (PRR and NDC).

```
measure =[ ...
     217 220 217 214 216 ...
     216 216 216 212 219 ...
     216 218 216 212 220 ...
     216 216 215 212 220 ...
     219 216 215 212 220 ...
     220 220 216 212 220]';

part = [ ...
     1 2 3 4 5 1 2 3 4 5 ...
     1 2 3 4 5 1 2 3 4 5 ...
     1 2 3 4 5 1 2 3 4 5]';

appraiser =[...
     1 1 1 1 1 ...
     1 1 1 1 1 ...
     1 1 1 1 1 ...
     2 2 2 2 2 ...
     2 2 2 2 2 ...
     2 2 2 2 2]';
```

11.33. **Additive R&R ANOVA for Measuring Impedance.** Repeat the analysis in Example 11.17 by assuming that parts and operators do not interact, that is, with an additive ANOVA model. Provide comparisons with results from gagerr, where the option ('model','linear') is used.

11.34. **Round and Wrinkled Peas.** Mendelian advocate Bateson (1913) provides table describing result of Mendel's peas experiment. The counts of round and wrinkled peas are provided for 10 plants.

Plant #	Round #	Wrinkled #	Total	Plant #	Round #	Wrinkled #	Total
1	45	12	57	6	26	6	32
2	27	8	35	7	88	24	112
3	24	7	31	8	22	10	32
4	19	10	29	9	28	6	34
5	32	11	43	10	25	7	32

(a) Test the hypothesis that the sample of 10 plants has homogeneous proportions, $H_0 : p_1 = p_2 = \cdots = p_{10}$. Report the p-value and discuss.
(b) The plants are F_2 generation with expected frequency of round peas equal to 3/4. How would you change the statistic in (a) to test $H_0 : p_1 = p_2 = \cdots = p_{10} = 3/4$?
Hint: When $\bar{p}$ is replaced by 3/4, the χ^2 statistic will have k degrees of freedom instead of $k - 1$. Why?

11.35. **Aberrant Crypt Foci.** Aberrant crypt foci (ACF) are abnormal collections of tube-like structures that are precursors to tumors (McLellan et al.,

1991). Researchers exposed 22 rats to a carcinogen and then counted the number of ACFs in the rat colons. There were three treatment groups based on time since first exposure to the carcinogen, as 6, 12, or 18 weeks.

```
ACFweek6  = [1 3 5 1 2 1 1]
ACFweek12 = [3 1 2 6 0 0 4 1]
ACFweek18 = [10 6 6 7 5 7 6]
```

Assume a Poisson model for the number of ACFs, where the intensity parameters λ_6, λ_{12}, and λ_{18} correspond to counts after 6, 12, and 18 weeks, respectively.

Test the hypothesis $H_0 : \lambda_6 = \lambda_{12} = \lambda_{18}$. Use $\alpha = 0.05$.

MATLAB AND WINBUGS FILES AND DATA SETS USED IN THIS CHAPTER

http://statbook.gatech.edu/Ch11.Anova/

anomct.m, anovarep.m, Barspher.m, C677T.m, cardiac.m, chrysanthemum.m, circularity.m, coagulationtimes.m, cochcdf1.m, crabs.m, Cochtest.m, dialysis.m, dorsalspines.m, dunn.m, epsGG.m, fanova.m, fibers.m, hair.m, insects.m, insulin.m, maternalbehavior.m, Mauspher.m, Mausphercnst.m, micePO2.m, nest1.m, nest2.m, nesta.m, nestAOV.m, nestedcomplex.m, nestnylon.m, nicardipine.m, orthosis.m, orthosisraw.m, pedometer1300.m, poissonmeans.m, powerANOVA.m, qtukey.m, RandR.m, RandR2.m, RandR3.m, ratsblocking.m, raynaud.m, screws.m, secchim.m, simvastatin.m, sphertest.m, ssizerepeated.m, tmcomptest.m, turnips.m, welchanovapower.m, welchexample.m, welchsimul.m, wolves.m

anovacoagulation.odc, dialysis.odc, insulin.odc, isoflavon.odc, simvastatin.odc, vortex.odc

arthritis.dat|mat, bees.mat, bolton.dat|mat, chairyoga.mat, clouds.txt, cochlear.xlsx, crabs.dat|mat, michelson.dat, deaf.xls, Ortho.dat, oxygen.dat, pmr1300.mat, porcinedata.xls, redclover.mat, Proliferation.xls, secchi.mat|xls, secchi.xls, silicone.dat, silicone1.dat, simvastatin.dat, twisting.dat|mat|xlsx, wolves.dat|mat

---

CHAPTER REFERENCES

Altman, D. G. (1991). *Practical Statistics for Medical Research*. Chapman and Hall, London.

Bartlett M. S. (1937). Properties of sufficiency and statistical tests. *Proc. R. Soc. Lond. A*, **160**, 268–282.

Bateson, W. (1913). *Mendel's Principles of Heredity*. Cambridge University Press, UK, 345 pp.

Bell, M. A. and Foster, S. A. (1994). Introduction to the evolutionary biology of the three-spine stickleback. In: Bell and Foster (eds.) *The Evolutionary Biology of the Threespine Stickleback*. Oxford University Press, Oxford, pp. 277–296.

Bolton, S. (1984). *Pharmaceutical Statistics - Practical and Clinical Applications*. Marcel Dekker, New York, p. 255.

Box, G. E. P. (1953). Non-normality and tests on variances. *Biometrika*, **40**, 318–335.

Box, G. E. P, Hunter, J. S., and Hunter, W. G. (2005). *Statistics for Experimenters II*, 2nd ed. Wiley, Hoboken, NJ. Brix-Christensen, V., Vestergaard, C., Andersen, S. K., Krog, J., Andersen, N. T., Larsson, A., Schmitz, O., and Tønnesen, E. (2005). Evidence that acute hyperinsulinaemia increases the cytokine content in essential organs after an endotoxin challenge in a porcine model. *Acta Anaesthesiol. Scand.*, **49**, 1429–1435.

Campbell, N. A. and Mahon, R. J. (1974). A multivariate study of variation in two species of rock crab of genus *Leptograpsus*. *Austral. J. Zool.*, **22**, 417–425.

Cochran W. G. (1941). The distribution of the largest of a set of estimated variances as a fraction of their total. *Ann. Eugen.*, **11**, 47–52.

Conover, W. J. (1999). *Practical Nonparametric Statistics*, 3rd ed. Wiley, New York.

Daugridas, J. T. and Ing, T. (1994). *Handbook of Dialysis*. Little, Brown, Boston.

Diggle, P. J., Liang, K. Y., and Zeger, S. L. (1994). *Analysis of Longitudinal Data*. Oxford University Press, Oxford.

Dunnett, C. W. (1955). A multiple comparison procedure for comparing several treatments with a control. *J. Am. Stat. Assoc.*, **50**, 1096–1121.

Ehlke, N. J. and Vellekson, D. J. (1999). Red clover variety trials. *Minnesota Agricultural Experiment Station, University of Minnesota*. Report ES-WW-7338.

Ehrstein, J. and Croarkin, M. C. (1984). Standard method for measuring resistivity of silicon wafers with an in-line four-point probe. Unpublished NIST data set. In: http://www.itl.nist.gov/div898/strd/anova/anova.html.

Fan, J. and Lin, S-K. (1998). Test of significance when data are curves. *J. Am. Stat. Assoc.*, **93**, 1007–1021.

Fisher, R. A. (1918). Correlation between relatives on the supposition of Mendelian inheritance. *Trans. R. Soc. Edinb.*, **52**, 399–433.

Fisher, R. A. (1921). Studies in crop variation. I. An examination of the yield of dressed grain from Broadbalk. *J. Agric. Sci.*, **11**, 107–135.

Halperin, M., Greenhouse, S. W., Cornfield, J., and Zalokar, J. (1955). Tables of percentage points for the studentized maximum absolute deviation in normal samples. *J. Am. Stat. Assoc.*, **50**, 185–195.

Houf, R. E. and Burman, D. B. (1988). Statistical analysis of power module thermal test equipment performance. *IEEE Trans. Components Hybrids Manuf. Technol.*, **11**, 516–520.

Johnson, R. A. (2000). *Miller and Freunds Probability and Statistics for Engineers*, 6th ed. Prentice Hall, Upper Saddle River, NJ.

Jolicoeur, P. (1959). Multivariate geographical variation in the wolf *Canis lupus L. Evolution*, **13**, 3, 283–299.

Kahan, A., Amor, B., Menkès, C. J., Weber, S., Guérin, F., and Degeorges, M. (1987). Nicardipine in the treatment of Raynaud's phenomenon: a randomized double-blind trial. *Angiology*, **38**, 333–337.

Kutner, M. H., Nachtsheim, C. J., Neter, J., and Li, W. (2004). *Applied Linear Statistical Models*, 5th ed. McGraw-Hill, Boston, MA.

Lanzen, J. L., Braun, R. D., Ong, A. L., and Dewhirst, M. W. (1998). Variability in blood flow and PO2 in tumors in response to carbogen breathing. *Int. J. Radiat. Oncol. Biol. Phys.*, **42**, 855–859.

Levene, H. (1960). In: Olkin et al. (eds.) *Contributions to Probability and Statistics: Essays in Honor of Harold Hotelling, I.* Stanford University Press, Palo Alto, pp. 278–292.

Liu, W. (1997). Sample size determination of Dunnett's procedure for comparing several treatments with a control. *J. Stat. Plann. Infer.*, **62**, 2, 255–261.

Marascuillo, L. A. and Serlin, R. C. (1988). *Statistical Methods for the Social and Behavioral Sciences*. Freeman, New York.

McLellan, E. A., Medline, A., and Bird, P. R. (1991). Dose response and proliferative characteristics of aberrant crypt foci: Putative preneoplastic lesions in rat colon. *Carcinogenesis*, **12**, 11, 2093–2098.

Michelson, A. A. (1927). Measurement of the velocity of light between Mount Wilson and Mount San Antonio, *Astrophys. J.*, **15**, 1–14.

Miller, A. J., Shaw, D. E., Veitch, L. G., and Smith, E. J. (1979). Analyzing the results of a cloud-seeding experiment in Tasmania. *Commun. Stat. Theor. M.*, **8**, 10, 1017–1047.

Montgomery, D. C. (1984). *Design and Analysis of Experiments*, 2nd edn. Wiley, New York.

Montgomery, D. C. (2005). *Statistical Control Theory*, 5th edn. Wiley, Hoboken, NJ.

Ott, E. R. (1967). Analysis of means – A graphical procedure. *Industrial Quality Control*, **24**, 101–109.

Ozturk, H., Durga, J., van de Rest, O., and Verhoef, P. (2005). The MTHFR 677C→T genotype modifies the relation of folate intake and status with plasma homocysteine in middle-aged and elderly people. *Ned. Tijdschr. Klin. Chem. Labgeneesk*, **30**, 208–217.

Park, W. (1932). Studies on the change in nectar concentration produced by the honeybee, *Apis mellifera*. Part I: Changes that occur between the flower and the hive. *Iowa Agric. Exp. Stat. Res. Bull.*, **151**.

Quenouille, M. H. (1953). *The Design and Analysis of Experiment*. Charles Griffin & Co., London, pp. 360.

Ramsay, J. and Silverman, B. (1997). *Functional Data Analysis*. Springer, New York.

Sokal, R. and Karten, I. (1964). Competition among genotypes in *Tribolium castaneum* at varying densities and gene frequencies (the black locus). *Genetics*, **49**, 195–211.

Št Lam R. U. E. (2010). Scrutiny of variance results for outliers: CochranŠs test optimized. *Anal. Chim. Acta*, 659, 68–84. `doi:10.1016/j.aca.2009.11.032`

Venables, W. N. and Ripley, B. D. (2002). *Modern Applied Statistics with S*, 4th ed. Springer, New York.

Walpole, R. E., Myers, R. H., Myers, S. L., and Ye, K. (2007). *Probability and Statistics for Engineers and Scientists*, 8th ed. Pearson-Prentice Hall, Upper Saddle River, NJ.

Welch, B. L. (1951). On the comparison of several mean values: An alternative approach. *Biometrika*, **38**, 330–336.

Wilson, M. C. and Shade, R. E. (1967). Relative attractiveness of various luminescent colors to the cereal leaf beetle and the meadow spittle-bug. *J. Econ. Entomol.*, **60**, 578–580.

Woodworth, G. G. (2004). *Biostatistics: A Bayesian Introduction*, Wiley, Hoboken, NJ, pp. 384.

Zar, J. H. (2010). *Biostatistical Analysis*, 5th ed. Pearson-Prentice-Hall, Englewood Cliffs, NJ, 944 pp.

Chapter 12
Models for Tables

The object of this present paper is to deal with this novel conception of what I have termed contingency, and to see its relation to our older notions of association and normal correlation. The great value of the idea of contingency for economic, social, and biometric statistics seems to me to lie in the fact that it frees us from the need of determining scales before classifying our attributes.

– Karl Pearson

WHAT IS COVERED IN THIS CHAPTER

• Contingency Tables and Testing for Independence in Categorical Data
• Measuring Association in Contingency Tables
• Three-Dimensional Tables
• Tables with Fixed Marginals: Fisher's Exact Test
• Combining Contingency Tables: Mantel–Haenszel Theory
• Paired Tables: McNemar and Liddell Tests
• Risk Differences, Risk Ratios, and Odds Ratios for Paired Experiments
• Stuart–Maxwell, Bowker, Garth, and Cochran's Tests

12.1 Introduction

The focus of this chapter is the analysis of tabular data. Although the measurements could be numerical, the tables summarize only the counts along the levels of two or more crossed factors according to which the data are tabulated. In this text, we go beyond the traditional introductory coverage and discuss topics such as three-dimensional tables, multiple tables (Mantel–Haenszel theory), paired tables (McNemar, Liddell, Stuart–Maxwell, and Bowker tests), and risk theory (risk differences, risk ratios, and odds ratios) for paired tables. The risk theory for 2×2 tables was already discussed in Chapter 10 in the context of comparing two population proportions.

The dominant statistical procedure in this chapter is testing for the independence of two cross-tabulated factors. In cases where the marginal counts are fixed before the sampling, the test for independence becomes the test for homogeneity of one factor across the levels of the other factor. Although the concepts of homogeneity and independence are different, the mechanics of the two tests are the same. Thus, sometimes it is important to know how an experiment was conducted and whether the table marginal counts were fixed prior to sampling.

The paired tables, like the paired t-test or repeated measure design, are preferred to ordinary or parallel tables whenever pairing is feasible. In paired tables, we would usually be interested in testing the agreement of proportions. The pairing allows for control of confounding by other factors and gives more precise estimators of variance of proportions and functions of the proportions.

12.2 Contingency Tables: Testing for Independence

To formulate a test for the independence of two crossed factors, we need to recall the definition of independence of two events and two random variables. Two events R and C are independent if the probability of their intersection is equal to the product of their individual probabilities,

$$\mathbb{P}(R \cap C) = \mathbb{P}(R) \cdot \mathbb{P}(C).$$

For random variables independence is defined using the independence of events. For example, two random variables X and Y are independent if the events $\{X \in I_x\}$ and $\{Y \in I_y\}$, where I_x and I_y are arbitrary intervals, are independent.

To motivate inference for tabulated data we also need a brief review of two-dimensional discrete random variables, discussed in Chapter. 5. A two-dimensional discrete random variable (X, Y), where $X \in \{x_1, \ldots, x_r\}$

and $Y \in \{y_1, \ldots, y_c\}$, is fully specified by its probability distribution, which is given in the form of a table:

	y_1	y_2	$\cdots$	y_c	Marginal
x_1	p_{11}	p_{12}		p_{1c}	$p_{1\cdot}$
x_2	p_{21}	p_{22}		p_{2c}	$p_{2\cdot}$
x_r	p_{r1}	p_{r2}		p_{rc}	$p_{r\cdot}$
Marginal	$p_{\cdot 1}$	$p_{\cdot 2}$		$p_{\cdot c}$	1

The two components X and Y in (X,Y) are independent if all cell probabilities are equal to the product of the associated marginal probabilities, that is, if $p_{ij} = \mathbb{P}(X = x_i, Y = y_j) = \mathbb{P}(X = x_i)\mathbb{P}(Y = y_j) = p_{i\cdot} \times p_{\cdot j}$ for each i, j. If there exists a cell (i,j) for which $p_{ij} \neq p_{i\cdot} \times p_{\cdot j}$, then X and Y are dependent. The marginal distributions for components X and Y are obtained by taking the sums of probabilities in the table, row-wise and column-wise,

respectively: $\dfrac{X \mid x_1 \ x_2 \ \ldots \ x_r}{p \mid p_{1\cdot} \ p_{2\cdot} \ \ldots \ p_{r\cdot}}$ and $\dfrac{Y \mid y_1 \ y_2 \ \ldots \ y_c}{p \mid p_{\cdot 1} \ p_{\cdot 2} \ \ldots \ p_{\cdot c}}$.

Example 12.1. **Dependence of Components in 2D Random Variable.** If (X,Y) is defined by

X \ Y	10	20	30	40
-1	0.1	0.2	0	0.05
0	0.2	0	0.05	0.1
1	0.1	0.1	0.05	0.05

then the marginals are $\dfrac{X \mid -1 \quad 0 \quad 1}{p \mid 0.35 \ 0.35 \ 0.3}$ and $\dfrac{Y \mid 10 \ 20 \ 30 \ 40}{p \mid 0.4 \ 0.3 \ 0.1 \ 0.2}$, and X and Y are dependent since we found a cell, for example, $(2,1)$, such that $0.2 = \mathbb{P}(X = 0, Y = 10) \neq \mathbb{P}(X = 0) \cdot \mathbb{P}(Y = 10) = 0.35 \cdot 0.4 = 0.14$. As we indicated, it is sufficient for one cell to violate the condition $p_{ij} = p_{i\cdot} \times p_{\cdot j}$ in order for X and Y to be dependent.

Instead of random variables and cell probabilities that provide intuition, we consider an empirical counterpart, a table of observed frequencies. The table is defined by the levels of two factors: R and C. The levels are not necessarily numerical but could be, and most often are, categorical, ordinal, or interval. For example, when assessing the possible dependence between gender (factor R) and personal income (factor C), the levels for R are categorical {male, female}, and for C are interval, say, $\{[0, 30K), [30K, 60K), [60K, 100K), \geq 100K\}$. In the table below, factor R has r levels coded as $1, \ldots, r$ and factor C has c levels coded as $1, \ldots, c$. A cell (i,j) is an intersection of the ith row and the jth column and contains n_{ij} observations. The sum of the ith row is denoted by $n_{i\cdot}$ while the sum of the jth column is denoted by $n_{\cdot j}$.

	1	2	$\cdots$	c	Total
1	n_{11}	n_{12}		n_{1c}	$n_{1\cdot}$
2	n_{21}	n_{22}		n_{2c}	$n_{2\cdot}$
r	n_{r1}	n_{r2}		n_{rc}	$n_{r\cdot}$
Total	$n_{\cdot 1}$	$n_{\cdot 2}$		$n_{\cdot c}$	$n_{\cdot\cdot}$

Denote the total number of observations $n_{\cdot\cdot} = \sum_{i=1}^{r} n_{i\cdot} = \sum_{j=1}^{c} n_{\cdot j}$ simply by n. The empirical probability of the cell (i,j) is $\frac{n_{ij}}{n}$, and the empirical marginal probabilities of levels i and j are $\frac{n_{i\cdot}}{n}$ and $\frac{n_{\cdot j}}{n}$, respectively.

When factors R and C are independent, the frequency in the cell (i,j) is *expected* to be $n \cdot p_{i\cdot} \cdot p_{\cdot j}$. This can be estimated by empirical frequencies

$$e_{ij} = n \times \frac{n_{i\cdot}}{n} \times \frac{n_{\cdot j}}{n},$$

that is, as the product of the total number of observations n and the corresponding empirical marginal probabilities. After simplification, the empirical frequency in the cell (i,j) for independent factors A and B is

$$e_{ij} = \frac{n_{i\cdot} \times n_{\cdot j}}{n}.$$

By construction, the table containing "independence" frequencies e_{ij} would have the same row and column totals as the table containing observed frequencies n_{ij}, that is,

	1	2	$\cdots$	c	Total
1	e_{11}	e_{12}		e_{1c}	$n_{1\cdot}$
2	e_{21}	e_{22}		e_{2c}	$n_{2\cdot}$
r	e_{r1}	e_{r2}		e_{rc}	$n_{r\cdot}$
Total	$n_{\cdot 1}$	$n_{\cdot 2}$		$n_{\cdot c}$	$n_{\cdot\cdot}$

To measure the deviation from independence of the factors, we compare n_{ij} and e_{ij} over all cells. There are several measures for discrepancy between observed and expected frequencies, the two most important being Pearson's χ^2,

$$\chi^2 = \sum_{i=1}^{r} \sum_{j=1}^{c} \frac{(n_{ij} - e_{ij})^2}{e_{ij}}, \tag{12.1}$$

and the *likelihood ratio* statistic G^2,

$$G^2 = 2 \sum_{i=1}^{r} \sum_{j=1}^{c} n_{ij} \log\left(\frac{n_{ij}}{e_{ij}}\right).$$

Both statistics χ^2 and G^2 are approximately distributed as chi-square with $(r-1) \times (c-1)$ degrees of freedom, and their large values are critical for H_0. The approximation is good when cell frequencies are not small. Informal requirements are that no empty cells should be present and that observed counts should not be less than 5 for at least 80% of the cells. We focus on χ^2 since the inference using statistic G^2 is similar.

Thus, for testing H_0: Factors R and C are independent, versus H_1: Factors R and C are dependent, the test statistic is χ^2 given in (12.1) and the test is summarized as

Null	Alternative	α-level rejection region	p-value (MATLAB)
H_0: R,C ind.	H_1: R,C dep.	$[\chi^2_{df,1-\alpha}, \infty)$	`1-chi2cdf(chi2, df)`

where $df = (r-1) \cdot (c-1)$.

When some cells have expected frequencies < 5, the Yates correction for continuity is sometimes used, in which case the statistic χ^2 gets the form

$$\chi^2 = \sum_{i=1}^{r} \sum_{j=1}^{c} \frac{(|n_{ij} - e_{ij}| - 0.5)^2}{e_{ij}}.$$

Although recommended by some practitioners, this correction is controversial. It was originally proposed by Yates (1934) to bring closer tail probabilities of this statistics to the hypergeometric tails for contingency tables with fixed marginals (discussed in Section 12.4).

We denoted by p_{ij} the population counterpart to $\hat{p}_{ij} = n_{ij}/n$ and by $p_{i\cdot}$ and $p_{\cdot j}$ the population counterparts of $\hat{p}_{i\cdot} = n_{i\cdot}/n$ and $\hat{p}_{\cdot j} = n_{\cdot j}/n$. In terms of p's, the independence hypothesis takes the form

$H_0 : p_{ij} = p_{i\cdot} \times p_{\cdot j}$ for all i,j versus $H_1 : p_{ij} \neq p_{i\cdot} \times p_{\cdot j}$ for at least one i,j.

The following MATLAB program, 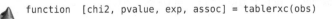 `tablerxc.m`, calculates expected frequencies, the value of the χ^2 statistic, and the associated p-value.

```
function [chi2, pvalue, exp, assoc] = tablerxc(obs)
```

```
% Contingency Table r x c for testing the probabilities
%
% Input:
%    obs - r x c matrix of observations.
%
% Output:
%    chi2 - statistic (approx distributed as chi-square
%                        with (r-1)(c-1) degrees of freedom)
%    pvalue  - p - value
%    exp - matrix of expected frequencies
%    asoc - structure containing association measures: phi,
%              C, and Cramer's V
% Example of use:
% [chi2,pvalue,exp,assoc]=tablerxc([6 14 17 9; 30 32 17 3])
%----------------------------------------------------------
  [r c]=size(obs);                %size of matrix
  n = sum(sum(obs));              %total s. size
  columns = sum(obs);             %col sums 1 x c
  rows = sum(obs')';              %row sums r x 1
  exp = rows * columns ./ n;      %[r x c] matrix
  chi2 = sum(sum((exp - obs).^2./exp ));
  df=(r-1)*(c-1);
  pvalue = 1- chi2cdf(chi2, df);
  % measures of association
  if (df == 1)
    assoc.phi = sqrt(chi2/n);
  end
  assoc.C = sqrt(chi2/(n + chi2));
  assoc.V = sqrt(chi2/(n * (min(r,c)-1)));
```

Example 12.2. **PAS and Streptomycin Cures for Pulmonary Tuberculosis.** Data released by the British Medical Research Council in 1950 (analyzed also in Armitage and Berry, 1994) concern the efficiency of para-amino-salicylic acid (PAS), streptomycin, and their combination in the treatment of pulmonary tuberculosis. Outcomes of sputum culture test performed on patients after the treatment are categorized as "Positive smear," "Negative smear and positive culture," and "Negative smear and negative culture."

The table below summarizes the findings on 273 treated patients with a TB diagnosis.

	Smear (+)	Smear (−) culture (+)	Smear (−) culture (−)	Total
PAS	56	30	13	99
Streptomycin	46	18	20	84
PAS and streptomycin	37	18	35	90
Total	139	66	68	273

There are two factors here: cure and sputum results. We will test the hypothesis of their independence at the level $\alpha = 0.05$.

```
D =[ ...
    56    30    13 ;  %99
    46    18    20 ;  %84
    37    18    35 ]; %90
%  139    66    68    273
  [chi2,pvalue,exp] = tablerxc(D)

%  chi2 = 17.6284
%  pvalue = 0.0015
%  exp =
%    50.4066   23.9341   24.6593
%    42.7692   20.3077   20.9231
%    45.8242   21.7582   22.4176
```

The p-value is 0.0015, which is significant. For $\alpha = 0.05$, the rejection region of the test comprises values larger than `chi2inv(1-0.05, (3-1)*(3-1))`, which is the interval $[9.4877, \infty)$. Since $17.6284 > 9.4877$, the hypothesis of independence is rejected.

Remark. The χ^2 statistic in (12.1) tests the hypothesis of independence if the sampling needed to construct the table is unrestricted. More precisely, only the total sample size n is preset in advance without any restriction on the marginal counts. If either row sums or column sums are preset in advance, the statistic in (12.1) tests for the homogeneity of one factor with respect to the other. Examples of such designs are given in Exercises 12.16 and 12.19.

12.2.1 Measuring Association in Contingency Tables

In a contingency table, statistic χ^2 tests the hypothesis of independence and in some sense quantifies the association between the two factors. However, as a measure, χ^2 is not normalized and depends on the sample size, while its distribution depends on the number of rows and columns. There are several measures that are calibrated to the interval $[0,1]$ and measure the strength of association in a way similar to R^2 in a regression context. We discuss three measures of association: the ϕ-coefficient, the contingency coefficient C, and Cramér's V coefficient.

ϕ-Coefficient. If a 2×2 contingency table classifying n elements produces statistic χ^2, then the ϕ-coefficient is defined as

$$\phi = \sqrt{\frac{\chi^2}{n}}.$$

Contingency Coefficient C**.** If an $r \times c$ contingency table classifying n elements produces statistic χ^2, then the contingency coefficient C is defined as

$$C = \sqrt{\frac{\chi^2}{\chi^2 + n}}.$$

Cramér's V **Coefficient.** If an $r \times c$ contingency table classifying n elements produces statistic χ^2, then Cramér's V coefficient is defined as

$$V = \sqrt{\frac{\chi^2}{n(k-1)}},$$

where k is the smaller of r and c, $k = \min\{r, c\}$. If the number of levels for any factor is 2, then Cramér's V becomes the ϕ-coefficient.

Example 12.3. **Thromboembolism and Contraceptive Use.** A data set, considered in more detail by Worchester (1971), contains a cross-classification of 174 subjects with respect to the presence of thromboembolism and contraceptive use as a risk factor. For a refinement of this data set by an additional factor (Smoking status), see Exercise 12.23.

	Contraceptive use	No contraceptive use	Total
Thromboembolism	26	32	58
Control	10	106	116
Total	36	138	174

Statistic χ^2 is 30.8913 with a p-value of $2.7289 \cdot 10^{-8}$, and the hypothesis of independence of thromboembolism and contraceptive usage is strongly rejected. For this table,

$$\phi = \sqrt{\frac{\chi^2}{n}} = \sqrt{\frac{30.8913}{174}} = 0.4214,$$

$$C = \sqrt{\frac{\chi^2}{\chi^2 + n}} = \sqrt{\frac{30.8913}{30.8913 + 174}} = 0.3883.$$

For 2×2 tables, ϕ and Cramér's V coincide. The fourth component in the output of tablerxc.m is a data structure with measures of association.

```
[chi2, pvalue, exp, stat]=tablerxc([26,32; 10,106])

%chi2 =    30.8913
%pvalue =  2.7289e-008
%
%exp =
%     12    46
%     24    92
%
%stat =
%    phi: 0.4214
%      C: 0.3883
%      V: 0.4214
```

12.2.2 *Power Analysis for Contingency Tables*

Power analysis for contingency tables involves evaluation of a noncentral χ^2 in which the noncentrality parameter λ depends on an appropriate effect size.

The traditional effect size in this context is Cohen's w, which is in fact equivalent to the ϕ-coefficient, $\phi = \sqrt{\chi^2/n}$, where n is the total table size. Unlike the ϕ-coefficient that is used for 2×2 tables only, w is used for arbitrary $r \times c$ tables and can exceed 1. In this case, $w^2 = (s-1)V^2$, where $s = \min\{r,c\}$ and V is Cramér's V. Effects $w = 0.1, 0.3$, and 0.5 correspond to a small, medium, and large size, respectively.

For prospective analyses, the noncentrality parameter λ is nw^2, and for retrospective analyses, λ is the observed χ^2. The power is

$$1 - \beta = 1 - ncx^2(\chi^2_{k,1-\alpha}, k, \lambda),$$

where $k = (r-1) \times (c-1)$ is the number of degrees of freedom and $\chi^2_{k,1-\alpha}$ is the $(1-\alpha)$-quantile of a χ^2_k distribution.

Example 12.4. **Power and Sample Size for a 2×6 Table.**
   (a) Find the power in a 2×6 contingency table, for $n = 180$ and $w = 0.3$ (medium effect).

```
w = 0.3; n = 180; k = (2-1)*(6-1); alpha = 0.05; lambda=n*w^2
pow = 1-ncx2cdf( chi2inv(1-alpha,k), k, lambda)    %0.8945
```

   (b) What sample size ensures the power of 95% in a contingency table 2×6, for an effect of $w = 0.3$ and $\alpha = 0.05$?

```
beta = 0.05; alpha = 0.05; k = (2-1)*(6-1); w=0.3;
pf = @(n)  ncx2cdf( chi2inv(1-alpha,k), k, n*w^2) - beta;
ssize = fzero(pf, 200)        %219.7793 approx 220
```

What is the meaning of effect size w in the context of contingency tables? Let $p_{ij}^{(0)}$ and $p_{ij}^{(1)}$ be the (i,j)th cell probabilities under H_0 and H_1 respectively. Then,

$$w^2 = \frac{\chi^2}{n} = \sum_{i,j} \frac{\left(p_{ij}^{(1)} - p_{ij}^{(0)}\right)^2}{p_{ij}^{(0)}}.$$

12.2.3 Cohen's Kappa

Cohen's kappa (Cohen, 1960) is a widely used descriptor of agreement between two testing procedures. This descriptor is motivated by calibrating the observed agreement by an agreement due to chance. If p_c is the proportion of agreement due to chance and p_o the proportion of observed agreement, then

$$\hat{\kappa} = \frac{p_o - p_c}{1 - p_c}. \tag{12.2}$$

For a paired table representing the results of n tests by two devices or ratings by two raters

	+	−	
+	a	b	$a+b$
−	c	d	$c+d$
	$a+c$	$b+d$	$n = a+b+c+d$

Cohen's kappa index is defined as

$$\hat{\kappa} = \frac{2(ad - bc)}{(a+b)(b+d) + (a+c)(c+d)}. \tag{12.3}$$

The expression in Eq. (12.2) is equivalent to that in Eq. (12.3) for $p_o = a/n + d/n$ and $p_c = (a+b)(a+c)/n^2 + (b+d)(c+d)/n^2$, respectively. The former is the observed agreement equal to the proportion of $(+,+)$ and $(-,-)$ outcomes, while the latter is the proportion of agreement when the results are independent within fixed marginal proportions, that is, an agreement due to chance.

There are no formal rules for judging $\hat{\kappa}$, but here is the standard:

$\hat\kappa$	Degree of agreement
<0.20	Poor
0.20–0.40	Fair
0.40–0.60	Moderate
0.60–0.80	Good
≥ 0.8	Very good

The MLE of κ is

$$\hat\kappa_{mle} = \frac{4(ad - bc) - (b - c)^2}{(2a + b + c)(2d + b + c)},$$

and it is obtained from (12.2) by taking $p_o = a/n + d/n$ and $p_c = P^2 + (P')^2$, where $P = (2a + b + c)/(2n)$ and $P' = (2d + b + c)/(2n)$ are, respectively, the MLEs of the prevalence of + and − in the population.

There are several approximations for the variance of $\hat\kappa$. Two standard estimators are the Block–Kraemer (BK) and Garner (G) approximations:

(1) (BK):

$$\mathbb{V}\text{ar}\,\hat\kappa \approx \frac{1 - \hat\kappa}{n}\left((1 - \hat\kappa)(1 - 2\hat\kappa) + \frac{\hat\kappa(2 - \hat\kappa)}{2P(1 - P')}\right);$$

(2) (G)

$$\mathbb{V}\text{ar}\,\hat\kappa \approx \frac{4}{(1 - p_c)^2 n^2 \left(\frac{1}{a+1} + \frac{1}{b+1} + \frac{1}{c+1} + \frac{1}{d+1}\right)}.$$

The sampling distribution of $\hat\kappa$ is asymptotically normal, and the approximation is satisfactory if n is not too small (e.g., $n > 30$) and κ is not too close to 1. This leads, in a standard manner, to approximate confidence intervals for κ, as

$$\hat\kappa \pm z_{1-\alpha/2}\sqrt{\mathbb{V}\text{ar}\,\hat\kappa}.$$

Example 12.5. **Two Sensors.** A company producing medical sensor A is applying for FDA approval of a new version B. Both sensors A and B are prone to errors, and a gold standard is absent. The FDA is requesting that the new sensor be comparable to the currently used one, and the company decides to include Cohen's $\hat\kappa$ statistic in the report.

The experiment consisted of 2,803 trials and resulted in a paired table [97 11; 6 2689], where 97 was the number of (+, +) outcomes and 2,689 the number of (−, −) outcomes. The code ◄ cohen.m finds Cohen's $\hat\kappa$ and 95%

confidence intervals for the population κ, based on the two estimators of
variance, Block-Kraemer and Garner.

```
data =[97 11;   6 2689]
%data =
% 97            11
% 6           2689
a=data(1,1); b=data(1,2); c=data(2,1); d=data(2,2);
apb = a+b; cpd = c+d; apc = a+c; bpd = b+d;
n = a + b + c + d;
%------------
p0 = (a + d)/n;                    %Observed agreement
pc = (apb*apc + cpd*bpd)/n^2; %Chance agreement
Pre  = (2*a + b + c)/(2 * n);  %Prevalence of +
pcc = Pre^2 + (1-Pre)^2;       %MLE of chance agreement
%------------
kappa = 2 * (a * d - b * c)/(apb*bpd + apc*cpd) %0.9163
%or kappa=(p0-pc)/(1-pc)

kappamle = (4 * (a*d - b*c) - (b -c)^2)/...
  ((2 * a + b + c) * (2 * d + b + c))  %0.9163
%or kappamle=(p0-pcc)/(1-pcc)
%------------
%Block-Kraemer variance estimator
varbk = (1- kappa)/n * ( (1- kappa)*(1-2* kappa) + ...
  (kappa * (2 -  kappa)/(2 * Pre * (1-Pre)))) %4.0731e-004
%Garner variance estimator
vargarner = 4/( (1- pc)^2 * n^2 * (1/(a+1) + 1/(b+1) + ...
    1/(c+1) + 1/(d + 1) ) )  %4.0971e-004
%------------
%Confidence intervals
[ kappa - 1.96 * sqrt(varbk) ...
        kappa + 1.96 * sqrt(varbk)]  %0.8767    0.9558
[ kappa - 1.96 * sqrt(vargarner) ...
    kappa + 1.96 * sqrt(vargarner)]  %0.8766    0.9560
```

Cohen's κ is estimated to be 91.63%, which represents very good agreement.

12.3 Three-Way Tables

A natural extension of two-way tables for testing the independence of two
factors are n-dimensional tables for testing the independence of n factors.
We will discuss a three-dimensional extension; the interested reader can
consult Zar (2010), Agresti (2002), or Fienberg (2000) for more detailed coverage. In three-dimensional tables the counts constitute three-dimensional
arrays characterized by rows, columns, and pages. We associate factors with
these three dimensions and consider row, column, and page factors (R, C,

and P). In the cell (i,j,k) there are n_{ijk} observations, $i = 1,\ldots,r$, $j = 1,\ldots,c$, and $k = 1,\ldots,p$.

Denote the total number of observations by n. The empirical probability of the cell (i,j,k) is n_{ijk}/n, and the empirical marginal probabilities of row i, column j, and page k are $n_{i..}/n$, $n_{.j.}/n$, and $n_{..k}/n$. The numerators are calculated as the sums over all indices replaced by dots. For example, $n_{.j.} = \sum_{i=1}^{r} \sum_{k=1}^{p} n_{ijk}$, $j = 1,\ldots,c$.

Complete Independence. We are interested in testing the hypothesis H_0 that the factors R, C, and P are independent. The alternative H_1 would be that the factors are not independent. Under H_0, the frequency in the cell (i,j,k) is *expected* to be

$$e_{ijk} = n \times \frac{n_{i..}}{n} \times \frac{n_{.j.}}{n} \times \frac{n_{..k}}{n},$$

as the product of the total number of observations n and the corresponding empirical marginal probabilities. After simplification, the expected frequency in the cell (i,j,k) becomes

$$e_{ijk} = \frac{n_{i..} \times n_{.j.} \times n_{..k}}{n^2}.$$

The number of free parameters is $(r-1) + (c-1) + (p-1)$. The test statistic is

$$\chi^2 = \sum_{i=1}^{r} \sum_{j=1}^{c} \sum_{k=1}^{p} \frac{(n_{ijk} - e_{ijk})^2}{e_{ijk}}, \tag{12.4}$$

and the *likelihood ratio* statistic is

$$G^2 = 2 \sum_{i=1}^{r} \sum_{j=1}^{c} \sum_{k=1}^{p} n_{ijk} \log\left(\frac{n_{ijk}}{e_{ijk}}\right).$$

When H_0 is true, both statistics χ^2 and G^2 follow approximately a χ^2-distribution with $rcp - 1 - (r-1) - (c-1) - (p-1) = rcp - r - c - p + 2$ degrees of freedom. Large values of χ^2 are critical for H_0.

Joint Independence. If H_0 is rejected, multiple alternatives are possible. For example, all three factors R, C, and P are mutually dependent, factors R and C are dependent but both are independent of factor P, and so on. To answer why H_0 was rejected, one needs to test the hypothesis whether each single factor is independent of the other two. This kind of independence is called *joint independence*. There are three such tests, and they are summarized in the following table:

Factor	e_{ijk}	df
R vs. (C,P)	$\frac{n_{i..} \times n_{.jk}}{n}$	$(r-1)(cp-1)$
C vs. (R,P)	$\frac{n_{.j.} \times n_{i.k}}{n}$	$(c-1)(rp-1)$
P vs. (R,C)	$\frac{n_{..k} \times n_{ij.}}{n}$	$(p-1)(rc-1)$

Here $n_{ij.} = \sum_k n_{ijk}$, $n_{.jk} = \sum_i n_{ijk}$, and $n_{i.k} = \sum_j n_{ijk}$. Also, as before, $n_{i..} = \sum_j \sum_k n_{ijk}$, $n_{.j.} = \sum_i \sum_k n_{ijk}$, and $n_{..k} = \sum_i \sum_j n_{ijk}$. The χ^2 statistic is calculated as in (12.4), and the degrees of freedom are given in the table. In particular, suppose that we want to test independence of P versus (R,C). The χ^2 from (12.4) is calculated using

$$e_{ijk} = n \times \frac{n_{ij.}}{n} \times \frac{n_{..k}}{n}.$$

The number of free parameters is $(rc-1) + (p-1)$, which leads to $rcp - 1 - (rc-1)(p-1) = (p-1)(rc-1)$ degrees of freedom for the χ^2 statistic in (12.4). Statistics and degrees of freedom for R versus C,P and C versus R,P are calculated similarly.

Conditional Independence. A very important kind of independence in three-way tables is conditional independence. Suppose that factors R and C are related and that R and P are also related. We are interested in checking the independence of factors C and P. Unconditionally the factors C and P will be related via their joint relation to factor R, but when accounted for R, the C and P may be independent. Such independence is called *conditional independence*.

There are three such tests, and they are summarized in the following table:

Factor	e_{ijk}	df
R vs. C given P	$\frac{n_{i.k} \times n_{.jk}}{n_{..k}}$	$p(r-1)(c-1)$
R vs. P given C	$\frac{n_{ij.} \times n_{.jk}}{n_{.j.}}$	$c(r-1)(p-1)$
C vs. P given R	$\frac{n_{ij.} \times n_{i.k}}{n_{i..}}$	$r(c-1)(p-1)$

We will explain how to test conditional independence of C and P given R, for other combinations the procedure is similar.

The the χ^2 statistic in (12.4) is calculated using

$$e_{ijk} = \frac{n \times (n_{ij.}/n) \times (n_{i.k}/n)}{n_{i..}/n} = \frac{n_{ij.} \times n_{i.k}}{n_{i..}}.$$

The number of free parameters is $(r-1) + r(c-1) + r(p-1)$ which leads to $rcp - 1 - (r-1) - r(c-1) - r(p-1) = r(c-1)(p-1)$ degrees of freedom.

Example 12.6. **Anolis Lizards of Bimini.** This well-known data set comes from the paper of Schoener (1968) and is also used in Fienberg (1970). The researcher was interested in structural habitat categories for Anolis lizards of Bimini: *sagrei* (brown anole) adult males versus *distichus* (trunk anole) adult and subadult males. The brown anole and trunk anole are medium-sized, fairly robust, "trunk-ground" lizards. They generally prefer the fairly open vegetation of disturbed sites, where they adopt a head-down, sit-and-wait posture and perch low on large trunks or fenceposts.

The researcher was interested in the preferences of these two species with respect to the perch height and diameter.

		A. sagrei Perch diameter		A. distichus Perch diameter	
		≤ 4	> 4	≤ 4	> 4
Perch height	> 4.75	32	11	61	41
(in feet)	≤ 4.75	86	35	73	70

The data is classified with respect to three dichotomous factors: Height at levels Low (≤ 4.75) and High (> 4.75), Diameter at levels Small (≤ 4) and Large (> 4), and Species with levels *A. sagrei* and *A. distichus*. Are these three factors independent? The null hypothesis is that the factors are independent and the alternative is that they are not. The MATLAB program ◢ tablerxcxp.m calculates the expected frequencies under mutual independence hypothesis and provides the χ^2 statistic, its degrees of freedom, and the p-value. This program also outputs the expected frequencies in the format that matches the input data. Here, Height is the row factor R, Diameter is the column factor C, and Species is the page factor P.

```
anolis = [32 11; 86 35];
anolis(:,:,2)=[61 41; 73 70];
[ch2 df pv exp]=tablerxcxp(anolis)
%ch2 = 23.9055
%df = 4
%pv =8.3434e-005
%exp(:,:,1) =
%   35.8233 22.3185
%   65.2231 40.6351
%exp(:,:,2) =
```

```
%  53.5165 33.3417
%  97.4370 60.7048
```

The hypothesis H_0 is rejected ($p = 8.3434 \cdot 10^{-5}$), and we infer that the three factors are not independent. However, this analysis does not fully explain the dependencies responsible for rejecting H_0. Are all three factors dependent, or maybe two of the factors are mutually dependent and the third is independent of both? When H_0 is rejected, we need a partial independence test, similar to pairwise comparisons in the case where the ANOVA hypothesis is rejected. The MATLAB program for this test is

 `partialrxcxp.m`.

```
[ch2 df pv exp]=partialrxcxp(anolis,'r_cp')
  %ch2 = 12.3028; df=3; pv = 0.0064, exp=...
[ch2 df pv exp]=partialrxcxp(anolis,'c_pr')
  %ch2 = 14.4498; df=3; pv = 0.0024, exp=...
[ch2 df pv exp]=partialrxcxp(anolis,'p_rc')
  %ch2 = 23.9792; df=3; pv =2.5231e-005, exp=...
```

Three tests are performed: (i) the row factor independent of column/page factors (`partial = 'r_cp'`), (ii) the column factor independent of row/page factors (`partial = 'c_rp'`), and (iii) the page factor independent of row/column factors (`partial = 'p_rc'`). It is evident from the output that all three tests produced significant χ^2, that is, each factor depends on the two others. Since Height was the row factor R, Diameter the column factor C, and Species the page factor P, we conclude that the strongest dependence is that of Species factor on the height and diameter of the perch, `partialrxcxp(anolis,'p_rc')`, with a p-value of 0.000025.

12.4 Contingency Tables with Fixed Marginals: Fisher's Exact Test

In his monograph, Fisher (1935) provides an example of a small 2×2 contingency table related to a tea-tasting experiment. It is about a woman who claimed to be able to judge whether tea or milk was poured in a cup first. The woman was given eight cups of tea, in four of which tea was poured first, and was told to guess which four had tea poured first. The contingency table for this design is

	Guess milk first	Guess tea first	Total
Milk first	x	$4 - x$	4
Tea first	$4 - x$	x	4
Column total	4	4	8

The number of correct guesses "Milk first" in the cell $(1,1)$, x, can take values 0, 1, 2, 3, or 4 with the probabilities $\frac{\binom{4}{x}\binom{4}{4-x}}{\binom{8}{4}}$, as in

```
hygepdf(0:4, 8,4,4)
% ans =   0.0143   0.2286   0.5143   0.2286   0.0143
```

These are hypergeometric probabilities, applicable here since the marginal counts are fixed.

For $x = 4$ the probability of obtaining this table by chance is 0.0143, while for $x = 3$ the probability of getting this or a more extreme table by chance is $0.2286 + 0.0143 = 0.2429$, and so on. Thus, the probability of the woman's guessing correctly, that is, if $x = 4$, would be less than 5%.

Suppose that in the table

	Column 1	Column 2	Row total
Row 1	a	b	$a+b$
Row 2	c	d	$c+d$
Column total	$a+c$	$b+d$	$n = a+b+c+d$

marginal counts $a+b$, $c+d$, $a+c$, and $b+d$ are fixed. Then a, b, c, and d are constrained by these marginals. We are interested if the probabilities that an observation will be in column 1 are the same for rows 1 and 2. Denote these probabilities as p_1 and p_2. The null hypothesis here is not the hypothesis of independence but the hypothesis of homogeneity, $H_0 : p_1 = p_2$. The test is close to the two-sample problem considered in Chapter 10; however, in this case the samples are dependent because of marginal constraints.

The statistic T to test H_0 is simply the number of observations in the cell $(1,1)$:

$$T = a.$$

If H_0 is true, then T has a hypergeometric distribution $\mathcal{HG}(n, a+b, a+c)$, that is,

$$\mathbb{P}(T = x) = \frac{\binom{a+b}{x} \cdot \binom{c+d}{a+c-x}}{\binom{n}{a+c}} = \frac{\binom{a+c}{x} \cdot \binom{b+d}{a+b-x}}{\binom{n}{a+b}}, \quad (12.5)$$

for $x = 0, 1, \ldots, \min\{a+b, a+c\}$.

This is an easy consequence of the fact that if $X \sim \mathcal{B}in(a+b, p_1)$ and $Y \sim \mathcal{B}in(c+d, p_2)$ are independent, then X given $Y = a+c$ has hypergeometric $\mathcal{HG}(n, a+b, a+c)$ distribution only if $p_1 = p_2$. Since the probabilities in $\mathcal{HG}(n,a+b,a+c)$ and $\mathcal{HG}(n,a+c,a+b)$ are identical, the probability $\mathbb{P}(T=x)$ in (12.5) has two equivalent forms.

The p-value against the one-sided alternative $H_1 : p_1 < p_2$ is hygecdf(a, n, a+b, a+c) and against the alternative $H_1 : p_1 > p_2$ is 1 - hygecdf(a-1, n, a+b, a+c). If the hypothesis is two-sided, then the p-value cannot be obtained by doubling one-sided p-value due to asymmetry of the hypergeometric distribution. The two-sided p-value is obtained as the sum of all probabilities hygepdf(x, n, a+b, a+c), $x = 0, 1, \ldots, \min\{a+b, a+c\}$ that are smaller than or equal to hygepdf(a, n, a+b, a+c).

Example 12.7. **Gender Balance.** There are 22 subjects enrolled in a clinical trial and 9 are females. Researchers plan to administer 11 portions of a drug and 11 placebos. Only 2 females are administered the drug. Are the proportions of males and females assigned to the drug significantly different? What are the p-values for one- and two-sided alternatives?

```
fisherexact.m
a = 2; b = 7; c = 9; d =4;
n = a + b + c + d;
T = a;
pval = hygecdf(T,n,a+c,a+b)  %H1: p1<p2    pval=0.0402
% or equivalently pval=hygecdf(T,n,a+b,a+c)
%
pa = hygepdf(T,n,a+c,a+b);
for  i = 1:min(a+b, a+c)+1
    p(i) = hygepdf(i-1,n,a+c,a+b) ;
end
pval2 = sum(p(p <= pa))   %H1: p1 ~= p2  pval2=0.0805
```

Since the one-sided p-value is less than 5%, we reject the hypothesis of homogeneity of adminstration of a drug versus placebo with respect to gender. For the two-sided alternative, we fail to reject H_0.

Example 12.8. **The Effect of Passive Smoking on Lung Cancer.** Lawal (2003) considers the following data originally published by Correa et al. (1983) on the effect of passive smoking on lung cancer. A total of 155 non-smoking ever-married females were tabulated by their lung cancer status and husband's smoking status.

Is the proportion of lung cancer cases homogeneous with respect to the husband's smoking status? Find the p-value for both one- and two-sided alternatives.

	Cancer status		
	Case	Control	Total
Spouse smoked	14	61	75
Spouse did not smoke	8	72	80
Total	22	133	155

Here $H_0 : p_1 = p_2$ and $H_1 : p_1 > p_2$ or $H_1 : p_1 \neq p_2$.

```
a = 14; b = 61; c = 8; d =72;
n = a + b + c + d;
T = a;
%H1: p1 > p2
pval = 1-hygecdf(T-1,n,a+c,a+b) %0.0941
%H1: p1 ~= p2
pa = hygepdf(T,n,a+c,a+b);
for  i = 1:min(a+b, a+c)+1
    p(i) = hygepdf(i-1,n,a+c,a+b) ;
end
pval2 = sum(p(p <= pa))          %0.1669
```

Thus, Fisher's exact test fails to reject the null hypothesis at 5% significance level.

Fisher's exact test remains valid for designs with random row totals, random column totals, or tables with random marginals as in the previous section. In this case the tests are conservative, and more powerful versions exist. A benefit of using Fisher's exact test is that it operates with small cell frequencies; for example, a 0 count in a table cell is a possibility. Since χ^2 or normal approximations assume large n and np_{ij}s preferably larger than 5, the reason for the popularity of Fisher's exact test is obvious.

Extensions of Fisher's exact test to tables of larger size are possible, but the methodology is considerably more complex. Freeman and Halton (1951) extended the test to 2×3 and 3×3 tables. Implementations in MATLAB can be found on MATLAB Central (◀ MyFisher33.m by Giuseppe Cardillo, File #15482).

12.5 Stratified Tables: Mantel–Haenszel Test

In Section 10.7, 2×2 tables were discussed in the context of comparing two proportions. In previous sections of this chapter we discussed independence and marginal homogeneity for 2×2 tables.

Here we discuss multiple 2×2 tables and inference from combined information. The Mantel–Haenszel methodology can be used in 2×2 tables to control for a variable that stratifies the data. This stratification leads to multiple tables, one for each level of controlled variable. The Mantel–Haenszel methodology can be used for (i) testing the conditional independence of two factors, (ii) measuring the degree of conditional association (risk ratios), or (iii) conducting meta-analysis of the conditional odds ratios. All conditioning is on the variable by which the tables are stratified. There are several other uses of the Mantel–Haenszel methodology such as in survival analysis (logrank test of Mantel, p. 818) and in depairing of McNemar's paired designs (p. 614).

12.5.1 *Testing Conditional Independence or Homogeneity*

Suppose that k independent classifications into a 2×2 table are observed. We could denote the ith such table by

a_i	b_i	$a_i + b_i$
c_i	d_i	$c_i + d_i$
$a_i + c_i$	$b_i + d_i$	n_i

The tables give counts broken down by binary levels of two factors, and the separate tables usually correspond to the levels of a third factor that needs to be controlled. Imagine that we want to test for the independence of two factors, say, political association (Democrat, Republican) and opinion about some social issue (Support, Oppose). The single contingency table may not be significant, but when controlled by gender (two tables, one for males, the other for females) or by age group, the dependence may turn out significant. Thus, multiple tables make inference more precise by controlling for an influential variable.

For each of k tables consider a cell at the position $(1,1)$, so-called *pivot* cell, with a_i counts. If the two tabulated factors are independent, then the counts a_i should be close to "expected" counts $e_i = (a_i + b_i)(a_i + c_i)/n_i$.

The test statistic measuring discrepancies in the pivot cell over all k tables is

$$\chi^2 = \frac{(|A - E| - 1/2)^2}{V}, \text{ where} \qquad (12.6)$$

$$A = \sum_{i=1}^{k} a_i, \ E = \sum_{i=1}^{k} e_i, \ V = \sum_{i=1}^{k} \frac{(a_i + b_i)(c_i + d_i)(a_i + c_i)(b_i + d_i)}{n_i^2(n_i - 1)},$$

which has an approximate χ^2-distribution with 1 degree of freedom when the hypothesis of independence/homogeneity is true. Large values of χ^2 are critical for H_0.

It is interesting that even for sparse individual tables the χ^2-approximation holds as long as the sum of row totals in all tables is larger than 20, say, $\sum_i(a_i + b_i), \sum_i(c_i + d_i) > 20$.

As in contingency tables, if the marginal sums are fixed in advance, the subsequent inference does not concern the independence; it concerns the homogeneity of one factor within the levels of the other factor. The following example tests for homogeneity of proportions of cancer incidence among smokers and nonsmokers stratified by populations in different cities.

Example 12.9. **Smoking and Cancer in Three Chinese Cities.** The three 2×2 tables provide classification of people from three Chinese cities, Zhengzhou, Taiyuan, and Nanchang, with respect to smoking habits and incidence of lung cancer (Liu, 1992).

	Zhengzhou			Taiyuan			Nanchang		
Cancer Diagnosis:	Yes	No	Total	Yes	No	Total	Yes	No	Total
Smoker	182	156	338	60	99	159	104	89	193
Nonsmoker	72	98	170	11	43	54	21	36	57
Total	254	254	508	71	142	213	125	125	250

We can apply the Mantel–Haenszel test to decide if the proportions of cancer incidence for smokers and nonsmokers coincide for the three cities: $H_0 : p_{1i} = p_{2i}$, where p_{1i} is the proportion of incidence of cancer among smokers in city i and p_{2i} is the proportion of incidence of cancer among nonsmokers in city i, $i = 1, 2, 3$. We use the two-sided alternative, $H_1 : p_{1i} \neq p_{2i}$, for some $i \in \{1, 2, 3\}$ and fix the type I error rate at $\alpha = 0.10$.

To compute χ^2 in (12.6), we find A, E, and V. From the tables, $A = \sum_i a_i = 182 + 60 + 104 = 346$. Also, $E = \sum_i e_i = 338 \cdot 254/508 + 159 \cdot 71/213 + 193 \cdot 125/250 = 169 + 53 + 96.5 = 318.5$.

$$
\begin{aligned}
V &= \sum_{i=1}^{k} \frac{(a_i + b_i)(c_i + d_i)(a_i + c_i)(b_i + d_i)}{n_i^2(n_i - 1)} \\
&= \frac{338 \cdot 254 \cdot 170 \cdot 254}{508^2 \cdot 507} + \frac{159 \cdot 71 \cdot 54 \cdot 142}{213^2 \cdot 212} + \frac{193 \cdot 125 \cdot 57 \cdot 125}{250^2 \cdot 249} \\
&= 28.33333 + 9 + 11.04518 = 48.37851.
\end{aligned}
$$

Therefore,

$$
\chi^2 = \frac{(|346 - 318.5| - 0.5)^2}{48.37851} = 15.0687.
$$

Because the statistic χ^2 is distributed approximately as χ_1^2, the p-value (via MATLAB m-file ◀ mantelhaenszel.m) is 0.0001.

```
  [chi2, pval] = mantelhaenszel([182 156; 72 98; ...
        60 99; 11 43; 104 89; 21 36])
%chi2 = 15.0687
%pval  = 1.0367e-004
```

In this case, there is clear evidence that the cancer rates are not homogeneous among the three cities.

Example 12.10. **Oral Contraceptive Use and Myocardial Infarction Risk.** Shapiroet al. (1979) investigated the effect of oral-contraceptive use on the risk of myocardial infarction. The possible link was investigated in 234 pre-menopausal women with a first infarction and 1,742 hospital controls. Results are summarized in five two-by-two tables stratified by age groups:

| Contraceptive | 25–29 y.o. | | 30–34 y.o. | | 35–39 y.o. | | 40–44 y.o. | | 45–49 y.o. | |
use	MI	Contr	MI	Contr	MI	Contr	MI	Contr	MI	Contr
Yes	4	62	9	33	4	26	6	9	6	5
No	2	224	12	390	33	330	65	362	93	301

If age stratification is ignored, then a cumulative table is

| Contraceptive | All ages | |
use	MI	Contr
Yes	29	135
No	205	1607

We are interested in testing for the independence of the contraceptive use and the incidence of MI.

If the significance level is set at 1%, the χ^2 analysis of the cumulative table fails to reject the hypothesis of independence.

```
[chi2, pval,~] = tablerxc([29 135; 205 1607])
% chi2 = 5.8443
% pval = 0.0156
```

By using the more appropriate Mantel–Haenszel test for the conditional independence (conditional on age group), we find that departure from the independence hypothesis is highly significant.

```
[chi2, pval] =  mantelhaenszel([4 62; 2 224;     9 33; 12 390; ...
    4 26; 33 330;    6 9; 65 362;     6 5;  93 301])
% chi2 = 32.7927
% pval = 1.0253e-08
```

This is an example of "perils of aggregation," where highly dependent factors, as indicated by Mantel–Haenszel test adequate here, may be not be found significantly dependent if the aggregate table was used.

Remark. (i) Breslow–Day statistic for testing overall homogeneity in multiple 2×2 tables is

$$\chi^2_{BD} = \sum_{i=1}^{k} \frac{n_i (a_i d_i - b_i c_i)^2}{(a_i + b_i)(c_i + d_i)(a_i + c_i)(b_i + d_i)},$$

which is distributed as χ^2_{k-1}. Unlike the Mantel–Haenszel statistic that remains valid for small counts in individual tables as long as cumulative counts are large, the Breslow–Day statistic requires large counts for each table.

(ii) It is possible to extend the Mantel–Haenszel theory for $r \times c$ tables. An overview and references can be found in Landis et al. (1998).

12.5.2 Odds Ratio from Stratified Tables

An overview of risk analysis for a 2×2 table was provided in Section 10.7, where we discussed risk difference, risk ratio, and odds ratio. Suppose that in a particular study the participating subjects are stratified to k groups according to some classifying feature/factor, and that each strata produces an independent 2×2 table. We are interested in combining the results from the individual tables for the risk analysis of all subjects. The goal is to eliminate confounding effect of the factor used for stratification.

In this section we discuss how to calculate the combined odds ratio. Let the rows in the table corresponding to ith group represent Types 1 and 2 and the columns Presence/Absence of a particular attribute:

	Presence	Absence	
Type 1	a_i	b_i	$a_i + b_i$
Type 2	c_i	d_i	$c_i + d_i$
	$a_i + c_i$	$b_i + d_i$	n_i

For example, Types could be case and control subjects and Presence/Absence could be related to a particular risk factor. The tables can be stratified according to some other risk factor or demographic feature that we would like to control, such as gender, age, smoking status, socioeconomic status, etc.

Then the proportion of subjects with the attribute present for Type 1 is $a_i/(a_i + b_i)$, and the proportion of subjects with the attribute present for Type 2 is $c_i/(c_i + d_i)$.

The observed odds ratio (for the types) in a single ith table is

$$\frac{a_i/b_i}{c_i/d_i} = \frac{a_i d_i}{b_i c_i}.$$

The combined odds ratio for all k strata is defined as

$$o_{mh} = \frac{\sum_i a_i d_i / n_i}{\sum_i b_i c_i / n_i}. \tag{12.7}$$

The expression o_{mh} was proposed in Mantel and Haenszel (1959) and represents the weighted average of individual odds ratios $\frac{a_i d_i}{b_i c_i}$ with weights proportional to $b_i c_i / n_i$ and summing up to 1.

An approximation to the sample variance of the log odds ratio o_{mh} is given by the Robins, Breslow, and Greenland (RBG) variance formula (Robins et al., 1986):

$$\widehat{\mathrm{Var}}\left(\log(o_{mh})\right) = \frac{\sum_i R_i P_i}{2R^2} + \frac{\sum_i P_i S_i + Q_i R_i}{2RS} + \frac{\sum_i S_i Q_i}{2S^2},$$

where $P_i = (a_i + d_i)/n_i$, $Q_i = (b_i + c_i)/n_i$, $R_i = a_i d_i / n_i$, $S_i = b_i c_i / n_i$, $R = \sum_i R_i$, and $S = \sum_i S_i$. If k is equal to 1, the RBG variance formula reduces to the familiar $(\frac{1}{a} + \frac{1}{b} + \frac{1}{c} + \frac{1}{d})$, (p. 471).

The $(1 - \alpha)100\%$ confidence interval for population odds ratio O is

$$\left[o_{mh} \exp\{-z_{1-\alpha/2}\, s\}, \ o_{mh} \exp\{z_{1-\alpha/2}\, s\}\right],$$

where $s = \sqrt{\widehat{\mathrm{Var}}\left(\log(o_{mh})\right)}$.

Another way of controlling confounding factors in risk calculations is by matching subjects according to all factors that we want to control. Some results on the inference from matched pair tables, or paired tables, are provided next.

12.6 Paired Tables: McNemar's Test

Another type of table commonly used in dental, opthalmology, and pharmacology trials is matched-pair table summarizing the designs in which interventions are applied to the same patient. For example, in randomized split-mouth trials comparing the effectiveness of tooth-specific interventions to prevent decay, one tooth in a subject is randomly selected to receive treatment A, while the contralateral tooth in the same subject receives treatment B. Another example is crossover trials testing the efficacy of drugs. In this design, a patient is randomly administered either treatment A or B in the first time period and then administered the other treatment in the second time period. The link between the split-mouth design and crossover trials is apparent – the tooth location in the split-mouth design is analogous to time in the crossover design.

The matched-pair design has statistical advantages. This design controls for many confounding factors because the control and test groups are subject to the same environment. Thus differences in outcomes between test and control groups are likely attributable to the treatment. Moreover, since control and test groups receive both interventions, matched-pair studies usually require no more than half the number of subjects to produce the same precision as parallel group studies.

McNemar's test is used for inference in paired tables. Although data for the McNemar test resemble contingency tables, the structure of the tables and the inference are quite different. For simplicity, assume that measurements at *Before* and *After* on the same subject result in *Positive* and *Negative* responses. From N subjects we obtain $2N$ responses, or N pairs of responses, organized as follows:

		After		Total
		Positive	Negative	
Before	Positive	A	B	$A + B$
	Negative	C	D	$C + D$
	Total	$A + C$	$B + D$	N

For example, A is the number of subjects (pairs of responses) where both *Before* and *After* resulted in a positive. More generally, *Before* and *After* could be any two different groups of matched subjects that produce binary responses.

The marginal sum $A + B$ is the number of positives in *Before*, and $A + C$ is the number of positives in *After*. The proportion of positives in *Before* is $\hat{p}_1 = \frac{A+B}{N}$, while the proportion of positives in *After* is $\hat{p}_2 = \frac{A+C}{N}$. Let the population counterparts of $\hat{p}_1$ and $\hat{p}_2$ be p_1 and p_2.

Any paired table can be "parallelized," but information about pairing is lost. The table above has the following parallelized form:

	Positive	Negative	Total
Before	$A + B$	$C + D$	N
After	$A + C$	$B + D$	N
Total	$2A + B + C$	$B + C + 2D$	$2N$

All inference regarding population p_1 and p_2 (risk differences, risk ratios, and odds ratios) can be conducted using the parallelized table; however, such inference does not take into account potentially substantial information contained in pairing. Omitting this information can influence decisions. Such errors of ignoring pairing, and treating paired data as parallel, are frequent in existing literature.

Next we discuss inference for risk differences, risk ratios, and odds ratios. These concepts were discussed in Chapter 10 for independent proportions or equivalently 2 × 2 parallel tables; here we take pairing into consideration.

12.7 Risk Differences, Risk Ratios, and Odds Ratios for Paired Tables

In the context of epidemiological studies when a single subject cannot be classified simultaneously as exposed and nonexposed, the measurements are taken on a matched pair of subjects, one exposed and one not exposed. All other characteristics should match or be as close as possible (gender, age, socioeconomic status, race, etc.)

		Non-exposed		Total
		Disease	No disease	
Exposed	Disease	A	B	$A + B$
	No disease	C	D	$C + D$
	Total	$A + C$	$B + D$	N

Table 12.1 Matched-pair design table typical for epidemiological studies.

In this table sample proportions of diseased among the exposed and unexposed are $\hat{p}_1 = \frac{A+B}{N}$ and $\hat{p}_2 = \frac{A+C}{N}$, respectively.

12.7.1 Risk Differences

The McNemar test examines the difference between the proportions that derive from the marginal sums and tries to infer if the two population proportions p_1 and p_2 differ significantly. The difference between this test and the test for two proportions from Chapter 10, page 468, is that in the paired tables the two proportions are *not independent*. Note that both sample proportions $\hat{p}_1 = (A + B)/N$ and $\hat{p}_2 = (A + C)/N$ depend on A from the upper left cell of the table. The inference about the difference of population counterparts p_1 and p_2 involves only entries B and C from the table, since A/N cancels.

Under the hypothesis $H_0 : p_1 = p_2$, both B and C are distributed as binomial $Bin(B + C, 0.5)$. In this case, $\mathbb{E}B = \frac{B+C}{2}$ and $\mathbb{V}ar\,(B) = \frac{B+C}{4}$. By the CLT,

$$Z = \frac{B - \mathbb{E}B}{\sqrt{\mathbb{V}ar\,(B)}} = \frac{B - (B+C)/2}{\sqrt{B+C}/2} = \frac{B - C}{\sqrt{B+C}}$$

has approximately standard normal $\mathcal{N}(0,1)$ distribution. Thus, after squaring we obtain the statistic

$$\chi^2 = Z^2 = \frac{(B-C)^2}{B+C},$$

which is approximately χ^2-distributed with 1 degree of freedom. Large values of χ^2 are critical for the two-sided alternative.

Note that if one-sided alternatives $H_1 : p_1 > p_2$ or $H_1 : p_1 < p_2$ are of interest, one should use statistic Z instead of χ^2. When $B + C$ is small, it is recommended to use a continuity correction in χ^2 as $\chi^2 = \frac{(|B-C|-1)^2}{B+C}$.

The confidence interval for the difference in proportions is

$$[\hat{p}_1 - \hat{p}_2 - z_{1-\alpha/2}\, s, \ \ \hat{p}_1 - \hat{p}_2 + z_{1-\alpha/2}\, s],$$

where s is the square root of s^2, and

$$s^2 = \frac{\hat{p}_1(1-\hat{p}_1)}{N} + \frac{\hat{p}_2(1-\hat{p}_2)}{N} - \frac{2(\hat{p}_{11}\hat{p}_{22} - \hat{p}_{12}\hat{p}_{21})}{N},$$

where, as before, $\hat{p}_1 = (A+B)/N$, $\hat{p}_2 = (A+C)/N$, $\hat{p}_{11} = A/N$, $\hat{p}_{12} = B/N$, $\hat{p}_{21} = C/N$, and $\hat{p}_{22} = D/N$. Note that the first two factors in the expression for s^2 are as in the case of independent proportions (e.g., page 10.6), while the third factor accounts for the dependence. Note also that this sample variance is equivalent to squared standard deviation in (10.13) when $\hat{p}'_1 = B/N$ and $\hat{p}'_2 = C/N$.

Remark. Since under the null hypothesis B is distributed as binomial $Bin(B + C, 0.5)$, exact inference is possible. A two sided p-value is the sum of binomial probabilities for B values at least as far from $(B + C)/2$ as observed. See Agresti (1992) for a survey of exact inferences for contingency tables.

12.7.2 Risk Ratios

An estimator of the population risk ratio (or relative risk) $R = p_1/p_2$ is defined as

$$r = \frac{\hat{p}_1}{\hat{p}_2} = \frac{A+B}{A+C},$$

with an estimator of variance for the $\log r$ as,

$$s_{\log r}^2 = \frac{1}{A+B} + \frac{1}{A+C} - \frac{2A}{(A+B)(A+C)}.$$

The $(1-\alpha) \times 100\%$ CI for the logarithm of population risk ratio is

$$\left[\log r - z_{1-\alpha/2}\, s_{\log r},\ \ \log r + z_{1-\alpha/2}\, s_{\log r} \right],$$

and the confidence interval on the population risk ratio is obtain by taking antilogs of the two bounds,

$$\left[r\, \exp\left\{ -z_{1-\alpha/2}\, s_{\log r} \right\},\ \ r\, \exp\left\{ z_{1-\alpha/2}\, s_{\log r} \right\} \right].$$

12.7.3 Odds Ratios

To conduct the inference on odds ratios in paired tables, we use the Mantel–Haenszel theory with parallel tables constructed from a paired table. Any table with N paired observations generates N Mantel–Haenszel tables

$$\begin{array}{|c|c|}\hline 1 & 0 \\\hline 0 & 0 \\\hline\end{array} \mapsto \begin{array}{|c|c|}\hline 1 & 1 \\\hline 0 & 0 \\\hline\end{array} \qquad \begin{array}{|c|c|}\hline 0 & 1 \\\hline 0 & 0 \\\hline\end{array} \mapsto \begin{array}{|c|c|}\hline 1 & 0 \\\hline 0 & 1 \\\hline\end{array} \qquad \begin{array}{|c|c|}\hline 0 & 0 \\\hline 1 & 0 \\\hline\end{array} \mapsto \begin{array}{|c|c|}\hline 0 & 1 \\\hline 1 & 0 \\\hline\end{array} \ \text{ and } \ \begin{array}{|c|c|}\hline 0 & 0 \\\hline 0 & 1 \\\hline\end{array} \mapsto \begin{array}{|c|c|}\hline 0 & 0 \\\hline 1 & 1 \\\hline\end{array}$$

where the black table is paired and the red is parallel.

		After	
		Positive	Negative
Before	Positive	•	•
	Negative	•	•

$\longrightarrow$

	Before	After
Positive	•	•
Negative	•	•

For example, **1** in the paired table $\begin{array}{|c|c|}\hline 1 & 0 \\\hline 0 & 0 \\\hline\end{array}$ where both *Before* and *After* are *Positive* is translated to 1 for each *Before* and *After* in a table $\begin{array}{|c|c|}\hline 1 & 1 \\\hline 0 & 0 \\\hline\end{array}$, contrasting *Positives* and *Negatives* for *Before* and *After*. From an analysis of Mantel–Haenszel odds ratio o_{mh} in (12.7) one can see that $\sum_i a_i d_i$ is B and $\sum_i b_i c_i$ is C. All $n_i = 2$ in (12.7) cancel.

Thus the estimator for population odds ratio O is

$$o = \frac{B}{C}.$$

To find the confidence interval on the $\log o$, we use the Miettinen's test-based method (Miettinen, 1976). First note that when $H_0 : O = 1$ is true, McNemar's χ^2 statistic can be expressed as

$$\chi^2 = \frac{(\log o - \log 1)^2}{\mathbb{V}\mathrm{ar}\,(\log o)},$$

in which the only unknown is $\mathbb{V}\mathrm{ar}\,[\log o]$, since $\log o = \log(B/C)$ and $\chi^2 = (|B - C| - 1)^2/(B + C)$ (McNemar) are easy to find. By solving for $\mathbb{V}\mathrm{ar}\,(\log o)$ and taking the solution as $\widehat{\mathbb{V}\mathrm{ar}}\,(\log o)$, we can approximate the $(1 - \alpha) \times 100\%$ confidence interval for $\log O$ as

$$[\log o - z_{1-\alpha/2}\, s, \;\; \log o + z_{1-\alpha/2}\, s],$$

where $s = \sqrt{\widehat{\mathbb{V}\mathrm{ar}}\,(\log o)}$. The confidence interval for the population O is derived from the above using antilogs.

Another formula for variance of o in a matched-pairs table case comes from the RBG variance estimator for the Mantel–Haenszel unmatched cases (page 610), $s = \sqrt{s_{\log o}^2} = \sqrt{1/B + 1/C}$. It can be seen that the estimator of $\log o$ is consistent, since when the number of tables goes to infinity, $s \to 0$.

MATLAB function ◀ mcnemart.m computes the estimators and confidence intervals for McNemar's layout.

```
function [stats] = mcnemart(matr, alpha)
%  matr is 2x2 table matrix [A B; C D].

if  nargin==1, alpha=0.05 ; end

A=matr(1,1); B= matr(1,2); C= matr(2,1); D=matr(2,2);
% If any entry is 0, add 0.5 to all cells
if( A*B*C*D==0 )
    matr = matr + 0.5;
end
N = A + B + C + D;    stats.N = N;

stats.chi2 = (B-C)^2/(B+C);  %mc nemar's chi2
stats.pval = 1- chi2cdf(stats.chi2, 1);
stats.h = stats.chi2 > chi2inv(1-alpha, 1); %H0 rejected?

p1 = (A + B)/N; stats.p1=p1;  % row prob
p2 = (A + C)/N; stats.p2=p2;  % column prob

p11 = A/N; p12 = B/N;  p21 = C/N;  p22 = D/N;

% risk difference
```

```
stats.rd = p1 - p2;
   delta = p11*p22 - p12*p21;
s = sqrt( p1*(1-p1)/N + p2*(1-p2)/N + 2*delta/N );
ssel = sqrt( ((B+C)*N  - (B-C)^2)/N^3 );
   lbd =  p1 - p2 - norminv(1-alpha/2)*s;
   ubd = p1 - p2 + norminv(1-alpha/2)*s;
stats.rdint = [lbd, ubd];
% risk ratio
stats.rr = (A+B)/(A+C);
varlrr = 1/(A+B) + 1/(A+C) - 2*A/((A+B)*(A+C));
   lblrr = log((A+B)/(A+C)) - norminv(1-alpha/2)*sqrt(varlrr) ;
   ublrr = log((A+B)/(A+C)) + norminv(1-alpha/2)*sqrt(varlrr) ;
stats.rrint = [exp(lblrr), exp(ublrr)];
% odds ratio
stats.or = B/C;
% miettinen approx for variance
s2lor1 = (log(B/C))^2 * (B+C)/((abs(B-C)-1)^2 + 0.0000001);
slor1 = sqrt(s2lor1);
   lblor1 = log(B/C) - norminv(1-alpha/2)*slor1 ;
   ublor1 = log(B/C) + norminv(1-alpha/2)*slor1 ;
stats.ormiett  = [exp(lblor1), exp(ublor1)];
% RBG approx for variance
slor2 = sqrt(1/B + 1/C);
lblor2 = log(B/C) - norminv(1-alpha/2)*slor2 ;
ublor2 = log(B/C) + norminv(1-alpha/2)*slor2 ;
rgblint = [lblor2, ublor2];
stats.orrgb  = [exp(lblor2), exp(ublor2)];
% Liddell's exact method
 df1l = 2*C +2; df2l = 2*B;
 F1 = finv(1-alpha/2, df1l, df2l);
 tl = B/(B+(C+1)*F1 );
 orexall = tl/(1-tl);
%
 df1u = 2*B +2; df2u = 2*C;
 F2 = finv(1-alpha/2, df1u, df2u);
 tu = (B+1)*F2/(C+(B+1)*F2 );
 orexalu = tu/(1-tu);
 stats.orliddell  = [orexall, orexalu];
```

Example 12.11. **Split-Mouth Trials for Dental Sealants.** Randomized split-mouth trials (RSM) are frequently used in dentistry to examine the effectiveness of preventive interventions that impact individual teeth as opposed to the whole mouth. For example, to examine the effectiveness of dental sealants in preventing caries, a permanent first molar is randomly chosen for the intervention, while its contralateral tooth serves as the control. Because the control and test teeth are subject to the same oral environment, this design controls for many confounding factors such as diet, tooth morphology, and oral hygiene habits. Thus, differences in outcomes between test and control teeth are likely attributable to the treatment. Due to this pairing, adequate power may be achieved with a smaller sample size than if the teeth were independent.

Forss and Halme (1998) report results of a split-mouth study that started in 1988 with 166 children with the goal of assessing tooth-sealant materials. Participants were children from Finland aged 5 to 14 years with a mean age 11 years. To be included in the study, children had to have a contralateral pair of newly erupted, sound, unsealed permanent first or second molar teeth.

Interventions on the occlusal surfaces of sound first or second permanent molars involved glass ionomer Fuji III sealant as a treatment and third-generation, resin-based, light-cured Delton sealant as a control. The results were recorded at the 7-year follow-up involving 97 children, as the dropout rate was 42%.

		Control (resin)		
		Caries	No caries	Total
Treatment (Fuji III)	Caries	8	15	23
	No caries	8	66	74
	Total	16	81	97

Risk differences, risk ratios, and odds ratios with respective confidence intervals are obtained by the MATLAB program ◢ mcnemart.m. Several approaches to confidence intervals for population odds ratio O (exact, approximate RBG, approximate Miettinen) are presented. The odds ratio o is 1.8750, the exact 95% confidence interval for O is orliddell = [0.7462, 5.1064], and two approximations are ormiett = [0.7003, 5.0198] and orrgb = [0.7950, 4.4224].

The complete output from mcnemart.m is

```
mcnemart([8, 15; 8, 66])
         N: 97
      chi2: 2.1304
      pval: 0.1444
         h: 0
        p1: 0.2371
        p2: 0.1649
        rd: 0.0722
     rdint: [-0.0545 0.1989]
        rr: 1.4375
     rrint: [0.8807 2.3464]
        or: 1.8750
   ormiett: [0.7003 5.0198]
     orrgb: [0.7950 4.4224]
  orliddell: [0.7462 5.1064]
```

Example 12.12. **Testing for Salmonella.** Large discrepancies are usually found when different ELISAs for the diagnosis of pig salmonellosis are compared. Mainar-Jaime et al. (2008) explored the diagnostic agreement of two commercial assays: (i) Salmonella Covalent Mix-ELISA (Svanovir) as test A, and (ii) Swine Salmonella Antibody Test Kit (HerdCheck) as test B, for the detection of antibodies to Salmonella spp. in slaughter pigs.

Two populations of pigs slaughtered in abattoirs from Saskatchewan, Canada, have been analysed:

Population 1: Animals from farms marketing $< 10{,}000$ pigs/year.
Population 2: Animals from farms marketing $\geq 10{,}000$ pigs/year.

	Population 1			Population 2		
	Test B +	Test B −	Total	Test B +	Test B −	Total
Test A +	11	16	27	2	5	7
Test A −	6	119	125	1	72	73
Total	17	135	152	3	77	80

From McNemar's χ^2 test it could be concluded that test A significantly differs from test B in the proportion positive at $\alpha = 0.05$ for Population 1 and in the aggregate table. The tests show no significant difference for Population 2.

Sample Size in McNemar's Test. Miettinen (1968) proposed a power analysis for McNemar's test based on a normal approximation for test statistics under alternative hypotheses.

Let $w = p_{12} + p_{21}$ be the total probability of a discording case. Let $0 < \Delta < w$ be the discrepancy we want to control. In these terms, McNemar's null hypothesis is

$$H_0 : p_{12} = p_{21} = w/2,$$

and the alternative is defined by

$$p_{12} = (w \pm \Delta)/2 \quad \text{and} \quad p_{21} = (w \mp \Delta)/2. \tag{12.8}$$

The sample size needed to reject H_0 with power $1 - \beta$ in α-level testing against the one-sided alternative, specified by the choice of w and Δ in (12.8), is

$$N = \frac{\left(z_{1-\alpha}\sqrt{w} + z_{1-\beta}\sqrt{w - \Delta^2(3+w)/(4w)}\right)^2}{\Delta^2}$$

$$1 - \beta = \Phi\left(\frac{|\Delta|\sqrt{N} - z_{1-\alpha}\sqrt{w}}{\sqrt{w - \Delta^2(3+w)/(4w)}}\right).$$

If the alternative is two sided, $1 - \alpha$ in normal quantiles is replaced by $1 - \alpha/2$.

Here are simple MATLAB @-functions implementing sample size and power calculations

```
%w = p12+p21  is the probability of discording
%     delta is discrepancy we want to capture.
%H0: p12=p21 = w/2   vs.  H1: |p12 - p21| = delta
%     p21=(w + delta)/2;   p12 = (w-delta)/2
%=================================
ssmn = @(alpha, beta, w, delta) ...
    (norminv(1-alpha)*sqrt(w)+norminv(1-beta)* ...
    sqrt(w - delta^2*(3 + w)/(4*w)) )^2/delta^2

power =@(alpha, N, w, delta) ...
    normcdf( (abs(delta)*sqrt(N)-norminv(1-alpha)*sqrt(w))/ ...
    sqrt(w - delta^2*(3 + w)/(4*w)) )
```

Example 12.13. **OPEN/CLOSED Cases.** This example is adapted from Berger and Sidik (2002). A regulatory agency sometimes checks the accuracy of analyses of a medical laboratory. The laboratory has information when it is checked. An experimenter thinks the laboratory is more careful when it knows that the results may be scrutinized. To confirm this the experimenter sends two samples from each person for antibody analysis. In one case, the sample is labeled as an OPEN case (part of a check); in the other case, the sample is labeled as CLOSED case (not part of a check). The experimenter also has a gold standard and knows whether the laboratory analysis is correct or incorrect. This is a matched pairs design and the one-sided alternative $H_1 : p_{12} < p_{21}$ is of interest. Here p_{12} is the probability that CLOSED/OPEN was correct/incorrect and p_{21} the probability that CLOSED/OPEN was incorrect/correct.

What total sample size, in number of pairs, is needed to reject H_0 with a power 0.8 in a 0.05-level test against the one-sided alternative, when $p_{12} = 0.1$ and $p_{21} = 0.2$?

From $p_{12} = 0.2$ and $p_{21} = 0.1$, we find $w = 0.3$ and $\Delta = 0.1$. Thus, `ssmn(0.05, 0.2, 0.3, 0.1)` % 179.6304, and a sample of size $N = 180$ pairs ensures the desired power. The exact power for $N = 180$ pairs is 0.8007, as `power(0.05, 180, 0.3, 0.1)` % 0.8007.

12.7.4 Liddell's Procedure

Paired proportions have traditionally been compared using McNemar's test but an exact alternative due to Liddell (1983) is preferable. In fact, some argue that McNemar test for testing odds ratios in paired tables should not be used when the exact test is available and readily implementable.

As in McNemar's setup, the B count in the table is treated as a realization of binomial variable $Bin(B + C, p)$. Under $H_0 : OR = 1$, the distribution of B is $Bin(B + C, 1/2)$.

The Liddell's exact test uses the link between binomial and F distributions. The test statistic $F = B/(C + 1)$ has an F-distribution with $2(C + 1)$ and $2B$ degrees of freedom. Confidence limits for OR are calculated as follows:

$$\left[\frac{B}{(C + 1) \times F_{2(C+1),2B,1-\alpha/2}}, \frac{B \times F_{2(B+1),2C,1-\alpha/2}}{C} \right]$$

Giuseppe Cardillo's function liddell.m is available on Matlab Central FileExchange (#22024) and this book's website.

*Example 12.14. **Schistosoma Mansoni**. Schistosoma mansoni is a parasite that is found in Africa, Caribbean, and parts of South America. Among human parasitic diseases in tropical and subtropical areas, schistosomiasis ranks second behind malaria in terms of socioeconomic and public health impact. Sleigh et al. (1982) compare results of one Bell and one Kato–Katz examination performed on each of 315 stool specimens from residents in an area in northeastern Brazil endemic for schistosomiasis mansoni. The following table, discussed also in Kirkwood and Sterne (2003), summarizes the findings:*

	Kato–Katz positive	Kato–Katz negative	Total
Bell positive	184	54	238
Bell negative	14	63	77
Total	198	117	315

Are the probabilities of detecting Schistosomiasis mansoni different for the two tests? Find the odds ratio and its 95% confidence interval using Liddell's procedure.

From the output of liddell.m we have

```
liddell([184 54; 14 63])
Liddell's exact test
Maximum likelihood estimate of OR =  3.8571
Exact 95% confidence interval = 2.1130 to 7.5193
F = 3.6000 p-value (two-sided)= 0.00000111
```

The odds ratio is estimated by $OR = 54/14 = 3.8571$, meaning that odds of a positive result by Kato-Katz test are almost 4 times the odds of a positive result by Bell's test. Liddell's exact method indicates that this odds ratio falls between 2.1130 and 7.5193 with confidence 95%.

12.7.5 Analyzing Discordant Pairs: Garth Test*

Assume that in a clinical assay the patients receiving two drugs A and B can receive them in order (A, B), that is, drug A first, or in order (B, A), that is, drug B first. The order of administration may be responsible for carryover effects and possible unwanted drug interactions affecting the efficacy.

The researchers are interested in drug effects but want to control for the order effects. Usually the drug effect is measured by some pharmacokinetic outcome, and for such continuous measures the ANOVA methodology is appropriate. When observations are dichotomous (Effect present – No effect present), Garth (1979) developed a test that is capable of both comparing drug efficiency and testing for order effect.

Consider the following paired table

		Drug B	
		Effect	No effect
Drug A	Effect	$\bullet$	n_A
	No effect	n_B	$\bullet$

where n_A (n_B) is the number of subjects for which drug A (B) produced an effect and drug B (A) did not.

As in the McNemar test, we focus only on discordant counts n_A and n_B and ignore the pairs where effects agree (marked as $\bullet$). Let $n_d = n_A + n_B$ be the number of discordant pairs, n_{AB} the number of discordant pairs with A as the first drug, and n_{BA} the number of discordant pairs with B as the first drug. Count n_A can be split as $n_A = n_{A,AB} + n_{A,BA}$, where $n_{A,AB}$ is the portion of n_A for the order (A, B). Analogously, $n_B = n_{B,AB} + n_{B,BA}$. Lastly, let n_1 be the number of subjects for which only the first drug was effective, and n_2 the number of subjects for which only the second drug was effective. Note that $n_1 = n_{A,AB} + n_{B,BA}$ and $n_2 = n_{A,BA} + n_{B,AB}$.

Using $n_{A,AB}$, $n_{B,AB}$, $n_{A,BA}$, and $n_{B,BA}$, we form two tables in which the rows relate to (i) the type of drug (A or B), and (ii) to the position (first or second) of the effective drug,

Table (i)	Order		
	(A,B)	(B,A)	
Effect with A	$n_{A,AB}$	$n_{A,BA}$	n_A
Effect with B	$n_{B,AB}$	$n_{B,BA}$	n_B
	n_{AB}	n_{BA}	n_d

Table (ii)	Order		
	(A,B)	(B,A)	
Effect with first drug	$n_{A,AB}$	$n_{B,BA}$	n_1
Effect with second drug	$n_{B,AB}$	$n_{A,BA}$	n_2
	n_{AB}	n_{BA}	n_d

Now the Fisher exact test applies to both tables. For table (i), the test is that there is no order effect, while for table (ii), the test is for no difference in drugs as their effect is concerned. For more discussion, see Everitt (1977).

Example 12.15. **Emetogenic drugs A and B.** In a study, patients received two emetogenic drugs A and B, and the drug effect was the presence of nausea/vomitting. The patients were randomized with respect to drugs' order. For 34 patients, 16 assigned to (A, B) and 18 assigned to (B, A) sequences, the results disagree. The table counts were $n_{A,AB} = 9, n_{A,BA} = 14, n_{B,AB} = 7$, and $n_{B,BA} = 4$.

(a) Interpret the table count $n_{A,AB}$ in the terms of this study.

(b) Test the hypothesis that there is no difference in drug effects in terms of causing nausea/vomitting.

(c) Test the hypothesis that there is no difference in effect with respect to sequence of drug administration.

Use Fisher exact tests and $\alpha = 0.05$.

Solution. (a) In this study $n_{A,AB} = 9$ means that for 9 patients out of 34 for which the results disagreed, and out of 16 assigned to sequence (A, B), 9 had positive response to the first drug administered. It also means that 9 patients, out of 23 positively responding to drug A, took drug A first.

```
%(b,c)
[h p]=fisherexactt([9 4; 7 14]) %for drug efficacy
 %H0: p1=p2 vs. H1: p1 =/= p2
 %h = 0
 %p =  0.0764
[h p]=fisherexactt([9 14; 7 4]) %for drug order
 H0: p1=p2 vs. H1: p1 =/= p2
 h = 0
 p = 0.2743
```

As the drug efficacy is of concern, neither the drugs nor the order of administration were significant at 5% level in two-sided testing. If we had applied the McNemar test on the discordant counts $n_A = n_{A,AB} + n_{A,BA} =$

23 and $n_B = n_{B,AB} + n_{B,BA} = 11$, the results would have been significant, with a p-value of 0.0396. This would have been an inappropriate conclusion in this context.

✏️

12.7.6 Multicategorical Paired Tables: Stuart–Maxwell Test*

A natural generalization of the McNemar test is a design in which there are N matched pairs where each response can be classified into $k > 2$ different categories. There are two possible generalizations of the McNemar test: tests for symmetry and tests for marginal homogeneity. For 2×2 tables these two tests coincide, but the tests are quite different for $k \times k$ tables, when $k > 2$.

12.7.6.1 Tests for Marginal Homogeneity

We will be interested in the equality of proportions for categories in the two populations that form the paired responses. This is equivalent to homogeneity of marginal proportions. For simplicity, we first discuss the case when $k = 3$.

Assume that paired responses come from *Before* and *After* and each response may result in one of three categories: *Positive*, *Neutral*, and *Negative*.

		After		
	Positive	Neutral	Negative	Total
Positive	n_{11}	n_{12}	n_{13}	$n_{1\cdot}$
Before Neutral	n_{21}	n_{22}	n_{23}	$n_{2\cdot}$
Negative	n_{31}	n_{32}	n_{33}	$n_{3\cdot}$
Total	$n_{\cdot 1}$	$n_{\cdot 2}$	$n_{\cdot 3}$	N

Notice that the table counts matched pairs; for example, n_{23} is the number of cases where *Before* resulted in neutral and *After* resulted in negative. Thus, again, we have a total of $2N$ responses organized into N matched pairs.

Denote by p_{ij} the population proportion of subjects that are classified as i (one of *Positive*, *Neutral*, or *Negative*) in *Before*, and as j in *After*.

Let $p_{i\cdot} = \sum_j p_{ij}$ and $p_{\cdot j} = \sum_i p_{ij}$ be the marginal probabilities of the three categories in *Before* and *After*, respectively. The null hypothesis of interest is that the population proportions of the three categories are the same for the two groups, *Before* and *After*,

$$H_0: \; p_{1\cdot} = p_{\cdot 1}, \; p_{2\cdot} = p_{\cdot 2}, \text{ and } \; p_{3\cdot} = p_{\cdot 3},$$

which is the hypothesis of marginal homogeneity. The alternative is any violation of H_0.

The test statistic is

$$\chi^2 = \frac{\bar{n}_{23}(n_{1.} - n_{.1})^2 + \bar{n}_{13}(n_{2.} - n_{.2})^2 + \bar{n}_{12}(n_{3.} - n_{.3})^2}{2(\bar{n}_{12}\bar{n}_{13} + \bar{n}_{12}\bar{n}_{23} + \bar{n}_{13}\bar{n}_{23})}, \qquad (12.9)$$

where $\bar{n}_{ij} = (n_{ij} + n_{ji})/2$.

The statistic χ^2 in (12.9) has a χ^2-distribution with 2 degrees of freedom, and its large values are critical.

Rule: Reject H_0 at significance level α if $\chi^2 > \chi^2_{2,1-\alpha}$, where $\chi^2_{2,1-\alpha}$ is a $1 - \alpha$ quantile of a χ^2-distribution with 2 degrees of freedom (`chi2inv(1-alpha, 2)`). The p-value of this test is `1-chi2cdf(chi2, 2)` for `chi2` as in (12.9).

Example 12.16. **Galton and ALW Features in Fingerprints.** In his influential book *Finger Prints*, Sir Francis Galton details the description and distributions of arch-loop-whorl (ALW) features of human fingerprints (Fig. 12.1). Some of his findings from 1892 are still in use today.

Tables 12.2a,b are from Galton (1892) and tabulate ALW features on forefingers in pairs of school children.

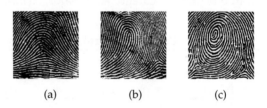

(a)   (b)   (c)

Fig. 12.1 (a) Arch, (b) loop, and (c) whorl features in a fingerprint.

In Table 12.2a, subjects A and B are paired at random from a large population of subjects. In Table 12.2b, subjects A and B are two brothers with randomized assignment of order (A, B).

Galton was interested in knowing if there was any influence of fraternity on the dependence of ALW features. Using Stuart–Maxwell test, we will test for the equality of marginal distributions, H_0: Population proportions of fingerprint features for individuals from groups A and B are the same.

The function ⟁ stuartmaxwell.m gives the following output:

```
%galtonalw.m
indeppairs = [5 12 8;  8 18  8; 9 20 13];
fratpairs  = [5 12 2;  4 42 15; 1 14 10];
```

Table 12.2 (a) Table with random pairing of children. (b) Table with fraternal pairing.

		A		
		Arch	Loop	Whorl
	Arch	5	12	8
B	Loop	8	18	8
	Whorl	9	20	13

(a)

		A		
		Arch	Loop	Whorl
	Arch	5	12	2
B	Loop	4	42	15
	Whorl	1	14	10

(b)

```
[stat] = stuartmaxwell(indeppairs)
%stat =
%stuartmaxwellchi2 = 5.8023
%stuartmaxwellpval = 0.0550
[stat] = stuartmaxwell(fratpairs)
%stat =
%stuartmaxwellchi2 = 4.2738
%stuartmaxwellpval = 0.1180
```

It is evident that for random pairing, the hypothesis of equal proportions H_0 almost rejected at 5% level (p-value 0.0550), while in the case of fraternal pairing the equality of proportions of fingerprint features is not rejected (p-value 0.1180). If treated as contingency tables, as originally intended by Galton, the hypothesis of independence between two groups A and B on the basis of fingerprint features is not rejected for random pairing (p-value 0.9520), while it is rejected for the case of fraternal pairing (p-value 0.0247).

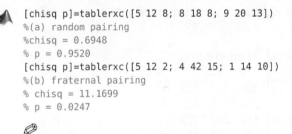

```
[chisq p]=tablerxc([5 12 8; 8 18 8; 9 20 13])
%(a) random pairing
%chisq = 0.6948
% p = 0.9520
[chisq p]=tablerxc([5 12 2; 4 42 15; 1 14 10])
%(b) fraternal pairing
% chisq = 11.1699
% p = 0.0247
```

When $k \geq 3$, an elegant representation of χ^2 statistic is possible in terms of matrices. Let n_{ij} be the entries in the paired table and let $n_{.j}$ and $n_{i.}$ be the column and row sums. Define vector $\boldsymbol{d} = (d_1, d_2, \ldots, d_{k-1})$ and matrix $\boldsymbol{S} = (s_{ij})_{k-1 \times k-1}$ as

$$d_i = n_{i.} - n_{.i},$$
$$s_{ii} = n_{i.} + n_{.i} - 2n_{ii}, \ i = 1, \ldots, k-1;$$
$$s_{ij} = -(n_{ij} + n_{ji}), \ i \neq j, \ i,j = 1, \ldots, k-1.$$

Then

$$\chi^2 = d' S^{-1} d$$

is the statistic for testing the hypothesis of marginal homogeneity

$$H_0 : p_{i.} = p_{.i},\ 1 \le i \le k$$

and has a χ^2-distribution with $k - 1$ degrees of freedom.

Bhapkar (1966) also proposed a test for marginal homogeneity. The test statistic is the similar to Stuart–Maxwell's, but the difference lies in the calculation of variance-covariance matrix. Bhapkar's χ^2 is calculated as

$$\chi^2 = d' B^{-1} d,$$

where

$$B = S - \frac{1}{n} dd'.$$

Here d and S are as in Stewart–Maxwell's procedure. Function stuartmax-well.m performs both Stuart–Maxwell and Bhapkar tests.

Although the Bhapkar and Stuart-Maxwell tests are asymptotically ($N \to \infty$) equivalent, the Bhapkar test is a more powerful alternative to the Stuart-Maxwell test for small N.

Example 12.17. **Unaided Distant Vision.** The table below, provided by Stuart (1953), presents visual acuity of 7,477 women. The women, aged 30 to 39, were employees in Britain's Royal Ordnance factories in 1943 to 1946. For each woman, the left and right eyes were classified into vision grades, from 1 (highest) to 4 (lowest).

Right eye	Left eye grade			
grade	1	2	3	4
1	1520	266	124	66
2	234	1512	432	78
3	117	362	1772	205
4	36	82	179	492

We will test whether the distribution of Grade score is homogeneous for the left and right eye, that is $H_0 : p_{i.} = p_{.i},\ i = 1,\ldots,4$.

```
eyes = [1520  266  124    66;      234 1512  432    78; ...
         117  362 1772   205;       36   82  179   492];
[stats] = stuartmaxwell(eyes)
stats =
 stuartmaxwellchi2: 11.9566
 stuartmaxwellpval: 0.0075
       bhapkarchi2: 11.9757
       bhapkarpval: 0.0075
```

As is evident from the output of `stuartmaxwell`, the hypothesis of symmetry for this table is rejected. Note also that, since $N = 7477$ is large, the Stuart–Maxwell and Bhapkar tests are practically indistinguishable. ✎

12.7.6.2 Tests for Symmetry

The hypothesis of table symmetry,

$$H_0 : p_{ij} = p_{ji}, \quad \text{for all } i, j,$$

is stronger than the hypothesis of marginal homogeneity. If the paired table is symmetric, the marginal homogeneity is satisfied,

$$H_0 \Rightarrow p_{\cdot i} = p_{i \cdot}, \ i = 1, 2, \ldots, k.$$

The opposite is not true: the marginal homogeneity does not imply symmetry, except for $k = 2$ (McNemar's test). If the hypothesis of symmetry is satisfied, then all conditional distributions are homogeneous. That is, for any fixed j,

$$\mathbb{P}(A \text{ in category } i | \ B \text{ in category } j) =$$
$$\mathbb{P}(B \text{ in category } i \ | \ A \text{ in category } j), \ i = 1, 2, \ldots, k.$$

Thus, the hypothesis of symmetry, or equivalently the equality of all conditional distributions, implies the marginal homogeneity hypothesis.

Bowker (1948) suggested generalization of McNemar's statistics for $k \times k$ tables as,

$$\chi^2 = \sum_{i<j} \frac{(n_{ij} - n_{ji})^2}{n_{ij} + n_{ji}},$$

which under the null hypothesis of symmetry has approximately χ^2 distribution with $k(k-1)/2$ degrees of freedom.

Edwards (1948) suggested a continuity correction for McNemar statistics that is applicable to Bowker's χ^2,

$$\chi^2_{corr} = \sum_{i<j} \frac{(|n_{ij} - n_{ji}| - 1)^2}{n_{ij} + n_{ji}},$$

while May and Johnson (2001) suggested

$$\chi^2_{mj} = \sum_{i<j} \frac{n \cdot (n_{ij} - n_{ji})^2}{n \cdot (n_{ij} + n_{ji}) - (n_{ij} - n_{ji})^2}.$$

Both modifications of Bowker test are distributed as chi-square with $k(k - 1)/2$ degrees of freedom.

Krampe and Kuhnt (2007) compared performance of the three symmetry tests and found that Bowker and May–Johnson approximations work well if the table counts are not small and that the Bowker-corrected χ^2 provides a conservative test. Function ◀ bowker.m performs the Bowker, Bowker-corrected, and May–Johnson's tests.

Example 12.18. **Noise and Quiet Traffic.** The research network "Quiet traffic" sponsored by the Bundesministerium für Bildung und Forschung, Germany, examined among other things the effect of traffic noise on humans. Kuhnt et al. (2004) provide data on 72 subjects exposed twice to road and rail noise, with four different noise intensities (40, 52, 70, and 82 [dB]). The subjects report the level of annoyance by the noise on a Likert scale with 5 levels (1 = very low, 2 = low, 3 = moderate, 4 = high, and 5 = very high).

		Second exposure				
		1	2	3	4	5
	1	51	28	3	0	0
First	2	15	68	40	5	1
exposure	3	0	29	77	21	1
	4	0	4	19	80	14
	5	0	1	5	26	88

The researches were interested in whether the subjects classify noise differently when they are exposed to it for the second time, that is, in testing the table symmetry.

```
        data=[ 51 28  3  0  0; ...
               15 68 40  5  1; ...
                0 29 77 21  1; ...
                0  4 19 80 14; ...
                0  1  5 26 88];
    stat = bowker(data)
stat =
               bowkerchi2:  15.1616
               bowkerpval:   0.1263
      bowkercorrectedchi2:  11.1814
      bowkercorrectedpval:   0.3436
           mayjohnsonchi2:  15.2448
           mayjohnsonpval:   0.1234
```

The hypothesis of symmetry is not rejected. Note that the Bowker-corrected test is conservative here. Several entries in the table are 0 or small, so the χ^2 approximation may not be satisfactory.

12.7.7 Cochran's Q Test*

If we observe a binary response, say, present and absent, for r subjects on c occasions, we might be interested in testing whether the presence rate changes over time. If $c = 2$, then we would perform McNemar's test by counting the instances $(0, 0)$, $(0,1)$, $(1, 0)$, and $(1, 1)$ for r subjects and forming a paired table of counts. When $c > 2$, then Cochran's Q test is used. Let $a_{ij} = 1$ if a success is observed at location (i,j), and 0 otherwise. Define row, column, and total means as

$$R_i = \frac{1}{c} \sum_{j=1}^{c} a_{ij},$$

$$C_j = \frac{1}{r} \sum_{i=1}^{r} a_{ij},$$

$$T = \frac{1}{rc} \sum_{i=1}^{r} \sum_{j=1}^{c} a_{ij}.$$

Then

$$Q = \frac{r^2(c-1)}{c} \times \frac{\sum_{j=1}^{c}(C_j - T)^2}{\sum_{i=1}^{r} R_i (1 - R_i)}$$

has χ^2 with $c - 1$ degrees of freedom.

The test for H_0 : no change in incidence of 1 over the repeated measures, is implemented by the function ◢ cochrant.m.

Example 12.19. **0–1 Image Homogeneity.** As an illustration of Cochran's test we generate a matrix of size 1024×1024 of Bernoulli random variables. The left-hand side submatrix 1024×512 is generated as i.i.d. Bernoullis $Ber(0.5)$, and the right-hand side submatrix of the same size is generated as i.i.d. Bernoulli $Ber(0.51)$. This visually undetectable difference (Fig. 12.2) is easily captured by Cochran's test.

```
a1 = binornd(1,0.5, [1024 512]);
a2 = binornd(1,0.51, [1024 512]);
a =[a1 a2];
[h Q p]=cochrant(a)
% h = 1
% Q = 1.1984e+003
% p = 1.1233e-004
```

Cochran's test is equivalent to McNemar's test when pairs $(1, 1)$,$(1, 0)$,$(0, 1)$, and $(0, 0)$ are counted and organized as a 2×2 table. Indeed if $c = 2$,

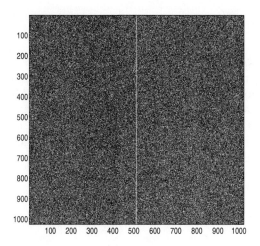

Fig. 12.2 Matrix 1024×1024 of Bernoullis. Submatrix 1024×512 on the left side is generated as i.i.d. Bernoulli $Ber(0.5)$, while the right-hand side submatrix of the same size is generated as i.i.d. Bernoulli $Ber(0.51)$.

```
b = binornd(1, 0.5, [200 2]);
[h Q p]=cochrant(b)
  %h = 0
  %Q = 0.8526
  %p = 0.3558
n11 = 0; n12=0; n21=0; n22=0;
for i = 1:r
    if ( b(i,1)+b(i,2) == 2)       n11=n11+1;
    elseif  ( b(i,1)+b(i,2) == 0) n22=n22+1;
    elseif ( b(i,1) > b(i,2) )     n12=n12+1;
    else  n21=n21+1;
    end
end
[h chi2 pval] = mcnemart([n11 n12; n21 n22])
  %h = 0
  %chi2 = 0.8526
  %pval = 0.3558
```

12.8 Exercises

12.1. **Amoebas and Intestinal Disease.** When an epidemic of severe intestinal disease occurred among workers in a plant in South Bend, Indiana, doctors said that the illness resulted from infection by the amoeba *Entamoeba histolytica*. There are actually two varieties of these amoebas, large and small, and the large ones were believed to be causing the disease. Doctors suspected that the presence of the small amoebas might help people resist infection by the large ones. To check on this, public health officials chose a random sample of 138 apparently healthy workers and determined if they were infected with either the large and/or small amoebas. The table below provides the resulting data (Cohen, 1973). Is the presence of the large variety independent of the presence of the small one? Test at 5% significance level.

Small variety	Large variety Present	Absent	Total
Present	12	23	35
Absent	35	68	103
Total	47	91	138

12.2. **Drinking and Smoking.** Alcohol and nicotine consumption during pregnancy are believed to be associated with certain characteristics in children. Since drinking and smoking behaviors may be related, it is important to understand the nature of this relationship when fully assessing their influence on children. In a study by Streissguth et al. (1984), 452 mothers were classified according to their alcohol intake prior to pregnancy recognition and their nicotine intake during pregnancy. The data are summarized in the following table:

Alcohol (ounces/day)	Nicotine (mg/day) None	1–15	16 or more
None	105	7	11
0.01–0.10	58	5	13
0.11–0.99	84	37	42
1.00 or more	57	16	17

(a) Calculate the column sums. In what way does the pattern of alcohol consumption vary with nicotine consumption?
(b) Calculate the row sums. In what way does the pattern of nicotine consumption vary with alcohol consumption?
(c) Formulate H_0 and H_1 for assessing whether or not alcohol consumption and nicotine consumption are independent.
(d) Compute the table of expected counts.
(e) Find the χ^2 statistic. Report the degrees of freedom and the p-value.
(f) What do you conclude from the analysis of this table?

12.3. **Aortic Valve Replacement and Bleeding Complications.** Généreux et al. (2014) sought to identify the incidence, predictors, and prognostic impact of bleeding complications (BC) after surgical aortic valve replacement (SAVR) compared with transcatheter aortic valve replacement

(TAVR). The TAVR procedure could be transfemoral (TF) or transapical (TA). Bleeding complications have been found to be the strongest independent predictor of 1-year patient mortality.

The study involved 657 severely symptomatic patients randomly assigned to SAVR, TAVR-TF, or TAVR-TA, and indicators of major bleeding complications have been recorded for each patient.

The data are given in the table:

	TAVR-TF	TAVR-TA	SAVR
BC	27	9	71
No BC	213	95	242
Total	240	104	313

Test the hypothesis that the incidence of BC is independent of AVR procedure types. Use $\alpha = 0.05$.

12.4. **Family Size.** A demographer surveys 1,000 randomly chosen American families and records their family sizes and family incomes:

	Family size
Family income	2 3 4 5 6 7
Low	145 81 57 22 9 8
Middle	151 73 71 33 13 10
High	124 60 80 42 13 8

Do the data provide sufficient evidence to conclude that family size and family income are statistically dependent?
(a) State the H_0 and H_1 hypotheses.
(b) Perform the test using $\alpha = 0.05$ and comment.

12.5. **Nightmares.** Over the years numerous studies have sought to characterize the nightmare sufferer. From these studies has emerged the stereotype of someone with high anxiety, low ego strength, feelings of inadequacy, and poorer-than-average physical health. A study by Hersen (1971) explored whether gender is independent of having frequent nightmares. Using Hersen's data summarized in the table below, test the hypothesis of independence at the level $\alpha = 0.05$.

	Men	Women
Nightmares often (at least once a month)	55	60
Nightmares seldom (less than once in a month)	105	132

12.6. **Independence of Segregation.** According to a mathematical model of inheritance, at a given meiosis, the probability of an allele at one locus passing to the gamete is independent of an allele at any locus on another chromosome passing to the gamete. This is usually referred to as the law of independent segregation of genes. Also, one allele from the pair at any locus passes to the gamete with probability equal to $1/2$.

Roberts et al. (1939) conducted an extensive set of experiments for testing independent segregation of genes in mice and rats. One of the results of a mating of the form $Aa\ Bb\ Dd \times aa\ bb\ dd$ is reported as:

	ab	aB	Ab	AB	Total
d	427	440	509	460	1836
D	494	467	462	475	1898
Total	921	907	971	935	3734

Using the χ^2 statistic, test for the independence of segregation at $\alpha = 5\%$. *Hint:* The expected number in each cell is $1/2 \times 1/2 \times 1/2 \times 3734 = 466.75$.

12.7. **Site of Corpus Luteum in Caesarean Births.** Williams (1921) observed that in Caesarean-section births, the corpus luteum was located in the right ovary 23 times and 16 times in the left for male children. For female children the numbers were 13 and 12, respectively. Test for the independence of ovary side and gender of a child.

12.8. **An Easy Grade?** A student wants to take a statistics course with a professor who is an easy grader. There are three professors scheduled to teach the course sections next semester. The student manages to obtain a random sample of grades given by the three professors this past year.

Observed	Prof A	Prof B	Prof C	Total
Grades A	10	12	28	50
Grades B	35	30	15	80
Grades C	15	30	25	70
Total	60	72	68	200

Using a significance level of 1%, test the hypothesis that a student's grade is independent of the professor.

12.9. **Importance of Bystanders.** When a group of people is confronted with an emergency, a diffusion-of-responsibility process can interfere with an individual's responsiveness. Darley and Latané (1968) asked subjects to participate in a discussion carried over an intercom. Aside from the experimenter to whom they were speaking, subjects thought that there were zero, one, or four other people (bystanders) also listening over the intercom. Part way through the discussion, the experimenter feigned serious illness and asked for help. Darley and Latané noted how often the subject sought help for the experimenter as a function of the number of supposed bystanders. The data are summarized in the table:

	Sought assistance	No assistance
No bystanders	11	2
One bystander	16	10
Four bystanders	4	9

What could Darley and Latané conclude from the results?

(a) State H_0 and H_1.

(b) Perform the test at a 5% significance level.

12.10. **Manual and Ocular Laterality: Is There a Link?** Tabular data on manual and ocular laterality in 413 students of University College, London, aged 18–24, were first reported by Woo (1928). Ocular laterality was measured by four tests: near point, visual acuity, spherical refractions, and general astigmatism. Manual laterality was measured by grip strength, first/last pull, lack of endurance, steadiness, and balancing tests. The data are summarized in the following table:

	Left-eyed	Ambiocular	Right-eyed	Total
Left-handed	34	62	28	124
Ambidexterous	27	28	20	75
Right-handed	57	105	52	214
Total	118	195	100	413

(a) Test the hypothesis that the ocular and manual dexterity are dependent. Report the p-value and discuss.

(b) Omit ambidexterous and ambiocular counts to test whether any side is significantly dominant for (i) manual or (ii) ocular laterality.

12.11. **Streptococcus Pyogenes and Tonsil Size.** Holmes and Williams (1954) classify 1,398 children aged 0–15 years according to their relative tonsil size and whether or not they were carriers of *Streptococcus pyogenes*.

	Present but not enlarged	Enlarged	Greatly enlarged	Total
Carriers	19	29	24	72
Noncarriers	497	560	269	1326
Total	516	589	293	1398

Is the size of tonsils dependent on the *Streptococcus pyogenes* carrier status?

If there is dependence, tonsil size could be regarded as the dependent variable, while the presence or absence of *Streptococcus pyogenes* is regarded as a possible explanatory factor. This distinction is in keeping with possible biological mechanisms: if there is a causal relationship between the two variables, it is almost certainly in the direction indicated rather than the reverse.

Test the hypothesis of independence at $\alpha = 0.05$ level.

12.12. **Psychosis in Adopted Children.** Numerous studies have been done to determine the etiology of schizophrenia. Such factors as biological, psychocultural, and sociocultural influences have been suggested as possible causes of this disorder. To test if schizophrenia has a hereditary component, researchers compared adopted children whose biological mothers are schizophrenic ("exposure") to adopted children whose biological mothers are normal ("nonexposure").

Furthermore, the child-rearing abilities of adoptive families have been assessed to determine if there is a relationship between those children who become psychotic and the type of family into which they are adopted. The families are classified as follows: healthy, moderately disturbed, and severely disturbed.

The following data are from an experiment described in Carson and Butcher (1992).

			Type of adoptive family			
	Healthy		Moderately disturbed		Severely disturbed	
Diagnosis	Exp	Nonexp	Exp	Nonexp	Exp	Nonexp
None	41	42	11	26	6	15
Psychotic	10	11	18	25	38	28

(a) For moderately disturbed families, find the risk difference for a child's psychosis with the mother's schizophrenia as a risk factor. Find a 95% confidence interval for the risk difference.

(b) For severely disturbed families, find the risk ratio and odds ratio for psychosis with the mother's schizophrenia as a risk factor. Find a 95% confidence interval for the risk ratio and odds ratio.

(c) Is a mother's schizophrenia a significant factor for a child's psychosis overall? Assess this using the Mantel–Haenszel test. Find the overall odds ratio and corresponding 95% confidence interval.

12.13. **More Perils of Aggregation: Berkeley Admission Data.** Examination of aggregate data on graduate admissions to the University of California, Berkeley, for fall 1973 shows a clear but misleading pattern of bias against female applicants (Bickel et al., 1975). For the six major graduate programs denoted here as A–F, a total of 4,526 students applied, 2,691 males and 1,835 females. Among 1,755 admitted students, 1,198 were males and 557 females.

(a) Using a 2×2 contingency table show that gender and admission are dependent. Also, show that the population proportions are significantly different (test for two proportions on page 465).

(b) The numbers in (a) brought up accusations of gender bias in Berkeley admissions. However, when stratified by department/program, the admission results were as follows:

		Admitted	Not admitted	Total
Program A	Men	512	313	825
	Women	89	19	108
Program B	Men	353	207	560
	Women	17	8	25
Program C	Men	120	205	325
	Women	202	391	593
Program D	Men	138	279	417
	Women	131	244	375
Program E	Men	53	138	191
	Women	94	299	393
Program F	Men	22	351	373
	Women	24	317	341

Using Mantel–Haenszel's test show that gender and admission status are not significantly related.

Remark. This exercise is not exactly Simpson's paradox, but it exemplifies perils of aggregation. In 4 out of 6 schools women had a higher admission rate. Yet, the overall admission rate for women was significantly lower.

12.14. **The Midtown Manhattan Study.** The data set below has been analyzed by many authors (Haberman, Goodman, Agresti, etc.) and comes from the study by Srole et al. (1962). The data cross-classifies 1,660 young New York residents with respect to two factors: mental health and parents' socioeconomic status (SES).

The mental health factor is classified by four categories: Well, Mild symptom formation, Moderate symptom formation, and Impaired. The parents' SES has six categories ranging from High (1) to Low (6).

Here is the table:

	Well	Mild	Moderate	Impaired
1 (High)	64	94	58	46
2	57	94	54	40
3	57	105	65	60
4	72	141	77	94
5	36	97	54	78
6 (Low)	21	71	54	71

(a) Test the hypothesis that mental health and parents' SES are independent factors at a 5% confidence level. You can use ◀ tablerxc.m code.

(b) The table below provides the expected frequencies. Explain what the expected frequencies are. Explain how the number **104.0807** from the table was obtained, and show your work.

	Well	Mild	Moderate	Impaired
1 (High)	48.4542	95.0145	57.1349	61.3964
2	45.3102	88.8494	53.4277	57.4127
3	53.0777	**104.0807**	62.5867	67.2548
4	71.0169	139.2578	83.7398	89.9855
5	49.0090	96.1024	57.7892	62.0994
6 (Low)	40.1319	78.6952	47.3217	50.8512

12.15. **Tonsillectomy and Hodgkin's Disease.** A study by Johnson and Johnson (1972) involved 85 patients with Hodgkin's disease. Each of these had a healthy sibling. In 26 of these pairs, both individuals had had tonsillectomies (T); in 37 pairs, both individuals had not had tonsillectomies (N); in 15 pairs, only the healthy individual had had a tonsillectomy; in 7 pairs, only the one with Hodgkin's disease had had a tonsillectomy.

	Healthy/T	Healthy/N	Total
Patient/T	26	15	41
Patient/N	7	37	44
Total	33	52	85

A goal of the study was to determine whether there was a link between the disease and having had a tonsillectomy: is the proportion of those who had tonsillectomies the same among those with Hodgkin's disease as among those who do not have it? Test at $\alpha = 5\%$,

12.16. **School Spirit at Duke.** Duke has always been known for its great school spirit and support of its athletic teams, as evidenced by the famous Cameron Crazies. One way that school enthusiasm is shown is by donning Duke paraphernalia including shirts, hats, shorts, and sweatshirts. A project in an introductory statistics class was to explore possible links between school spirit, measured by the number of students wearing paraphernalia, and some other attributes. It was hypothesized that men would wear Duke clothes more frequently than women. The data were collected on the Bryan Center walkway starting at a random hour on five different days. Each day 100 men and 100 women were tallied, with results shown in the table below:

	Duke paraphenalia	No Duke paraphenalia	Total
Male	131	369	500
Female	52	448	500
Total	183	817	1000

Test the hypothesis that population of male and female Duke students are homogeneous with respect to wearing Duke paraphernalia?

12.17. **Two Halloween Questions with Easy Answers.** A study was designed to test whether or not aggression is a function of anonymity. The study was conducted as a field experiment on Halloween (Fraser, 1974); 300 children were observed unobtrusively as they made their rounds. Of these 300 children, 173 wore masks that completely covered their faces, while 127 wore no masks. It was found that 101 children in the masked group displayed aggressive or antisocial behavior versus 36 children in the unmasked group.

(a) Are anonymity and aggression independent? Use $\alpha = 0.01$.

(b) If p_1 is the (population) proportion for aggressive behavior for subjects wearing a mask and p_2 is the proportion for subjects not wearing a mask, find a 95% confidence interval for the odds ratio:

$$\frac{p_1/(1 - p_1)}{p_2/(1 - p_2)}.$$

12.18. **Runners and Heart Attack.** The influence of running on preventing heart attacks has been studied by a local runners club. The following two-way table classifies 350 people as runners or nonrunners, and whether or not they have had a heart attack. The factors are runner status and history of heart attack.

	Heart attack	No heart attack	Total
Runner	12	112	124
Nonrunner	36	190	226
Total	48	302	350

(a) Test for the independence of factors. Use $\alpha = 0.05$.

(b) Explain in words (in terms of this problem) what constitutes errors of the first and second kind in the testing.

12.19. **Perceptions of Dangers of Smoking.** A poll was conducted to determine if perceptions of the hazards of smoking were dependent on whether or not the person smoked. One hundred smokers and one hundred nonsmokers were randomly selected and surveyed. The results are given below:

Smoking is:	Very dangerous	Dangerous	Somewhat dangerous	Not dangerous
Smokers	21 (35.5)	29 (30)	29 ( )	21 ( )
Nonsmokers	50 (35.5)	31 ( )	11 ( )	8 ( )

Test the hypothesis that perception of the dangers of smoking is homogeneous with respect to smoking status at $\alpha = 0.05$. Three theoretical frequencies in parentheses are already calculated.

12.20. **Red Dye No. 2.** Fienberg (1980) discusses an experiment in which the food additive Red Dye No. 2 was fed to two groups of rats at various dosages. Some rats died during the experiment, which lasted 131 weeks, and the remainder were sacrificed at the end of the 131st week. All rats were examined for tumors.

Age of death	0–131 weeks		Terminal sacrifice	
Dosage	Low	High	Low	High
Tumor present	4	7	0	7
Tumor absent	26	16	14	14

Test the mutual independence of the three factors. Report the p-value and discuss.

12.21. **Cyclaneusma Needlecast.** Cyclaneusma needlecast is a plant disease that affect pines. The affected trees have yellow needles that later become more of a tan color, with darker brown transverse bands developing on the needle surfaces. It is difficult to control this disease because spore production and infection can take place whenever temperatures are above freezing and needles are wet.

Data in the table below describe an assessment of the outbreak of needlecast disease in two geographic locations of the western Italian Alps in September 2010.

The affected species were Scots, Austrian White, and Swiss Mountain pines, which are randomly sampled from the two locations.

Cyclaneusma needlecast disease data					
	Location A		Location B		
	Needlecast disease		Needlecast disease		
Type of pine	Yes	No	Yes	No	Totals
Scots	4	21	3	27	55
Austrian White	17	40	14	67	138
Swiss Mountain	61	43	70	52	226

(a) Are Species, Locations, and Disease mutually independent?
(b) Is Location independent from (Disease, Species)?
(c) Is Disease independent from Location given the Species?

12.22. **Leukoplakia.** Data (Table 12.3) from Hamerle and Tutz (1980) come from a study on leukoplakia, which is a clinical term used to describe patches of keratosis visible as adherent white patches on the membranes of the oral cavity. Although most leukoplakia patches are benign and considered not dangerous, sometimes they are coexistent with oral cancer. Often cancers on the floor of the mouth, beneath the tongue, occur next to areas of leukoplakia.

The objective is to explore the association between the disease and smoking. The data on this association are stratified by alcohol consumption level, also considered to be a risk factor.

		Leukoplakia	
		Yes	No
Alcohol	Smoker		
No	Yes	26	10
	No	8	8
$(0g, 40g]$	Yes	38	8
	No	43	24
$(40g, 80g]$	Yes	4	1
	No	14	17
$> 80g$	Yes	1	0
	No	3	7

Table 12.3 Contingency table for oral leukoplakia.

(a) Using Haenszel–Mantel procedure test the hypothesis that smoking and the disease are associated. Use $\alpha = 0.05$.
(b) Aggregate over alcohol consumption levels into a single 2×2 table on smoking versus disease status. Does the test for association agree with the decision from (a)?

12.23. **Thromboembolism, Smoking, and Contraceptive Use.** The data presented in the table below were first analyzed by Worcester (1971) and then re-analyzed by Bishop et al. (1975, p. 112). It was also discussed in Example 12.3 in an aggregate form. The first response variable is the presence or absence of thromboembolism. Variables 2 and 3 are both stimuli variables, being the use or nonuse of oral contraception and smoking or nonsmoking, respectively.

Thromboembolism data from Worcester (1971)					
	Smoker		Nonsmoker		
	Contraceptive user		Contraceptive user		
Type of patient	Yes	No	Yes	No	Totals
Thromboembolism	14	7	12	25	58
Control	2	22	8	84	116

(a) Identify the three factors and test the hypothesis that they are mutually independent.
(b) If the hypothesis in (a) is rejected, test the hypothesis that the factor Type of patient is independent of the pair of factors Smoker and Contraceptive use.
(c) Using Mantel–Haenszel methodology, test the hypothesis that the probabilities of thromboembolism are equal for the contraceptive users or non-users by accounting for the smoking status.

12.24. **ADA Polymorphism, Age and Gender.** ADA (Adenosine Deaminase) is a gene that plays a critical role in the immune system. ADA controls the cellular level of adenosine, directly affecting immune response as well as metabolic rate. ADA*1 is the dominant form of the gene, and ADA*2 homozygous causes death in early embryonic stages. However, if individuals are heterozygous with ADA*2, they have higher levels of circulating adenosine, which can have a number of repercussions. This polymorphism can be looked into further by identifying risk factors. The population proportion of homozygous ADA*1 is 0.8575

Napolioni and Lucarini (2010) report data on 884 unrelated healthy individuals participating in a study, looking for genetic frequencies.

Age	Males	
	ADA*1/1	ADA*1/2
<71	115	16
80–85	111	32
86–90	58	6
>90	59	3

Age	Females	
	ADA*1/1	ADA*1/2
<71	109	32
80-85	120	17
86-90	82	16
>90	91	17

(a) Determine if the genotype is independent from age grouping for both male and female. Conduct separate tests at $\alpha = 0.05$.

(b) Recode the data from both tables so you can check if the proportion of homozygous subjects is homogeneous for males and females, stratified by the age level. Use Mantel–Haenszel's test at the 5% level.

(c) Aggregate observed counts in (b) across the age groups to determine if gender and genotype are independent.

12.25. **Meta Analysis of Amantadine for Treatment of Influenza.** The flu (influenza) can be caused by many different viruses. The drug amantadine is FDA-approved for treatment and chemoprophylaxis of the influenza A virus infections among adults and children aged one year and older. Data on five randomized controlled trials of amantadine for preventing influenza (Table 12.4) and analysis of its efficacy were presented by Jefferson et al. (2002). The outcome is presence/absence of influenza-like-lllness (ILI).

For us, it is of interest to explore the dependence of factors Treatment (levels: Drug/Placebo) and ILI Status (levels: ILI/No ILI). Each of the

Trial	Drug (n/N)	Placebo (n/N)
Calmander (1968)	33/47	31/47
Oker-Blom (1970)	62/141	88/152
Schapira (1971)	49/157	39/140
Pettersson (1980)	66/95	69/97
Quarless (1981)	42/107	44/99

Table 12.4 Five trials of amantadine for prevention of influenza. Outcome is the proportion of influenca-like-illnesses (ILI), n cases out of N subjects.

five studies can be presented in a form of 2×2 contingency table. For example, Oker-Blom (1970) data can be tabulated as

	Drug	Placebo	Total
ILI	62	88	150
No ILI	79	64	143
Total	141	152	293

(a) For Oker-Blom data, test independence of factors Treatment and ILI Status using the χ^2-test. You can use `tablerxc.m` and report the p-value. What is your conclusion?

(b) Repeat the analysis in (a) for the remaining 4 studies. For which study the p-value is the smallest.

(c) Sum up the 5 tables into a single table. Conduct the analysis the same as in (a). Are the factors significantly dependent?

(d) Conduct simultaneous analysis of all 5 tables using Mantel–Haenszel test. Report the p-value.

(e) In one paragraph, summarize your findings in (a)–(d).

12.26. **UPSIT.** Le (2009) provides data on a matched case-control study conducted in order to evaluate the cumulative effects of acrylate and methacrylate on olfactory function. Cases were defined as subjects scoring below the 10th percentile on the UPSIT (University of Pennsylvania Smell Identification Test).

	Cases	
Controls	Exposed	Unexposed
Exposed	25	22
Unexposed	9	21

(a) Test to compare the cases versus controls. State your hypotheses clearly and conduct the test. Use $\alpha = 0.05$.

(b) In the table above there are 154 subjects in total, 77 cases and 77 controls. A tabulation that ignores pairing/matching is

	Cases	Controls	Total
Exposed	34	47	81
Unexposed	43	30	73
Total	77	77	154

Repeat the test from (a), without the information on matching. Compare the results and discuss.

12.27. **Mycosis Fungoides and Cutting Oils.** In a case-control study reported by Cohen (1977) researchers wanted to examine the association between mycosis fungoides, a type of lymphoma that begins in the skin and eventually spreads to internal organs, and a history of employment in an industrial environment with exposure to cutting oils, a risk factor. After matching 54 subjects with the disease (Cases) with 54 subjects without the disease (Controls), the history of exposure to risk factors was investigated.

		Controls		
		History of exposure	No history of exposure	Total
	History of exposure	16	13	29
Cases	No history of exposure	3	22	25
	Total	19	35	54

(a) The Cases seem to be more likely to be exposed to the risk factor than the Controls. Explore if this is statistically significant, at the 5% level.
(b) Fully "parallelize" the table, that is, find b, c, and d:

	Cases	Controls
History of exposure	$16 + 13$	b
No history of exposure	c	d

For the table in this form explore if the factors Mycosis and Exposure are significantly dependent. Use $\alpha = 0.05$.

12.28. **H. pylori and ELISA Revisited.** In Exercise 4.16 the researchers also compared performance of oral fluid ELISA (OD=0.3) and serum ELISA in the cohort of 81 subjects. We are interested in comparing the specificities of the two tests. From the information given in Exercise 4.16, and the fact that 3 control subjects tested positive with both ELISA tests, distill a single paired table:

		Oral fluid ELISA		
		$-$	$+$	Total
Serum ELISA	$-$			
	$+$		3	
	Total			47

Because we are interested in the specificities, note that this table involves only 47 control subjects. Using ◀ mcnemart.m compare the population specificities of the two tests.

12.29. **Hepatic Arterial Infusion.** Allen-Mersh et al. (1994), as well as Spiegel-halter et al. (2004), reported the results of a trial in which patients under-going therapy for liver metastasis were randomized to receive it either systematically (as is standardly done) or via hepatic arterial infusion (HAI). Of 51 randomized to HAI, 44 died, and of 49 randomized to sys-tematic therapy, 46 died. Estimate the log odds ratio and find a 95% confidence interval.

12.30. **Vaccine Efficacy Study.** Consider the data from a vaccine efficacy study (Chan, 1998). In a randomized clinical trial of 30 subjects, 15 were inocu-lated with a recombinant DNA influenza vaccine, and the remaining 15 were inoculated with a placebo. Twelve of the 15 subjects in the placebo group (80%) eventually became infected with influenza, whereas for the vaccine group, only 7 of the 15 subjects (47%) became infected. Suppose that p_1 is the probability of infection for the vaccine group and p_2 is the probability of infection for the placebo group.
What is the one-sided p-value for testing the null hypothesis $H_0 : p_1 = p_2$ obtained by Fisher's exact test?

12.31. **Marriages in Surinam.** Speckmann (1965) and Lawal (2003) provide data on the religious affiliations of husbands and wives in 264 marriages in Surinam.

		Wife		
	Christian	Muslim	Hindu	Total
Christian	17	1	7	25
Husband Muslim	1	66	6	73
Hindu	9	6	151	166
Total	27	73	164	264

Here Hindu counts combine Sanatan Dharm and Arya Samaj affiliations. Are the population proportions of husbands who are Christian, Mus-lim, or Hindu equal to the proportions of wives who belong to those three religious groups? Note that the data are paired and that the Stuart–Maxwell test is appropriate.

12.32. **CYP1A1 Polymorphysm and Risk of Leukemia.** Researchers (Ma et al, 2002) were interested in the association between the genotypic frequen-cies of the cytochrome P450 1A1 polymorphism and the risk of child-hood leukemia. Data were collected as part of the Northern California Childhood Leukemia Study. Matched case-control observations were ab-stracted from a large number of genetic/environmental variables that potentially influence the risk of childhood leukemia. The cases are chil-dren ages 0 to 14 years old with newly diagnosed leukemia (1995 to 1999) obtained from major hospitals in the San Francisco Bay Area. Compari-son with California State Cancer Registry data shows that > 90% of the eligible children were ascertained. The control children were randomly selected from birth certificate records and matched to cases with respect to sex, age, race, and county of birth. The CYP1A1/leukemia data consisting of 175 matched pairs of acute lymphoblastic leukemia cases and their controls (117 concordant and 58 discordant pairs) are given in Table 12.5.

	Control: AA*	Control: AG	Control: GG	Total
Case: AA*	103	26	2	131
Case: AG	23	14	2	39
Case: GG	1	4	0	5
Total	127	44	4	175

Table 12.5 The observed numbers of matched pairs by case-control status and CYP1A1 genotypes

(a) Using Stuart–Maxwell and Bhapkar procedures, test the marginal homogeneity of genotypes for case and control children.
(b) Using Bowker's procedure, test for the symmetry of the table. Use $\alpha = 0.05$.

12.33. **Left-/Right-Side Tooth Decay.** Tomizawa et al. (2006) give tabular data on number of decayed teeth in 349 men aged 18–39. Counts are made on left and right side of mouth in each subject. The data were collected in a dental clinic in Sapporo City, Japan, from 2001 to 2005.

		Right		
	0–4	5–8	9+	Total
0–4	118	37	2	157
Left 5–8	21	87	23	131
9+	2	11	48	61
Total	141	135	73	349

Using Stuart–Maxwell and Bhapkar procedure test the marginal homogeneity. Also test for the symmetry using Bowker procedure. Use $\alpha = 0.05$.

MATLAB FILES USED IN THIS CHAPTER

http://statbook.gatech.edu/Ch12.Tables/

annoyance.m, anole.m, berkeleyadmissions.m, bowker.m, cochranappl.m,
cochrant.m, cohen.m, concoef.m, conditionalrxcxp.m, fisherexact.m,
fisherexactt.m, galtonalw.m, hindu.m, jointrxcxp.m, kappa1.m,
liddell.m, mantelhaenszel.m, mcnemart.m, myfisher33.m, oesophageal.m,
partialrxcxp.m, psychosis.m, PAS.m, stuartmaxwell.m, table2x2.m,
tablerxc.m, tablerxcxp.m, unmatch.m

CHAPTER REFERENCES

Agresti A. (1992). A survey of exact inference for contingency tables. *Stat. Sci.*, **7**, 1, 131–153.

Agresti, A. (2002). *Categorical Data Analysis*, 2nd ed. Wiley, New York.

Allen-Mersh T. G., Earlam, S., Fordy, C., Abrams, K., and Houghton, J. (1994). Quality of life and survival with continuous hepatic-artery floxuridine infusion for colorectal liver metastases. *Lancet*, **344**, 8932, 1255–1260.

Armitage, P. and Berry, G. (1994). *Stat. Meth. Med. Res.*, 3rd ed. Blackwell, Oxford.

Berger, R. and Sidik, K, (2002). Exact unconditional tests for a 2 x 2 matched-pairs design. *Statistical Methods in Medical Research*, **12**, 2, 91–108.

Bhapkar V. P. (1966). A note on the equivalence of two test criteria for hypotheses in categorical data. *J. Amer. Stat. Assoc.*, **61**, 228–235.

Bickel, P. J., Hammel, E. A., and O'Connell, J. W. (1975). Sex bias in graduate admissions: data from Berkeley. *Science, New Series*, **187**, 4175, 398–404.

Bishop, Y. M. M., Fienberg, S. E., and Holand, P. W. (1975). *Discrete Multivariate Analysis: Theory and Practice*. MIT Press, Cambridge, MA.

Chan, I. (1998). Exact tests of equivalence and efficacy with a non-zero lower bound for comparative studies. *Stat. Med.*, **17**, 1403–1413.

Cohen, J. (1960). A coefficient of agreement for nominal scales. *Educ. Psychol. Meas.*, **20**, 37–46.

Cohen, J. E. (1973). Independence of Amoebas. In *Statistics by Example: Weighing Chances*, edited by F. Mosteller, R. S. Pieters, W. H. Kruskal, G. R. Rising, and R. F. Link, with the assistance of R. Carlson and M. Zelinka. Addison-Wesley, Reading, MA.

Cohen, S. R. (1977). Mycosis fungoides: clinicopathologic relationships, survival, and therapy in 54 patients, with observation on occupation as a new prognostic factor. MS Thesis at Yale University School of Medicine, New Haven, CT.

Correa, P., Pickle, L. W., Fontham, E., Lin, Y., and Haenszel, W. (1983). Passive smoking and lung cancer. *Lancet*, **2**, 8350, 595–597.

Darley, J. M. and Latané, B. (1968). Bystander intervention in emergencies: diffusion of responsibility. *J. Personal. Soc. Psychol.*, **8**, 4, 377–383.

Everitt, B. S. (1977). *The Analysis of Contingency Tables*, Chapman & Hall, London.

Fienberg, S. E. (1970). The analysis of multidimensional contingency tables. *Ecology*, **51**, 419–433.

Fienberg, S. E. (1980). *The Analysis of Cross-classified Categorical Data*. MIT Press, Cambridge, MA.

Fisher, R . A. (1935). *The Design of Experiments* (8th ed. 1966). Oliver and Boyd, Edinburgh.

Forss, H. and Halme, E. (1998). Retention of a glass ionomer cement and a resin-based fissure sealant and effect on carious outcome after 7 years. *Community Dent. Oral Epidemiol.*, **26**, 1, 21–25.

Fraser, S. C. (1974). Deindividuation: effects of anonymity on aggression in children. Unpublished technical report. University of Southern California, Los Angeles.

Freeman, G. H. and Halton, J. H. (1951). Note on an exact treatment of contingency goodness-of-fit and other problems of significance. *Biometrika*, **38**, 141–149.

Galton, F. (1892). *Finger Prints*, Macmillan, London.

Gart, J. J. (1969). An exact test for comparing matched proportions in cross-over designs. *Biometrika*, **56**, 75–80.

Généreux, P., Cohen, D. J., Williams, M. R., Mack, M., Kodali, S. K., Svensson, L. G., Kirtane, A. J., Xu, K., McAndrew, T. C., Makkar, R., Smith, C. R., and Leon, M. B. (2014). Bleeding complications after surgical aortic valve replacement compared with transcatheter aortic valve replacement: insights from the PARTNER I Trial (Placement of Aortic Transcatheter Valve). *J. Amer. Coll. Cardiol.*, **63**, 11, 1100–1109.

Holmes, M. C., and Williams, R. E. O. (1954). The distribution of carriers of *Streptococcus pyogenes* among 2413 healthy children. *J. Hyg. Camb.*, **52**, 165–179.

Jefferson, T. O., Demicheli, V., Deeks, J. J., and Rivetti, D. (2002). Amantadine and rimantadine for preventing and treating influenza A in adults. *Cochrane Database Syst. Rev.* 2002 (4): CD001169. [PubMed]

Johnson, S. and Johnson, R. (1972). Tonsillectomy history in Hodgkin's disease. *New Engl. J. Med.*, **287**, 1122–1125.

Hamerle, A. K. P. and Tutz, G. (1980). Kategoriale Reaktionen in multifaktoriellen Versuchsplanen und mehrdimensionale Zusammenhangsanalysen. *Arch Psychol. (Frankf)*, 53–68.

Hersen, M. (1971). Personality characteristics of nightmare sufferers. *J. Nerv. Ment. Dis.*, **153**, 29–31.

Kirkwood, B. and Sterne, J. (2003). *Essential Medical Statistics*, 2nd ed. Blackwell, Oxford.

Krampe, A. and Kuhnt, S. (2007). Bowker's test for symmetry and modifications within algebraic framework. *Comput. Stat. Data Anal.*, **51**, 9, 4124–4142.

Kuhnt, S., Schürmann, C., and Griefahn, B. (2004). Annoyance from multiple transportation noise: Statistical models and outlier detection. *Meth. Inform. Med.*, **5**, 510–515.

Landis. J. R., Sharp, T. J., Kuritz, S. J., and Koch, G. G. (1998). Mantel-Haenszel methods. In *Encyclopedia of Biostatistics*. Wiley, New York.

Lawal, H. B. (2003). *Categorical Data Analysis with SAS and SPSS Applications*. Erlbaum, Mahwah.

Le, C. (2009). *Health and Numbers: A Problems-Based Introduction to Biostatistics*, 3rd ed. Wiley, Hoboken, NJ.

Liddell, F. D. K. (1983). Simplified exact analysis of case-referent studies; matched pairs; dichotomous exposure. *J. Epidemiol. Commun. Health*, **37**, 82–84.

Ma, X., Buffler. P. A., Selvin, S., Matthay, K. K., Wiencke, J. L., and Reynolds, P. (2002). Daycare attendance and risk of childhood acute lymphoblastic leukemia. *Br. J. Cancer*, **86**, 1419–1424.

Mainar-Jaime, R. C., Atashparvar, N., Chirino-Trejo, M., and Blasco, J. M. (2008). Accuracy of two commercial enzyme-linked immunosorbent assays for the detection of antibodies to *Salmonella spp.* in slaughter pigs from Canada. *Prevent. Vet. Med.*, **85**, 41–51.

Mantel, N. and Haenszel, W. (1959). Statistical aspects of the analysis of data from retro-spective studies of disease. *J. Natl. Cancer Inst.*, **22**, 719–748.

Maxwell, A. E. (1970). Comparing the classification of subjects by two independent judges. *Br. J. Psychiatry*, **116**, 651–655.

May, W. L. and Johnson, W. D. (2001). Symmetry in square contingency tables: tests of hypotheses and confidence interval construction. *J. Biopharm. Stat.*, **11**, 23–33.

Miettinen, O. S. (1976). Estimability and estimation in case-referent studies. *Am. J. Epidemiol.*, **103**, 226–235.

Miettinen, O. S. (1968). The matched pairs design in the case of all-or-none responses. *Biometrics*, **24**, 339–352.

Napolioni, V. and Lucarini, N. (2010). Gender-specific association of ADA genetic poly-morphism with human longevity. *Biogerontology*, **11**, 4, 457–462.

Pearson, K. (1904). On the theory of contingency and its relation to association and normal correlation. In *Draper's Company Research Memoirs, Biometric Series I*, Dulau, London.

Roberts, E., Dawson, W. M., and Madden, M. (1939). Observed and theoretical ratios in Mendelian inheritance. *Biometrika*, **31**, 56–66.

Robins, J., Breslow, N., and Greenland, S. (1986). Estimators of the Mantel–Haenszel variance consistent in both sparse data and large-strata limiting models. *Biometrics*, **42**, 311–323.

Schoener, T. W. (1968). The Anolis lizards of Bimini: resource partitioning in a complex fauna. *Ecology*, **48**, 704–726.

Shapiro, S., Slone, D., Rosenberg, L., Kaufmann, D. W., Stolley, P. D., and Mietinnen, O. S. (1979). Oral-contraceptive use in relation to myocardial infarction. *Lancet*, **313**, 8119, 743–747.

Sleigh, A., Hoff, R., Mott, K., Barreto, M., de Paiva, T., Pedrosa, J., and Sherlock, I. (1982). Comparison of filtration staining (Bell) and thick smear (Kato) for the detection and quantitation of Schistosoma mansoni eggs in faeces. *Trop. Med. Hyg.*, **76**, 3, 403–406.

Speckmann, J. D. (1965). Marriage and kinship among the Indians in Surinam. *Assen: VanGorgum*, 32–33.

Spiegelhalter, D. J., Abrams, K. R., and Myles, J. P. (2004). *Bayesian Approaches to Clinical Trials and Health Care Evaluation.* Wiley, New York.

Srole, L., Langner, T., Michael, S., Opler, M., and Rennie, T. (1962). *Mental Health in the Metropolis: The Midtown Manhattan Study*, vol. 1. McGraw-Hill, New York.

Streissguth, A. P., Martin, D. C., Barr, H. M., Sandman, B. M., Kirchner, G. L., and Darby, B. L. (1984). Intrauterine alcohol and nicotine exposure: attention and reaction time in 4-year-old children. *Dev. Psychol.*, **20**, 533–541.

Stuart, A. (1953). The estimation and comparison of strengths of association in contin-gency tables. *Biometrika*, **40**, 105–110.

Tomizawa, S., Miyamoto, N., and Iwamoto, M. (2006). Linear column-parameter symme-try model for square contingency tables: application to decayed teeth data. *Biom. Lett.*, **43**, 91–98.

Williams, J. W. (1921). A critical analysis of twenty-one years' experience with cesarean section. *Johns Hopkins Hosp. Bull.*, **32**, 173–184.

Woo, T. L. (1928). Dextrality and sinistrality of hand and eye: second memoir. *Biometrika*, **20A**, 1/2, 79–148.

Worchester, J. (1971). The relative odds in 2^3 contingency table. *Am. J. Epidemiol.*, **93**, 145–149.

Yates, F. (1934). Contingency table involving small numbers and the χ^2 test. Supplement to *J. Roy. Stat. Soc.*, **1**, 2, 217–235.

Zar, J. H. (2010). *Biostatistical Analysis*, 5th ed. Pearson-Prentice-Hall, Englewood Cliffs, NJ.

Chapter 13
Correlation

The invalid assumption that correlation implies cause is probably among the two or three most serious and common errors of human reasoning.

– Stephen Jay Gould

WHAT IS COVERED IN THIS CHAPTER

- Calculating the Pearson Coefficient of Correlation, Conditional Correlation
- Testing the Null Hypothesis of No Correlation
- General Test for Correlation and Confidence Intervals, Fisher z-Transformation
- Inference for Two or More Correlation Coefficients
- Nonparametric Correlation Measures: Spearman's and Kendall's Correlation Coefficient

13.1 Introduction

Collins English Dictionary[1] defines *correlation* as a *mutual or reciprocal relationship between two or more things, the act or process of correlating or the state*

---

[1] correlation. (n.d.). Collins English Dictionary - Complete & Unabridged 10th Edition. Retrieved April 16, 2017 from Dictionary.com website http://www.dictionary.com/browse/correlation

of being correlated, and the extent of correspondence between the ordering of two variables.

Statistically, correlation is a measure of the particular affinity between two sets of comparable measurements. It is often incorrectly believed that the notions of correlation and statistical dependence and causality coincide. According to an anecdote, the number of drownings at a particular large beach in one season could be positively and significantly correlated with the number of ice-creams sold at the beach during the same period of time. Of course, nobody would argue that the relationship is causal. Purchasing an ice-cream at the beach does not increase the risk of drowning; the positive correlation is caused by a latent or lurking variable, the number of visitors at the beach.

> Correlation is, in informal terms, a constrained dependence. For example, the common measure of correlation, the Pearson coefficient of correlation, quantifies the strength and direction of the *linear* relationship between two variables.

If measurements are correlated, then they are dependent, but not necessarily vice versa. A simple example involves points on a unit circle. One selects n angles φ_i, $i = 1,\ldots,n$, uniformly from $[0,2\pi]$. The angles determine n points on the unit circle, $(x_i,y_i) = (\cos\varphi_i,\sin\varphi_i)$, $i = 1,\ldots,n$. Although $x = (x_1,\ldots,x_n)$ and $y = (y_1,\ldots,y_n)$ are functionally dependent, since $x_i^2 + y_i^2 = 1$, their coefficient of correlation is 0 or very close to 0. See also Exercise 13.1. Only for a normal distribution do the notions of correlation and independence coincide: uncorrelated normally distributed measurements are independent.

13.2 The Pearson Coefficient of Correlation

In Chapter 2, where we discussed the summaries of multidimensional data we also mentioned correlations between the variables and connected the coefficient of correlation with scatterplots and a descriptive statistical methodology. In this chapter we will take an inferential point of view on correlations. For developing the tests and confidence intervals we will assume that the data come from normal distributions for which the population coefficient of correlation is ρ.

Suppose that pairs $(X_1,Y_1),(X_2,Y_2),\ldots,(X_n,Y_n)$ are observed, and assume that Xs and Ys come from normal $\mathcal{N}(0,\sigma_X^2)$ and $\mathcal{N}(0,\sigma_Y^2)$ distributions and that the correlation coefficient between X and Y is ρ_{XY}, or simply ρ. We are interested in making an inference about ρ when n pairs (X_i,Y_i) are observed.

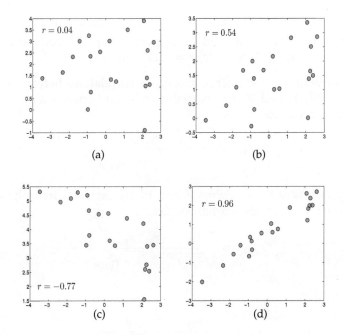

Fig. 13.1 A simulated example showing scatterplots with correlated components (a) $r = 0.04$, (b) $r = 0.54$, (c) $r = -0.77$, and (d) $r = 0.96$.

From the observations, the following sums of squares are formed: $S_{xx} = \sum_{i=1}^{n}(X_i - \overline{X})^2$, $S_{yy} = \sum_{i=1}^{n}(Y_i - \overline{Y})^2$, and $S_{xy} = \sum_{i=1}^{n}(X_i - \overline{X})(Y_i - \overline{Y})$. Then, an estimator of ρ is

$$r = \frac{S_{xy}}{\sqrt{S_{xx}\, S_{yy}}}.$$

An alternative expression for r is

$$r = \frac{\sum_i X_i Y_i - n\overline{X}\,\overline{Y}}{\sqrt{\left(\sum_i X_i^2 - n(\overline{X})^2\right)\left(\sum_i Y_i^2 - n(\overline{Y})^2\right)}}.$$

This estimator is an MLE for ρ, and it is asymptotically unbiased, that is, $\lim_{n\to\infty} \mathbb{E} r_n = \rho$.

Sample correlation coefficient r is always in $[-1,1]$. To see this, consider the Cauchy–Schwartz inequality that states: For any two vectors $(u_1, u_2, \ldots, u_n)$ and $(v_1, v_2, \ldots, v_n)$, it holds that $(\sum_i u_i v_i)^2 \le \sum_i u_i^2 \sum_i v_i^2$.

Substitutions $u_i = X_i - \overline{X}$ and $v_i = Y_i - \overline{Y}$ prove the inequality $r^2 \leq 1$. Figure 13.1 shows simulated correlated data (green dots) with correlation coefficients (a) $r = 0.04$, (b) $r = 0.54$, (c) $r = -0.77$, and (d) $r = 0.96$.

Partial Correlation. Let r_{xy}, r_{xz}, and r_{yz} be correlations between the pairs of variables X, Y, and Z. The partial correlation between X and Y when effect of Z is accounted for is

$$r_{xy.z} = \frac{r_{xy} - r_{xz}r_{yz}}{\sqrt{1 - r_{xz}^2}\sqrt{1 - r_{yz}^2}}.$$

When the effects of two variables Z and W are excluded from the correlation between X and Y, the partial coefficient is

$$r_{xy.zw} = \frac{r_{xy.w} - r_{xz.w}r_{yz.w}}{\sqrt{1 - r_{xz.w}^2}\sqrt{1 - r_{yz.w}^2}}$$

or, equivalently,

$$r_{xy.zw} = \frac{r_{xy.z} - r_{xw.z}r_{yw.z}}{\sqrt{1 - r_{xw.z}^2}\sqrt{1 - r_{yw.z}^2}},$$

depending on the order of exclusion of W and Z.

Remark. The partial coefficient of correlation $r_{xy.z}$ referred to a correlation between X and Y when the effects of Z are removed from *both* X and Y. If we wanted the correlation between X and Y when the effect of Z is removed only from Y, then we should use semi-partial correlation,

$$r_{x(y.z)} = \frac{r_{xy} - r_{xz}r_{yz}}{\sqrt{1 - r_{yz}^2}} = r_{xy.z}\sqrt{1 - r_{xz}^2}.$$

Semi-partial correlations are utilized in multiple regressions for variable selection.

13.2.1 Inference About ρ

Assume that pairs $(X_i, Y_i), i = 1, \ldots, n$ are sampled from a bivariate normal distribution $\mathcal{N}_2(\mu_X, \mu_Y, \sigma_X^2, \sigma_Y^2, \rho)$ with density as in (6.1). We refer to ρ as the population coefficient of correlation. To test $H_0 : \rho = 0$ against one of the alternatives $\rho >, \neq, < 0$, the t-statistic is used:

$$t = r\sqrt{\frac{n-2}{1-r^2}},$$

which has a t-distribution with $df = n - 2$ degrees of freedom.

Alternative	α-level rejection region	p-value
$H_1 : \rho > 0$	$[t_{df,1-\alpha}, \infty)$	1-tcdf(t, df)
$H_1 : \rho \neq 0$	$(-\infty, t_{df,\alpha/2}] \cup [t_{df,1-\alpha/2}, \infty)$	2*tcdf(-abs(t), df)
$H_1 : \rho < 0$	$(-\infty, t_{df,\alpha}]$	tcdf(t, df)

Example 13.1. **Rats and Maze.** The table below gives the number of times X rats ran through a maze and the time Y it took them to run through the maze on their last trial.

Rat	Trials (X)	Time (Y)
1	8	10.9
2	9	8.6
3	6	11.4
4	5	13.6
5	3	10.3
6	6	11.7
7	3	10.7
8	2	14.8

(a) Find r.

(b) Is the maze learning significant? Test the hypothesis that the population correlation coefficient ρ is 0, versus the alternative that ρ is negative.
$\sum_i X_i = 42$, $\sum_i X_i^2 = 264$, $\sum_i Y_i = 92$, $\sum_i Y_i^2 = 1084.2$, $\sum_i X_i Y_i = 463.8$, $\overline{X} = 5.25$, $\overline{Y} = 11.5$.

$$r = \frac{463.8 - 8 \cdot 5.25 \cdot 11.5}{\sqrt{(264 - 8 \cdot 5.25^2)(1084.2 - 8 \cdot 11.5^2)}} = \frac{-19.2}{\sqrt{43.5 \cdot 26.2}} = -0.5687.$$

For testing $H_0 : \rho = 0$ versus $H_1 : \rho < 0$, we find

$$t = r\sqrt{\frac{n-2}{1-r^2}} = -1.6937.$$

For $\alpha = 0.05$ the critical value is $t_{6,0.05} = -1.9432$ and the null hypothesis is not rejected.

Remark. Sample size is critical for our decision. If for the same $r = -0.5687$, the sample size n were 30, then $t = -3.6588$, and H_0 would be rejected, since $t < t_{28,0.05} = -1.7011$.

In MATLAB:

```
X = [8 9 6 5 3 6 3 2];
Y = [10.9 8.6 11.4 13.6 10.3 11.7 10.7 14.8];
n=8;
cxy=cov(X,Y)
  % cxy =
  %       6.2143    -2.7429
  %      -2.7429     3.7429
rxy=cxy(1,2)/sqrt(cxy(1,1)*cxy(2,2))
  %rxy =    -0.5687
tstat = rxy * sqrt(n-2)/sqrt(1-rxy^2)
  %tstat =  -1.6937
pval = tcdf(tstat, n-2)
  %pval =     0.0706
  %n=8, p-val = 7% > 5% do not reject H_0
  %=============
n=30;
tstat = rxy * sqrt(n-2)/sqrt(1-rxy^2)
  %tstat =-3.6588
pval = tcdf(tstat, n-2)
  % pval = 5.2031e-04
```

When partial correlations are of interest, the test is based on the statistic $t = r\sqrt{\frac{n-q-2}{1-r^2}}$, which has $df = n - q - 2$ degrees of freedom. Here q is the number of other variables accounted for.

Example 13.2. **Mg-CaO Data.** In many ways magnesium (symbol Mg) is very similar to calcium (Ca), and determination of the quantity of one may be influenced by the presence of the other. Extremely small amounts of calcium, especially in the presence of much magnesium, as in magnesites and fused magnesia, cannot be determined satisfactorily by direct precipitation as an oxalate. Hazel and Eglof (1946) proposed a new method, denoted as method B. Its performance was compared to the traditional alcohol method (treatment with a mixture of ethyl alcohol, methyl alcohol and sulfuric acid), denoted here as method A.

The column y is obtained by using the method A, while the column z is obtained by a Hazel–Eglof method B. Column x is the exact amount of CaO present in MgO + CaO, India magnesite, and magnesite 104. Part of this data (hazel.dat|mat) was analyzed also by Youden (1951, p.44).

	x	y	z
	CaO Present	CaO Found Method A	CaO Found Method B
	4.0	3.7	3.9
	8.0	7.8	8.1
	8.5	7.7	8.9
	9.1	8.4	9.2
	9.7	9.4	10.0
	9.9	9.5	9.9
	12.5	12.1	12.4
	16.0	15.6	16.0
	20.0	19.8	19.8
	25.0	24.5	25.0
	26.8	26.1	27.1
	31.0	31.1	31.1
	36.0	35.5	35.8
	40.0	39.4	40.1
	40.0	39.5	40.1

Find and test the significance of $r_{yz.x}$.

```
%hazel.m
hazel=[...
  4.0    3.7     3.9
  8.0    7.8     8.1
...
 40.0   39.4    40.1
 40.0   39.5    40.1 ];
%or just load 'hazel.dat|mat'
x=hazel(:,1); y=hazel(:,2); z=hazel(:,3);   n=length(x);
mycor = @(x,y) (x(:)-mean(x(:)))'*(y(:)-mean(y(:)))/ ...
          sqrt( (x(:)-mean(x(:)))'*(x(:)-mean(x(:)) )*...
                (y(:)-mean(y(:)))'*(y(:)-mean(y(:)))));
rxy=mycor(x,y); rxz = mycor(x,z); ryz=mycor(y,z);
%    0.9998          0.9999          0.9996
ryz_x = (ryz - rxy * rxz)/sqrt((1-rxy^2)*(1-rxz^2)) %-0.3861
t = ryz_x * sqrt(n-2-1)/sqrt(1-ryz_x^2)             %-1.4498
pval = 2 * tcdf(-abs(t), n-2-1)                     % 0.1727
```

Thus, the extreme correlation between y and z (0.9996, p-value of about $1.6 \cdot 10^{-21}$) becomes insignificant (p-value of 0.1727), even negative, when variable x is taken into account. In the context of the problem, when accounted for the true content of CaO, the methods A and B are not significantly correlated.

13.2.1.1 Confidence Intervals for ρ

Confidence intervals for the population coefficient of correlation ρ are obtained not by using direct measurements, but in a transformed domain. This transformation is known as Fisher's z-transformation and is introduced next.

Let $(X_{11}, X_{21}), \ldots, (X_{1n}, X_{2n})$ be a sample from a bivariate normal distribution $\mathcal{N}_2(\mu_1, \mu_2, \sigma_1^2, \sigma_2^2, \rho)$, and $\overline{X}_i = \frac{1}{n} \sum_{j=1}^n X_{ij}$, $i = 1, 2$.

The Pearson coefficient of linear correlation

$$r = \frac{\sum_{i=1}^n (X_{1i} - \overline{X}_1)(X_{2i} - \overline{X}_2)}{\left[\sum_{i=1}^n (X_{1i} - \overline{X}_1)^2 \cdot \sum_{i=1}^n (X_{2i} - \overline{X}_2)^2 \right]^{1/2}}$$

has a complicated distribution involving special functions, (e.g., Anderson, 1984, page 113). However, it is well known that the asymptotic distribution for r is normal, $\mathcal{N}(\rho, \frac{(1-\rho^2)^2}{n})$. Since the variance is a function of the mean $[\sigma^2(\rho) = \frac{(1-\rho^2)^2}{n}$, see equation (6.4)], the transformation defined as

$$\vartheta(\rho) = \int \frac{c}{\sigma(\rho)} d\rho$$

$$= \int \frac{c\sqrt{n}}{1 - \rho^2} d\rho$$

$$= \frac{c\sqrt{n}}{2} \int \left(\frac{1}{1-\rho} + \frac{1}{1+\rho} \right) d\rho$$

$$= \frac{c\sqrt{n}}{2} \log \left(\frac{1+\rho}{1-\rho} \right) + k$$

stabilizes the variance. This is known as Fisher's z-transformation for the correlation coefficient. Typical choices for c and k are $c = 1/\sqrt{n}$ and $k = 0$.

Assume that r and ρ are mapped to w and ζ as

$$w = \frac{1}{2} \log \left(\frac{1+r}{1-r} \right) = \text{arctanh } r, \quad \zeta = \frac{1}{2} \log \left(\frac{1+\rho}{1-\rho} \right) = \text{arctanh } \rho.$$

The distribution of w is approximately normal, $\mathcal{N}(\zeta, \frac{1}{n-3})$, and this approximation is quite accurate when ρ^2/n^2 is small and n is as low as 20.

The inverse z-transformation is

$$r = \tanh(w) = \frac{\exp\{2w\} - 1}{\exp\{2w\} + 1}.$$

The use of Fisher's z-transformation is illustrated on finding the confidence intervals for ρ and testing hypotheses about ρ.

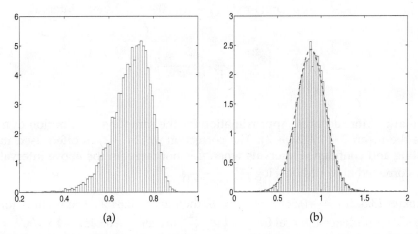

Fig. 13.2 (a) Simulational run of 10,000 rs from a bivariate population having a theoretical $\rho = \sqrt{2}/2$. (b) The same rs transformed into ws with the normal approximation superimposed.

To exemplify the above-stated transformations, we generated $n = 30$ pairs of normally distributed random samples with theoretical correlation $\sqrt{2}/2$. This was done by generating two i.i.d. normal samples a and b of length 30 and taking the transformation $x_1 = a + b$, $x_2 = b$. The sample correlation coefficient r was found. This was repeated $M = 10,000$ times. A histogram of 10,000 sample correlation coefficients is shown in Figure 13.2a. A histogram of the z-transformed rs is shown in Figure 13.2b with a superimposed normal approximation $\mathcal{N}(\text{arctanh}(\sqrt{2}/2), 1/(30 - 3))$.

The sampling distribution for w is approximately normal with mean $\zeta = \frac{1}{2}\log\frac{1+\rho}{1-\rho}$ and variance $1/(n-3)$. This approximation is satisfactory when the number of pairs exceeds 20. Thus, the $(1-\alpha)100\%$ confidence interval for $\zeta = \frac{1}{2}\log\frac{1+\rho}{1-\rho}$ is

$$[w_L, w_U] = \left[w - \frac{z_{1-\alpha/2}}{\sqrt{n-3}}, \ w + \frac{z_{1-\alpha/2}}{\sqrt{n-3}} \right],$$

where $w = \frac{1}{2} \log \frac{1+r}{1-r}$.

Since $r = \frac{e^{2w}-1}{e^{2w}+1}$, an approximate $(1-\alpha)100\%$ confidence interval for ρ is

$$[r_L, r_U] = \left[\frac{e^{2w_L}-1}{e^{2w_L}+1}, \ \frac{e^{2w_U}-1}{e^{2w_U}+1} \right].$$

Remark. More accurate approximation for the sampling distribution of w has the mean $\zeta + \rho/(2n-2)$. The correction $\rho/(2n-2)$ is often used in testing and confidence intervals when n is not large. In the above interval the corrected w would be $\frac{1}{2} \log \frac{1+r}{1-r} - \frac{r}{2(n-1)}$.

Example 13.3. If $r = -0.5687$ and $n = 8$, then $w = -0.6456$. The bounds for the 95% confidence interval for $\zeta = \frac{1}{2} \log \frac{1+\rho}{1-\rho}$ are $w_L = -0.6456 - 1.96/\sqrt{5} = -1.522$ and $w_U = -0.6456 + 1.96/\sqrt{5} = 0.2309$. The confidence interval for ρ is obtained by back-transforming w_L and w_U using $r = \frac{e^{2w}-1}{e^{2w}+1}$. The result is $[-0.9091, 0.2269]$.

🖉

13.2.1.2 Test for $\rho = \rho_0$

We saw that when $\rho_0 = 0$, this test is a t-test with a statistic given as $t = r\sqrt{\frac{n-2}{1-r^2}}$. When $\rho_0 \neq 0$, then the test of $H_0 : \rho = \rho_0$ versus $H_1 : \rho >, \neq, < \rho_0$ does not have a simple generalization. An approximate test is based on a normal approximation. Under $H_0 : \rho = \rho_0$, the test statistic

$$z = \frac{\sqrt{n-3}}{2} \left[\log \left(\frac{1+r}{1-r} \right) - \log \left(\frac{1+\rho_0}{1-\rho_0} \right) \right],$$

has an approximately standard normal distribution. This implies the following testing rules:

Alternative	α-level rejection region	p-value
$H_1 : \rho > \rho_0$	$[z_{1-\alpha}, \infty)$	`1-normcdf(z)`
$H_1 : \rho \neq \rho_0$	$(-\infty, z_{\alpha/2}] \cup [z_{1-\alpha/2}, \infty)$	`2*normcdf(-abs(z))`
$H_1 : \rho < \rho_0$	$(-\infty, z_\alpha]$	`normcdf(z)`

Example 13.4. **Interparticular Spacing in Nanoprisms.** There is much interest in knowing and understanding how nanoparticles interact optically. One reason is the use of nanoparticles as plasmon rulers. Plasmon rulers are beneficial to the field of biomedical engineering as they allow researchers to measure changes and differences in DNA or cells at the nano level. This is promising for diagnostics, especially with respect to genetic disorders, which could be potentially identified by data from a plasmon ruler. In order to create a basis for this idea, research must be done to determine the effect of different interparticle spacing on the maximum wavelength of absorbance of the particles. While a linear correlation between measured separation and wavelength is not strong, researchers have found that the correlation between the reciprocal of separation, recsep=1/separation, and the logarithm of a wavelength, logwl=log(wavelength), is strong. The data from the lab of Dr. Mostafa El-Sayed, Georgia Tech, are given in the table below, as well as in ⌨ nanoprism.dat.

recsep	logwl	recsep	logwl	recsep	logwl
2.9370	0.0694	2.9284	0.0433	2.9212	0.0331
2.9196	0.0288	2.9149	0.0221	2.9106	0.0121
2.9047	0.0080	2.9047	0.0069	2.9031	0.0049
2.9320	0.0714	2.9154	0.0427	2.9165	0.0336
2.9085	0.0292	2.9090	0.0240	2.9047	0.0197
2.9058	0.0052	2.9025	0.0104	2.9025	0.0087
2.8976	0.0078	2.8971	0.0065	2.8976	0.0062

(a) Test the hypothesis that the population coefficient of correlation for the transformed measures is $\rho = 0.96$ against the alternative $\rho < 0.96$. Use $\alpha = 0.05$.

(b) Find a 95% confidence interval for ρ.

From the MATLAB code below, we see that $r = 0.9246$ and the p-value for the test is 0.0832. Thus, at a 5% significance level $H_0 : \rho = 0.96$ is not rejected. The 95% confidence interval for ρ in this case is $[0.8203, 0.9694]$.

```
%nanoprism.m
load 'nanoprism.dat'
recsep = nanoprism(:,1);
logwl = nanoprism(:,2);
r = corr(recsep, logwl)     %0.9246
n = length(logwl);
```

```
fisherz = @(x) atanh(x); invfisherz = @(x) tanh(x);
%fisher z and inverse transformations as pure functions
%Forming z for testing H0: rho = rho0:
z = (fisherz(r) - fisherz(0.96))/(1/sqrt(n-3))    %-1.3840
pval = normcdf(z)  %0.0832

%95% confidence interval
invfisherz([fisherz(r) - norminv(0.975)/sqrt(n-3) ,...
            fisherz(r) + norminv(0.975)/sqrt(n-3)])
   %0.8203    0.9694
```

Remark. Inference for population ρ using Fisher z transform is valid when samples sizes are larger than $n = 5$. When $n < 15$, it is recommended to correct r as $r^* = r\left[1 + \frac{1+r^2}{2(n-3)}\right]$.

If $n > 15$, the distribution of random variable

$$t = \frac{(r - \rho)\sqrt{n-2}}{\sqrt{(1-\rho^2)(1-r^2)}}$$

is well approximated with t-distribution with $n - 2$ degrees of freedom. This fact is sometimes used to devise an alternative test for testing $H_0 : \rho = \rho_0$ versus one-sided or two-sided alternatives, in a standard way.

If $n > 200$, one can approximate distribution of r directly as normal $\mathcal{N}\left(\rho, \frac{1-\rho^2}{n-1}\right)$.

13.2.1.3 Test for the Equality of Two Correlation Coefficients

In some testing scenarios we might be interested to know if the correlation coefficients from two bivariate populations, ρ_1 and ρ_2, are equal. From the first population the pairs $(X_i^{(1)}, Y_i^{(1)})$, $i = 1,\ldots,n_1$ are observed. Analogously, from the second population the pairs $(X_i^{(2)}, Y_i^{(2)})$, $i = 1,\ldots,n_2$ are observed and sample correlations r_1 and r_2 are calculated. The populations are assumed normal and independent, but components X and Y within each population might be correlated.

The test statistic for testing $H_0 : \rho_1 = \rho_2$ versus $H_1 : \rho_1 >, \neq, < \rho_2$ is expressed in terms of Fisher's z-transformations of the sample correlations r_1 and r_2:

$$z = \frac{w_1 - w_2}{\sqrt{\frac{1}{n_1-3} + \frac{1}{n_2-3}}},$$

where $w_i = \frac{1}{2} \log \frac{1+r_i}{1-r_i}$, $i = 1,2$ and n_1 and n_2 are the number of pairs in the first and second sample, respectively.

Alternative	α-level rejection region	p-value
$H_1 : \rho_1 > \rho_2$	$[z_{1-\alpha}, \infty)$	`1-normcdf(z)`
$H_1 : \rho_1 \neq \rho_2$	$(-\infty, z_{\alpha/2}] \cup [z_{1-\alpha/2}, \infty)$	`2*normcdf(-abs(z))`
$H_1 : \rho_1 < \rho_2$	$(-\infty, z_{\alpha}]$	`normcdf(z)`

If H_0 is not rejected, then one may be interested in pooling the two sample estimators r_1 and r_2. This is done in the domain of z-transformed values as

$$w_p = \frac{(n_1 - 3)w_1 + (n_2 - 3)w_2}{n_1 + n_2 - 6}$$

and inverting w_p to r_p via

$$r_p = \frac{1 - \exp\{2w_p\}}{1 + \exp\{2w_p\}}.$$

Example 13.5. **Swallowtail Butterflies.** The following data were extracted from a larger study by Brower (1959) on a speciation in a group of swallowtail butterflies. Morphological measurements are (X – length of eighth tergile, Y – length of superuncus)$\times 8$, in mm.

Species	X	Y	X	Y	X	Y	X	Y
Papilio	24	14	21	15	20	17.5	21.5	16.5
multicaudatus	21.5	16	25.5	16	25.5	17.5	28.5	16.5
	23.5	15	22	15.5	22.5	17.5	20.5	19
	21	13.5	19.5	19	26	18	23	17
	21	18	21	17	20.5	16	22.5	15.5
Papilio	20	11.5	21.5	11	18.5	10	20	11
rutulus	19	11	20.5	11	19.5	11	19	10.5
	21.5	11	20	11.5	21.5	10	20.5	12
	20	10.5	21.5	12.5	17.5	12	21	12.5
	21	11.5	21	12	19	10.5	19	11
	18	11.5	21.5	10.5	23	11	22.5	11.5
	19	13	22.5	14	21	12.5	19.5	12.5

The observed correlation coefficients are $r_1 = -0.1120$ (for *P. multicaudatus*) and $r_2 = 0.1757$ (for *P. rutulus*). We are interested if the corresponding population correlation coefficients ρ_1 and ρ_2 are significantly different.

The Fisher z-transformations of r_1 and r_2 are $w_1 = -0.1125$ and $w_2 = 0.1776$. The test statistic is $z = \frac{-0.1125 - 0.1776}{\sqrt{1/17 + 1/25}} = -0.9228$. For this value of

z the p-value against the two-sided alternative is 0.3561, and the null hypothesis of the equality of population correlations is not rejected. Here is a MATLAB session for the exercise above.

```
PapilioM=[24,   14;    21,   15;    20,   17.5;   21.5, 16.5; ...
          21.5, 16;    25.5, 16;    25.5, 17.5;   28.5, 16.5; ...
          23.5, 15;    22,   15.5;  22.5, 17.5;   20.5, 19;    ...
          21,   13.5;  19.5, 19;    26,   18;      23,   17;   ...
          21,   18;    21,   17;    20.5, 16;      22.5, 15.5];

PapilioR=[20,   11.5;  21.5, 11;    18.5, 10;     20,   11;    ...
          19,   11;    20.5, 11;    19.5, 11;     19,   10.5; ...
          21.5, 11;    20,   11.5;  21.5, 10;     20.5, 12;    ...
          20,   10.5;  21.5, 12.5;  17.5, 12;     21,   12.5; ...
          21,   11.5;  21,   12;    19,   10.5;   19,   11;    ...
          18,   11.5;  21.5, 10.5;  23,   11;     22.5, 11.5; ...
          19,   13;    22.5, 14;    21,   12.5;   19.5, 12.5];
PapilioMX=PapilioM(:,1); % X_m
PapilioMY=PapilioM(:,2); % Y_m

PapilioRX=PapilioR(:,1); % X_r
PapilioRY=PapilioR(:,2); % Y_r

n1=length(PapilioMX);
n2=length(PapilioRX);

r1=corr(PapilioMX, PapilioMY); % -0.1120
r2=corr(PapilioRX, PapilioRY); %  0.1757
%test for rho1 = 0
   pval1 = 2* tcdf(-abs(r1*sqrt(n1-3)/sqrt(1-r1^2)), n1-3);
   % 0.6480
   %test for rho2 = 0
   pval2 = 2* tcdf(-abs(r2*sqrt(n2-3)/sqrt(1-r2^2)), n2-3);
   %0.3806

fisherz = @(x)  1/2*log( (1+x)/(1-x) );
%Fisher z transformation as pure function
w1 = fisherz(r1);  %-0.1125
w2 = fisherz(r2);  %0.1776

%test for rho1 = rho2  vs. rho1 ~= rho2
z = (w1 - w2)/sqrt(1/(n1-3) + 1/(n2-3)) %-0.9228
pval = 2 * normcdf(-abs(z)) %0.3561
```

13.2.1.4 Testing the Equality of Several Correlation Coefficients

We are interested in testing $H_0 : \rho_1 = \rho_2 = \cdots = \rho_k$.

Let r_i be the correlation coefficient based on n_i pairs, $i = 1, \ldots, k$, and let $w_i = \frac{1}{2} \log \frac{1+r_i}{1-r_i}$ be its Fisher transformation. Define

$$N = (n_1 - 3) + (n_2 - 3) + \cdots + (n_k - 3) = \sum_{i=1}^{k} n_i - 3k, \quad \text{and}$$

$$\overline{w} = \frac{(n_1 - 3)w_1 + (n_2 - 3)w_2 + \cdots + (n_k - 3)w_k}{N}.$$

Then the statistic

$$\chi^2 = (n_1 - 3)w_1^2 + (n_2 - 3)w_2^2 + \cdots + (n_k - 3)w_k^2 - N(\overline{w})^2$$

has a χ^2-distribution with $k - 1$ degrees of freedom. Large values of χ^2 are critical for H_0.

Example 13.6. **Correlations in Fisher's Iris Data.** In testing whether the correlations between sepal and petal lengths differ for the species: setosa, versicolor, and virginica, the p-value found was close to 0.

```
load fisheriris
%Correlations between sepal and petal lengths
%for species setosa, versicolor and virginica
n1=50; n2=50; n3=50; k=3; N=n1+n2+n3-3*k;
r =[ corr(meas(1:50, 1), meas(1:50, 3)), ...
     corr(meas(51:100, 1), meas(51:100, 3)), ...
     corr(meas(101:150, 1), meas(101:150, 3)) ]
                     %0.2672    0.7540    0.8642
fisherz = @(r)   1/2 * log( (1+r)./(1-r) );
       w=fisherz(r)
                     %0.2738    0.9823    1.3098
wbar =    [n1-3   n2-3 n3-3]/N * w'        %0.8553
chi2 =    [n1-3,  n2-3, n3-3]*(w.^2)'- N*wbar^2  %26.3581
pval = 1-chi2cdf(chi2, k-1)          %1.8897e-006
```

13.2.1.5 Power and Sample Size in Inference About Correlations

Power and sample size computations for testing correlations use the fact that Fisher's z-transformed sample correlation coefficients have an approximately normal distribution, as we have seen before.

The statistic $t = r\sqrt{\frac{n-2}{1-r^2}}$, which has a t-distribution with $df = n - 2$ degrees of freedom, determines the critical points of the rejection region as

$$r^* = \sqrt{\frac{t_{n-2,1-\alpha}^2}{t_{n-2,1-\alpha}^2 + (n - 2)}},$$

where α is replaced by $\alpha/2$ for the two-sided alternative, and where the sign of the square root is negative if $H_1 : \rho < 0$.

Let z-transformations of r and r^* be w and w^*, respectively. Then the power, as a function of w, is approximately

$$1 - \beta = \Phi\left((w - w^*)\sqrt{n - 3}\right), \tag{13.1}$$

with $(w - w^*)$ replaced by $(w^* - w)$ if $H_1 : \rho < 0$.

The calculated power is retrospective, that is, the power for $\rho = r$, where r is observed. For prospective studies, w in (13.1) is replaced by z-transform of ρ_1 from $H_1 : \rho = \rho_1$.

The sample size needed to achieve a power of $1 - \beta$, in a test of level α against the one-sided alternative, is approximately,

$$n = \left(\frac{z_{1-\alpha} + z_{1-\beta}}{w}\right)^2 + 3.$$

Here the effect size w is z-transformation of ρ_1 from the specific alternative, $H_1 : \rho = \rho_1$, and $z_{1-\alpha}$ and $z_{1-\beta}$ are quantiles of a standard normal distribution. When the alternative is two-sided, $z_{1-\alpha}$ is replaced by $z_{1-\alpha/2}$.

Example 13.7. **Power for Rats and Mazes.** In Example 13.1 it was found that $r = -0.5687$. For $n = 8$, the null hypothesis of no correlation was not rejected. It was also discussed that if, for the same observed r, the sample size were $n = 30$, the null hypothesis would be rejected with a p-value of $5.2031e - 04$.

What would be the retrospective power of this test in a 5% testing against the alternative $H_1 : \rho < 0$?

If w^* is taken with the negative sign, and $(w - w^*)$ in (13.1) replaced by $(w^* - w)$, the approximate power is found to exceed 95%:

```
r=-0.5687;
n=30;
rstar = -sqrt( tinv(1-0.05,n-2)^2/( (n-2) + tinv(1-0.05,n-2)^2 )) %-0.3061
fisherz = @(x) atanh(x);
wstar = fisherz(rstar); w = fisherz(r);
power = normcdf((wstar-w)*sqrt(n-3)) %0.9565
```

Example 13.8. **Sample Size in Testing the Correlation.** One wishes to determine the size of a sample sufficient to reject $H_0 : \rho = 0$ with a power of $1 - \beta = 0.90$ in a test of level $\alpha = 0.05$ whenever $\rho \geq 0.4$. Here we take $H_1 : \rho = 0.4$ and find that a sample of size 51 will be necessary:

```
fisherz = @(x) atanh(x);
w  = fisherz(0.4) % 0.4236
n =((norminv(0.95)+norminv(0.90))/w)^2 + 3  %50.7152
```

If $H_0 : \rho = 0$ is to be rejected whenever $|\rho| \geq 0.4$, then a sample of size 62 is needed:

```
n =((norminv(0.975)+norminv(0.90))/w)^2 + 3   %61.5442
```

13.2.1.6 Multiple Correlation Coefficient

Consider three variables X_1, X_2, and Y. The correlation between Y and the pair X_1, X_2 is measured by the multiple correlation coefficient

$$R_{y.x_1x_2} = \sqrt{\frac{r^2_{x_1y} - 2r_{x_1x_2} \cdot r_{x_1y} \cdot r_{x_2y} + r^2_{x_2y}}{1 - r^2_{x_1x_2}}} \; .$$

This correlation is significant if

$$F = \frac{R^2_{y.x_1x_2}/2}{\left(1 - R^2_{y.x_1x_2}\right)/(n-3)}$$

is large. Here n is the number of triplets (X_1, X_2, Y), and statistic F has an F-distribution with $2, n-3$ degrees of freedom.

The general case $R = R_{y.x_1x_2...x_k}$ is analogous. Statistic

$$F = \frac{R^2/k}{(1 - R^2)/(n-k-1)}$$

has an $F_{k,n-k-1}$-distribution. We will see in the next chapter (page 687) that R^2 represents the coefficient of determination and testing its significance is equivalent to testing the significance of the multiple regression of Y on $X_1, \ldots, X_k$.

13.2.2 Bayesian Inference for Correlation Coefficients

To conduct a Bayesian inference on a correlation coefficient, a bivariate normal distribution for the data is assumed and Wishart's prior is placed on the inverse of the covariance matrix. Recall that the Wishart distribution is a multivariate counterpart of a gamma distribution (more precisely, of a χ^2-distribution) and the model is in fact a multivariate analogue of a gamma prior on the normal precision parameter.

To illustrate this Bayesian model, a bivariate normal sample of size $n = 46$ is generated. For this sample Pearson's coefficient of correlation was found to be $r = 0.9908$, with sample variances of $s^2_x = 8.1088$ and $s^2_y = 35.0266$.

. WinBUGS code ⚡corr.odc, as given below, is run, and Bayes' estimators for the population correlation ρ and component variances σ_x^2 and σ_y^2 are obtained as 0.9901, 8.114, and 34.99. These values are close to the classical estimators since the priors are noninformative. The hyperparameters of Wishart's prior are matrix W and degrees of freedom df, and low degrees of freedom, df=3, make this prior "vague."

As an exercise, compute a classical 95% confidence interval for ρ (page 656) and compare it with the 95% credible set [0.9823, 0.9952]. Are the intervals similar?

```
model{
for( i in 1:nn){
    y[i,1:2] ~ dmnorm( mu[i,], Tau[,] )
    mu[i,1] ~ dnorm(mu.x, tau1)
    mu[i,2] ~ dnorm(mu.y, tau2)
    }
mu.x ~ dnorm(0, 0.0001)
mu.y ~ dnorm(0, 0.0001)
tau1 ~ dgamma(0.001, 0.001)
tau2 ~ dgamma(0.001, 0.001)
Tau[1:2,1:2] ~ dwish( W[,], df )
df <- 3
Sigma[1:2, 1:2]  <- inverse(Tau[,])
rho <- Sigma[1,2]/sqrt(Sigma[1,1]*Sigma[2,2])
}

DATA
list( nn=46, y = structure(.Data = c( 0.5674,    -1.6458,
    -0.4656,    -4.8531,
     1.5253 ,   -2.1200,
     ...
     8.3435,    14.7693,
    11.0151,    18.8988,
     8.9949,    15.3797,
    10.5287,    18.1455), .Dim=c(46,2)) ,
W = structure(.Data = c(1,0,0,1),.Dim=c(2,2) )   )

INIT
list(Tau = structure(.Data = c(1,1,1,1), .Dim = c(2,2)),
  mu.x = 1, mu.y = 1, tau1=1, tau2=1)
```

	mean	sd	MC error	val2.5pc	median	val97.5pc	start	sample
Sigma[1,1]	8.114	1.757	0.01462	5.361	7.881	12.21	1001	100000
Sigma[1,2]	16.68	3.625	0.03035	11.0	16.2	25.13	1001	100000
Sigma[2,1]	16.68	3.625	0.03035	11.0	16.2	25.13	1001	100000
Sigma[2,2]	34.99	7.570	0.06331	23.15	33.98	52.63	1001	100000
rho	0.9901	0.0033	3.287E-5	0.9823	0.9906	0.9952	1001	100000

13.3 Spearman's Coefficient of Correlation

Charles Edward Spearman was a late bloomer, academically speaking. He received his Ph.D. at the age of 48, after having served as an officer in the British army for 15 years. He is most famous in the field of psychology, where he theorized that "general intelligence" was a function of a comprehensive mental competence rather than a collection of multifaceted mental abilities. His theories eventually led to the development of factor analysis.

Spearman (1904) proposed the rank correlation coefficient long before statistics became a scientific discipline. For bivariate data, an observation has two coupled components (X, Y) that may or may not be related to each other. Let $\rho = \mathrm{Corr}(X, Y)$ represent the unknown correlation between two components. In a sample of n, let $R_1, \dots, R_n$ denote the ranks for the first component X and $S_1, \dots, S_n$ denote the ranks for Y. For example, if $x_1 = x_{(3)}$ is the third smallest value from $x_1, \dots, x_n$ and $y_1 = y_{(5)}$ is the fifth smallest value from $y_1, \dots, y_n$, then $(R_1, S_1) = (3, 5)$. The Spearman coefficient of correlation is simply Pearson's coefficient of correlation the corresponding ranks,

$$\hat{\rho} = \frac{\sum_{i=1}^{n}(R_i - \overline{R})(S_i - \overline{S})}{\sqrt{\sum_{i=1}^{n}(R_i - \overline{R})^2 \cdot \sum_{i=1}^{n}(S_i - \overline{S})^2}}. \tag{13.2}$$

This expression can be simplified. From (13.2), $\overline{R} = \overline{S} = (n+1)/2$ and $\sum(R_i - \overline{R})^2 = \sum(S_i - \overline{S})^2 = n\mathrm{Var}\,(R_i) = n(n^2 - 1)/12$. Define D as the difference between ranks, i.e., $D_i = R_i - S_i$. With $\overline{R} = \overline{S}$, we can see that

$$D_i = (R_i - \overline{R}) - (S_i - \overline{S})$$

and

$$\sum_{i=1}^{n} D_i^2 = \sum_{i=1}^{n}(R_i - \overline{R})^2 + \sum_{i=1}^{n}(S_i - \overline{S})^2 - 2\sum_{i=1}^{n}(R_i - \overline{R})(S_i - \overline{S}),$$

i.e.,

$$\sum_{i=1}^{n}(R_i - \overline{R})(S_i - \overline{S}) = \frac{n(n^2 - 1)}{12} - \frac{1}{2}\sum_{i=1}^{n} D_i^2.$$

By dividing both sides of the equation by $\sqrt{\sum_{i=1}^{n}(R_i - \overline{R})^2 \cdot \sum_{i=1}^{n}(S_i - \overline{S})^2} = \sum_{i=1}^{n}(R_i - \overline{R})^2 = n(n^2 - 1)/12$, we obtain

$$\hat{\rho} = 1 - \frac{6\sum_{i=1}^{n} D_i^2}{n(n^2 - 1)}. \tag{13.3}$$

Consistent with Pearson's coefficient of correlation, Spearman's coefficient of correlation also ranges between -1 and 1. If there is perfect agreement, meaning all the differences are 0, then $\hat{\rho} = 1$. The scenario that maximizes $\sum D_i^2$ occurs when ranks are perfectly opposite: $R_i = n - S_i + 1$. In this case $\sum D_i^2 = n(n^2 - 1)/3$, and $\hat{\rho} = -1$.

If the sample is large enough, then Spearman's statistic can be approximated using the normal distribution. It was shown that if $n > 10$, then

$$Z = (\hat{\rho} - \rho)\sqrt{n - 1} \sim \mathcal{N}(0,1).$$

If Spearman's correlation $\hat{\rho}$ is given, an approximation for Pearson's coefficient of correlation is

$$r \approx 2\sin\frac{\pi\hat{\rho}}{6}.$$

Example 13.9. **Tread Wear for Tires.** Stichler et al. (1953) provide a list of tread wear for tires, each tire measured by two methods based on (a) weight loss and (b) groove wear.

Weight	Groove	Weight	Groove
45.9	35.7	41.9	39.2
37.5	31.1	33.4	28.1
31.0	24.0	30.5	28.7
30.9	25.9	31.9	23.3
30.4	23.1	27.3	23.7
20.4	20.9	24.5	16.1
20.9	19.9	18.9	15.2
13.7	11.5	11.4	11.2

For this data, $\hat{\rho} = 0.9265$. Note that if we opt for the parametric measure of correlation, the Pearson coefficient is 0.948.

Ties in the Data: The statistics in (13.2) and (13.3) are not designed for paired data that include tied measurements. If ties exist in the data, a simple adjustment should be made. Define $u' = \sum u(u^2 - 1)/12$ and $v' = \sum v(v^2 - 1)/12$ where the us and vs are the ranks for X and Y adjusted (e.g., averaged) for ties. Then

$$\hat{\rho}' = \frac{n(n^2 - 1) - 6\sum_{i=1}^{n} D_i^2 - 6(u' + v')}{\{[n(n^2 - 1) - 12u'][n(n^2 - 1) - 12v']\}^{1/2}},$$

and it holds that, for large n,

$$Z = (\hat{\rho}' - \rho)\sqrt{n-1} \sim \mathcal{N}(0,1).$$

The MATLAB function `corr(x,y,'type','Spearman')` computes the Spearman correlation coefficient for column vectors x and y.

13.4 Kendall's Tau

M. G. Kendall formalized an alternative measure of dependence among ranked data, originally proposed and used in the nineteenth century, by analyzing "concordant" and "discordant" pairs in a bivariate sample.

From (X_i, Y_i), $i = 1,\ldots,n$, one can choose $\binom{n}{2}$ different pairs. The pair $(X_i, Y_i), (X_j, Y_j)$ is concordant if either $X_i < X_j$ and $Y_i < Y_j$ or $X_i > X_j$ and $Y_i > Y_j$. The pair is called discordant if either $X_i < X_j$ and $Y_i > Y_j$ or $X_i > X_j$ and $Y_i < Y_j$. For example, the pair $(2,4)$ and $(1,-1)$ is concordant, while the pair $(-2,4)$ and $(1,-1)$ is discordant.

Let n_C and n_D be the number of concordant and discordant pairs respectively, among all possible $\binom{n}{2}$ pairs. Kendall's $\hat{\tau}$ statistic (Kendall, 1938) is defined as

$$\hat{\tau} = \frac{n_C - n_D}{\binom{n}{2}} = \frac{2S}{n(n-1)} \tag{13.4}$$

for $S = n_C - n_D$ and no ties in the samples present.

Coefficient $\hat{\tau}$ from (13.4) can also be represented as

$$\hat{\tau} = 1 - \frac{4Q}{n(n-1)}, \quad Q = \sum_{i=1}^{n-1} \sum_{j=i+1}^{n} \mathbf{1}(r_i > r_j),$$

where r_is are defined via ranks of the second sample corresponding to the ordered ranks of the first sample, $\{1,2,\ldots,n\}$, as

$$\begin{pmatrix} 1 & 2 & \ldots & n \\ r_1 & r_2 & \ldots & r_n \end{pmatrix}.$$

In this notation, Q is the number of inversions among the pairs of ranks of the second sample corresponding to the ordered ranks of the first sample.

Quantities Q and S are connected, as

$$S = \frac{n(n-1)}{2} - 2Q.$$

The population τ is the probability of a pair being concordant, minus the probability of a pair being discordant. If (X_1, Y_1) and (X_2, Y_2) is a pair with the same distribution as (X, Y), then

$$\tau = \mathbb{P}\left[(X_1 - X_2)(Y_1 - Y_2) > 0\right] - \mathbb{P}\left[(X_1 - X_2)(Y_1 - Y_2) < 0\right].$$

The population τ is also a standard correlation between two random variables, $\text{sign}(X_1 - X_2)$ and $\text{sign}(Y_1 - Y_2)$.

When $n > 10$, a normal approximation can be applied for S and $\hat{\tau}$. Using this normal approximation, we will find the confidence interval for the population τ and test the hypothesis that the rank correlation is significant.

The sample variance of S and $\hat{\tau}$ when no ties are present are approximately

$$\mathbb{V}\text{ar}\, S = \frac{n(n-1)(2n+5)}{18} \quad \text{and} \quad \mathbb{V}\text{ar}\,\hat{\tau} = \frac{2(2n+5)}{9n(n-1)}.$$

With the presence of ties, the expressions for sample variance are more complicated, but the expressions above can serve as approximations if the number of ties in not excessive. Then the $(1 - \alpha)100\%$ confidence interval for τ is

$$\left[\hat{\tau} - z_{1-\alpha/2}\sqrt{\frac{2(2n+5)}{9n(n-1)}}, \quad \hat{\tau} + z_{1-\alpha/2}\sqrt{\frac{2(2n+5)}{9n(n-1)}}\right] \cap [-1,1].$$

To test the hypothesis $H_0 : \tau = 0$ versus one-sided or two-sided alternatives we will use a z-test with statistic

$$Z = \hat{\tau}\sqrt{\frac{9n(n-1)}{2(2n+5)}},$$

that under H_0 has a standard normal distribution. Since $\hat{\tau}$ is derived from S, and S is an integer, we can get more precise p-values using continuity corrections. For example, a test of H_0 against the two-sided alternative $H_1 : \tau \neq 0$ would have p-value, in terms of MATLAB, as

```
p  = 2 * min( normcdf(3*(S+1/2)*sqrt(2)/sqrt(n*(n-1)*(2*n + 5))),...
            1-normcdf(3*(S-1/2)*sqrt(2)/sqrt(n*(n-1)*(2*n + 5))) )
```

In the case of numerical data, Kendall's and Pearson's coefficients of correlation could be linked. Here we can approximate r as

$$r \approx \sin\frac{\hat{\tau}\pi}{2}.$$

Example 13.10. **Prevention of Vitreous Loss.** Limbal incisions were made in rabbit eyes to mirror the initial steps of lens extraction. The vitreous body loses water when the eye is open and decreases in weight, as reported by

Galin et al. (1971). The results had implications in the context of cataract surgery. The authors measured the vitreous body weight for each eye of 15 New Zealand albino rabbits. One eye had been open for 5 minutes (y), while the other served as a control (x). The measurements of vitreous weight (in mg) are provided next:

Rabbit #	1	2	3	4	5	6	7	8
Control eye (x)	1848	1532	1460	1947	1810	1718	1686	1617
Open eye (y)	1738	1440	1388	1756	1692	1629	1583	1499
Rabbit #	9	10	11	12	13	14	15	
Control eye (x)	1724	1873	1928	2226	1708	1605	1822	
Open eye (y)	1596	1794	1785	2044	1602	1491	1702	

In the code ◢ rabbits.m we found no ties, $n_C = 100$, $n_D = 5$, $\hat{\tau} = 0.9048$, and a 95% confidence interval of $[0.5276, 1.0000]$.

```
%rabbits.m
x=[1848 1532 1460 1947 1810 1718 1686 1617 ...
          1724 1873 1928 2226 1708 1605 1822];
y=[1738 1440 1388 1756 1692 1629 1583 1499 ...
          1596 1794 1785 2044 1602 1491 1702];
n=15;
%nc-number of concordant pairs,%nd - discordant %nt - ties
nc = 0; nd = 0; nt=0;
for i = 1:n-1
  for j = i+1:n
    if       sign(x(i) - x(j)) * sign(y(i) - y(j))  > 0   nc=nc+1;
    elseif sign(x(i) - x(j)) * sign(y(i) - y(j))  < 0   nd=nd+1;
    else     nt=nt+1;
    end
  end
end
tauhat = (nc-nd)/(nc+nd)                 %0.9048
%Sample variance of tau
tauvar = 2*(2*n+5)/(9*n*(n-1))           %0.0370
%95% CI for population tau
[max(0,tauhat - norminv(0.975)*sqrt(tauvar)) ...
 min(1,tauhat + norminv(0.975)*sqrt(tauvar)) ]%[0.5276    1.0000]
S=nc-nd    %95
%p-value in testing H0: tau=0 vs. H_1: tau != 0.
pval=2*(1-normcdf(abs(tauhat), 0, sqrt(tauvar)))  %2.5853e-06
```

Remark. The generalized coefficient of correlation is defined as

$$\rho = \frac{\sum_{i<j} c_{ij}(X) c_{ij}(Y)}{\left(\sum_{i<j} c_{ij}^2(X) \times \sum_{i<j} c_{ij}^2(X) \right)^{1/2}}, \qquad (13.5)$$

for $X = (X_1,\ldots,X_n)$ and $Y = (Y_1,\ldots,Y_n)$. Then Pearson's, Spearman's, and Kendall's coefficients of correlation are recovered for $c_{ij}(X) = X_j - X_i$, $c_{ij}(X) = R_j - R_i = r(X_j) - r(X_i)$, and $c_{ij}(X) = \text{sign}(X_j - X_i)$, respectively.

13.5 *Cum hoc ergo propter hoc*

We conclude this chapter with a discussion on the misuses of correlations. The fallacy that correlation implies causation is summarized by Gould's quote at the beginning of the chapter (Latin *Cum hoc ergo propter hoc* meaning "With this, therefore because of this"). We already mentioned the "link" between ice-cream sold on the beach and the number of drowning accidents, but the correlation fallacy causes more serious damage to science. Spurious correlations are often misused in medical and health science and attributed to causations. The number of published studies with *voodoo* causations, often conflicted from study to study, is stunning.

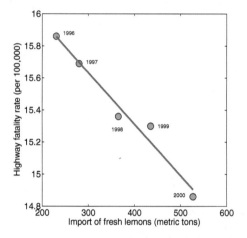

Fig. 13.3 Fresh lemons imported to the United States from Mexico (in metric tons; US Department of Agriculture) and total US highway fatality rate (per 100,000; US NHTSA, DOT HS 810 780).

As an extreme case of spurious correlation we give an example, popular among bloggers on the Web, involving data on imports of fresh lemons from Mexico (1996–2000) and US highway fatality rates (1996–2000), Figure 13.3. The correlation is $r = -0.986$ and is highly significant ($p < 0.0002$), even with sample size $n = 5$. Some bloggers provided "causal links" citing less expensive car air-fresheners that make drivers happy or slower traffic caused by trucks from Mexico transporting lemons. See also Exercise 13.14 for an interesting example of spurious correlation.

There are two possible errors in correlation inference caused by grouping data. The first one is if two disparate groups are combined. It could be that for each group there may not be correlation, but when the groups are combined, the correlation may be significant and, of course, spurious. Figure 13.4 illustrates this point. Observations represented by red circles (group 1, $r = 0.0643$), as well as the pairs represented by blue circles (group 2, $r = 0.0079$), show no significant correlation. However, when the groups are combined, the correlation increases to $r = 0.7031$, and it is significant with a p-value of 1.5×10^{-5}. Details can be found in ◢ spur.m.

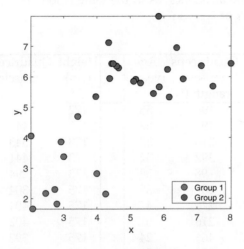

Fig. 13.4 Spurious correlation when two groups of uncorrelated pairs are combined.

The second error is more subtle. Often, repeated bivariate measurements are considered as independent and an artificial correlation due to a blocking factor is introduced. For example, if for 15 subjects one measures weight (X) and skinfold thickness (Y) before and after a diet and combines the measurements, then due to the "increased sample size" a significance of correlation between X and Y is likely to be found.

13.6 Exercises

13.1. **Correlation between Uniforms and Their Squares.** Generate 10,000 uniform random numbers between -1 and 1 in the form of a vector x.

Demonstrate that y=x.^2 has a small correlation with x, regardless of their perfect functional relationship. Explain why.

13.2. **Muscle Strength of "Ethanol Abusers."** It is estimated that 10% of European and North American adults, and up to one-third of acute hospital admissions, are alcoholics. Obviously, the high proportion of alcoholics in the hospitalized population imposes severe financial constraints on health authorities and emphasizes the need for primary caretakers to focus on minimizing alcohol misuse. A staggering two-thirds of chronic ethanol abusers have skeletal muscle myopathy (Martin et al., 1985; Worden, 1976).

Hickish et al. (1989) provide height, quadriceps muscle strength, and age data in 41 male alcoholics, as in the table below. The data are available as alcos.xls or alcos.ascii.

Height (cm)	Quadriceps muscle strength (N)	Age (years)	Height (cm)	Quadriceps muscle strength (N)	Age (years)
155	196	55	172	147	32
159	196	62	173	441	39
159	216	53	173	343	28
160	392	32	173	441	40
160	98	58	173	294	53
161	387	39	175	304	27
162	270	47	175	404	28
162	216	61	175	402	34
166	466	24	175	392	53
167	294	50	175	196	37
167	491	35	176	368	51
168	137	65	177	441	49
168	343	41	177	368	48
168	74	65	177	412	32
170	304	55	178	392	49
171	294	47	178	540	41
172	294	31	178	417	42
172	343	38	178	324	55
172	147	31	179	270	32
172	319	39	180	368	34
172	466	53			

(a) Find the sample correlation between Height and Strength, r_{HS}. Test the hypothesis that the population correlation coefficient between Height and Strength ρ_{HS} is significantly positive at the level $\alpha = 0.01$.

(b) Since an increase in Age is expected to decrease the Strength (negative correlation), find the correlation between Height and Strength when

Age is accounted for; that is, find $r_{HS.A}$. Test the hypothesis that $\rho_{HS.A}$ is positive at the level $\alpha = 0.01$.

(c) Find an approximate 95% confidence interval for ρ_{HS}.

13.3. **Vending Machine and Pharmacy Errors.** Mr. Joseph Bentley, the owner of a pharmacy store, wants to remove the Coke vending machine standing in front of his store because he believes the vending machine influences the number of errors the store employees make. More precisely, as more Coke is sold outside his store, more errors are made. He provided the following data:

Errors made	5	3	10	9	5	7	8	4
Coke sold	112	100	220	250	100	200	160	100

Find the coefficient of correlation. Comment on why this correlation is high. Is there a causation; are Coke sales alone influencing the pharmacy employees?

13.4. **Vending Machine and Pharmacy Errors Revisited.** Refer to Exercise 13.3. In addition to Errors and Coke, Mr. Bentley provided the count of people that pass by his store (and the vending machine):

Errors made	5	3	10	9	5	7	8	4
Coke	112	100	220	250	100	200	160	100
People	10000	6000	17000	20000	9000	15000	14000	8000

Find the coefficient of correlation between Errors and Coke sales while accounting for the number of people and comment.

13.5. **Education, Socioeconomic Status, and Infant Mortality.** On the basis of $n = 28$ records it was found that the education level of the mother (X) was negatively correlated with infant mortality (Y) as $r_{xy} = -0.6$. Socioeconomic status (Z) is a variable that was not taken into account. Z is correlated with X and Y as $r_{xz} = 0.65$ and $r_{yz} = -0.7$. Find $r_{xy.z}$ and test for the significance of the population counterpart $\rho_{xy.z}$.

13.6. **Corn Yields and Rainfall.** The following table published by Misner (1928) has been analyzed by Ezekiel and Fox (1959).

X: rainfall measurements in inches, in the six states, from 1890 to 1927. Year 1 in the data below corresponds to 1890.

Y: yearly corn yield in bushels per acre, in six Corn Belt states (Iowa, Illinois, Nebraska, Missouri, Indiana, and Ohio).

year	X	Y	year	X	Y
1	9.6	24.5	20	12.0	32.3
2	12.9	33.7	21	9.3	34.9
3	9.9	27.9	22	7.7	30.1
4	8.7	27.5	23	11.0	36.9
5	6.8	21.7	24	6.9	26.8
6	12.5	31.9	25	9.5	30.5
7	13.0	36.8	26	16.5	33.3
8	10.1	29.9	27	9.3	29.7

9	10.1	30.2	28	9.4	35.0
10	10.1	32.0	29	8.7	29.9
11	10.8	34.0	30	9.5	35.2
12	7.8	19.4	31	11.6	38.3
13	16.2	36.0	32	12.1	35.2
14	14.1	30.2	33	8.0	35.5
15	10.6	32.4	34	10.7	36.7
16	10.0	36.4	35	13.9	26.8
17	11.5	36.9	36	11.3	38.0
18	13.6	31.5	37	11.6	31.7
19	12.1	30.5	38	10.4	32.6

Find the sample correlation coefficient r and a 95% confidence interval for the population coefficient ρ.

13.7. **Drosophilæ.** Sokoloff (1966) reported the correlation between body weight and wing length in *Drosophila pseudoobscura* as 0.52 in a sample of $n_1 = 39$ at the Grand Canyon, and as 0.67 in a sample of $n_2 = 20$ at Flagstaff, Arizona. Do the correlations in these two populations differ significantly? Use $\alpha = 0.05$.

13.8. **Confidence Interval for the Difference of Two Correlation Coefficients.** Using the results on testing the equality of two correlation coefficients, develop a $(1 - \alpha)100\%$ confidence interval for their difference.

13.9. **Oxygen Intake.** The human body takes in more oxygen when exercising than when it is at rest, and to deliver the oxygen to the muscles, the heart must beat faster. Heart rate is easy to measure, but the measurement of oxygen uptake requires elaborate equipment. If oxygen uptake (VO2) is strongly correlated with heart rate (HR) under a particular set of exercise conditions, then its predicted, rather than measured, values could be used for various research purposes.[2]

HR	VO2	HR	VO2
94	0.473	108	1.403
96	0.753	110	1.499
95	0.929	113	1.529
95	0.939	113	1.599
94	0.832	118	1.749
95	0.983	115	1.746
94	1.049	121	1.897
104	1.178	127	2.040
104	1.176	135	2.231
106	1.292		

Find the sample correlation r and calculate a 95% confidence interval for its population counterpart, ρ.

---

[2] Data provided by Paul Waldsmith from experiments conducted in Don Corrigan's lab at Purdue University, West Lafayette, Indiana.

13.10. **Obesity and Pain.** Khimich (1997) found that a pain threshold increases in obese subjects and increases with age. Obesity is measured as the percentage over ideal weight (X). The response to pain is measured by using the threshold of the nociceptive flexion reflex (Y), which is a measure of the pricking pain sensation in an individual. Measurements X and Y are considered to be normal. We are interested in an inference about the correlation between X and Y. The following data were obtained:

X	89	90	75	30	51	75	62	45	90	20
Y	2	3	4	4.5	5.5	7	9	13	15	14

(a) Using results $n = 10$, $\sum_i X_i Y_i = 4{,}461.5$, $\sum_i X_i = 627$, $\sum_i X_i^2 = 45{,}141$, $\sum_i Y_i = 77$, $\sum_i Y_i^2 = 799.5$, calculate the Pearson coefficient of correlation, r.
(b) Test the hypothesis that the population coefficient of correlation, ρ, is 0, against the one-sided alternative $H_1 : \rho < 0$. Use $\alpha = 0.05$.
(c) Let the age Z (in years) of the individuals from the table be as follows (in the corresponding order): 20 18 23 19 44 51 36 47 60 55. Find the partial coefficient of correlation $r_{xy.z}$ if $r_{xz} = -0.2089$ and $r_{yz} = 0.8627$.
(d) Find a 95% confidence interval for ρ.

13.11. **Connection between Partial and Multiple Coefficients of Correlation.** Show

$$R_{x.yz}^2 = 1 - (1 - r_{xz}^2)(1 - r_{xy.z}^2) = 1 - (1 - r_{xy}^2)(1 - r_{xz.y}^2).$$

13.12. **Swallowtail Butterflies Continued.** In Example 13.5 correlations between morphological measurements, X – length of eighth tergile and Y – length of superuncus, were compared for species *Papilio multicaudatus* and *Papilio rutulus*, and no significant differences were found.
Brower (1959) also provides the same morphological measurements for *Papilio glaucus*, the Eastern tiger swallowtail:

Species	X	Y	X	Y	X	Y	X	Y
Papilio	17.5	9	19	9	18	10	19	8
glaucus	19	8.5	21	8.5	19	9	21	9.5
	19	9.5	17.5	9	23.5	10.5	16	9.5
	19.5	10	17	7	16	8	18	9
	18.5	8.5	19	8.5	16.5	8.5	18.5	9.5
	17.5	9.5	19.5	8	18.5	8	19	9
	20.5	9	17	9.5	18.5	9	16.5	9
	17.5	9	19	9	18.5	7.5	18.5	8.5
	17.5	9	16.5	11	23	9	22	8
	18	10	21.5	9	18.5	9	20.5	8
	21	10	21	8	22.5	9.5	18	10

(a) Find the observed correlation coefficient r_3 for *P. glaucus*.

(b) If population correlation coefficients for the three species *P. multicaudatus*, *P. rutulus*, and *P. glaucus*, are ρ_1, ρ_2, and ρ_3, respectively, test the hypothesis $H_0 : \rho_1 = \rho_2 = \rho_3$. Report the p-value and discuss.

13.13. **Van Der Waerden's Rank Correlation.** Let $R_x = (R_1, \ldots, R_n)$ and $R_y = (R_1', \ldots, R_n')$ be the vectors of ranks for $X_1, \ldots, X_n$ and $Y_1, \ldots, Y_n$. Van Der Waerden's rank correlation coefficient is defined as

$$R = \frac{\sum_{i=1}^n z_{R_i/(n+1)} \times z_{R_i'/(n+1)}}{\sum_{i=1}^n z_{i/(n+1)}^2}.$$

Here, for example, $z_{R_i/(n+1)}$ denotes the $R_i/(n+1)$th quantile of a standard normal distribution.

When the rank correlation is absent (H_0), the quantity

$$S = \sum_{i=1}^n z_{R_i/(n+1)} \times z_{R_i'/(n+1)}$$

is approximately normal with $\mathbb{E}S = 0$ and $\mathbb{V}\text{ar}\, S = \frac{1}{n-1} \left[\sum_{i=1}^n z_{i/(n+1)}^2 \right]^2$ whenever $n > 10$.

(a) Using data on rabbit eyes from Exercise 13.10, find Van Der Waerden's rank correlation coefficient R.

(b) For the same data test the hypothesis that the population Van Der Waerden's rank correlation is 0 versus the two-sided alternative. Use asymptotic distribution of S.

13.14. **Age of Miss America and the Murders by Hot Objects.** An interesting example of spurious correlation is found between the ages of the crowned Miss America and the number of homicides by steam, hot vapors, and hot objects (for the United States, as reported by the CDC), for years 1999–2007:

Year	1999	2000	2001	2002	2003	2004	2005	2006	2007
Age of Miss America	24	24	24	21	22	21	24	22	20
Number of Deaths	7	7	7	3	4	3	8	4	2

(a) Plot the two data sets on the same graph, using Year as the x-axis;

(b) Test the significance of this correlation and report the p-value.

MATLAB AND WINBUGS FILES AND DATA SETS USED IN THIS CHAPTER

http://statbook.gatech.edu/Ch13.Corr/

   corroncircle.m, corrs.m, errorscoke.m, fisherzsimu.m, hazel.m, histo.m, iriscorr.m, lemon.m, nanoprism.m, ObesityPain.m, rabbits.m, spur.m, sputious2.m, vanderwaerdencorr.m, variouscorrs.m

corr.odc

hazel.dat|mat, nanoprism.dat

CHAPTER REFERENCES

Anderson, T. W., (1984). *An Introduction to Multivariate Statistical Analysis,* 2nd ed., Wiley, New York.

Arvin, D. V. and Spaeth, R. (1998). Trends in Indiana's water use, 1986–1996. Indiana Department of Natural Resources Special Report, No. 1.

Brower, L. P. (1959). Speciation in butterflies of the *Papilio glaucus* group. I: Morphological relationships and hybridizations. *Evolution,* **13,** 40–63.

Ezekiel, M. and Fox, K. A. (1959). *Methods of Correlation and Regression Analysis.* Wiley, New York.

Galin, M. A., Robbins, R., and Obstbaum, S. (1971). Prevention of vitreous loss. *Br. J. Ophthal.,* **55,** 533–537.

Hickish, T., Colston, K., Bland, J. M., and Maxwell J. D. (1989). Vitamin D deficiency and muscle strength in male alcoholics. *Clin. Sci.,* **77,** 171–176.

Kendall, M. (1938). A new measure of rank correlation. *Biometrika,* **30,** 1–2, 81–89.

Khimich, S. (1997). Level of sensitivity of pain in patients with obesity. *Acta Chir. Hung.,* **36,** 166–167.

Martin, R., Ward, K., Slavin, G., Levi, J., and Peters, T. J. (1985). Alcoholic skeletal myopathy, a clinical and pathological study. *Q. J. Med.,* **55,** 233–251.

Misner, E. G. (1928). Studies of the relationship of weather to the production and price of farm products. I: Corn. Mimeographed publication. Cornell University, March 1928.

Sokoloff, A. (1966). Morphological variation in natural and experimental populations of *Drosophila pseudoobscura* and *Drosophila persimilis. Evolution,* **20,** 49–71.

Spearman, C. (1904). General intelligence: objectively determined and measured. *Am. J. Psychol.,* **15,** 201–293.

Stichler, R. G., Richey, G. G., and Mandel, J. (1953). Measurement of treadware of commercial tires. *Rubber Age,* **73**, 2.

Worden R. E. (1976). Pattern of muscle and nerve pathology in alcoholism. *NY Acad. Sci.,* **273**, 351–359.

Chapter 14
Regression

The experiments showed further that the mean filial regression towards mediocrity was directly proportional to the parental deviation from it.

– Francis Galton (1886)

WHAT IS COVERED IN THIS CHAPTER

- Ordinary Linear Regression
- Confidence Intervals and Hypothesis Tests for Parameters and Responses in Linear Regression
- Multiple Regression, Matrix Formulation
- Model Selection and Assessment in Multiple Regression
- Power Analysis in Regression
- Regression Nonlinear in Predictors, Errors-in-Variables Regression
- Analysis of Covariance (ANCOVA)

14.1 Introduction

The rather curious name *regression* was given to a statistical methodology by British scientist Sir Francis Galton, who analyzed the heights of sons and the average heights of their parents. From his observations, Galton concluded that sons of very tall (or short) parents were generally taller (or shorter) than average, but not as tall (or short) as their parents. The re-sults were published in 1886 under the title *Regression Towards Mediocrity in Hereditary Stature.* In due course of time the word *regression* became syn-onymous with the statistical study of the functional relationship between two or more variables. The data set illustrating Galton's finding and used by Pearson is given in ⌨ pearson.dat. The scatterplot and regression fits are analyzed in ◢ galton.m and summarized in Figure 14.1. The circles corre-spond to pairs of father–son heights, the black line is the line $y = x$, the red line is the regression line, and the green line is the regression line con-strained to pass through the origin. Galton's findings can be summarized by the observation that the slope of the regression (red) line was signifi-cantly smaller than the slope of the 45° line.

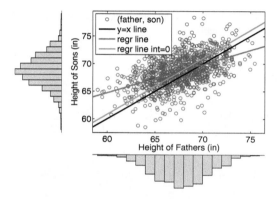

Fig. 14.1 Galton's father–son height data (used by Pearson). The *circles* correspond to pairs of father–son heights, the *black line* is the line $y = x$, the *red line* is the regression line, and the *green line* is the regression line constrained to pass through the origin.

Usually the response variable y is "regressed" on several predictors or covariates, $x_1, \dots, x_k$, and this raises many interesting questions involving the model choice and fit, collinearity among the predictors, and others. When we have a single predictor x and a linear relation between y and x, the regression is called a simple linear regression.

14.2 Simple Linear Regression

Assume that we observed n pairs $(x_1, y_1), \dots, (x_n, y_n)$, and each observation y_i can be modeled as a linear function of x_i, plus an error,

$$y_i = \beta_0 + \beta_1 x_i + \epsilon_i, \quad i = 1, \dots, n.$$

Here β_0 and β_1 are the population intercept and slope parameters, respectively, and ϵ_i is the error. We assume that the errors are not correlated and have mean 0 and variance σ^2; thus, $\mathbb{E}y_i = \beta_0 + \beta_1 x_i$ and $\mathbb{V}\text{ar}\, y_i = \sigma^2$. The goal is to estimate this linear model, that is, estimate β_0, β_1, and σ^2 from the n observed pairs. To put our discussion in context, we consider a study of factors affecting patterns of insulin-dependent diabetes mellitus in children.

Example 14.1. **Diabetes Mellitus in Children.** Diabetes mellitus is a condition characterized by hyperglycemia resulting from the body's inability to use blood glucose for energy. In type 1 diabetes, the pancreas no longer makes insulin, and therefore blood glucose cannot enter the cells to be used for energy.

The objective of this study was to investigate the dependence of the level of serum C-peptide on various other factors in order to understand the patterns of residual insulin secretion. C-peptide is a protein produced by the beta cells of the pancreas whenever insulin is made. Thus, the level of C-peptide in the blood is an index of insulin production.

The part of the data from Sockett et al. (1987), discussed in the context of statistical modeling by Hastie and Tibshirani (1990), is given next. The response measurement is the logarithm of C-peptide concentration (pmol/ml) at the time of diagnosis, and the predictor is the base deficit, a measure of acidity.

Deficit (x)	−8.1	−16.1	−0.9	−7.8	−29.0	−19.2	−18.9	−10.6	−2.8	−25.0	−3.1
Log C-peptide (y)	4.8	4.1	5.2	5.5	5	3.4	3.4	4.9	5.6	3.7	3.9

Deficit (x)	−7.8	−13.9	−4.5	−11.6	−2.1	−2.0	−9.0	−11.2	−0.2	−6.1	−1
Log C-peptide (y)	4.5	4.8	4.9	3.0	4.6	4.8	5.5	4.5	5.3	4.7	6.6

Deficit (x)	−3.6	−8.2	−0.5	−2.0	−1.6	−11.9	−0.7	−1.2	−14.3	−0.8	−16.8
Log C-peptide (y)	5.1	3.9	5.7	5.1	5.2	3.7	4.9	4.8	4.4	5.2	5.1

Deficit (x)	−5.1	−9.5	−17.0	−3.3	−0.7	−3.3	−13.6	−1.9	−10.0	−13.5
Log C-peptide (y)	4.6	3.9	5.1	5.1	6.0	4.9	4.1	4.6	4.9	5.1

We will follow this example in MATLAB as an annotated step-by-step code/output of ◀ cpeptide.m. For more sophisticated analysis, MATLAB has quite advanced built-in regression tools, regress, regstats, robustfit, stepwise, and many other more-or-less specialized fitting and diagnostic tools.

After importing the data, we specify p, which is the number of parameters, rename the variables, and find the sample size.

```
Deficit =[-8.1 -16.1 -0.9 -7.8 -29.0 -19.2 -18.9 -10.6 -2.8...
  -25.0 -3.1 -7.8 -13.9 -4.5 -11.6 -2.1 -2.0 -9.0 -11.2 -0.2...
  -6.1 -1 -3.6 -8.2 -0.5 -2.0 -1.6 -11.9 -0.7 -1.2 -14.3 -0.8...
  -16.8 -5.1 -9.5 -17.0 -3.3 -0.7 -3.3 -13.6 -1.9 -10.0 -13.5];
```

```
logCpeptide =[ 4.8 4.1 5.2 5.5 5 3.4 3.4 4.9 5.6 3.7 3.9 ...
        4.5 4.8 4.9 3.0 4.6 4.8 5.5 4.5 5.3 4.7 6.6 5.1 3.9 ...
        5.7 5.1 5.2 3.7 4.9 4.8 4.4 5.2 5.1 4.6 3.9 5.1 5.1 ...
        6.0 4.9 4.1 4.6 4.9 5.1];
%%%%%%%%%%%%%%%%%%%%%%%%%%%%%%%%%%%%%%%%%
p = 2; %number of parameters, (beta0, beta1)
        %"Deficit" measurement is "x",   "logCpeptide" is "y".
x = Deficit' ;       %as a column vector
y = logCpeptide' ;   %as a column vector
n = length(x);
```

It is of interest to express the log C-peptide (variable y) as a linear function of alkaline deficiency (variable x), and the population model $y = \beta_0 + \beta_1 x + \epsilon$ is postulated. Finding estimators for β_0 and β_1 is an exercise in calculus – finding the extrema of a function of two variables. The following derivation is known as the least-squares method, which is a broad mathematical methodology for approximate solutions of overdetermined systems, first described by Gauss at the end of the eighteenth century. The best regression line minimizes the sum of squares of errors:

$$L = \sum_{i=1}^{n} \epsilon_i^2 = \sum_{i=1}^{n} (y_i - (\beta_0 + \beta_1 x_i))^2.$$

When pairs (x_i, y_i) are considered fixed, L is a function of β_0 and β_1 only. Minimizing L amounts to solving the so-called *normal* equations

$$\frac{\partial L}{\partial \beta_0} = -2 \sum_{i=1}^{n} [y_i - \beta_0 - \beta_1 x_i] = 0 \quad \text{and}$$

$$\frac{\partial L}{\partial \beta_1} = -2 \sum_{i=1}^{n} [x_i y_i - \beta_0 x_i - \beta_1 x_i^2] = 0,$$

that is,

$$n\beta_0 + \beta_1 \sum_{i=1}^{n} x_i = \sum_{i=1}^{n} y_i \quad \text{and}$$

$$\beta_0 \sum_{i=1}^{n} x_i + \beta_1 \sum_{i=1}^{n} x_i^2 = \sum_{i=1}^{n} x_i y_i. \tag{14.1}$$

Let $\bar{x} = \frac{1}{n} \sum_{i=1}^{n} x_i$ and $\bar{y} = \frac{1}{n} \sum_{i=1}^{n} y_i$ be the sample means of predictor values and the responses. If

$$S_{xy} = \sum_{i=1}^{n} (x_i - \bar{x})(y_i - \bar{y}) = \sum_{i=1}^{n} y_i (x_i - \bar{x}) = \sum_{i=1}^{n} x_i y_i - n\bar{x}\,\bar{y},$$

$$S_{xx} = \sum_{i=1}^{n}(x_i - \bar{x})^2 = \sum_{i=1}^{n}x_i^2 - n\bar{x}^2,$$

and

$$S_{yy} = \sum_{i=1}^{n}(y_i - \bar{y})^2 = \sum_{i=1}^{n}y_i^2 - n\bar{y}^2,$$

then the values for β_0 and β_1 minimizing L or, equivalently, solving the normal equations (14.1) are

$$\hat{\beta}_1 = \frac{S_{xy}}{S_{xx}} \quad \text{and} \quad \hat{\beta}_0 = \bar{y} - \hat{\beta}_1\bar{x}.$$

We will simplify the notation by denoting $\hat{\beta}_0$ by b_0 and $\hat{\beta}_1$ by b_1. Thus, the fitted regression equation is

$$\hat{y} = b_0 + b_1 x, \quad \text{with}$$

$$b_1 = \frac{S_{xy}}{S_{xx}} \quad \text{and} \quad b_0 = \bar{y} - b_1\bar{x}.$$

Note that the estimator of slope b_1 is connected with the sample correlation coefficient r_{XY},

$$b_1 = \frac{S_{xy}}{S_{xx}} = \frac{S_{xy}}{\sqrt{S_{xx}}\sqrt{S_{yy}}} \times \frac{\sqrt{S_{yy}}}{\sqrt{S_{xx}}} = r_{XY}\frac{\sqrt{S_{yy}}}{\sqrt{S_{xx}}}.$$

For values $x = x_i$, the fits $\hat{y}_i$ are obtained as

$$\hat{y}_i = b_0 + b_1 x_i,$$

with the residuals $e_i = y_i - \hat{y}_i$. The residuals are the most important diagnostic modality in regression. They explain how well the predicted data $\hat{y}_i$ fit the observations, and if the fit is not good, residuals indicate what caused the problem.

```
%Sums of Squares
SXX = sum( (x - mean(x)).^2 )   %SXX=2.1310e+003
SYY = sum( (y - mean(y)).^2 )   %SYY=21.807
SXY = sum( (x - mean(x)).* (y - mean(y)) ) %SXY=105.3477
%estimators of coefficients beta1 and beta0
b1 = SXY/SXX                    %0.0494
b0 = mean(y) - b1 * mean(x)     %5.1494
```

```
 % predictions
yhat = b0 + b1 * x;
 %residuals
res = y - yhat;
```

We found that `yhat=5.1494+0.0494*x`. Figure 14.2 shows a scatterplot of log C-peptide level (y) against alkaline deficiency (x) with superimposed regression fit $b_0 + b_1 x$.

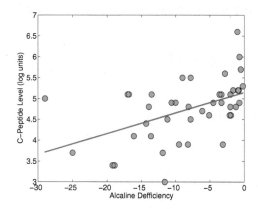

Fig. 14.2 Scatterplot of log C-peptide level (y) against alkaline deficiency (x). The regression fit (*red*) is $\hat{y} = 5.1494 + 0.0494x$.

The sum of squared residuals, $\sum_{i=1}^{n} e_i^2$, is denoted by SSE.

One can show that $SSE = S_{yy} - b_1 S_{xy}$ and that $\mathbb{E}(SSE) = (n - 2)\sigma^2$. Thus, the mean square error $MSE = SSE/(n - 2)$ is an unbiased estimator of error variance σ^2. Recall the fundamental ANOVA identity $SST = SSTr + SSE$. In regression terms, the fundamental ANOVA identity has the form

$$SST = SSR + SSE,$$

where $SST = S_{yy}$, $SSR = b_1 S_{xy}$, and $SSE = \sum_{i=1}^{n} e_i^2$. Since

$$\mathbb{E}SSR = \sigma^2 + \beta_1^2 S_{xx},$$

SSR has an associated 1 degree of freedom and the regression mean sum of squares MSR is $SSR/1 = SSR$.

The statistic MSR becomes an unbiased estimator of variance σ^2 when $\beta_1 = 0$. Thus, if $H_0 : \beta_1 = 0$ is true, one should have $F = MSR/MSE$ close to 1, since under H_0 both MSR and MSE estimate the same quantity, σ^2. Under H_0, the statistic $F = MSR/MSE$ has an F-distribution with 1 and $n - 2$ degrees of freedom.

Large values of F indicate that there is a contribution of β_1 in MSR, and discrepancy from H_0 can be assessed using an F-test. The sums of squares, degrees of freedom, mean squares, F-statistic and p-value associated with observed F are customarily summarized in an ANOVA table:

Source	DF	SS	MS	F	p-value
Regression	1	SSR	$MSR = SSR$	$F = \frac{MSR}{MSE}$	$\mathbb{P}(F_{1,n-2} > F)$
Error	$n-2$	SSE	$MSE = \frac{SSE}{n-2}$		
Total	$n-1$	SST			

The p-value is associated with testing of H_0, which essentially states that covariate x does not influence the response y, and the same fit can be obtained by just taking $\bar{y}$ as the model for y_is.

```
%ANOVA Identity
SST = sum((y - mean(y)).^2)     %this is also SYY
SSR = sum((yhat - mean(y)).^2)  %5.2079
SSE = sum((y - yhat).^2)        %=sum(res.^2), 16.599
% forming F and testing the adequacy of linear regression
MSR = SSR/(p - 1)   %5.2079
MSE = SSE/(n - p)   %estimator of variance, 0.4049
s = sqrt(MSE)       %0.6363
F = MSR/MSE         %12.8637
pvalue = 1-fcdf(F, p-1, n-p)
%testing H_0: regression has beta1=0,
%that is, there is no need for linear fit, p-val = 0.00088412
```

The preceding calculations are arranged in the ANOVA table:

Source	DF	SS	MS	F	p-value
Regression	1	5.2079	5.2079	12.8637	0.0009
Error	41	16.5990	0.4049		
Total	42	21.8070			

Figure 14.3a shows a plot of residuals $y_i - \hat{y}_i$ against x_i. A normalized histogram of residuals with a superimposed normal distribution $\mathcal{N}(0, 0.6363^2)$, is given in Figure 14.3b.

The quantity R^2, called the *coefficient of determination*, is defined as

$$R^2 = \frac{SSR}{SST} = 1 - \frac{SSE}{SST}.$$

The R^2 in this context coincides with the square of the correlation coefficient between $(x_1, \ldots, x_n)$ and $(y_1, \ldots, y_n)$. However, the representation of R^2 via the ratio SSR/SST is more illuminating. In words, R^2 explains what proportion of the total variability (SST) encountered in observations is explained or accounted for by the regression (SSR). Thus, a high R^2 is desirable in any regression. Note that $F = \frac{MSR}{MSE} = \frac{(n-2)R^2}{1-R^2}$ and the F test is

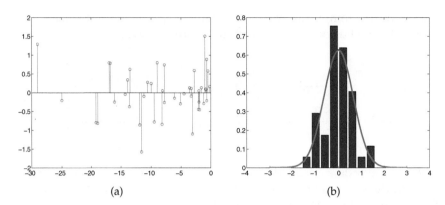

(a)   (b)

Fig. 14.3 (a) Plot of residuals $e_i = y_i - \hat{y}_i$ against x. (b) Normalized histogram of residuals with superimposed normal distribution $\mathcal{N}(0, s^2)$, with s estimated as $\sqrt{MSE} = 0.6363$.

equivalent to testing that the population correlation coefficient ρ (the correlation between X's and Y's) is significant.

The adjusted R^2 (Ezekiel, 1930), defined as

$$R^2_{adj} = 1 - \frac{n-1}{n-p} \cdot \frac{SSE}{SST} = 1 - \frac{MSE}{MST},$$

where $MST = SST/(n-1) = s^2_y$, is important in cases with multiple predictors ($p > 2$), since it penalizes inclusion of predictors in the model.

```
%  Other measures of goodness of fit
R2 = SSR/SST       %0.2388
R2adj = 1 - (n-1)/(n-p)* SSE/SST     %0.2203
```

Often, instead of regressing y_i on x_i, one regresses y_i on $x_i - \bar{x}$ as

$$y_i = \beta_0^* + \beta_1(x_i - \bar{x}).$$

This is beneficial for several reasons. In practice, we calculate only the estimator b_1. Since the fitted line contains the point $(\bar{x}, \bar{y})$, the intercept β_0^* is estimated by $\bar{y}$, and our regression fit is

$$\hat{y}_i = \bar{y} + b_1(x_i - \bar{x}).$$

In the Bayesian context, estimating β_0^* and β_1 is more stable and efficient than estimating β_0 and β_1 directly, since $\bar{y}$ and b_1 are uncorrelated.

Estimators b_0 and b_1 are unbiased estimators of population's β_0 and β_1. We will show that they are unbiased and that their variance is intimately connected with the variance of responses, σ^2:

$$\mathbb{E}b_1 = \beta_1 \quad \text{and} \quad \mathbb{V}\text{ar}\, b_1 = \frac{\sigma^2}{S_{xx}},$$

$$\mathbb{E}b_0 = \beta_0 \quad \text{and} \quad \mathbb{V}\text{ar}\, b_0 = \sigma^2 \left(\frac{1}{n} + \frac{(\bar{x})^2}{S_{xx}}\right).$$

Here is the rationale:

$$\mathbb{E}b_1 = \mathbb{E}\frac{S_{xy}}{S_{xx}} = \frac{1}{S_{xx}}\mathbb{E}\sum_{i=1}^{n} y_i(x_i - \bar{x})$$

$$= \frac{1}{S_{xx}}\mathbb{E}\sum_{i=1}^{n}(\beta_0^* + \beta_1(x_i - \bar{x}) + \epsilon_i)(x_i - \bar{x})$$

$$= \frac{1}{S_{xx}}\left[\sum_{i=1}^{n}\beta_0^*(x_i - \bar{x}) + \sum_{i=1}^{n}\beta_1(x_i - \bar{x})^2 + \mathbb{E}\sum_{i=1}^{n}\epsilon_i(x_i - \bar{x})\right]$$

$$= \frac{1}{S_{xx}}[0 + \beta_1 S_{xx} + 0] = \beta_1.$$

$$\mathbb{V}\text{ar}\, b_1 = \mathbb{V}\text{ar}\frac{S_{xy}}{S_{xx}} = \frac{1}{S_{xx}^2}\sum_{i=1}^{n}\mathbb{V}\text{ar}\,(y_i(x_i - \bar{x}))$$

$$= \frac{1}{S_{xx}^2}\sum_{i=1}^{n}\sigma^2(x_i - \bar{x})^2 = \frac{\sigma^2}{S_{xx}}.$$

Since $b_0 = \bar{y} - b_1\bar{x}$,

$$\mathbb{E}b_0 = \mathbb{E}(\bar{y} - b_1\bar{x}) = \beta_0 + \beta_1\bar{x} - \beta_1\bar{x} = \beta_0$$

and

$$\mathbb{V}\text{ar}\, b_0 = \mathbb{V}\text{ar}\,\bar{y} + \mathbb{V}\text{ar}\,(b_1\bar{x}) - 2\,\mathbb{C}\text{ov}(\bar{y}, b_1\bar{x})$$

$$= \frac{\sigma^2}{n} + (\bar{x})^2\frac{\sigma^2}{S_{xx}} - 2 \cdot 0 = \sigma^2\left[\frac{1}{n} + \frac{(\bar{x})^2}{S_{xx}}\right].$$

An alternative expression for $\mathbb{V}\text{ar}\, b_0$ is $\sigma^2\frac{\overline{x^2}}{S_{xx}}$, for $\overline{x^2} = \frac{1}{n}\sum_{i=1}^{n}x_i^2$. Sample counterparts of $\mathbb{V}\text{ar}\, b_0$ and $\mathbb{V}\text{ar}\, b_1$ will be needed for the inference in subsequent sections; they are obtained by plugging in the MSE in place of σ^2.

The covariance between b_0 and b_1 is

$$\mathrm{Cov}(b_0, b_1) = \mathrm{Cov}(\bar{y} - b_1 \cdot \bar{x}, b_1) = \mathrm{Cov}(\bar{y}, b_1) - \bar{x} \cdot \mathrm{Var}(b_1) = -\bar{x} \cdot \frac{\sigma^2}{S_{xx}},$$

since $\mathrm{Cov}(\bar{y}, b_1) = 0$. The correlation between b_0 and b_1 is then readily found as

$$\mathrm{Corr}(b_0, b_1) = -\frac{\bar{x}}{\sqrt{\frac{1}{n} \sum_{i=1}^{n} x_i^2}}.$$

In MATLAB the estimator of σ and sample standard deviations of estimators b_0 and b_1 from Example 14.1 are as follows:

```
% s, sb0, and sb1
 s = sqrt(MSE)   %s = 0.6363
%Standard errors of parameter estimators
sb1 = s/sqrt(SXX)    %sb1 = 0.0138
sb0 = s * sqrt(1/n + (mean(x))^2/SXX )   %sb0 = 0.1484
```

14.3 Inference in Simple Linear Regression

To find the estimators of regression parameters and calculate their expectations and variances, we do not need the distributional properties of errors, except that they are independent, have a mean 0, and a variance that does not vary with x. However, to test the hypotheses about the population intercept and slope, and to find confidence intervals, we need to assume that the errors ϵ_i are i.i.d. normal. In practice, the residual analysis is conducted to verify whether the normality assumption is justified.

14.3.1 Inference about the Slope Parameter

For a given constant β_{10}, the test for

$$H_0 : \beta_1 = \beta_{10}$$

relies on the statistic

$$t = \frac{b_1 - \beta_{10}}{\sqrt{s^2 / S_{xx}}},$$

where $s^2 = MSE$. This statistic under H_0 has a t-distribution with $n - 2$ degrees of freedom, and testing is done as follows:

Alternative	α-level rejection region	p-value (MATLAB)
$H_1 : \beta_1 > \beta_{10}$	$[t_{n-2,1-\alpha}, \infty)$	1-tcdf(t,n-2)
$H_1 : \beta_1 \neq \beta_{10}$	$(-\infty, t_{n-2,\alpha/2}] \cup [t_{n-2,1-\alpha/2}, \infty)$	2*tcdf(-abs(t),n-2)
$H_1 : \beta_1 < \beta_{10}$	$(-\infty, t_{n-2,\alpha}]$	tcdf(t,n-2)

The distribution of the test statistic is derived from a linear representation of b_1 as

$$b_1 = \sum_{i=1}^{n} a_i y_i, \quad a_i = \frac{x_i - \overline{x}}{S_{xx}}.$$

Under H_0, $b_1 \sim \mathcal{N}(\beta_{10}, \sigma^2/S_{xx})$. Thus,

$$t = \frac{b_1 - \beta_{10}}{\sqrt{s^2/S_{xx}}} = \frac{b_1 - \beta_{10}}{\sqrt{\sigma^2/S_{xx}}} \times \frac{\sigma}{s} = \frac{\frac{b_1 - \beta_{10}}{\sqrt{\sigma^2/S_{xx}}}}{\sqrt{\frac{SSE}{(n-2)\sigma^2}}},$$

which, by definition, has a t_{n-2}-distribution, as $Z/\sqrt{\frac{\chi_{n-2}^2}{n-2}}$. We also used the fact that $s^2 = MSE = SSE/(n-2)$.

The $(1-\alpha)100\%$ confidence interval for β_1 is

$$\left[b_1 - t_{n-2,1-\alpha/2} \frac{s}{\sqrt{S_{xx}}}, \; b_1 + t_{n-2,1-\alpha/2} \frac{s}{\sqrt{S_{xx}}} \right].$$

14.3.2 Inference about the Intercept Parameter

For a given constant β_{00}, the test for

$$H_0 : \beta_0 = \beta_{00}$$

relies on the statistic

$$t = \frac{b_0 - \beta_{00}}{s\sqrt{\frac{1}{n} + \frac{(\overline{x})^2}{S_{xx}}}}.$$

Under H_0 this statistic has a t-distribution with $n - 2$ degrees of freedom, and testing is done as follows:

Alternative	α-level rejection region	p-value (MATLAB)
$H_1 : \beta_0 > \beta_{00}$	$[t_{n-2,1-\alpha}, \infty)$	1-tcdf(t,n-2)
$H_1 : \beta_0 \neq \beta_{00}$	$(-\infty, t_{n-2,\alpha/2}] \cup [t_{n-2,1-\alpha/2}, \infty)$	2*tcdf(-abs(t),n-2)
$H_1 : \beta_0 < \beta_{00}$	$(-\infty, t_{n-2,\alpha}]$	tcdf(t,n-2)

This is based on the representation of b_0 as $b_0 = \bar{y} - b_1\bar{x}$ and under H_0 $b_0 \sim \mathcal{N}\left(\beta_{00}, \sigma^2\left(\frac{1}{n} + \frac{(\bar{x})^2}{S_{xx}}\right)\right)$. Thus,

$$t = \frac{b_0 - \beta_{00}}{s\sqrt{1/n + (\bar{x})^2/S_{xx}}} = \frac{b_1 - \beta_{00}}{\sigma\sqrt{1/n + (\bar{x})^2/S_{xx}}} \times \frac{\sigma}{s} = \frac{\frac{b_0 - \beta_{00}}{\sigma\sqrt{1/n + (\bar{x})^2/S_{xx}}}}{\sqrt{\frac{SSE}{(n-2)\sigma^2}}},$$

which by definition has a t_{n-2}-distribution, as $Z/\sqrt{\frac{\chi^2_{n-2}}{n-2}}$.

The $(1 - \alpha)100\%$ confidence interval for β_0 is

$$\left[b_0 - t_{n-2,1-\alpha/2}\, s\sqrt{\frac{1}{n} + \frac{(\bar{x})^2}{S_{xx}}}, \; b_0 + t_{n-2,1-\alpha/2}\, s\sqrt{\frac{1}{n} + \frac{(\bar{x})^2}{S_{xx}}}\right].$$

```
% Are the coefficients equal to 0?
t1 = b1/sb1  %3.5866
pb1 = 2 * (1-tcdf(abs(t1),n-p) ) %8.8412e-004
t0 = b0/sb0  %34.6927
pb0 = 2 * (1-tcdf(abs(t0),n-p) ) %0

% Test H_0: beta1 = 0.04 vs. H_1: beta1 > 0.04
tst1 = (b1 - 0.04)/sb1          %0.6846
ptst1 = 1 - tcdf( tst1, n-p )    %0.2487
% Test H_0: beta0 = 5.8 vs. H_1: beta0 < 5.8
tst2 = (b0 - 5.8)/sb0           %-4.3836
ptst2 = tcdf(tst2, n-p )         %3.9668e-005
%%%%%%%%%%%%%%%%%%%%%%%%%%%%%%%%%%%%%
% Find 95% CI for beta1
[b1 - tinv(0.975, n-p)*sb1,  b1 + tinv(0.975, n-p)*sb1]
%  0.0216    0.0773
% Find 99% CI for beta0
[b0 - tinv(0.995, n-p)*sb0,  b0 + tinv(0.995, n-p)*sb0]
%  4.7484    5.5503
```

14.3.3 Inference about the Variance

Testing $H_0 : \sigma^2 = \sigma_0^2$ relies on the statistic $\chi^2 = \frac{(n-2)MSE}{\sigma_0^2} = \frac{SSE}{\sigma_0^2}$. This statistic under H_0 has a χ^2-distribution with $n - 2$ degrees of freedom and testing is done as follows:

Alternative	α-level rejection region	p-value (MATLAB)
$H_1 : \sigma^2 < \sigma_0^2$	$[0, \chi_{n-2,\alpha}^2]$	`chi2cdf(chi2,n-2)`
$H_1 : \sigma^2 \neq \sigma_0^2$	$[0, \chi_{n-2,\alpha/2}^2] \cup [\chi_{n-2,1-\alpha/2}^2, \infty)$	`2*chi2cdf(ch,n-2)`
$H_1 : \sigma^2 > \sigma_0^2$	$[\chi_{n-2,1-\alpha}^2, \infty)$	`1-chi2cdf(chi2,n-2)`

where `chi2` is the test statistic and `ch=min(chi2,1/chi2)`.
The $(1 - \alpha)100\%$ confidence interval for σ^2 is

$$\left[\frac{SSE}{\chi_{n-2,1-\alpha/2}^2}, \frac{SSE}{\chi_{n-2,\alpha/2}^2} \right].$$

The following MATLAB script tests $H_0 : \sigma^2 = 0.5$ versus $H_1 : \sigma^2 < 0.5$ and finds a 95% confidence interval for σ^2. As is evident, H_0 is not rejected (p-value 0.1981), and the interval is $[0.2741, 0.6583]$.

```
% Test H_0: sigma2 = 0.5 vs. H_1: sigma2 < 0.5
ch2 = SSE/0.5     %33.1981
ptst3 = chi2cdf(ch2, n-p)     %0.1981
% Find 95% CI for sigma2
[SSE/chi2inv(0.975, n-p),   SSE/chi2inv(0.025, n-p)]
%    0.2741    0.6583
```

14.3.4 Inference about the Mean Response

Suppose that the regression $\hat{y} = b_0 + b_1 x$ has been found and that we are interested in making an inference about the response $y_m = \mathbb{E}(y|x = x^*) = \beta_0 + \beta_1 x^*$. The statistic for y_m is $\hat{y}_m = b_0 + b_1 x^*$, and it is a random variable since both b_0 and b_1 are random variables.

The $\hat{y}_m$ is an unbiased estimator of y_m, $\mathbb{E}\hat{y}_m = E(b_0 + b_1 x^*) = \beta_0 + \beta_1 x^* = y_m$, as expected. The variance of $\hat{y}_m$ is obtained from representation $\hat{y}_m = b_0 + b_1 x^* = \bar{y} + b_1(x^* - \bar{x})$ and the fact that the correlation between $\bar{y}$ and b_1 is zero:

$$\mathbb{V}\mathrm{ar}\,\hat{y}_m = \sigma^2 \left(\frac{1}{n} + \frac{(x^* - \bar{x})^2}{S_{xx}} \right).$$

Thus,

$$\hat{y}_m \sim \mathcal{N} \left(\beta_0 + \beta_1 x^*, \sigma^2 \left(\frac{1}{n} + \frac{(x^* - \bar{x})^2}{S_{xx}} \right) \right),$$

from which we develop the inference.

The test

$$H_0 : y_m = y_0$$

relies on the statistic

$$t = \frac{\hat{y}_m - y_0}{s \sqrt{\frac{1}{n} + \frac{(x^* - \bar{x})^2}{S_{xx}}}}.$$

This statistic under H_0 has a t-distribution with $n - 2$ degrees of freedom and testing is done as in the cases of β_0 and β_1.

The $(1 - \alpha)100\%$ confidence interval for $y_m = \beta_0 + \beta_1 x^*$ is

$$\left[\hat{y}_m - t_{n-2,1-\alpha/2}\, s \sqrt{\frac{1}{n} + \frac{(x^* - \bar{x})^2}{S_{xx}}}\,,\ \hat{y}_m + t_{n-2,1-\alpha/2}\, s \sqrt{\frac{1}{n} + \frac{(x^* - \bar{x})^2}{S_{xx}}} \right].$$

14.3.5 Inference about a New Response

Suppose that the regression $\hat{y} = b_0 + b_1 x$ has been established and that we are interested in predicting the response $\hat{y}_{pred}$ for a new observation, corresponding to a covariate $x = x^*$. Given the value $x = x^*$, the difference between the inference about the mean response y_m discussed in the previous section and the inference about an individual outcome y_{pred} is substantial. As in the previous subsection, $\hat{y}_{pred} = b_0 + b_1 x^*$, and the mean of $\hat{y}_{pred}$ is $\mathbb{E}(\hat{y}_{pred}) = \beta_0 + \beta_1 x^* = y_{pred}$, which is in fact equal to y_m.

Where y_{pred} and y_m differ is in their variability. The variability of $\hat{y}_{pred}$ has two sources, first, the variance of the distribution of ys for $x = x^*$, which is σ^2, and, second, the variance of sampling distribution for $b_0 + b_1 x^*$, which is $\sigma^2 \left(\frac{1}{n} + \frac{(x^* - \bar{x})^2}{S_{xx}} \right)$. Thus, $\mathbb{V}\mathrm{ar}(\hat{y}_{pred}) = MSE + \mathbb{V}\mathrm{ar}(\hat{y}_m)$.

The distribution for $\hat{y}_{pred}$ is normal,

$$\hat{y}_{pred} \sim \mathcal{N}\left(\beta_0 + \beta_1 x^*, \ \sigma^2 \left(1 + \frac{1}{n} + \frac{(x^* - \bar{x})^2}{S_{xx}}\right)\right),$$

and the subsequent inference is based on this distribution.

The test

$$H_0 : y_{pred} = y_0$$

relies on the statistic

$$t = \frac{\hat{y}_{pred} - y_0}{s\sqrt{1 + \frac{1}{n} + \frac{(x^* - \bar{x})^2}{S_{xx}}}}.$$

This statistic under H_0 has a t-distribution with $n - 2$ degrees of freedom, which implies the inference.

The $(1 - \alpha)100\%$ confidence interval for y_{pred} is

$$\left[\hat{y}_{pred} - t_{n-2,1-\alpha/2}\, s\sqrt{1 + \frac{1}{n} + \frac{(x^* - \bar{x})^2}{S_{xx}}}, \ \hat{y}_{pred} + t_{n-2,1-\alpha/2}\, s\sqrt{1 + \frac{1}{n} + \frac{(x^* - \bar{x})^2}{S_{xx}}}\right].$$

```
% predicting y for the new observation x, CI and PI
newx = -5; %Deficit = -5
y_newx = b0 + b1 * newx  % 4.9022
sym = s * sqrt(1/n + (mean(x) - newx)^2/SXX )
  %st.dev. for mean response, sym = 0.1063
syp = s * sqrt(1 + 1/n + (mean(x) - newx)^2/SXX )
  %st.dev. for the prediction syp = 0.6451
alpha = 0.05;
  %mean response interval
lbym = y_newx - tinv(1-alpha/2, n-p) * sym;
rbym = y_newx + tinv(1-alpha/2, n-p) * sym;
  % prediction interval
lbyp = y_newx - tinv(1-alpha/2, n-p) * syp;
rbyp = y_newx + tinv(1-alpha/2, n-p) * syp;
  % the intervals
[lbym rbym]   % 4.6875    5.1168
[lbyp rbyp]   % 3.5994    6.2050
```

Remark. Suppose that for $x = x^*$, instead of a single new response, we anticipate m new responses and wish to find the prediction interval for their average. The prediction interval in this case is obtained by replacing $\sqrt{1 + \frac{1}{n} + \frac{(x^* - \bar{x})^2}{S_{xx}}}$ with $\sqrt{\frac{1}{m} + \frac{1}{n} + \frac{(x^* - \bar{x})^2}{S_{xx}}}$.

Next, we will find Bayesian estimators of the regression parameters in the same example, *Diabetes Mellitus in Children*, by using WinBUGS. On page 688 we mentioned that taking $x_i - \bar{x}$ as a predictor instead of x_i is beneficial in the Bayesian context. From such a parametrization of regression,

$$y_i = \beta_0^* + \beta_1(x_i - \bar{x}) + \epsilon_i,$$

the traditional intercept β_0 is then obtained as $\beta_0^* - \beta_1 \bar{x}$.

```
model{
 for (i in 1:ntotal){
 y[i] ~ dnorm( mui[i], tau )
 mui[i] <-  bb.0 + b.1 *(x[i] - mean(x[]))
 yres[i] <- y[i] - mui[i]
 }
 bb.0 ~ dnorm(0, 0.0001)
 b.0 <- bb.0 - b.1 * mean(x[])
 b.1 ~ dnorm(0, 0.0001)
 tau ~ dgamma(0.001, 0.001)
 s <- 1/sqrt(tau)
 }


DATA
list(ntotal=43,
y = c(4.8, 4.1, 5.2, 5.5, 5.0, 3.4, 3.4, 4.9, 5.6, 3.7,
      3.9, 4.5, 4.8, 4.9, 3.0, 4.6, 4.8, 5.5, 4.5, 5.3,
      4.7, 6.6, 5.1, 3.9, 5.7, 5.1, 5.2, 3.7, 4.9, 4.8,
      4.4, 5.2, 5.1, 4.6, 3.9, 5.1, 5.1, 6.0, 4.9, 4.1,
      4.6, 4.9, 5.1),
x = c(-8.1, -16.1, -0.9, -7.8, -29.0, -19.2, -18.9, -10.6,
      -2.8, -25.0, -3.1, -7.8, -13.9, -4.5,  -11.6, -2.1,
      -2.0, -9.0, -11.2, -0.2, -6.1, -1.0, -3.6, -8.2,
      -0.5, -2.0, -1.6, -11.9, -0.7, -1.2, -14.3, -0.8,
      -16.8, -5.1, -9.5, -17.0, -3.3, -0.7, -3.3, -13.6,
      -1.9, -10.0, -13.5))

INITS
list(bb.0 = 0, b.1 = 0, tau=1)
```

The output is given in the table below. It contains Bayesian estimators `b.0` for β_0 and `b.1` for β_1. In the least-squares regression we found that $b_1 = S_{xy}/S_{xx} = 0.0494$, $b_0 = \bar{y} - b_1 \cdot \bar{x} = 5.1494$, and $s = \sqrt{MSE} = 0.6363$. Since priors were noninformative, we expect that the Bayes estimators will be close to the classical. Indeed that is the case: `b.0 = 5.149`, `b.1 = 0.0494`, and `s = 0.6481`.

The classical standard errors of estimators for β_0 and β_1 are sb0 = 0.1484 and sb1 = 0.0138, while the corresponding Bayesian estimators are 0.1525 and 0.01418. The classical 95% confidence interval for β_1 was found to be [0.0216,0.0773]. The Bayesian 95% credible set for β_1 is [0.02139,0.07733], as is evident from val2.5pc and val97.5pc in the output below:

	mean	sd	MC error	val2.5pc	median	val97.5pc	start	sample
b.0	5.149	0.1525	3.117E-4	4.848	5.149	5.449	2001	200000
b.1	0.0494	0.0141	3.072E-5	0.02139	0.04944	0.07733	2001	200000
s	0.6481	0.0734	1.771E-4	0.5236	0.6415	0.811	2001	200000
yres[1]	0.05111	0.09944	2.175E-4	-0.1444	0.05125	0.2472	2001	200000
yres[2]	-0.2537	0.1502	3.459E-4	-0.5499	-0.2533	0.0418	2001	200000
yres[3]	0.09544	0.1431	2.925E-4	-0.1861	0.09505	0.378	2001	200000
yres[4]	0.7363	0.09957	2.167E-4	0.5406	0.7364	0.9325	2001	200000
...								
yres[41]	-0.4552	0.1333	2.727E-4	-0.7173	-0.4555	-0.1919	2001	200000
yres[42]	0.245	0.1028	2.314E-4	0.04251	0.2451	0.4475	2001	200000
yres[43]	0.6179	0.125	2.879E-4	0.3718	0.6179	0.8632	2001	200000

Thus, the Bayesian approach to regression estimation is quite close to the classical when the priors on β_0 and β_1 and the precision $\tau = 1/\sigma^2$ are noninformative.

Example 14.2. **Hubble Regression.** Hubble's constant (H) is one of the most important numbers in cosmology because it is instrumental in estimating the size and age of the universe. This long-sought number indicates the rate at which the universe is expanding, from the primordial "Big Bang." The Hubble constant can be used to determine the intrinsic brightness and masses of stars in nearby galaxies, examine those same properties in more distant galaxies and galaxy clusters, deduce the amount of dark matter present in the universe, obtain the scale size of faraway galaxy clusters, and serve as a test for theoretical cosmological models.

In 1929, Edwin Hubble, a distinguished American astronomer, investigated the relationship between the distance of a galaxy from the Earth and the velocity with which it appears to be receding. Galaxies appear to be moving away from us no matter which direction we look. This is thought to be the result of the "Big Bang." Hubble hoped to provide some knowledge about how the universe was formed and what might happen in the future. The data collected included distances (megaparsecs[1]) to $n = 24$ galaxies and their recessional velocities (km/sec).

Hubble's law is as follows: Recessional velocity = H × distance, where H is Hubble's constant (units of H are [km/sec/Mpc]). By working backwards in time, the galaxies appear to meet in the same place. Thus 1/H can be used to estimate the time since the Big Bang, a measure of the age of the universe.

---

[1] 1 parsec = 3.26 light years

Distance in megaparsecs ([Mpc])	0.032	0.034	0.214	0.263	0.275	0.275
	0.45	0.5	0.5	0.63	0.8	0.9
	0.9	0.9	0.9	1.0	1.1	1.1
	1.4	1.7	2.0	2.0	2.0	2.0
Recessional velocity ([km/sec])	170	290	−130	−70	−185	−220
	200	290	270	200	300	−30
	650	150	500	920	450	500
	500	960	500	850	800	1090

A regression analysis seems appropriate; however, there is no intercept term in Hubble's law. Can you verify that the constant term of the regression analysis is not significantly different than 0 at any *reasonable*[2] level of α. Find the 95% confidence interval for the slope β_1, also known as Hubble's constant H, from the given data.

The age of the universe as predicted by Hubble (in years) is about 2.3 billion years.

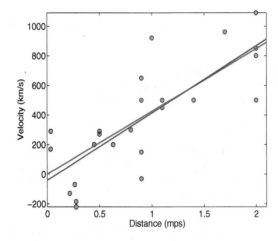

Fig. 14.4 Hubble's data and regression fits. The *blue line* is an unconstrained regression (with intercept fitted), and the *red line* is a no-intercept fit. The slope for the no-intercept fit is $b_1 = 423.9373$ (=H).

```
%H = 423.9373
secinyear =60*60*24*365   %31536000
kminmps = 3.08568025 * 10^19;
age = 1/H  *  kminmps/secinyear   %2.3080e+009
```

Modern measurements put H at approx. 70, thus predicting the age of the universe to about 14 billion years. Figure 14.4 showing Hubble's data and regression fits is generated by ◀ hubble.m.

---

[2] Reasonable here means level α not larger than 0.10.

14.4 Calibration

Often in regression problems we need to make an inference about the predictor x when a response y is observed or assumed. This typically arises in the context of instrument calibration, which gives the name to the statistical methodology used to solve this kind of problems.

A naïve solution, sometimes referred as the *reverse method*, is to reverse the roles of x and y, fit a regression $\hat{x} = c_0 + c_1 y$, and apply the results from Section 14.3.5. The problem with this approach is that it assumes that measurements y_i are observed without error, while x_is are observed with error, which may not be the case since x's are typically controlled. More seriously, when y_i's are random, the values $y^*, \bar{y}$, and S_{yy} in

$$s_{x_0} = s_x \sqrt{1 + \frac{1}{n} + \frac{(y^* - \bar{y})^2}{S_{yy}}}, \quad s_x = \sqrt{\frac{S_{xx}}{n-2}},$$

are not fixed constants but random, unlike $x^*, \bar{x}$, and S_{xx} in a reverse counterpart $s\sqrt{1 + \frac{1}{n} + \frac{(x^* - \bar{x})^2}{S_{xx}}}$ from Section 14.3.5.

The standard method uses the original regression $\hat{y} = b_0 + b_1 x$ and, for the response y^*, predicts x^* as

$$\hat{x}^* = \frac{1}{b_1}(y^* - b_0).$$

This method is called *inverse method* because the linear equation is inverted, In this case the expectation and variance of $\hat{x}^*$ are approximated as

$$\mathbb{E}\,\hat{x}^* \approx x^* + \frac{\sigma^2}{\beta_1^2 S_{xx}}(x^* - \bar{x}), \quad \text{Var}\,\hat{x}^* \approx \frac{\sigma^2}{\beta_1^2}\left(1 + \frac{1}{n} + \frac{(x^* - \bar{x})^2}{S_{xx}}\right).$$

The $(1 - \alpha)100\%$ prediction interval for x^* is

$$\left[\hat{x}^* - t_{n-2,1-\alpha/2}\frac{s}{b_1}\sqrt{1 + \frac{1}{n} + \frac{(\hat{x}^* - \bar{x})^2}{S_{xx}}},\ \hat{x}^* + t_{n-2,1-\alpha/2}\frac{s}{b_1}\sqrt{1 + \frac{1}{n} + \frac{(\hat{x}^* - \bar{x})^2}{S_{xx}}}\right].$$

If for a single x^*, $y_1^*, y_2^*, \ldots, y_m^*$ are observed, then y^* from above is replaced by the mean $\bar{y}^*$, $\hat{x}^* = \frac{1}{b_1}(\bar{y}^* - b_0)$, and the prediction interval becomes

$$\left[\hat{x}^* - t_{n-2,1-\alpha/2} \frac{s}{b_1} \sqrt{\frac{1}{m} + \frac{1}{n} + \frac{(\hat{x}^* - \overline{x})^2}{S_{xx}}} , \hat{x}^* + t_{n-2,1-\alpha/2} \frac{s}{b_1} \sqrt{\frac{1}{m} + \frac{1}{n} + \frac{(\hat{x}^* - \overline{x})^2}{S_{xx}}} \right].$$

Example 14.3. **Concentration of Caprolactone.** Thonnard (2006) analyzes data on 10 solutions of caprolactone in the solvent tetrahydrofuran. As a control, a solution without the caprolactone is also provided.

Each of these 11 solutions was injected in a gaschromatograph three times. The measures from each injection are recorded in a form of surface, which translates to the estimated concentration. The known solution concentrations x for the 10 solutions are paired with the surface readings y, three for each concentration. The no caprolactone solution results in three pairs $(0, 0)$. Thus, the data presented in Table 14.1 consist of 33 (x, y) pairs:

Observations	Concentration x (in (g/l))	Surface Measures y
1–3	9.71	24.276, 24.083, 24.276
4–6	8.52	20.206, 20.199, 20.223
7–9	7.96	19.773, 19.759, 19.765
10–12	6.82	16.743, 16.587, 16.744
13–15	5.85	15.081, 15.121, 15.274
16–18	4.95	12.636, 12.641, 12.682
19–21	3.91	9.869, 9.906, 9.883
22–24	2.98	7.624, 7.592, 7.585
25–27	2.07	4.638, 4.666, 4.649
28–30	1.02	2.860, 2.859, 2.896
31–33	0.00	0.00, 0.00, 0.00

Table 14.1 Concentration and surface of caprolactone in the solution.

Suppose that we have a new solution, for which we do not know the concentration of caprolactone, x^*. After injecting this unknown solution three times in the gaschromatograph, three observations of y are obtained, $y_1^* = 1.582, y_2^* = 1.793$, and $y_3^* = 1.787$. We will estimate x^* and provide the 95% prediction interval.

MATLAB script ◢ caprolactone.m finds the prediction $\hat{x}^* = 0.6329$ and the 95% prediction interval for x^* as $[0.4047, 0.8611]$.

```
data=[
    9.71  24.276
    9.71  24.083
    9.71  24.276
    8.52  20.206
    8.52  20.199
    8.52  20.223
    ...
```

```
    1.02    2.860
    1.02    2.859
    1.02    2.896
    0.00    0.000
    0.00    0.000
    0.00    0.000];
 x = data(:,1); y=data(:,2);
 n=length(x)  %33
SXX = sum( (x - mean(x)).^2 ) %305.8026
SXY = sum( (x - mean(x)).* (y - mean(y)) ) %749.4972
b1 = SXY/SXX %2.4509
b0 = mean(y) - b1 * mean(x)  %0.1694
s=sqrt(sum(  (y - (b0+b1*x)).^2 )/(n-2)) %0.4216
%
ystars=[1.582  1.793 1.787]; m=length(ystars);
ystar = mean(ystars);
xbar=mean(x);
xstar = 1/b1 * (ystar - b0)    %0.6329
LB=xstar - tinv(0.975, n-2) * s/b1 * sqrt(1/m+1/n+(xstar-xbar)^2/SXX);
UB=xstar + tinv(0.975, n-2) * s/b1 * sqrt(1/m+1/n+(xstar-xbar)^2/SXX);
[LB xstar UB] %[0.4047 0.6329 0.8611]
```

This was the *inverse* method. Compare this solution with the *reverse* method in which x and y are flipped and the prediction interval from Section 14.3.5 is used. ✐

14.5 Testing the Equality of Two Slopes*

Let (x_{1i}, y_{1i}), $i = 1, \ldots, n_1$ and (x_{2i}, y_{2i}), $i = 1, \ldots, n_2$, be pairs of measurements obtained from two groups, and for each group the regression is estimated as

$$y_{1i} = b_{0(1)} + b_{1(1)} x_{1i} + e_{i(1)}, i = 1, \ldots, n_1, \text{ and}$$
$$y_{2i} = b_{0(2)} + b_{1(2)} x_{2i} + e_{i(2)}, i = 1, \ldots, n_2,$$

where in groups $i = 1, 2$ the statistics $b_{0(i)}$ and $b_{1(i)}$ are estimators of the respective population parameters, intercepts $\beta_{0(i)}$, and slopes $\beta_{1(i)}$. We are interested in testing the equality of the population slopes,

$$H_0 : \beta_{1(1)} = \beta_{1(2)},$$

against the one- or two-sided alternatives.
    The test statistic is

$$t = \frac{b_{1(1)} - b_{1(2)}}{s.e.(b_{1(1)} - b_{1(2)})}, \tag{14.2}$$

where the standard error of the difference $b_{1(1)} - b_{1(2)}$ is

$$s.e.(b_{1(1)} - b_{1(2)}) = \sqrt{s^2 \left[\frac{1}{S_{xx(1)}} + \frac{1}{S_{xx(2)}} \right]},$$

and s^2 is the pooled estimator of variance,

$$s^2 = \frac{SSE_1 + SSE_2}{n_1 + n_2 - 4}.$$

Statistic t in (14.2) has a t-distribution with $n_1 + n_2 - 4$ degrees of freedom and, in addition to testing, could be used for a $(1 - \alpha)$ 100% confidence interval for $\beta_{1(1)} - \beta_{1(2)}$,

$$[(b_{1(1)} - b_{1(2)}) \mp t_{n_1+n_2-4,1-\alpha/2} \times s.e.(b_{1(1)} - b_{1(2)})].$$

Example 14.4. **Cadmium Poisoning.** Chronic cadmium poisoning is an insidious disease associated with the development of emphysema and the excretion in the urine of a characteristic protein of low molecular weight. The first signs of chronic cadmium poisoning become apparent following a latent interval after exposure has ended. Respiratory functions deteriorate faster with in age. The data set featured in Armitage and Berry (1994) gives ages (in years) and vital capacity (in liters) for 84 men working in the cadmium industry, cadmium.dat|mat|xlsx. The observations with flag exposure equal to 0 denote persons unexposed to cadmium oxide fumes, while flags 1 and 2 correspond to exposed persons. The purpose of the study was to assess the degree of influence of exposure to respiratory functions. Since respiratory functions are influenced by age, regardless of exposure, age as a covariate needs to be taken into account. Thus, the suggested methodology is to test the equality of the slopes in group regressions of vital capacity to age:

$$H_0 : \beta_{1(exposed)} = \beta_{1(unexposed)} \quad \text{versus} \quad H_1 : \beta_{1(exposed)} < \beta_{1(unexposed)}.$$

The research hypothesis is that the regression in the exposed group is "steeper," that is, the vital capacity decays significantly faster with age. This corresponds to a smaller slope parameter for the exposed group since in this case the slopes are negative (Figure 14.5). The inference is supported by the following MATLAB code.

```
xlsread vitalcapacity.xlsx;
twos = ans;
  x1 = twos( twos(:,3) > 0, 1);   y1 = twos( twos(:,3) > 0, 2);
  x2 = twos( twos(:,3) ==0, 1);   y2 = twos( twos(:,3) ==0, 2);
  n1=length(x1); n2 = length(x2);
SXX1 = sum((x1 - mean(x1)).^2)   %4.3974e+003
```

```
SXX2 = sum((x2 - mean(x2)).^2)   %6.1972e+003
SYY1 = sum((y1 - mean(y1)).^2)   %26.5812
SYY2 = sum((y2 - mean(y2)).^2)   %20.6067
SXY1 = sum((x1 - mean(x1)).*(y1 - mean(y1)))  %-236.3850
SXY2 = sum((x2 - mean(x2)).*(y2 - mean(y2)))  %-189.7116
b1_1 = SXY1/SXX1    %-0.0538
b1_2 = SXY2/SXX2    %-0.0306
SSE1 = SYY1 - (SXY1)^2/SXX1   %13.8741
SSE2 = SYY2 - (SXY2)^2/SXX2   %14.7991
s2 = (SSE1 + SSE2)/(n1 + n2 - 4)   %0.3584
s = sqrt(s2)   %0.5987
seb1b2 = s * sqrt( 1/SXX1 + 1/SXX2 )  %0.0118
t = (b1_1 - b1_2)/seb1b2     %-1.9606
pval = tcdf(t, n1 + n2 - 4)    %0.0267
```

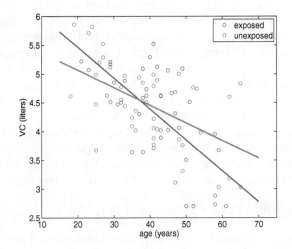

Fig. 14.5 Samples from exposed (*red*) and unexposed (*green*) groups with fitted regression lines. The slopes of the two regressions are significantly different with a *p*-value smaller than 3%.

Thus, the hypothesis of equality of slopes is rejected with a *p*-value of 2.67%.

Remark. Since the distribution of *t*-statistic is calculated under H_0, which assumes parallel regression lines, the more natural estimator s22, in place of s2, takes into account this fact. The number of degrees of freedom in the *t*-statistic changes to $n_1 + n_2 - 3$. The changes in the inference are minimal, as evidenced from the MATLAB code accounting for s22:

```
%s22 accounts for equality of slopes:
s22 = (SYY1 + SYY2 - ...
 (SXY1 + SXY2)^2/(SXX1 + SXX2))/(n1 + n2 - 3) %0.3710
s = sqrt(s22)   %0.6091
seb1b2 = s * sqrt( 1/SXX1 + 1/SXX2 )  %0.0120
```

```
t = (b1_1 - b1_2)/seb1b2        %-1.9270
pval = tcdf(t, n1 + n2 - 3)     %0.0287
```

For the case of the confidence interval, estimator s2 and $n_1 + n_2 - 4$ degrees of freedom should be used.

✎

14.6 Multiple Regression

It is often the case that in an experiment leading to regression analysis more than a single covariate is available. For example, in Chapter 2, page 32, we discussed an experiment in which two indexes of the amount of body fat (Siri and Brozek indexes) were calculated from the body density measure. In addition, a variety of body measurements, including weight, height, adiposity, neck, chest, abdomen, hip, thigh, knee, ankle, biceps, forearm, and wrist, were recorded. Recall that the body density measure is complicated and potentially unpleasant, since it is taken by submerging the subject in water. Therefore, it is of interest to ask whether the Brozek index can be well predicted using the nonintrusive measurements.

If $x_1, x_2, \ldots, x_k$ are variables, covariates, or predictors, and we have n joint measurements of covariates and the response, $x_{i1}, x_{i2}, \ldots, x_{ik}, y_i$, $i = 1, 2, \ldots, n$, then multiple regression expresses the response as a linear combination of covariates, plus an intercept and an additive error:

$$y_i = \beta_0 + \beta_1 x_{i1} + \beta_2 x_{i2} + \cdots + \beta_k x_{ik} + \epsilon_i, i = 1, \ldots, n.$$

The errors ϵ_i are assumed to be independent and normal with mean 0 and constant variance σ^2. We will denote by k the number of covariates, but the number of parameters in the model is $p = k + 1$ because the intercept β_0 should be added. To avoid confusion, we will mostly use p to represent the number of parameters in all expressions that involve dimensions and derived statistics.

As in the case of a single predictor, we will be interested in estimating and testing the coefficients, error variance, and mean and prediction responses. However, multiple regression brings several new challenges when compared to a simple regression. The two main challenges are (i) the possible presence of *multicollinearity* among the covariates, that is, covariates being correlated among themselves, and (ii) a multitude of possible models and the need to find the *best model* by identifying the "best" subset of predictors. A synonym for multiple regression is multivariable regression. Sometimes the multiple regression is wrongly termed multivariate regression; this terminology is reserved for the case where the response y is multivariate, which is not the case here.

14.6.1 Matrix Notation

The regression equations for all n observations $(x_{i1}, x_{i2}, \ldots, x_{ik}, y_i)$, $i = 1, \ldots, n$, can be written as

$$y_1 = \beta_0 + \beta_1 x_{11} + \beta_2 x_{12} + \cdots + \beta_k x_{1k} + \epsilon_1,$$
$$y_2 = \beta_0 + \beta_1 x_{21} + \beta_2 x_{22} + \cdots + \beta_k x_{2k} + \epsilon_2,$$

$$\vdots$$

$$y_n = \beta_0 + \beta_1 x_{n1} + \beta_2 x_{n2} + \cdots + \beta_k x_{nk} + \epsilon_n,$$

and also in convenient matrix form as

$$y = X\beta + \epsilon,$$

where

$$y = \begin{bmatrix} y_1 \\ y_2 \\ \vdots \\ y_n \end{bmatrix},$$

$$X = \begin{bmatrix} x_1 \\ x_2 \\ \vdots \\ x_n \end{bmatrix} = \begin{bmatrix} 1 & x_{11} & x_{12} & \cdots & x_{1k} \\ 1 & x_{21} & x_{22} & \cdots & x_{2k} \\ \vdots & & \vdots & & \vdots \\ 1 & x_{n1} & x_{n2} & \cdots & x_{nk} \end{bmatrix}, \quad \beta = \begin{bmatrix} \beta_0 \\ \beta_1 \\ \vdots \\ \beta_k \end{bmatrix}, \quad \text{and} \quad \epsilon = \begin{bmatrix} \epsilon_1 \\ \epsilon_2 \\ \vdots \\ \epsilon_n \end{bmatrix}.$$

Note that y and ϵ are $n \times 1$ vectors, X is an $n \times p$ matrix, and β is a $p \times 1$ vector. Here $p = k + 1$. To find the least-squares estimator of β, one minimizes the sum of squares:

$$\sum_{i=1}^{n} (y_i - (\beta_0 + \beta_1 x_{i1} + \cdots + \beta_k x_{ik}))^2.$$

The minimizing solution

$$b = \begin{bmatrix} b_0 \\ b_1 \\ \vdots \\ b_k \end{bmatrix}$$

satisfies the system (normal equations)

$$X'Xb = X'y, \tag{14.3}$$

and the least-squares estimator of β is

$$b = (X'X)^{-1}X'y.$$

The fitted values are obtained as

$$\hat{y} = Xb = X(X'X)^{-1}X'y,$$

and the residuals are

$$e = y - \hat{y} = y - Xb = y - X(X'X)^{-1}X'y = (I - X(X'X)^{-1}X')y,$$

where I is an $n \times n$ identity matrix.

The matrix $H = X(X'X)^{-1}X'$ that appears in the expressions for fitted values and residuals is important in this context; it is called the *hat* matrix. In terms of the hat matrix H,

$$\hat{y} = Hy, \quad \text{and} \quad e = (I - H)y.$$

Matrices H and $I - H$ are projection matrices, and the n-dimensional vector y is projected to $\hat{y}$ by H and to the residual vector e by $I - H$. Any projection matrix A is *idempotent*, which means that $A^2 = A$. Simply put, a projection of a projection will be the same as the original projection. Geometrically, vectors $\hat{y}$ and e are orthogonal since the product of their projection matrices is 0. Indeed, because H is idempotent, $H(I - H) = H - H^2 = H - H = 0$.

The errors ϵ_i are independent and the variance of vector e is $\sigma^2 I$.

We will illustrate some concepts from multiple regression using dataset fat.dat; all calculations are part of MATLAB script fatregdiag.m:

```
load 'fat.dat'
    casen = fat(:,1);    %case number
    broz  = fat(:,2);    %dependent variable
    siri  = fat(:,3);    %function of densi
    densi = fat(:,4);    %an intrusive measure
%below are the predictors (except ffwei)
age     = fat(:,5);         weight  = fat(:,6);    height = fat(:,7);
adiposi = fat(:,8);  %adiposity is BMI index=weight/height^2
ffwei   = fat(:,9);  %fat free weight, excluded from predictors
                     % since it involves body fat and brozek
neck    = fat(:,10);    chest  = fat(:,11);    abdomen = fat(:,12);
hip     = fat(:,13);    thigh  = fat(:,14);    knee    = fat(:,15);
ankle   = fat(:,16);    biceps = fat(:,17);    forearm = fat(:,18);
wrist   = fat(:,19);

vecones = ones(size(broz)); % necessary for the intercept

 disp('=========================================================')
```

```
disp('        p = 15, 14 variables + intercept')
disp('=========================================================')

Z =[age  weight  height  adiposi   neck    chest    abdomen ...
     hip  thigh   knee    ankle    biceps  forearm  wrist];
X =[vecones Z];
Y = broz
%   X is design matrix, n x p where n is the number of subjects
%   and p is the number of parameters, or number of predictors+1.
%   varnames = ['intercept=0' 'age=1' 'weight=2' 'height=3'
%   'adiposi=4' 'neck=5' 'chest=6' 'abdomen=7' 'hip=8' 'thigh=9'
%   'knee=10' 'ankle=11''biceps=12' 'forearm=13' 'wrist=14'];
[n, p] = size(X)
b = inv(X' * X) * X'* Y;
H = X * inv(X' * X) * X';
max(max(H * H - H));   %0 since H is projection matrix
Yhat = H * Y;          %or Yhat  = X * b;
%--------------------------------------------------------
```

14.6.2 Sums of Squares and an ANOVA Table

Sums of squares, SST, SSR, and SSE, for multiple regression have simple expressions in matrix notation. Here we introduce matrix J, which is an $n \times n$ matrix in which each element is 1. The total sum of squares can be calculated as

$$SST = y'y - \frac{1}{n}y'Jy = y'\left(I - \frac{1}{n}J\right)y.$$

The error sum of squares is $SSE = e'e = y'(I - H)'(I - H)y = y'(I - H)y$ because $I - H$ is a symmetric projection matrix.

By taking the difference, we obtain

$$SSR = SST - SSE = y'\left(H - \frac{1}{n}J\right)y.$$

The number of degrees of freedom for SST, SSR, and SSE are $n - 1$, $p - 1$, and $n - p$, respectively. Thus, the multiple regression ANOVA table is as follows:

Source	DF	SS	MS	F	p-value
Regression	$p - 1$	SSR	$MSR = \frac{SSR}{p-1}$	$F = \frac{MSR}{MSE}$	$\mathbb{P}(F_{p-1,n-p} > F)$
Error	$n - p$	SSE	$MSE = \frac{SSE}{n-p}$		
Total	$n - 1$	SST			

where large values of F are critical for H_0, which states that the covariates $x_1, \ldots, x_k$ do not influence the response y. Formally, the null hypothesis is

$$H_0 : \beta_1 = \beta_2 = \cdots = \beta_k = 0,$$

while the alternative is that at least one β_i, $i = 1, \ldots, k$ is not 0.

As in the simple regression, R^2 is the coefficient of determination,

$$R^2 = \frac{SSR}{SST} = 1 - \frac{SSE}{SST}.$$

Adding more variables to a regression always increases R^2, even when the added covariates have nothing to do with the experiment. If two models have comparable R^2s, then, according to Ockham's razor,[3] the simpler model should be preferred, and adding new variables to the regression should be penalized. One way to achieve this is via an adjusted coefficient of determination,

$$R^2_{adj} = 1 - \frac{n-1}{n-p} \frac{SSE}{SST},$$

which is one of the criteria for comparing models.

The estimator of the error variance σ^2 is MSE. We can find the confidence intervals of the components of e and $\hat{y}$ using, respectively, the diagonal elements of covariance matrices $MSE \times (I - H)$ and $MSE \times H$.

```
%Sums of Squares
J=ones(n); I = eye(n);
SSR = Y' * (H - 1/n * J) * Y;
SSE = Y' * (I - H) * Y;
SST = Y' * (I - 1/n * J) * Y;
MSR = SSR/(p-1) %806.7607
MSE = SSE/(n-p) %15.9678
F = MSR/MSE      %50.5243
pval = 1-fcdf(F, p-1, n-p)         %0
Rsq = 1 - SSE/SST                  %0.7490
Rsqadj = 1 - (n-1)/(n-p) * SSE/SST  %0.7342
s = sqrt(MSE)                      %3.9960
%-------------------------------------------------------
```

---

[3] *Pluralitas non est ponenda sine neccesitate*, which translates into English as "Plurality should not be posited without necessity" (William of Ockham, 1287–1347).

14.6.3 Inference About Regression Parameters and Responses

The covariance matrix for a vector of estimators of regression coefficients b is equal to

$$s_b^2 = MSE \times (X'X)^{-1}.$$

Its $(j+1)$st diagonal element is an estimator of variance for b_j, and off-diagonal elements at the position $(j+1, k+1)$ are estimators of covariances between b_j and b_k, where $j, k = 0, \ldots, p-1$. Note that here we count the rows/columns of $p \times p$ matrix $(X'X)^{-1}$ from 1 to p, while the corresponding indices in b run from 0 to $p-1$.

When finding a confidence interval or testing a hypothesis about a particular β_j, $j = 0, \ldots p-1$, we use b_j and s_{b_j} (square root of $(j+1)$st diagonal element of s_b^2) in the same way as in the univariate regression, only this time the test statistic t has $n - p$ degrees of freedom instead of $n - 2$. Several subsequent MATLAB scripts are excerpts from the file ◀ fatregdiag.m:

```
sig2 = MSE * inv(X' * X);% covariances among b's
sb=sqrt(diag(sig2));
tstats = b./sb;
pvals = 2 * tcdf(-abs(tstats), n-p);
 disp('-------------------------------------')
 disp('    var#        t         pval        ')
 disp('-------------------------------------')
 [ (0:p-1)'   tstats    pvals    ]
 %-----------------------------------------
 %        var#        t         pval
 %-----------------------------------------
 %      0        -0.9430     0.3467
 %      1.0000    1.8942     0.0594
 %      2.0000   -1.6298     0.1045
 %
 %      ...
 %     12.0000    0.9280     0.3543
 %     13.0000    2.3243     0.0210
 %     14.0000   -2.9784     0.0032
```

Note that in the fat example, the intercept is not significant ($p = 0.3467$), nor is the coefficient for variable #12 (biceps) ($p = 0.3543$), while the coefficient for variable #13 (forearm) is significant ($p = 0.0210$).

The regression response $y_m = x^*b$, evaluated at $x^* = (1 \ x_1^* \ x_2^* \ \ldots \ x_k^*)$, has sample variance

$$s_{y_m}^2 = (x^*)s_b^2(x^*)'.$$

For a prediction, the variance is, as in univariate regression, $s_{y_p}^2 = s_{y_m}^2 + MSE$. For the inference about regression response, the t-statistic with $n - p$

degrees of freedom is used. As an example, assume that for a "new" person with covariates $x_h = (1\ 38\ 191\ 72\ 26\ 41\ 104\ 95\ 101.5\ 66\ 39\ 24\ 31\ 30\ 18.5)$, a prediction of the Brozek index is needed. The model gives a prediction of 19.5143, and the variances for mean response and individual response are estimated below. This is sufficient to calculate confidence intervals for the mean and individual responses.

```
%-------------------------------------------------
% predicting mean and individual responses
% with 95% confidence/prediction intervals
Xh=[1   38   191  72    26    41   104  ...
       95   101.5  66   39    24   31   30    18.5];
   Yh = Xh * b   %19.5143
   sig2h = MSE *  Xh  * inv(X' * X)   * Xh';
   sig2hpre = MSE * (1 + Xh  * inv(X' * X)   * Xh');
   sigh = sqrt(sig2h);
   sighpre = sqrt(sig2hpre);
   %95% CI's on the mean and individual responses
   [Yh-tinv(0.975, n-p)*sigh, Yh+tinv(0.975, n-p)*sigh]
       %[17.4347,   21.5940]
   [Yh-tinv(0.975, n-p)*sighpre, Yh+tinv(0.975, n-p)*sighpre]
       %[11.3721   27.6566]
```

A Noninformative Bayesian Approach. A Bayesian inference in multiple linear regression is based on a prior on (β, σ^2) and the likelihood

$$
f(y|\beta,\sigma^2) = \left(\frac{1}{\sqrt{2\pi\sigma^2}}\right)^n \exp\left\{-\frac{1}{2\sigma^2}(y - X\beta)'(y - X\beta)\right\}
$$
$$
= \left(\frac{1}{\sqrt{2\pi\sigma^2}}\right)^n \exp\left\{-\frac{1}{2\sigma^2}\left[(y - \hat{y})'(y - \hat{y}) + (\beta - b)'(X'X)(\beta - b)\right]\right\},
$$

where $b = (X'X)^{-1}X'y$ is the least squares estimator of β, and $\hat{y} = Xb$.
    With noninformative prior

$$
\pi(\beta,\sigma^2) \propto \frac{1}{\sigma^2},
$$

the posterior for $(n - p)\frac{s^2}{\sigma^2}$ is χ^2_{n-p}, while the posterior for β is multivariate normal $\mathcal{MVN}_p(b,\sigma^2(X'X)^{-1})$. The marginal posterior for β, when σ^2 is integrated out, is multivariate t with location b. Thus, the Bayes estimator for β coincides with b, which is also traditional estimator (MLE). The Bayes estimator of σ^2 is $\frac{n-p}{n-p-2}s^2$, for $s^2 = \frac{1}{n-p}(y - \hat{y})'(y - \hat{y})$, and $n > p + 2$.
    The $(1 - \alpha)100\%$ HPD set for β is

$$
\left\{\beta \,\middle|\, (\beta - b)'(X'X)(\beta - b) \leq p\,s^2\,F_{1-\alpha,p,n-p}\right\},
$$

where $F_{1-\alpha,p,n-p}$ is the $(1-\alpha)$ quantile of an $F_{p,n-p}$-distribution.

For any particular component β_j,

$$t = \frac{\beta_j - b_j}{s\sqrt{c_{jj}}},$$

has t_{n-p} posterior distribution. Here c_{jj} is $j+1$st diagonal element of $(XX)^{-1}$.

A more general prior on (β,σ^2) is multivariate normal-inverse gamma $(\mathcal{NIG})$. This conjugate prior is capable of incorporating various prior information about the parameters and the prior $\pi(\beta,\sigma^2) \propto \frac{1}{\sigma^2}$ can be obtained as a limiting case of a $\mathcal{NIG}$ prior. Chapter 9 of O'Hagan (1994) provides an excellent description of the resulting Bayesian inference.

14.7 Diagnostics in Multiple Regression

Three important deficiencies in a multiple linear model can be diagnosed: (i) the presence of outliers, (ii) the nonconstant error variance, and (iii) a possible suboptimal model selection. Although the exposition level of these diagnostic methods exceeds the level in introductory coverage of regression, multiple regression modeling is important in practice and provides an important step to understanding more sophisticated nonlinear models, such as generalized linear models. For this reason, we provide a basic overview of residual and influence analysis, as well as an assessment of multicollinearity and choice of model. For readers interested in a more comprehensive treatment of multivariable linear models, the book by Rawlings et al. (2001) is a comprehensive resource.

14.7.1 Residual Analysis and Influence

In setting the regression model, we made several assumptions about population errors (independence, zero-mean, constant variance, normality). The residuals, defined as $e_i = y_i - \hat{y}_i$, can be thought of as observed errors if the model is correct and should confirm our assumptions. For this reason, the residuals are examined graphically (histograms, plots against fits $\hat{y}$, against particular predictors, or, when it makes sense, against their order). When individual data points fail the residual check, we may suspect outliers in an otherwise correct model. However, if the residual analysis shows systematic deviations (trends, nonconstant variance), the model should be questioned. An interesting example of importance of residual analysis was constructed by Anscombe (1973), see Exercise 14.11.

Next we discus the ordinary residuals and three modifications more appropriate for the statistical analysis.

The ordinary residuals $e_i = y_i - \hat{y}_i$ are components of $(I - H)y$. The *leverages* h_{ii} are diagonal elements of the hat matrix H. These are important descriptors of design matrix X and explain how far x_i is from $\bar{x}$. All leverages are bounded $1/n \le h_{ii} \le 1$, and their sum is $\sum h_{ii} = p$, the number of regression parameters. Although the errors in the regression are independent and with the same variance, the residuals are correlated and with different sample variances $s^2(1 - h_{ii})$, where $s^2 = MSE$.

The *studentized residual* is the ordinary residual divided by its standard deviation

$$r_i = \frac{e_i}{s\sqrt{1 - h_{ii}}},$$

and recalls the t-statistic. Such residuals are scale-free comparable, and values outside the interval $[-2.5, 2.5]$ are potential outliers. Sometimes these residuals are called *internally studentized*, since the standard deviation $s = \sqrt{MSE}$ depends on the ith observation.

Externally studentized residuals, also called R-Student residuals, are measures of influence of the ith observation (y_i, x_i) on the ith residual. Instead of s, the residuals are studentized by an *external standard deviation*,

$$s_{-i} = \sqrt{\frac{(n-p)s^2 - e_i^2/(1 - h_{ii})}{n - p - 1}}. \tag{14.4}$$

This external estimate of σ comes from the model fitted without the ith observation; however, the refitting is not necessary due to a simple expression in (14.4). Externally studentized residuals

$$t_i = \frac{e_i}{s_{-i}\sqrt{1 - h_{ii}}},$$

can be tested, since they are t distributed with $n - p - 1$ degrees of freedom. Of course, if multiple residuals are tested simultaneously, then it should be done in the spirit of multiple hypothesis testing (Section 9.9).

PRESS (acronym for prediction sum of squares; Allen, 1974) residuals $e_{i,-i} = e_i/(1 - h_{ii})$ also remove the impact of the ith observation (y_i, x_i) on the fit at x_i. This is a cross-validatory residual that measures how a model built without using the ith observation would predict the ith response.

For model assessment, the statistic *PRESS* is useful. It is defined as a sum of squares of *PRESS* residuals,

$$PRESS = \sum_i e_{i,-i}^2,$$

and used in defining the *prediction R^2*,

$$R^2_{pred} = 1 - \frac{PRESS}{SST}.$$

Here SST is the total sum of squares $\sum_i (y_i - \bar{y})^2$. The ordinary R^2 is defined as $1 - SSE/SST$, and in R^2_{pred} the SSE is replaced by $PRESS$. Since the *average prediction error* is defined as $\sqrt{PRESS/n}$, good models should have a small $PRESS$.

The following table summarizes standard types of residuals used in residual analysis:

1. Ordinary residuals	$e_i = y_i - \hat{y}_i$
2. Studentized residuals	$r_i = \dfrac{e_i}{s\sqrt{1-h_{ii}}}$
3. Externally studentized residuals	$t_i = \dfrac{e_i}{s_{-i}\sqrt{1-h_{ii}}}$
4. Prediction sum of squares residuals (PRESS)	$e_{i,-i} = \dfrac{e_i}{1-h_{ii}}$

DFBETAS stands for *difference in betas*. It measures the influence of the ith observation on β_j:

$$DFBETAS_{ij} = \frac{b_j - b_{j(-i)}}{s_{-i}\sqrt{c_{jj}}},$$

where b_j is the estimator of β_j, $b_{j(-i)}$ is the estimator of β_j when the ith observation is excluded, and c_{jj} is the $(j+1)$st diagonal element in $(X'X)^{-1}$. Large *DFBETAS* may indicate which predictor might be influential. The recommended threshold is $2/\sqrt{n}$. The black boxes in Figure 14.8 are at combinations $(\beta_0 - \beta_{14}$ vs. indices of observations) for which `abs(DFBetas>2/sqrt(n))`.

Since several *DFBETAS* can be large, it is useful to pay attention to those corresponding to large *DFFITS*. The *DFFITS* measure the influence of the ith observation on the prediction $\hat{y}_i$:

$$DFFITS_i = \frac{\hat{y}_i - \hat{y}_{i,-i}}{s_{(-i)}\sqrt{h_{ii}}} = \sqrt{\frac{h_{ii}}{1-h_{ii}}} \cdot \frac{e_i}{s_{-i}\sqrt{1-h_{ii}}}.$$

The value $\hat{y}_{i,-i}$ is the prediction of y_i on the basis of a regression fit without the ith observation. The observation is considered influential if its *DFFITS* value exceeds $2\sqrt{p/n}$.

A measure related computationally to *DFFITS* is Cook's distance, D_i:

$$D_i = (DFFITS_i)^2 \cdot \frac{s^2_{-i}}{ps^2}.$$

Cook's distance measures the effect of the ith observation on the whole vector $b = \hat{\beta}$. An observation is deemed influential if its Cook's distance exceeds $4/n$.

Figure 14.7 shows ordinary residuals plotted against predicted values $\hat{y}$. The radii of circles are proportional to |Dffits| in panel (a) and to Cook's distance in panel (b).

Influential observations are not necessarily outliers and should not be eliminated from a model only on the basis of their influence. Such observations can be identified by their influence on a predicted value. One often finds predictions of the ith response $\hat{y}_{i,-i}$ in a regression in which the ith case (y_i, x_i) is omitted (Figure 14.6).

```
%prediction of y_i with ith observation  removed
%hat y_i(-i)
ind = 1:n;
Yhati = [];
for i = 1:n
    indi = find(ind ~= i);
    Yi = Y(indi);
    Xi=X(indi,:);
    bi = inv(Xi' * Xi) * Xi'* Yi;
    Yhatii = X(i,:) * bi;
    Yhati =[Yhati; Yhatii];
end
Yhati %prediction of y_i without i-th observation
%------------------------------------------------
```

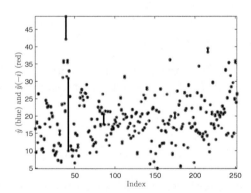

Fig. 14.6 Predicted responses $\hat{y}_i$ (*blue*) and predicted responses $\hat{y}_{i,-i}$ (*red*). Large changes in prediction signify an influential observation.

```
%============ residual analysis==================
hii = diag(H);                  %leverages
resid = (I - H)*Y;              %ordinary residuals
sresid = sqrt(MSE .* (1-hii));
stresid = resid./sresid %studentized residuals
%----------studentized deleted residuals---
di = Y - Yhati; %or  di  = resid./(1-hii)
%di is also called PRESS residual
PRESS =sum(di.^2)
R2pred = 1-PRESS/SST %R^2 predictive

sminusi = sqrt(((n-p)*MSE*ones(n,1) -...
    resid.^2./(1-hii))/(n-p-1));    %stdev(-i)
ti  = resid ./(sminusi .* sqrt(1-hii))
% externally studentized residuals
% outliers based on leverage = hii
outli=hii/mean(hii);
 find(outli > 3)
   % 31 36 39 41 42 86 159 175 206

%influential observations
Dffits = ti .* sqrt( hii ./(1-hii)) %influ ith to ith
find(abs(Dffits) > 2 * sqrt(p/n));
% 31 39 42 82 86 128 140 175 207 216 221 231 250
%
CooksD = resid.^2 .* (hii./(1-hii).^2)/(p * MSE)
% influence if ith to all;
find(CooksD > 4/n) %find influential
%31 39 42 82 86 128 175 207 216 221 250
```

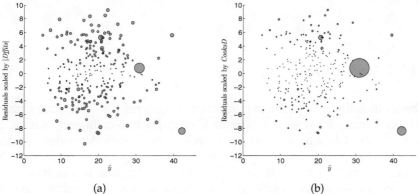

(a)   (b)

Fig. 14.7 Ordinary residuals plotted against predicted values $\hat{y}$. The radii of circles are proportional to |Dffits| in panel (a) and to Cook's distance in panel (b). The observations with the largest circles are in both cases the 42nd and the 39th.

```
%DFBETAS- influence if ith obs on jth coefficient
```

```
    cii = diag(inv(X' * X));
    DFBetas =[];
 for i = 1:n
    indi = find(ind ~= i);
    Yi = Y(indi);
    Xi=X(indi,:);
    bi  = inv(Xi' * Xi) * Xi'* Yi;
    Hi = Xi * inv(Xi' * Xi)   * Xi';
    SSEi  = Yi' * (eye(n-1) - Hi) * Yi;
    MSEi  = SSEi./(n-p-1);
    DFBetasi = (b - bi)./sqrt(MSEi .* cii) ;
    DFBetas = [DFBetas; DFBetasi'];
 end
```

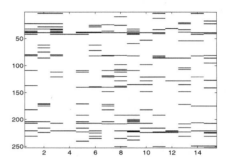

Fig. 14.8 DFBETAS: The x-axis enumerates $\beta_0 - \beta_{14}$, while on the y-axis are plotted the indices of observations. The *black boxes* are at combinations for which abs(DFBetas)>2/sqrt(n)).

14.7.2 Multicollinearity

The multicollinearity problem in regression concerns the correlation among the predictors. Suppose that matrix X contains two collinear columns x_i and $x_j = 2x_i$. Obviously covariates x_i and x_j are linearly dependent, and x_j does not bring any new information about the response. This collinearity makes matrix X not of full rank and $X'X$ singular, that is, not invertible, and normal equations (14.3) have no unique solution. In reality, if multicollinearity is present, then matrix $X'X$ is not singular, but near-singular, in the sense that its determinant is close to 0, making the inversion of $X'X$, and consequently the solution $\hat{\beta}$, numerically unstable. This happens when either two or more variables are highly correlated or when a variable has a small variance (and, in a sense, is correlated to the intercept).

In addition to numerical instability in computing the least squares estimators, the multicollinearity can lead to inferential inconsistencies. Multicollinear predictors tend to have t-statistics for their corresponding coefficients shrunk toward 0, leading to the possibility of having a significant regression, as assessed by statistic F, in which none of the individual predictors turns out significant.

There are many measures of multicollinearity discussed in the literature. We will discuss the *condition number*, *condition indexes*, and *multicollinearity index*, which are global measures, and the *variance inflation factor*, which is linked to a particular predictor.

Let $\mathbf{Z}_{n \times k}$ be a matrix obtained from the design matrix $\mathbf{X}$ by omitting the vector of ones, Z = X(:,2:end). Let $\lambda_1 \geq \lambda_2 \geq \cdots \geq \lambda_k$ be ordered eigenvalues of correlation matrix of the predictors, in MATLAB, corr(Z). The condition number (Belsley et al., 1980) is defined as the square root of the ratio of the largest and smallest eigenvalues,

$$K = \sqrt{\frac{\lambda_1}{\lambda_k}}.$$

Values for K starting at around 10 are concerning, values between 30 and 100 influence the results, and values over 100 indicate a serious collinearity problem.

The condition indexes for corr(Z) defined as

$$K_j = \sqrt{\frac{\lambda_1}{\lambda_j}}, \, j = 1,\ldots,k,$$

are indicative of possible presence of multiple near-linear relationships among the predictors. For example, if three condition indexes are large, say exceed 30, this would mean that among predictors there are three competing near-linear dependencies independent of each other.

The multicollinearity index (*MCI*) suggested by Thisted (1980) is defined as

$$MCI = \sum_{j=1}^{k} \left(\frac{\lambda_k}{\lambda_j} \right)^2$$

Here, λ_k is the smallest eigenvalue of corr(Z). The *MCI* falls between 1 and k, it is equal to 1 for exactly collinear variables, and to k for orthogonal variables. The values of *MCI* close to 1 indicate high collinearity, values larger than 2 are not alarming.

The variance inflation factor (*VIF*) explains the extent of correlation of a particular variable x_j to the rest of predictors. It is defined as

$$VIF_j = \frac{1}{1 - R_j^2},$$

where R_j^2 is the coefficient of determination in regression of x_j to the rest of predictors. The name *variance inflation factor* comes from an alternate expression for variance of b_j,

$$\mathbb{V}\text{ar}(b_j) = \frac{1}{1 - R_j^2} \times \frac{\sigma^2}{\sum_{i=1}^{n}(x_{ij} - \bar{x}_j)^2}, \quad j = 1, \dots, p-1,$$

where x_{ij} are elements of design matrix X without the column of 1's corresponding to the intercept. It is obvious that when R_j approaches 1, the variance of b_j explodes. VIFs exceeding 10 are considered serious.

Computation of VIF's is simple. One finds the inverse of the correlation matrix for the predictors, The diagonal elements of this inverse are the VIFs, `diag(inv(corr(X(:,2:end)))))`, where X is the design matrix.

As a global measure of multicollinearity an average VIF is used,

$$\overline{VIF} = \frac{1}{k}\sum_{j=1}^{k} VIF_j, \quad k = p-1.$$

Values of $\overline{VIF}$ substantially larger than 1 indicate a multicollinearity problem.

A note of caution: VIF diagnostic sometimes can miss a problem since the intercept is not included in the analysis.

Multicollinearity can be diminished by excluding problematic variables causing the collinearity in the first place. Alternatively, groups of variables can be combined and merged into a single variable.

Another way to diminish multicolinearity is to keep all variables but "condition" matrix $X'X$ by adding kI, for some $k > 0$, to the normal equations. This is known as ridge regression (Hoerl and Kennard, 1970). There is a tradeoff: the solutions of $(X'X + kI)\hat{\beta} = X'y$ are more stable, but some bias is introduced.

```
%
    Z = X(:,2:end);
    RXX = corr(Z);
lambdas = eig(RXX);
K =  sqrt(max(lambdas)/min(lambdas))  % 19.3713
Ki = sqrt(max(lambdas)./lambdas);
Ki'
% 1.0000    2.4975    2.9039    3.6422    3.8436    5.2447    5.4643
% 5.8257    6.9182    8.1185   19.3713   12.4119   13.7500   10.6779
MCI = sum( (min(lambdas)./lambdas).^2 )    % 1.5853
VIF = diag (inv(RXX));
VIF'
% 2.2509   33.7869    2.2566   16.1634    4.4307   10.6846   13.3467
```

```
% 15.1583    7.9615    4.8288    1.9455    3.6745    2.1934    3.3796
```

Alternatively, for most of the above calculations one can use MATLAB's regstats or diagnostics.m:

```
s = regstats(Y,Z,'linear','all');
[index,res,stud_res,lev,DFFITS1,Cooks_D,DFBETAS]=diagnostics(Y,Z);
```

14.7.3 Variable Selection in Regression

Model selection involves finding a subset of predictors from a large number of potential predictors that is optimal in some sense.

We defined the coefficient of determination R^2 as the proportion of model-explained variability, and it seems natural to choose a model that maximizes R^2. It turns out that this is not a good idea since the maximum will always be achieved by that model that has the maximal number of parameters. It is a fact that R^2 increases when even a random or unrelated predictor is included.

The adjusted R^2 penalizes the inclusion of new variables and represents a better criterion for choosing a model. However, with p parameters there would be 2^p possible candidate models, and even for moderate p this could be a formidable number.

There are two mirroring procedures, forward selection and backward selection, that are routinely employed in cases where checking all possible models is infeasible. Forward selection proceeds in the following way:

STEP 1. Start with the intercept-only model. Choose the predictor that has the largest R^2 among the models with a single variable. Call this variable x_1.

STEP 2. Assume that the model already has m variables, $x_1,\ldots,x_m$, for some $m \geq 1$. Select the variable x_{m+1} that gives the maximal increase to R^2 and refit the model.

STEP 3. Denote by $SSR(x_1,\ldots,x_m)$ the regression sum of squares for a regression fitted with variables $x_1,\ldots,x_m$. Then $R(x_{m+1}|x_1,\ldots,x_m) = SSR(x_1,\ldots,x_{m+1}) - SSR(x_1,\ldots,x_m)$ is the contribution of the $(m+1)$st variable and it is considered significant if

$$R(x_{m+1}|x_1,\ldots,x_m)/MSE(x_1,\ldots,x_{m+1}) > F_{1,n-m-1,1-\alpha}, \qquad (14.5)$$

where $MSE(x_1,\ldots,x_{m+1})$ the mean square error for a regression fitted with variables $x_1,\ldots,x_{m+1}$. If relation (14.5) is satisfied, then variable x_{m+1} is included in the model. Increase m by one and go to **STEP 2**.

If relation (14.5) is not satisfied, then the contribution of x_{m+1} is not significant, in which case go to **STEP 4**.

STEP 4. Stop with the model that has m variables. **END**

The MSE in (14.5) was estimated from the full model. Note that the forward selection algorithm is "greedy" and chooses the single best improving variable at each step. This, of course, may not lead to the optimal model since in reality variable x_1, which is the best for one-variable models, may not be included in the best two-variable model.

Backward stepwise regression starts with the full model and removes variables with insignificant contributions to R^2. Seldom do these two approaches end with the same candidate model.

MATLAB's Statistics Toolbox has two functions for stepwise regression: `stepwisefit`, a function that proceeds automatically from a specified initial model and entrance/exit tolerances, and `stepwise`, an interactive tool that allows you to explore individual steps in a process.

An additional criterion for the goodness of a model is the Mallows C_p. This criterion evaluates a proposed model with k variables and $p = k + 1$ parameters. The Mallows C_p is calculated as

$$C_p = (n - p)\frac{s^2}{\hat{\sigma}^2} - n + 2p,$$

where s^2 is the *MSE* of the candidate model and $\hat{\sigma}^2$ is an estimator of σ^2, usually taken to be the best available estimate. The *MSE* of the full model is typically used as $\hat{\sigma}^2$.

A common misinterpretation is that in C_p, p is referred to as the number of predictors instead of parameters. This is correct only for models without the intercept (or when 1 from the vector of ones in the design matrix is declared as a predictor).

Adequate models should have a small C_p that is close to p. Typically, a plot of C_p against p for all models is made. The "southwesternmost" points close to the line $C_p = p$ correspond to adequate models. The C_p criterion can also be employed in forward and backward variable selection as a stopping rule.

14.7.4 Bayesian Model Selection in Multiple Regression

Next, we revisit fat.dat with some Bayesian analyses. Four competing models are compared using the Laud–Ibrahim predictive criterion, LI. Models with smaller LI are favored.

Laud and Ibrahim (1995) argue that agreement of model-simulated predictions and original data should be used as a criterion for model selection. If for y_i responses $\hat{y}_{i,new}$ are hypothetical replications according to the posterior predictive distribution of competing model parameters, then

$$\mathrm{LI} = \sum_{i=1}^{n} (\mathbb{E}\hat{y}_{i,new} - y_i)^2 + \mathbb{V}\mathrm{ar}\,(\hat{y}_{i,new})$$

measures the discrepancy between the observed and model-predicted data. A smaller LI is better. The file ✣ fat.odc performs a Laud–Ibrahim Bayesian model selection and prefers model #2 of the four models analyzed.

```
#fat.odc
model{
for(j in 1:N ){
# four competing models
mu[1, j] <- b1[1] + b1[2] *age[j] + b1[3]*wei[j] + b1[4]*hei[j] +
             b1[5]*adip[j] + b1[6]*neck[j] + b1[7]*chest[j] +
             b1[8]*abd[j] + b1[9]*hip[j] + b1[10]*thigh[j] +
             b1[11]*knee[j]+b1[12]*ankle[j]+b1[13]*biceps[j] +
             b1[14]*forea[j] + b1[15]*wrist[j]
mu[2, j] <- b2[1]+b2[2]*wei[j]+b2[3]*adip[j]+b2[4]*abd[j]
mu[3, j] <- b3[1]+b3[2]*adip[j]
mu[4, j] <- b4[1]*wei[j]+b4[2]*abd[j]+b4[3]*abd[j]+b4[4]*wrist[j]
                }
#LI - Laud-Ibrahim Predictive Criterion. LI-smaller-better
for(i in 1:4 ){
    tau[i] ~ dgamma(2,32)
    LI[i] <- sqrt( sum(D2[i,]) + pow(sd(broz.new[i,]),2))
# data sets 1-4 for different models
                for (j in 1:N) {
                broz2[i,j] <- broz[j]
                broz2[i,j]  ~ dnorm(mu[i,j],tau[i])
                broz.new[i,j] ~ dnorm(mu[i,j],tau[i])
                D2[i,j] <- pow(broz[j]-broz.new[i,j],2)
                           }
                }
# Compare predictive criteria between models i and j
# Comp[i,j] is 1 when LI[i]<LI[j], i-th model better.
for (i in 1:3) { for (j in i+1:4)
                   {Comp[i,j] <- step(LI[j]-LI[i])}}
# priors
   for (j in 1:15)  { b1[j] ~ dnorm(0,0.001)}
   for(j in 1:4)    { b2[j] ~ dnorm(0,0.001)
                      b4[j] ~ dnorm(0,0.001)}
   for(j in 1:2)    { b3[j] ~ dnorm(0,0.001)}
}

#DATA 1:  Load this first
list(N = 252)

# DATA2: Then load the variables

broz[] age[] wei[]  hei[]   ...  biceps[] forea[] wrist[]
12.6 23 154.25 67.75 ...    32.0  27.4  17.1
23.4 38.5 93.6 83.00 ...    30.5  28.9  18.2
...248 lines deleted...
25.3 72 190.75 70.50 ...    30.5  29.4  19.8
30.7 74 207.50 70.00 ...    33.7  30.0  20.9
END
```

```
#the line behind 'END' has to be empty

# INITS (initialize by loading tau's
#            and generating the rest
list(tau=c(1,1,1,1))
```

The output is given in the table below. Note that even though the posterior mean of LI[4] is smaller that that of LI[2], it is the posterior median that matters. Model #2 is more frequently selected as the best compared to model #4.

	mean	sd	MC error	val2.5pc	median	val97.5pc	start	sample
LI[1]	186.5	209.2	17.63	84.0	104.6	910.5	1001	20000
LI[2]	96.58	23.46	1.924	85.08	93.14	131.2	1001	20000
LI[3]	119.6	5.301	0.03587	109.4	119.5	130.2	1001	20000
LI[4]	94.3	4.221	0.03596	86.33	94.2	103.0	1001	20000
Comp[1,2]	0.2974	0.4571	0.02785	0.0	0.0	1.0	1001	20000
Comp[1,3]	0.5844	0.4928	0.03939	0.0	1.0	1.0	1001	20000
Comp[1,4]	0.3261	0.4688	0.03008	0.0	0.0	1.0	1001	20000
Comp[2,3]	0.9725	0.1637	0.01344	0.0	1.0	1.0	1001	20000
Comp[2,4]	0.5611	0.4963	0.01003	0.0	1.0	1.0	1001	20000
Comp[3,4]	5.0E-5	0.007071	4.98E-5	0.0	0.0	0.0	1001	20000

More comprehensive treatment of Bayesian approaches in linear regression can be found in O'Hagan (1994) and Ntzoufras (2009).

14.8 Sample Size in Regression

The evaluation of power in a regression with $p - 1$ variables and p parameters (an intercept is present) requires specification of a significance level and precision. Suppose that we want a power such that a total sample size of $n = 61$ would make $R^2 = 0.2$ significant for $\alpha = 0.05$ and the number of predictor variables $p - 1 = 3$. Cohen's effect size here is defined as $f^2 = R^2/(1 - R^2)$. Unlike the ANOVA where the values of $f^2 \approx 0.01$ corresponded to small, $f^2 \approx 0.0625$ to medium, and $f^2 \approx 0.16$ to large effects, in regression the values of $f^2 \approx 0.02$ corresponded to small, $f^2 \approx 0.15$ to medium, and $f^2 \approx 0.35$ to large effects. Note that from $f^2 = R^2/(1 - R^2)$ one gets $R^2 = f^2/(1 + f^2)$, which can be used to check the adequacy of the elicited/required effect size.

The power, similar to ANOVA, is found using the noncentral F-distribution,

$$1 - \beta = \mathbb{P}(F^{nc}(p - 1, n - p, \lambda) > F^{-1}(1 - \alpha, p - 1, n - p)), \qquad (14.6)$$

where $\lambda = nf^2$ is the noncentrality parameter.

Example 14.5. **Power Analysis in Regression.** For $p = 4$, $R^2 = 0.2$, that is, $f^2 = 0.25$ (a medium-to-large effect), and a sample size of $n = 61$, one gets $\lambda = 61 \times 0.25 = 15.25$, and a power of approximately 90%.

```
p=4; n=61; lam=15.25;
1-ncfcdf( finv(1-0.05, p-1, n-p), p-1, n-p, lam)
% ans =   0.9014
```

14.9 Linear Regression That Is Nonlinear in Predictors

In linear regression, "linear" concerns the parameters, not the predictors. For instance,

$$\frac{1}{y_i} = \beta_0 + \frac{\beta_1}{x_i} + \epsilon_i, \ \ i = 1,\dots,n,$$

and

$$y_i = \epsilon_i \times \exp\{\beta_0 + \beta_1 x_{1i} + \beta_2 x_{2i}\}, \ \ i = 1,\dots,n,$$

are examples of a linear regression. There are many functions where x and y can be linearized by an obvious transformation of x or y or both; however, one needs to be mindful that in such transformations the normality and homoscedasticity of errors is often compromised. In such a case, fitting a regression is simply an optimization task without natural inferential support. Bayesian solutions involving MCMC are generally more informative regarding the inference (Bayes estimators, credible sets, predictions).

An example where errors are not affected by the transformation of variables is a polynomial relationship between x and y postulated as

$$y_i = \beta_0 + \beta_1 x_i + \beta_2 x_i^2 + \cdots + \beta_k x_i^k + \epsilon_i, \ \ i = 1,\dots,n,$$

which is in fact a linear regression. Simply, the k predictors are $x_{1i} = x_i$, $x_{2i} = x_i^2, \dots, x_{ki} = x_i^k$, and estimating the polynomial relationship is straightforward. The example below is a research problem in which a quadratic relationship is used.

Example 14.6. **Von Willebrand Factor.** Von Willebrand disease is a bleeding disorder caused by a defect or deficiency of a blood clotting protein called the von Willebrand factor. This glue-like protein, produced by the cells that line blood vessel walls, interacts with blood cells called platelets to form a plug that prevents bleeding. In order to understand the differential bonding mechanics underlying von Willebrand-type bleeding disorders, researchers at Georgia Tech studied the interactions between the wild-type platelet GPIba molecule (receptor) and wild-type von Willebrand factor (ligand).

The mean stop time rolling parameter was calculated from frame-by-frame rolling velocity data collected at 250 frames per second. Mean stop time indicates the amount of time a cell spends stopped, so it is analogous to the bond lifetime. This parameter, being an indicator for how long a bond is stopped before the platelet moves again, can be used to assess the bond lifetime and off-rate (Yago et al., 2008).

For the purpose of exploring interactions between the force and the mean stop times, Ficoll 6% is added to increase the viscosity.

Data are courtesy of Dr. Leslie Coburn. The mat file ▣ coburn.mat contains the structure coburn with data fields coburn.fxssy, where x = 0, 6 is a code for Ficoll absence/presence and y = 1, 2, 4, ..., 256 denotes the shear stress (in dyn/cm^2). For example, coburn.f0ss16 is a 243 × 1 vector of mean stop times obtained with no Ficoll, under a shear stress of 16 dyn/cm^2.

Shear stress	2	4	8	16	32	64	128	256
Shear number	1	2	3	4	5	6	7	8
Mean stop time	f0ss2	f0ss4	f0ss8	f0ss16	f0ss32	f0ss64	f0ss128	f0ss256
Sample size	26	57	157	243	256	185	62	14

We fit a regression on the logarithm (base 2) of mean stop time log2mst, with no Ficoll present, as a quadratic function of share stress number $\log_2(\text{dyn/cm}^2)$. This regression is linear in parameters with two predictors, shearn and the squared shear, shearn2=shearn^2.

The regression fit is

$$\text{log2mst} = -6.2423 + 0.8532 \text{ shear} - 0.0978 \text{ shear2}.$$

The regression is significant ($F = 63.9650, p = 0$); however, its predictive power is rather weak, with $R^2 = 0.1137$.

Figure 14.9 is plotted by the script ◢ coburnreg.m.

```
%coburnreg.mat
load 'coburn.mat';
mst=[coburn.f0ss2;  coburn.f0ss4;  coburn.f0ss8;   coburn.f0ss16; ...
     coburn.f0ss32; coburn.f0ss64; coburn.f0ss128; coburn.f0ss256];

shearn = [1 * ones( 26,1); 2 * ones( 57,1); 3 * ones(157,1); ...
          4 * ones(243,1); 5 * ones(256,1); 6 * ones(185,1); ...
          7 * ones( 62,1); 8 * ones( 14,1)];

shearn2 = shearn.^2; %quadratic term

%design matrix
X = [ones(length(shearn),1) shearn  shearn2];
[b,bint,res,resint,stats] = regress(log2(mst), X) ;
```

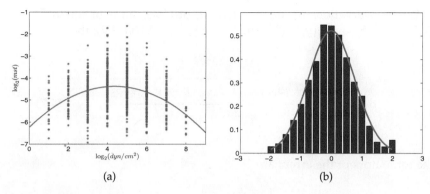

Fig. 14.9 (a) Quadratic regression on log mean stop time. (b) Residuals fitted with normal density.

```
        %b0=-6.2423,  b1=0.8532,  b2 =-0.0978
stats   %R2,      F,          p,          sigma2
        %0.1137   63.9650     0.0000      0.5856
```

The R^2 is about 11%, which is small, but represents improvement if only simple linear regression with covariate sharen were used. Note that residuals are approximately normal (Figure 14.9b), as expected.

14.10 Errors-in-Variables Linear Regression*

Assume that in the context of regression both responses Y and covariates X are measured with error. This is a frequent scenario in research labs in which it would be inappropriate to apply standard linear regression, which assumes that covariates are designed and constant.

This scenario in which covariates are observed with error is called errors-in-variables (EIV) linear regression. There are several formulations for EIV regression (Fuller, 2006). For pairs from a bivariate normal distribution (x_i, y_i), $i = 1, \ldots, n$, the EIV regression model is

$$y_i \sim \mathcal{N}(\beta_0 + \beta_1 \xi_i, \sigma_y^2)$$
$$x_i \sim \mathcal{N}(\xi_i, \sigma_x^2).$$

In an equivalent form, the regression is $\mathbb{E}y_i = \beta_0 + \beta_1 \mathbb{E}x_i = \beta_0 + \beta_1 \xi_i$, $i = 1, \ldots, n$, and the inference on parameters β_0 and β_1 is made conditionally on ξ_i. To make the model identifiable, parameter $\eta = \frac{\sigma_x^2}{\sigma_y^2}$ is assumed known. Note that we do not need to know individual variances, just their ratio.

If the observations (x_i, y_i), $i = 1, \ldots, n$, produce sums of squares $S_{xx} = \sum_i (x_i - \bar{x})^2$, $S_{yy} = \sum_i (y_i - \bar{y})^2$, and $S_{xy} = \sum_i (x_i - \bar{x})(y_i - \bar{y})$, then

$$\hat{\beta}_1 = \frac{-(S_{xx} - \eta S_{yy}) + \sqrt{(S_{xx} - \eta S_{yy})^2 + 4\eta S_{xy}^2}}{2\eta S_{xy}}, \qquad (14.7)$$

$$\hat{\beta}_0 = \bar{y} - \hat{\beta}_1 \bar{x}.$$

The estimators of the errors are

$$\hat{\sigma}_x^2 = \frac{1}{2n} \times \frac{\eta}{1 + \eta \hat{\beta}_1^2} \sum_{i=1}^{n} \left(y_i - (\hat{\beta}_0 + \hat{\beta}_1 x_i) \right)^2,$$

$$\hat{\sigma}_y^2 = \frac{\hat{\sigma}_x^2}{\eta}.$$

When $\eta = 1$, meaning variances of errors are the same, the solution in (14.7) coincides with a numerical least-squares minimization (so called orthogonal regression). However, for $\eta = 0$, meaning no errors in covariates, $\sigma_x^2 = 0$, we are back to the standard regression where $\hat{\beta}_1 = \frac{S_{xy}}{S_{xx}}$ (Exercise 14.24).

Example 14.7. **Predicting Albumin from Plasma Volume.** Griffin et al. (1945) reported plasma volume (in cc) and circulating albumin (in g) for $n = 58$ healthy males. Both quantities were measured with error, and it was assumed that the variance of plasma measurement exceeded the variance of circulating albumin by a factor of 200. Using EIV regression, establish an equation that would be used to predict circulating albumin from plasma volume. The data are given in ⬛ circalbumin.dat, where the first column contains plasma volume and the second the measured albumin. The script ▲ errorinvar.m calculates the following EIV regression equation: $\mathbb{E} y_i = 0.0521 \cdot \mathbb{E} x_i - 13.1619$. This straight line is plotted in red in Figure 14.10. For comparison, the standard regression is $\mathbb{E} y_i = 0.0494 x_i - 5.7871$, and it is plotted in black. Parameter η was set to 200, and the variances are estimated as $s_x^2 = 4852.8$ and $s_y^2 = 24.2641$.

✎

14.11 Analysis of Covariance

Analysis of covariance (ANCOVA) is a linear model that includes two types of predictors: quantitative, like regression, and categorical, like ANOVA. It

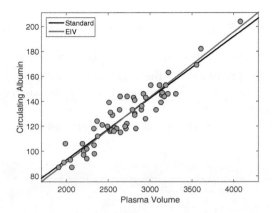

Fig. 14.10 EIV regression (*red*) and standard regression (*black*) for `circalbumin` data.

can be formulated in quite general terms, but we will discuss the case of a single predictor of each kind.

The quantitative variable x is linearly connected with the response, and for a fixed categorical variable, the problem is exactly regression. However, for a fixed value of x, the model is ANOVA with treatments/groups defined by the categorical variable.

The rationale behind the merging of the two models is that in a range of experiments, modeling the problem as regression only or as ANOVA only may be inadequate. By introducing a quantitative covariate to ANOVA, or equivalently groups/treatments to regression, we may better account for the variability in data and produce better modeling and prediction results.

We will analyze two illustrative examples. In the first example, which will be solved in MATLAB, the efficacies of two drugs for lowering blood pressure are compared by analyzing the drop in blood pressure after taking the drug. However, the initial blood pressure measured before the drug is taken should be taken into account since a drop of 50, for example, is not the same if the initial pressure was 90 as opposed to 180.

In the second example, which will be solved in WinBUGS, the measured response is the strength of synthetic fiber, and the two covariates are the fiber's diameter (quantitative) and the machine on which the fiber was produced (categorical with three levels).

We assume the model

$$y_{ij} = \mu + \alpha_i + \beta(x_{ij} - \bar{x}) + \epsilon_{ij},\ i = 1, \ldots, a;\ j = 1, \ldots, n, \qquad (14.8)$$

where a is the number of levels/treatments, n is a common sample size within each level, and $\bar{x} = \frac{1}{an} \sum_{i,j} x_{ij}$ is the overall mean of xs. The errors ϵ_i

are assumed to be independent normal with mean 0 and constant variance σ^2.

For practical reasons the covariates x_{ij} are centered as $x_{ij} - \bar{x}$ in order to simplify the expressions for the estimators. Let $\bar{x}_i = \frac{1}{n} \sum_j x_{ij}$ be the ith treatment mean for the xs. The means $\bar{y}$ and $\bar{y}_i$ are defined analogously, $\bar{y} = \frac{1}{an} \sum_{i,j} y_{ij}$ and $\bar{y}_i = \frac{1}{n} \sum_j y_{ij}$.

In ANCOVA we calculate the sums of squares and mixed-product sums as

$$
\begin{aligned}
S_{xx} &= \sum_{i=1}^a \sum_{j=1}^n (x_{ij} - \bar{x})^2 \\
S_{xy} &= \sum_{i=1}^a \sum_{j=1}^n (x_{ij} - \bar{x})(y_{ij} - \bar{y}) \\
S_{yy} &= \sum_{i=1}^a \sum_{j=1}^n (y_{ij} - \bar{y})^2 \\
T_{xx} &= \sum_{i=1}^a (\bar{x}_i - \bar{x})^2 \\
T_{xy} &= \sum_{i=1}^a (\bar{x}_i - \bar{x})(\bar{y}_i - \bar{y}) \\
T_{yy} &= \sum_{i=1}^a (\bar{y}_i - \bar{y})^2 \\
Q_{xx} &= \sum_{i=1}^a \sum_{j=1}^n (x_{ij} - \bar{x}_i)^2 = S_{xx} - T_{xx} \\
Q_{xy} &= \sum_{i=1}^a \sum_{j=1}^n (x_{ij} - \bar{x}_i)(y_{ij} - \bar{y}_i) = S_{xy} - T_{xy} \\
Q_{yy} &= \sum_{i=1}^a \sum_{j=1}^n (y_{ij} - \bar{y}_i)^2 = S_{yy} - T_{yy}
\end{aligned}
$$

We are interested in finding estimators for the parameters in model (14.8), the common mean μ, treatment effects α_i, regression slope β, and the variance of the error σ^2.

The estimators are $\hat{\mu} = \bar{y}$, $b = \hat{\beta} = Q_{xy}/Q_{xx}$, and $\hat{\alpha}_i = \bar{y}_i - \bar{y} - b(\bar{x}_i - \bar{x})$. The estimator of the variance, σ^2, is $s^2 = MSE = SSE/(a(n-1)-1)$, where $SSE = Q_{yy} - Q_{xy}^2/Q_{xx}$.

If there are no treatment effects, that is, if all $\alpha_i = 0$, then the model is a plain regression and

$$ y_{ij} = \mu + \beta(x_{ij} - \bar{x}) + \epsilon_{ij}, \ i = 1,\ldots,a; \ j = 1,\ldots,n. $$

In this reduced case the error sum of squares is $SSE' = S_{yy} - S_{xy}^2/S_{xx}$, with $an - 2$ degrees of freedom. Thus, the test $H_0 : \alpha_i = 0$ is based on an F-statistic,

$$ F = \frac{(SSE' - SSE)/(a-1)}{SSE/(a(n-1)-1)}, $$

that has an F-distribution with $a - 1$ and $a(n-1) - 1$ degrees of freedom.

The test for regression $H_0 : \beta = 0$ is based on the statistic

$$ F = \frac{Q_{xy}^2/Q_{xx}}{SSE/(a(n-1)-1)}, $$

which has an F-distribution with 1 and $a(n-1)-1$ degrees of freedom.
  Next we provide a MATLAB solution for a simple ANCOVA layout.

Example 14.8. **Kodlin's Blood Pressure Experiment.** Kodlin (1951) reported
an experiment that compared two substances for lowering blood pressure,
denoted as substances A and B. Two groups of animals are randomized to
the two substances and a decrease in pressure is recorded. The initial pressure is recorded. The data for the blood pressure experiment appear in the
table below. Compare the two substances by accounting for the possible effect of the initial pressure on the decrease in pressure. Discuss the results of
the ANCOVA analysis and compare them with those obtained by ignoring
the potential effect of the initial pressure.

Substance A			Substance B		
Animal	Decrease	Initial	Animal	Decrease	Initial
1	45	135	11	34	90
2	45	125	12	55	135
3	20	125	13	50	130
4	50	130	14	45	115
5	25	105	15	30	110
6	37	130	16	45	140
7	50	140	17	45	130
8	20	93	18	23	95
9	25	110	19	40	90
10	15	100	20	35	105

We will use $\alpha = 0.05$ for all significance assessments.

```
%kodlin.m
%x = Initial; y = Decrease; g=1 for 'A', g=2 for 'B'
x = [135 125 125 130 105 130 140  93 110 100 ...
      90 135 130 115 110 140 130  95  90 105];
y = [45 45 20 50 25 37 50 20 25 15 ...
      34 55 50 45 30 45 45 23 40 35];
g = [1 1 1 1 1 1 1 1 1 1 2 2 2 2 2 2 2 2 2 2];

a = 2; n = 10;

x1=x(g==1);   x2=x(g==2);
y1=y(g==1);   y2=y(g==2);
x1b=mean(x1); x2b=mean(x2);
y1b=mean(y1); y2b=mean(y2);
xb = mean(x); yb = mean(y);

SXX = sum( (x - xb).^2 );
SXY = sum( (x - xb).* (y - yb) );
SYY = sum( (y - yb).^2 );

QXX = sum( (x1-x1b).^2 + (x2-x2b).^2  );
```

```
QXY = sum( (x1-x1b).* (y1 - y1b) + ...
    (x2-x2b).*(y2 - y2b)  );
QYY = sum( (y1-y1b).^2 + (y2-y2b).^2  );
% estimators of model parameters mu, alpha_i, beta
mu = yb           %mu = 36.7
b = QXY/QXX       %b = 0.5175
alpha = [y1b y2b] - [yb yb] - b*([x1b x2b]-[xb xb])
                  %alpha =[-4.8714    4.8714]
TXX = SXX - QXX;
TXY = SXY - QXY;
TYY = SYY - QYY;

SSE = QYY - QXY^2/QXX;
MSE = SSE/(a * (n-1) - 1 )            %MSE = 60.6542
SSEp = SYY - SXY^2/SXX;

% F-test for testing $H_0: alpha_i = 0 (all i)
F1 = ((SSEp - SSE)/(a-1)) /MSE        %F1 = 7.6321
pvalF1 = 1- fcdf(F1,a-1, a*(n-1) - 1) %pvalF1=0.0133

% F-test for testing $H_0: beta = 0
F2 = (QXY^2/QXX)/MSE    %F2 = 24.5668
pvalF2 = 1 - fcdf(F2, 1, a * (n-1)-1 )
                       %pvalF2 = 0.00012
```

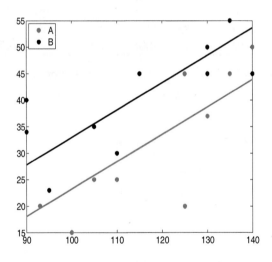

Fig. 14.11 Scatterplot of decrease (y) against the initial value (x) for substances A and B.

Note that both null hypotheses are found to be significant, but the testing of the equality of treatments seems to be the more important hypothesis in this example. The conclusion is that substances A and B are significantly different at $\alpha = 5\%$ significance level (p-value = 0.0133). Figure 14.11 plots the decrease (y) against the initial value (x) for substances A and B. The ANCOVA fits $\hat{y} = 36.7 \pm 4.8714 + 0.5175 \cdot (x - 116.65)$ are superimposed.

If we conducted the ANOVA test by ignoring the covariates, the test would find no differences among the drugs (p-value = 0.2025). Likewise, if both treatments were lumped together and the response regressed on x, the regression would be significant but with a p-value more than eight times larger than in ANCOVA.

```
oX = [ones(size(x'))  x'];
[b, bint, r, rint, stats] = regress(y', oX);
stats    % 0.4599    15.3268    0.0010    83.0023
[p, table, statan] = anoval(y, g);
p        %0.2025
```

This example shows how important accounting for sensible predictors is and how selecting the right statistical procedure is critical in modeling and decision making.

✎

MATLAB has a built-in function `aoctool` that, when invoked as `aoctool` (x,y,g), opens a front-end suite with various modeling and graphing capabilities. The output `stats` from `aoctool` can be imported into `multcompare` for a subsequent pairwise comparison analysis.

14.11.1 Sample Size in ANCOVA

ANCOVA can be thought of as an ANOVA for the transformed observations,

$$y_{ij}^* = y_{ij} - \beta(x_{ij} - \overline{x}), \quad i = 1, 2, \ldots, a; \ j = 1, \ldots, n.$$

where, as before, a denotes the number of groups/treatments, and n is the common group size. The total sample size is $N = an$.

Thus the power analysis in ANCOVA is related to that of ANOVA with a small modification. Namely, for a single covariate the number of degrees of freedom in the denominator of F statistic becomes $(n-1)a - 1$ instead of $(n-1)a$. More generally, if k covariates are employed, the number of degrees of freedom would reduce to $(n-1)a - k$.

With this modification power analysis proceeds as it was described in Section 11.8.

Example 14.9. **Kodlin's Experiment Revisited.** If we are to repeat Kodlin's experiment described in Example 14.8 and wanted to be able to find medium effect $f^2 = 0.25^2$ with the power of 85% in $\alpha = 0.05$ level testing, we will use (11.5). Here we have a single covariate, $a = 2$ treatments/groups, and $n = 10$ observations per group. The power for such experiment would be 0.1872.

```
a=2; n=10; f2=0.25^2; c=1;
df1=a-1;    df2=(n-1)*a - c;   lam=a*n*f2;
1-ncfcdf( finv(0.95, df1, df2), df1, df2, lam) %ans =0.1842
```

Thus, the experimenters "were lucky" that the observed effect in the experiment was large enough to produce a significant result (p-value was 0.0133).

If we were to design a new experiment and require a power of at least 85%, for the medium effect $f^2 = 0.25^2$, and $\alpha = 0.05$, the required sample size would be 73 per treatment.

```
a=2;   f2=0.25^2;
f= @(n) 1-ncfcdf( finv(0.95, a-1, (n-1)*a - 1),...
                  a-1, (n-1)*a - 1, a*n*f2)-0.85
ssize=fzero(f,100)   %ssize =72.8073
```

✎

14.11.2 Bayesian Approach to ANCOVA

Bayesian approach implemented via WinBUGS is conceptually simple: the ANCOVA model is stated in its direct form, constraints on the ANOVA coefficients imposed (either STZ or CR), and the priors on all free parameters are elicited and set. Unlike the classical approach where more than one covariate in the regression part and more than one factor in the ANOVA part significantly increase the complexity of ANCOVA model, the Bayesian approach via WinBUGS remains simple and straightforward.

The following example solved in WinBUGS illustrates handling AN-COVA in a Bayesian manner.

Example 14.10. **ANCOVA Fibers.** Three machines produce monofilament synthetic fiber for medical use (surgery, implants, devices, etc.). The measured response is the strength y (in pounds, lb.) and a covariate is the diameter x (in inches/1000).

```
#ancovafibers.odc
model{
for (i in 1:ntotal){
y[i] ~ dnorm( mui[i], tau )
mui[i] <-  mu + alpha[g[i]] + beta1 *(x[i] - mean(x[]))
}
#alpha[1] <- 0.0;                 #CR  constraints
 alpha[1] <- -sum( alpha[2:a] );#STZ constraints

mu ~ dnorm(0, 0.0001)
beta1 ~ dnorm(0, 0.0001)
beta0 <- mu - beta1 * mean(x[])
```

```
alpha[2] ~ dnorm(0, 0.0001)
alpha[3] ~ dnorm(0, 0.0001)
tau ~ dgamma(0.001, 0.001)
var <- 1/tau
}

DATA
list(ntotal=15, a=3,
y = c(36, 41, 39, 42, 49,
      40, 48, 39, 45, 44,
      35, 37, 42, 34, 32),
x = c(20, 25, 24, 25, 32,
      22, 28, 22, 30, 28,
      21, 23, 26, 21, 15),
g = c( 1,  1,  1,  1,  1,
       2,  2,  2,  2,  2,
       3,  3,  3,  3,  3)  )

INITS
list(mu=1, alpha=c(NA, 0, 0), beta1=0, tau=1)
```

The output from WinBUGS is

	mean	sd	MC error	val2.5pc	median	val97.5pc	start	sample
alpha[1]	0.182	0.6618	0.001848	−1.129	0.1833	1.494	10001	100000
alpha[2]	1.218	0.6888	0.003279	−0.1576	1.217	2.596	10001	100000
alpha[3]	−1.4	0.743	0.003757	−2.876	−1.399	0.07178	10001	100000
beta0	17.18	3.086	0.01394	11.01	17.18	23.31	10001	100000
beta1	0.9547	0.1266	5.634E-4	0.7019	0.9543	1.207	10001	100000
mu	40.2	0.4564	0.001372	39.29	40.2	41.11	10001	100000
var	3.118	1.671	0.007652	1.276	2.716	7.348	10001	100000

The regression equations corresponding to the three treatments (machines) are

$$\hat{y}_{1i} = 17.18 + 0.182 + 0.9547 \cdot x_i,$$
$$\hat{y}_{2i} = 17.18 + 1.218 + 0.9547 \cdot x_i, \text{ and}$$
$$\hat{y}_{3i} = 17.18 - 1.4 + 0.9547 \cdot x_i.$$

Note that all three 95% credible sets for `alpha` contain 0, while the credible set for the slope `beta1` does not contain 0. This analysis of credible sets is not a formal Bayesian testing, but it agrees with the output from

```
y = [36 41 39 42 49   40 48 39 45 44   35 37 42 34 32];
x = [20 25 24 25 32   22 28 22 30 28   21 23 26 21 15];
g = [1  1   1  1  1    2  2  2  2  2    3  3  3  3  3];
aoctool(x, y, g) %parallel lines option
```

where the p-value corresponding to ANOVA part is 0.11808. The test for the regression slope is significant with a p-value of 4.2645e-06.

✎

14.12 Exercises

14.1. **Regression with Three Points.** Points $(x_1, y_1) = (1,1)$, $(x_2, y_2) = (2,2)$, and $(x_3, y_3) = (3,2)$ are given.
(a) Find (by hand calculation) the regression line that best fits the points, and sketch a scatterplot of points with the superimposed fit.
(b) What are the predictions at 1, 2, and 3 (i.e, $\hat{y}_1$, $\hat{y}_2$, and $\hat{y}_3$)?
(c) What are SST, SSR, and SSE?
(d) Find an estimator of variance $\hat{\sigma}^2$.

14.2. **Age and IVF Success Rate.** The highly publicized (recent TV reports) in vitro fertilization (IVF) success cases for women in their late fifties all involve donors' eggs. If the egg is the woman's own, the story is quite different.
IVF, an assisted reproductive technology (ART) procedure, involves extracting a woman's eggs, fertilizing the eggs in the laboratory, and then transferring the resulting embryos to the woman's uterus through the cervix. Fertilization involves a specialized technique known as intracytoplasmic sperm injection (ICSI).
The table below shows the live-birth success rate per transfer rate from a woman's own eggs, by age of recipient. The data are for the year 1999, published by the CDC at http://www.cdc.gov/art/ARTReports.htm.

Age (x)	24	25	26	27	28	29	30	31	32	33	34	35
Percentage (y)	38.7	38.6	38.9	41.4	39.7	41.1	38.7	37.6	36.3	36.9	35.7	33.8

Age (x)	36	37	38	39	40	41	42	43	44	45	46
Percentage (y)	33.2	30.1	27.8	22.7	21.3	15.4	11.2	9.2	5.4	3.0	1.6

Select the ages in the range 33–46, which shows an almost linear decay of the success rate (Figure 14.12). For the selected ages fit the linear model $\hat{y} = b_0 + b_1 x$. Would the quadratic relationship $\hat{y} = b_0 + b_1 x + b_2 x^2$ be more appropriate than the linear?

14.3. **Sharp Dissection and Severity of Postoperative Adhesions.** Postoperative adhesions are formed after surgical cardiac and great vessel procedures as part of the healing process. Scar tissue makes reentry complex and increases the rate of iatrogenic lesions. Currently, as reoperations are needed in 10% to 20% of heart surgeries, various methods have been investigated to prevent or decrease the severity of postoperative adhesions.
The surgical time spent in the adhesiolysis procedure and amounts of sharp dissection are informative summaries to to predict the severity of pericardial adhesions, as reported by Lopes et al. (2009). For example, the authors reported a linear relationship between the logarithm of the amount of sharp dissection lasd and severity score sesco assessed by a

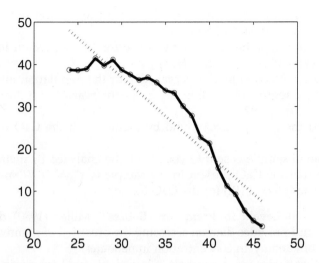

Fig. 14.12 Success rate (in %) versus age (in years).

standard categorization. Use the data in MATLAB format to answer (a)–(e).

```
shdiss = [14  108  39  311   24  112  104 ...
          382   42  74   67  145   21   93 ...
           75  381  36   36   73  239   35  69];
lasd = log(shdiss);

sesco = [6  12   9  18    7  12  12 ...
        17   8  11  14   15   7  12 ...
        13  18   9   9   10  16   7  10];
```

(a) Write down the linear relationship between lasd and sesco.
(b) What is R^2 here and what does it represent?
(c) Test the hypothesis $H_0 : \beta_0 = 0$ versus the alternative $H_1 : \beta_0 < 0$. Use $\alpha = 0.05$. The critical cut points are provided at the back of the problem, or, alternatively, you can report the p-value.
(d) Find a 95% confidence interval for the population slope β_1.
(e) For lasd=4 predict the severity score. Find a 99% confidence interval for the mean response.

14.4. **Mg–CaO Data Revisited.** Consider the data on CaO measurements discussed in Example 13.2. In data matrix ▦ hazel.dat|mat there are columns y and z corresponding to two ways to assess CaO (methods A and B), as well as a column x that gives the exact CaO amount.
(a) Fit two linear regressions. The first regression should express CaO found (A) (y) as a linear function of CaO present (x); that is, find the

equation $y = b_0 + b_1 x$. Then find the second regression as $z = d_0 + d_1 x$. Discuss the adequacy of regression fits.

(b) Test the hypothesis about the slope for the regression for (A) $H_0 : \beta_1 = 1$ versus the alternative $H_1 : \beta_1 < 1$.

(c) Find a 95% confidence interval (CI) for the population intercepts for each of the regressions. Is 0 in any of the intervals? What does it mean if it is and if it is not?

(d) Find the average CaO found by method A if the CaO present was $x^* = 15$.

(e) A small sample is sent to your lab to be analyzed by method A. It is known that the CaO present in the sample is $x^* = 15$. What is the 95% prediction interval (PI) for the CaO found?

14.5. **Kanamycin Levels in Premature Babies.** Miller (1980) describes a project and provides data on assessing the precision of noninvasive measuring of kanamycin concentration in neonates.

Premature babies are susceptible to infections, and kanamycin (an aminoglycoside) is used for the treatment of sepsis. Since kanamycin is ineffective at low doses and potentially harmful at high doses, it is necessary to constantly monitor its levels in a premature baby's body during treatment. The standard procedure for measuring serum kanamycin levels is to take blood samples from a heel. Unfortunately, due to frequent blood sampling, neonates are left with badly bruised heels.

Kanamycin is routinely administered through an umbilical catheter. An alternative procedure for measuring serum kanamycin would be to reverse the flow in the catheter and draw a blood sample from it. The concern about this noninvasive method is that the blood drawn from the point close to an infusion may have an elevated level of kanamycin compared to blood samples from more distant points in the body.

In a carefully designed experimental setup, blood samples from 20 babies were obtained simultaneously from an umbilical catheter and a heel venapuncture (using a heelstick). If the agreement is satisfactory, physicians would be willing to use the catheter values instead of heelstick values.

Here are the data:

Baby	Heelstick	Catheter	Baby	Heelstick	Catheter
1	23.0	25.2	11	26.4	24.8
2	33.2	26.0	12	21.8	26.8
3	16.6	16.3	13	14.9	15.4
4	26.3	27.2	14	17.4	14.9
5	20.0	23.2	15	20.0	18.1
6	20.0	18.1	16	13.2	16.3
7	20.6	22.2	17	28.4	31.3
8	18.9	17.2	18	25.9	31.2
9	17.8	18.8	19	18.9	18.0
10	20.0	16.4	20	13.8	15.6

(a) Model the Heelstick responses with Catheter as the predictor in a linear regression.

(b) Are there any unusual observations? Does regression improve when unusual observations are removed from the analysis?

(c) Test $H_0 : \beta_1 = 1$ versus $H_1 : \beta_1 < 1$ at the level $\alpha = 0.05$.

(d) Find a 97% confidence interval for the population intercept.

(e) Find 95% confidence and prediction intervals for the regression response when cath = 20.

(f) Using WinBUGS, estimate the parameters in a Bayesian regression with noninformative priors. Compare Bayesian and the least-squares solutions.

14.6. **Degradation of Scaffolds.** In an experiment conducted at the Georgia Tech/ Emory Center for the Engineering of Living Tissues, the goal was to find a suitable biomechanical replacement for cartilage, better known as tissue engineered cartilage. There are many factors (dimensional or mechanical) at which the cartilage scaffold is tested to assess whether it is a viable replacement. One of the problems is the degradation of scaffolds as the tissue grows, which affects all of the experimental metrics. The experimental data collected comprise a tissue growth experiment in which no cells were added, thus approximating the degradation of the scaffold over a sequence of 8 days. The dynamic shear summaries capture two physical phenomena, the modulus or the construct's ability to resist deformation under load and the frequency at which the modulus was evaluated. This modulus provides a measure of the extent of interconnectivity within the fibrous scaffold.

The table below contains moduli, in 1000s, for the frequency $f = 1$ over 8 days, each day represented by three independent measurements.

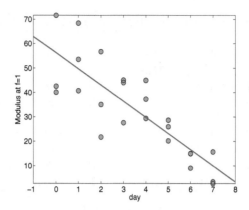

Fig. 14.13 Moduli measured against time for frequency $f = 1$.

Day	Mod at $f = 1$	Day	Mod at $f = 1$
0	42.520	4	44.929
0	71.590	4	29.348
0	40.063	4	37.259
1	68.397	5	28.625
1	53.527	5	25.956
1	40.676	5	20.179
2	21.724	6	14.994
2	35.032	6	9.051
2	56.687	6	14.923
3	44.029	7	2.692
3	45.058	7	15.688
3	27.579	7	3.420

The data and the code are available in the file ◄ degradation.m.

(a) Use the moduli for frequency $f = 1$. Write down the linear regression model: mod1 $= b_0 + b_1 \cdot$ day, where b_0 and b_1 are estimators of the population intercept and slope (Figure 14.13). What is R^2 for your regression?

(b) Test the hypothesis that the population intercept is equal to 100 versus the alternative that it is smaller than 100.

(c) Find a 96% confidence interval for the population slope.

(d) For day $= 5.5$, find the prediction of the modulus. What is the standard deviation for this predicted value?

14.7. **Glucosis in Lactococcus Lactis.** The data set ▢ Lactis.dat is courtesy of Dr. Eberhard Voit at the Georgia Institute of Technology and is an excerpt from a larger collection of data dealing with glycolysis in the bacterium *Lactococcus lactis* MG1363 (which is involved in essentially all yogurts, cheeses, etc.). The experiment was conducted in aerobic conditions with cell suspension in a 50-mM KPi buffer with a pH of 6.5, and 20 mM

[6-13C] glucose. There are four columns in the file 💾 Lactis.dat: time, glycose, lactate, and acetate.

The levels of extracelular metabolites lactate and acetate are monitored over time. After time $t = 5$, the level of lactate stabilizes around 30, while the level of acetate shows a linearly increasing trend in time.

(a) Plot lactate level against time. Take a subset of lactate levels for times > 5 (86 observations) and find basic descriptive statistics.

(b) Check the normality of the subset data. Test the hypothesis that the mean lactate level for $t > 5$ is 30 against the two-sided alternative.

(c) Select acetate levels for $t > 5$. Fit a linear relationship of acetate level against time and show that the linear model is justified by providing and discussing the ANOVA table.

14.8. **Weight and Latency in Rats.** Data consisting of rat body weight (grams) and latency to seizure (minutes) are given for 15 rats (adapted from Kleinbaum et al., 1987):

Rat #	Weight	Latency	Rat #	Weight	Latency	Rat #	Weight	Latency
1	348	1.80	2	372	1.95	3	378	2.90
4	390	2.30	5	392	1.10	6	395	2.50
7	400	1.30	8	409	2.00	9	413	1.70
10	415	2.00	11	423	2.95	12	428	2.25
13	464	3.05	14	468	3.70	15	470	3.62

It is of interest to regress the latency (y) to weight (x).

(a) Test the hypothesis $H_0 : \beta_0 = 0$ against the two-sided alternative. Use $\alpha = 0.05$.

(b) Test the hypothesis $H_0 : \beta_1 = 0.02$ against the alternative $H_0 : \beta_1 < 0.02$. Use $\alpha = 0.05$.

(c) Find a 95% confidence interval for the slope β_1.

(d) For weight $wei = 410$ find the mean latency response, $\hat{y}_m$. Test the hypothesis $H_0 : \hat{y}_m = 3$ versus the alternative $H_1 : \hat{y}_m < 3$. Test the same hypothesis for the predicted response $\hat{y}_{pred}$. In both tests use $\alpha = 0.05$.

14.9. **Rinderpest Virus in Rabbits.** Temperatures (temp) were recorded in a rabbit at various times (time) after the rabbit was inoculated with rinderpest virus (modified from Carter and Mitchell, 1958). Rinderpest (RP) is an infectious viral disease of cattle, domestic buffalo, and some species of wildlife; it is commonly referred to as cattle plague. It is characterized by fever, oral erosions, diarrhea, lymphoid necrosis, and high mortality.

Time after injection	Temperature
(time in hrs)	(temp in ° F)
24	102.8
32	104.5
48	106.5
56	107.0
70	105.1
72	103.9
80	103.2
96	102.1

(a) Demonstrate that a linear regression with one predictor (time) gives an insignificant F-statistic and relatively low R^2.

(b) Include time2 (squared time) as the second predictor, making the regression quadratic in variables, but still linear in coefficients. Show that this regression is significant and has a larger R^2.

(c) Find the 90% confidence interval for the coefficient of the quadratic term and test the hypothesis that the intercept is equal to 100 versus the one-sided alternative that it is smaller than 100. Use $\alpha = 0.05$.

14.10. **Hemodilution.** Clark et al. (1975) examined the fat filtration characteristics of a packed polyester-and-wool filter used in arterial lines during clinical hemodilution. They collected data on the filter's recovery of solids for ten patients who underwent surgery. The table below shows removal rates of lipids and cholesterol:

	Removal rates, mg/kg/L $\cdot 10^{-2}$	
Patient	Lipids (x)	Cholesterol (y)
1	3.81	1.90
2	2.10	1.03
3	0.79	0.44
4	1.99	1.18
5	1.03	0.62
6	2.07	1.29
7	0.74	0.39
8	3.88	2.30
9	1.43	0.93
10	0.41	0.29

(a) Fit a regression line to the data, with cholesterol as the response variable and lipids as the covariate. Discuss the adequacy of the proposed linear fit.

(b) What test is the resulting regression p-value referring to? State H_0 and H_1.

(c) Find the 95% confidence interval for the population intercept β_0.

(d) Test the hypothesis that the population slope β_1 is equal to 2/3 versus the one-sided alternative that it is less than 2/3. Use $\alpha = 0.05$.

(e) Predict the cholesterol rate for lipids at the level $1.65\ \mathrm{mg/kg/L} \times 10^{-2}$. Find the 95% confidence interval for the mean response and prediction interval for individual response.

14.11. **Anscombe's Data Sets.** A celebrated classic example of the role of residual analysis and statistical graphics in statistical modeling was created by Anscombe (1973). He constructed four different data sets (x_i, y_i), $i = 1, \dots, 11$, that share the descriptive statistics necessary to establish a linear regression fit $\hat{y} = \hat{\beta}_0 + \hat{\beta}_1 x$.

A linear model is appropriate for data set 1; the scatterplots and residual analysis suggest that data sets 2–4 seem not to be amenable to linear modeling.

	Set 1										
x	10	8	13	9	11	14	6	4	12	7	5
y	8.04	6.95	7.58	8.81	8.33	9.96	7.24	4.26	10.84	4.82	5.68
	Set 2										
x	10	8	13	9	11	14	6	4	12	7	5
y	9.14	8.14	8.74	8.77	9.26	8.10	6.13	3.10	9.13	7.26	4.74
	Set 3										
x	10	8	13	9	11	14	6	4	12	7	5
y	7.46	6.77	12.74	7.11	7.81	8.84	6.08	5.39	8.15	6.42	5.73
	Set 4										
x	8	8	8	8	8	8	8	19	8	8	8
y	6.58	5.76	7.71	8.84	8.47	7.04	5.25	12.50	5.56	7.91	6.89

(a) Using MATLAB fit the regression line for the four sets and provide four ANOVA tables. What statistics are the same? What statistics differ?

(b) Plot the residuals against the fitted values. Discuss the appropriateness of the regressions.

14.12. **Potato Leafhopper.** Potato leafhopper (*Empoasca fabae* is found throughout much of the United States east of the Rocky Mountains. It feeds on nearly 200 kinds of plants. Feeding and egg laying cause the infested plant damage. Eggs are deposited in the midrib or larger veins of the leaves, or in the petioles or stems. The leaves turn yellow, or sometimes pink or purple, and become wilted or stunted.

The length of the developmental period (in days) of the potato leafhopper, from egg to adult seems to be dependent on temperature (Kouskolekas and Decker, 1966). The original data were weighted means, but for the purpose of this exercise we consider them as though they were single observed values.

Temp °F	59.8	67.6	70.0	70.4	74.0	75.3	78.0	80.4	81.4	83.2	88.4	91.4	92.5
Development (days)	58.1	27.3	26.8	26.3	19.1	19.0	16.5	15.9	14.8	14.2	14.4	14.6	15.3

(a) Find a 98% confidence interval for population slope β_1.

(b) Test the hypothesis that the intercept is equal to 60 against the alternative that it is larger than 60. Take $\alpha = 0.01$.

(c) What is the 96% confidence interval for the mean response (mean number of days) if the temperature is 85°F?

14.13. **Force Sensor Calibration.** Your friend is conducting an experiment to measure the grasping force a robot arm exerts on an object. To accomplish this, he puts a force sensor on an object so that when the robot arm grasps it, he can measure the force. The force sensor maps change in electric pressure (in Volts) to force (in Newtons).

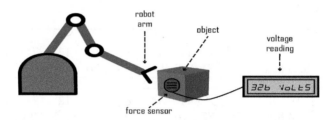

Fig. 14.14 The robot arm grasps on an object that has a force sensor. The force sensor outputs voltage readings which can be mapped to force.

The more you push on the force sensor, the higher is the electric pressure, and the higher is the voltage reading. A setup is shown in Figure 14.14. Before starting the experiment, your friend runs into a problem: he lost the force sensor data sheet! To calibrate the force sensor, he applies known weights to the force sensor and records the voltage output. His measurements are shown below:

Force (x)	0	1	2	3	5	7	10	15	20	25
Voltage (y)	326	375	403	438	555	646	799	1005	1223	1383

Force (x)	30	60	65	70	80	100	110	130	150	170
Voltage (y)	1653	2207	2250	2400	2767	3287	3727	4513	5170	5417

(a) Given the data he collected, fit a linear regression with force as the independent, controlled variable and voltage as the response.

(b) When the robot grasps the object, your friend records a voltage of 850 Volts. Using the regression from (a) estimate what this voltage corresponds to in force? Denote this force by x^* and its estimator by $\hat{x}^*$. Find a 95% CI for x^*.

(c) Flip now the roles of x and y and fit a linear regression in which x (force) is predicted by y (voltage). Predict now the force x^* for a voltage of 850 and compare it with the prediction in (b). Compare 95% CI's as well.

(d) Although the prediction in part (c) looks "more natural," it is not formally correct; the prediction as in (b) is recommended. Can you guess why? State your arguments in one or two sentences.

14.14. **Determining Average Slope.** One way to learn about the function of the kidney is to study the rate at which it produces and consumes different substances. An important quantity is the rate at which oxygen is consumed, since this is related to the kidney's workload. Another item of interest is the rate at which the kidney reabsorbs ionic sodium from the urine. This activity, known as sodium pumping, requires energy.

Data 🖻 kidneyouabain.dat|mat|xslx analyzed by Hyde (1980) contains measurements from 8 kidneys treated by the drug ouabain. Ouabain, from Somali waabaayo – "arrow poison," is g-strophanthin, a poisonous cardiac glycoside. This drug inhibits sodium pumping in kidneys.

The first column is the number of the kidney, the second column provides dry weight in milligrams multiplied by 10, the third column shows sodium reabsorption in $\mu Eq/min$, multiplied by 100. The last column is oxygen consumption in $\mu moles/min$, multiplied by 1000. The measurements are taken over 10-minute intervals.

Although both measurements have errors, the investigators felt that sodium measurements are more precise and should be taken as the covariate, and oxygen measurements as the response.

The average slope in regressions is of interest; in fact researchers were interested in estimating the reciprocal of the slope, which measures sodium pumping efficiency of the kidney.

(a) Find the slopes $b_{1,i}, i = 1, \ldots, 8$, in regressions of oxygen consumption (as y) to sodium reabsorption (as x), one for each kidney.

(b) Find a weighted average $b_1 = \sum_{i=1}^{8} w_i b_{1,i}$ where the weights w_i are proportional to $1/Sxx_i$ and sum up to 1. Here Sxx_i is $\sum_{k=1}^{6} (x_k - \bar{x})^2$ for the ith kidney. What is the variance of b_1?

(c) Using Bayesian hierarchical model, estimate an overall slope. Use noninformative priors. Compare this slope and its variance to results from (b).

Hint: In (b) the weights are inverse proportional to slope variances. The variance of weighted average slope is $\sigma^2 / \sum_{i=1}^{8} Sxx_i$. For part (c) consult 🐀Rats form Examples Vol I provided with the distribution of Win-BUGS/OpenBUGS.

14.15. **Cross-validating a Bayesian Regression.** In this exercise covariates x_1 and x_2 are simulated as

```
x1 = rand(1, 40);   x2 = floor(10 * rand(1,40)) + 1;
```

and the response variable y is obtained as

```
y = 2 + 6 * x1 - 0.5 * x2 + 0.8*randn(size(x1));
```

Write a WinBUGS program that selects 20 triples (x_1, x_2, y) to train the linear regression model $\hat{y} = b_0 + b_1 x_1 + b_2 x_2$ and then uses the remaining 20 triples to evaluate the model by comparing the original responses y_i, $i = 21, \ldots, 40$, with regression-predicted values $\hat{y}_i$, $i = 21, \ldots, 40$. The comparison involves calculating the MSE, the mean of $(y_i - \hat{y}_i)^2$, $i = 21, \ldots, 40$.

This is an example of how a cross-validation methodology is often employed to assess statistical models.

How do the Bayesian estimators of β_0, β_1, β_2, and σ compare to the "true" values 2, 6, -0.5, and 0.8?

14.16. **Taste of Cheese.** As cheddar cheese matures, a variety of chemical processes take place. The taste of mature cheese is related to the concentration of several chemicals in the final product. In a study of cheddar cheese from LaTrobe Valley of Victoria, Australia, samples of cheese were analyzed for their chemical composition and were subjected to taste tests. The table below presents data (from the experiments of G. T. Lloyd and E. H. Ramshaw, CISRO Food Research, Victoria, Australia, analyzed in Moore and McCabe, 2006) for one type of cheese manufacturing process. *Taste* is the response variable of interest. The taste scores were obtained by combining scores from several tasters. Three of the chemicals whose concentrations were measured are *acetic acid, hydrogen sulfide,* and *lactic acid*. For acetic acid and hydrogen sulfide, log transformations were taken.

Taste	Acetic	H2S	Lactic	Taste	Acetic	H2S	Lactic
12.3	4.54	3.13	0.86	20.9	5.16	5.04	1.53
39.0	5.37	5.44	1.57	47.9	5.76	7.59	1.81
5.6	4.66	3.81	0.99	25.9	5.70	7.60	1.09
37.3	5.89	8.73	1.29	21.9	6.08	7.97	1.78
18.1	4.90	3.85	1.29	21.0	5.24	4.17	1.58
34.9	5.74	6.14	1.68	57.2	6.45	7.91	1.90
0.7	4.48	3.00	1.06	25.9	5.24	4.94	1.30
54.9	6.15	6.75	1.52	40.9	6.37	9.59	1.74
15.9	4.79	3.91	1.16	6.4	5.41	4.70	1.49
18.0	5.25	6.17	1.63	38.9	5.44	9.06	1.99
14.0	4.56	4.95	1.15	15.2	5.30	5.22	1.33
32.0	5.46	9.24	1.44	56.7	5.86	10.20	2.01
16.8	5.37	3.66	1.31	11.6	6.04	3.22	1.46
26.5	6.46	6.92	1.72	0.7	5.33	3.91	1.25
13.4	5.80	6.69	1.08	5.5	6.18	4.79	1.25

(a) Find the equation in multiple linear regression that predicts Taste using Acetic, H2S, and Lactic as covariates.

(b) For Acetic = 5, H2S = 8, and Lactic = 2, estimate the regression response $\hat{Y}_h$ and find standard deviations for the mean and individual responses. [Ans. 43.37, 6.243, 11.886]

(c) Find the 98% confidence interval for the intercept β_0.

(d) Construct an ANOVA table.

(e) Find ordinary, studentized, and studentized deleted residuals for the observation Y_8.

(f) Find $DFFITS_8$ and $COOKSD_8$.

(g) Find $DFBETAS_8$ on the Lactic coefficient.

(h) Find and discuss the VIF.

14.17. **Slowing the Progression of Arthritis.** Arthritis is caused by the breakdown of collagen in joint cartilage by the enzyme MMP-13. The antibiotic doxycycline is a general inhibitor of MMPs and, by inhibiting the activity of MMP-13, is an effective method of slowing the progression of arthritis. At present, doxycycline is used to treat both rheumatiod arthritis and osteoarthritis. The rabbit's HIG-82 synovial cell line was used to model arthritis. MMP-13 was prepared by adding PMA, which guarantees its presence, and APMA, which activates it. The enzyme MMP-13 was mixed with a quenched substrate. When the enzyme cleaves the substrate, it fluoresces. This fluorescence was used to measure the amount of MMP-13 activity. Doxycycline, which decreases the amount of enzyme activity, was added in increasing concentrations: 0, 25, 50, 75, 100, and 200 micromols. The decrease in the fluorescence produced by the cleaved substrate when doxycycline was present was used to measure the decrease in activity of the enzyme. The same experiment was performed on three different plates, and the data was normalized.

The data set 🖵 arthritis1.mat can be found on the book's website. The data file contains 72 observations (rows); the first column represents doxycycline concentration (0, 25, 50, 75, 100, and 200), the second column is the fluorescence response, and the third column is the plate number.

(a) Fit the linear regression where fluorescence is the response and doxycycline concentration is the predictor. Predict the fluorescence if the doxycycline concentration is equal to 125.

(b) Since three plates are present, run aoctool with doxycycline concentration as x, fluorescence as y, and plate number as g. Is there a significant difference between the plates?

14.18. *Drosophila* **Offspring Prediction.** In a Genetics Lab at Duke University students[4] were involved in a project aimed to predict the sizes of *Drosophila* offspring population.

For 22 different mating bottles, a number of yeast **flakes** and a number of **virgin** flies was set and recorded. After several **days** the number of **offsprings** was counted.

For all bottles, the genotype of flies, temperature, and the number of males (3 per bottle), were kept the same. The bottles were attended in the same manner and the count was made at 4:00 pm for each bottle. The

---

[4] David Lee and Craig Cook, Project in STAT110, ISDS, Duke University, Spring 1996.

earliest counting day was after 2 weeks of setting the bottle, since virgin flies need at least 13-14 days to reproduce after mating. Data is available in the file 💻 drosophila.dat|mat|xlsx.

Bottle	Flakes	Virgins	Days	Offsprings	Bottle	Flakes	Virgins	Days	Offsprings
1	8	11	23	75	12	3	1	19	4
2	2	6	15	13	13	7	5	16	14
3	5	9	20	48	14	10	4	21	35
4	9	7	17	32	15	8	12	22	72
5	10	15	18	67	16	11	12	24	85
6	6	9	17	39	17	3	11	22	51
7	5	6	25	60	18	10	6	16	20
8	4	10	20	45	19	3	8	21	50
9	8	7	16	30	20	7	9	21	61
10	6	11	23	75	21	5	7	24	48
11	9	10	18	61	22	8	10	19	51

Although counts of offsprings are integers and Poisson regression is appropriate (to be covered in Chapter 15), since the average offspring count is about 50, normal approximation is very good and the multiple linear regression is adequate.

(a) Regress offsprings on flakes, virgins, and days. Write down the equation that links the number of offsprings to predictors.

(b) What is the overall R^2? A new bottle is set with 10 flakes and 10 virgins. How many offsprings is predicted if the count is made after 22 days?

(c) Run MATLAB's procedure stepwise. Is a smaller model recommended?

(d) Variable virgins enters during stepwise procedure first and accounts for more than 66% of total variability in the response. How many offsprings do you predict using univariate regression with only this variable as the predictor (the same value as in (b): virgins=10)? Find a 95% Prediction Interval for the number of offsprings in this case. Is the offspring prediction from (b) contained in this interval?

14.19. **Insulin on Opossum Liver.** Corkill (1932) provides data on the influence of insulin on opossum liver. In the experimental setup the 20 animals (common gray Australian opossums – *Trichosurus*) fasted for 24 or 36 hours. Ten animals, four from the 24-hour fasting group and six from the 36-hour fasting group, were injected with insulin, while the remaining ten animals served as controls, that is, they received no insulin. After 3 to 4 hours liver glycogen and blood sugar were measured. The weights of the animals were recorded as well.

The goal of the study was to explore the deposition of liver glycogen after the insulin regimen in opossums. In rabbits and cats, for example, it was previously found that insulin induced significant glycogen storage.

This study found a slight depletion of liver glycogen after the insulin treatment.

Our goal is to model the liver glycogen based on weight, level of blood sugar, insulin indicator, and fasting regime.

Is the insulin indicator (0 no, 1 yes) an important covariate in the model?

Animal	Weight	Liver glycogen	Blood sugar	Fasting period	Insulin
1	1502	1.80	0.124	24	0
2	1345	0.95	0.115	24	0
3	1425	1.12	0.128	24	0
4	1650	1.05	0.110	24	0
5	1520	0.45	0.052	24	1
6	1300	0.48	0.050	24	1
7	1250	0.75	0.045	24	1
8	1620	0.60	0.040	24	1
9	1725	0.76	0.130	36	0
10	1450	0.51	0.112	36	0
11	1800	0.48	0.105	36	0
12	1685	0.34	0.121	36	0
13	1560	0.38	0.116	36	0
14	1650	0.45	0.108	36	0
15	1650	0.65	0.032	36	1
16	1575	0.28	0.025	36	1
17	1260	0.10	0.045	36	1
18	1485	0.26	0.050	36	1
19	1520	0.18	0.030	36	1
20	1616	0.30	0.028	36	1

14.20. **Prostate Cancer Data.** This data set comes from the study by Stamey et al. (1989) that examined the relationship between the level of serum prostate specific antigen (Yang polyclonal radioimmunoassay) and a number of histological and morphometric measures in 97 patients who were about to receive a radical prostatectomy. The data are organized as data structure prost with first 8 fields (prost.lcavol - prost.pgg45) as predictors, and the 9th field (prost.lpsa) as the response.

(a) Load the data into MATLAB and run procedure stepwise. Write down the regression equation suggested by stepwise.

(b) Mr. Smith (a new patient) has response $y = 2.3$ and covariates:

$$x_1 = 1.4, x_2 = 3.7, x_3 = 65, x_4 = 0.1, x_5 = 0, x_6 = -0.16, x_7 = 7, \text{ and } x_8 = 30.$$

How close to the measured response $y = 2.3$ does the regression from (a) predict y for Mr. Smith? Denote this prediction by $\hat{y}_p$. Calculate the residual $r = \hat{y}_p - y$. *Hint:* In calculating $\hat{y}_p$ you should use only covariates x_i suggested by stepwise procedure.

(c) The best, in an R^2 sense, single predictor for y is x_1 – the logarithm of the cancer volume. Fit the univariate regression using x_1 as the predictor.

x_1 prost.lcavol	Logarithm of cancer volume	
x_2 prost.lweight	Logarithm of prostate weight	
x_3 prost.age	Patient's age	
x_4 prost.lbph	Logarithm of benign prostatic hyperplasia amount	
x_5 prost.svi	Seminal vesicle invasion, 0 – no, 1 – yes.	
x_6 prost.lcp	Logarithm of capsular penetration	
x_7 prost.gleason	Gleason score	
x_8 prost.pgg45	Percentage Gleason scores 4 or 5	
y prost.lpsa	Logarithm of prostate specific antigen	

Table 14.2 Fields in structure file prost. First 8 fields are predictors, and the last is the response to be modeled.

What is $\hat{y}_p$ for Mr. Smith based on this univariate regression? Find a 95% prediction interval for y_p. Is $y = 2.3$ in the interval?

14.21. **Assessing the Volume in Shortleaf Pine.** *Pinus echinata* (shortleaf pine) forests provided innumerable railroad ties for our nation's expanding railroad network in the late 19th and early 20th century. The wood is now used for general construction, exterior and interior finishing, and pulpwood.

Volume is the most widely used measure of wood quantity and is often estimated in standing trees for the assessment of economic value or commercial utilization potential. Volume is usually estimated from such measurements as diameter and merchantable height. The proposed equation for volume is

$$V = \alpha_0 D^{\alpha_1} H^{\alpha_2} \eta, \tag{14.9}$$

where D is the diameter at breast height (1.3 m) and H is the merchantable height. Parameters α_0, α_1, and α_2 depend on the tree species while η is a multiplicative error with lognormal distribution with parameters $\mu = 0$ and σ^2.

Bruce and Schumacher (1935) provided data on 70 shortleaf pine trees consisting of D (inches), H (ft), and V (cu ft). The dataset is in file
🖳 shortleaf.dat.

(a) Apply logarithmic transformation on both sides of equation (14.9) and by linear regression, estimate α_0, α_1, and α_2 and σ^2.

(b) For $D = 15$ in and $H = 85$ ft estimate the volume and find the 95% confidence interval for the mean response.

(c) If you are to select a single best predictor for $\log V$, which one would you choose?

14.22. **Hocking–Pendleton Data.** This popular data set was constructed by Hocking and Pendelton (1982) to illustrate that an influential observation may not be outlier, and that an outlier may not be influential. The data, given in file 📋hockpend.dat|mat, is organized as a matrix of size 26×4; the predictors x_1, x_2, and x_3 are the first three columns, and the response y is the fourth column.

(a) Fit the linear regression model with the three covariates, report the parameter estimates and R^2.

(b) Is the multicolinearity problem here?

(c) Is any of the 26 observations influential (in the sense of DFFITS, or Cook's Distance)?

(d) Is any of the 26 observations potential outlier (in the sense of a large studentized residual)?

(e) Using forward variable selection propose a possibly simpler model.

14.23. **Squids.** Data, analyzed by Freund and Wilson (1998), were obtained on 22 squids. The dependent variable y is the weight of the squid in pounds. The predictor variables represent measurements on the beak or mouth of the squid. The data are provided in the file squids.csv|dat|xlsx|mat

Column 1 Observation	–
Column 2 Rostral length in inches	x_1
Column 3 Wing length in inches	x_2
Column 4 Rostral to notch length	x_3
Column 5 Notch to wing length	x_4
Column 6 Width in inches	x_5
Column 7 Weight of the squid in pounds	y

Scientists wanted to know how useful beak measurements are in predicting the weight of the squid. Answering this question was important in the study of sizes of squid eaten by sharks and tuna, since the beak is indigestible.

(a) Using multiple linear regression, estimate a linear model that expresses the squid weight y using the predictors $x_1 - x_5$

(b) What y is predicted for $x_1 = 1.52, x_2 = 1.12$, $x_3 = 0.622$, $x_4 = 0.917$, and $x_5 = 0.324$? What is the 95% confidence interval for the mean response?

(c) Which of the 22 observations are influential? Answer by using either DFFITS or Cook's distance.

(d) Is the multicolinearity a problem here? Explain.

(e) Using the forward selection procedure propose a more compact model.

(f) Redo (a) and (b) in Bayesian fashion using noninformative priors. A shell-ODC with the data is provided as squids.odc. Are the results of classical and Bayesian analyses similar?

14.24. **Slope in EIV Regression.** Show that the EIV regression slope in (14.7) tends to S_{xy}/S_{xx} when $\eta \to 0$. [*Hint:* Apply L'Hôpital's rule.]

14.25. **Interparticular Spacing and Wavelength in Nanoprisms 2.** In the context of Example 13.4, let $x = (\text{separation})^{-1}$ and $y = \log(\text{wavelength})$.

The part of MATLAB regression output for 📖 nonoprism.dat data set is given below.

```
[b, bint, r, rint, stats] = regress(y,[ones(size(x)) x])

%b =
%    -4.7182
%     1.6289
%
%bint =
%    -5.6578    -3.7787
%     1.3061     1.9516
%
%r =
%     0.0037
%    -0.0084
%     ...
%     0.0058
%     0.0046
%
%rint =
%    -0.0101     0.0176
%    -0.0232     0.0064
%     ...
%    -0.0096     0.0213
%    -0.0110     0.0203
%
%stats =
%     0.8545   111.5735     0.0000     0.0001
```

Using information contained in this output, answer the following questions:

(a) What is the regression equation linking x and y?

(b) Predict $y = \log(\text{wavelength})$ for $x = 2.9$. What is the *wavelength* for such x?

(c) What is R^2 here and how is it interpreted? What is the F-statistic here? Is it significant?

(d) The 95% confidence interval for the population slope β_1 is [1.3061, 1.9516]. Using information in this output construct a 99% confidence interval for β_1.

14.26. **Kodlin's Experiment Revisited.** Estimate ANCOVA parameters for Kodlin's blood pressure experiment (Example 14.8) in a Bayesian fashion using WinBUGS/OpenBUGS. Use noninformative priors. Compare classical parameter estimators from Example 14.8 with the Bayesian counterparts.

MATLAB AND WINBUGS FILES AND DATA SETS USED IN THIS CHAPTER

http://statbook.gatech.edu/Ch14.Reg/

adhesions.m, ancovafibers.m, caprolactone.m, coburnreg.m, cpeptide.m, degradation.m, diabetes.m, diagnostics.m, dissection.m, errorinvar.m, fatreg.m, fatreg1.m, fatregdiag.m, galton.m, hemo.m, histn.m, hubble.m, invitro.m, kanamycin.m, kodlin.m, myeb.m, oldfaithful.m, pedometer1.m, ratwei.m, silverzinc.m, tastecheese.m, vitalcapacity.m, vonneumann.m

ancovafibers.odc, fat.odc, mellitus.odc, regressionpred.odc, vortex.odc

adhesion.xls, alcos.dat|xls, arthritis1.dat|mat, bmp2.dat|mat|xlsx, circalbumin.dat, coburn.mat, Cpaptide.dat|mat, Cpeptideext.dat|mat, drosophila.dat|mat|xlsx, fat.dat|xlsx, galton.dat, galtoncompact.dat, kanamycin.dat, kidneyouabain.dat|mat|xlsx, Lactis.dat, nanoprism.dat, pearson.dat, pmr1.mat, prost.mat, prostate.dat, ranunculus.xlsx, shortleaf.dat, silverzinc.dat|mat, vitalcapacity.xlsx

---

CHAPTER REFERENCES

Allen, D. M. (1974). The relationship between variable selection and data augmentation and a method for prediction. *Technometrics*, **16**, 125–127.

Anscombe, F. (1973). Graphs in statistical analysis. *Am. Stat.*, **27**, 17–21.

Armitage. P. and Berry, G. (1994). *Statistical Methods in Medical Research.* Blackwell Science, London.

Belsley, D. A., Kuh, E., and Welsch, R. E. (1980). *Regression Diagnostics: Identifying Influential Data and Sources of Collinearity.* Wiley, New York.

Bruce, D. and Schumacher, F. X. (1935) *Forest Mensuration.* American Forestry Series. McGraw Hill, New York.

Carter, G. and Mitchell, C. (1958). Methods for adapting the virus of rinderpest to rabbits. *Science*, **128**, 252–253.

Clark, R., Margraf, H., and Beauchamp, R. (1975). Fat and solid filtration in clinical perfusions. *Surgery*, **77**, 216–224.

Corkill, B. (1932). The influence of insulin on the liver glycogen of the common grey australian "opossum" (Trichosurus). *J. Physiol.*, **75**, 1, 29–32. PMCID: PMC1394507.

Ezekiel, M. (1930). *Methods of Correlation Analysis.* Wiley, New York.

Freund, R. J. and Wilson, W. J. (1998). *Regression Analysis: Statistical Modeling of a Response Variable.* Academic Press, San Diego, CA.

Galton, F. (1886). Regression towards mediocrity in hereditary stature. *J. Anthropol. Inst. Great Br. Ireland,* **15**, 246–263.

Griffin, G. E., Abbott, W. E., Pride, M. P., Runtwyler, E., Mautz, F. R., and Griffith, L. (1945). Available (thiocyanate) volume total circulating plasma proteins in normal adults, *Ann. Surg.,* **121**, 3, 352–360.

Hastie, T. and Tibshirani, R. (1990). *Generalized Additive Models.* Chapman & Hall, London.

Hocking, R. R. and Pendleton, O. J. (1982). The regression dilemma. *Comm. Stat. Theory Methods,* **12**, 497–527.

Hoerl, A. E. and Kennard, R. (1970). Ridge regression: Biased estimation for nonorthogonal problems. *Technometrics,* **12**, 55–67.

Hyde, J. (1980). Determining an average slope. In *Biostatistics Casebook.* Miller, R. G. Jr., Effron, B., Brown, B. W. Jr., and Moses, L. E. (eds.), 171–189, Wiley, New York.

Kleinbaum, D. G., Kupper, L. L., and Muller, K. E. (1987). *Applied Regression Analysis and Other Multivariable Methods.* PWS-Kent, Boston.

Kodlin, D. (1951). An application of the analysis of covariance in pharmacology. *Arch. Int. Pharmacodyn. Ther.,* **87**, 1–2, 207–211.

Kouskolekas, C. and Decker, G. (1966). The effect of temperature on the rate of development of the potato leafhopper, *Empoasca fabae* (Homoptera: Cicadelidae). *Ann. Entomol. Soc. Am.,* **59**, 292–298.

Laud, P. and Ibrahim, J. (1995). Predictive model selection. *J. R. Stat. Soc. Ser. B,* **57**, 1, 247–262.

Lopes, J. B., Dallan, L. A. O., Moreira, L. P. F., Carreiro, M. C., Rodrigues, F. L. B., Mendes, P. C., and Stol, N. A. G. (2009). New quantitative variables to measure postoperative pericardial adhesions. Useful tools in experimental research. *Acta Cir. Bras.,* **24**, São Paulo.

Moore, D. and McCabe, G. (2006). *Introduction to the Practice of Statistics,* 5th ed. Freeman, San Francisco.

Ntzoufras, I. (2009). *Bayesian Modeling Using WinBUGS.* Wiley, Hoboken, NJ.

O'Hagan, A. (1994). *Bayesian Inference.* Kendall's Advanced Theory of Statistics, Vol. 2B. Arnold, London.

Rawlings, J. O., Pantula, S., and Dickey, D. (2001). *Applied Regression Analysis: A Research Tool.* Springer Texts in Statistics. Springer, New York.

Sockett, E. B., Daneman, D., Clarson, C., and Ehrich, R. M. (1987). Factors affecting and patterns of residual insulin secretion during the first year of type I (insulin dependent) diabetes mellitus in children. *Diabetes,* **30**, 453–459.

Stamey, T. A., Kabalin, J. N., McNeal, J. E., Johnstone, I. M., Freiha, F., Redwine, E. A. and Yang, N. (1989). Prostate specific antigen in the diagnosis and treatment of adenocarcinoma of the prostate: II. Radical prostatectomy treated patients. *J. Urol.,* **141**, 5, 1076–1083.

Thisted, R. A. (1980). Comment on Smith and Campbell (1980). *J. Amer. Stat. Assoc.,* **75**, 81–86.

Thonnard, M. (2006). Confidence Intervals in Inverse Regression, MS Thesis. Technische Universiteit Eindhoven, Netherlands.

Yago, T., Lou, J., Wu, T., Yang, J., Miner, J. J., Coburn, L., Lopez, L. A., Cruz, M. A., Dong, J.-F., McIntire, L. V., McEver, R. P., and Zhu, C. (2008). Platelet glycoprotein Iba forms catch bonds with human WT vWF but not with type 2B von Willebrand disease vWF. *J. Clin. Invest.,* **118**, 9, 3195–3207.

Chapter 15
Regression for Binary and Count Data

There are 10 types of people in the world, those who can read binary, and those who can't.

– Anonymous

15.1 Introduction

Traditional simple or multiple linear regression assumes a normally distributed response centered at a linear combination of the predictors. For example, in simple regression, the response y_i is modeled as normal $\mathcal{N}(\beta_0 + \beta_1 x_i, \sigma^2)$, where the expectation, conditional on covariate x_i, is a linear function of x_i.

For some regression scenarios this model is inadequate because the response is not normally distributed. The response could be categorical, for example, with two or more categories ("disease present–disease absent," "survived–died," "low–medium–high," etc.) or be integer valued ("number with the disease," "number of failures," etc.), and yet a response may still depend on a covariate or a vector of covariates, x. In this chapter we discuss logistic and Poisson regressions that are appropriate models for binary and counting responses.

In logistic regression, the responses are binary, coded, without loss of generality, as 0 and 1 (or as 1 and 2 in WinBUGS). In Poisson regression, the responses are nonnegative integers well modeled by a Poisson distribution in which the rate λ depends on one or more covariates. The covariates enter the model in a linear fashion; however, their connection with the expected response is nonlinear.

Both logistic and Poisson regressions are examples of a wide class of models called generalized linear models (GLMs). The term generalized linear model refers to models introduced by Nelder and Wedderburn (1972) and popularized by the monograph of McCullagh and Nelder (1982, second edition 1989). In a canonical GLM model, the response variable y_i is assumed to follow a distribution from the class of distributions called the exponential family, with mean μ, which is assumed to depend on covariates via their linear combination. The exponential family is a rich family of distributions and includes almost all important distributions (normal, Bernoulli, binomial, Poisson, gamma, etc.). This link between the mean μ and covariates can be nonlinear, but the distribution of y_i depends on covariates only via their linear combination. Linear regression is a special case of GLMs for normally distributed responses, in which the mean is directly modeled by a linear combination of covariates.

15.2 Logistic Regression

Assume that a response y_i, depending on a covariate x_i, is categorical and can take two possible values. Examples of such responses include male–female, sick–healthy, alive–dead, pass–fail, success–failure, win–loss, etc. One usually assigns provisional numerical values to the responses, say 0 and 1, mainly to simplify notation. Our interest is in modeling the probability of response $y = 1$ given the observed covariates.

Classical least-squares regression $y_i = \beta_0 + \beta_1 x_i + \epsilon_i$ is clearly inadequate since for unbounded x_i the linear term $\beta_0 + \beta_1 x_i$ is unbounded as well. In addition, the residuals have only two possible values, and the variance of y_i is not free of x_i.

We assume that y is Bernoulli distributed with $\mathbb{E}y = \mathbb{P}(y = 1) = p$. Since $\mathbb{E}(y_i | X = x_i) = \mathbb{P}(y_i = 1 | X = x_i) = p_i$ is a number between 0 and 1, it is reasonable to model p_i as $F(\beta_0 + \beta_1 x_i)$ for some probability CDF F. Equivalently,

$$F^{-1}(p) = \beta_0 + \beta_1 x.$$

In principle, any monotone cumulative distribution function F can provide a link between the probability p and the covariate(s), but the most used distributions are logistic, normal, and complementary log-log, leading to logistic, probit, and clog-log regressions. The most popular among the three is the logistic regression because its coefficients, measuring the impact of the predictors on the binary response y, have convenient interpretations via the log odds of the events $\{y = 1\}$. In logistic regression, $F^{-1}(p) = \log \frac{p}{1-p}$ is called the *logit* and denoted as $\text{logit}(p)$.

15.2.1 Fitting Logistic Regression

The basic statistical model for logistic regression is

$$y_i \sim Ber(p_i),$$

$$\text{logit}(p_i) = \log \frac{p_i}{1 - p_i} = \beta_0 + \beta_1 x_i, \quad i = 1,\ldots,n, \tag{15.1}$$

when the responses are Bernoulli, 0 or 1.

When multiple measurements correspond to the same covariate, it is convenient to express the responses as binomial counts. That is, n_i responses corresponding to covariate x_i are grouped together and y_i is the number of responses equal to 1:

$$y_i \sim Bin(n_i, p_i),$$

$$\text{logit}(p_i) = \log \frac{p_i}{1 - p_i} = \beta_0 + \beta_1 x_i, \quad i = 1,\ldots,k, \quad \sum_{i=1}^{k} n_i = n. \tag{15.2}$$

In principle, it is always possible to express the binomial response model via Bernoulli responses by repeating y_i times the response 1 and $n_i - y_i$ times the response 0 for the same covariate x_i. The converse is not possible in general, especially if y_i depends on a continuous covariate. Assessing the goodness of fit for models of type (15.1) is a well-known problem since the asymptotic distributional results do not hold even for arbitrarily large sample sizes.

How does one estimate parameters β_0 and β_1 in logistic model (15.1) or (15.2)? The traditional least-squares algorithm that is utilized in linear regression is not applicable. Estimating the model coefficients amounts to solving a nonlinear equation, and this is done by an iterative procedure. The algorithm for a single predictor is illustrated and implemented in the m-file ◀ logisticmle.m. There, the Newton–Raphson method is used to solve non-linear likelihood equations and calculate coefficients b_0 and b_1 as estimators of population parameters β_0 and β_1. Details regarding the background and convergence of methods for estimating model parameters are beyond the scope of this text and can be found in McCullagh and Nelder (1989).

Once the parameters β_0 and β_1 are estimated, the probability $p_i = \mathbb{P}(y = 1|x = x_i)$ is obtained as

$$\hat{p}_i = \frac{\exp\{b_0 + b_1 x_i\}}{1 + \exp\{b_0 + b_1 x_i\}} = \frac{1}{1 + \exp\{-b_0 - b_1 x_i\}}.$$

In addition to the nature of response y_i, there is another key difference between ordinary and logistic regressions. For linear regression, the variance does not depend on the mean $\beta_0 + \beta_1 x_i$; it is constant for all x_i. This is one of the assumptions for linear regression. In logistic regression, the variance is not constant; it is a function of the mean. From (15.1), $\mathbb{E}y_i = p_i$ and $\mathbb{V}\mathrm{ar}\, y_i = \mathbb{E}y_i(1 - \mathbb{E}y_i)$.

If $p - 1$ covariates ($p \geq 2$ parameters) are available, as is often the case, then

$$X_i'b = \ell_i = b_0 + b_1 x_{i1} + b_2 x_{i2} + \cdots + b_{p-1} x_{i,p-1}$$

replaces $b_0 + b_1 x_i$, where X_i' is the ith row of a design matrix X of size $n \times p$, and $b = (b_0 \;\; b_1 \;\; \ldots \;\; b_{p-1})'$. Now,

$$\hat{p}_i = \frac{\exp\{X_i'b\}}{1 + \exp\{X_i'b\}} = \frac{1}{1 + \exp\{-X_i'b\}},$$

with b maximizing the log-likelihood

$$\log L(\beta) = \ell(\beta) = \sum_{i=1}^{n} y_i \cdot (X_i'\beta) - \sum_{i=1}^{n} \log\left(1 + \exp\{X_i'\beta\}\right).$$

Example 15.1. **Caesarean-Section Infections.** A Caesarean-section, or C-section, is major abdominal surgery, so mothers who undergo C-sections are more likely to have an infection, excessive bleeding, blood clots, more postpartum pain, a longer hospital stay, and a significantly longer recovery. The data in this example comes from Munich's *Klinikum Großharden* (Fahrmeir and Tutz, 1996) and concerns infections in births by C-section. The response variable of interest is the occurrence or nonoccurrence of infection. Three covariates, each at two levels, were considered as important for the occurrence of infection:

noplan – C-section delivery was planned (0) or not planned (1);

riskfac – risk factors for the mother, such as diabetes, overweight, previous C-section birth, etc., are present (1) or not present (0); and

antibio – antibiotics as a prophylaxis are given (1) or not given (0). Table 15.1 provides the results:

Table 15.1 Caesarean-section delivery data.

		Planned			No plan	
		Infection			Infection	
	Yes	No	Total	Yes	No	Total
Antibiotics						
Risk factor yes	1	17	18	11	87	98
Risk factor no	0	2	2	0	0	0
No antibiotics						
Risk factor yes	28	30	58	23	3	26
Risk factor no	8	32	40	0	9	9

Here is the MATLAB code that uses built-in functions `glmfit` and `glmval` to fit and present the model:

```
infection = [ 1 11   0   0 28 23  8   0];
total =      [18 98   2   0 58 26 40   9];
proportion = infection./total;
noplan =    [ 0  1  0  1  0  1  0  1];
riskfac =   [ 1  1  0  0  1  1  0  0];
antibio =   [ 1  1  1  1  0  0  0  0];
[b,dev,stats] = glmfit([noplan' riskfac' antibio'],...
              [infection' total'],'binomial','logit');
logitFit = ...
   glmval(b,[noplan' riskfac' antibio'],'logit');
```

The resulting additive model (with no interactions) is

$$\log \frac{\mathbb{P}(\text{infection})}{\mathbb{P}(\text{no infection})} = \beta_0 + \beta_1 \cdot \text{noplan} + \beta_2 \cdot \text{riskfac} + \beta_3 \cdot \text{antibio}$$

with estimators of βs as

b_0	b_1	b_2	b_3
−1.8926	1.0720	2.0299	−3.2544

The interpretation of the estimators for β coefficients is illuminating if we look at the odds ratio $\dfrac{\mathbb{P}(\text{infection})}{\mathbb{P}(\text{no infection})}$:

$$\frac{\mathbb{P}(\text{infection})}{\mathbb{P}(\text{no infection})} = \exp(\beta_0) \cdot \exp(\beta_1 \text{ noplan}) \cdot \exp(\beta_2 \text{ riskfac}) \cdot \exp(\beta_3 \text{ antibio}).$$

For example, when `antibio=1`, that is, when antibiotics are given, the estimated odds of infection $\mathbb{P}(\text{infection})/\mathbb{P}(\text{no infection})$ increase by the factor $\exp(-3.25) = 0.0388$, that is, the odds decrease 25.79 times. Of course, these statements are valid only if the model is accurate. Other competing models, such as probit or clog-log, may result in different changes in risk ratios.

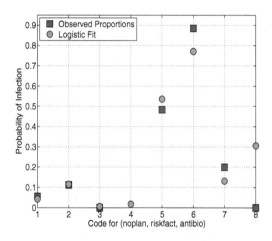

Fig. 15.1 Caesarean delivery infection predictions. For a triple (noplan, riskfac, antibio), the numbers on the x-axis code as follows: **1** = (0, 1, 1), **2** = (1, 1, 1), **3** = (0, 0, 1), **4** = (1, 0, 1), **5** = (0, 1, 0), **6** = (1, 1, 0), **7** = (0, 0, 0), and **8** = (1, 0, 0). *Blue squares* are the observed relative frequencies and *green circles* are the model-predicted probabilities of infection. Note that point **4** does not have an observed proportion.

The m-function ◀ `logisticmle.m` also gives standard errors for estimators of βs. Table 15.2 provides t-values, that is, ratios of coefficients and their standard deviations, for testing if the coefficients are significantly different from 0. These are known as Wald's Z statistics, since they are approximately normal.

The *deviance* (page 761) of this model as a measure of goodness of fit is distributed as χ^2 with 3 degrees of freedom. The number of degrees of freedom is calculated as 7 (the number of groups with observations) minus 4 (four estimated parameters $\beta_0 - \beta_3$). Since the deviance is found significant,

```
dev = 10.9967;
```

Table 15.2 t-ratios (Wald's Z statistic) for the estimators $b = \hat{\beta}$.

	b	s_b	t
Intercept	−1.8926	0.4124	−4.5893
noplan	1.0720	0.4253	2.5203
riskfac	2.0299	0.4553	4.4588
antibio	−3.2544	0.4812	−6.7624

```
pval = 1 - chi2cdf(dev, 7-4)    % 0.0117
```

the fit of this model is inadequate. To improve the fit, one may include the interactions.

One may ask why the regression model was needed in the first place. The probabilities of interest could be predicted by relative frequencies. For example, in the case (noplan = 0, riskfac = 1, antibio = 1), the relative frequency of infection was $1/18 = 0.0556$, just slightly larger than the model-predicted $\hat{p} = 0.0424$. There are two benefits in using regression. First, the model is able to predict probabilities in the cases where no patients are present, such as for (noplan = 1, riskfac = 0, antibio = 1). Second, the predictions for the cases where $y = 1$ is not observed are "borrowing strength" from other data and are not modeled individually. For example, zero as an estimator in the case (noplan = 1, riskfac = 0, antibio = 0) is not reasonable; the model-based estimator $\hat{p} = 0.3056$ is more realistic. Figure 15.1 compares observed and model-predicted infection rates. For a triple of covariates (noplan, riskfac, antibio), the numbers on the x-axis code as follows: **1** = (0, 1, 1), **2** = (1, 1, 1), **3** = (0, 0, 1), **4** = (1, 0, 1), **5** = (0, 1, 0), **6** = (1, 1, 0), **7** = (0, 0, 0), and **8** = (1, 0, 0). Note that point **4** does not have an observed proportion; however, the model-predicted proportion can be found. For computational aspects refer to file ◢ caesarean.m.

Next, we provide a Bayesian solution to this example and compare the model fit with the classical fit above. The comparisons are summarized in Table 15.1.

C-SECTION INFECTIONS

```
model{
  for(i in 1:N){
    inf[i] ~ dbin(p[i],total[i])
    logit(p[i]) <- beta0 + beta1*noplan[i] +
                   beta2*riskfac[i] + beta3*antibio[i]
  }
  beta0 ~dnorm(0, 0.00001)
  beta1 ~dnorm(0, 0.00001)
  beta2 ~dnorm(0, 0.00001)
  beta3 ~dnorm(0, 0.00001)
}

DATA
```

```
list(inf=c(1, 11, 0, 0, 28, 23, 8, 0),
     total = c(18, 98, 2, 0, 58, 26, 40, 9),
     noplan = c(0,1,0,1,0,1,0,1),
     riskfac = c(1,1, 0, 0, 1,1, 0, 0),
     antibio =c(1,1,1,1,0,0,0,0), N=8)
```

INITS

```
list(beta0 =0, beta1=0, beta2=0, beta3=0)
```

	mean	sd	MC error	val2.5pc	median	val97.5pc	start	sample
beta0	−1.964	0.4258	0.001468	−2.853	−1.945	−1.183	1001	1000000
beta1	1.111	0.4339	8.857E-4	0.2851	1.102	1.986	1001	1000000
beta2	2.104	0.4681	0.00159	1.226	2.09	3.066	1001	1000000
beta3	−3.335	0.4915	9.756E-4	−4.337	−3.322	−2.411	1001	1000000
deviance	32.24	2.873	0.00566	28.67	31.59	39.49	1001	1000000

Table 15.3 Comparison of classical and noninformative Bayes estimators $b = \hat{\beta}$, with estimators of standard deviations.

	b	s_b	$\hat{\beta}_B$	$\hat{\sigma}_B$
Intercept	−1.8926	0.4124	−1.964	0.4258
noplan	1.0720	0.4253	1.111	0.4339
riskfac	2.0299	0.4553	2.104	0.4681
antibio	−3.2544	0.4812	−3.335	0.4915

15.2.2 Assessing the Logistic Regression Fit

The measures for assessing the goodness of linear regression fit that we covered in Chapter 14, R^2, F, MSE, etc., are not appropriate for logistic regression. As in the case of linear regression, there is a range of measures for assessing the performance of logistic regression, and we will briefly discuss a few.

The significance of model parameters $\beta_0, \beta_1, \ldots, \beta_{p-1}$ is tested by the so-called Wald's test. One finds the statistic $Z_i = \frac{b_i}{s(b_i)}$ that has an approximate normal distribution if the coefficient β_i is 0. Equivalently, $W_i = \frac{b_i^2}{s^2(b_i)}$ with an approximate χ^2-distribution with 1 degree of freedom can be used. Large values of $|Z_i|$ or W_i are critical for $H_0 : \beta_i = 0$.

The sample variances $s^2(b_i)$ are diagonal elements of $(X'VX)^{-1}$, where

$$X = \begin{bmatrix} 1 & x_{11} & x_{12} & \cdots & x_{1,p-1} \\ 1 & x_{21} & x_{22} & \cdots & x_{2,p-1} \\ & & \cdots & & \\ 1 & x_{n1} & x_{n2} & \cdots & x_{n,p-1} \end{bmatrix}$$

is the design matrix and

$$V = \begin{bmatrix} \hat{p}_1(1-\hat{p}_1) & 0 & \cdots & 0 \\ 0 & \hat{p}_2(1-\hat{p}_2) & \cdots & 0 \\ & & \cdots & \\ 0 & 0 & \cdots & \hat{p}_n(1-\hat{p}_n) \end{bmatrix}.$$

The customary measure for goodness of fit is *deviance*, defined as

$$D = -2\log \frac{\text{likelihood of the fitted model}}{\text{likelihood of the saturated model}}.$$

For the logistic regression in (15.2), where y_i is the number of 1s and $n_i - y_i$ is the number of 0s in class i, the likelihood is $L = \prod_{i=1}^{k} p_i^{y_i}(1-p_i)^{n_i-y_i}$ and the deviance is

$$D = -2\sum_{1=1}^{k} \left\{ y_i \log\left(\frac{\hat{y}_i}{y_i}\right) + (n_i - y_i)\log\left(\frac{n_i - \hat{y}_i}{n_i - y_i}\right) \right\},$$

where $\hat{y}_i = n_i \hat{p}_i$ is the model fit for y_i. The saturated model estimates p_i as $\hat{p}_i = y_i/n_i$ and $\hat{y}_i = y_i$, providing the fit that matches the observations.

The deviance statistic in this case has a χ^2-distribution with $k - p$ degrees of freedom, where k is the number of classes/groups and p is the number of parameters in the model. Recall that in the previous example the deviance of the model was distributed as χ^2 with $k - p = 7 - 4 = 3$ degrees of freedom.

For both Bernoulli and binomial observations, the mean and variance depend on a single parameter, p. When the mean is well fitted, the variance may be underfitted (overdispersion in data) or overfitted (underdispersion in data). The ratio D/df is often used to indicate over- or underdispersion in the data.

The traditional χ^2-statistic for the goodness of fit in model (15.2) is defined as

$$\chi^2 = \sum_{i=1}^{k} \left[\frac{(y_i - \hat{y}_i)^2}{\hat{y}_i} + \frac{(y_i - \hat{y}_i)^2}{n_i - \hat{y}_i} \right],$$

where n_i is the number of observations in class i, $i = 1,\ldots,k$. This statistic has an approx. χ^2-distribution with $k - p$ degrees of freedom.

Goodness of fit measure G is defined as the difference of deviance between the null model (intercept-only model) and the model under consideration. G has a χ^2-distribution with $p - 1$ degrees of freedom, and small values of G are critical, suggesting that the deviance did not improve significantly by adding covariates.

The logistic model can always be expressed in terms of Bernoulli outcomes, where y_i is 0 or 1, as in (15.1). Then $k = n$, $n_i = 1$, the likelihood for the saturated model, is $\prod_{i=1}^{n} y_i^{y_i}(1 - y_i)^{1-y_i} = 1$ (we assume that $0^0 = 1$), and the deviance for the Bernoulli representation becomes

$$D = -2 \sum_{i=1}^{n} [y_i \log \hat{p}_i + (1 - y_i) \log(1 - \hat{p}_i)].$$

Statistic D does not follow any specific distribution, regardless of the sample size. Likewise, the Pearson χ^2 becomes

$$\chi^2 = \sum_{i=1}^{n} \frac{(y_i - \hat{p}_i)^2}{\hat{p}_i(1 - \hat{p}_i)} \tag{15.3}$$

in model (15.1) and does not follow any specific distribution, either.

To further evaluate the model, several types of residuals are available. Deviance residuals are defined as

$$r_i^D = \text{sign}(y_i - \hat{y}_i) \sqrt{2 \left\{ y_i \log \left(\frac{y_i}{\hat{y}_i} \right) + (n_i - y_i) \log \left(\frac{n_i - y_i}{n_i - \hat{y}_i} \right) \right\}}, \, i = 1, \ldots, k,$$

for model (15.2) and

$$r_i^D = \text{sign}(y_i - \hat{p}_i) \sqrt{2 y_i \log \hat{p}_i + (1 - y_i) \log(1 - \hat{p}_i)}, \, i = 1, \ldots, n,$$

for model (15.1). The deviance D is decomposed to the sum of squares of deviance residuals in an ANOVA-like fashion as $D = \sum (r_i^D)^2$. The squared residual $(r_i^D)^2$ measures the contribution of the ith case to the deviance.

Deviance residuals can be plotted against the order of sampling to explore for possible trends and outliers. Also useful for checking the model are half-normal plots where the ordered absolute values r_i^D are plotted against the normal quantiles $\Phi^{-1} \left(\frac{i+n-1/8}{2n+1/2} \right)$. These kinds of plots are an extension of Atkinson's (1985) half-normal plots in regular linear regression models. Deviation from a straight line in a half-normal plot indicates model inadequacy.

For the model in (15.1), the Pearson residual is defined as

$$r_i^{pea} = \frac{y_i - \hat{p}_i}{\sqrt{\hat{p}_i(1 - \hat{p}_i)}},$$

and the sum of squares of r_i^{pea} constitutes Pearson's χ^2 statistic,

$$\sum_{i=1}^{n} (r_i^{pea})^2 = \sum_{i=1}^{n} \frac{(y_i - \hat{p}_i)^2}{\hat{p}_i(1 - \hat{p}_i)},$$

as in (15.3). This statistic represents a discrepancy measure; however, as we mentioned, it does not follow the χ^2-distribution, even asymptotically.

In the case of continuous covariates, large n and small n_i, Hosmer and Lemeshow proposed a χ^2-statistic based on the grouping of predicted values $\hat{p}_i$. All $\hat{p}_i$ are ordered and divided into g approximately equal groups, usually 10. For 10 groups, sample deciles of ordered $\hat{p}_i$ can be used.

The Hosmer–Lemeshow statistic is

$$\chi^2_{HL} = \sum_{i=1}^{g} \frac{(n_i - n\bar{p}_i)^2}{n\bar{p}_i},$$

where g is the number of groups, n_i is the number of cases in the ith group, and $\bar{p}_i$ is the average of model (predicted) probabilities for the cases in the ith group. The χ^2_{hl} statistic is compared to χ^2_{g-2} quantiles, and small p-values indicate that the fit is poor. In the case of ties, that is, when there are blocks of items with the same predicted probability $\hat{p}$, the blocks are not split but assigned to one of the two groups that share the block. The details of the algorithm can be found in Hosmer and Lemeshow (1989).

In the case of linear regression, R^2, as a proportion of model-explained variability in observations, has a strong intuitive appeal in assessing the regression fit. In the case of logistic regression, there is no such intuitive R^2. However, there are several proposals of R^2-like measures, called pseudo-R^2. Most of them are defined in terms of model likelihood or log-likelihood. The model likelihood and log-likelihood are calculated using the logit model,

$$\ell_i = b_0 + b_1 x_{i1} + \cdots + b_{p-1} x_{i,p-1},$$

$$LL_p = LL(b_0, \ldots, b_{p-1}) = \sum_{i=1}^{n} (y_i \times \ell_i - \log(1 + \exp\{\ell_i\})),$$

and the model likelihood is $L_p = \exp\{LL_p\}$. The null model is fitted without covariates, and

$$\ell_0 = b_0,$$

$$LL_{null} = LL(b_0) = \sum_{i=1}^{n} (y_i \times \ell_0 - \log(1 + \exp\{\ell_0\})).$$

The null model likelihood is $L_{null} = \exp\{LL_{null}\}$.

By analogy to linear regression, $R^2 = \frac{SSR}{SST} = \frac{SST - SSE}{SST}$,

$$R^2_{mf} = \frac{LL_{null} - LL_p}{LL_{null}} = 1 - \frac{LL_p}{LL_{null}},$$

defines McFadden's pseudo-R^2. Some other counterparts of R^2 are

Cox–Snell: $\quad R^2_{cs} = 1 - \left[\frac{L_{null}}{L_p}\right]^{2/n}$;

Nagelkerke: $\quad R^2_n = \dfrac{1 - \left[\frac{L_{null}}{L_p}\right]^{2/n}}{1 - (L_{null})^{2/n}}$;

Effron: $\quad R^2_e = 1 - \dfrac{\sum_{i=1}^n (y_i - \hat{p}_i)^2}{\sum_{i=1}^n (y_i - \bar{y})^2}, \quad \bar{y} = \dfrac{\sum_{i=1}^n y_i}{n}.$

Example 15.2. **Arrhythmia.** Patients who undergo coronary artery bypass graft (CABG) surgery have an approximately 19% to 40% chance of developing atrial fibrillation (AF). AF is a quivering, chaotic motion in the upper chambers of the heart, known as the atria. AF can lead to the formation of blood clots, causing greater in-hospital mortality, strokes, and longer hospital stays. While this can be alleviated with drugs, it is very expensive and sometimes dangerous if not warranted. Ideally, several risk factors that would indicate an increased risk of developing AF in this population could save lives and money by indicating which patients need pharmacological intervention. Researchers began collecting data from CABG patients during their hospital stay such as demographics like age and sex, as well as heart rate, cholesterol, operation time, etc. Then the researchers recorded which patients developed AF during their hospital stay. The goal was to evaluate the probability of AF given the measured demographic and risk factors.

The data set ⬛arrhythmia.dat, courtesy of Dr. Matthew C. Wiggins, contains the following variables:

Y	Fibrillation
X_1	Age
X_2	Aortic cross clamp time
X_3	Cardiopulmonary bypass time
X_4	Intensive care unit (ICU) time
X_5	Average heart rate
X_6	Left ventricle ejection fraction
X_7	Anamnesis of hypertension
X_8	Gender [1 - female; 0 - male]
X_9	Anamnesis of diabetes
X_{10}	Previous MI

The MATLAB script ◀arrhythmia.m provides a logistic regression fit. The script calculates deviance and several goodness-of-fit measures.

```
load 'Arrhythmia.mat'
Y = Arrhythmia(:,1);
X = Arrhythmia(:,2:11);     %Design matrix n x (p-1) without
                            %vector 1 (intercept)
Xdes =[ones(size(Y)) X];   %with the intercept: n x p
n = length(Y); %number of subjects
alpha = 0.05;  %alpha for CIs

[b, dev, stats]=glmfit(X,Y, 'binomial','link','logit')

lin = Xdes * b  %linear predictor, n x 1 vector
```

Figure 15.2 shows observed arrhythmia responses (0 or 1) with their logistic fit.

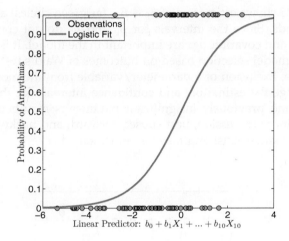

Fig. 15.2 Arrhythmia responses 0 or 1 with their logistic fit. The abscise axis is the linear predictor `lin`.

With the linear predictor, fitted probabilities for $\{Y_i = 1\}$ are given as $\hat{p}_i$. The estimators of the βs with their standard deviations and p-values for the Wald test are given next. The intercept is significantly nonzero (0.0158), and the variable x1 (*age*) is strongly significant (0.0005). This agrees with the inference based on confidence intervals; only the intervals for β_0 and β_1 do not contain 0, or, equivalently, the intervals for the odds ratio, $\exp\{\beta_0\}$ and $\exp\{\beta_1\}$, do not contain 1.

```
phat = exp(lin)./(1 + exp(lin));
V = diag( phat .* (1 - phat) );
sqrtV = diag( sqrt(phat .* (1 - phat) ))
sb = sqrt(   diag( inv( Xdes' * V * Xdes ) ) )
```

```
% inv( Xdes' * V * Xdes ) is stats.covb
% Wald tests for parameters beta
z = b./sb   %tests for beta_i = 0, i=0,...,p-1
pvals = 2 * normcdf(-abs(z))    %p-values
%[0.0158;  0.0005;  0.3007;  0.2803;  0.1347;  0.8061
% 0.4217;  0.3810;  0.6762;  0.0842;  0.5942]
%(1-alpha)*100% CI for betas
 CIs = [b - norminv(1 - alpha/2) * sb , b + norminv(1-alpha/2) * sb]
 %(1-alpha)*100% CIs for odds ratios
 CIsOR = exp([b-norminv(1-alpha/2)*sb , b+norminv(1-alpha/2)*sb])
%     0.0000    0.1281
%     1.0697    1.2711
%     0.9781    1.0744
%
%     ...
%     0.8628   10.3273
%     0.4004    4.9453
```

Figure 15.3 shows estimators of $\beta_0 - \beta_{10}$ (as green circles) and 95% confidence bounds. Since the intervals for β_0 and β_1 do not contain 0, both the intercept and covariate *age* are important in the model. It is tempting to do variable/model selection based on outcomes of Wald's test – but this is not advisable. Exclusion of a parameter/variable from the model will necessarily change the estimators and confidence intervals for the remaining parameters and previously insignificant parameters may become significant. As in linear regression, best subset, forward, and backward variable selection procedures exist and may be implemented.

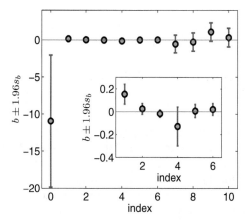

Fig. 15.3 Estimators of $\beta_0 - \beta_{10}$ are shown as *green circles*, and 95% confidence intervals are given. For comparison, the intervals for $\beta_1 - \beta_6$ are shown separately on a different scale.

Next, we find the log-likelihoods for the model and null model. The model deviance is 78.2515, while the difference of deviances between the models is 26.1949. This would be a basis for a likelihood ratio test if the response were grouped. Since in a Bernoulli setup the distributions of deviance and G are not χ^2, the testing needs to be done by one of the response-grouping methods, such as the Hosmer–Lemeshaw method.

```
%Log-likelihood
loglik = sum( Y  .* lin - log( 1 + exp(lin) ))   %-39.1258
%fitting null model.
[b0, dev0, stats0] = glmfit(zeros(size(Y)),Y,'binomial','link','logit')
 %b0=-0.6381,  dev0=104.4464, stats=... (structure)
loglik0 = sum( Y  .* b0(1) - log(1 + exp(b0(1))) )  %-52.2232
%
G = -2 * (loglik0 -  loglik)  % 26.1949
dev0 - dev %26.1949, the same as G, difference of deviances
%
%model deviance
devi = -2 * sum( Y  .* log( phat + eps) + (1-Y ).*log(1 - phat + eps) )
                 %78.2515,   directly
dev              %78.2515,   glmfit output
-2 * loglik    %78.2515,   as a link between loglik and deviance
```

Several measures correspond to R^2 in the linear regression context: Mac-Fadden's pseudo-R^2, Cox–Snell R^2, Nagelkerke R^2, and Effron's R^2. All measures fall between 0.25 and 0.4.

```
%McFadden Pseudo R^2, equivalent expressions
mcfadden  = -2*(loglik0-loglik)/(-2*loglik0) %0.2508
1 - loglik/loglik0 %0.2508
%
coxsnell =1-(exp(loglik0)/exp(loglik))^(2/n) %0.2763
%
nagelkerke=(1-(exp(loglik0)/exp(loglik))^(2/n))/...
    (1-exp(loglik0)^(2/n))      %0.3813
%
effron=1-sum((Y-phat).^2)/sum((Y-sum(Y)/n).^2) %0.2811
```

Next we find several types of residuals: ordinary, Pearson, deviance, and Anscombe.

```
ro = Y  - phat;  %Ordinary residuals

%Deviance Residuals
rdev = sign(Y  - phat) .* sqrt(-2 * Y .* log(phat+eps) - ...
                    2*(1 - Y) .* log(1 - phat+eps));
%Anscombe Residuals
ransc = (betainc(Y,2/3,2/3) - betainc(phat,2/3,2/3) ) .* ...
                   ( phat   .* (1-phat) + eps).^(1/6);
%
% Model deviance is recovered as
%the sum of squared dev. residuals
sum(rdev.^2)  %78.2515
```

Figure 15.4 shows four kinds of residuals (ordinary, Pearson, deviance, and Anscombe), plotted against $\hat{p}$.

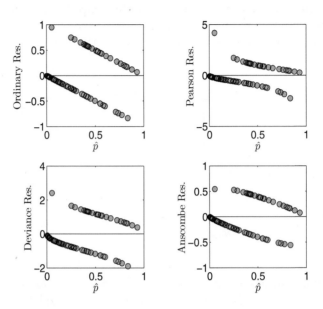

Fig. 15.4 Ordinary, Pearson, deviance, and Anscombe residuals plotted against $\hat{p}$.

If the model is adequate, the smoothed residuals should result in a function close to 0. Figure 15.5 shows Pearson's residuals smoothed by a loess smoothing method (◀ loess.m).

Influential and outlying observations can be detected with a plot of absolute values of residuals against half-normal quantiles. Figure 15.6 was produced by the script below and shows a half-normal plot. The upper and lower bounds (in red) show an empirical 95% confidence interval and were obtained by simulation. The sample of size 19 was obtained from Bernoulli $\mathcal{B}er(\hat{p})$, where $\hat{p}$ is the model fit, and then the minimum, mean, and maximum of the absolute residuals of the simulated values were plotted.

```
k = 1:n;
q = norminv((k + n - 1/8)./(2 * n + 1/2));
plot( q, sort(abs(rdev)),  'k-','LineWidth',1.5);

% Simulated Envelope
rand('state',1)
env =[];
for i = 1:19
    surrogate = binornd(1, phat);
    rdevsu = sign(surrogate - phat).*sqrt(- 2*surrogate .* ...
    log(phat+eps)- 2*(1 - surrogate) .* log(1 - phat+eps) );
    env = [env   sort(abs(rdevsu))];
```

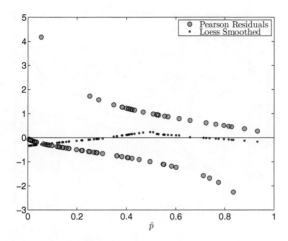

Fig. 15.5 Pearson's residuals (*green circles*) smoothed. The *red circles* show the result of smoothing.

```
end
envel=[min(env'); mean(env' ); max(env' )]';
hold on
plot( q , envel(:,1), 'r-');
plot( q , envel(:,2), 'g-');
plot( q , envel(:,3), 'r-');
xlabel('Half-Normal Quantiles','Interpreter','LaTeX')
ylabel('Abs. Dev. Residuals','Interpreter','LaTeX')
h=legend('Abs. Residuals','Simul. $95%$ CI','Simul. Mean',2)
set(h,'Interpreter','LaTeX')
axis tight
```

To predict the mean response for a new observation, we selected a "new person" with specific covariate values. For this person the estimator for $\mathbb{P}(Y = 1)$ is 0.3179, and 0 for a single future response. A single future response is in fact a classification problem: individuals with a specific set of covariates are classified as either 0 or 1.

```
%Probability of  Y=1 for a new observation
Xh =[1  72  81  130  15  78  43  1   0   0   1]' ;
% responses for a new person
pXh = exp(Xh' *  b)/(1 + exp(Xh' * b) ) %0.3179
%(1-alpha) * 100% CI
ppXh = Xh' * b  %-0.7633
s2pXp = Xh' * inv( Xdes' * V * Xdes ) * Xh  %0.5115
spXh = sqrt(s2pXp)    %0.7152
% confidence interval on the linear part
li = [ppXh-norminv(1-alpha/2)*spXh  ...
  ppXh+norminv(1-alpha/2)*spXh]  %-2.1651    0.6385
% transformation to the CI for the mean response
exp(li)./(1 + exp(li)) %0.1029    0.6544
```

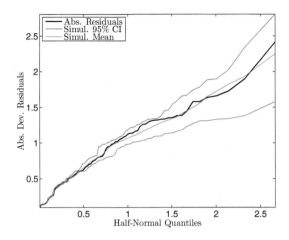

Fig. 15.6 Half-normal plot for deviance residuals.

```
%Predicting single future observation
cutoff = sum(Y)/n %0.3457
%
pXh > cutoff   %Ynew = 0
```

Next, we provide a Bayesian solution to the Arrhythmia logistic model (Arrhythmia.odc) and compare classical and Bayesian model parameters.

```
model{
  eps <- 0.00001
    for(i in 1:N){
    Y[i] ~ dbern(p[i])
    logit(p[i]) <- beta[1] + beta[2] * X1[i]+
             beta[3] * X2[i] + beta[4] * X3[i]+ beta[5] * X4[i] +
             beta[6] * X5[i] + beta[7] * X6[i]  + beta[8] * X7[i] +
             beta[9] * X8[i] + beta[10] * X9[i] + beta[11] * X10[i]
devres[i] <- 2*Y[i]* log(Y[i]/p[i] +eps) +
                 2*(1 - Y[i])*log((1-Y[i])/(1-p[i])+eps)
  }
  for(j in 1:11){
beta[j] ~ dnorm(0, 0.0001)
        }
dev  <-  sum(devres[])
}

DATA + INITS (see Arrhythmia.odc)
```

The classical and Bayesian model parameters are shown in the table:

	$\hat{\beta}_0$	$\hat{\beta}_1$	$\hat{\beta}_2$	$\hat{\beta}_3$	$\hat{\beta}_4$	$\hat{\beta}_5$	$\hat{\beta}_6$	$\hat{\beta}_7$	$\hat{\beta}_8$	$\hat{\beta}_9$	$\hat{\beta}_{10}$	Dev.
Classical	−10.95	0.1536	0.0248	−0.0168	−0.1295	0.0071	0.0207	−0.5377	−0.2638	1.0936	0.3416	78.25
Bayes	−13.15	0.1863	0.0335	−0.0236	−0.1541	0.0081	0.0025	−0.6419	−0.3157	1.313	0.4027	89.89

15.2.3 *Probit and Complementary Log-Log Links*

We have seen that for logistic regression,

$$\hat{p}_i = F(\ell_i) = \frac{\exp\{\ell_i\}}{1 + \exp\{\ell_i\}},$$

where $\ell_i = b_0 + b_1 x_{i1} + \cdots + b_{p-1} x_{i,p-1}$ is the linear part of the model.
    A *probit* regression uses a normal distribution instead,

$$\hat{p}_i = \Phi(\ell_i),$$

while for the complementary log-log, the extreme value (Gumbel type I for the minimum) distribution

$$F(x) = 1 - \exp\{-\exp\{x\}\}$$

is used.
    The complementary log-log link interprets the regression coefficients in terms of the hazard ratio rather than the log odds ratio. It is defined as

$$\text{clog-log} = \log(-\log(1 - p)).$$

The clog-log regression is typically used when the outcome $\{y = 1\}$ is rare. See also Remark on page 775 on a link between clog-log and Poisson regressions. Unlike the logit and probit links, the clog-log link is asymmetric, that is, clog-log$(p) \neq -$clog-log$(1 - p)$.
    Probit models are popular in a bioassay context. A disadvantage of probit models is that the link Φ^{-1} does not have an explicit expression, although approximations and numerical algorithms for its calculation are readily available.
    Once the linear part ℓ_i in a probit or clog-log model is fitted, the probabilities are estimated as

$$\hat{p}_i = \Phi(\ell_i) \quad \text{or} \quad \hat{p}_i = 1 - \exp(-\exp(\ell_i)),$$

respectively. In MATLAB, the probit and complementary log-log links are optional arguments, `'link','probit'` or `'link','cloglog'`.

Example 15.3. **Bliss Data.** In his 1935 paper, Bliss provides a table showing a number of flour beetles killed after 5 hours of exposure to gaseous carbon disulfide at various concentrations. This data set has since been used extensively by statisticians to illustrate and compare models for binary and binomial data.

Table 15.4 Bliss beetle data.

Dose $(\log_{10} CS_2\ mgl^{-1})$	Number of Beetles	Number Killed
1.6907	59	6
1.7242	60	13
1.7552	62	18
1.7842	56	28
1.8113	63	52
1.8369	59	53
1.8610	62	61
1.8839	60	60

The following Bayesian model is applied on the Bliss data, and a probit fit is provided (bliss.odc).

```
model{
    for( i in 1 : N ) {
  y[i] ~ dbin(p[i],n[i])
        probit(p[i]) <- alpha.star + beta * (x[i] - mean(x[]))
   yhat[i] <- n[i] * p[i]
     }
    alpha <- alpha.star - beta * mean(x[])
    beta ~ dnorm(0.0,0.001)
    alpha.star ~ dnorm(0.0,0.001)
}

DATA
list( x = c(1.6907, 1.7242, 1.7552, 1.7842,
               1.8113, 1.8369, 1.8610, 1.8839),
  n = c(59, 60, 62, 56, 63, 59, 62, 60),
  y = c(6, 13, 18, 28, 52, 53, 61, 60), N = 8)
INITS
list(alpha.star=0, beta=0)
```

	mean	sd	MC error	val2.5pc	median	val97.5pc	start	sample
alpha	−35.03	2.652	0.01837	−40.35	−35.01	−29.98	1001	100000
alpha.star	0.4461	0.07724	5.435E-4	0.2938	0.4461	0.5973	1001	100000
beta	19.78	1.491	0.0104	16.94	19.77	22.78	1001	100000
yhat[1]	3.445	1.018	0.006083	1.757	3.336	5.725	1001	100000
yhat[2]	10.76	1.69	0.009674	7.643	10.7	14.26	1001	100000
yhat[3]	23.48	1.896	0.01095	19.77	23.47	27.2	1001	100000
yhat[4]	33.81	1.597	0.01072	30.62	33.83	36.85	1001	100000
yhat[5]	49.59	1.623	0.01208	46.28	49.63	52.64	1001	100000
yhat[6]	53.26	1.158	0.008777	50.8	53.33	55.33	1001	100000
yhat[7]	59.59	0.7477	0.00561	57.91	59.68	60.82	1001	100000
yhat[8]	59.17	0.3694	0.002721	58.28	59.23	59.71	1001	100000

If instead of `probit`, the clog-log was used, as `cloglog(p[i]) <- alpha.star + beta * (x[i] - mean(x[]))`, then the coefficients are

	mean	sd	MC error	val2.5pc	median	val97.5pc	start	sample
alpha	-39.73	3.216	0.02195	-46.24	-39.66	-33.61	1001	100000
beta	22.13	1.786	0.01214	18.73	22.09	25.74	1001	100000

For comparisons we take a look at the classical solution (◢ `beetleBliss2.m`). Figure 15.7 shows three binary regressions (logit, probit and clog-log) fitting the Bliss data.

```
disp('Logistic Regression 2: Bliss Beetle Data')
lw = 2.5;
set(0, 'DefaultAxesFontSize', 16);
fs = 15;
msize = 10;
beetle=[...
1.6907  6 59; 1.7242 13 60; 1.7552 18 62; 1.7842 28 56;...
1.8113 52 63; 1.8369 53 59; 1.8610 61 62; 1.8839 60 60];
%%%%%%%%%%%%%%%%%%
xi = beetle(:,1);  yi=beetle(:,2); ni=beetle(:,3);
figure(1)
[b, dev, stats] = glmfit(xi,[yi ni],'binomial','link','logit');
[b1, dev1, stats1] = glmfit(xi,[yi ni],'binomial','link','probit');
[b2, dev2, stats2] = glmfit(xi,[yi ni],'binomial','link','comploglog');

xs = 1.5:0.01:2.0;
ys  = glmval(b, xs, 'logit');
y1s = glmval(b1, xs, 'probit');
y2s = glmval(b2, xs, 'comploglog');
% Plot
plot(xs,ys,'r-','LineWidth',lw)
hold on
plot(xs,y1s,'k--','LineWidth',lw)
plot(xs,y2s,'-.','LineWidth',lw)
plot(xi, yi./ni, 'o','MarkerSize',msize,...
      'MarkerEdgeColor','k','MarkerFaceColor','g')
axis([1.5 2.0 0 1])
grid on
xlabel('Log concentration')
ylabel('Proportion killed')
legend('Obs. proportions','Logit','Probit','Clog-log',2)
```

The table below compares the coefficients of the linear part of the three models. Note that classical and Bayesian results are close because the priors in the Bayesian model are noninformative.

	Classical			Bayes		
Link	Logit	Probit	Clog-log	Logit	Probit	Clog-log
Intercept	-60.72	-34.94	-39.57	-60.78	-35.03	-39.73
Slope	34.27	19.73	22.04	34.31	19.78	22.13

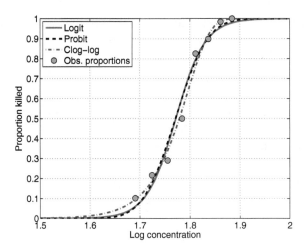

Fig. 15.7 Bliss data (*green* dots). Regression fit with logit link *red*, probit link *black*, and clog-log link *blue*.

15.3 Poisson Regression

Poisson regression models the counts $y = \{0,1,2,3,\dots\}$ of rare events in a large number of trials. Typical examples are unusual adverse events, accidents, incidence of a rare disease, device failures during a particular time interval, etc. Recall that Poisson random variable $Y \sim \mathcal{P}oi(\lambda)$ has the probability mass function

$$f(y) = \mathbb{P}(Y = y) = \frac{\lambda^y}{y!} \exp\{-\lambda\}, \quad y = 0,1,2,3,\dots,$$

with both mean and variance equal to the rate parameter $\lambda > 0$.

Suppose that n counts of y_i, $i = 1,\dots,n$ are observed and that each count corresponds to a particular value of a covariate x_i, $i = 1,\dots,n$. A typical Poisson regression can be formulated as follows:

$$y_i \sim \mathcal{P}oi(\lambda_i), \tag{15.4}$$
$$\log(\lambda_i) = \beta_0 + \beta_1 x_i, \quad i = 1,\dots,n,$$

although other relations between λ_i and the linear part $\beta_0 + \beta_1 x_i$ are possible as long as λ_i remains positive. More generally, λ_i can be linked to a linear expression containing $p - 1$ covariates and p parameters as

$$\log(\lambda_i) = \beta_0 + \beta_1 x_{i1} + \beta_2 x_{i2} + \dots + \beta_{p-1} x_{i,p-1}, i = 1,\dots,n.$$

In terms of model (15.4) the Poisson rate λ_i is the expectation, and its logarithm can be expressed as $\log \mathbb{E}(y_i|X = x_i) = \beta_0 + \beta_1 x_i$. When the covariate x_i gets a unit increment, $x_i + 1$, then

$$\log \mathbb{E}(y_i|X = x_i + 1) = \beta_0 + \beta_1 x_i + \beta_1 = \log \mathbb{E}(y_i|X = x_i) + \beta_1.$$

Thus, parameter β_1 can be interpreted as the increment to log rate when the covariate gets an increment of 1. Equivalently, $\exp\{\beta_1\}$ is the ratio of rates,

$$\exp\{\beta_1\} = \frac{\mathbb{E}(y_i|x_i + 1)}{\mathbb{E}(y_i|x_i)}.$$

The model-assessed mean response is $\hat{y}_i = \exp\{b_0 + b_1 x_i\}$, where b_0 and b_1 are the estimators of β_0 and β_1. Strictly speaking, the model predicts the rate $\hat{\lambda}_i$, but the rate is interpreted as the expected response.

Remark. If a Poisson regression model

$$y_i \sim \mathcal{P}oi(\lambda_i), \; \lambda_i = \exp\{\beta_0 + \beta_1 x_{i1} + \cdots + \beta_k x_{ik}\} = \exp\{\ell_i\}, \; i = 1, \ldots, n,$$

is dichotomized as

$$y_i^* = \begin{cases} 1, & y_i > 0 \\ 0, & y_i = 0 \end{cases},$$

then an adequate model for y_i^* is a binary regression with the clog-log link. Indeed,

$$\mathbb{P}(y_i^* = 1) = 1 - \mathbb{P}(y_i^* = 0) = 1 - \exp\{-\lambda_i\} = 1 - \exp\{-\exp(\ell_i)\}.$$

The deviance of the model, D, is defined as

$$D = 2 \sum_{i=1}^{n} \left(y_i \log \frac{y_i}{\hat{y}_i} - (y_i - \hat{y}_i) \right),$$

where $y_i \log y_i = 0$ if $y_i = 0$. As in logistic regression, the deviance is a measure of goodness of fit of a model and for a Poisson model has a χ^2-distribution with $n - p$ degrees of freedom.

Deviance residuals, defined as

$$r_i^{dev} = \text{sign}(y_i - \hat{y}_i) \times \sqrt{2 y_i \log \frac{y_i}{\hat{y}_i} - 2(y_i - \hat{y}_i)},$$

satisfy $D = \sum_{i=1}^{n} \left(r_i^{dev} \right)^2$. Note that the sum of squares of the deviance residuals simplifies to $D = 2\sum_{i=1}^{n} y_i \log(y_i/\hat{y}_i)$, since in the Poisson regression $\sum_{i=1}^{n}(y_i - \hat{y}_i) = 0$.

Pearson's residuals are defined as

$$r_i^{pea} = \frac{y_i - \hat{y}_i}{\sqrt{\hat{y}_i}}.$$

Then the Pearson goodness-of-model-fit statistic $\chi^2 = \sum_{i=1}^{n}(r_i^{pea})^2$ also has a χ^2-distribution with $n - p$ degrees of freedom. Although homoscedastic, those residuals are asymmetric.

Freedman–Tukey residuals are defined as

$$r_i^{ft} = \sqrt{y_i} + \sqrt{y_i + 1} - \sqrt{4\hat{y}_i + 1}$$

and Anscombe residuals (closest to normality) as

$$r_i^{a} = \frac{3}{2} \times \frac{y_i^{2/3} - \hat{y}_i^{2/3}}{\hat{y}_i^{1/6}}.$$

The pseudo-R^2 for the Poisson regression is defined as $1 - D/D_0$. Here D is deviance of the model in question, and D_0 is deviance of the intercept-only model,

$$D_0 = 2\sum_{i=1}^{n} y_i \log(y_i/\bar{y}),$$

where $\bar{y}$ is the sample mean of the observations.

Some additional diagnostic tools are exemplified in the following case study (◢ ihga.m):

Example 15.4. **Case Study: Danish IHGA Data.** In an experiment conducted in the 1980s (Hendriksen et al., 1984), 572 elderly people living in a number of villages in Denmark were randomized, 287 to a control (C) group who received standard care, and 285 to an experimental group who received standard care plus IHGA: a kind of preventive assessment in which each person's medical and social needs were assessed and acted upon individually. The important outcome was the number of hospitalizations during the 3-year life of the study.

```
% IHGA
% data
x0 = 0 * ones(287,1);   x1 = 1 * ones(285,1);
    %covariate 0-no intervention, 1- intervention
y0 = [0*ones(138,1); 1*ones(77,1); 2*ones(46,1);...
```

Table 15.5 Distribution of number of hospitalizations in IHGA study.

Group	# of hospitalizations 0 1 2 3 4 5 6 7	n	Mean	Variance
Control	138 77 46 12 8 4 0 2	287	0.944	1.54
Treatment	147 83 37 13 3 1 1 0	285	0.768	1.02

```
     3*ones(12,1); 4 * ones(8,1);  5*ones(4,1); 7*ones(2,1)];
y1 = [0*ones(147,1); 1*ones(83,1); 2*ones(37,1);...
     3*ones(13,1); 4 * ones(3,1);  5*ones(1,1); 6*ones(1,1)];
     %response # of hospitalizations
x =[x0; x1];   y=[y0; y1];
xdes = [ones(size(y)) x];
[n p] = size(xdes)

[b dev stats] = glmfit(x,y,'poisson','link','log')
yhat = glmval(b, x,'log') %model predicted responses

% Pearson residuals
rpea  = (y - yhat)./sqrt(yhat);
% deviance residuals
rdev = sign(y - yhat) .* sqrt(-2*y.*log(yhat./(y + eps))-2*(y - yhat));
% Friedman-Tukey residuals
rft = sqrt(y) + sqrt(y + 1) - sqrt(4 * yhat + 1)
% Anscombe residuals
ransc = 3/2 * (y.^(2/3) - yhat.^(2/3) )./(yhat.^(1/6))
```

Figure 15.8 shows four our types of residuals in Poisson regression fit of IHGA data: Pearson, deviance, Friedman–Tukey, and Anscombe. The residuals are plotted against responses y.

```
loglik = sum(y  .* log(yhat+eps) - yhat  - log(factorial(y)));
%
[b0, dev0, stats0] = glmfit(zeros(size(y)),y,'poisson','link','log')
yhat0 = glmval(b0, zeros(size(y)),'log');
loglik0  = sum( y .* log(yhat0 + eps) - yhat0 - log(factorial(y)))

G  = -2 * (loglik0 -  loglik) %LR test, nested model chi2  5.1711
dev0 - dev % the same as G, difference of deviances        5.1711
pval = 1-chi2cdf(G,1)     %0.0230
```

Under H_0 stating that the model is null (model with an intercept and no covariates), the statistic G will have $df = p - 1$ degrees of freedom, in our case $df = 1$. Since this test is significant ($p = 0.0230$), the covariate contributes significantly to the model.

Below are several ways to express the deviance of the model.

```
%log-likelihood for saturated model
loglliksat = sum(y.*log(y+eps)-y-log(factorial(y))) %-338.1663
m2LL = -2 * sum( y  .* log(yhat./(y + eps)) ) %819.8369
deviance = sum(rdev.^2) %819.8369
dev %819.8369    from glmfit
-2*(loglik - loglliksat) % 819.8369
```

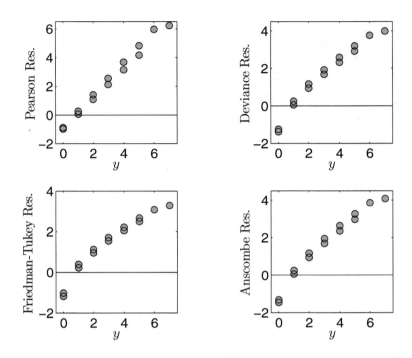

Fig. 15.8 (a) Four types of residuals in a Poisson regression fit of IHGA data: Pearson, deviance, Friedman–Tukey, and Anscombe. The residuals are plotted against responses y.

The following is a Bayesian model fit in WinBUGS (geriatric.odc).

```
model
{
for (i in 1:n)
    {
       y[i] ~ dpois(lambda[i] )
       log( lambda[i]) <- beta.0 + beta.1 * x[i]
    }
 beta.0 ~ dnorm(0, 0.0001)
 beta.1 ~ dnorm(0, 0.0001)
 lambda.C <- exp(beta.0)
 lambda.E <- exp(beta.0 + beta.1 )
 diff <- lambda.E - lambda.C
 meffect  <- exp( beta.1 )
 }


DATA
list( y = c(0, 0, 0, 0, 0, 0, 0, 0, 0, 0, 0, 0, 0, 0, 0, 0, 0,
0, 0, 0, 0, 0, 0, 0, 0, 0, 0, 0, 0, 0, 0, 0, 0, 0, 0, 0, 0,
                     ...
4, 4, 4, 4, 4, 4, 4, 4, 5, 5, 5, 5, 7, 7, 0, 0, 0, 0, 0, 0, 0,
0, 0, 0, 0, 0, 0, 0, 0, 0, 0, 0, 0, 0, 0, 0, 0, 0, 0, 0, 0, 0,
                     ...
2, 2, 2, 2, 2, 2, 2, 2, 3, 3, 3, 3, 3, 3, 3, 3, 3, 3, 3, 3, 3,
4, 4, 4, 5, 6 ),
x = c(0, 0, 0, 0, 0, 0, 0, 0, 0, 0, 0, 0, 0, 0, 0, 0, 0, 0, 0,
0, 0, 0, 0, 0, 0, 0, 0, 0, 0, 0, 0, 0, 0, 0, 0, 0, 0, 0, 0, 0,
                     ...
0, 0, 0, 0, 0, 0, 0, 0, 0, 0, 0, 0, 0, 0, 0, 0, 0, 0, 0, 0, 0,
0, 0, 0, 0, 0, 0, 0, 0, 0, 0, 0, 0, 0, 1, 1, 1, 1, 1, 1, 1, 1,
                     ...
1, 1, 1, 1, 1, 1, 1, 1, 1, 1, 1, 1, 1, 1, 1, 1, 1, 1, 1, 1, 1,
1, 1, 1, 1, 1 ), n = 572 )

INITS
list( beta.0 = 0.0, beta.1 = 0.0 )
```

	mean	sd	MC error	val2.5pc	median	val97.5pc	start	sample
beta.0	-0.05915	0.06093	2.279E-4	-0.1801	-0.05861	0.05881	1001	100000
beta.1	-0.2072	0.09096	3.374E-4	-0.3861	-0.2071	-0.02923	1001	100000
deviance	1498.0	2.012	0.006664	1496.0	1498.0	1504.0	1001	100000
diff	-0.1764	0.07737	2.872E-4	-0.3285	-0.1763	-0.02493	1001	100000
lambda.C	0.9443	0.05749	2.137E-4	0.8352	0.9431	1.061	1001	100000
lambda.T	0.7679	0.05188	1.784E-4	0.6693	0.7668	0.8732	1001	100000
meffect	0.8162	0.07437	2.76E-4	0.6797	0.8129	0.9712	1001	100000

Example 15.5. **Cellular Differentiation Data.** In a biomedical study of the immunoactivating ability of the agents TNF (tumor necrosis factor) and IFN

(interferon) to induce cell differentiation, the number of cells that exhibited markers of differentiation after exposure to TNF or IFN was recorded (Piergorsch et al., 1988; Fahrmeir and Tutz, 1994). At each of the 16 dose combinations of TNF/IFN, 200 cells were examined. The number y of differentiating cells corresponding to a TNF/IFN combination are given in Table 15.6.

Table 15.6 Cellular differentiation data.

Number cells diff	Dose of TNF (U/ml)	Dose of IFN (U/ml)	Number cells diff	Dose of TNF (U/ml)	Dose of IFN (U/ml)
11	0	0	31	10	0
18	0	4	68	10	4
20	0	20	69	10	20
39	0	100	128	10	100
22	1	0	102	100	0
38	1	4	171	100	4
52	1	20	180	100	20
69	1	100	193	100	100

The suggested model is Poisson with the form

$$\lambda = \mathbb{E}(y|\text{TNF}, \text{IFN}) = \exp\{\beta_0 + \beta_1\, \text{TNF} + \beta_2\, \text{IFN} + \beta_3\, \text{TNF}\times\text{IFN}\}.$$

From

```
load 'celular.dat'
number = celular(:,1);
TNF =    celular(:,2);
IFN =    celular(:,3);
[b, dev, stats] = glmfit([TNF IFN  TNF.*IFN],number,...
     'poisson','link','log')
```

the estimators for β_0–β_3 are

b_0	b_1	b_2	b_3
3.43463	0.01553	0.00895	−0.0000567

Since $b_3 < 0.0001$, it is tempting to drop the interaction term. However, since the standard error of b_3 is $s.e.(b_3) = 0.000013484$, Wald's $Z = b_3/s.e.(b_3)$ statistic is -4.2050, suggesting that the term TNF×IFN might be significant. However, the overdispersion parameter, which theoretically should be stats.s=1, is estimated as stats.s=3.42566, and the Wald statistic should be adjusted to $Z' = -0.0000567/(3.42566 \times 0.000013484) = 1.2275$. Since the p-value is 2 * normcdf(-1.2275) = 0.2196, after all, the interaction term turns out not to be significant and the additive model could be fit:

```
[b, dev, stats] = glmfit([TNF IFN],number,...
     'poisson','link','log')
```

which gives estimates $b_0 = 3.57311$, $b_1 = 0.01314$, and $b_3 = 0.00585$. Details can be found in ◀ celular.m.

The additive model was also fit in a Bayesian manner.

```
model
{
for (i in 1:n)
    {
      numbercells[i] ~ dpois(lambda[i])
      lambda[i] <- exp(beta0 + beta1 * tnf[i]  + beta2 * ifn[i])
    }
beta0 ~ dnorm(0, 0.00001)
beta1 ~ dnorm(0, 0.00001)
beta2 ~ dnorm(0, 0.00001)
}


DATA
list(n=16,
numbercells = c(11,18,20,39,22,38,52,69,31,68,69,128,102,171,180,193),
tnf = c(0,0,0,0,    1,1,1,1,    10,10,10,10,    100,100,100,100),
ifn = c(0,4,20,100,    0,2,20,100,    0,4,20,100,    0,4,20,100 ) )


INITS
list(b0=0, b1=0, b2=0)
```

	mean	sd	MC error	val2.5pc	median	val97.5pc	start	sample
b0	3.573	0.05139	0.001267	3.473	3.574	3.672	1001	100000
b1	0.01313	5.921E-4	1.18E-5	0.01197	0.01314	0.01431	1001	100000
b2	0.00585	6.399E-4	1.142E-5	0.004585	0.005855	0.007086	1001	100000

Note that, because of the noninformative priors on β_0, β_1, and β_3, the Bayesian estimators almost coincide with the MLEs from glmfit.

✐

15.4 Log-linear Models

Poisson regression can be employed in the context of contingency tables (Chapter 12). The logarithm of the table cell count is modeled in a linear fashion.

In an $r \times c$ contingency table, let the probability of a cell (i, j) be p_{ij}. If n subjects are cross-tabulated, let n_{ij} be the number of subjects classified in the cell (i, j). The count n_{ij} is realization of random variable N_{ij}. We assume that the sample size n is random, since in that case the cell frequency N_{ij} is a Poisson random variable with the intensity μ_{ij}. If the sample size n is fixed, then the N_{ij}s are realizations of a multinomial random variable. In Fisher's exact test context (page 602), we saw that if in addition the marginal counts are fixed, the N_{ij}s are hypergeometric random variables.

Then the expected table is

	1	2	$\cdots$	c	Total
1	np_{11}	np_{12}		np_{1c}	$np_{1.}$
2	np_{21}	np_{22}		np_{2c}	$np_{2.}$
r	np_{r1}	np_{r2}		np_{rc}	$np_{r.}$
Total	$np_{.1}$	$np_{.2}$		$np_{.c}$	n

Note that both n_{ij} and $e_{ij} = n_i. \times n_{.j}/n$ involve observations and empirical marginal probabilities defined by the observed marginal frequencies. If we denote $\mu_{ij} = \mathbb{E} N_{ij} = np_{ij}$, then both n_{ij} and e_{ij} are estimators of μ_{ij}; the first is unconstrained and the second is constrained by the assumption of independence of factors, $p_{ij} = p_i.p_{.j}$. This point is important since e_{ij} are "expected" under independence once the table is observed and fixed, while μ_{ij} are expectations of the random variables N_{ij}.

The log-linear model for the expected frequency μ_{ij} is given as

$$\log \mu_{ij} = \lambda_0 + \lambda_i^R + \lambda_j^C + (\lambda^{RC})_{ij}, \quad i = 1,\ldots,r; j = 1,\ldots,c,$$

where λ_i^R and λ^C are contributions by row and column, respectively, and λ_{ij}^{RC} is a row–column interaction term.

This model is called a *saturated log-linear model* and is similar to the two-factor ANOVA model. Note that an unconstrained model has $(1 + r + c + r \times c)$ free parameters but only $r \times c$ observations, n_{ij}. To make this over-parameterized model identifiable, constraints on parameters are imposed. A standard choice is STZ: $\sum \lambda_i^R = 0$, $\sum_j \lambda_j^C = 0$, $\sum_i(\lambda^{RC})_{ij} = \sum_j(\lambda^{RC})_{ij} = 0$. Thus, with the constraints we have $1 - (r - 1) + (c - 1) + (r - 1)(c - 1) = r \times c$ free parameters, which is equal to the number of observations, and the model gives a perfect fit for the observed frequencies. This is the reason why this model is called *saturated*.

The hypothesis of independence in a contingency table that was discussed Chapter 12 has simple form:

$$H_0 : \lambda_{ij}^{RC} = 0, \quad i = 2,\ldots,r; \; j = 2,\ldots,c.$$

Under H_0 the log-linear model becomes additive:

$$\log \mu_{ij} = \lambda_0 + \lambda_i^R + \lambda_j^C, \; i = 1,\ldots,r; \; j = 1,\ldots,c.$$

The MLEs of components in the log-linear model (not derived here, but see Agresti, 2002) are

$$\hat{\lambda}_0 = \frac{\sum_{i,j} \log n_{ij}}{rc},$$

$$\hat{\lambda}_i^R = \frac{\sum_j \log n_{ij}}{c} - \hat{\lambda}_0,$$

$$\hat{\lambda}_j^C = \frac{\sum_i \log n_{ij}}{r} - \hat{\lambda}_0,$$

$$\hat{\lambda}_{ij}^{RC} = \log n_{ij} - (\hat{\lambda}_0 + \hat{\lambda}_i^R + \hat{\lambda}_j^C).$$

If any n_{ij} is equal to 0, then *all* entries in the table are replaced by $n_{ij} + 0.5$.

Traditional analysis involves testing that particular λ components are equal to 0. One approach, often implemented in statistical software, would be to find the variance of $\hat{\lambda}$, $\mathbb{V}\mathrm{ar}\,(\hat{\lambda})$, using a hypergeometric model, and then use the statistic $(\hat{\lambda})^2 / \mathbb{V}\mathrm{ar}\,(\hat{\lambda})$ that has a χ^2-distribution with one degree of freedom (λ here is any of λ_0, λ_i^R, λ_j^C, or λ_{ij}^{RC}). Large values of this statistic are critical for H_0.

Next, we focus on the Bayesian analysis of a log-linear model.

Example 15.6. **Log-linear Model and Bystanders.** In the psychological experiment of Exercise 12.9, seeking assistance (help) was dependent on a subject's perception of the number of bystanders. The resulting chi-square statistic $\chi^2 = 7.908$ was significant with a *p*-value of about 2%. We revisit this exercise and provide a Bayesian solution using a log-linear approach. The WinBUGS program 🖳bystanders.odc is used to conduct statistical inference.

```
model{
    for (i in 1:r) for (j in 1:c)  n[i,j] ~ dpois(mu[i,j]);
    log(mu[i,j]) <- lambda0+ lambdaR[i]+lambdaC[j]+lambdaRC[i,j]
                    }

    lambda0 ~ dnorm(0,prec)

    lambdaR[1] <- 0
    lambdaC[1] <- 0
    for (i in 2:r) { lambdaR[i] ~ dnorm(0,prec) }
    for (i in 2:c) { lambdaC[i] ~ dnorm(0,prec) }

    for (j in 1 : c) {  lambdaRC[1, j] <- 0 }
    for (i in 2 : r) {  lambdaRC[i, 1] <- 0;
    for (j in 2 : c) {  lambdaRC[i, j] ~ dnorm(0, prec)}
}

DATA
list(r=3,c=2,prec=0.0001,
    n=structure(.Data=c(11,2,16,10,4,9),.Dim=c(3,2)))
```

```
INITS
list(lambda0 = 0,lambdaR = c(NA,0,0), lambdaC=c(NA,0),
lambdaRC = structure(.Data = c(NA, NA, NA,0,NA,0),
                      .Dim = c(3,2)) )
```

	mean	sd	MC error	val2.5pc	median	val97.5pc	start	sample
lambda0	2.351	0.3075	0.003347	1.707	2.366	2.908	2001	100000
lambdaC[2]	−1.922	0.8446	0.01376	−3.796	−1.85	−0.481	2001	100000
lambdaR[2]	0.3895	0.3977	0.004085	−0.372	0.3842	1.186	2001	100000
lambdaR[3]	−1.089	0.6145	0.006468	−2.375	−1.061	0.03655	2001	100000
lambdaRC[2,2]	1.435	0.9372	0.01451	−0.2453	1.38	3.465	2001	100000
lambdaRC[3,2]	2.801	1.051	0.01653	0.9045	2.744	5.041	2001	100000

The hypothesis of independence is assessed by testing that all λ_{RC} are equal to 0. In this case lambdaRC[1,1], lambdaRC[1,2], and lambdaRC[2,1] are set to 0 because of identifiability constraints. The 95% credible set for interaction lambdaRC[2,2] contains 0 – therefore, this interaction is not significant. However, lambdaRC[3,2] is significantly positive since the credible set [0.9045, 5.041] does not contain 0.

Example 15.7. **Upton–Fingleton Square.** An example from Upton and Fingleton (1985) concerns finding directional trends in spatial count data. The simple count data set is given as follows:

0	0	0	1	0
3	3	5	2	0
2	3	6	2	5
1	5	4	6	7
6	2	4	3	4

We are interested in testing if there are any north–south or east–west trends present in the spatial pattern. The idea of Upton and Fingleton was to establish a Poisson regression where the response was the intensity n_{ij} in the location (i,j) and the covariates are x-coordinate, related to the east–west trend, and y-coordinate, related to the north–south trend.

Thus, the observed frequencies n_{ij} will be modeled as

$$\mathbb{E}(n_{ij}) = \exp\{\beta_0 + \beta_1 x_i + \beta_2 y_j\}, \quad i,j = 1,\dots,5,$$

where $x_i = i$ and $y_j = j$. A significant β_1 or β_2 in a well-fitting Poisson regression will indicate the presence of corresponding trends (see UptonFingleton.m for details).

15.5 Exercises

15.1. **Blood Pressure and Heart Disease.** This example is based on data given by Cornfield (1962). A sample of male residents of Framingham, Massachusetts, aged 40–59, were chosen. During a 6-year follow-up period they were classified according to several factors, including blood pressure and whether they had developed coronary heart disease. The results for these two variables are given in the table below. The covariate for blood pressure represents an interval, for example, 122 stands for blood pressure in the interval 117–126.

Blood pressure	No disease	Disease	Total
112 (<117)	156	3	159
122 (117–126)	252	17	269
132 (127–136)	284	12	296
142 (137–146)	271	16	287
152 (147–156)	139	12	151
162 (157–166)	85	8	93
177 (167–186)	99	16	115
197 (> 186)	43	8	51

Using logistic regression, estimate the probability of disease for a person with an average blood pressure equal to 158.

15.2. **Blood Pressure and Heart Disease in WinBUGS.** Use data from the previous exercise. Parameter p represents the probability of developing coronary disease and can be estimated as

$$\hat{p} = \frac{e^{b_0 + b_1 BP}}{1 + e^{b_0 + b_1 BP}},$$

where b_0 and b_1 are Bayes estimators of β_0 and β_1 obtained by WinBUGS. Use noninfromative priors on β_0 and β_1.
(a) What are the values b_0 and b_1? Provide plots of posterior densities for b_0 and b_1.
(b) What are the 95% credible intervals for β_0 and β_1?
(c) Estimate the probability of disease for a person with an average blood pressure of 158.

15.3. **Sex of Diamond-backed Terrapins and Incubation Temperature.** Temperature-dependent sex determination, observed in some reptiles and fish, is a type of environmental sex determination in which the temperatures experienced during embryonic development determine the sex of the offspring. Below are data on the relationship between the ratio of male/female diamond-backed terrapins (*Malaclemys terrapin*) and incubation temperature, as reported by Burke and Calichio (2014):

Temp in °C	Male	Female
24	21	0
26	39	7
28	34	19
30	2	36
32	0	29

Predict the probability of a female terrapin for a temperature of 29°C.

15.4. **Health Promotion.** Students at the University of the Best in England (UBE) investigated the use of a health promotion video in a doctor's surgery. Covariates **Age** and **Amount of Weekly Exercise** for a sample of 30 men were obtained, and each man was asked a series of questions on the video. On the basis of the responses to these questions the psychologist simply recorded whether the promotion video was **Effective** or **Not effective**. The collected data are provided in Table 15.7.

Number	Age	Exercise	Code of response	Response
1	27	3	0	Not effective
2	26	5	0	Not effective
3	28	10	1	Effective
4	40	4	1	Effective
5	19	3	0	Not effective
6	36	14	0	Not effective
7	41	5	1	Effective
8	27	5	0	Not effective
9	33	1	0	Not effective
10	34	2	1	Effective
11	51	8	1	Effective
12	33	21	1	Effective
13	23	12	0	Not effective
14	41	19	1	Effective
15	38	2	0	Not effective
16	25	9	0	Not effective
17	40	6	0	Not effective
18	36	25	1	Effective
19	40	13	0	Not effective
20	39	3	0	Not effective
21	45	10	1	Effective
22	39	5	0	Not effective
23	40	2	1	Effective
24	20	1	0	Not effective
25	47	9	1	Effective
26	31	17	1	Effective
27	37	7	1	Effective
28	30	0	0	Not effective
29	32	13	1	Effective
30	25	10	0	Not effective

Table 15.7 Age, exercise, and effectiveness of the video.

(a) Use the fitted model to estimate the probability that a male of age 40 who exercises 10 hours a week would find the video "effective."
Comment: The linear predictor is

$$-7.498 + 0.17465 \text{ Age} + 0.16324 \text{ Exercise.}$$

The positive coefficient with covariate Age means that older subjects tend to respond "video is effective" with higher probabilities. Similarly, the positive coefficient with predictor Exercise indicates that increasing the values of Exercise also increases the probabilities for the response "video is effective."
(b) Comment on the fit of the model based on deviance and residuals.

15.5. **IOP.** Laser refractive surgery often decreases Intraocular Pressure (IOP) and may lead to hypotony (clinically significant low IOP that may lead to corneal decompensation, accelerated cataract formation, maculopathy, and discomfort).
An investigator wished to determine whether the post-operative IOP in patients after laser refractive surgery was related to the residual thickness of the cornea. In a sample of 140 patients who had undergone laser surgery, post-operative IOP and the thickness of the cornea were measured. The data set is provided in 📖 iop2.dat|mat which consists of two columns, (1) indicator of low IOP (IOP < 10) and (2) central corneal thickness (in micrometers).

(a) Fit the logistic regression with cornea thickness as the predictor of incidence of low IOP.
(b) For a person who had a refractive surgery with residual thickness of cornea of 480 micrometers, what is the risk of a low IOP?
(c) Compare deviances for two links: logit (as in (a)), probit and comploglog. Which link provides the best fit? (*Hint:* Deviances are in the output of glmfit).

15.6. **PONV.** Despite advances over the past decade, including the advent of 5-HT_3 receptor antagonists, combination therapy, and multimodal strategies, postoperative nausea and vomiting (PONV) remains a serious and frequent adverse event associated with surgery and anesthesia. PONV can be very distressing for patients, can lead to medical complications, and impose economic burdens. A meta-analysis of several studies gives rates of 37% for nausea and 20% for vomiting in patients undergoing general anesthesia. However, indiscriminate prophylaxis is not recommended (the "prevent-or-cure" dilemma).
There are considerable variations in the reported incidence of PONV, which can be attributed to a number of factors. Risk factors for PONV can be divided into patient risk factors, procedural risk factors, anesthetic risk factors, and postoperative risk factors. The main and well-

understood risk factors are gender, history of motion sickness/PONV, smoking status, and use of postoperative opioids.

A data set 🖥 PONV.xls or PONV.mat (courtesy of Jelena Velickovic, MD anesthesiologist from Belgrade) contains records of 916 patients consisting of some demographic, anamnetic, clinical, and procedural variables. Several variables are of interest to be modeled and controlled: manifestation of PONV from 0 to 2 hours after surgery, PONV from 2 to 24 hours, and PONV from 0 to 24 hours (PONV0to2, PONV2to24, and PONV0to24). Three score variables (SinclairScore, ApfelScore, and LelaScore) summarize the relevant demographic and clinical information prior to surgery with the goal of predicting PONV.

The starter file ponv.m gives a basic explanation of variables and helps to read all the variables into MATLAB.

Fit a logistic model for predicting PONV0to24 based on a modified Sinclair Score defined as MSS = SinclairScore + 1/20 * LelaScore. What is the probability of PONV0to24 for a person with MSS = 1.3?

15.7. **Mannose-6-phosphate Isomerase.** McDonald (1985) counted allele frequencies at the mannose-6-phosphate isomerase (Mpi) locus in the amphipod crustacean *Megalorchestia californiana*, which lives on sandy beaches of the Pacific coast of North America. There were two common alleles, Mpi90 and Mpi100. The latitude of each collection location, the number of each allele, and the proportion of the Mpi100 allele are shown here:

Location	Latitude	Mpi90	Mpi100	Prop Mpi100
Port Townsend, WA	48.1	47	139	0.748
Neskowin, OR	45.2	177	241	0.577
Siuslaw River, OR	44.0	1087	1183	0.521
Umpqua River, OR	43.7	187	175	0.483
Coos Bay, OR	43.5	397	671	0.628
San Francisco, CA	37.8	40	14	0.259
Carmel, CA	36.6	39	17	0.304
Santa Barbara, CA	34.3	30	0	0.000

Estimate the probability of seeing allele Mpi100, taking into account latitude as a covariate, using logistic regression.

15.8. **Arthritis Treatment Data.** The data were obtained from Koch and Edwards (1988) for a double-blind clinical trial investigating a new treatment for rheumatoid arthritis. In this data set, there were 84 subjects of different ages who received an active or placebo treatment for their arthritis pain, and the subsequent extent of improvement was recorded as marked, some, or none. The dependent variable **improve** was an ordinal categorical observation with three categories (0 for none, 1 for some, and 2 for marked). The three explanatory variables were **treatment** (1 for active or 0 for placebo), **gender** (1 for male, 2 for female), and **age**

(recorded as a continuous variable). The data in 🖳 arthritis2.dat is organized as a matrix, with 84 rows corresponding to subjects and 5 columns
containing ID number, treatment, gender, age, and improvement status:

$$
\begin{array}{ccccc}
57 & 1 & 1 & 27 & 1 \\
9 & 0 & 1 & 37 & 0 \\
46 & 1 & 1 & 29 & 0 \\
\multicolumn{5}{c}{\cdots} \\
15 & 0 & 2 & 66 & 1 \\
10 & 2 & 74 & 2 \\
71 & 0 & 2 & 68 & 1
\end{array}
$$

Dichotomize the variable **improve** as improve01=improve>0; and fit the binary regression with **improve01** as a response and **treatment, gender,**
and **age** as covariates. Use the three links logit, probit, and comploglog
and compare the models by comparing the deviances.

15.9. **Third-degree Burns.** The data for this exercise, discussed in Fan et al.
(1995), refer to $n = 435$ adults who were treated for third-degree burns by
the University of Southern California General Hospital Burn Center. The
patients were grouped according to the area of third-degree burns on the
body. For each midpoint of the groupings "log(area +1)," the number of
patients in the corresponding group who survived and the number who
died from the burns was recorded:

Log(area+1)	Survived	Died
1.35	13	0
1.60	19	0
1.75	67	2
1.85	45	5
1.95	71	8
2.05	50	20
2.15	35	31
2.25	7	49
2.35	1	12

(a) Fit the logistic regression on the probability of death due to third-
degree burns with the covariate log(area+1).
(b) Using your model, estimate the probability of survival for a person
for which log(area + 1) equals 2.

15.10. **Diabetes Data.** The data repository of Andrews and Herzberg (1985)
features a data set containing measures of blood glucose, insulin levels,
relative weights, and clinical classifications of 145 subjects in 🖳 diabetes.dat:

1 0.81	80	356	124	55 1
2 0.95	97	289	117	76 1
3 0.94	105	319	143	105 1
...				
143 0.90	213	1025	29	209 3
144 1.11	328	1246	124	442 3
145 0.74	346	1568	15	253 3

The columns represent the following variables:

Variable	Meaning
relwt	Relative weight
glufast	Fasting plasma glucose
glutest	Test plasma glucose
instest	Plasma insulin during test
sspg	Steady-state plasma glucose
group	Clinical group: (3) overt diabetic; (2) chem. diabetic; (1) normal

From the variable group form the variable dia and set it to 1 if diabetes is present (group=1,2) and 0 if the subject has no diabetes (group=1). Find the regression of the five other variables on dia.

15.11. **Remission Ratios over Time.** A clinical trial on a new anticancer agent produced the following remission ratios for 40 patients on trial at each of the six stages of the trial:

9/40 14/40 22/40 29/40 33/40 35/40

Fit the logistic model for the probability of remission if the stages are measured in equal time units. Give the probability that a new patient on this regiment will be in remission at stage 4 and discuss how this probability compares to 29/40.

15.12. **Death of Sprayed Flour Beetles.** Hewlett (1974) and Morgan (2000) provide data that have been considered by many researchers in bioassay theory. The data consist of a quantal bioassay for pyrethrum in which the mortality of adult flour beetles (*Tribolium castaneum*) was measured over time under four dose levels. The columns are cumulative numbers of dead adult flour beetles exposed initially to pyrethrum, a well-known plant-based insecticide. Mixed with oil, the pyrethrum was sprayed at the given dosages over small experimental areas in which the groups of beetles were confined but allowed to move freely. The beetles were fed during the experiment in order to eliminate the effect of natural mortality.

Dose(mg/cm^2)		0.20		0.32		0.50		0.80	
Sex		M	F	M	F	M	F	M	F
Day	1	3	0	7	1	5	0	4	2
	2	14	2	17	6	13	4	14	9
	3	24	6	28	17	24	10	22	24
	4	31	14	44	27	39	16	36	33
	5	35	23	47	32	43	19	44	36
	6	38	26	49	33	45	20	46	40
	7	40	26	50	33	46	21	47	41
	8	41	26	50	34	47	25	47	42
	9	41	26	50	34	47	25	47	42
	10	41	26	50	34	47	25	48	43
	11	41	26	50	34	47	25	48	43
	12	42	26	50	34	47	26	48	43
	13	43	26	50	34	47	27	48	43
Group size		144	152	69	81	54	44	50	47

Using WinBUGS, model the proportion of dead beetles using sex, dosage, and day as covariates. The data set in WinBUGS format can be found in tribolium.odc.

15.13. **Mortality in Swiss White Mice.** An experiment concerning the influence of diet on the rate of *Salmonella enteritidis* infection (mouse typhoid) among W-Swiss mice was conducted by Schneider and Webster (1945). Two diets, one of whole wheat and whole dried milk (coded 100) and the other synthetic (coded 191), are compared against two doses of bacilli, 50K and 500K. The experimental results are summarized in the following table.

	Dose 50K		Dose 500K	
Diet	Number of mice	Number of surviving	Number of mice	Number of surviving
100	293	194	144	54
191	296	120	141	25

The authors conclude that diet is able to condition natural resistance, but one of the factors was the genetic constitution of the mice employed.
Model the probability of survival using an additive logistic regression with the covariates Diet and Dose.

15.14. **Kyphosis Data.** The measurements in kyphosis.dat are from Hastie and Tibshirani (1990, p. 301) and were collected on 83 patients undergoing corrective spinal surgery (Bell et al., 1994). The objective was to determine the important risk factors for kyphosis, or the forward flexion of the spine at least 40 degrees from vertical following surgery. The covariates are age in months, the vertebrae level at which the surgery started, and the number of vertebrae involved.

The data set 🖳 kyphosis.dat has four columns:

$$
\begin{array}{llll}
71 & 5 & 3 & 0 \\
158 & 14 & 3 & 0 \\
128 & 5 & 4 & 1 \\
& \cdots & & \\
120 & 13 & 2 & 0 \\
42 & 6 & 7 & 1 \\
36 & 13 & 4 & 0
\end{array}
$$

where the first three columns are age, start, and number, and where the fourth column is binary indicator for the presence of kyphosis, 0 if absent and 1 if present.

Using logistic regression, model the probability of present kyphosis given the risk factors age, start, and number.

15.15. **Prostate Cancer.** The prostate cancer clinical trial data of Byar and Green (1980) is given in the file 🖳 prostatecanc.dat with a description of the variables in 🖳 prostatecanc.txt or ◀ prostatecanc.m. There are 475 observations with 12 measured covariates. The response is the stage (3 or 4) of a patient assessed by a physician.

Propose a model for predicting the stage by a subset of predictors.

15.16. **Pediculosis Capitis.** An outbreak of *Pediculosis capitis* is being investigated in a girls' school containing 291 pupils. Of 130 children who live in a nearby housing estate, 18 were infested, and of 161 who live elsewhere, 37 were infested. Thus, the school girls are stratified by the housing attribute into two groups: (A) the nearby housing estate and (B) elsewhere.

(a) Test the hypothesis that the population proportions of infested girls for groups A and B are the same.

(b) Run a logistic regression that predicts the probability of a girl being infested including the predictor housing that takes value 1 if the girl is from group A and 0 if she is from group B. All you need are the sample sizes from groups A and B and the corresponding incidences of infestation. You might need to recode the data and represent the summarized data as 291 individual cases containing the incidence of infestation and housing status.

(c) A sample of 26 girls from group A and 34 from group B are randomly selected for more detailed modeling analysis. The instances of infestation (0-no, 1-yes), housing (A=1, B=0), family income (in thousands), family size, and girl's age are recorded. The data are available in the file 🖳 lice.xls. Propose the logistic model that predicts the probability that a girl who is infested will possess some or all of the predictors (housing, income, size, age). This is an open-ended question, and you are expected to defend your proposed model.

(c.1) According to your model, what is the probability of a girl being infested if housing is A, family income is 74, family size is 4, and age is 12? Of course, you will use only the values of the predictors included in your model.

(c.2) If the family size is increased by 1 and all other covariates remain the same, how much do the odds of infestation change?

(d) The 55 affected girls were divided randomly into two groups of 29 and 26. The first group received a standard local application and the second group a new local application. The efficacy of each was measured by clearance of the infestation after one application. By this measure the standard application failed in ten cases and the new application in five. Is the new treatment more effective? This part may mimic the methodology in (a).

15.17. **Finney Data.** In a controlled experiment to study the effect of the rate and volume of air inspired on a transient reflex vasoconstriction in the skin of the digits, 39 tests under various combinations of rate and volume of air inspired were conducted (Finney, 1947). The end point of each test was whether or not vasoconstriction occurred.

Volume	Rate	Response	Volume	Rate	Response	Volume	Rate	Response
3.70	0.825	Constrict	3.50	1.09	Constrict	1.80	1.50	Constrict
1.25	2.50	Constrict	0.75	1.50	Constrict	0.95	1.90	No constrict
0.80	3.20	Constrict	0.70	3.50	Constrict	1.90	0.95	Constrict
0.60	0.75	No constrict	1.10	1.70	No constrict	1.60	0.40	No constrict
0.90	0.75	No constrict	0.90	0.45	No constrict	2.70	0.75	Constrict
0.80	0.57	No constrict	0.55	2.75	No constrict	2.35	0.03	No constrict
0.60	3.00	No constrict	1.40	2.33	Constrict	1.10	1.83	No constrict
0.75	3.75	Constrict	2.30	1.64	Constrict	1.10	2.20	Constrict
3.20	1.60	Constrict	0.85	1.415	Constrict	1.20	2.00	Constrict
1.70	1.06	No constrict	1.80	1.80	Constrict	0.80	3.33	Constrict
0.40	2.00	No constrict	0.95	1.36	No constrict	0.95	1.90	No constrict
1.35	1.35	No constrict	1.50	1.36	No constrict	0.75	1.90	No constrict
1.60	1.78	Constrict	0.60	1.50	No constrict	1.30	1.625	Constrict

Model the probability of vasoconstriction as a function of two covariates, Volume and Rates. Use MATLAB and compare the results with WinBUGS output. The data in MATLAB and WinBUGS formats is provided in finney.dat.

15.18. **Diagnosing Sagittal Synostosis.** (Courtesy of Dr. Marcus Walker.) In early human development, the skull is made up of many different bones, and the gaps between these bones are called cranial sutures. During the first few years after birth, these cranial sutures allow the bones to grow and the skull to expand to make room for the growing brain. These bones naturally grow closer together and fuse to form one solid skull for protection, but if any of these cranial sutures fuse too early while the brain is still rapidly growing, a condition known as *craniosynostosis* occurs that results in skull deformation and constriction of the developing brain. Craniosynostosis occurs in approximately 1 in every 2,000 chil-

dren can lead to developmental disabilities, blindness, and even death if
left untreated. One specific type of craniosynostosis, called sagittal synostosis, is characterized by a premature fusion of the suture that runs
from the front to the back of the skull. This has been observed to cause
an elongation of the skull and opening of two sutures that run down the
sides of the skull. Using measurements from a CT scan of the volume
inside the skull at different areas and the distances between the bones
in the cranial sutures, it is possible to diagnose sagittal synostosis. Data

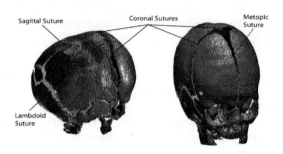

Fig. 15.9 Types of suture in early human development.

set `sagittal.dat|mat|xlsx` contains measurements on 60 children, 20
with diagnosed sagittal synostosis and 40 normal controls (Walker, 2013;
Credits to Dr. Barbara Boyan, Dr. Zvi Schwartz and Dr. Chris Hermann)
The variables are described in the following table:

Column	Variable	Meaning
1	y	1 - Synostosis, 0 - Normal
2	x_1	Percent of volume in the front of the skull
3	x_2	Percent of sagittal suture that is open
4	x_3	Tangent distance in the left coronal suture
5	x_4	Tangent distance in the right coronal suture

(a) Model y by a logistic regression that uses x_1 and x_2 as predictors.
Write down the model. Predict the probability of synostosis for a child
with measured $x_1 = 20$ and $x_2 = 61$.
(b) If the predicted probability exceeds 0.5, decide $\hat{y} = 1$. What is the total
number of errors made by your model, that is, how many predicted $\hat{y}_i$'s
differ from the corresponding observed $\hat{y}_i$?
(c) Show that the logistic model with all four variables $x_1, \ldots, x_4$ as predictors makes no errors; that is, all $\hat{y}_i$ are equal to y_i.

15.19. **Shocks.** An experiment was conducted (Dalziel et al., 1941) to assess the
effect of small electrical currents on farm animals, with the eventual goal
of understanding the effects of high-voltage power lines on livestock. The

experiment was carried out with seven cows using six shock intensities, 0, 1, 2, 3, 4, and 5 milliamps (shocks on the order of 15 milliamps are painful for many humans). Each cow was given 30 shocks, 5 at each intensity, in random order. The entire experiment was then repeated, so each cow received a total of 60 shocks. For each shock, the response of mouth movement was either present or absent. The data as quoted give the total number of responses, out of 70 trials, at each shock level. We ignore cow differences and differences between blocks (experiments).

Current (milliamps) x	Number of responses y	Number of trials n	Proportion of responses p
0	0	70	0.000
1	9	70	0.129
2	21	70	0.300
3	47	70	0.671
4	60	70	0.857
5	63	70	0.900

Using logistic regression and noninformative priors on its parameters, estimate the proportion of responses after a shock of 2.5 milliamps. Find 95% credible set for the population proportion.

15.20. **Separating Wild and Mutant** C. *elegans.* C. *elegans,* a soil dwelling nematode, is a highly studied multicellular organism that offers several experimental advantages such as a short life span, ease of culture, and transparency. Synapses in C. *elegans* can be imaged with fluorescent gene reporters such as GFP. San Miguel Delgadillo, et al. (2012) monitored presynaptic sites in motorneurons through a GFP-tagged protein located in synaptic vesicles. From these images, descriptors for the morphology of synaptic sites can be extracted to provide a phenotypic profile of the animal.

Researchers were faced with the problem of differentiating wild type and mutant C. *elegans* populations based on the very subtle phenotypic differences that can be presented at presynaptic sites, as quantified through 74 descriptors obtained through quantitative analysis of fluorescent images.

The data ⬛ celegans.dat|mat|xlsx (courtesy of Hang Lu Lab at Georgia Tech) contains 1,584 rows (observations) and 76 columns. The first 74 columns contain descriptors for synapse morphology (features), the 75th column is type of mutant (1–7, 0 is wild), and the last column contains an indicator of mutation (0 – wild, 1 – mutant).

Denote the matrix of features by X and the indicator of mutation by y. The features X_2, X_{20}, X_{37}, X_{51}, and X_{56} are selected as the most predictive columns in X for determination of mutation.

(a) Fit the logistic regression

$$\text{logit}(\mathbb{P}(y=1)) = \beta_0 + \beta_1 X_2 + \beta_2 X_{20} + \beta_3 X_{37} + \beta_4 X_{51} + \beta_5 X_{56} + \epsilon.$$

(b) Using the fit from (a) estimate the probability that an observed C. $elegans$ is a mutant if the recorded features were $X_2 = 8.5$, $X_{20} = 12.5$, $X_{37} = 600$, $X_{51} = 90$, and $X_{56} = 0.5$, respectively.
(c) Find the deviance and McFadden pseudo-R^2 for the model in (a).
(d) Describe in words (one paragraph) how you would classify C. $elegans$ to mutant and wild using logistic regression. According to your description, how would you classify the C. $elegans$ from (b)?

15.21. **Ants.**   The data set 🖼 ants.csv|dat|mat|xlsx, discussed in Gotelli and Ellison (2002), provides the ant species richness (number of ant species) found in 64-square-meter sampling grids in 22 forests (coded as 1) and 22 bogs (coded as 2) surrounding the forests in Connecticut, Massachusetts, and Vermont. The sites span $3°$ of latitude in New England. There are 44 observations on four variables (columns in data set): Ants – number of species, Habitat – forests (1) and bogs (2), Latitude, and Elevation – in meters above sea level.
(a) Using Poisson regression, model the number of ant species (Ants) with covariates Habitat and Elevation. Report the model coefficients and deviance.
(b) For a sampling grid unit located in a forest at the elevation of 100 m how many species the model from (a) predicts?
(c) Do the calculations from (a) and (b) using Open/WinBUGS with non-informative priors on all parameters. For the model coefficients and the prediction report 95% credible sets.

15.22. **Sharp Dissection and Postoperative Adhesions Revisited.** In Exercise 14.3 we fitted a linear relationship between the logarithm of the amount of sharp dissection lasd (predictor) and severity score sesco (response). Criticize this linear model. Model this relationship using Poisson regression and graphically compare the linear and Poisson fits.

15.23. **Airfreight breakage.**   A substance used in biological and medical research is shipped by air freight to users in cartons of 1,000 ampules. The data below, involving ten shipments, were collected on the number of times a carton was transferred from one aircraft to another over the shipment route (X) and the number of ampules found to be broken upon arrival (Y).

X	1	0	2	0	3	1	0	1	2	0
Y	16	9	17	12	22	13	8	15	19	11

Using WinBUGS, fit Y by Poisson regression, with X as a covariate. According to your model, how many packages will be broken if the number of shipment routes is $X = 4$?

15.24. **Body Fat Affecting Accuracy of Heart Rate Monitors.** In the course
Problems in Biomedical Engineering I at Georgia Tech, a team of students
investigated whether the readings of heart rate from a chest strap monitor (Polar T31) were influenced by the subject's percentage of body fat.
Hand counts facilitated by a stethoscope served as the gold standard.
The absolute differences between device and hand counts (AD) were regressed on body fat (BF) measurements. The measurements for 28 subjects are provided below:

Subj.	BF	AD	Subj.	BF	AD	Subj.	BF	AD	Subj.	BF	AD
1	17.8	4	8	18.8	1	15	25.1	3	22	24.1	6
2	13.2	3	9	13.4	0	16	18.3	2	23	12.9	2
3	7.7	3	10	39.4	7	17	16.9	3	24	30.1	6
4	11.8	1	11	6.8	1	18	27.8	6	25	17.1	4
5	23.9	0	12	25.0	6	19	36.0	5	26	18.4	4
6	27.2	0	13	19.9	0	20	31.9	1	27	14.6	4
7	27.6	0	14	23.0	9	21	17.4	2	28	26.8	3

A significant nonconstant representation of AD as a function of BF can be
translated as a significant influence of percent body fat on the accuracy
of the device.
(a) Why is linear regression not adequate here?
(b) Fit the Poisson regression $AD \sim \mathcal{P}oi(\exp\{b_0 + b_1 BF\})$. Is the slope b_1
from the linear part significantly positive?
(c) Using WinBUGS find the 95% credible set for the slope b_1 in the linear
part of a Bayesian Poisson regression model. Use non-informative priors.

15.25. **Micronuclei Assay.** The micronuclei (MN) assay procedure involves
breaking the DNA of lymphocytes in a blood sample with a powerful
dose of radiation, then measuring the efficiency of its ability to repair
itself. Micronuclei are fragments of DNA that have not healed back into
either of the two daughter nuclei after irradiation. The MN assay entails
scoring the number of micronuclei; the higher the number, the less efficient is a person's DNA repair system. The dose response of the number
of micronuclei in cytokinesis-blocked lymphocytes after in-vitro irradiation of whole blood with X-rays in the dose range 0–4 Gy was studied
by Thierens et al. (1991). The data provided in table are from one patient
(male, 54 y.o.) and represent the frequency of micronuclei numbers for
six levels of radiation, 0, 0.5, 1, 2, 3, and 4 Gy.

	Number of micronuclei						
	0	1	2	3	4	5	6
0	976	21	3	0	0	0	0
0.5	936	61	3	0	0	0	0
Dose 1	895	94	11	0	0	0	0
(in Gy) 2	760	207	32	1	0	0	0
3	583	302	97	12	6	0	0
4	485	319	147	35	11	2	1

(a) Fit a Poisson regression in which the number of micronuclei is the response (y) and dose is a covariate (x). Plot Poisson intensity λ as a function of dose.

(b) What is the average number of micronuclei for dose of 3.5 Gy?

(c) Simulate 1,000 micronuclei counts for λ corresponding to dose 3.5 Gy. Summarize the simulation output.

Hint: Use MATLAB's `glmfit` with `'poisson'` distribution and `'log'` link, This will estimate the regression coefficients needed for (b–c). For plotting, use `glmval`.

The data need to be appropriately recoded and one way to do so is shown below:

```
y=[ zeros(976,1); ones(21,1); 2*ones(3,1); ...
        zeros(936,1); ones(61,1); 2*ones(3,1); ...
        zeros(895,1); ones(94,1); 2*ones(11,1); ...
        zeros(760,1); ones(207,1); 2*ones(32,1); 3*ones(1,1);...
        zeros(583,1); ones(302,1); 2*ones(97,1); 3*ones(12,1); 4*ones(6,1);...
        zeros(485,1); ones(319,1); 2*ones(147,1); 3*ones(35,1); 4*ones(11,1);...
            5*ones(2,1); 6*ones(1,1)];

x =[zeros(976+21+3,1); ...
        0.5 * ones(936+61+3,1);...
        ones(895+94+11,1);...
        2*ones(760+207+32+1,1);...
        3*ones(583+302+97+12+6,1);...
        4*ones(485+319+147+35+11+2+1,1)];
```

15.26. **Miller Lumber Company Customer Survey.** Kutner et al. (2005) analyze a data set from a survey of customers of the Miller Lumber Company. The response is the total number of customers (in a representative 2-week period) coming from a tract of a metropolitan area within 10 miles from the store. The covariates include five variables concerning the tracts: number of housing units, average income in dollars, average housing unit age in years, distance to nearest competitor in miles, and distance to store in miles. Fit and assess a Poisson regression model for the number of customers as predicted by the covariates. The data are in ◢ `lumber.m`.

15.27. **SO$_2$, NO$_2$, and Hospital Admissions.** Fan and Chen (1999) discuss a public health data set consisting of daily measurements of pollutants and other environmental factors in Hong Kong between January 1, 1994 and December 31, 1995. The association between levels of pollutants and the number of daily hospital admissions for circulation and respiratory problems is of particular interest.

The data file 📄 hospitaladmissions.dat consists of six columns: (1) year, (2) month, (3) day in month, (4) concentration of sulfur dioxide SO_2, (5) concentration of pollutant nitrogen NO_2, and (6) daily number of hospital admissions.

(a) Using logistic regression, determine how the probability of a high level of sulfur dioxide (with values $> 20\,\mu g/m^3$) is associated with the level of pollutant nitrogen NO_2.

(b) Using a Poisson regression model, explore how the expected number of hospital admissions varies with the level of NO_2.

(c) Suppose that on a particular day the level of NO_2 was measured at 100. Estimate the probability of a high level of sulfur dioxide in (a) and the expected number of hospital admissions in (b).

15.28. **Kidney Stones.** Charig et al. (1986) provide data on the success rates of two methods of treating kidney stones: open surgery methods and percutaneous nephrolithotomy. There are two predictors: size of stone and method. Size of stone is set at two levels: $< 2\,$cm in diameter, coded as Small, and $> 2\,$cm in diameter, coded as Large. The two methods are coded as A (open surgery) and B (percutaneous nephrolithotomy). The outcome of interest is the outcome of the treatment (Success, Failure).

Count	Size	Method	Outcome
81	Small	A	Success
6	Small	A	Failure
234	Small	B	Success
36	Small	B	Failure
192	Large	A	Success
71	Large	A	Failure
55	Large	B	Success
25	Large	B	Failure

There are four combinations of the covariates: Small A, Small B, Large A, and Large B. Find the relative frequencies of the outcome "Success" and compare them with the model-predicted probabilities using logistic regression.

Show that these data hide Simpson's paradox.

MATLAB AND WINBUGS FILES AND DATA SETS USED IN THIS CHAPTER

http://statbook.gatech.edu/Ch15.Logistic/

arrhythmia.m, arthritis2.m, beetleBliss1.m, beetleBliss2.m, bumpus.m, caesarean.m, celular.m, counterr.m, dmdreg.m, ihga.m, kyphosis.m, logisticmle.m, logisticSeed.m, lumber.m, outbreak.m, ponv.m, prostatecanc.m, sagittal.m, UptonFingleton.m

accidentssimple.odc, arrhythmia.odc, beetles.odc, bliss.odc, bystanders.odc, caesarean.odc, celldifferentiation.odc, errors1.odc, geriatric.odc, microdamage.odc, raynaud.odc, remission.odc, tribolium.odc, tromboembolism.odc

ants.dat, arrhythmia.mat|xlsx, arrhythmiadata.m, arthritis2.dat, birthweight.dat, bumpus.mat, cardiac.mat|txt, celegans.dat|mat|xlsx, celular.dat, diabetes.dat, dmd.dat|mat, finney.dat, hospitaladmissions.dat, ihgadat.m, kyphosis.dat|txt, lowbwt.dat, microdamage.dat|mat, outbreak.mat, pima.dat, PONV.mat|xls, programm.mat, prostatecanc.dat, sagittal.mat|xlsx, tribolium.mat

---

CHAPTER REFERENCES

Andrews, D. F. and Herzberg, A. M. (1985). *Data: A Collection of Problems from Many Fields for the Student and Research Worker*. Springer, New York.

Atkinson, A. C. (1985). *Plots, Transformations and Regression: An Introduction to Graphical Methods of Diagnostic Regression Analysis*. Oxford University Press, New York.

Bell, D. F., Walker, J. L., O'Connor, G., and Tibshirani, R. (1994). Spinal deformity after multiple-level cervical laminectomy in children. *Spine*, **19**, 4, 406–411.

Bliss, C. I. (1935). The calculation of the dose-mortality curve. *Ann. Appl. Biol.*, **22**, 134–167.

Burke, R. L. and Calichio, A. M. (2014). Temperature-dependent sex determination in the diamond-backed terrapin (*Malaclemys terrapin*). *J. Herpetol.*, **48**, 4, 466–470.

Byar, D. P. and Green, S. B. (1980). The choice of treatment for cancer patients based on covariate information: application to prostate cancer. *Bull. Cancer (Paris)*, **67**, 477–488.

Charig, C. R., Webb, D. R., Payne, S. R., and Wickham, J. A. E. (1986). Comparison of treatment of renal calculi by open surgery, percutaneous nephrolithotomy, and extracorpeal shockwave lithotripsy. *Br. Med. J.*, **292**, 879–882.

Cornfield, J. (1962). Joint dependence of risk of coronary heart disease on serum cholesterol and systolic blood pressure: a discriminant function analysis. *Federat. Proc.*, **21**, 58–61.

Dalziel, C. F., Lagen, J. B., and J. L. Thurston, J. L. (1941). Electric shocks. *Trans. IEEE*, **60**, 1073–1079.

Fahrmeir, L. and Tutz, G. (1994). *Multivariate Statistical Modelling Based on Generalized Linear Models*. Springer, New York.

Fan, J. and Chen, J. (1999). One-step local quasi-likelihood estimation. *J. R. Stat. Soc. Ser. B*, **61**, 4, 927–943.

Fan, J., Heckman, N. E., and Wand, M. P. (1995). Local polynomial kernel regression for generalized linear models and quasi-likelihood functions. *J. Am. Stat. Assoc.*, **90**, 429, 141–150.

Finney, D. J. (1947). The estimation from individual records of the relationship between dose and quantal response. *Biometrika*, **34**, 320–334.

Gotelli, N. J. and Ellison, A. M. (2002). Biogeography at a regional scale: determinants of ant species density in bogs and forests of New England. *Ecology*, **83**, 1604–1609.

Hastie, T. and Tibshirani, R. (1990). *Generalized Additive Models*. Chapman & Hall, London.

Hendriksen, C., Lund, E., and Stromgard, E. (1984). Consequences of assessment and intervention among elderly people: a 3 year randomised, controlled trial. *Br. Med. J.*, **289**, 1522–1544.

Hewlett, P. S. (1974). Time from dosage to death in beetles Tribolium castaneum, treated with pyrethrins or DDT, and its bearing on dose-mortality relations. *J. Stored Prod. Res.*, **10**, 27–41.

Hosmer, D. and Lemeshow, S. (1989). *Applied Logistic Regression* (2nd edn. 2000). Wiley, New York.

Koch, G. and Edwards, S. (1988). Clinical efficiency trials with categorical data. In: Peace, K. E. (ed) *Biopharmaceutical Statistics for Drug Development*. Dekker, New York, pp. 403–451.

Kutner, M. H., Nachtsheim, C. J., Neter, J., and Li, W. (2005). *Applied Linear Statistical Models*. McGraw-Hill/Irwin, New York.

McCullagh, P. and Nelder, J. A. (1989). *Generalized Linear Models*, 2nd edn. Monographs on Statistics and Applied Probability. Chapman & Hall/CRC, Boca Raton.

Morgan, B. J. T. (2000). *Applied Stochastic Modelling*. Arnold Texts in Statistics. Oxford University Press, New York.

Piergorsch, W. W., Weinberg, C. R., and Margolin, B. H. (1988). Exploring simple independent action in multifactor table of proportions. *Biometrics*, **44**, 595–603.

San Miguel Delgadillo, A., Crane, M., Kurshan, P., Kang Shen, K., and Lu, H. (2012). High Throughput Genetic Screens of *C. Elegans* with Microfluidics and Computer Vision. Proceedings of 2012 AIChE Annual Meeting, p. 507, Pittsburgh.

Schneider, H. A. and Webster, L. T. (1945). Nutrition of the host and natural resistance to infection. I: The effect of diet on the response of several genotypes of musculus to salmonella enteritidis infection. *J. Exp. Med.*, **81**, 4, 359–384.

Thierens, H., Vral, A., and de Ridder, L. (1991). Biological dosimetry using the micronucleus assay for lymphocytes: interindividual differences in dose response. *Health Phys.*, **61**, 5, 623–630.

Upton, G. J. G. and Fingleton B. (1985). *Spatial Data Analysis by Example*, vol. 1. Wiley, Chichester.

Walker, M. (2013). *Cranial Suture Quantification and Intracranial Volume Asymmetry Analysis for Craniosynostosis Predictive Modeling and Assessment*. Master's Thesis at EE/BME, Georgia Institute of Technology.

Chapter 16

Inference for Censored Data and Survival Analysis

The first condition of progress is the removal of censorship.

– George Bernard Shaw

WHAT IS COVERED IN THIS CHAPTER

- Parametric Models for Time-to-Event Data
- Kaplan–Meier Estimator, Mantel's Logrank Test
- Cox Proportional Hazards Model
- Bayesian Approaches

16.1 Introduction

Survival analysis models the survival times of a group of subjects, usually with some kind of medical condition, and generates a survival curve, which shows how many of the subjects are "alive" or survive over time.

What makes survival analysis different from standard regression methodology is the presence of censored observations; in addition, some subjects may leave the study and will be lost to follow-up. Such subjects were known to have survived for some amount of time (up until the time we last saw them), but we do not know how much longer they might ultimately have survived. Several methods have been developed for using this "at least this long" information to finding unbiased survival curve estimates, the most popular being the nonparametric method of Kaplan and Meier.

An observation is said to be *censored* if we know only that it is less than (or greater than) a certain known value. For instance, in clinical trials, one could be interested in patients' survival times. Survival time is defined as the length of time between diagnosis and death, although other "start" events, such as surgery, and other "end" events, such as relapse of disease, increase in tumor size beyond a particular threshold, rejection of a transplant, etc., are commonly used. Because of many constraints, trials cannot be run until the endpoints are observed for all patients. Just because for a particular subject the time to endpoint is not fully observed, partial information is still available: the patient survived up to the end of the observational period, and this should be incorporated into the analysis. Such observations are called *right censored*. Observations can also be *left censored*, for example, an assay may have a detection threshold. In order to utilize information contained in censored observations, special methods of analysis are required.

Engineers are often interested in the reliability of various devices, and most of the methodology from survival analysis is applicable in device reliability analyses. There are of course important differences. In the reliability of multicomponent systems an important aspect is optimization of the number and position of components. Analogous considerations with organs or parts of organs as components in living systems (animals or humans) are impossible. Methods such as "accelerated life testing" commonly used in engineering reliability are inappropriate when dealing with human subjects.

However, in comparing the lifetimes of subjects in clinical trials involving different treatments (humans, animals) or different engineering interventions (systems, medical devices), the methodology that deals with censored observations is shared.

16.2 Definitions

Let T be a continuous random variable with CDF $F(t)$ representing a lifetime. The survival (or survivor) function is the tail probability for T, expressed as $S(t) = 1 - F(t)$, $t > 0$. The function $S(t)$ gives the probability of surviving up to time t, that is,

$$S(t) = \mathbb{P}(T > t).$$

The hazard function or hazard rate is defined as

$$h(t) = \frac{f(t)}{S(t)} = \frac{f(t)}{1 - F(t)},$$

when T has a density $f(t)$. Note that $S'(t) = -f(t)$. It is insightful to represent $h(t)$ in limit terms,

$$\frac{S(t) - S(t + \Delta t)}{\Delta t} \times \frac{1}{S(t)} = \frac{F(t + \Delta t) - F(t)}{\Delta t} \times \frac{1}{S(t)}$$

$$= \frac{\mathbb{P}(t < T \leq t + \Delta t)}{\Delta t\, \mathbb{P}(T > t)}$$

$$= \frac{\mathbb{P}(T \leq t + \Delta t | T > t)}{\Delta t},$$

when $\Delta t \to 0$. It represents an instantaneous probability that an event that was not observed up to time t will be observed before $t + \Delta t$, when $\Delta t \to 0$.

Cumulative hazard is defined as

$$H(t) = \int_0^t h(s)ds.$$

Both hazard and cumulative hazard uniquely determine the distribution of lifetime F,

$$F(t) = 1 - \exp\left\{ -\int_0^t h(s)ds \right\} = 1 - \exp\{-H(t)\}.$$

Cumulative hazard can also be connected to the survival function as $H(t) = -\log S(t)$, or

$$S(t) = \exp\{-H(t)\}. \tag{16.1}$$

Example 16.1. **Constant Hazard.** The hazard function for an exponential distribution with density $f(t) = \lambda e^{-\lambda t}, t \geq 0, \lambda > 0$ is constant in time, $h(t) = \lambda$.

📝

Example 16.2. **Linear Hazard.** Identify distributions for which the hazard rate is linear, $h(t) = a + bt, t \geq 0$.

Since

$$H(t) = \int_0^t h(u)du = at + \frac{bt^2}{2},$$

according to (16.1),

$$S(t) = 1 - F(t) = \exp\left\{-at - \frac{bt^2}{2}\right\},$$

and

$$f(t) = (a + bt)\exp\left\{-at - \frac{bt^2}{2}\right\}.$$

This represents a density if $b \geq 0$. For $a = 0$ the above is the Rayleigh distribution, and for $b = 0$ it is the exponential.

Example 16.3. **Weibull's Hazard.** The hazard rate for a one-parameter Weibull distribution with CDF $F(t) = 1 - \exp\{-t^\gamma\}$ and density $f(t) = \gamma t^{(\gamma-1)}\exp\{-t^\gamma\}$, $t \geq 0, \gamma > 0$ is $h(t) = \gamma t^{\gamma-1}$. The parameter γ is called a *shape* parameter. Depending on the shape parameter γ, the hazard function $h(t)$ could model various types of survival analyses.

The two-parameter version of Weibull that was introduced in Section 5.5.9 is used more frequently. It is defined as $F(t) = 1 - \exp\{-\lambda t^\gamma\}$, with density $f(t) = \lambda\gamma t^{\gamma-1}\exp\{-\lambda t^\gamma\}$, $t \geq 0, \gamma > 0, \lambda > 0$. The parameter λ is called a *rate* parameter. In this case $h(t) = \lambda\gamma t^{\gamma-1}$. Figure 16.1 shows hazard functions for different values of shape parameters γ. For $\gamma = 1$, the hazard is constant (Weibull distribution becomes exponential), for $\gamma < 1$, the hazard is decreasing, and for $\gamma > 1$, it is increasing.

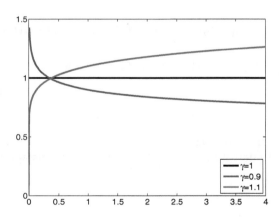

Fig. 16.1 Hazard function of two parameter Weibull distribution with $\lambda = 1$ and $\gamma = 0.9, 1$, and 1.1.

Two important summaries in the parametric case (where the survival distribution is specified up to a parameter) are mean residual life (mrl) and median life, defined, respectively, as

$$\text{mrl}(t) = \frac{\int_t^\infty S(x)dx}{S(t)},$$
$$t_{0.5}: \ S(t_{0.5}) = 0.5.$$

The hazard rate $h(t)$ and mean residual life $\text{mrl}(t)$ are connected,

$$h(t) = \frac{1 + (\text{mrl}(t))'}{\text{mrl}(t)}.$$

Example 16.4. **Exponential Mean Residual Life.** For an exponential lifetime, the expected lifetime and mrl coincide. Indeed, $\mathbb{E}T = 1/\lambda$, while

$$\text{mrl}(t) = \frac{\int_t^\infty e^{-\lambda x}dx}{e^{-\lambda t}} = \frac{1}{\lambda}.$$

At first glance this looks like a paradox; the expected total lifetime $\mathbb{E}T$ is equal to the residual lifetime $\text{mrt}(t)$ regardless of t. This is an example of the "inspection paradox" that follows from the memoryless property of the exponential distribution, page 201.

The median lifetime of an exponential distribution is $t_{0.5} = (\log 2)/\lambda$.

✐

Example 16.5. **Estimation Parameters in Weibull Distribution by Regression.** Recall that the CDF of a two-parameter Weibull distribution is

$$F(x) = 1 - e^{-\lambda x^r},$$

where λ is rate parameter and r is the shape parameter. Given a sample $X_1, X_2, \ldots, X_n$, the goal is to estimate λ and r.

The MLE equations do not have solutions in closed form and numerical approximations are used. Simpler and often superior estimation is based on a combination of method of moments and linear regression.

Note that

$$1 - F(x) = e^{-\lambda x^r}$$
$$\log(1 - F(x)) = -\lambda x^r$$
$$\log\left(\frac{1}{1 - F(x)}\right) = \lambda x^r.$$

After taking the logarithm, the equation becomes,

$$\log\left(\log\left(\frac{1}{1 - F(x)}\right)\right) = \log(\lambda) + r \log x.$$

If $X_{(i)}$ is ith order statistic from $X_1, \ldots, X_n$ coming from distribution with a CDF F, it holds that

$$\mathbb{E}F(X_{(i)}) = \frac{i}{n+1}.$$

This is true for any distribution, since $F(X_{(i)}) = U_{(i)}$ where $U_{(i)}$ is the ith order statistic from uniform $\mathcal{U}(0,1)$ distribution, and $\mathbb{E}U_{(i)} = i/(n+1)$.

The idea is to calculate regression responses y_i using $\mathbb{E}F(X_{(i)})$ in place of $F(X_{(i)})$,

$$y_i = \log\left(\log\left(\frac{1}{1 - \mathbb{E}F(X_{(i)})}\right)\right) = \log\left(\log\left(\frac{n+1}{n+1-i}\right)\right),$$

and regress the responses against ordered observations $x_i = X_{(i)}$.

If b_0 and b_1 are estimated intercept and slope in this regression, then

$$\hat{\lambda} = e^{b_0} \quad \text{and} \quad \hat{r} = b_1.$$

The following MATLAB script simulates 2,000 observations from Weibull $\mathcal{W}ei(2,3)$ distribution. The estimation of parameters r and λ is done by regression, and estimators are compared results from built in MATLAB's weblfit.

```
%estimweibull.m
s= RandStream('mt19937ar','Seed',0);
RandStream.setGlobalStream(s);
%
lambda=3;
r=2;
% Simulate 2000 Wei(r=2, lambda=3)
n = 2000;
xx = wblrnd( lambda^(-1/r), r,[n 1]);
%In MATLAB's parametrization of Weibull distribution
% the scale parameter is  lambda^(-1/r)
x = sort(xx);        i=(1:n)';
y = log(log((n+1)./(n +1 - i)));
```

```
b = regress(y, [ones(size(x)) log(x)]);
lambdahat = exp(b(1))                    %3.0676
rhat = b(2)                              %2.0371
%
%Built in Weibull fits
parmhat = wblfit(x);
lambdahat = parmhat(1)^(-parmhat(2))  %3.1063
rhat = parmhat(2)                        %2.0538
```

16.3 Inference with Censored Observations

We will consider the case of right-censored data (the most common type of censoring) and two approaches: a parametric approach, in which the survival function $S(t)$ would have a specific functional form, and a non-parametric approach, in which no such functional form is assumed.

16.3.1 Parametric Approach

In the parametric approach the models depend on the parameters, and the parameters are estimated by taking into account both uncensored and censored observations. We will show how to find an MLE in the general case and illustrate it on an exponential lifetime distribution.

Let (t_i, δ_i), $i = 1, \ldots, n$ be observations of a lifetime T for n individuals, with $\delta \in \{0, 1\}$ indicating fully observed and censored lifetimes, and let k observations be fully observed while $n - k$ are censored. Suppose that the underlying lifetime T has a density $f(t|\theta)$ with survival function $S(t|\theta)$. Then the likelihood is

$$L(\theta|t_1, \ldots, t_n) = \prod_{i=1}^{n} (f(t_i|\theta)^{1-\delta_i} \times (S(t_i|\theta))^{\delta_i} = \prod_{i=1}^{k} f(t_i|\theta) \times \prod_{i=k+1}^{n} S(t_i|\theta).$$

Since $h(t_i|\theta) \times (S(t_i|\theta)) = f(t_i|\theta)$, then

$$L(\theta|t_1, \ldots, t_n) = \prod_{i=1}^{n} \left(h(t_i|\theta)^{1-\delta_i} \times S(t_i|\theta) \right). \qquad (16.2)$$

Example 16.6. **MLE for Censored Exponential Lifetimes.** We will show that for an exponential lifetime in the presence of right-censoring, the MLE for

λ is

$$\hat{\lambda} = \frac{k}{\sum_{i=1}^{n} t_i},$$
(16.3)

where k is the number of noncensored data points and $\sum_{i=1}^{n} t_i$ is the sum of *all* observed and censored times. From (16.2), the likelihood is $L = \lambda^k \exp\{-\lambda \sum_{i=1}^{n} t_i\}$. By taking the log and differentiating, we get the MLE as the solution to $\frac{k}{\lambda} - \sum_{i=1}^{n} t_i = 0$.

The variance of the MLE $\hat{\lambda}$ is $k/(\sum_{i=1}^{n} t_i)^2$ and can be used to find the confidence interval for λ (Exercise 16.2).

Example 16.7. **Immunoperoxidase and BC.** Data analyzed in Sedmak et al. (1989) and also in Klein and Moeschberger (2003) represent times to death (in months) for breast cancer patients with different immunohistochemical responses. Out of 45 patients in the study, 9 were immunoperoxidase positive while the remaining 36 were negative (+ denotes censored time).

Immunoperoxidase negative
19, 25, 30, 34, 37, 46, 47, 51, 56, 57, 61, 66, 67, 74, 78, 86, 122+, 123+, 130+, 130+, 133+, 134+, 136+, 141+, 143+, 148+, 151+, 152+, 153+, 154+, 156+, 162+, 164+, 165+, 182+, 189+
Immunoperoxidase positive
22, 23, 38, 42, 73, 77, 89, 115, 144+

Assume that lifetimes are exponentially distributed and that rates λ_1 (for Immunoperoxidase negative) and λ_2 (for Immunoperoxidase positive) are to be estimated. The following MATLAB code finds MLEs of λ_1 and λ_2, first directly by using (16.3) and then by using MATLAB's built-in function `mle` with option `'censoring'`.

```
ImmPeroxNeg=[...
19, 25, 30, 34, 37, 46, 47, 51, 56, 57, 61, 66, 67, 74, 78, 86,...
122, 123, 130, 130, 133, 134, 136, 141, 143, 148, 151, 152,...
153, 154, 156, 162, 164, 165, 182, 189];
CensorIPN=[0,0,0,0,0,0,0,0,0,0,0,0,0,0,0,0,...
  1,1,1,1,1,1,1,1,1,1,1,1,1,1,1,1,1,1,1,1,1];

ImmPeroxPos=[...
22, 23, 38, 42, 73, 77, 89, 115, 144];
CensorIPP=[0,0,0,0,0,0,0,0,1];

%number of observed (non-censored)
k1 = sum(1-CensorIPN)    %16
k2 = sum(1-CensorIPP)    %8

% MLEs of rate lambda for 2 samples.
hatlam1 = k1/sum(ImmPeroxNeg)  %0.0042
hatlam2 = k2/sum(ImmPeroxPos)  %0.0128

[reclambdahat1 lamci1] = mle(ImmPeroxNeg, ...
    'distribution','exponential','censoring',CensorIPN)
   % 237.6250
   % 153.6769  415.7289
[reclambdahat2 lamci2] = mle(ImmPeroxPos, ...
    'distribution','exponential','censoring',CensorIPP)
   % 77.8750
   % 43.1959   180.3793
%(MATLAB parametrization) scale to rate
lambdahat1 = 1/reclambdahat1  %0.0042
lambdahat2 = 1/reclambdahat2  %0.0128
```

As is indicated by the code, the patients who are immunoperoxidase positive are at increased risk since the rate $\hat{\lambda}_2 = 0.0128$ exceeds $\hat{\lambda}_1 = 0.0042$.

In MATLAB, dfittool, normfit, wblfit, and other commands for fitting parametric distributions can be applied to censored data by specifying the censoring vector at the input.

16.3.2 Nonparametric Approach: Kaplan–Meier or Product–Limit Estimator

Assume that individuals in the study are assessed at discrete time instances $t_1, t_2, \ldots, t_k$, which may not be equally spaced. Typically, the times t_i are selected when failures occur. If we want to calculate the probability of survival up to time t_i, then by the chain rule of conditional probabilities and their Markovian property,

$\hat{S}(t_i) = \mathbb{P}(\text{ surviving to time } t_i\ ) = \mathbb{P}(\text{ survived up to time } t_1\ )$

$\times\ \mathbb{P}(\text{ surviving to time } t_2\ |\ \text{ survived up to time } t_1\ )$

$\times\ \mathbb{P}(\text{ surviving to time } t_3\ |\ \text{ survived up to time } t_2\ )$

$\cdots$

$\times\ \mathbb{P}(\text{ surviving to time } t_i\ |\ \text{ survived up to time } t_{i-1}\ ).$

It is assumed that $t_0 = 0$.

Suppose that r_i subjects are at risk at time t_{i-1} and are not censored at time t_{i-1}. In the ith interval $[t_{i-1}, t_i)$ among these r_i subjects d_i have an event, ℓ_i are censored, and r_{i+1} survive. The r_{i+1} subjects will be at risk at the beginning of the $(i+1)$th time interval $[t_i, t_{i+1})$, that is, at time t_i. Thus, $r_i = d_i + \ell_i + r_{i+1}$. We can estimate the probability of survival up to time t_i, given that one survived up to time t_{i-1}, as $1 - d_i/(r_{i+1} + d_i + \ell_i) = 1 - d_i/r_i$.

The ℓ_i subjects censored at time t_i do not contribute to the survival function for times $t > t_i$.

$$\hat{S}(t) = \left(1 - \frac{d_1}{r_1}\right) \times \left(1 - \frac{d_2}{r_2}\right) \times \cdots \times \left(1 - \frac{d_i}{r_i}\right)$$

$$= \prod_{t_i \leq t} \left(1 - \frac{d_i}{r_i}\right), \quad \text{for } t \geq t_1;$$

$$\hat{S}(t) = 1, \quad \text{for } t < t_1.$$

This is the celebrated Kaplan–Meier or product-limit estimator (Kaplan and Meier, 1958). This result has been one of the most influential developments in the past century in statistics; the paper by Kaplan and Meier is the most cited paper in the field of statistics (Stigler, 1994).

For uncensored observations, the Kaplan–Meier estimator is identical to the complement of the empirical CDF. The difference occurs when there is a censored observation – then the Kaplan–Meier estimator takes the "weight" normally assigned to that observation and distributes it evenly among all observed values to the right of the censored observation. This is intuitive because we know that the true value of the censored observation must be somewhere to the right of the censored value, but information about what the exact value should be is lacking. Thus all observed values larger than the censored observation are treated in the same way.

The variance of Kaplan–Meier estimator is estimated by Greenwood's formula (Greenwood, 1926):

$$\tau_{\hat{S}}^2(t) = \left(\hat{S}(t)\right)^2 \times \sum_{t_i \leq t} \frac{d_i}{r_i(r_i - d_i)}.$$

The pointwise confidence intervals (for a fixed time t_*) for the survival function $S(t_*)$ can be found in several ways. The most popular confidence intervals are

linear

$$\left[\hat{S}(t_*) - z_{1-\alpha/2}\, \tau_S(t_*),\ \hat{S}(t_*) + z_{1-\alpha/2}\, \tau_S(t_*)\right],$$

log-transformed

$$\left[(\hat{S}(t_*))\, \exp\left\{\frac{-z_{1-\alpha/2}\, \tau_S(t_*)}{\hat{S}(t_*)}\right\},\ (\hat{S}(t_*))\, \exp\left\{\frac{z_{1-\alpha/2}\, \tau_S(t_*)}{\hat{S}(t_*)}\right\}\right],$$

and *log-log-transformed*

$$\left[(\hat{S}(t_*))^{v},\ (\hat{S}(t_*))^{1/v}\right],\ \ v = \exp\left\{\frac{z_{1-\alpha/2}\, \tau_S(t_*)}{\hat{S}(t_*)\, |\log \hat{S}(t_*)|}\right\}.$$

Although not centered at $\hat{S}(t_*)$, the log- and log-log-transformed intervals are considered superior to the linear. This is because $\hat{S}(t_*)$ is not well approximated by a normal distribution, especially when $S(t_*)$ is close to 0 or 1.

The pointwise confidence intervals given above differ from simultaneous confidence bounds on $S(t)$ for which the confidence of $1 - \alpha$ means that the probability that *any* part of the curve $S(t)$ will fall outside the bounds does not exceed α. Such general bounds are naturally wider than those generated by pointwise confidence intervals, since the overall confidence is controlled. Two important types of confidence bands are Nair's equal precision bands and the Hall–Wellner bands. Description of these bounds are beyond the scope of this text; see Klein and Moeschberger (2003, p. 109), for further discussion and implementation. The bounds computed in MATLAB's `[f,t,flo,fup]=ecdf(...)` also return lower and upper confidence bounds for the CDF. These bounds are calculated using Greenwood's formula and are not simultaneous confidence bounds.

The Kaplan–Meier estimator also provides an estimator for the cumulative hazard $H(t)$ as

$$\hat{H}(t) = -\log\left(\hat{S}(t)\right).$$

Better small-sample performance in estimating the cumulative hazard can be achieved by the Nelson–Aalen estimator,

$$\tilde{H}(t) = \begin{cases} 0, & \text{for } t \leq t_1 \\ \sum_{t_i < t} d_i / r_i, & \text{for } t > t_1, \end{cases}$$

with an estimated variance $\sigma_{\tilde{H}}^2(t) = \sum_{t_i < t} d_i / r_i^2$. By using $\tilde{H}(t)$ and $\sigma_{\tilde{H}}^2(t)$, pointwise confidence intervals on $H(t)$ can be obtained.

Example 16.8. **Catheter Complications in Peritoneal Dialysis.** The following example is from Chadha et al. (2000). The authors studied a sample of 36 pediatric patients undergoing acute peritoneal dialysis through Cook catheters. They wished to examine how long these catheters performed properly. They noted the date of complication (either occlusion, leakage, exit-site infection, or peritonitis).

Half of the subjects had no complications before the catheter was removed. Reasons for removal of the catheter in this group of patients were that the patient recovered ($n = 4$), the patient died ($n = 9$), or the catheter was changed to a different type electively ($n = 5$). If the catheter was removed prior to complications, that represented a censored observation, because they knew that the catheter remained complication free at least until the time of removal.

Day	At Risk, r_i	Censored, ℓ_i	Fail, d_i	$1 - \frac{d_i}{r_i}$	KM
1	36	8	2	$1 - 2/36 = 0.944$	0.9444
2	$36 - 8 - 2 = 26$	2	2	$1 - 2/26 = 0.92$	$0.92 \cdot 0.944 = 0.8718$
3	$26 - 2 - 2 = 22$	1	2	$1 - 2/22 = 0.91$	$0.91 \cdot 0.872 = 0.7925$
4	$22 - 1 - 2 = 19$	1	1	$1 - 1/19 = 0.95$	$0.95 \cdot 0.793 = 0.7508$
5	$19 - 1 - 1 = 17$	6	3	$1 - 3/17 = 0.82$	0.6183
6	$17 - 6 - 3 = 8$	0	2	$1 - 2/8 = 0.75$	0.4637
7	$8 - 0 - 2 = 6$	0	1	$1 - 1/6 = 0.83$	0.3865
10	$6 - 0 - 1 = 5$	0	2	$1 - 2/5 = 0.60$	0.2319
12	$5 - 0 - 2 = 3$	0	2	$1 - 2/3 = 0.33$	0.0773
13	$3 - 0 - 2 = 1$	0	1	$1 - 1/1 = 0.00$	0.0000

MATLAB script ◀ chada.m finds the Kaplan–Meier estimator and generates Figure 16.2. plots

```
%chada.m
times=[1,1,1,1,1,1,1,1,1,1,2,2,2,2,...
       3,3,3,4,4,5,5,5,5,5,5,5,5,5,...
       6,6,7,10,10,12,12,13];
censored =[1,1,1,1,1,1,1,1,0,0,1,1,...
           0,0,1,0,0,1,0,1,1,1,1,1,...
           1,0,0,0,0,0,0,0,0,0,0,0];

% Calculate and plot KM estimator
ple(times, censored)
```

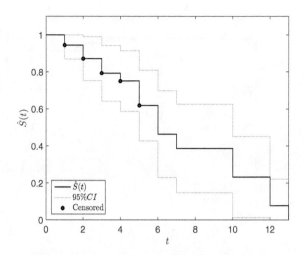

Fig. 16.2 Kaplan–Meier estimator for Catheter Complications data.

Example 16.9. **Strength of Weathered Cord.** Data from Crowder et al. (1991) lists strength measurements (in coded units) for 48 pieces of weathered cord. Seven of the pieces of cord were damaged and yielded strength measurements that are considered right-censored. That is, because the damaged cord was taken off the test, we know only the lower limit of its strength. In the MATLAB code below, the vector data represents the strength measurements, and the vector censor indicates (with a zero) if the corresponding observation in data is censored.

```
data=[36.3, 41.7, 43.9, 49.9, 50.1, 50.8, 51.9, 52.1, 52.3, 52.3,...
      52.4, 52.6, 52.7, 53.1, 53.6, 53.6, 53.9, 53.9, 54.1, 54.6,...
      54.8, 54.8, 55.1, 55.4, 55.9, 56.0, 56.1, 56.5, 56.9, 57.1,...
      57.1, 57.3, 57.7, 57.8, 58.1, 58.9, 59.0, 59.1, 59.6, 60.4,...
      60.7,      26.8, 29.6, 33.4, 35.0, 40.0, 41.9, 42.5];
censor=[zeros(1,41), ones(1,7)];
[table] = ple(data, censor)
```

The table below shows how the Kaplan–Meier estimator is calculated for the first 16 measurements, which includes 7 censored observations. Figure 16.3 shows the estimated survival function for the cord strength data.

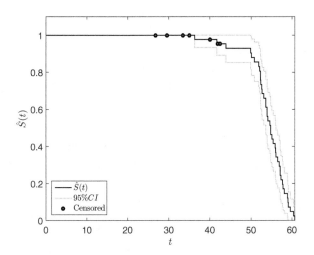

Fig. 16.3 Kaplan–Meier estimator cord strength (in coded units).

t_i	r_i	d_i	ℓ_i	$1 - \frac{d_i}{r_i}$	$\hat{S}(t_i)$
0	48	0	4	1.0000	1.0000
36.3	44	1	1	0.9773	0.9773
41.7	42	1	2	0.9762	0.9540
43.9	39	1	0	0.9744	0.9295
49.9	38	1	0	0.9737	0.9051
50.1	37	1	0	0.9730	0.8806
50.8	36	1	0	0.9722	0.8562
51.9	35	1	0	0.9714	0.8317
52.1	34	1	0	0.9706	0.8072
52.3	33	2	0	0.9394	0.7583
52.4	31	1	0	0.9677	0.7338
$\vdots$	$\vdots$	$\vdots$	$\vdots$	$\vdots$	$\vdots$
57.8	8	1	0	0.8750	0.1712
58.1	7	1	0	0.8571	0.1468
58.9	6	1	0	0.8333	0.1223
59.0	5	1	0	0.8000	0.0978
59.1	4	1	0	0.7500	0.0734
59.6	3	1	0	0.6667	0.0489
60.4	2	1	0	0.5000	0.0245
60.7	1	1	0	0	0

16.3.3 *Comparing Survival Curves*

In clinical trials it is often important to compare survival curves calculated for cohorts undergoing different treatments. Often one is interested in comparing the new treatment to the existing one or to a placebo. In comparing two survival curves, we are testing whether the corresponding hazard functions $h_1(t)$ and $h_2(t)$ coincide:

$$H_0 : h_1(t) = h_2(t) \qquad \text{versus} \qquad H_1 : h_1(t) >, \neq, < h_2(t).$$

The simplest comparison involves exponential lifetime distributions where the comparison between survival/hazard functions is simply a comparison of constant rate parameters. The statistic is calculated using logarithms of hazard rates as

$$Z = \frac{\log \lambda_1 - \log \lambda_2}{\sqrt{1/k_1 + 1/k_2}},$$

where k_1 and k_2 are numbers of observed (uncensored) survival times in the two comparison groups. Now, the inference relies on the fact that statistic Z is approximately standard normal.

Example 16.10. **Comparing the Rates.** In Example 16.7 the rate for immunoperoxidase-positive patients was larger than that of immunoperoxidase-negative patients. Was this difference significant?

```
z = (log(lambdahat1)-log(lambdahat2))/sqrt(1/k1+1/k2)   %-2.5763
p = normcdf(z)  %0.0050
```

As is evident from the code, the hazard rate λ_2 (and, in the case of exponential distribution, hazard function) for the immunoperoxidase-positive patients is significantly larger than the rate for negative patients, λ_1, with a p-value of half a percent.

🖎

Logrank Test. The logrank test compares survival functions in a nonparametric fashion. It was proposed by Mantel (1966) and applies Haenszel–Mantel theory on survival data in the form of 2×2 tables.

Let (r_{11}, d_{11}), $(r_{12}, d_{12}), \ldots, (r_{1k}, d_{1k})$ be the number of people at risk and the number of people who died at times $t_{11}, t_{12}, \ldots, t_{1k}$ in the first cohort, and (r_{21}, d_{21}), $(r_{22}, d_{22}), \ldots, (r_{2m}, d_{2m})$ be the number of people at risk and the number of people who died at times $t_{21}, t_{22}, \ldots, t_{1m}$ in the second cohort. We merge the two data sets together with the corresponding times. Thus, there will be $D = k + m$ time points if there are no ties, and each time point corresponds to a death from either the first or second cohort. For example, if times of events in the first sample are 1, 4, and 10 and in the second 2, 3, 7, and 8, then in the merged data sets the times will be 1, 2, 3, 4, 7, 8, and 10.

For a time t_i from the merged data set, let r_{1i} and r_{2i} correspond to the number of subjects at risk in cohorts 1 and 2, respectively, and let $r_i = r_{1i} + r_{2i}$ be the number of subjects at risk in the combined sample. Analogously, let d_{1i}, d_{2i}, and $d_i = d_{1i} + d_{2i}$ be the number of events at time t_i.

Then, if $H_0 : h_1(t) = h_2(t)$ is true, d_{1i} has a hypergeometric distribution with parameters (r_i, d_i, r_{1i}).

	Event	No event	At risk
Treatment 1	d_{1i}	$r_{1i} - d_{1i}$	r_{1i}
Treatment 2	d_{2i}	$r_{2i} - d_{2i}$	r_{2i}
Merged	d_i	$r_i - d_i$	r_i

Since $d_{1i} \sim \mathcal{HG}(r_i, d_i, r_{1i})$, the expectation and variance of d_{1i} are

$$
\mathbb{E}d_{1i} = r_{1i} \times \frac{d_i}{r_i},
$$

$$
\mathbb{Var}\,(d_{1i}) = \frac{r_{i1}}{r_i}\left(1 - \frac{r_{i1}}{r_i}\right)\left(\frac{r_i - d_i}{r_i - 1}\right)d_i.
$$

Note that in the terminology of the Kaplan–Meier estimator, the number of subjects with no event at time t_i is equal to $r_i - d_i = r_{i+1} + \ell_i$, where ℓ_i is the number of subjects censored in the time interval (t_{i-1}, t_i) and r_{i+1} is the number of subjects at risk at the beginning of the subsequent interval (t_i, t_{i+1}).

The test statistic for testing $H_0 : h_1(t) = h_2(t)$ against the two-sided alternative $H_1 : h_1(t) \neq h_2(t)$ is

$$
\chi^2 = \frac{\left(\sum_{i=1}^{D}(d_{1i} - \mathbb{E}(d_{1i}))\right)^2}{\sum_{i=1}^{D} \mathbb{Var}\,(d_{1i})}, \tag{16.4}
$$

which has a χ^2-distribution with 1 degree of freedom. The continuity correction 0.5 can be added to the numerator of the χ^2-statistic as $(|\sum_{i=1}^{D}(d_{1i} - \mathbb{E}(d_{1i}))| - 0.5)^2$ when the sample size is small. If the statistic is calculated as chi2, then its large values are critical and the p-value of the test is equal to 1-chi2cdf(chi2,1).

If the alternative is one-sided, $H_1 : h_1(t) < h_2(t)$ or $H_1 : h_1(t) > h_2(t)$, then the preferable statistic is

$$Z = \frac{\sum_{i=1}^{D} (d_{1i} - \mathbb{E}(d_{1i}))}{\sqrt{\sum_{i=1}^{D} \mathrm{Var}\,(d_{1i})}}$$

and the p-values are `normcdf(Z)` and `1-normcdf(Z)`, respectively. A more general statistic is of the form

$$Z = \frac{\sum_{i=1}^{D} W(t_i)\,(d_{1i} - \mathbb{E}(d_{1i}))}{\sqrt{\sum_{i=1}^{D} W^2(t_i)\mathrm{Var}\,(d_{1i})}},$$

where $W(t_i) = 1$ (as above), $W(t_i) = r_i$ (Gehan's statistic), and $W(t_i) = \sqrt{r_i}$ (Tarone–Ware statistic).

Example 16.11. **Mantel's Logrank Step-by-Step.** To illustrate the logrank test, we consider a simple example. Consider two trials A and B with outcomes 4.5, 7, 7, 8.9+, 9.1, and 6.1+, 7, 9, 10+.

At combined event times t_i the tables

	Event	No event	At risk
A	d_{1i}	$r_{1i} - d_{1i}$	r_{1i}
B	d_{2i}	$r_{2i} - d_{2i}$	r_{2i}
	d_i	$r_i - d_i$	r_i

need to be formed. Organize the data as follows:

A	4.5		7	7	8.9+	9.1
B		6.1+		7	9	10+

In the merged sample, there are four times with events: 4.5, 7, 9, and 9.1, leading to four tables. At time 4.5, five in group A are at risk, one with the event. In group B, four are at risk, none with the event. For time 7, four are at risk and two with the event in A, and three are at risk and one with the event in B. Likewise for times 9 and 9.1. The four tables are

$$\begin{array}{c|cc} A & 1 & 4 & 5 \\ B & 0 & 4 & 4 \\ \hline & 1 & 8 & 9 \end{array} \quad \begin{array}{c|cc} A & 2 & 2 & 4 \\ B & 1 & 2 & 3 \\ \hline & 3 & 4 & 7 \end{array} \quad \begin{array}{c|cc} A & 0 & 1 & 1 \\ B & 1 & 1 & 2 \\ \hline & 1 & 2 & 3 \end{array} \quad \text{and} \quad \begin{array}{c|cc} A & 1 & 0 & 1 \\ B & 0 & 1 & 1 \\ \hline & 1 & 1 & 2 \end{array}.$$

Since $d_{11} \sim \mathcal{HG}(9,1,5)$, the expectation and variance of d_{11} are respectively $\mathbb{E}d_{11} = 5 \times \frac{1}{9} = \frac{5}{9}$, and $\mathrm{Var}\,(d_{11}) = \frac{5}{9}\left(1 - \frac{5}{9}\right)\left(\frac{9-1}{9-1}\right) \times 1 = \frac{20}{81}$. Thus, observed events for group A, their expectations, and variances are:

d_{1i}	1	2	0	1
$\mathbb{E}d_{1i}$	5/9	12/7	1/3	1/2
$\mathrm{Var}\,(d_{1i})$	20/81	24/49	2/9	1/4

Equation in (16.4) leads to $\chi^2 = 0.6653$. Under the null hypothesis this statistic has a χ^2-distribution with one degree of freedom, so the p-value is 0.4147 (`1-chi2cdf(0.6653,1)`).

Notice that ◀ `manteltwo.m` produces the same result:

```
tA = [4.5  7  7    8.9  9.1];      cA = [0 0 0 1 0];
tB = [6.1  7  9  10];              cB = [1 0 0 1];
[p ch2]= manteltwo(tA, cA, tB, cB)
% p  =    0.4147
% ch2 =  0.6653
```

Example 16.12. **Histiocytic Lymphoma.**   The data (from McKelvey et al., 1976; Armitage and Berry, 1994) given below are survival times (in days) since entry to a trial by patients with diffuse histiocytic lymphoma. Two cohorts of patients are considered: (1) with stage III and (2) with stage IV of the disease. The observations with + are censored.

Stage	6	19	32	42	42	43+	94	126+	169+	207	211+
III	227+	253	255+	270+	310+	316+	335+	346+			
Stage	4	6	10	11	11	11	13	17	20	20	21
IV	22	24	24	29	30	30	31	33	34	35	39
	40	41+	43+	45	46	50	56	61+	61+	63	68
	82	85	88	89	90	93	104	110	134	137	160+
	169	171	173	175	184	201	222	235+	247+	260+	284+
	290+	291+	302+	304+	341+	345+					

Using a logrank test, we will assess the equality of the two survival curves. The function manteltwo.m calculates χ^2 statistic and the corresponding *p*-value:

```
load 'YourDataPath\limphoma.mat'
tA = limphoma.survive(limphoma.stage == 1);
         cA = limphoma.censored(limphoma.stage == 1);
tB = limphoma.survive(limphoma.stage == 2);
         cB = limphoma.censored(limphoma.stage == 2);
 [p chi2] = manteltwo(tA, cA, tB, cB)
 % p = 0.0096
 % chi2 = 6.7097
```

The hypothesis of equality of survival curves in this case is rejected, *p*-value is 0.0096. Kaplan–Meier estimators of the two survival functions are shown in Figure 16.4. MATLAB Central contains several functions conducting logrank test. Function logrank.m (Cardillo, 2008) is an example.

16.4 The Cox Proportional Hazards Model

We often need to take into account that survival is influenced by one or more covariates, which may be categorical (e.g., the kind of treatment a

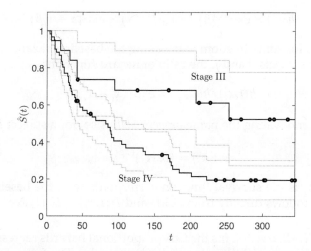

Fig. 16.4 Kaplan–Meier estimators of the two survival functions. The groups correspond to stage III and stage IV cohorts of patients.

patient received) or continuous (e.g., the patient's age, weight, or drug dosage). For simple situations involving a single factor with just two values (e.g., drug versus placebo), we discussed a method for comparing the survival curves for the two groups of subjects. For more complicated situations, a regression-type model that incorporates the effect of each predictor on the shape of the survival curve is needed.

Assume that the log hazard for subject i can be modeled via a linear relationship:

$$\log h(t, x_i) = \beta_0 + \beta_1 x_{1,i} + \cdots + \beta_p x_{p,i},$$

where $x_i = x_{1,i}, \ldots, x_{p,i}$ is p-dimensional vector of covariates associated with subject i. The Cox model assumes that β_0 is a log baseline hazard, $\log h_0(t) = \beta_0$, namely the log hazard for a "person" for whom all covariates are 0 (Cox, 1972; Cox and Oakes, 1984). Alternatively, we can set the baseline hazard to correspond to a typical person for whom all covariates are averages of covariates from all subjects in the study. For the Cox model,

$$\log h(t, x_i) = \log h_0(t) + \beta_1 x_{1,i} + \cdots + \beta_p x_{p,i},$$

or, equivalently,

$$h(t, x_i) = h_0(t) \times \exp\{\beta_1 x_{1,i} + \cdots + \beta_p x_{p,i}\} = h_0(t) \times \exp\{x_i'\beta\}.$$

Inclusion of an intercept would lead to nonidentifiability because

$$h_0(t) \times \exp\{x_i'\beta\} = (h_0(t)e^{-\alpha}) \times \exp\{\alpha + x_i'\beta\}.$$

This allows for some freedom in choosing the baseline hazard.

For two subjects, i and j, the ratio of hazard functions

$$h(t,x_i)/h(t,x_j) = \exp\{(x_i' - x_j')\beta\}$$

is free of t, motivating the name *proportional*. Also, for a subject i,

$$S(t,x_i) = (S_0(t))^{\exp\{x_i'\beta\}},$$

where $S_0(t)$ is the survival function corresponding to the baseline hazard $h_0(t)$. This follows directly from (16.1) and $H(t,x_i) = H_0(t)\exp\{x_i'\beta\}$.

In MATLAB, `coxphfit` fits the Cox proportional hazards regression model, which relates survival times to predictor variables. The following example uses `coxphfit` to fit Cox's proportional hazards model.

Example 16.13. **Mayo Clinic Trial in PBC.** Primary biliary cirrhosis (PBC) is a rare but fatal chronic liver disease of unknown cause, with a prevalence of about $1/20,000$. The primary pathologic event appears to be the destruction of interlobular bile ducts, which may be mediated by immunologic mechanisms.

The PBC data set available at StatLib is an excerpt from the Mayo Clinic trial in PBC of the liver conducted between 1974 and 1984. From a total of 424 patients that met eligibility criteria, 312 PBC patients participated in the double-blind, randomized, placebo-controlled trial of the drug D-penicillamine. Details of the trial can be found in Markus et al. (1989).

Survival statuses were recorded for as many patients as possible until July 1986. By that date, 125 of the 312 patients had died and 187 were censored.

The variables contained in the data set 📊 pbc.xls|dat are described in the following table:

```
casen = pbc(:,1);         %case number 1-312
lived = pbc(:,2);         %days lived (from registration to study date)
indicatord = pbc(:,3);    %0 censored, 1 death
treatment = pbc(:,4);     %1 - D-Penicillamine, 2 - Placebo
age = pbc(:,5);           %age in years
gender = pbc(:,6);        %0 male, 1 female
ascites= pbc(:,7);        %0 no, 1 yes
hepatomegaly=pbc(:,8);    %0 no, 1 yes
spiders = pbc(:,9);       %0 no, 1 yes
edema = pbc(:,10);        %0 no, 0.5 yes/no therapy, 1 yes/therapy
bilirubin = pbc(:,11);    %bilirubin [mg/dl]
cholesterol = pbc(:,12);  %cholesterol [mg/dl]
albumin = pbc(:,13);      %albumin [gm/dl]
ucopper =pbc(:,14);       %urine copper [mg/day]
aphosp =pbc(:,15);        %alcaline phosphatase [U/liter]
```

```
sgot = pbc(:,16);         %SGOT [U/ml]
trig =pbc(:,17);          %triglycerides [mg/dl]
platelet = pbc(:,18);     %# platelet count [#/mm^3]/1000
prothro = pbc(:,19);      %prothrombin time [sec]
histage = pbc(:,20);      %hystologic stage [1,2,3,4]
```

To illustrate the CPH model, in this example we selected four predictors and formed a design matrix X as

```
X = [treatment age gender edema];
```

The treatment has two values, 1 for treatment by D-penicillamine and 2 for placebo. The variable edema takes three values: 0 if no edema is present, 0.5 when edema is present but no diuretic therapy was given or edema resolved with diuretic therapy, and 1 if edema is present despite adminis-tration of diuretic therapy.

The variable lived is the lifetime observed or censored, a censoring vec-tor is 1-indicatord, and a baseline hazard is taken to be a hazard for which all covariates are set to 0.

```
[b,logL,H,stats] = coxphfit(X,lived,...
  'censoring',1-indicatord,'baseline',0);
```

The output H is a two-column matrix as a discretized cumulative hazard estimate. The first column of H contains values from the vector lived, while the second column contains the estimated baseline cumulative hazard eval-uated at lived.

To illustrate the model, we selected two subjects from the study to find survival curves corresponding to their covariates. Subject #100 is a 51-year-old male with no edema who received placebo while subject #275 is a 38-year-old female with no edema who received D-penicillamine treatment.

```
X(100,:) %2.0000     51.4689   0         0
%          placebo;  51 y.o.;  male;     no edema;
X(275,:) %1.0000     38.3162   1.0000    0
% D-Penicillamine;   38 y.o.;  female; no edema;
```

First, we find cumulative hazards at the mean values of predictors, as well as for subjects #100 and #275, as

$$H(t,\bar{x}) = H_0(t) \times \exp\{\beta_1 \bar{x}_1 + \cdots + \beta_4 \bar{x}_4\},$$
$$H(t,x_i) = H_0(t) \times \exp\{\beta_1 x_{1,i} + \cdots + \beta_4 x_{4,i}\}, \quad i = 100,\ 275.$$

```
Hmean(:,2)     = H(:,2) .* exp(mean(X)*b);  %c.haz. average
Hsubj100(:,2) = H(:,2) .* exp(X(100,:)*b); %subject #100
Hsubj275(:,2) = H(:,2) .* exp(X(275,:)*b); %subject #275
```

Here the estimators of coefficients $\beta_1, \ldots, \beta_4$ are

```
b'
%     0.0831     0.0324   -0.3940     2.2424
```

Note that the `treatment` coefficient $0.0831 > 0$ indicates that, given all other covariates fixed, the placebo increases the risk over the treatment. Also note that `age` and `edema` statuses also increase the risk, while the risk for female subjects is smaller.

Next, from cumulative hazards we find survival functions

```
Smean      = exp(-Hmean(:,2));
Ssubj100   = exp(-Hsubj100(:,2));
Ssubj275   = exp(-Hsubj275(:,2));
```

The subsequent commands plot the survival curves for an "average" subject (blue), as well as for subjects #100 (black) and #275 (red); see Figure 16.5.

```
stairs(H(:,1),Smean,'b-','linewidth',2)
hold on
stairs(H(:,1),Ssubj100,'k-')
stairs(H(:,1),Ssubj275,'r-')
xlabel('$t$ (days)','Interpreter','LaTeX')
ylabel('$\hat S(t)$','Interpreter','LaTeX')
legend('average subject','subject #100', 'subject #275', 3)
axis tight
```

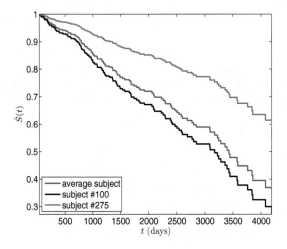

Fig. 16.5 Cox model for survival curves for a "subject" with average covariates, subject #100 (51-year-old male on placebo, no edema), and subject #275 (38-year-old female on treatment, no edema).

16.5 Bayesian Approach

We will focus on parametric models in which the lifetime distributions are specified up to unknown parameters. The unknown parameters will be assigned prior distributions and the inference will proceed in a Bayesian fashion. Nonparametric Bayesian modeling of survival data is possible; however, the methodology is advanced and beyond the scope of this text. For a comprehensive coverage, see Ibrahim et al. (2001).

Let survival time T have distribution $f(t|\theta)$, where θ is unknown parameter. For $t_1, \ldots, t_k$ observed and $t_{k+1}, \ldots, t_n$ censored times, the likelihood is

$$L(\theta|t_1, \ldots, t_n) = \prod_{i=1}^{k} f(t_i|\theta) \times \prod_{i=k+1}^{n} S(t_i|\theta).$$

If the prior on θ is $\pi(\theta)$, then the posterior is

$$\pi(\theta|t_1, \ldots, t_n) \propto L(\theta|t_1, \ldots, t_n) \times \pi(\theta).$$

The Bayesian estimator of hazard is

$$\hat{h}_B(t) = \int h(t|\theta)\pi(\theta|t_1, \ldots, t_n)d\theta$$

and the survival function is

$$\hat{S}_B(t) = \int S(t|\theta)\pi(\theta|t_1, \ldots, t_n)d\theta.$$

Example 16.14. **Exponential Lifetimes with Gamma Prior.** In Example 16.6 we found that for the exponential lifetime in the presence of censoring, the likelihood was $L(\lambda) = \lambda^k \exp\{-\lambda \sum_{i=1}^{n} t_i\}$, where k is the number of uncensored data and $\sum_{i=1}^{n} t_i$ is the sum of *all* observed and censored times. The resulting MLE for λ was

$$\hat{\lambda} = k / \sum_{i=1}^{n} t_i.$$

If a gamma $\mathcal{G}a(\alpha, \beta)$ prior on λ is adopted, $\pi(\lambda) \propto \lambda^{\alpha-1}\exp\{-\beta\lambda\}$, then the conjugacy of the exponential/gamma pair (page 341) leads to the posterior

$$\pi(\lambda|t_1, \ldots, t_n) \propto \lambda^{k+\alpha-1}\exp\{-(\beta + \sum_{i=1}^{n} t_i)\,\lambda\},$$

from which the Bayes estimator of λ is the posterior mean,

$$\hat{\lambda}_B = \frac{k+\alpha}{\beta + \sum_{i=1}^{n} t_i}.$$

The posterior predictive distribution of future failure time t_{n+1} is

$$f(t_{n+1}|t_1,\ldots,t_n) = \int_0^\infty \lambda e^{-\lambda t_{n+1}} \times \pi(\lambda|t_1,\ldots,t_n)d\lambda$$

$$= \frac{(k+\alpha)(\beta + \sum_{i=1}^{n} t_i)^{\alpha+k}}{(\beta + \sum_{i=1}^{n} t_i + t_{n+1})^{\alpha+k+1}}, \quad t_{n+1} > 0.$$

This distribution is known as an inverse beta distribution.

The Bayes estimator of hazard function coincides with the Bayes estimator of λ,

$$\hat{h}_B(t) = \frac{k+\alpha}{\beta + \sum_{i=1}^{n} t_i},$$

while the Bayes estimator of the survival function is

$$\hat{S}_B(t) = \left(1 + \frac{t}{\beta + \sum_{i=1}^{n} t_i}\right)^{-(k+\alpha)}.$$

The expression for $\hat{S}_B(t)$ can be derived from the moment-generating function of a gamma distribution.

When the posterior distribution is intractable, one can use WinBUGS.

16.6 Survival Analysis in WinBUGS

WinBUGS uses two arrays to define censored observations: observed (uncensored) times and censored times. For example, an input such as

```
list(times = c(0.5, NA, 1, 2, 6, NA, NA),
t.censored = c(0,  0.9, 0, 0, 0,  9, 12))
```

corresponds to times $\{0.5, 0.9+, 1, 2, 6, 9+, 12+\}$.

In WinBUGS, direct time-to-event modeling is possible with exponential, Weibull, gamma, and log-normal densities.

There is a multiplier I that is used to implement censoring. For example, if Weibull dweib(r, mu) observations are on the input, the multiplier is an indicator that the observation exceeded time t.censored[i]:

```
t[i] ~ dweib(r, mu) I(t.censored[i],)
```

For uncensored data one sets `t.censored[i] = 0`. The above describes right-censoring. Left-censored observations are modeled using the multiplier `I(,t.censored[i])`.

Example 16.15. **Bayesian Immunoperoxidase and BC.** In Example 16.7 MLEs and confidence intervals on λ_1 and λ_2 were found. In this example we find Bayes estimators and credible sets.

Before discussing the results, note that when observations are censored, their values are unknown parameters in the Bayesian model and predictions can be found. Because they are treated as parameters, all censored observations need to be initialized, and the initial values should exceed the censoring times. Here is the WinBUGS program that estimates λ_1 and λ_2.

```
model{
for(i in 1:n1)  {
   ImmPeroxNeg[i] ~ dexp(lam1) I(CensorIPN[i], )
   }
for(i in 1:n2)  {
   ImmPeroxPos[i] ~ dexp(lam2) I(CensorIPP[i], )
   }
lam1 ~ dgamma(0.001, 0.001)
lam2 ~ dgamma(0.001, 0.001)
}

DATA

list( n1 = 36, n2 = 9,
ImmPeroxNeg=c(19, 25, 30, 34, 37, 46, 47, 51,
56, 57, 61, 66, 67, 74, 78, 86,
NA, NA, NA, NA, NA,  NA, NA, NA, NA, NA,
NA, NA, NA, NA, NA,  NA, NA, NA, NA, NA),
CensorIPN = c(0,0,0,0,0,0,0,0,0,0,0,0,0,0,0,0,
122, 123, 130, 130, 133, 134, 136, 141, 143, 148,
151, 152, 153, 154, 156, 162, 164, 165, 182, 189),
ImmPeroxPos = c(22, 23, 38, 42, 73, 77, 89, 115, NA),
CensorIPP= c(0,0,0,0,0,0,0,0,144))

INITS

list(lam1=1, lam2 = 1,
ImmPeroxNeg=c(NA, NA, NA, NA, NA, NA, NA, NA,
                         NA, NA, NA, NA, NA, NA, NA, NA,
 200, 200, 200, 200, 200, 200, 200, 200, 200, 200,
 200, 200, 200, 200, 200, 200, 200, 200, 200, 200),
ImmPeroxPos = c(NA, NA, NA, NA, NA, NA, NA, NA, 200)   )
```

	mean	sd	MCrror	val2.5pc	median	val97.5pc	start	sample
ImmPeroxNeg[17]	374.8	270.8	0.4261	128.0	289.7	1107.0	1001	500000
ImmPeroxNeg[18]	376.2	270.6	0.4568	129.0	291.2	1107.0	1001	500000
ImmPeroxNeg[19]	383.2	270.6	0.4342	136.0	298.0	1115.0	1001	500000
	...							
ImmPeroxNeg[35]	435.1	270.2	0.4319	188.1	349.9	1166.0	1001	500000
ImmPeroxNeg[36]	442.3	270.6	0.4272	195.0	357.1	1173.0	1001	500000
ImmPeroxPos[9]	233.1	102.8	0.1730	146.0	200.5	508.2	1001	500000
lam1	0.0042	0.0010	2.849E-6	0.0024	0.0041	0.0065	1001	500000
lam2	0.0128	0.0045	7.496E-6	0.0055	0.0123	0.0231	1001	500000

The Bayes estimator of λ_1 is $\hat{\lambda}_{1,B} = 0.004211$, and the 95% credible set is $[0.002404, 0.006505]$. The Bayes estimator and interval are close to their classical counterparts (Exercise 16.2).

Example 16.16. **Smoking Cessation Experiment.** The data set for this example comes from a clinical trial discussed in Banerjee and Carlin (2004). A number of smokers entered into a smoking cessation study, and 263 of them quit. These 263 quitters were monitored and checked to see if and when they relapsed. RelapseT is the time to relapse; it is either observed or censored, with the censoring indicator contained in the vector censored.time. The independent covariates are Age (age of the individual), AgeStart (age when he/she started smoking), SexF (Female=1, Male=0), SIUC (whether the individual received an intervention or not), and F10Cigs (the average number of cigarettes smoked per day).

A logistic distribution is constrained to a nonnegative domain to model RelapseT. The parameters of dlogis(mu,tau) are the mean mu, which depends on the linear combination of covariates, and tau, which is a rate parameter. The standard deviation is $\pi/(\sqrt{3}\tau) \approx 1.8138/\tau$.

```
model {
for (i in 1:N)
     {
        RelapseT[i] ~ dlogis(mu[i],tau) I(censored.time[i],)
        mu[i]   <-  beta[1] + beta[2] * Age[i] + beta[3] * AgeStart[i] +
        beta[4] * SexF[i] + beta[5] * SIUC[i] + beta[6] * F10Cigs[i]
        }
for( j in 1:6){
beta[j] ~ dnorm(0, 0.01)
                    }
tau ~ dgamma(1,0.01)
meanT <- mean(mu[])
sigma <- 1.8138/tau     #1.8138 ~ pi/sqrt(3)


# Evaluate Survival Curve for a Subject with covariates:
Ag <- 50; AgSt <- 18; SxF <- 0; S <- 1; Cigs <- 20;
fmu    <-  beta[1] + beta[2] * Ag  + beta[3] * AgSt +
               beta[4] * SxF  + beta[5] * S   + beta[6] *  Cigs
```

```
for(i in 1:100) {
    time[i]  <-  i/10
    Surv[i] <- 1/(1 + exp(tau*(time[i] - fmu)))
    }
}
# Data and Inits omitted (see Smoking.odc)
```

	mean	sd	MCrror	val2.5pc	median	val97.5pc	start	sample
beta[1]	2.686	2.289	0.1254	-2.026	2.701	7.284	1001	100000
beta[2]	0.07817	0.03844	0.002078	0.009544	0.07815	0.1509	1001	100000
beta[3]	-0.07785	0.07222	0.003745	-0.2181	-0.07965	0.06802	1001	100000
beta[4]	-0.9555	0.5382	0.009515	-2.02	-0.9494	0.08981	1001	100000
beta[5]	1.666	0.5993	0.01482	0.4886	1.672	2.817	1001	100000
beta[6]	-0.02347	0.02459	9.21E-4	-0.07072	-0.02394	0.0281	1001	100000
meanT	5.314	0.3382	0.00604	4.712	5.292	6.034	1001	100000
sigma	3.445	0.345	0.005367	2.841	3.421	4.187	1001	100000
tau	0.5317	0.05256	8.15E-4	0.4332	0.5302	0.6384	1001	100000
Surv[1]	0.9628	0.01129	5.199E-4	0.9366	0.9642	0.9808	1001	100000
Surv[2]	0.9609	0.01173	5.459E-4	0.9336	0.9623	0.9797	1001	100000
	...							
Surv[99]	0.1402	0.05249	0.003402	0.06123	0.1334	0.2645	1001	100000
Surv[100]	0.1342	0.05118	0.003311	0.05765	0.1274	0.2557	1001	100000

The Surv values (ordinate, mean, standard deviation, median, and quantiles) are exported to MATLAB as data file ▦ smokingoutbugs.mat. The file smokingbugs.m reads in the data and plots the posterior estimator of the survival curve (Figure 16.6). Note that the survival curve is $S(t|\mu,\tau) = 1/(1 + \exp\{\tau(t - \mu)\})$. The posterior distribution of $S(t|\mu,\tau)$ is understood as a distribution of a function of μ and τ for t fixed.

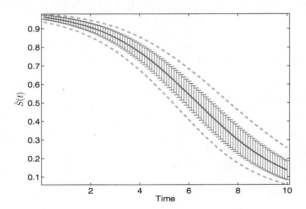

Fig. 16.6 Bayesian estimator of survival curve. The *green* bands are the 0.025 and 0.975 percentiles of the posterior distribution for $S(t)$, while the *blue* errorbars have a size of posterior standard deviations.

Example 16.17. **Duration of Remissions in Acute Leukemia.** A data set analyzed by Freireich et al. (1963), and subsequently by many authors, comes from a trial of 42 leukemia patients (under age of 20) treated in 11 US hospitals.

The effect of 6-mercaptopurine (6-MP) therapy on the duration of remissions induced by adrenal corticosteroids has been studied as a model for testing new agents. Some patients were treated with 6-MP and the rest were controls. The trial was designed as matched pairs. The matching was done with respect to remission status (partial = 1, complete = 2). Randomization to 6-MP or control arms was done within a pair. Patients were followed until leukemia relapsed or until the end of the study. Overall survival was not significantly different for the two treatment programs since patients maintained on placebo were treated with 6-MP when relapse occurred.

#	Status	Contr	6-MP	#	Status	Contr	6-MP	#	Status	Contr	6-MP
1	1	1	10	8	2	11	34+	15	2	8	17+
2	2	22	7	9	2	8	32+	16	1	23	35+
3	2	3	32+	10	2	12	25+	17	1	5	6
4	2	12	23	11	2	2	11+	18	2	11	13
5	2	8	22	12	1	5	20+	19	2	4	9+
6	1	17	6	13	2	4	19+	20	2	1	6+
7	2	2	16	14	2	15	6	21	2	8	10+

```
#Duration of Steroid-induced Remissions in Acute Leukemia
model {
for (i in 1:n) {
            log(mu[i]) <- b0+ b1[treat[i]] + b2[status[i]]
            t[i] ~ dweib(r,mu[i])  I(t.cen[i],)
            S[i]  <- exp(-mu[i]*pow(t[i],r));
            f[i]  <- mu[i]*r*pow(t[i],r-1)*S[i]
            Lik[i] <- pow(f[i],1-delta[i])*pow(S[i],delta[i]);
            logLik[i] <- log(Lik[i])
            }

b0 ~ dnorm(0,0.00001)
b1[1] <- 0
b1[2] ~ dnorm(0,0.001)
b2[1] <- 0
b2[2] ~ dnorm(0,0.001)
r ~ dgamma(0.01,0.01)

     Dev <-    -2*sum(logLik[]) #deviance

}

DATA

list(n=42,
t = c(1, 22, 3, 12, 8, 17, 2, 11, 8, 12,
        2, 5, 4, 15, 8, 23, 5, 11, 4, 1, 8,
```

```
        10, 7, NA, 23, 22, 6, 16, NA, NA, NA,
        NA, NA, NA, 6, NA, NA, 6, 13, NA, NA, NA),
t.cen = c(0,0,0,0,0,0,0,0,0,0,0,0,0,0,0,
            0,0,0,0,0,0,0,0,0,32,0,0,0,0,
        34,32,25,11,20,19,0,17,35,0,0,9,6,10),
treat = c(1,1,1,1,1, 1,1,1,1,1, 1,1,1,1,1, 1,1,1,1,1,1,
            2,2,2,2,2, 2,2,2,2,2, 2,2,2,2,2, 2,2,2,2,2,2),
status = c(1,2,2,2,2,1,2,2,2,2,2,1,2,2,2,1,1,2,2,2,2,2,
            1,2,2,2,2,1,2,2,2,2,2,1,2,2,2,1,1,2,2,2,2,2),
delta = c(0,0,0,0,0, 0,0,0,0,0, 0,0,0,0,0, 0,0,0,0,0,0,
            0,0,1,0,0, 0,0,1,1,1, 1,1,1,0,1, 1,0,0,1,1,1) )
```

INITS

```
list(r=1,b0 = 0, b1=c(NA,0),b2=c(NA,0),
t = c(NA,NA,NA,NA,NA,NA,NA,NA,NA,NA,NA,NA,NA,NA,NA,
        NA,NA,NA,NA,NA,NA,NA,NA,32,NA,NA,NA,NA,34,32,
        25,11,20,19,NA,17,35,NA,NA,9,6,10))
```

	mean	sd	MCrror	val2.5pc	median	val97.5pc	start	sample
Dev	241.2	7.527	0.02825	228.5	240.6	257.8	1001	200000
b0	−3.209	0.6867	0.009883	−4.636	−3.181	−1.949	1001	200000
b1[2]	−1.772	0.4241	0.003255	−2.629	−1.76	−0.9696	1001	200000
b2[2]	0.1084	0.4306	0.002543	−0.6967	0.09381	0.9995	1001	200000
r	1.37	0.2069	0.003338	0.9877	1.361	1.798	1001	200000
t[24]	57.34	29.86	0.1341	32.62	48.55	132.9	1001	200000
t[29]	58.96	29.15	0.1345	34.6	50.17	134.7	1001	200000
...								
t[41]	37.21	30.96	0.1128	7.161	29.06	115.5	1001	200000
t[42]	39.68	31.09	0.1177	10.95	31.45	117.3	1001	200000

Example 16.18. **Photocarcinogenicity.** Grieve (1987) and Dellaportas and Smith (1993) explored photocarcinogenicity in four treatment groups with 20 rats in each treatment group. Treatment 3 is the test drug, and the others are some type of control. Response is time to death or censoring time.

We will find Bayes' estimators for median survival times for the four treatments. The survival times are modeled as Weibull $\mathcal{W}ei(r,\mu_i)$, $i = 1,\ldots,4$. The WinBUGS code specifies the censoring and priors on the Weibull model. The data part of the code provides observed and censored times. Posterior densities for median survival times are given in Figure 16.7.

```
model{
for(i in 1 : M) {
        for(j in 1 : N) {
t[i, j] ~ dweib(r, mu[i])I(t.cen[i, j],)
                            }
    mu[i] <- exp(beta[i])
```

```
     beta[i] ~  dnorm(0.0, 0.001)
     median[i] <-  pow(log(2) * exp(-beta[i]), 1/r)
                  }
r ~ dexp(0.001)
}
```

DATA

```
list( t = structure(.Data = c(12, 1, 21, 25, 11, 26, 27, 30, 13,
12, 21, 20, 23, 25, 23, 29, 35, NA, 31, 36,      32, 27, 23, 12,
18, NA, NA, 38, 29, 30, NA, 32, NA, NA, NA, NA, 25, 30,
37, 27,     22, 26, NA, 28, 19, 15, 12, 35, 35, 10, 22, 18, NA,
12, NA, NA, 31, 24, 37, 29,     27, 18, 22, 13, 18, 29, 28, NA,
16, 22, 26, 19, NA, NA, 17, 28, 26, 12, 17, 26), .Dim = c(4,20)),
t.cen = structure(.Data = c(0, 0, 0, 0, 0, 0, 0, 0, 0, 0, 0, 0,
0, 0, 0, 0, 0, 40, 0, 0, 0, 0, 0, 0, 40, 40, 0, 0, 0, 40, 0, 40,
40, 40, 40, 0, 0, 0, 0, 0, 0, 10, 0, 0, 0, 0, 0, 0, 0, 0, 0, 24, 0,
40, 40, 0, 0, 0, 0, 0, 0, 0, 0, 0, 0, 20, 0, 0, 0, 0, 29, 10, 0,
0, 0, 0, 0, 0), .Dim = c(4, 20)), M = 4, N = 20)
```

INITS

```
list( r=1, beta=c(0,0,0,0) )  #generate the rest
```

	mean	sd	MCrror	val2.5pc	median	val97.5pc	start	sample
median[1]	23.82	1.977	0.02301	20.20	23.74	27.96	1001	100000
median[2]	35.11	3.459	0.01650	29.23	34.79	42.82	1001	100000
median[3]	26.80	2.401	0.02169	22.46	26.66	31.94	1001	100000
median[4]	21.38	1.845	0.01446	18.09	21.26	25.34	1001	100000

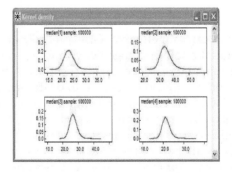

Fig. 16.7 Posterior densities of median survival times median[1]–median[4].

16.7 Exercises

16.1. **Simulation of Censoring.**

```
y = exprnd(10,50,1); % Random failure times exponential(10)
d = exprnd(20,50,1); % Drop-out times exponential(20)
t = min(y,d);        % Observe the minimum of these times
censored = (y>d);    % Observe whether the subject failed
```

Using MATLAB's `ecdf` calculate and plot empirical CDF and confidence bounds for arguments `t` and `censored`.

16.2. **Immunoperoxidase.** In the context of Example 16.7 find a confidence interval for λ_1, the rate parameter for immunoperoxidase-negative patients. Use the fact that for MLE $\hat{\lambda}_1$,

$$\frac{\hat{\lambda}_1 - \lambda_1}{\hat{\lambda}/\sqrt{k_1}} \text{ or}$$
$$\sqrt{k_1}\,(\log \hat{\lambda}_1 - \log \lambda_1),$$

both have an approximately standard normal distribution. Here, k_1 is the number of non-censored life-times.

16.3. **Expected Lifetime.** Let T be a lifetime with survival function $S(t)$. Using integration by parts in the definition of $\mathbb{E}T$, show

$$\mathbb{E}T = \int_0^\infty S(t)dt.$$

16.4. **Rayleigh Lifetimes.** A lifetime T would have Rayleigh distribution with scale parameter θ if its CDF is given by

$$F(t) = 1 - \exp\left\{-\frac{t^2}{2\theta}\right\}, \ t \geq 0.$$

(a) Find the PDF, survival function $S(t)$, hazard $h(t)$, and cumulative hazard $H(t)$.
(b) Assume that lifetimes $T_1 = 10, T_2 = 8, T_3 = 6$ and $T_4 = 10$ have been observed. Find the MLE of θ. If $\mathbb{E}T^2 = 2\theta$, show that the corresponding moment matching estimator coincides with MLE. Evaluate the MLE for the four observed values.
(c) Find the theoretical median life t_{med}. Is it close to observed median life $T_{med} = (8 + 10)/2 = 9$?

16.5. **Log-logistic Lifetimes.** A lifetime T would have log-logistic $\mathcal{LL}(\alpha, \beta)$ distribution if its PDF and CDF are given by

$$f(t|\alpha,\beta) = \frac{\frac{\alpha}{\beta}\left(\frac{t}{\beta}\right)^{\alpha-1}}{\left(1+\left(\frac{t}{\beta}\right)^{\alpha}\right)^2}, \ t \geq 0,$$

and

$$F(t|\alpha,\beta) = 1 - \frac{1}{1+\left(\frac{t}{\beta}\right)^{\alpha}} = \frac{t^{\alpha}}{\beta^{\alpha}+t^{\alpha}}, \ t \geq 0.$$

Here the parameters α (shape) and β (scale) are both positive.
If T has a log-logistic distribution with shape parameter α and scale parameter β, then $Y = \log(T)$ has a logistic distribution with location parameter $\log(\beta)$ and scale parameter $1/\alpha$.
The kth raw moment of T is given by

$$\mathbb{E}T^k = \beta^k\frac{k\pi/\alpha}{\sin(k\pi/\alpha)}, \ k < \alpha.$$

(a) Find the survival function $S(t)$, hazard $h(t)$, and cumulative hazard $H(t)$.
(b) Assume that $\alpha = 2$ and that lifetimes $T_1 = 4, T_2 = 2, T_3 = 2$, and $T_4 = 7$ from $\mathcal{LL}(2,\beta)$ have been observed. Find the moment matching estimator of β. Evaluate the estimator for the four observed values.
(c) Find t_p, the p-quantile for T. Estimate the median life $t_{1/2}$ and quartiles $t_{1/4}$ and $t_{3/4}$ for the given data.

HINT. For cumulative hazard no need to integrate. Since in (b) $\alpha = 2$ only the first moments can be matched.

16.6. **Weathered Cord Data Parametric Fit.** Using one of MATLAB's built-in functions, mle or evfit, fit the EV distribution to the data in Example 16.9. The PDF and CDF of EV distribution (Gumbel type I distribution for the minimum) are

$$f(t) = \frac{1}{b}\exp\left(\frac{t-a}{b}\right)\exp\left(-\exp\left(\frac{t-a}{b}\right)\right),$$

$$F(t) = 1 - \exp\left(-\exp\left(\frac{t-a}{b}\right)\right), \qquad t \in \mathbb{R}.$$

Plot the fitted survival function $S(t|\hat{a},\hat{b})$ over the Kaplan–Meier estimator from Example 16.9. Comment on the agreement of the parametric and nonparametric estimators of the survival function.
Hint: Estimators of a and b for EV distribution are theta(1) and theta(2) in

```
% Strength of Weathered Cord Data
T=[36.3,41.7, ... ,41.9,42.5];
censor=[zeros(1,41),ones(1,7)];
[theta,thetaCI] = mle(T, 'distribution', 'ev', 'censoring', censor)
%  or: [theta thetaCI] = evfit(T,0.05,censor)
```

16.7. **Cumulative Hazards and Maxwell.** The components E_1, E_2, and E_3 are connected as in the graph below.

The components E_1 and E_2 have cumulative hazards

$$H(t) = \begin{cases} \lambda t, & 0 \le t < 1 \\ \lambda t^2, & t \ge 1 \end{cases}$$

with $\lambda_1 = 2$ and $\lambda_2 = 0.5$, respectively. The component E_3 has a lifetime T_3 with the Maxwell distribution for which the PDF and CDF are

$$f(t) = \sqrt{\frac{2}{\pi}} \frac{t^2}{\sigma^2} \exp\left\{-\frac{t^2}{2\sigma^2}\right\}, \text{ and}$$

$$F(x) = 2\Phi\left(\frac{t}{\sigma}\right) - 1 - \sqrt{\frac{2}{\pi}} \frac{t}{\sigma} \exp\left\{-\frac{t^2}{2\sigma^2}\right\}.$$

The expectation of T_3 is $\mathbb{E}T_3 = 2\sigma\sqrt{2/\pi}$. Unlike the λ's, the parameter σ is not known in advance. From the previous experiments with components identical to E_3, the following survival times are available: 1, 3, 4, 1, and 1.
Using the above information, find the probability of the system being operational up to time (a) $t = 0.5$, (b) $t = 2$.

16.8. **A System with Log-Burr Lifetimes.** In the network from Exercise 16.7 all three components are identical and have a lifetime T given by the log-Burr CDF:

$$F(t) = \frac{\left(1 - e^{\lambda t}\right)^2}{1 + \left(1 - e^{\lambda t}\right)^2} = 1 - \frac{1}{1 + \left(1 - e^{\lambda t}\right)^2}, \quad t \ge 0, \lambda > 0.$$

The parameter λ is not known; however, it could be estimated. In previous experiments involving $n = 6$ of such components the following lifetimes were obtained: $T_1 = 3, T_2 = 5, T_3 = 3, T_4 = 6, T_5 = 2$, and $T_6 = 5$.
(a) If it is known that for this distribution

$$\mathbb{E}T = \frac{\pi}{4\lambda},$$

estimate λ by the moment-matching approach.
(b) What is the probability that the network is operational at time $t = 4$?

16.9. **Censored Rayleigh.** The lifetime (in hours) of a certain sensor has Rayleigh distribution, with survival function

$$S(t) = \exp\left\{ -\frac{1}{2}\lambda t^2 \right\}, \; \lambda > 0.$$

Twelve sensors are placed under test for 100 hours, and the following failure times are recorded 23, 40, 41, 67, 69, 72, 84, 84, 88, 100+, 100+. Here + denotes a censored time.
(a) If failure times $t_1, \ldots, t_r$ are observed, and $t_{r+1}^+, \ldots, t_n^+$ are censored, show that the MLE of λ is

$$\hat{\lambda} = \frac{2r}{\sum_{i=1}^{r} t_i + \sum_{i=r+1}^{n} t_i^{+2}}.$$

Evaluate the MLE for the given data. Consult Example 16.6.
(b) Calculate and plot the Kaplan–Meier estimator and superimpose $S(t)$ evaluated at $\hat{\lambda}$.

16.10. **MLE for Equally Censored Data.** A cohort of n subjects is monitored in the time interval $[0, T]$, where T is fixed in advance. Suppose that r failures are observed (r can be any number from 0 to n) at times $t_1, t_2, \ldots, t_r \leq T$. There are $(n - r)$ subjects that survived the entire period $[0, T]$, and their failure times are not observed.
Suppose that $f(t)$ is the density of a lifetime. The likelihood is

$$L = C \prod_{i=1}^{r} f(t_i) \, (1 - F(T))^{n-r}$$

for some normalizing constant C.
(a) Express the likelihood L for the exponential lifetime distribution, that is, $f(t) = \lambda e^{-\lambda t}, \; t \geq 0$, and $F(t) = 1 - e^{-\lambda t}, \; t \geq 0$.
(b) Take the log of the likelihood, $\ell = \log(L)$.
(c) Find the derivative of ℓ with respect to λ. Set the derivative to 0 and solve for λ. If the solution $\hat{\lambda}$ maximizes ℓ, ($\ell''(\hat{\lambda}) < 0$), then $\hat{\lambda}$ is the MLE of λ.

(d) Show that the MLE is

$$\hat{\lambda}_{mle} = \frac{r}{\sum_{i=1}^{r} t_i + (n-r)T}.$$

(e) If in the interval $[0,8]$ four subjects failed at times $t_1 = 2$, $t_2 = 5/2$, $t_3 = 4$, and $t_4 = 5$, and two subjects survived without failure, find the estimator λ, assuming that the lifetime has an exponential distribution. Check your calculations with

```
data = [2  2.5  4  5  8  8];
cens = [0  0    0  0  1  1];
lamrec = mle(data, 'distribution','exponential','censoring',cens)
lammle = 1/lamrec   %MATLAB uses reciprocal parametrization
```

(f) What is the MLE of λ if the unobserved failure times are ignored, that is, only four observed failure times, t_1, t_2, t_3, and t_4, are used?

16.11. **Cancer of Tongue.** Sickle-Santanello et al. (1988) analyze data on 80 males diagnosed with cancer of the tongue. Data are provided in the file 📄 tongue.dat. The columns in the dataset are as follows:
(i) Tumor DNA profile (1=aneuploid tumor, 2 = diploid tumor);
(ii) Time to death or on-study time (in weeks); and
(iii) Censoring indicator (1 = censored, 0 = observed)
(a) Calculate and plot Kaplan–Meier estimators for the two types of tumor.
(b) Using Mantel's logrank procedure ◀ manteltwo.m, test the hypothesis that the two survival functions do not differ significantly.
(c) Fit the Cox proportional hazard regression with tumor profile as the covariate. What is the 95% CI for the slope β? Compare the results with the conclusions from (b).

16.12. **Rayleigh and Bayes.** It was observed that in clinical studies dealing with cancer survival times follow Rayleigh distribution with pdf

$$f(x) = 2\lambda t e^{-\lambda t^2}, \quad t \geq 0, \, \lambda > 0.$$

(a) Show that the hazard function is linear.
(b) Find the mean survival time as a function of λ.
(c) For $t_1, \ldots, t_k$ observed and $t_{k+1}, \ldots, t_n$ censored times, show that the likelihood is proportional to

$$L \propto \lambda^k \exp\{-\lambda \sum_{i=1}^{n} t_i^2\}.$$

If the prior on λ is gamma $\mathcal{G}(\alpha, \beta)$, show that the posterior is gamma $\mathcal{G}(\alpha + k, \beta + \sum_{i=1}^{n} t_i^2)$.

(d) Show that the Bayes estimators of the hazard and survival functions are

$$\hat{h}_B(t) = \frac{2(k+\alpha)t}{\beta + \sum_{i=1}^{n} t_i^2} \quad \text{and} \quad \hat{S}_B(t) = \left(1 + \frac{t^2}{\beta + \sum_{i=1}^{n} t_i^2}\right)^{-(k+\alpha)}.$$

16.13. **Experiment at Strangeways Lab.** Pike (1966) provides data from an experiment conducted by Glücksmann and Cherry at Strangeways Laboratory, Cambridge, UK, in which two groups of female rats were exposed to carcinogen DMBA, and the number of days to death from vaginal cancer was recorded. The two groups were distinguished by pretreatment regime.

Group 1	143	164	188	188	190	192	206	209	213	216	216*
	220	227	230	234	244*	246	265	304			
Group 2	142	156	163	198	204*	205	232	232	233	233	233
	233	239	240	261	280	280	296	296	323	344*	

The * on two entries in each group indicate that the observation was censored.

(a) Find and plot Kaplan–Meier estimators of survival functions for the two groups.

(b) Using the logrank procedure, test the hypothesis that the two survival functions are the same at significance level $\alpha = 0.05$.

16.14. **Exercise Stress Test.** Campbell and Swinscow (2009) describe an experiment in which 20 patients, 10 of normal weight and 10 severely overweight, underwent an exercise stress test. The patients had to lift a progressively increasing load for up to 12 minutes, but they were allowed to stop earlier if they could do no more. On two occasions the equipment failed before 12 minutes. The times (in minutes) achieved were:

Normal weigh	2, 4, 6, 8, 8**, 9, 10, 12*, 12*, 12*
Overweight	1, 3, 4, 5, 6, 7, 7**, 9, 11, 12*

Here * means that the end of test was reached, and ** stands for equipment failure.

Use the logrank test to compare these two groups. Report the p-value.

16.15. **Western White Clematis.** Muenchow (1986) tested whether male or female flowers (of *Clematis ligusticifolia*) were equally attractive to insects. The data in table represent waiting times (in minutes), which includes censored data (observations with +).

Male flowers					Female flowers				
1	6	11	19	54	1	8	28	43	90+
1	7	14	19	61	2	9	29	56	94+
2	7	14	27	68	4	14	29	57	96
2	8	14	27	69	4	15	29	59	96+
4	8	16	30	70	5	18	30	67	100+
4	8	16	31	83	6	18	32	71	102+
5	9	17	35	95	7	19	35	75	105+
5	9	17	36	102+	7	23	35	75+	
6	9	18	40	104+	8	23	37	78+	
6	11	19	43		8	26	39	81	

Compare survival functions for male and female flowers.

16.16. **Gastric Cancer Data.** Stablein et al. (1981) provide data on 90 patients affected by locally advanced, nonresectable gastric carcinoma. The patients are randomized to two treatments: chemotherapy alone (coded as 0) and chemotherapy plus radiation (coded as 1). Survival time is reported in days, with censoring indicator (1 = censored, 0 = death).
Data are provided in file: gastric.cls|dat|xlsx where the column 1 is the treatment code, column 2 is the survival time, and column 3 is censoring indicator:

```
1 17 0
1 42 0
1 44 0
...
0 1512 1
0 1519 1
```

Is the treatment type significantly affecting the survival time? Use the logrank test and Cox proportional hazards regression to answer this question.

16.17. **Duration of Remissions in Acute Leukemia.** In Example 16.17 times to remission in leukemia patients (controls and treated with 6-MP drug) were analyzed using a Bayesian model. For this data compare the two survival functions, for 6-MP treatment and control, using the logrank test. Ignore matching by variable status. You can use the function ◀ manteltwo.m.

16.18. **Time to Second Birth.** The Medical Birth Registry of Norway was established in 1967 and contains information on all births in Norway since that time. The data set 📖 secondbirth.dat|xlsx distilled from the registry provides the time between first and second births for a selection of 53,558 women.[1]

It is of interest to explore whether this time was possibly affected if the first child died within one year of its birth.

The data set contains the following variables (as columns)

age	Age of mother at first birth (in years)
sex	Sex of first child (1 = boy, 2 = girl)
death	First child died within one year (0 = no, 1 = yes)
time	Time from first birth to second birth or censoring (in days)
status	Censoring indicator (0 = censored, 1 = birth)

(a) Fit the Cox proportional hazard regression with variables age, sex, and death as covariates. What is the 95% CI for the parameter β_3 corresponding to variable death? Is the variable sex significant?

(b) Plot survival functions for death = 0 and death = 1.

(c) Using Mantel's procedure ◀ manteltwo.m, test the hypothesis that the two survival functions are not significantly different.

16.19. **Rats on Three Diets.** The data is taken from the study by King et al. (1979). The researchers studied influence of dietary fat, food type and amount, and the dietary antioxidant butylated hydroxytoluene (BHT) on tumor induction and tumor growth by 7,12-dimethyl-benz[a]anthracene. The study was to determine whether ingestion of polyunsaturated fat decreased or antagonized the inhibitory action of the antioxidant in comparison to diets that either contained equivalent amounts of a saturated fat or were very low in fat. The data provided in 🖳 rat.csv|dat|mat|xlsx consist of the tumor-free time (in days) in 90 rats on three different diets (column 1), censoring indicator (column 2), and a diet code (column 3). The three diets are coded according to the fat content: 1 stands for a low fat diet, 2 for a saturated fat diet, and 3 for a unsaturated fat diet.

(a) Fit the Weibull distribution for each of the three diets. Are the confidence intervals for the Weibull parameters overlapping?

(b) Find the Kaplan–Meier estimators for each of the three diets and superimpose the Weibull survival functions fitted in (a).

(c) Compare survival functions for the low fat and saturated fat diets. Use Mantel's logrank test with $\alpha = 0.05$.

16.20. **Dukes' C Colorectal Cancer and Diet Treatment.** Colorectal cancer is a common cause of death. In the advanced stage of disease, when the disease is first diagnosed in many patients, surgery is the only treatment. Cytotoxic drugs, when given as an adjunct to surgery, do not prevent relapse and do not increase the survival in patients with advanced disease.

---

[1] The Medical Birth Registry of Norway is acknowledged for allowing the usage of the data and Dr. Stein Emil Vollset for providing the data.

Interest has been shown, at least by patients, in a nutritional approach to treatment, where diet plays a critical role in the disease management program.

In a controlled clinical trial, McIllmurray and Turkie (1987) evaluated the diet treatment in patients with Dukes' C colorectal cancer, because the residual tumour mass is small after operation, the relapse rate is high, and no other effective treatment is available. The diet treatment consisted of linolenic acid, an oil extract of the seed from the evening primrose plant *Onagraceae Oenothera biennis* and vitamin E.

The data for the treatment and control patients are given below:

Treatment	Survival time (months)
Linoleic acid ($n_1 = 25$)	1+, 5+, 6, 6, 9+, 10, 10, 10+, 12, 12, 12, 12, 12+, 13+, 15+, 16+, 20+, 24, 24+, 27+, 32, 34+, 36+, 36+, 44+
Control ($n_2 = 24$)	3+, 6, 6, 6, 6, 8, 8, 12, 12, 12+, 15+, 16+, 18+, 18+, 20, 22+, 24, 28+, 28+, 28+, 30, 30+, 33+, 42

(a) Estimate the Weibull parameters for the two groups and superimpose the resulting $\hat{S}(t)$ on the corresponding Kaplan–Meier estimators.

(b) For both treatment (linoleic acid) and control groups, find times at which 50% of the patients are surviving. Use $\hat{S}(t)$ from (a).

(c) Using Mantel's logrank procedure manteltwo.m, test the hypothesis that the two survival functions are the same.

A starter MATLAB file with data, dukes.m, is provided.

MATLAB AND WINBUGS FILES AND DATA SETS USED IN THIS CHAPTER

http://statbook.gatech.edu/Ch16.Survival/

chada.m, cordband.m, coxreadmissions.m, ImmunoPerox.m, KMmuenchow.m, leukemiadirect.m, leukemiaremission.m, lifetables.m, limphomaLR.m, logrank.m, logrank1.m, MacDonaldCancer.m, matneltwo.m, Muenchow.m, PBC.m, ple.m, secondbrth.m, simulation1.m, simulation2.m, smokingbugs.m, tongue.m, weibullsim.m

ibrahim1.odc, ibrahim2.odc, Immunoperoxidase.odc, Leukemia.odc, photocar.odc, Smoking.odc

gehan.dat, KMmuenchow.txt, limphoma.mat, pbc.xls, pbcdata.dat, prostatecanc.dat, secondbirth.dat|xlsx, smokingoutbugs.mat, tongue.dat|xlsx

---

CHAPTER REFERENCES

Armitage, P. and Berry, G. (1994). *Statistical Methods in Medical Research*, 3rd edn. Blackwell, London.

Banerjee, S. and Carlin, B. P. (2004). Parametric spatial cure rate models for interval-censored time-to-relapse data. *Biometrics*, **60**, 268–275.

Bhattacharjee, A., Richards, W. G., Staunton, J., Li, C., Monti, S., Vasa, P., Ladd, C., Beheshti, J., Bueno, R., Gillette, M., Loda, M., Weber, G., Mark, E. J., Lander, E. S., Wong, W., Johnson, B. E., Golub, T. R., Sugarbaker, D. J., and Meyerson, M. (2001). Classification of human lung carcinomas by mRNA expression profiling reveals distinct adenocarcinoma subclasses. *Proc. Natl. Acad. Sci. USA*, **98**, 24, 13790–13795.

Campbell, M. J. and Swinscow, T. D. V. (2009). *Statistics at Square One*, 11th ed. Wiley-Blackwell, London.

Cardillo, G. (2008). LogRank: comparing survival curves of two groups using the log rank test. http://www.mathworks.com/matlabcentral/fileexchange/22317.

Cardillo, G. (2008). KMPLOT: Kaplan–Meier estimation of the survival function. http://www.mathworks.com/matlabcentral/fileefileexchange/22293

Chadha, V., Warady, B. A., Blowey, D. L., Simckes, A. M., and Alon, U. S. (2000). Tenckhoff catheters prove superior to Cook catheters in pediatric acute peritoneal dialysis. *Am. J. Kidney Dis.*, **35**, 6, 1111–1116.

Cox, D. R. (1972). Regression models and life tables (with discussion). *J. R. Stat. Soc. Ser. B*, **34**, 187–220.

Cox, D. R. and Oakes, D. (1984). *Analysis of Survival Data*. Chapman & Hall, London.

Crowder, M. J., Kimber, A. C., Smith, R. L., and Sweeting, T. J. (1991). *Analysis of Reliability Data*. Chapman and Hall, London

Dellaportas, P. and Smith, A. F. M. (1993). Bayesian inference for generalized linear and proportional hazards models via Gibbs sampling. *Appl. Stat.*, **42**, 443–460.

Freireich, E., Gehan, E., Frei, E. III, Schroeder, L., Wolman, I., Anbari, R., Burgert, E., Mills, S., Pinkel, D., Selawry, O., Moon, J., Gendel, B., Spurr, C., Storrs, R., Haurani, F., Hoogstraten, B., and Lee, S. (1963). The effect of 6 mercaptopurine on the duration of steroid-induced remissions in acute leukemia. *Blood*, **21**, 6, 699–716.

Glasser, M. (1967). Exponential survival with covariance. *J. Am. Stat. Assoc.*, **62**, 561–568.

Grieve, A. P. (1987). Applications of Bayesian software: two examples. *Statistician*, 36, 283–288.

Greenwood, M. Jr. (1926). The natural duration of cancer. *Reports of Public Health and Medical Subjects*, **33**. His Majesty's Stationary Office, London.

Ibrahim, J. G., Chen, M. H., and Sinha, D. (2001). *Bayesian Survival Analysis*. Springer, New York.

Kaplan, E. L. and Meier, P. (1958). Nonparametric estimation from incomplete observations. *J. Am. Stat. Assoc.*, **53**, 457–481.

King, M. M., Bailey, D. M., Gibson, D. D., Pitha, J. V., and McCay, P. B. (1979). Incidence and growth of mammary tumors induced by 7,12-dimethylbenz[a]anthracene as related to the dietary content of fat and antioxidant. *J. Natl. Cancer Inst.*, **63**, 657–663.

Klein, J. P. and Moeschberger, M. L. (2003). *Survival Analysis: Techniques for Censored and Truncated Data*, 2nd ed. Springer, New York.

MacDonald, E. J. (1963). The epidemiology of melanoma. *Ann. N.Y. Acad. Sci.*, **100**, 4–17.

Mantel, N. (1966). Evaluation of survival data and two new rank order statistics arising in its consideration. *Cancer Chemother. Rep.*, **50** , 3, 163-170.

Markus, B. H., Dickson, E. R., Grambsch, P. M., Fleming, T. R., Mazzaferro, V., Klintmalm, G. B., Wiesner, R. H., Van Thiel, D. H., and Starzl, T. E. (1989). Efficiency of liver transplantation in patients with primary biliary cirrhosis. *New Engl. J. Med.*, **320**, 26, 1709–1713.

McIllmurray, M. B. and Turkie, W. (1987). Controlled trial of linoleic acid in Dukes's C colorectal cancer. *Br. Med. J. (Clin. Res. Ed.)*, **294**, 1260, correction in **295**, 475.

McKelvey, E. M., Gottlieb, J. A., Wilson, H. E., Haut, A., Talley, R. W., Stephens, R., Lane, M., Gamble, J. F., Jones, S. E., Grozea, P. N., Gutterman, J., Coltman, C., and Moon, T. E. (1976). Hydroxyldaunomycin (Adriamycin) combination chemotherapy in malignant lymphoma. *Cancer*, **38**, 4, 1484–1493.

Muenchow, G. (1986). Ecological use of failure time analysis. *Ecology*, **67**, 246–250.

Pike, M. C. (1966). A method of analysis of a certain class of experiments in carcinogenesis. *Biometrics*, **22**, 1, 142–161. http://www.jstor.org/stable/2528221

Sedmak, D. D., Meineke, T. A., Knechtges, D. S., and Anderson, J. (1989). Prognostic significance of cytokeratin-positive breast cancer metastases. *Mod. Pathol.*, **2**, 516–520.

Sickle-Santanello, B. J., Farrar, W. B., DeCenzo, J. F., Keyhani-Rofagha, S., Klein, J., Pearl, D., Laufman, H., and O'Toole R. V. (1988). Technical and statistical improvements for flow cytometric DNA analysis of paraffin-embedded tissue. *Cytometry*, **9**, 594–599.

Stablein, D. M., Carter, W. H., Novak, J. W. (1981). Analysis of survival data with nonproportional hazard functions. *Control. Clin. Trials*, **2**, 2, 149–159.

Chapter 17
Goodness-of-Fit Tests

Although Fisher was fond of pointing out the difficulties of assuming the correct prior distribution $p(\theta)$, he did not disdain to make a prodigious leap of faith in his selection of $f(x|\theta)$.

– Richard A. Tapia and James R. Thompson

WHAT IS COVERED IN THIS CHAPTER

- Quantile–Quantile Plots
- Pearson's χ^2-Test
- Kolmogorov–Smirnov Goodness-of-Fit Test
- Smirnov's Two-Sample Test
- Cramér–von Mises and Watson's Tests
- Rosenblatt's Test
- Moran's Test
- Testing Departures from Normality

17.1 Introduction

In traditional exposition of inferential statistics usually the family of distributions that model the data is given or assumed. Typically, the family is generated by a single distribution known up to a parameter or vector of parameters. Then the analysis proceeds to estimate or test the parameters which results in narrowing the family to a specific distribution. Often the distribution is tacitly assumed (Poisson, binomial, normal, exponential, etc.) and the interest is only in the parameters specifying the location or spread of such distributions. All this was done in Chapters 7 through 11.

Should we be more critical? Before the model or a family of models describing data is "assumed known," we should think how to select the model and test its adequacy.

Goodness-of-fit tests are procedures that test whether the distribution of a sample conforms to some fixed-in-advance distribution. We already saw Q–Q plots in Chapter 5 where the samples were compared to some theoretical distributions but in a descriptive fashion, without formal inference. In this chapter we discuss the celebrated Pearson's χ^2-test, the Kolmogorov–Smirnov (K-S) test, and several other inferential procedures for goodness of fit.

17.2 Probability Plots

Probability plots are graphical tools used to evaluate the agreement of the observed distribution to a postulated theoretical distribution. Two kinds of plots are traditionally used: Q–Q (quantile-quantile) plots and P–P (probability-probability) plots.

The Q–Q plots are more robust and considerably more popular, so we discuss them first.

17.2.1 Q–Q Plots

Quantile–quantile, or Q–Q, plots are a popular and informal diagnostic tool for assessing the distribution of data or comparing two distributions. There are two kinds of Q–Q plots, one that compares an empirical distribution with a theoretical one and another that compares two empirical distributions.

We will explain the plots with an example. Suppose that we generated some random sample X from a uniform $\mathcal{U}(-10,10)$ distribution of size $n = 200$. Suppose, for a moment, that we do not know the distribution of this data and want to check for normality by a Q–Q plot. We generate quantiles of a normal distribution for n equally spaced probabilities between 0 and 1, starting at $0.5/n$ and ending at $1 - 0.5/n$. The empirical quantiles for X are simply the elements of a sorted sample. If the empirical distribution matches the theoretical, the Q–Q plot is close to a straight line.

```
n=200;
X = unifrnd(-10, 10, [1 n]);
```

```
q=((0.5/n):1/n:(1-0.5/n));
qY = norminv(q);
qZ = unifinv(q);
qX = sort(X);
figure(1); plot(qX, qY, '*')
figure(2); plot(qX, qZ, '*')
```

The Q–Q plot is given in Figure 17.1a. Note that the ends of the Q–Q plot curve up and down from the straight line, indicating that the sample is not normal.

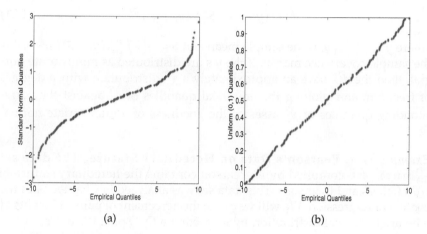

Fig. 17.1 (a) Q–Q plot of a uniform sample against normal quantiles. (b) Q–Q plot of a uniform sample against uniform quantiles.

If instead the statement `qY = norminv(q)` is replaced by `qY = unifinv(q)`, then the Q–Q plot indicates that the match is good and that the sample is consistent with uniform distribution (Figure 17.1b). Note that while the moments (mean and variance) of the sample and theoretical distributions may differ, matching the family of distributions is important. If the mean and variance differ, the straight line in the plot is shifted and has a different slope. Sometimes the straight line passing through the first and third quartiles is added to Q–Q plots for reference.

It has been part of the general knowledge that Q–Q plots can be misleading. Given a data set, the Q–Q plots may look approximately linear in the central part for many candidate distributions. A formal justification of why Q–Q plots cannot be used to develop a satisfactory GOF test is provided in Brown et al (2004).

MATLAB has built-in functions `qqplot`, `normplot`, `probplot`, and `cdfplot` for a variety of visualization and diagnostic tasks.

In our example we used p_i-quantiles for $p_i = (i - 0.5)/n$, $i = 1,\ldots,n$. In general, one can use $p_i = (i - c)/(n - 2c + 1)$ for any $c \in [0,1]$, and popular choices are $c = 1/3, c = 1/2$ (our choice), and $c = 1$. The choice $c = 0.3175$ is recommended if the distributions are long-tailed, since the extreme points on a Q–Q plot are more stable. This choice of c approximates the medians of order statistics from a uniform distribution on (0,1) and has invariance property; $F^{-1}(p_i)$ are medians of order statistics from F.

Multivariate extensions are possible. A simple procedure that can be used to assess multivariate normality is described next.

Given multivariate data $y_i = (y_{i1}, y_{i2}, \ldots, y_{ip})'$, $i = 1, \ldots, n$ one finds

$$d_i^2 = (y_i - \overline{y})' \, S^{-1} \, (y_i - \overline{y}), \tag{17.1}$$

where $\overline{y} = \frac{1}{n} \sum_{i=1}^{n} y_i$ is the sample mean and $S = \frac{1}{n-1} \sum_{i=1}^{n} (y_i - \overline{y})(y_i - \overline{y})'$ is the sample covariance matrix. If the y_is are distributed as multivariate normal, then the d_i^2s have an approximately a χ^2-distribution with p degrees of freedom, and plotting the empirical quantiles of d_i^2 against the corresponding quantiles of χ_p^2 assesses the goodness of a multivariate normal fit.

Example 17.1. **Pearson's Data on Hereditary Stature.** The data set pearson.dat compiled by K. Pearson contains the hereditary stature of 1078 fathers and their sons. This data set was analyzed in Chapter 14 in the context of re-gression. We will visualize the agreement of pairs of heights to a bivariate normal distribution by inspecting a Q–Q plot. We will calculate and sort d^2s in (17.1) and plot them against quantiles of a χ^2_2- distribution (Figure 17.2).

```
%qqbivariatenormal.m
load 'pearson.dat'
S = cov(pearson)
[m n] = size(pearson)    %[1078 2]
mp = repmat(mean(pearson), m, 1);
d2 = diag((pearson - mp) * inv(S)  * (pearson - mp)');

% chi-square (2 df) quantiles
p = 0.5/m:1/m:1;
quanch2 = chi2inv(p,2);

figure(1)
loglog(quanch2, sort(d2),'o','Markersize',msize,...
    'MarkerEdgeColor','k', 'MarkerFaceColor','g')
hold on
loglog([0.0002 50],[0.0002 50],'r','linewidth',lw)
axis([10^-4, 100, 10^-4, 100])
xlabel('Quantiles of chi^2_2 ')
ylabel('Sorted d^2')
```

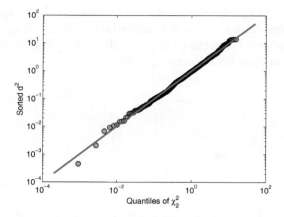

Fig. 17.2 Sorted values of d^2 plotted against the χ_2^2 quantiles. The linearity of the plot indicates a good fit of Pearson's data to a bivariate normal distribution.

17.2.2 P–P Plots

A P–P plot compares the empirical CDF of a variable with a proposed theoretical cumulative distribution function F. Recall that the empirical CDF, denoted by $F_n(x)$, is defined as the proportion of all observations less than or equal to x, so for ith order statistic $F_n(x_{(i)}) = i/n$. The P–P plot is constructed by first sorting n observations in increasing order: $x_{(1)} \le x_{(2)} \le \cdots \le x_{(n)}$. Then the ith ordered value is represented on the plot by the point whose ordinate is $F(x_{(i)})$ and whose abscissa is i/n.

Similar to the Q–Q plots, the P–P plots are used to visualize how close the selected theoretical distribution fits the data. Some disadvantages of P–P plots when compared to Q–Q plots are: (i) location and scale of the theoretical distribution have to be specified for P–P plots, and (ii) P–P plots are less sensitive to extreme observations, which are most influential in the model fit.

P–P plots are more discriminatory in the regions of high probability density. For example, if you compare a data distribution with a particular normal distribution, differences in the middle of the two distributions are more apparent on a P–P plot than on a Q–Q plot.

Example 17.2. **P–P and Q–Q Plots Compared.** In ◀ ppqq.m we generated 100 observations from a t-distribution with 7 degrees of freedom. We are interested in checking whether the distribution is normal by inspecting both P–P and Q–Q plots. The P–P plots require exact theoretical distribution parameters, so for the normal distribution we use the mean and standard deviation used to generate the observations. As is evident from Figure 17.3a, the P–P plot is not indicating a systematic deviation from the normality.

However, the Q–Q plot, which does not require parameter specification, is sensitive to observations from the tails, which in fact distinguishes normal and t-distributions. From the Q–Q plot in Figure 17.3b we see that extreme observations deviate from the straight line.

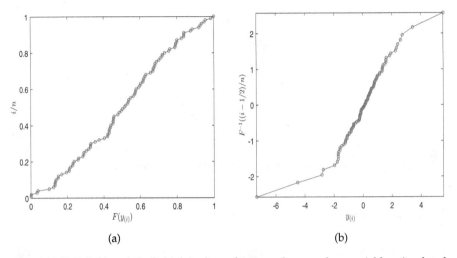

(a)   (b)

Fig. 17.3 P–P (left) and Q–Q (right) plots of 100 random random variables simulated from t_7-distribution, contrasted against the normal distribution.

17.2.3 Poissonness Plots

Poissonness plots were developed by Hoaglin (1980) and consist of assessing the goodness of fit of observed frequencies to frequencies corresponding to Poisson-distributed data. Suppose that frequencies $n_i, i = 0, 1, \ldots, k$ corresponding to realizations $0, 1, \ldots, k$ are observed and that for $i > k$ all $n_i = 0$. Let $N = \sum_{i=0}^{k} n_i$.

Then, the frequency n_i and the theoretical counterpart $e_i = N \times \frac{\lambda^i}{i!} \exp \{-\lambda\}$ should be close, if the data are consistent with a Poisson distribution,

$$n_i \sim N \times \frac{\lambda^i}{i!} \exp\{-\lambda\}.$$

By taking the logarithms of both sides one gets

$$\log n_i + \log(i!) - \log N \sim \log \lambda \cdot i - \lambda.$$

Thus, the plot of $\log n_i + \log(i!) - \log N$ against integer i should be linear with slope $\log \lambda$ if the frequencies are consistent with a Poisson $\mathcal{P}oi(\lambda)$ distribution. Such a plot is called a *poissonness plot*.

Example 17.3. **Poissonness.** The script ◢ poissonness.m illustrates this. $N = 50{,}000$ Poisson random variables with $\lambda = 4$ are generated, observed frequencies are calculated, and a poissonness plot is made (Figure 17.4).

```
rand('state',2)
N = 50000;  xx = poissrnd(4, 1, N); %simulate N Poisson(4) rv's.
% observed frequencies
ni =    [sum(xx==0) sum(xx==1) sum(xx==2) sum(xx==3)  sum(xx==4) ...
            sum(xx==5) sum(xx==6) sum(xx==7) sum(xx==8)  sum(xx==9) ...
        sum(xx==10) sum(xx==11) sum(xx==12) sum(xx==13)  sum(xx==14)]' ;

i=(0:14)'
poissonness = log(ni) + log(factorial(i)) - log(N);

plot(i, poissonness,'o')
xlabel('i')
ylabel('log(n_i) + log(i!) - log(N)')
hold on
plot(i, log(4)*i - 4,'r:') %theoretical line with lambda=4
[beta] = regress(poissonness, [ones(size(i)) i])

% beta =
%    -4.0591
%     1.4006
%
%     Slope in the linear fit is 1.4006
%     lambda can be estimated as  exp(1.4006) = 4.0576,
%     also, as negative intercept, 4.0509
```

17.3 Pearson's Chi-Square Test

Pearson's χ^2-test (Pearson, 1900) is the first formally formulated testing procedure in the history of statistical inference. In his 1984 *Science* article entitled "Trial by number," Ian Hacking states that the goodness-of-fit chi-square test introduced by Karl Pearson *ushered in a new kind of decision making* and places it among the top 20 discoveries since 1900, considering all branches of science and technology.

Suppose that $X_1, X_2, \ldots, X_n$ is a sample from an unknown distribution with CDF $F_X(x)$. We are interested in testing that this distribution is equal

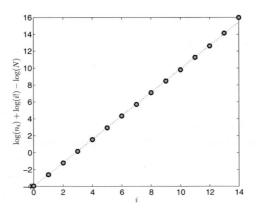

Fig. 17.4 Poissonness plot for simulated example involving $N = 50{,}000$ Poisson $\mathcal{P}oi(4)$ random variates. The slope of this plot estimates $\log \lambda$, while the intercept estimates $-\lambda$. The theoretical line, $\log(4)i - 4$ (in red), fits the points $(i, \log(n_i) + \log(i!) - \log(N))$ well.

to some specific distribution F_0, i.e., in testing the goodness-of-fit hypothesis $H_0 : F_X(x) = F_0(x)$.

Suppose that the domain (a,b) of the distribution F_0 is split into r nonoverlapping intervals, $I_1 = (a, x_1]$, $I_2 = (x_1, x_2]$ ... $I_r = (x_{r-1}, b)$. Such intervals have probabilities $p_1 = F_0(x_1) - F_0(a)$, $p_2 = F_0(x_2) - F_0(x_1)$, ..., $p_r = F_0(b) - F_0(x_{r-1})$, under the assumption that H_0 is true. Of course it is understood that the observations belong to the domain of F_0; if this is not the case, the null hypothesis is automatically rejected.

Let $n_1, n_2, \ldots, n_r$ be the observed frequencies of observations falling in the intervals $I_1, I_2, \ldots, I_r$. In this notation, n_1 is the count of observations from the sample $X_1, \ldots, X_n$ that fall in the interval I_1. Of course, $n_1 + \cdots + n_r = n$, since the intervals partition the domain of the sample.

The discrepancy between observed frequencies n_i and frequencies under F_0, np_i is the rationale for forming the statistic

$$\chi^2 = \sum_{i=1}^{r} \frac{(n_i - np_i)^2}{np_i}, \tag{17.2}$$

which has an approximate χ^2-distribution with $r - 1$ degrees of freedom. Alternative representations include

$$\chi^2 = \sum_{i=1}^{r} \frac{n_i^2}{np_i} - n \quad \text{and} \quad \chi^2 = n \left[\sum_{i=1}^{r} \left(\frac{\hat{p}_i}{p_i} \right) \hat{p}_i - 1 \right], \tag{17.3}$$

where $\hat{p}_i = n_i / n$.

In some cases, the distribution under H_0 is not *fully* specified; for example, one might conjecture that the data is exponentially distributed without knowing the exact value of λ. In this case, the unknown parameter can be estimated from the sample.

If k parameters needed to fully specify F_0 are estimated from the sample, the χ^2-statistic in (17.2) has a χ^2-distribution with $r - k - 1$ degrees of freedom. A degree of freedom in the test statistic is lost for each estimated parameter.

Firm recommendations on how to select the intervals or even the number of intervals for a χ^2-test do not exist. For a continuous distribution function one may make the intervals approximately equal in probability. Practitioners may want to arrange interval selection so that all $np_i > 1$ and that at least 80% of the np_is exceed 5. The rule of thumb is $n \geq 10$, $r \geq 3$, $n^2/r \geq 10$, $np_i \geq 0.25$.

Remark. Pearson's χ^2-test is sometimes referred to as "poorness-of-fit" test. This attribute is appropriate since the test can be decisive only if the model does not fit the data well. If the postulated distribution fits the data well, then it may not be unique, and possibly other distributions could be found that would pass the test as well. In terms of testing hypotheses, not rejecting H_0 does not mean that the distribution specified in the null hypothesis is the only distribution consistent with the data.

When some of the expected frequencies are less than 5, or when the number of classes is small, then the χ^2-test with Yates (Yates, 1934) corrections is recommended,

$$\chi^2 = \sum_{i=1}^{r} \frac{(|n_i - np_i| - 0.5)^2}{np_i},$$

which, under H_0, has a χ^2-distribution with $r - 1$ degrees of freedom.

Sometimes, instead of χ^2, the Freeman–Tukey statistic is used. This statistic, defined as

$$FT = 4 \sum_{i=1}^{r} \left(\sqrt{n_i} - \sqrt{np_i} \right)^2,$$

which under H_0 has approximately a χ^2 distribution with $r - 1$ degrees of freedom, is sometimes preferred to Pearson's χ^2-statistic.

Example 17.4. **Weldon's 26,306 Rolls of 12 Dice.** Pearson (1900) discusses Weldon's data (table below) as the main illustration for his test. Raphael Weldon, an evolutionary biologist and founder of biometry, rolled 12 dice simultaneously and recorded the number of times a 5 or a 6 was rolled. In his letter to Galton, dated February 2, 1894, Weldon provided the data and asked for Galton's opinion about their validity. Three contemporary British statisticians, Pearson, Edgeworth, and Yule, have also considered

Weldon's data. An interesting historic account can be found in Kemp and
Kemp (1991).

The results from 26,306 rolls of 12 dice are summarized in the following
table:

No. of dice resulting in ⚁ or ⚄ when 12 dice are rolled	Observed frequency
0	185
1	1149
2	3265
3	5475
4	6114
5	5194
6	3067
7	1331
8	403
9	105
10	14
11	4
12	0

In Pearson (1900), the value of χ^2-statistic was quoted as 43.9, which is
slightly different than the correct value. The discrepancy is probably due to
the use of expected frequencies rounded to the nearest integer and due to
the accumulation of rounding errors when calculating χ^2.

To find the expected frequencies, recall the binomial distribution. The
number of times ⚁ or ⚄ was rolled with 12 dice is binomial: $Bin(12,1/3)$.
To find the expected frequencies, multiply the total number of rolls 26,306
by the expected (theoretical) probabilities obtained from $Bin(12,1/3)$, as in
the following MATLAB output:

```
%weldon.m
obsfreq = [ 185  1149  3265  5475  6114  5194 ...
            3067  1331   403   105    14     4   0];

n = sum(obsfreq) %26306
expected = n * binopdf(0:12, 12, 1/3)

%expected =
%  1.0e+003 *
%
%  0.2027  1.2165  3.3454  5.5756  6.2726  5.0180  2.9272
%  1.2545  0.3920  0.0871  0.0131  0.0012  0.0000

chisqs = (obsfreq - expected).^2./expected
%chisqs =
%  1.5539  3.7450  1.9306  1.8155  4.0082  6.1695  6.6772
%  4.6635  0.3067  3.6701  0.0665  6.6562  0.0495
```

```
chi2 = sum(chisqs)
%chi2 =   41.3122

pval=1 - chi2cdf(chi2, 13-1)
     % pval = 4.3449e-005
crit = chi2inv(0.95, 13-1)
     % crit = 21.0261
```

As is evident from the MATLAB output, the observations are not supporting the fact that the dice were fair (*p*-value of 4.3449e-05); see also Figure 17.5.

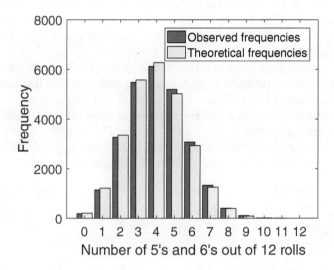

Fig. 17.5 Bar plot of observed frequencies and theoretical frequencies. Although the
graphs appear close, the large sample size makes the discrepancy of this size very unlikely. It is almost certain that one or more of Weldon's 12 dice were not well balanced.

Although the observed frequencies are fairly close to the theoretical,
given the large sample size $n = 26306$, the test exhibits excessive power and
rejects H_0 with a *p*-value smaller than 0.00005.

Example 17.5. **Is the Process Poisson?** The Poisson process is one of the
most important stochastic models. For example, random arrival times of
patients to a clinic are well modeled by a Poisson process. This means that
in any interval of time, say $[0,t]$, the number of arrivals is Poisson with
parameter λt. For such Poisson processes, the interarrival times follow an
exponential distribution with density $f(t) = \lambda e^{-\lambda t}$, $t \geq 0, \lambda > 0$. It is often
of interest to establish the Poissonity of a process since many theoretical

results available for such processes are ubiquitous in queueing theory and various engineering applications.

The interarrival times of the arrival process were recorded, and it was observed that $n = 109$ recorded times could be categorized as follows:

Interval	$0 \leq T < 1$	$1 \leq T < 2$	$2 \leq T < 3$	$3 \leq T < 4$	$4 \leq T < 5$	$5 \leq T < 6$	$T \geq 6$
Frequency	34	20	16	15	9	7	8

A simple calculation determined that the sample mean was $\overline{T} = 5/2$. Test the hypothesis that the process, with the interarrival times described above, is Poisson, at level $\alpha = 0.05$.

Given this data, one should test the hypothesis that the interarrival times are exponential. The density $f(t) = \lambda e^{-\lambda t}$, $t \geq 0$, $\lambda > 0$ corresponds to the CDF $F(t) = 1 - e^{-\lambda t}$, $t \geq 0$, $\lambda > 0$, and the theoretical probability of an interval $[a, b]$ is $F(b) - F(a)$.

However, we first need to estimate the parameter λ in order to calculate the the the theoretical probabilities. A standard estimator for the parameter λ is the reciprocal of the sample mean, $\hat{\lambda} = 1/\overline{T} = 1/(5/2) = 0.4$.

With this $\hat{\lambda}$ the theoretical frequencies of intervals $[3, 4]$ and $[4, 5]$, for example, are $np_4 = 109 \cdot (F(4) - F(3)) = 109 \cdot (1 - e^{-0.4 \cdot 4} - (1 - e^{-0.4 \cdot 3})) = 109 \cdot (e^{-0.4 \cdot 3} - e^{-0.4 \cdot 4}) = 0.099298 \cdot 109 = 10.823$ and $np_5 = 0.066561 \cdot 109 = 7.255$, respectively. Overall, the χ^2 statistic is equal to

$$\frac{(34 - 35.9335)^2}{35.935} + \frac{(20 - 24.089)^2}{24.089} + \frac{(16 - 16.147)^2}{16.147} + \frac{(15 - 10.823)^2}{10.823} +$$
$$\frac{(9 - 7.255)^2}{7.255} + \frac{(7 - 4.863)^2}{4.863} + \frac{(8 - 9.888)^2}{9.888} = 4.13.$$

The number of degrees of freedom is $df = 7 - 1 - 1 = 5$ and the 95% quantile for χ_5^2 is chi2inv(0.95,5)=11.071. Thus, we do not reject the hypothesis that the interarrival times are exponential, that is, the observed process is consistent with a Poisson process.

How should the number of classes be selected? For discrete distributions, the discrete values usually define the classes. For continuous distributions, both the number and sizes of classes can vary. In the previous example, for $n = 109$ we selected $r = 7$. A popular recommendation, influenced by Mann and Wald (1942), is to take r as a multiple of $n^{2/5}$ in the case of equiprobable classes. Although asymptotically optimal from the power-of-test considerations, the growth proportional to $n^{2/5}$ leads to a large number of classes for large n, inflating the variance of the χ^2 statistic, which affects even the nominal significance level of the test. Thus, in practice, the number of classes should increase with sample size but seldom to exceed 15.

Example 17.6. **Simulation of Exponentials.** A sample of $n = 1000$ exponential $\mathcal{E}(1/2)$ random variates is generated. For a moment, we pretend that the generating distribution is unknown. Using MATLAB's built-in function chi2gof, we test the consistency of the generated data with an exponential distribution with the rate λ estimated from the sample. Note that MATLAB parameterizes the exponential distribution with the reciprocal of rate parameter, which is estimated by sample mean.

```
X = exprnd(2, [1, 1000]);
[h,p,stats] = ...
chi2gof(X,'cdf',@(z)expcdf(z,mean(X)),'nparams',1,'nbins',7)
%
%h = 0
%p = 0.6220
%stats  =    chi2stat: 2.6271
%            df: 4
%            edges: [1x7 double]
%            O: [590 258 96 34 14 8]
%            E: [1x6 double]
```

The sample is consistent with the exponential distribution with p-value of 0.6220. Note that the number of intervals selected by MATLAB is 6, not the requested 7. This is because the upper tail intervals with a low expected count (< 5) are merged.

✒

Example 17.7. **Wrinkled Peas.** Mendel crossed peas that were heterozygotes for smooth/wrinkled, where smooth is dominant. The expected ratio in the offspring is 3 smooth to 1 wrinkled. He observed 423 smooth and 133 wrinkled peas. The expected frequency of smooth is calculated by multiplying the sample size (556) by the expected proportion (3/4) to yield 417. The same is done for wrinkled to yield 139. The number of degrees of freedom when an extrinsic hypothesis is used is the number of values of the nominal variable minus one. In this case, there are two values (smooth and wrinkled), so there is one degree of freedom.

```
chisq = (556/4 - 133)^2/(556/4) + (556*3/4 - 423)^2/(556*3/4)
    %chisq = 0.3453
1 - chi2cdf(chisq, 2-1)
    %ans = 0.5568
chisqy = (abs(556/4 - 133)-0.5)^2/(556/4) + ...
    (abs(556*3/4 - 423)-0.5)^2/(556*3/4) %with Yates correction
    %chisq = 0.2902
1-chi2cdf(chisqy, 2-1)
    %ans = 0.5901
```

We conclude that the theoretical odds 3:1 in favor of smooth peas are consistent with the observations at level $\alpha = 0.05$.

✒

Example 17.8. **Horse-Kick Fatalities.** During the latter part of the nineteenth century, Prussian officials gathered information on the hazards that horses posed to cavalry soldiers. Fatal accidents for 10 cavalry corps were collected over a period of 20 years (Preussischen Statistik). The number of fatalities due to kicks, x, was recorded for each year and each corps. The table below shows the distribution of x for these 200 "corps-years."

Number of deaths, x	Observed number of corps-years in which x fatalities occurred
0	109
1	65
2	22
3	3
≥ 4	1
	200

Altogether there were 122 fatalities $[109(0) +65(1) + 22(2) +3(3) + 1(4)]$, meaning that the observed fatality *rate* was 122/200, or 0.61 fatalities per corps-year. Von Bortkiewicz (1898) proposed a Poisson model for X with a mean of $c = 0.61$. The table below shows the observed and expected frequencies corresponding to $x = 0,1,2,\ldots$, etc. The expected frequencies are

$$ n \times \frac{0.61^i}{i!} \exp\{-0.61\}, $$

for $n = 200$ and $i = 0,1,2$, and 3. We put together all values ≥ 4 as a single class; this will ensure that the sum of the theoretical probabilities is equal to 1. For example, the expected frequencies np_i are

```
npi = 200 * [poisspdf(0:3, 0.61)  1-poisscdf(3, 0.61)]
   %npi = 108.6702   66.2888   20.2181   4.1110   0.7119
```

i	Fatalities	Observed number of corps-years, n_i	Expected number of corps-years, e_i
1	0	109	108.6702
2	1	65	66.2888
3	2	22	20.2181
4	3	3	4.1110
5	≥ 4	1	0.7119
		200	200

Now we calculate the statistic $\chi^2 = \sum_{i=1}^{5} (n_i - np_i)^2/(np_i)$ and find the p-value for the test and rejection region. Note that the number of degrees of freedom is $df = 5 - 1 - 1$ since $\lambda = 0.61$ was estimated from the data.

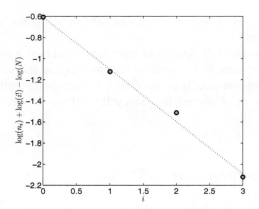

Fig. 17.6 Poissonness plot for von Bortkiewicz's data. The theoretical line, $\log(0.61)i - 0.61$, is shown in *red*. In this plot frequencies corresponding to $x = 3$ and $x \geq 4$ are grouped.

```
ni = [109 65 22 3 1]
   % ni = 109     65     22     3     1
 ch2 = sum( (ni-npi).^2 ./npi )
   % ch2 = 0.5999
 1-chi2cdf(0.5999, 5-1-1)
   % ans = 0.8965,  pvalue
 chi2inv(0.95, 5-1-1)
   % ans = 7.8147, critical value
   % rejection region [7.8147, infinity)
```

The Poisson distribution in H_0 is consistent with the data at the level $\alpha = 0.05$. Clearly the agreement between the observed and expected frequencies is remarkable; see also the poissonness plot in Figure 17.6.

Although very popular, Pearson's χ^2-test has several serious shortcomings. The most important is the dependence of statistic χ^2 on sample size, as in (17.3). In addition, for continuous distribution there is a degree of arbitrariness in setting the frequency intervals.

All this implies that χ^2 may not be well calibrated to measure closeness of empirical and theoretical distributions. For example, under H_0, the variance of χ^2-statistic is

$$\text{Var}_{H_0}(\chi^2) = \frac{1}{n}\left(2(n-1)(r-1) - r^2 + \sum_{i=1}^{r}\frac{1}{p_i}\right).$$

When r is large, some of the theoretical cell probabilities may be quite small, which inflates the variance and makes it much larger than the nominal $2(r-1)$. Another example is overpowering, as in the case of Weldon's data in Example 17.4. However, with smaller sample sizes, the χ^2-test may lack adequate power.

Kolmogorov–Smirnov family of tests described next have the advantage that they are genuinely distribution free as long as the underlying distribution is continuous and fully specified. Kolmogorov–Smirnov tests have no sample size or observed frequency limitations inherent for Pearson's χ^2-test.

17.4 Kolmogorov–Smirnov Tests

The first measure of goodness of fit for general distributions was derived by Kolmogorov (1933). Andrei Nikolaevich Kolmogorov the most accomplished and celebrated Russian mathematician of all time, made fundamental contributions to probability theory, including a test statistic for distribution functions, some of which are named after him. Nikolai Vasilyevich Smirnov another Russian mathematician, extended Kolmogorov's results to two samples.

17.4.1 Kolmogorov's Test

Let $X_1, X_2, \ldots, X_n$ be a sample from a population with a continuous, but unknown, CDF F. As in (2.2), let $F_n(x)$ be the empirical CDF based on $X_1, X_2, \ldots, X_n$.

We are interested in testing the hypothesis

$$H_0 : \text{(For all } x) \; F(x) = F_0(x) \quad \text{versus} \quad H_1 : \text{(There exists } x) \; F(x) \neq F_0(x),$$

where $F_0(x)$ is a fully specified continuous distribution.

The test statistic D_n is

$$D_n = \sup_{-\infty < x < \infty} |F_n(x) - F_0(x)|, \tag{17.4}$$

where sup stands for supremum, the smallest upper bound for a set.

This statistic is calculated from the sample as

$$D_n = \max \left\{ \max_{1 \le i \le n} \left(\frac{i}{n} - F_0(X_{(i)}) \right), \; \max_{1 \le i \le n} \left(F_0(X_{(i)}) - \frac{i-1}{n} \right) \right\}.$$

The first maximum in the curly brackets is sometimes denoted as D_n^+ and the second as D_n^-, so $D_n = \max\{D_n^+, D_n^-\}$. When hypothesis H_0 is true, Kolmogorov (1933) showed that the statistic $\sqrt{n}D_n$ is approximately distributed as

$$\lim_{n \to \infty} \mathbb{P}(\sqrt{n}D_n \leq x) = K(x) = 1 - 2\sum_{k=1}^{\infty} (-1)^{k-1} e^{-2k^2 x^2}, \; x \geq 0,$$

which allows one to approximate critical regions and p-values for this test.

The MATLAB file ◀ kscdf.m calculates the CDF K. Since large values of D_n are critical for H_0, the p-value of the test is approximated as

$$p \approx 1 - K(\sqrt{n}D_n),$$

or in MATLAB as `1-kscdf(sqrt(n)*Dn)`.

In practice, most K-S-type tests are two-sided, testing whether or not F is equal to F_0, the distribution postulated by H_0. Alternatively, one might test to see if the distribution is larger or smaller than a hypothesized F_0 (e.g., Kvam and Vidakovic, 2007). We remark that $D_n \neq \max_{1 \leq i \leq n} |F_0(X_{(i)}) - \frac{i}{n}|$. The usage of $\max_{1 \leq i \leq n} |F_0(X_{(i)}) - \frac{i}{n}|$ as D_n is a common error, probably influenced by (17.4).

Example 17.9. **Kolmogorov Test and 30 Exponentials.** To illustrate the Kolmogorov test, we simulate 30 observations from an exponential distribution with $\lambda = 1/2$.

```
% rand('state', 0);
% n = 30;   i = 1:n;
% x = exprnd(1/2,[1,n]); x = sort(x);
x = [...
    0.0256    0.0334    0.0407    0.0434    0.0562    0.0575...
    0.0984    0.1034    0.1166    0.1358    0.1518    0.2427...
    0.2497    0.2523    0.3608    0.3921    0.4052    0.4455...
    0.4511    0.5208    0.6506    0.7324    0.7979    0.8077...
    0.8079    0.8679    0.9870    1.4246    1.9949    2.309];
n = length(x); i = 1:n;
distances = [i./n - expcdf(x, 1/2); expcdf(x, 1/2) - (i-1)./n ];
Dn = max(max(distances))             %0.1048
pval = 1 - kscdf(sqrt(n)*Dn)         %0.8966
```

The p-value is 0.8966 and H_0 is not rejected. In other words, the sample is consistent with the hypothesis of the population's exponential distribution $\mathcal{E}(1/2)$.

The CDF $K(x)$ is an approximate distribution for $\sqrt{n}D_n$, and for small values of n it may not be precise. Better approximations use a continuity correction:

$$p \approx 1 - K\left(\sqrt{n}D_n + \frac{1}{6\sqrt{n}}\right).$$

This approximation, known as Bolshev's correction, is satisfactory for $n \geq$ 20.

```
1 - kscdf(sqrt(n) * Dn + 1/(6 * sqrt(n)) ) %0.8582
```

MATLAB has a built-in function, kstest, that produces similar output.

```
% form theoretical cdf
t = 0:0.01:20;
y = expcdf(t,1/2);
cdf = [t' y'];
[decision, pval, KSstat, critValue] = kstest(x, cdf, 0.01, 0)
        %decision =     0
        %pval   =  0.8626
        %KSstat =  0.1048
        %critValue =  0.2899
```

Figure 17.7 shows the empirical and theoretical distribution in this example, and it is produced by the code below.

```
%Plot
xx = 0:0.01:4;
plo = cdfplot(x); set(plo,'LineWidth',2);
hold on
plot(xx,expcdf(xx, 1/2),'r-','LineWidth',2);
legend('Empirical','Theoretical Exponential',...
        'Location','SE')
```

Note that in the two calculations the values of D_n statistic coincide but the p-values differ. This is because kstest uses a different approximation to the p-value.

✎

The Kolmogorov test has advantages over exact tests based on the χ^2 goodness-of-fit statistic, which depend on an adequate sample size and proper interval assignments for the approximations to be valid. A shortcoming of the Kolmogorov test is that the F_0 CDF in H_0 must be fully specified. That is, if location, scale, or shape parameters are estimated from the data, the critical region of the Kolmogorov test is no longer valid. An example is Lilliefors' test for departures from normality when the null distribution is not fully specified.

17.4.2 Smirnov's Test to Compare Two Distributions

Russian mathematician Nikolai Smirnov (Smirnov, 1939) extended the Kolmogorov test to compare two distributions based on independent samples

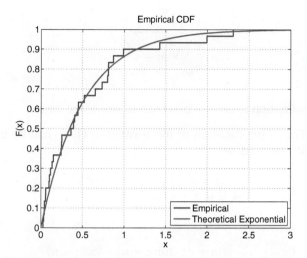

Fig. 17.7 Empirical and theoretical distributions in the simulated example.

from each population. Let $X_1, X_2, \ldots, X_m$ and $Y_1, Y_2, \ldots, Y_n$ be two independent samples from populations with unknown CDFs F_X and G_Y. Let $F_m(x)$ and $G_n(x)$ be the corresponding empirical distribution functions.

We would like to test

$$H_0 : (\text{For all } x) \; F_X(x) = G_Y(x) \quad \text{versus} \quad H_1 : (\text{There exists } x) \; F_X(x) \neq G_Y(x),$$

We will use an analogue of the Kolmogorov statistic

$$D_{m,n} = \max \left\{ \max_{1 \leq i \leq m} \left\{ \frac{i}{m} - G_n(X_{(i)}) \right\}, \; \max_{1 \leq j \leq n} \left\{ \frac{j}{n} - F_m(Y_{(j)}) \right\} \right\}.$$

The limiting distribution for $D_{m,n}$ can be expressed by Kolmogorov's CDF K as

$$\lim_{m,n \to \infty} \mathbb{P} \left(\sqrt{\frac{mn}{m+n}} D_{m,n} \leq x \right) = K(x),$$

and the p-value for the test is approximated as

$$p \approx 1 - K\left(\sqrt{\frac{mn}{m+n}} D_{m,n}\right). \qquad (17.5)$$

This approximation is good when both m and n are large.

Remark. The approximation of the p-value in (17.5) can be improved by continuity corrections as

$$p \approx 1 - K\left(\sqrt{\frac{mn}{m+n}}\left(D_{m,n} + \frac{|m-n|}{6mn} + b_{m,n}\right)\right),$$

where

$$b_{m,n} = \frac{(m+n)\,(\min(m,n) - \gcd(m,n))}{2mn(m+n+\gcd(m,n))}$$

and $\gcd(m,n)$ is the greatest common divisor of m and n. This approximation is satisfactory if $m, n > 20$.

If $m = n$, exact p-values can be found by using ranks and combinatorial calculations. This has a practical value only when this common sample size is small.

Example 17.10. **Normal and t.** To illustrate Smirnov's test, we simulate $m = 39$ observations from a normal $\mathcal{N}(-1, 2^2)$ distribution and $n = 35$ observations from t-distribution with 10 degrees of freedom, Figure 17.8. The hypothesis of equality of distributions is rejected at the level $\alpha = 0.05$ since the p-values are approximately 2%.

Notice that the corrected p-value 0.0196 is close to that in `kstest2`, and the differences are due to different approximation formulas.

```
x =[...
    -5.75    -3.89    -3.69    -3.68    -3.54    -2.59 ...
    -2.53    -2.40    -2.39    -2.27    -2.02    -1.72 ...
    -1.64    -1.54    -1.27    -1.11    -1.08    -1.00 ...
    -0.88    -0.84    -0.47    -0.36    -0.29    -0.24 ...
    -0.20    -0.19     0.02     0.15     0.25     0.45 ...
     0.51     0.72     0.74     0.96     1.25     1.33 ...
     1.49     2.39     2.59 ];

y=[...
    -2.72    -2.18    -1.31    -1.17    -1.00    -0.94 ...
    -0.78    -0.65    -0.63    -0.52    -0.40    -0.37 ...
    -0.30    -0.21    -0.19    -0.11    -0.05     0.12 ...
     0.14     0.25     0.27     0.35     0.45     0.48 ...
     0.48     0.60     0.71     0.76     0.79     1.01 ...
     1.10     1.10     1.12     1.36     2.03 ];
```

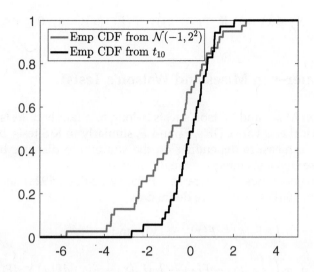

Fig. 17.8 Empirical distributions for the simulated example.

```
m = length(x); n=length(y);
i = 1:m; j = 1:n;
distances = [i./m - empiricalcdf(x, y), j./n - empiricalcdf(y, x)];
Dmn = max(max(distances))    %0.3414
z = sqrt(m * n/(m + n))*Dmn %1.4662
pval1 = 1 - kscdf(z)      %0.0271
%
bmn = (m+n)/(2*m*n) * (min(m,n) - gcd(m, n))/(m + n + gcd(m,n)) %0.0123
zz =   sqrt(m * n/(m + n))*(Dmn + abs(m-n)/(6*m*n) +  bmn )    %1.5211
pval2 = 1 - kscdf(zz) %0.0196
% MATLAB's built in function
[h,p,k] = kstest2(x,y)
   % 1
   % 0.0202
   % 0.3414
```

Before discussing Cramér–von Mises, Watson, and Moran's tests, we note that any goodness-of-fit null hypothesis $H_0 : F = F_0$, where F_0 is a fixed continuous distribution, can be restated in terms of a uniform distribution. This is a simple consequence of the following result:

Result. Let random variable X have a continuous distribution F. Then, $F(X)$ has a uniform distribution on $[0,1]$.

This important fact has a simple proof, since the continuity of F ensures the existence of F^{-1}:

$$\mathbb{P}(F(X) \leq x) = \mathbb{P}(X \leq F^{-1}(x)) = F(F^{-1}(x)) = x, \ 0 \leq x \leq 1.$$

17.5 Cramér–von Mises and Watson's Tests*

Cramér–von Mises and Watson's tests belong to a family of tests that compare empirical and true CDF's, F_n and F, similarly to K-S tests, but instead of a distance measure depending on the supremum distance between F_n and F, these two tests integrate the difference.

Cramér–von Mises' W_n (Cramér, 1928; von Mises, 1928) and Watson's U_n^2 (Watson, 1961) statistics are defined as

$$W_n^2 = n \int_{\mathbb{R}} [F_n(x) - F(x)]^2 dF(x), \ \text{and}$$

$$U_n^2 = n \int_{\mathbb{R}} \left\{ F_n(x) - F(x) - \int_{\mathbb{R}} [F_n(x) - F(x)] dF(x) \right\}^2 dF(x).$$

We are interested in testing $H_0 : F = F_0$ versus $H_1 : F \neq F_0$ if the sample $x_1, \dots, x_n$ is observed.

If $x_1' \leq x_2' \leq \cdots \leq x_n'$ is the ordered sample, then statistics W_n^2 and U_n^2 are calculated as

$$W_n^2 = \frac{1}{12n} + \sum_{i=1}^{n} \left(y_i - \frac{2i-1}{2n} \right)^2, \ \text{and}$$

$$U_n^2 = \frac{1}{12n} + \sum_{i=1}^{n} \left(y_i - \frac{2i-1}{2n} - \bar{y} + 1/2 \right)^2,$$

where $y_i = F_0(x_i')$ and $\bar{y} = \frac{1}{n} \sum_i y_i$. Note that the two statistics are connected,

$$U_n^2 = W_n^2 - n \left(\bar{y} - \frac{1}{2} \right)^2.$$

Precise tables for assessing significance of W_n^2 and U_n^2 are available from various sources (e.g., Pearson and Hartley, 1972). We will use Tiku's χ^2 approximation (Tiku, 1965) that is quite precise even for small values of n.

The statistic $\chi^2 = \frac{W_n^2 - a_1}{b_1}$ has approximately a χ^2-distribution with k_1 degrees of freedom for

$$a_1 = \frac{48n^2 - 137n + 87}{30C}, \ b_1 = \frac{C}{84n(4n-3)}, \ \text{and} \ k_1 = \frac{98n(4n-3)^3}{5C^2},$$

with $C = 32n^2 - 61n + 30$.

The statistic $\chi^2 = \frac{U_n^2 - a_2}{b_2}$ has approximately a χ^2-distribution with k_2 degrees of freedom for

$$a_2 = \frac{3n - 8}{60(2n - 3)}, \quad b_2 = \frac{2n - 3}{84n}, \quad \text{and} \quad k_2 = \frac{49n(n - 1)}{5(2n - 3)^2}.$$

In both cases large values of χ^2 are critical. The test is significant, that is, $H_0 : F(x) = F_0(x)$ is rejected, if $\chi^2 > \chi^2_{k,1-\alpha}$, where k is either k_1 or k_2 depending on the test used.

The Tiku approximations are precise when W_n and U_n are not small. Small values of these statistics are not interesting for the inference, since in this case respective p-values are close to 1.

The limiting CDF of W_n involves special functions

$$W(x) = \lim_{n \to \infty} \mathbb{P}(W_n \leq x)$$

$$= \frac{1}{\sqrt{2x}} \sum_{i=1}^{\infty} \frac{\Gamma(i + 1/2)}{\Gamma(1/2)\Gamma(i + 1)} \sqrt{4i + 1} \exp\left\{ -\frac{(4i + 1)^2}{16x} \right\}$$

$$\times \left\{ I_{-1/4}\left[\frac{(4i + 1)^2}{16x} \right] - I_{1/4}\left[\frac{(4i + 1)^2}{16x} \right] \right\}, \qquad (17.6)$$

where $I_a(x)$ is modified Bessel function of the first kind implemented in MATLAB as besseli. This CDF is calculated as MATLAB funcion cvmcdf.

If $n < 40$, corrected values $(W_n^2)'$ and $(U_n^2)'$ in place of W_n^2 and U_n^2 are recommended (Stephens, 1974):

$$(W_n^2)' = \left(W_n^2 - \frac{4}{10n} + \frac{6}{10n^2} \right)\left(1 + \frac{1}{n} \right), \quad \text{and}$$

$$(U_n^2)' = \left(U_n^2 - \frac{1}{10n} + \frac{1}{10n^2} \right)\left(1 + \frac{8}{10n} \right).$$

We illustrate Cramér–von Mises and Watson's tests on the data taken from Exercise 5.32.

Example 17.11. **Insulin Sensitivity Distribution.** In Exercise 5.32 a data set diabetes.xlsx was analyzed. The 8th column contained measurements of insulin sensitivity for 232 subjects, denoted here by X. This column was extracted to a separate data set diasi.dat. We are interested in testing whether the measurements X are consistent with gamma $\mathcal{G}a(3, 3/10)$ distribution at the significance level $\alpha = 0.05$. Is the data consistent with χ^2_{10} model?

Here we will use Cramér–von Mises and Watson's goodness-of-fit tests with their respective χ^2 approximations for the critical points.

```
%insens.m
alpha=0.05
load('diasi.dat')
x = sort(diasi);
n= length(x);
%model X ~ Ga(3, 3/10), 3/10 rate parameter.
y=gamcdf(x', 3, 10/3);                              %10/3 scale parameter
i1n = ( 2*(1:n) - 1 )/(2*n);
%Cramer-von Mises' W2
      w2 = 1/(12 * n) + sum( (y - i1n).^2 )                    %0.1793
      a1 = (48*n^2 - 137*n + 87)/(30*(32*n^2 - 61*n + 30))     %0.0498
      b1 = (32*n^2 - 61*n + 30)/(84*n*(4*n - 3))               %0.0948
      k1 = (98*n*(4*n - 3)^3)/(5*(32*n^2 - 61*n + 30)^2)       %1.2333
   w2crit = a1 + b1* chi2inv(1-alpha, k1)                      %0.4662
    chi2 = (w2 - a1)/b1                                        %1.3667
    pval1 = 1 - chi2cdf(chi2, k1)                              %0.3056
%Watson's U2
  u2=w2 - n*(mean(y)-1/2)^2                                    %0.1134
%or directly
  u2 = sum( (y - i1n - mean(y) + 1/2).^2 )+ 1/(12*n)           %0.1134
        a2= (3*n - 8)/(60*(2*n - 3))                           %0.0249
        b2=(2*n - 3)/(84*n)                                    %0.0237
        k2 = 49*n*(n-1)/(5*(2*n - 3)^2)                        %2.4713
    u2crit = a2 + b2* chi2inv(1-alpha, k2)                     %0.1875
    chi2 = (u2 - a2)/b2                                        %3.7413
    pval2 = 1 - chi2cdf(chi2, k2)                              %0.2152
```

Thus, insulin sensitivity measurements are consistent with gamma $\mathcal{G}a(3, 3/10)$ distribution with respect both Cramér–von Mises (approx p-value 0.3056) and Watson (approx p-value 0.2152) tests. The mean of this distribution is 10, so it was natural to check χ^2_{10} as a potential model. As an exercise, check that the model χ^2_{10} would be rejected by the tests, with p-values of 0.0123 and 1.2×10^{-6}, respectively.

Remark. Anderson and Darling (1952, 1954) proposed a modification of Cramér–von Mises statistic

$$A_n^2 = n \int_R [F_n(x) - F(x)]^2 \, \frac{1}{F(x)(1 - F(x))} \, dF(x). \tag{17.7}$$

The weight function $(F(x)(1 - F(x)))^{-1}$ puts more emphasis in the tails of F, compared to unweighed Cramér–von Mises statistic W_n^2.

At first glance the calculation of A_n^2 may look uninviting. However, it can be shown that

$$A_n^2 = -n - \frac{1}{n} \sum_{i=1}^{n} (2i - 1) \left[\log u_i + \log(1 - u_{n+1-i}) \right] \tag{17.8}$$

for $u_i = F(X_{(i)})$. Under the null hypothesis, $u_i = F_0(X_{(i)}) = U_{(i)}$, where $U_{(i)}$ is the ith order statistics in a sample from uniform $\mathcal{U}(0,1)$ distribution. Large values of A_n^2 are critical for H_0. The problem is the distribution of A_n^2. In applications the tabulated quantiles of the limiting statistic $A^2 = \lim_{n \to \infty} A_n^2$ are usually used. See also Exercise 17.24 for simulational approximations of quantiles and test p-value.

17.6 Rosenblatt's Test*

Rosenblatt (1952) extended Cramér–von Mises' test to compare two distributions, in the same spirit as Smirnov's test extended the Kolmogorov test. In many situations Rosenblatt's test may be more sensitive compared to Smirnov's test. Let, as in Section 17.4.2, $X_1, \ldots, X_m$ and $Y_1, \ldots, Y_n$ be two samples with cumulative distributions F and G respectively. It is of interest to test the hypothesis of homogeneity, $H_0 : F(x) = G(x)$.

Define

$$W_{m,n} = \frac{mn}{m+n} \int_{\mathbb{R}} [F_m(x) - G_n(x)]^2 \, dH_{m+n}(x),$$

where F_m and G_n are empirical CDF's based on samples $X_1, \ldots, X_m$ and $Y_1, \ldots, Y_n$, while H_{m+n} is an empirical CDF based on the combined sample $X_1, \ldots, X_m, Y_1, \ldots, Y_n$. The statistic $W_{m,n}$ can be expressed using only the ranks of X's and Y's in the combined sample,

$$W_{m,n} = \frac{1}{m+n} \left[\frac{1}{6} + \frac{1}{n} \sum_{i=1}^{m} (R_i - i)^2 + \frac{1}{m} \sum_{j=1}^{n} (S_j - j)^2 \right] - \frac{2mn}{3(m+n)}.$$

Here R_i and S_j are ranks of $X_{(i)}$ and $Y_{(j)}$ in the ordered combined sample. The statistic $W_{m,n}$ has the same limiting distribution ($m, n \to \infty$, $m/n \to$ const) as that of W_n in (13.5), implemented by ◀ cvmcdf.m. Large values of $W_{m,n}$ are critical for H_0.

When m, n are small, a correction

$$W_{m,n}' = \frac{1}{6} + \frac{W_{m,n} - \mathbb{E}W_{m,n}}{\sqrt{45 \, \mathbb{V}\mathrm{ar}\,(W_{m,n})}},$$

where

$$\mathbb{E}W_{m,n} = \frac{1}{6}\left(1 + \frac{1}{m+n}\right), \text{ and}$$

$$\mathbb{V}\mathrm{ar}\, W_{m,n} = \frac{1}{45}\left(1 + \frac{1}{m+n}\right)\left[1 + \frac{1}{m+n} - \frac{3}{4}\left(\frac{1}{m} + \frac{1}{n}\right)\right],$$

is used. The statistic $W'_{m,n}$ has the same limiting distribution as $W_{m,n}$ and can be used for m, n as small as 7.

Example 17.12. **Normal and t Revisited.** In Example 17.12 we illustrated Smirnov's test to test equality of distributions of two simulated samples. In this example we illustrate Rosenblatt's test using the same data set. The following excerpt is from MATLAB file ⬥ rosenblatt.m:

```
% From rosenblatt.m
x =[ -5.75     -3.89     -3.69     -3.68     -3.54     -2.59 ...
... deleted ...
     1.49     2.39     2.59 ];
y=[   -2.72     -2.18     -1.31     -1.17     -1.00     -0.94 ...
... deleted ...
     1.10     1.10     1.12     1.36     2.03 ];
m = length(x); n=length(y);
i = 1:m; j = 1:n;
z = [x y]; %combined sample
rz = ranks(z);
Ri = rz(1:m);  Sj=rz(m+1:m+n);
Wmn = 1/(m+n)*(1/6 + 1/n * sum( Ri - (1:m)).^2) + ...
      1/m * sum( (Sj - (1:n) ).^2 )) - 2 * m * n/(3*(m+n)) %0.6715
 pval1 = 1-cvmcdf(Wmn)                          %0.0149
%Test with corrections
EWmn = 1/6 * (1 + 1/(m+n))                      %0.1689
VarWmn = 1/45 * (1+1/(m+n) )*(1+ 1/(m+n) - ...
                3/4 * (1/m + 1/n))              %0.0219

Wprimemn = 1/6 + (Wmn-EWmn)/sqrt(45*VarWmn)  %0.6728
 pval2 = 1-cvmcdf(Wprimemn)                     %0.0148
```

Consistent with Smirnov's test, Rosenblatt's test rejects the hypothesis of homogeneity ($H_0 : F = G$) with p-value of 0.0149. The p-value for corrected statistic $W'_{m,n}$ is 0.0148, close to that of $W_{m,n}$. The correction is insignificant in this case but recommended when m, n are small. To compare the performance of Smirnov's and Rosenblatt's tests, we simulated 20,000 data sets as in this example, 39 observations from normal $\mathcal{N}(-1, 2^2)$ distribution and 35 observations from t-distribution with 10 degrees of freedom. The Smirnov test rejected the homogeneity hypothesis 84.67% of the time, while Rosenblatt's test rejected the null hypothesis 85.27% of the time. For both tests, corrections were used for increased precision. The simulations and results are given in MATLAB file ⬥ smirrosen.m.
✐

17.7 Moran's Test*

The result on page 865 stated the equivalence of testing the hypothesis $H_0 : F = F_0$ for observed $x_1, \ldots, x_n$, and testing for the uniformity of $y_1 = F_0(x_1), y_2 = F_0(x_2), \ldots, y_n = F_0(x_n)$, as long as the CDF F_0 is continuous.

Moran's test is based on spacings between ranked $y_1 \leq y_2 \leq \cdots \leq y_n$ which are, under H_0, an ordered sample form uniform $\mathcal{U}(0,1)$ distribution.

Define spacings

$$d_i = y_{(i+1)} - y_{(i)}, \quad i = 0, 1, 2, \ldots, n,$$

where $y_{(1)} \leq y_{(2)} \leq \cdots \leq y_{(n)}$ is the order statistic of $y_1, \ldots, y_n$ and $y_{(0)} = 0$ and $y_{(n+1)} = 1$. Then Moran's statistic

$$M_n = \sqrt{n} \left(\frac{n}{2} \sum_{i=0}^{n} d_i^2 - 1 \right)$$

has an approximately standard normal distribution, (Moran, 1947, 1951). This approximation is adequate if $n > 30$.

The hypothesis $H_0 : F = F_0$ is rejected if $M_n > z_\alpha$. The p-value is `1-normcdf(M)`.

Since Moran's M_n is minimized by identical spacing $d_i = 1/(n+1)$, small values of M_n may indicate nonrandom data.

Moran's test is complementary to other goodness-of-fit tests since it is sensitive to data clustering and anomalous spacing. The alternatives, consisting of long-tailed densities that remain undetected by Pearson's χ^2 or K-S-type tests, may be detected by Moran's test. The following example demonstrates that Moran's test is superior in detecting a long-tailed alternative compared to Kolmogorov's test.

Example 17.13. **Moran versus K-S.** In the following MATLAB script we repeated 1,000 times the following: (i) a sample of size 200 was generated from a t-distribution with 4 degrees of freedom, and (ii) the sample was tested for standard normality using Moran's and Kolmogorov's tests. Moran's test rejected H_0 203 times while Kolmogorov's test rejected H_0 122 times. The exact number of times H_0 was rejected varies slightly depending on the random number seed; in this example we used a combined multiple recursive generator: `stream=RandStream('mrg32k3a');`

```
%moran.m
n=200;
stream = RandStream('mrg32k3a');
```

```
RandStream.setDefaultStream(stream);
pvalmoran =[]; pvalks=[];
for i = 1:1000
  x =  trnd(4,[n 1]);           %simulate t_4
  y=normcdf(x,0,1);             %H0: F=N(0,1)
  yy =[0; sort(y); 1];
  Sd = sum(diff(yy,1).^2);
  Mn =sqrt(n)*(n * Sd/2 - 1);  %Moran's Stat
  p = 1-normcdf(Mn);
  pvalmoran = [pvalmoran p];
  [h pv dn] = kstest(x, [x normcdf(x, 0, 1)]);
  pvalks = [pvalks pv];
end
%Number of times H_0 was rejected in 1000 runs
sum(pvalmoran < 0.05)          %203
sum(pvalks < 0.05)             %122
```

This example emphasizes one of the shortcomings of K-S goodness-of-fit tests: they are sensitive in the middle of the distribution where the CDF is steep but not in the tails. The difference between t and normal distributions is mostly in the tails and Moran's spacings showed superior sensitivity.

✎

Remark. Moran's statistic is a special case of Kimball's result (Kimball, 1947). The statistic $K_{n,\alpha} = \frac{\sum_{i=0}^n d_i^\alpha - \mu_{n,\alpha}}{\sigma_{n,\alpha}}$ has a standard normal limiting distribution. Here $\mu_{n,\alpha} = \Gamma(\alpha + 1)/n^{\alpha-1}$, and $\sigma_{n,\alpha}^2 = [\Gamma(2\alpha + 1) - (\alpha^2 + 1)\Gamma^2(\alpha + 1)]/n^{2\alpha-1}$. One can show that $K_{n,2} = M_n$.

Darling (1953) proved that the sum of the log-spacings, properly normalized, has normal limiting distribution, as well.

17.8 Departures from Normality

Several tests are available specifically for the normal distribution. The Jarque–Bera test (Jarque and Bera, 1980) is a goodness-of-fit measure of departure from normality, based on the sample kurtosis and skewness. The test statistic is defined as

$$\chi_{JB}^2 = \frac{n}{6}\left(\gamma_n^2 + \frac{(\kappa_n - 3)^2}{4}\right),$$

where n is the sample size, γ_n is the sample skewness,

$$\gamma_n = \frac{\frac{1}{n}\sum_{i=1}^{n}(X_i - \overline{X})^3}{\left(\frac{1}{n}\sum_{i=1}^{n}(X_i - \overline{X})^2\right)^{3/2}},$$

and κ_n is the sample kurtosis,

$$\kappa_n = \frac{\frac{1}{n}\sum_{i=1}^{n}(X_i - \overline{X})^4}{\left(\frac{1}{n}\sum_{i=1}^{n}(X_i - \overline{X})^2\right)^2},$$

as on page 20.

Note that the first four moments are used jointly for the calculation of χ^2_{JB}. The statistic χ^2_{JB} has an asymptotic chi-square distribution with 2 degrees of freedom and measure of deviation from the skewness and kurtosis of normal distribution ($Sk = 0$ and $\kappa = 3$). This test was originally proposed by Bowman and Shenton (1975) who derived the χ^2_2 asymptotics. Jarque and Bera built on those results by exploring the power and finite sample properties.

In general, the approximation by a χ^2-distribution is crude even for large sample sizes since the convergence is slow. A case of an inadequate approximation is discussed in Example 17.14.

For this reason, the χ^2_2 approximation should not be used for sample sizes below 300. For small sample sizes one can use approximate p-values obtained by simulation, as in ◀ jarqueberatest.m

In MATLAB, the built-in Jarque–Bera test is called jbtest(x,alpha).

Example 17.14. **PCB in Yolks of Pelican Eggs.**   A well-known data set for testing agreement with a normal distribution comes from Risebrough (1972) who explored concentrations of polychlorinated biphenyl (PCB) in yolk lipids of pelican eggs. For $n = 65$ pelicans from Anacapa Island (northwest of Los Angeles), the concentrations of PCB were as follows:

452	184	115	315	139	177	214	356	166	246	177	289	175
296	205	324	260	188	208	109	204	89	320	256	138	198
191	193	316	122	305	203	396	250	230	214	46	256	204
150	218	261	143	229	173	132	175	236	220	212	119	144
147	171	216	232	216	164	185	216	199	236	237	206	87

Using the Jarque–Bera procedure, test the hypothesis that PCB concentrations are consistent with the normal distribution at the level $\alpha = 0.05$.

```
%anacapa1.m
anacapa=[452    184  %  <60 observations deleted>   237    206   87];
n=length(anacapa);    alpha=0.05;
skew = skewness(anacapa)            %0.7084
kurt = kurtosis(anacapa)            %4.2337
```

```
chi2jb = n/6 * (skew^2 + (kurt-3)^2/4)   %9.5588
approxp=1- chi2cdf(chi2jb, 2)            %0.0084
crit1 = chi2inv(1-alpha,2)              %5.9915
%more precise test by simulation...
[h stat pval crit] = jarqueberatest(anacapa)
%h = 1;   stat =9.5588;  pval = 0.0174; crit =5.2363
```

The sample size of 65 was not large enough to give an acceptable χ^2_2 approximation to the distribution of χ^2_{JB} statistic. The p-value found was 0.0084 suggesting decisive rejection of H_0. It turns out that this p-value is more than two times smaller than the exact p-value; see Figure 17.9a,b.

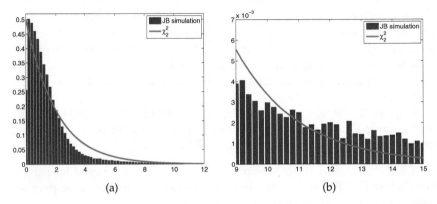

(a)   (b)

Fig. 17.9 (a) Empirical distribution of null distribution of JB statistic and its χ^2_2 approximation. Note inadequate approximation for the sample size of $n = 65$ as in this example. (b) Detail of tail behavior explaining why the approximation yielded incorrect p-value of 0.0084, when the correct p-value is close to 0.0174.

A better approximation is obtained by `jarqueberatest.m`, which finds the p-value as 0.0174. This is still significant at 5% level. The decisions by relying on χ^2_2 approximation would be wrong if the significance level was set to 1%.

✐

Kolmogorov–Smirnov test with Lilliefors' correction (Lilliefors, 1967) for departures from normality is a modification of the Kolmogorov's test, where the theoretical distribution is normal but not fully specified, as the Kolmogorov test requires. When the sample mean and variance are used to specify the null distribution, the usual K-S quantiles become conservative and simulations are used for their better approximation.

The MATLAB command for Kolmogorov's test with Lilliefors' correction is `[h,p,ksstat,critval] = lillietest(x, alpha)`, where the data to be tested are in vector x.

17.9 Ellimination of Unknown Parameters by Transformations

Instead of plugging in the estimators of unknown parameters to make the null distribution-specific, in some cases it is possible to transform observations and obtain parameter-free distributions. This freedom is "paid" by lowering the sample size, but only slightly. We demonstrate how this can be done in testing for exponentiality and normality.

Testing for Exponentiality. Let the null hypothesis be that a sample $X_1, X_2, \ldots, X_n$ is exponential $\mathcal{E}(\lambda)$ with rate parameter λ unspecified.

Sort the sample and define scaled spacings $d_i = (n + 1 - i)(X_{(i)} - X_{(i-1)})$, $i = 1, 2, \ldots, n$ with $X_{(0)} = 0$. Using the scaled spacings d_i, define

$$Y_j = \frac{\sum_{i=1}^{j} d_i}{\sum_{i=1}^{n} d_i}, \quad j = 1, 2, \ldots, n - 1.$$

The values $Y_1, \ldots, Y_{n-1}$ are i.i.d. uniform $\mathcal{U}(0,1)$. A test for exponentiality of $X_1, \ldots, X_n$ is equivalent to a test for uniformity of $Y_1, \ldots, Y_{n-1}$, which can be conducted by methods described in previous sections of this chapter.

Testing for Normality. Let the null hypothesis be that $X_1, X_2, \ldots, X_n$ came from normal $\mathcal{N}(\mu, \sigma^2)$ distribution, with parameters μ and σ^2 unknown. Fix any k, $1 \leq k \leq n$, and define

$$m_k = \frac{1}{n + \sqrt{n}} \sum_{i=1}^{n} X_i + \frac{1}{1 + \sqrt{n}} X_k.$$

Define now

$$Y_i = \begin{cases} m_k - X_i, & i = 1, 2, \ldots, k - 1 \\ m_k - X_{i+1}, & i = k, \ldots, n - 1 \end{cases}$$

Then Y_i, $i = 1, \ldots, n - 1$ are i.i.d. normal $\mathcal{N}(0, \sigma^2)$. Note that we could have eliminated unknown location parameter μ simply by subtracting the sample mean, $Y_i = X_i - \overline{X}$, but the resulting Y_i's would not be independent.

To eliminate σ^2, we transform Y_i as

$$T_j = \frac{Y_j}{\sqrt{\frac{1}{n-j-1} \sum_{k=j+1}^{n-1} Y_k^2}}, \quad j = 1, 2, \ldots, n - 2.$$

The T_j's are independent, t-distributed with $n - j - 1$ degrees of freedom. Finally, if F_{n-j-1} is the CDF of t-distribution with $n - j - 1$ degrees of freedom, the variables

$$U_j = F_{n-j-1}(T_j), \quad j = 1, \ldots, n-2$$

are distributed as i.i.d. uniform $\mathcal{U}(0,1)$.

Testing for the normality of $X_1, \ldots, X_n$ is now equivalent to testing for the uniformity of $U_1, \ldots, U_{n-2}$, which again can be done by using the tests discussed in previous sections.

These two cases of parameter elimination are illustrated in MATLAB script ◀ elimination.m. Two samples of size 200, one from exponential $\mathcal{E}(\lambda)$ distribution and the other from normal $\mathcal{N}(\mu, \sigma^2)$ distribution, are transformed to uniform $\mathcal{U}(0,1)$ distribution of respective sizes 199 and 198. To make simulation of the samples specific, we took $\lambda = 2.5$, $\mu = 3$, and $\sigma^2 = 2^2$, but these values have not been used in the transformations.

17.10 Exercises

17.1. **Q–Q Plot for $\sqrt{2\chi^2}$.** Simulate $N = 10,000$ χ^2 random variables with $k = 40$ degrees of freedom. Demonstrate empirically that

$$Z = \sqrt{2\chi^2} - \sqrt{2k-1}$$

is approximately standard normal $\mathcal{N}(0,1)$ by plotting a histogram and Q–Q plot.

Hint: `df=40; zs=sqrt(2*chi2rnd(df,[1 10000]))-sqrt(2*df-1);`

17.2. **Not at All Like Me.** You have a theory that if you ask subjects to sort one-sentence characteristics of people (e.g., "I like to get up early in the morning") into five groups ranging from *not at all like me* (group 1) to *very much like me* (group 5), the percentage falling in each of the five groups will be approximately 10, 20, 40, 20, and 10. You have one of your friends sort 50 statements, and you obtain the following data: 8, 9, 21, 8, and 4. Do these data support your hypothesis at $\alpha = 5\%$?

17.3. **Different Expressions for χ^2.** Derive expressions in (17.3) from the expression in (17.2).

17.4. **Cell Counts.** A student takes the blood cell count of five random blood samples from a larger volume of solution to determine if it is well mixed. She expects the cell counts to be distributed uniformly. The data is given below:

Sample blood cell count	Expected blood cell count
35	27.2
20	27.2
25	27.2
25	27.2
31	27.2
136	136

Can she depend on the results to be uniformly distributed? Perform a chi-square test with $\alpha = 0.05$.

17.5. **GSS Data.** Below is part of the data from the 1984 and 1990 General Social Survey, conducted annually by the National Opinion Research Center. Random samples of 1,473 persons in 1984 and 899 persons in 1990 were taken using multistage cluster sampling. One of the questions in a 67-question-long questionnaire was: *Do you think most people would try to take advantage of you if they get a chance, or would they try to be fair?*

	1984 survey	1990 survey
1. Would take advantage of you	507	325
2. Would try to be fair	913	515
3. Depends	47	53
4. No answer	6	6

Assuming that the 1984 frequencies are theoretical, using Pearson's χ^2 test, explore how the 1990 frequencies agree. Use $\alpha = 0.05$. State your findings clearly.

17.6. **Strokes on "Black Monday."** In a long-term study of heart disease, the day of the week on which 63 seemingly healthy men died was recorded. These men had no history of disease and died suddenly.

Day of week	Mon.	Tue.	Wed.	Thu.	Fri.	Sat.	Sun.
Deaths	22	7	6	13	5	4	6

(a) Test the hypothesis that these men were just as likely to die on one day as on any other. Use $\alpha = 0.05$.
(b) Explain in words what constitutes the error of the second kind in the testing from (a).

17.7. **Monocytes among Blood Cells Revisited.** In Exercise 7.12 we assumed Poisson and Binomial models to estimate parameters λ and p. We also found expected frequencies for the number of monocytes in 100 blood cells, for both models.
For which model is the χ^2 statistic smaller? Compare the p-values.

17.8. **Benford's Law.** Benford's law (Benford, 1938; Hill, 1998) concerns the
relative frequencies of leading digits of various data sets, numerical tables, accounting data, etc. Benford's law, also called the first digit law,
states that in numbers from many sources, the leading digit is much
more often a 1 than it is any other digit (specifically about 30% of the
time). Furthermore, the higher the digit, the less likely it is to occur as
the leading digit of a number. This applies to figures related to the natural world or figures of social significance, be it numbers taken from
electricity bills, newspaper articles, street addresses, stock prices, population numbers, death rates, areas or lengths of rivers, or physical and
mathematical constants.

More precisely, the Benford law states that the leading digit n, ($n = 1,\ldots,9$) occurs with probability $P(n) = \log_{10}(n + 1) - \log_{10}(n)$, or as in
the table below [probabilities $P(n)$ rounded to 4 decimal places]:

Digit n	1	2	3	4	5	6	7	8	9
$P(n)$	0.3010	0.1761	0.1249	0.0969	0.0792	0.0669	0.0580	0.0512	0.0458

An oft-cited data set is the distribution of the leading digit of all 309
numbers appearing in a particular issue of *Reader's Digest*.

Digit	1	2	3	4	5	6	7	8	9
Count	103	57	38	23	20	23	17	15	13

At level $\alpha = 0.05$, test the hypothesis that the observed distribution of
leading digits in *Reader's Digest* numbers is consistent with Benford's
law.

Hint: To speed up the calculation, some theoretical frequencies rounded
to two decimal places are provided:

93.01	54.41	•	29.94	24.47	•	•	15.82	14.15

17.9. **Cholestyramine.** In a type II coronary intervention study, patients with
type II hyperlipoproteinemia (a condition characterized by high levels of
beta-lipoproteins in the blood) and coronary heart disease were assigned
at random to a daily dose of 24 g of cholestyramine or to a placebo.
Cholestyramine removes bile acids from the body by forming insoluble
complexes. In the process, plasma cholesterol is converted to bile acid to
normalize acid levels, which ultimately lowers plasma cholesterol concentrations.

After 5 years, the number of vascular lesions were counted on each patient's angiogram. The data in table below are reported by Barnhart and
Sampson (1995):

Number of	Number of patients	
lesions	Cholestyramine group	Placebo group
0	5	2
1	4	4
2	6	6
3	5	4
4	7	6
5	8	9
6	6	7
7	6	5
8	7	2
9	2	4
≥ 10	4	8
Total	60	57

You have a choice between Poisson and negative binomial distributions, as possible models for the number of lesions in the cholestyramine group. Using the χ^2-test check which distribution better fits the data (e.g., results in smaller chi-square statistic).

For the negative binomial use $r = 4$ and $p = 0.44$, and for the Poisson use $\lambda = 4.6$.

17.10. **Simulational Exercise.**

(a) Generate a sample of size $n = 216$ from a standard normal distribution.

(b) Select intervals by cutting the range of simulated values by points $-2.7, -2.2, -2, -1.7, -1.5, -1.2, -1, -0.8, -0.5, -0.3, 0, 0.2, 0.4, 0.9, 1, 1.4, 1.6, 1.9, 2, 2.5,$ and 2.8.

(c) Using a χ^2-test confirm the normality of the sample.

(d) Repeat the previous test if your sample is contaminated by a Cauchy $Ca(0,1)$ distribution in the following way: `0.95 * normal_sample + 0.05 * cauchy_sample`.

17.11. **Black Wednesday.** This exercise concerns the analysis of 5-year data on association of temporal factors (day-of-the-week) to suicides in the United States. The number of suicides (in thousands) provided in the table below was pooled from the Multiple Cause of Death Files from the years 2000 to 2004.

Is the probability distribution of suicide in the US uniform over days of the week?

Day of the week	Suicides (in thousands)
Sunday	14.5
Monday	18.8
Tuesday	17.6
Wednesday	32.9
Thursday	14.6
Friday	14.7
Saturday	18.9
Total	132.0

(a) Using Pearson's χ^2 procedure, test the hypothesis that the probability distribution is discrete uniform, that is that the probability of a suicide on any given day of the week is $1/7$. Use $\alpha = 0.05$ as significance level.
(b) The data suggest an unusual *Black Wednesday* effect, and researchers tried to explain the phenomenon (Kposowa and D'Auria, 2010). Test the hypothesis that the probability distribution of a suicide on a particular weekday is

$$p_{SU} = p_{MO} = p_{TU} = p_{TH} = p_{FR} = p_{SA} = \frac{1}{8}, \quad p_{WE} = \frac{1}{4}.$$

For both (a) and (b) find p-values and compare them to 0.05-significance level.

Hint: Retain data as presented, in units of thousands. Using the actual numbers (e.g., for Monday the exact number was 18,824) would extensively overpower the test and a distribution with a satisfactory fit would be unreasonably close to the empirical distribution (relative frequencies).

17.12. **Deathbed Scenes.** Can some people postpone their death until after a special event takes place? It is believed that famous people do so with respect to their birthdays to which they attach some importance. A study by Philips (1972) seems to be consistent with this notion. Philips obtained data[1] on the months of birth and death of 1,251 famous Americans; the deaths were classified by the time period between the birth dates and death dates as shown in the following table:

B	e	f	o	r	e	Birth	A	f	t	e	r
6	5	4	3	2	1	month	1	2	3	4	5
90	100	87	96	101	86	119	118	121	114	113	106

---

[1] 348 were people listed in *Four Hundred Notable Americans* and 903 are listed as the foremost families in three volumes of *Who Was Who* for the years 1951–1960, 1943–1950, and 1897–1942.

(a) Clearly formulate the statistical question based on the observations above.

(b) Provide a solution for the question formulated in (a). Use $\alpha = 0.05$.

17.13. **Grouping in a Vervet Monkey Troop.** Struhsaker (1965) recorded, individual by individual, the composition of sleeping groups of wild vervet monkeys (*Cercopithecus aethiops*) in East Africa. A particular large troop was observed for 22 nights.

Size	1	2	3	4	5	6	7	8	9	10	11	≥12
Freq	19	14	19	11	7	7	3	2	3	1	2	1

It is well known that the sizes of groups among humans in various social settings (shoppers, playgroups, pedestrians) follows a truncated Poisson distribution:

$$P(X = k) = \frac{\lambda^k \exp\{-\lambda\}}{k!(1 - \exp\{-\lambda\})}, \quad k = 1,2,3,\ldots.$$

Using a χ^2 goodness-of-fit test, demonstrate that the truncated Poisson is not a good model for this data at the significance level $\alpha = 0.05$. To find the theoretical frequencies, λ can be estimated by the method of moments using an average size of $\overline{X} = 3.74$ and the equation $\lambda/(1 - \exp\{-\lambda\}) = 3.74$. In MATLAB `fzero(@(lam) lam - 3.74*(1-exp(-lam)), 3)` gives $\hat{\lambda} = 3.642$.

Struhsaker further argues that in this context the truncated negative binomial gives a reasonably good fit. Check this.

17.14. **Crossing Mushrooms.** In a botany experiment the results of crossing two hybrids of a species of mushrooms (*Agaricus bisporus*) gave observed frequencies of $120, 53, 36$, and 15. Do these results disagree with theoretical frequencies that specify a 9:3:3:1 ratio? Use $\alpha = 0.05$.

17.15. **Renner's Honey Data Revisited.** In Example 6.11 we argued that a lognormal distribution with parameters $\mu = -0.6084$ and $\sigma^2 = 1.0040^2$ provides a good fit for Renner's honey data. Find χ^2-statistics and the p-value of the fit. State your decision at level $\alpha = 0.05$. The midintervals in the first column of data set 📇 renner.mat|dat, $0.125, 0.375, 0.625, \ldots, 7.125$ correspond to the intervals $(0,0.25], (0.25,0.5], (0.5,0.75], \ldots, (7,7.25]$.

17.16. **Beta-Geometric Distribution, Smoking and Pregnancy.** Beta-Geometric distribution is defined in the following way. Given π, X is geometric $\mathcal{Geom}(\pi)$,

$$\mathbb{P}(X = m|\pi) = (1 - \pi)^{m-1}\pi, \quad m = 1,2,\ldots.$$

When π has beta distribution $\mathcal{Be}(a,b)$, the marginal (unconditional) distribution for X is beta-geometric

$$\mathbb{P}(X = m) = \frac{B(a+1, b+m-1)}{B(a,b)}, \quad m = 1, 2, \dots$$

where $B(a,b)$ is beta function, in MATLAB `beta(a,b)`. The beta-geometric distribution has a mean $\mathbb{E}X = \frac{a+b-1}{a-1}$. Higher moments involve special functions.

Weinberg and Gladen (1986) provide data in which the times (in menstrual cycles) taken by the couples who are attempting to conceive until pregnancy results. The data is retrospective, starting from the pregnancy in each case. This fecundity data set features 586 women contributing to a total of 1,844 cycles.

Cycles	1	2	3	4	5	6	7	8	9	10	11	12	> 12
Smokers	29	16	17	4	3	9	4	5	1	1	1	3	7
Nonsmokers	198	107	55	38	18	22	7	9	5	3	6	6	12

The authors fitted this data with beta-geometric distributions and found a satisfactory fit.

Using the chi-square goodness of fit test, confirm Weinberg and Gladen's findings for Nonsmokers only ($n = 486$). Use $a = 3.5$ and $b = 5$ at the level $\alpha = 0.05$.

Hint: Observed frequency for $m = 3$ is 55. The theoretical frequency is

```
betageom  = @(n, a, b) beta(a+1, b+n-1)/beta(a,b);
486*betageom(3, 3.5, 5)  %60.1858
```

17.17. Deviations From 90%-Circularity. Refer to Exercise 2.29 for the description of the experiment. The data set `dcirc.dat|mat` contains deviations from 90% of circularity parameter, for static condition after 2 days of growth. The researchers believe that this deviation is distributed as beta with parameters 3/2 and 5.

Using Kolmogorov's procedure test the hypothesis that the distribution of deviations from 90% circularity is indeed consistent with $\mathcal{B}e(3/2, 5)$. State your decision at the $\alpha = 0.05$ significance level. Report the p-value.

17.18. PCB in Yolks of Pelican Eggs. In Example 17.14 we tested the hypothesis that PCB concentrations are consistent with the normal distribution at the level $\alpha = 0.05$ using Jarque–Bera test. It was found that data significantly deviate from normality.

For the same data test for the normality using (a) Pearson's χ^2-test and (b) Lilliefors test. Compare the results.

Suppose that the null distribution is fully specified as normal $\mathcal{N}(200, 70^2)$. Test for the null distribution using (c) Kolmogorov's test, (d) Cramér–von Mises' test, and (e) Moran's test.

17.19. Number of Leaves per Whorl in *Ceratophyllum demersum.* After spending a year with Karl Pearson in London, biometrician Raymond Pearl published his work on mathematical and statistical modeling of growth of an aquatic plant, *Ceratophyllum demersum* (Pearl, 1907, p. 17). Among several interesting data sets, Pearl gives the table of frequencies of numbers of leaves per whorl (Series I, Plant 4)

Leaves	6	7	8	9	10	11	Total
Whorls	17	35	71	77	58	5	263

(a) Are the observed frequencies consistent with binomial $Bin(11, p)$ distribution at 5% significance level when p is estimated from the data?
(b) If $p = 0.777$ is considered to be known in advance, that is, no estimation from the given data was made, would the decision be the same as in (a)?

17.20. **From 1998–2002 US National Health Interview Survey (NHIS).** Stansfield and Carlton (2007) analyzed data in sibships of size 2 and 3 from the 1998–2002 US National Health Interview Survey (NHIS).
For the 25,468 sibships of size 2, among all 50,936 children 10 years of age and younger in the NHIS data set (the youngest age cohort available), they found that 51.38% were boys (B) and 48.62% were girls (G); the B/G ratio was 1.0566, or about 106 boys for every 100 girls. The number of boys is given in the following table:

Number of boys	0	1	2
Observed sibships	5,844	13,079	6,545

For the 7,541 sibships of size 3 from the same NHIS data set the number of boys among the first two children is given in the following table:

Number of boys	0	1	2
Observed sibships	1,941	3,393	2,207

(a) If the number of boys among the first two children in sibships of size 3 is binomial $Bin(2, 0.515)$, find what theoretical frequencies are expected in the table for sibships of size 3. Are the observed and theoretical frequencies close? Comment.
Hint: Find the binomial probabilities and multiply them by $n = 7,541$.
(b) For the 25,468 sibships of size 2, test the hypothesis that the probability of a boy is 1/2 versus the one-sided alternative, that is,

$$H_0 : p = 0.5 \quad \text{versus} \quad H_1 : p > 0.5,$$

at the level $\alpha = 0.05$.
Hint: Be careful about the n here. The rejection region of this test is $RR = [1.645, \infty)$.

17.21. **Neuron Fires Revisited.** The 989 firing times of a cell culture of neurons have been recorded. The records are time instances when a neuron sends a signal to another linked neuron (a spike) and the largest time instance is 1001.

The count of firings in 200 consecutive time intervals of length 5 (= 1000/200) time units is as follows:

2 7 5	5 1 6 8 4 6 4 5 7	7 5 4 7 4 4 5	9 6 5 6 6 5
5 6 4 10 7 8 8 2 9 5 4 4	4 3 8 3 2 7 6	7 5 6 6 4 6
5 7 3	5 6 5 5 2 7 7 6 4	8 8 9 7 3 3 3	5 6 6 4 6 4
5 4 4	5 2 3 5 1 4 4 3 2 10 4 7 2 1 7 9	4 3 6 8 6 5
2 4 4	3 7 2 2 6 1 3 7 6	6 4 7 5 5 8 7	3 5 5 4 5 6
10 3 6	9 5 7 2 8 4 4 2 3	4 3 3 3 5 4 2	2 7 4 5 4 5
3 5 3	5 5 7 2 5 5 4 8 4	3 5 7 4 4 8 5	4 3 6 4 6 4
1 7 9	4 2 5 4 4 4 4 7 2	4 4 5 2 4 7 5 12 1 5 4 6 7

Are the data consistent with a Poisson distribution?

17.22. Cloudiness in Greenwich. Pearse (1928) cites data describing the degree of cloudiness in Greenwich, UK, for days in the month of July between 1890 and 1904.

Cloudiness	0	1	2	3	4	5	6	7	8	9	10	Total
Days	320	129	74	68	45	45	55	65	90	148	676	1715

Assume that the measure of cloudiness can be modeled as a continuous random variable $C = 10 \cdot X$, where X has a beta $Be(\alpha,\beta)$ distribution. Find the best-fitting parameters α and β and assess the goodness of fit.

17.23. Distance between Spiral Reversals in Cotton Fibers. Tippett (1941) discusses a frequency histogram for a data set consisting of intervals and counts for a distance between spiral reversals in cotton fiber (Figure 17.10). The distances are given in units of mm^{-2}, and the sample size was 1,117.

Distance	[0,2.5)	[2.5,4.5)	[4.5,6.5)	[6.5,8.5)	[8.5,10.5)
Number	7	48	100	106	84
	[10.5,12.5)	[12.5,16.5)	[16.5,20.5)	[20.5,24.5)	[24.5,28.5)
	72	136	94	78	69
	[28.5,32.5)	[32.5,36.5)	[36.5,40.5)	[40.5,50.5)	[50.5,60.5)
	53	45	36	69	40
	[60.5,70.5)	[70.5,80.5)	[80.5,90.5)	[90.5,∞)	Total
	31	21	17	11	1117

Using Pearson's χ^2 criterion and significance level $\alpha = 0.05$, test the hypothesis that the data are consistent with the χ^2-distribution. Use the midpoints of intervals to estimate the mean of a theoretical χ^2-distribution. Recall that the mean in this case is equal to the number of degrees of freedom.

Fig. 17.10 Structure of spiral reversals in cotton fiber.

17.24. **Lin–Mudholkar Test.** Lin and Mudholkar (1980) proposed a test for normality that is based on the following fact: the mean and variance in a random sample are independent, if and only if, the parent population is normal. See also Exercise 6.13. The Lin–Mudholkar test is sensitive only to departures from normality due to skewness. test The procedure has little power in detecting non-normal symmetrical distributions. For example, uniform distributions or bimodal symmetric distributions will pass this test easily. Therefore, other tests should also be consulted if there is a concern for these types of departure from normality. MATLAB script ◢ LinMudholkar.m implements the test.

(a) Show that Lin–Mudholkar's test rejects the hypothesis of symmetry for data in Example 17.14, page 873.

(b) Generate two vectors:

```
x1=[normrnd(2,1,[1000 1]); normrnd(-2,1,[1000 1])];
x2=[normrnd(2,1,[1000 1]); normrnd(-2,1.15,[1000 1])];
```

The distributions of x1 and x2 are bimodal, thus clearly non-gaussian. Show that x1 easily passes Lin–Mudholkar's test while x2 fails to pass at 5% significance level. Can you explain why?

Anderson–Darling by Simulation. In this exercise you are asked to approximate quantiles of statistic A_n^2 given in (17.8) by simulation. Follow these steps:

- Take n equal to the size of the original sample.
- Take B runs (say $B = 10,000$ or more), and for each run $b = 1, \dots, B$:
- Simulate n uniforms $U_i, i = 1, \dots, n$, find their order statistics $u_i = U_{(i)}$, by (17.8) calculate $A_{n,b}^2$, and save it.
- The empirical distribution, based on B simulated $A_{n,b}^2$, produces an approximation to the quantiles and test p-value.

For example, for observed statistics A_n^2, the test p-value is approximated by $\#\{A_{n,b}^2 \geq A_n^2\}/B$. The quantiles can be approximated by sample quantiles of $A_{n,b}^2$. When n is large, this simulation approximates the limiting distribution, that of A^2.

The following 54 observations were generated from gamma $\mathcal{G}a(2,3)$ distribution. Here 3 is the rate parameter.

1.2346	0.1205	0.2018	0.3438	0.6359
0.8481	0.9760	1.0241	0.6305	0.1686
1.9662	0.8579	1.7217	0.2149	1.4038
0.4389	0.5844	0.4281	1.0587	1.0276
0.8505	0.8577	0.4915	0.2164	0.1886
1.0571	0.3483	0.1317	0.0610	0.7597
0.3718	0.4449	0.5733	0.5834	0.7581
0.2152	0.9859	0.1797	0.3645	0.7342
0.7709	0.3281	1.5775	0.6292	0.8452
0.7634	0.2071	0.6350	0.6514	0.5634
0.3921	0.7281	1.8689	0.8891	

(a) Find an approximation to $0.90, 0.95,$ and 0.99 quantiles for A_{54}^2. Check against Marsaglia and Marsaglia (2004) corresponding A^2 quantiles, accurate up to 20 decimal places, $1.93295783274159373 04, 2.4923671600494096176$ and 3.8781250216053948842.

(b) For the simulated data test the hypothesis

$$H_0 : F(x) \text{ is CDF of gamma } \mathcal{G}a(2,3) \quad \text{vs.} \quad H_1 = H_0^c.$$

Use approximate p-values, based on simulation in (a).

(c) As in (b), test the hypothesis

$$H_0 : F(x) \text{ is CDF of exponential } \mathcal{E}(3/2) \quad \text{vs.} \quad H_1 = H_0^c.$$

MATLAB FILES AND DATA SETS USED IN THIS CHAPTER

http://statbook.gatech.edu/Ch17.Goodness/

airconditions.m, anacapa.m, anacapa1.m, anacapapict.m, betageometric.m, consolidator.m, cvmcdf.m, elimination.m, empiricalcdf.m, examples.m, horsekicks.m, insens.m, jarqueberatest.m, kolsm.m, ks2cdf.m, kscdf.m, kscdfcensor.m, ksexample.m, mises.m, mises2.m, moran.m, mrjones.m, neuronfiresqq.m, neurongof.m, nomega2.m, nomega2cdf.m, notatalllikeme.m, omega2cdf.m, perk.m, plotsimulqq.m, poissonness.m, ppqq.m, qqbivariatenormal.m, qqnorm.m, qqplotsGOF.m, RANDmillion.m, rosenblatt.m, shapirowilksimulation.m, smirnov.m, smirnov2.m, smirrosen.m, swdist.m, tippett.m, watson.m, watson1.m, weldon.m

dcirc.dat|mat, diasi.dat, mrjones.mat, neuronfires.mat, RAND1Millrandomdigits.txt, renner.dat|mat

---

CHAPTER REFERENCES

Anderson, T. W. and Darling, D. A. (1952). Asymptotic theory of certain "goodness-of-fit" criteria based on stochastic processes. *Annals of Mathematical Statistics*, **23**, 193–212. doi:10.1214/aoms/1177729437.

Anderson, T. W. and Darling, D. A. (1954). A Test of Goodness-of-Fit. *J. Am. Stat. Assoc.*, **49**, 765–769. doi:10.2307/2281537.

Barnhart H. X. and Sampson A. R. (1995). Multiple population models for multivariate random length data–with applications in clinical trials. *Biometrics*, **51**, 1, 195–204.

Benford, F. (1938). The law of anomalous numbers. *Proc. Am. Philos. Soc.*, **78**, 551–572.

Bowman, K. O. and Shenton, L. R. (1975). Omnibus contours for normality based on $\sqrt{b_1}$ and b_2. *Biometrika*, **62**, 243–250.

Brown, L., DasGupta, A., Marden, J., and Politis, D. (2004). Characterizations, Sub and Resampling, and Goodness of Fit. In: Lecture Notes-Monograph Series, Vol. 45, A Festschrift for Herman Rubin, 180–206.

Cramér, H. (1928). On the composition of elementary errors. *Skand. Aktuarietidskr.*, **11**, 13–74; 141–180.

Darling, D. A. (1953). On a class of problems related to the random division of an interval. *Ann. Math. Statist.*, **24**, 239–253.

Hill, T. (1998). The first digit phenomenon. *Am. Sci.*, **86**, 358–363.

Hoaglin, D. C. (1980). A poissonness plot. *Am. Stat.*, **34**, 146–149.

Jarque, C. and Bera, A. (1980). Efficient tests for normality, homoscedasticity and serial independence of regression residuals. *Econ. Lett.*, **6**, 3, 255–259.

Kemp, W. A. and Kemp, D. C. (1991). Weldon's dice data revisited. *Am. Stat.*, **45**, 3, 216–222.

Kimball, B. F. (1947). Some basic theorems for developing tests of fit for the case of the non-parametric probability distribution function. *Ann. Math. Statist.*,**18**, 4, 540–548.

Kolmogorov, A. N. (1933). Sulla determinazione empirica di una legge di distributione. *Gior. Ist. Ital. Attuari*, **4**, 83–91.

Kposowa, A. J. and D'Auria, S. (2010). Association of temporal factors and suicides in the United States, 2000–2004 *Soc. Psychiatry Psychiatr. Epidemiol.*, **45**, 4, 433–445. doi: 10.1007/s00127-009-0082-9, PMCID: PMC2834764

Kvam, P. and Vidakovic, B. (2007). *Nonparametric Statistics with Applications to Science and Engineering*. Wiley, Hoboken.

Lilliefors, H. W. (1967). On the Komogorov–Smirnov test for normality with mean and variance unknown. *J. Am. Stat. Assoc.*, **62**, 399–402.

Lin, C.-C. and Mudholkar, G. S. (1980). A simple test for normality against asymmetric alternatives. *Biometrika*, **67**, 455–461.

Mann, H. B. and Wald, A. (1942). On the choice of the number of class intervals in the application of the chi square test. *Ann. Math. Statist.*, **13** , 3, 306–317.

Marsaglia, G. and Marsaglia, J. (2004). Evaluating Anderson-Darling Statistics. *J. Stat. Softw.*, **9**, 2, 1–5.

Moran, P. A. P. (1947). The random division of an interval – Part I. *J. Roy. Statist. Soc. Ser. B*, **9**, 92–98.

Moran, P. A. P. (1951). The random division of an interval – Part II. *J. Roy. Statist. Soc. Ser. B*, **13**, 147–150.

Pearl, R. (1907). Variation and differentiation in *Ceratophyllum*. *Carnegie Inst. Wash. Publ.*, **58**, 1–136.

Pearse, G. E. (1928). On corrections for the moment-coefficients of frequency distributions. *Biometrika*, **20** A, 314–355.

Pearson, K. (1900). On the criterion that a given system of deviations from the probable in the case of a correlated system of variables is such that it can be reasonably supposed to have arisen from random sampling. *Philos. Mag. Ser.*, **5**, 50, 157–175.

Pearson, E. S. and Hartley, H. O. (1972). *Biometrika Tables for Statisticians. Volume II* Cambridge University Press, London.

Phillips, D. P. (1972). Deathday and birthday: an unexpected connection. In: Tanur, J. M., ed. *Statistics: a guide to the unknown*. Holden-Day, San Francisco, 52–65.

Risebrough, R. W. (1972). Effects of environmental pollutants upon animals other than man. *Proc. Sixth Berkeley Symp. on Math. Statist. Prob.*, vol. 6, 443–453.

Rosenblatt, M. (1952). Limit theorems associated with variants of the von Mises statistic. *Annals of Mathematical Statistics*, **23**, 617–623.

Smirnov, N. (1939). On the estimation of the discrepancy between empirical curves of distribution for two independent samples. *Bull. Math. Univ. Moscou*, **2**, 3–14.

Stansfield, W. D. and Carlton, M. A. (2007). Human sex ratios and sex distribution in sibships of size 2. *Hum. Biol.*, **79**, 255–260.

Stephens, M. A. (1974). EDF Statistics for goodness of fit and some comparisons. *Journal of the American Statistical Association*, **69**, 347, 730–737.

Struhsaker, T. T. (1965). Behavior of the vervet monkey (*Cercopithecus aethiops*). PhD dissertation. University of California-Berkeley.

Tiku, M. L. (1965). Chi-square approximation for the distribution of goodness-of-fit statistics U_N^2 and W_N^2. *Biometrika*, **52**, 3/4, 630-633

von Bortkiewicz, L. (1898). Das Gesetz der kleinen Zahlen. Teubner, Leipzig.

von Mises, R. E. (1928). *Wahrscheinlichkeit, Statistik und Wahrheit*. Julius Springer, Vienna.

Watson, G. S. (1961). Goodness-of-fit tests on a circle. *Biometrika*, **48**, 1-2, 109–114.

Weinberg, C. R. and Gladen, B. C. (1986). The beta-geometric distribution applied to comparative fecundability studies, *Biometrics*, **42**, 552

Yates, F. (1934). Contingency table involving small numbers and the χ^2 test. *J. R. Stat. Soc.*, **1** (suppl.), 2, 217–235.

Chapter 18
Distribution-Free Methods

Assumptions are the termites of relationships.

– Henry Winkler

18.1 Introduction

Most of the methods we have covered until now are based on parametric assumptions; the data are assumed to follow some well-known family of distributions, such as normal, exponential, or Poisson. Each of these distributions is indexed by one or more parameters (e.g., the normal distribution has μ and σ^2), and at least one is presumed unknown and must be inferred. However, with complex experiments and messy sampling plans, the generated data might not conform to any well-known distribution. In the case where the experimenter is not sure about the underlying distribution of the data, statistical techniques that can be applied regardless of the true distribution of the data are needed. These techniques are called *distribution-free* or *nonparametric*. To quote statistician James V. Bradley (1968, p. 15):

> The terms nonparametric and distribution-free are not synonymous, and neither term provides an entirely satisfactory description of the class of statistics to which they are intended to refer. . . . . Popular usage, however, has equated the terms. . . . Roughly speaking, a nonparametric test is one which makes no hypothesis about the value of a parameter in a statistical density function, whereas a distribution-free test is one which makes no assumptions about the precise form of the sampled population.

Many basic statistical procedures, such as the *t*-test, test for Pearson's correlation coefficient, and analysis of variance, assume that the sampled data come from a normal distribution or that the sample size is sufficiently large for the CLT to make the normality assumption reasonable.

In this chapter, we will revisit the testing of means, the two-sample problem, ANOVA, and the repeated measures design, without making strict distributional assumptions about the data. The material in the next chapter dealing with goodness-of-fit tests is also considered a nonparametric methodology. For a more comprehensive account on the theory and use of nonparametric methods in engineering research, we direct the reader to Kvam and Vidakovic (2007).

The following table gives nonparametric counterparts to a few of the most popular inferential procedures. The acronym WSiRT stands for Wilcoxon's signed-rank test, while WSuRT stands for Wilcoxon's sum-rank test.

Parametric	Nonparametric
One-sample *t*-test for the location	Sign test, WSiRT
Paired *t*-test	Sign test, WSiRT
Two-sample *t*-test	WSuRT, Wilcoxon–Mann–Whitney
One-way ANOVA	Kruskal–Wallis test
Block design ANOVA	Friedman test

18.2 Sign Test

We start with the simplest nonparametric procedure, the sign test. It was used in an informal way as early as 1710 (Arbuthnot, 1710).

Suppose we are interested in testing the hypothesis H_0 that a population with a continuous CDF has a median m_0 against one of the alternatives $H_1 : med > \neq < m_0$.

We assign a $+$ sign when $X_i > m_0$, i.e., when the difference $X_i - m_0$ is positive, and a $-$ sign when $X_i < m_0$. The case $X_i = m_0$ (a tie) is theoretically impossible for continuous distributions, although when observations are coded with limited precision, ties are quite possible and most of the time present. Henceforth we will assume an ideal situation in which no ties occur.

If m_0 is the median and if H_0 is true, then from the definition of the median and continuity of CDF, $\mathbb{P}(X_i > m_0) = \mathbb{P}(X_i < m_0) = \frac{1}{2}$. Thus, the statistic T representing the total number of $+$ is equal to

$$T = \sum_{i=1}^{n} \mathbf{1}(X_i > m_0)$$

and has a binomial distribution with parameters n and $1/2$.

Let the level of the test, α, be specified. When the alternative is $H_1 : med > m_0$, the critical values of T are integers greater than or equal to k_α, which is defined as the smallest integer for which the relationship

$$\sum_{k=k_\alpha}^{n} \binom{n}{k} \left(\frac{1}{2}\right)^n < \alpha$$

holds.

Likewise, if the alternative is $H_1 : med < m_0$, the critical values of T are integers less than or equal to k'_α, which is defined as the largest integer for which the relationship

$$\sum_{k=0}^{k'_\alpha} \binom{n}{k} \left(\frac{1}{2}\right)^n < \alpha$$

holds.

If the alternative is two-sided, namely $H_1 : med \neq m_0$, the critical values of T are integers less than or equal to $k'_{\alpha/2}$ and greater than or equal to $k_{\alpha/2}$, which are defined, respectively, as the largest and smallest integers for which the inequalities

$$\sum_{k=0}^{k'_{\alpha/2}} \binom{n}{k} \left(\frac{1}{2}\right)^n < \alpha/2, \quad \text{and} \quad \sum_{k=k_{\alpha/2}}^{n} \binom{n}{k} \left(\frac{1}{2}\right)^n < \alpha/2$$

hold.

If the value T is observed, then in testing against the alternative $H_1 : med > m_0$, large values of T are critical and the p-value is $p = \sum_{i=T}^{n} \binom{n}{i} 2^{-n} = \sum_{i=0}^{n-T} \binom{n}{i} 2^{-n}$. When testing against the alternative $H_1 : med < m_0$, small values of T are critical and the p-value is $p = \sum_{i=0}^{T} \binom{n}{i} 2^{-n}$. When the alternative is two-sided, the p-value is $p = 2\sum_{i=0}^{T'} \binom{n}{i} 2^{-n}$ for $T' = \min\{T, n - T\}$.

Alternative	p-value
$H_1 : med > m_0$	$\sum_{i=T}^{n} \binom{n}{i} 2^{-n}$
$H_1 : med \neq m_0$	$2\sum_{i=0}^{T'} \binom{n}{i} 2^{-n}$ for $T' = \min\{T, n - T\}$
$H_1 : med < m_0$	$\sum_{i=0}^{T} \binom{n}{i} 2^{-n}$

Consider now the two-paired-sample case, and suppose that the samples $X_1, \ldots, X_n$ and $Y_1, \ldots, Y_n$ are observed. We are interested in knowing whether the population median of the differences $X_1 - Y_1, X_2 - Y_2, \ldots, X_n - Y_n$ is equal to 0. In this case, $T = \sum_{i=1}^{n} \mathbf{1}(X_i > Y_i)$, is the total number of strictly positive differences.

Although it is true that the hypothesis of equality of means is equivalent to the hypothesis that the mean of the differences is 0, for the medians an analogous statement is not true in general. More precisely, if for samples X and Y, $D = X - Y$, then $med(D)$ may not be equal to $med(X) - med(Y)$. Thus, with the sign test, we are not testing the equality of medians, but whether the median of the differences is 0.

Ties complicate the calculations but can be handled. Even when observations come from a continuous distribution, ties appear due to limited precision in the application part. There are several ways of dealing with ties:

(i) Ignore them. If there are s ties, use only the "untied" observations. Of course, the sample size drops to $n - s$.

(ii) Assign the winning sign to tied pairs. For example, if there are two minuses, two ties, and six pluses, consider the two ties as pluses.

(iii) Randomize. For each tie, flip a coin and assign a plus if the coin lands heads and minus if the coin lands tails.

In script ◢ signtst.m conducting the sign test (not to be mixed with MATLAB's built in signtest), the options for handling ties are: I, C, and R, for policies described in (i)–(iii).

Example 18.1. **TCDD Levels.** Many Vietnam veterans have dangerously high levels of the dioxin 2,3,7,8-TCDD in their blood and fat tissue as a result of their exposure to the defoliant Agent Orange. A study published in *Chemosphere* (vol. 20, 1990) reported on the TCDD levels of 20 Massachusetts Vietnam veterans who had possibly been exposed to Agent Orange. The amounts of TCDD (measured in parts per trillion) in blood plasma and fat tissue drawn from each veteran are shown in the table below:

TCDD in plasma	TCDD in fat tissue
2.5 3.1 2.1 4.6 1.6	4.9 5.9 4.4 4.6 1.4
3.5 3.1 1.8 7.2 1.8	3.5 7.0 4.2 7.7 1.8
6.8 3.0 36.0 20.0 2.0	10.0 5.5 41.0 11.0 2.5
4.7 6.9 3.3 2.5 4.1	4.4 7.0 2.9 2.3 2.5

Is there sufficient evidence of a difference between the distributions of TCDD levels in plasma and fat tissue for Vietnam veterans exposed to Agent Orange? Use the sign test and report the p-value.

```
tcddpla=[2.5  3.1   2.1   4.6    1.6 ...
         3.5  3.1   1.8   7.2    1.8 ...
         6.8  3.0  36.0  20.0    2.0 ...
         4.7  6.9   3.3   2.5    4.1];
tcddfat=[4.9  5.9   4.4   4.6    1.4 ...
         3.5  7.0   4.2   7.7    1.8 ...
        10.0  5.5  41.0  11.0    2.5 ...
         4.4  7.0   2.9   2.3    2.5];
% ignore ties
[pvae, pvaa, n, plusses, ties] = signtst(tcddpla, tcddfat)
%pvae = 0.3323
%pvaa = 0.3320
%n = 17
%plusses = 6
%ties = 3

% randomize ties
[pvae, pvaa, n, plusses, ties] = signtst(tcddpla, tcddfat,'R')
% pvae = 0.2632
% take the conservative, least favorable approach
[pvae, pvaa, n, plusses, ties] = signtst(tcddpla, tcddfat,'C')
% pvae = 0.1153
```

Overall, the sign test failed to find significant differences between the distributions of TCDD levels irrespective of the ties handling policy. The p-values in this example are for the two-sided alternative. The sensible one-sided alternative can be easily accommodated by reporting half of the two-sided p-values, although the care is needed.

Compare these results with MATLAB's built-in function signtest.

Remark. The sign test is discussed for its historic significance and its simplicity. For applications this test is suboptimal to other nonparametric procedures, such as Wilcoxon's signed-rank test covered in the next section. Nevertheless, the relative efficiency of sign test, compared to the t-test is $2/\pi \approx 63.7\%$ when the data come from the normal distribution. Informally, this means that for normal data, the testing precision of sign test achieved with 1000 observations will require 637 observations for the t-test.

Many distribution-free procedures are based on how observations within the sample are *ranked* compared to either a parameter θ or to another sample. Recall (Section 2.5) that ranks of a sample $X_1, X_2, \ldots, X_n$ are defined as indices of the ordered sample

$$r(X_1), r(X_2), \ldots, r(X_n).$$

Suppose that a random sample $X_1, \ldots, X_n$ from a continuous distribution F is ranked and that $R_i = r(X_i)$, $i = 1, \ldots, n$ are the ranks. Ranks R_i are random variables with discrete uniform distribution (page 171). The properties of integer sums lead to the following properties for ranks:

$$\mathbb{E}(R_i) = \sum_{j=1}^{n} \frac{j}{n} = \frac{n+1}{2},$$

$$\mathbb{E}(R_i^2) = \sum_{j=1}^{n} \frac{j^2}{n} = \frac{n(n+1)(2n+1)}{6n} = \frac{(n+1)(2n+1)}{6}$$

$$\mathbb{V}\mathrm{ar}\,(R_i) = \frac{n^2 - 1}{12}.$$

These relationships follow from the fact that for a random sample, ranks are distributed as discrete uniform, namely, for any i,

$$\mathbb{P}(R_i = j) = \frac{1}{n}, \quad 1 \le j \le n.$$

As we discussed in Chapter 2 (page 25), script ◢ ranks.m outputs ranks for an input sample:

```
ranks([2 1 7 1 15 9])
%ans = 3.0000   1.5000   4.0000   1.5000   6.0000   5.0000
```

Several statistical procedures described next are based on ranks.

18.3 Wilcoxon Signed-Rank Test

More powerful than the sign test is Wilcoxon's signed-rank test (Wilcoxon, 1945), where, in addition to signs, the corresponding ranks are taken into account.

Let the paired sample (X_i, Y_i), $i = 1, \ldots, n$ be observed and let $D_i = X_i - Y_i$, $i = 1, \ldots, n$ be the differences. In a two-sample problem, we are interested in testing that the true mean of the differences is 0. The only assumption is that under H_0 the distribution of the differences D_i, $i = 1, \ldots, n$ is symmetric about 0.

It is also possible to consider a one-sample scenario in which testing the hypothesis about the median *med* is of interest. Here $H_0 : med = m_0$ is tested versus the one- or two-sided alternative. Then observations X_i, $i = 1, \ldots, n$ are compared to m_0, and the differences are $D_i = X_i - m_0$, $i = 1, \ldots, n$. In this case, no assumption on the symmetry of X_i is needed. If X_i's have symmetric distribution, the word *median* may be replaced with *mean*.

For the WSiRT test, the absolute values of the differences $(|D_1|, |D_2|, \ldots, |D_n|)$ are ranked. Let $r(|D_1|), r(|D_2|), \ldots, r(|D_n|)$ be the ranks of the differences.

Under H_0, the expectations of the sum of positive differences and the sum of negative differences should be equal. Define

$$W^+ = \sum_{i=1}^{n} S_i \, r(|D_i|)$$

and

$$W^- = \sum_{i=1}^{n} (1 - S_i) \, r(|D_i|),$$

where $S_i = 1$ if $D_i > 0$ and $S_i = 0$ if $D_i < 0$. Cases where $D_i = 0$ are ties and are ignored. Thus, $W^+ + W^-$ is the sum of all ranks, and in the case of no ties, it is equal to $\sum_{i=1}^{n} i = n(n+1)/2$. The statistic for the WSiRT is the difference between the ranks of positive differences and ranks of negative differences:

$$W = W^+ - W^- = 2 \sum_{i=1}^{n} r(|D_i|) S_i - n(n+1)/2.$$

Rule: For the WSiRT, it is suggested that a large-sample approximation should be used for W. If the samples are from the same population, the differences should be well mixed, and the sum of the ranks of positive differences should be close to the sum of the ranks of negative differences. Thus, in this case, $\mathbb{E}(W) = 0$ and $\mathbb{V}\mathrm{ar}\,(W) = \sum_i (r(|D_i|)^2) = \sum_i i^2 = n(n+1)(2n+1)/6$ under H_0 and no ties in differences. The statistic

$$Z = \frac{W}{\sqrt{\mathbb{V}\mathrm{ar}\,(W)}}$$

has an approximately standard normal distribution, so `normcdf` can be used to evaluate the p-values of the observed statistic W with respect to a particular alternative.

Equivalently, the WSiRT can be based on the sum of the ranks of positive differences only (or, equivalently, the sum of the ranks of negative differences only). In that case, under H_0, $\mathbb{E}W^+ = n(n+1)/4$ and $\mathbb{V}\mathrm{ar}\,(W^+) = n(n+1)(2n+1)/24$, leading to

$$Z = \frac{W^+ - n(n+1)/4}{\sqrt{n(n+1)(2n+1)/24}},$$

which also has an approximately standard normal distribution.

A function that performs WSiRT is wsirt.m. The preamble of this file is given below:

```
function  [W, Z, p] = wsirt( data1, data2, alt )
% ---------------------------------------------------------
%   WILCOXON SIGNED RANK TEST, AN APPROXIMATION
%   Input:  data1, data2 - first and second sample
%           data2 can be a scalar if the test is one sample
%           alt - code for alternative hypothesis;
%                 -1  for mu1<mu2; 0 for mu1 ~= mu2; and 1 for mu1>mu2
%   Output: W - sum of all signed ranks
%           Z - standardized R but adjusted for the ties
%           p - p-value for testing the null hypothesis
%                against the alternative specified by the input alt
%   Examples of use:
%    >  dat1=[1 3 2 4 3 5 5 4 2 3 4 3 1 7 6 6 5 4 5 8 7];
%    >  dat2=[2 5 4 3 4 3 2 2 1 2 3 2 3 4 3 2 3 4 4 3 5];
%    >  [W, Zstat, pval] = wsirt(dat1, dat2, 1)
%
%   When using wsirt as one sample test in testing
%   H0: mu=mu0 vs.  H1:mu~= mu0
%    >  mu0 = 3.5;
%    >  dat1=[1 3 2 4 3 5 5 4 2 3 4 3 1 7 6 6 5 4 5 8 7];
%    > [W, Zstat, pval] = wsirt(dat1,mu0, 0)
%
%   wsirt.m comes with function ranks.m (ranking procedure)
%---------------------------------------------------------
```

Example 18.2. **Identical Twins.** This data set was discussed in Conover (1999). Twelve pairs of identical twins underwent psychological tests to measure the amount of aggressiveness in each person's personality. We are interested in comparing the twins to each other to see if the first-born twin tends to be more aggressive than the other. The results are as follows (the higher score indicates more aggressiveness).

First-born twin, X_i	86	71	77	68	91	72	77	91	70	71	88	87
Second-born twin, Y_i	88	77	76	64	96	72	65	90	65	80	81	72

The hypotheses are: H_0 : the mean aggressiveness scores for the two twins are the same, that is, $\mathbb{E}(X_i) = \mathbb{E}(Y_i)$, and H_1 : the first-born twin tends to be more aggressive than the other, that is, $\mathbb{E}(X_i) > \mathbb{E}(Y_i)$. The WSiRT is appropriate if we assume that $D_i = X_i - Y_i$ are independent and symmetric.

Below is the output of several tests wsirt, wsirtexa, and signrank. The null hypothesis is not rejected by wsirt; the p-value is 0.2382. The function wsirtexa performs exact test (p-value 0.2324), while signrank is MATLAB's built in function (p-value 0.2378).

```
%Aggressiveness Score in Twins
fb = [86 71 77 68 91 72 77 91 70 71 88 87];
```

```
sb = [88 77 76 64 96 72 65 90 65 80 81 72];
[~, ~, p] = wsirt(fb, sb, 1 )
% p = 0.2382
% Exact test
[~,   p] = wsirtexa(fb, sb, 1)
% p = 0.2324  % exact for n<100
% MATLAB's built in
[p,~] = signrank(fb, sb,'tail','right')
% p = 0.2378
```

Note that all tests have approximately the same p-values, and failed to reject H_0.

The WSiRT can be used to test the hypothesis of location $H_0 : \mu = \mu_0$ using a single sample, as in a one-sample t-test. The differences in the WSiRT are $X_1 - \mu_0, X_2 - \mu_0, \ldots, X_n - \mu_0$ instead of $X_1 - Y_1, X_2 - Y_2, \ldots, X_n - Y_n$, as in the two-sample WSiRT. However, the assumption of symmetry of distribution of X's is needed; without this assumption, μ_0 in H_0 is the median and not the mean.

Example 18.3. **WSiRT for the Moon Illusion.** In the Moon Illusion (Example 9.5), we tested $H_0 : \mu = 1$ against $H_1 : \mu > 1$. Here is the WSiRT version of this test.

```
moon = [1.73 1.06 2.03 1.40 0.95 1.13 1.41 1.73 1.63 1.56];
mu0 = 1;
mu0vec = mu0 * ones(size(moon));
[w, z, p]=wsirt(moon, mu0vec, 1)
%w = 53
%z = 2.7029
%p = 0.0040
```

Compared to the t-test where the p-value was found to be pval = 9.9885e-04, the WSiRT still rejects H_0 even though the p-value is higher, p = 0.004.

If the data are normal and WSiRT is used instead of the optimal t-test, then the relative efficiency is 95%. This efficiency is quite high: for a precision achieved with WSiRT using 100 observations, the t-test would require 95 observations.

18.4 Wilcoxon Sum-Rank and Mann–Whitney Tests

The Wilcoxon sum-rank (WSuRT) and Mann-Whitney (MW) tests are equivalent and often referred together as Wilcoxon-Mann-Whitney (WMW) test. Here we will discuss only the former, WSuRT.

The WSuRT is often used in place of a two-sample t-test when the populations being compared are independent, but possibly not normally distributed. An example of the sort of data for which this test could be used is responses on a Likert scale (e.g., 1 = much worse, 2 = worse, 3 = no change, 4 = better, 5 = much better). It would be inappropriate to use the t-test for such data because of their ordinal nature.

The WSuRT tells us more generally whether the groups are homogeneous or if one group is "better" than the other. More generally, the basic null hypothesis of the WSuRT is that the two populations are equal. That is, $H_0 : F_X(x) = F_Y(x)$. When stated in this way, this test assumes that the shapes of the distributions are similar, which is not a stringent assumption.

Let $X = X_1, \ldots, X_{n_1}$ and $Y = Y_1, \ldots, Y_{n_2}$ be two samples of sizes n_1 and n_2, respectively, from the populations that we want to compare. Assume that the samples are put together and that $n = n_1 + n_2$ ranks are assigned to their concatenation. The test statistic W_n is the sum of ranks (1 to n) corresponding to the first sample, X. For example, if $X_1 = 1, X_2 = 13, X_3 = 7, X_4 = 9$, and $Y_1 = 2, Y_2 = 0, Y_3 = 18$, then the value of W_n is $2 + 4 + 5 + 6 = 17$.

If the two populations have the same distribution, then the sum of the ranks of the first sample and those in the second sample should be close, relative to their sample sizes. The WSuRT statistic is

$$W_n = \sum_{i=1}^{n} i S_i(X, Y),$$

where $S_i(X, Y)$ is an indicator function defined as 1 if the ith ranked observation is from the first sample and 0 if the observation is from the second sample.

For example, for $X_1 = 1, X_2 = 13, X_3 = 7, X_4 = 9$ and $Y_1 = 2, Y_2 = 0, Y_3 = 18$, $S_1 = 0, S_2 = 1, S_3 = 0, S_4 = 1, S_5 = 1, S_6 = 1, S_7 = 0$. Thus

$$W_n = 1 \times 0 + 2 \times 1 + 3 \times 0 + 4 \times 1 + 5 \times 1 + 6 \times 1 + 7 \times 0 = 2 + 4 + 5 + 6 = 17.$$

If there are no ties, then under H_0

$$\mathbb{E}(W_n) = \frac{n_1(n+1)}{2} \quad \text{and} \quad \mathbb{Var}(W_n) = \frac{n_1 n_2(n+1)}{12}.$$

The statistic W_n achieves its minimum when the first sample is entirely smaller than the second, and its maximum when the opposite occurs:

$$\min W_n = \sum_{i=1}^{n_1} i = \frac{n_1(n_1+1)}{2}, \quad \max W_n = \sum_{i=n-n_1+1}^{n} i = \frac{n_1(2n - n_1 + 1)}{2}.$$

For the statistic W_n a normal approximation holds:

$$W_n \sim \mathcal{N}\left(\frac{n_1(n+1)}{2}, \frac{n_1 n_2(n+1)}{12}\right).$$

A better approximation is

$$\mathbb{P}(W_n \le w) \approx \Phi(x) + \phi(x)(x^3 - 3x)\frac{n_1^2 + n_2^2 + n_1 n_2 + n}{20 n_1 n_2(n+1)},$$

where $\phi(x)$ and $\Phi(x)$ are the PDF and CDF of a standard normal distribution, respectively, and $x = (w - \mathbb{E}(W_n) + 0.5)/\sqrt{\mathrm{Var}(W_n)}$. This approximation is satisfactory for $n_1 > 5$ and $n_2 > 5$ if there are no ties.

The MATLAB function that performs WSuRT is ◀ wsurt.m. The preamble of this function is given below:

```
function  [W, Z, p] = wsurt( data1, data2, alt )
%  -------------------------------------------------------------
%  WILCOXON SUM RANK TEST
%  Input:   data1, data2 - first and second sample
%           alt - code for alternative hypothesis;
%                  -1  mu1<m2; 0 mu1 ~= m2; and 1 mu1>mu2
%  Output: W - sum of the ranks for the first sample. If
%              there is no ties, the standardization by EW &
%              Var W allows using standard normal quantiles
%              as long as sample sizes are larger than 15-20.
%           Z - standardized W but adjusted for the ties
%           p - p-value for testing equality of distributions
%              (equality of locations) against the alternative
%              specified by input "alt"
%  Example of use:
%  >  dat1=[1 3 2 4 3 5 5 4 2 3 4 3 1 7 6 6 5 4 5 8 7 3 3 4];
%  >  dat2=[2 5 4 3 4 3 2 2 1 2 3 2 3 4 3 2 3 4 4 3 5];
%  >  [sumranks1, zstat, pval] = wsurt(dat1, dat2, 1)
%
%  Comes with function ranks.m (ranking procedure)
%-------------------------------------------------------------
```

Example 18.4. **Nanoscale Probes.** The development of intracellular nanoscale probes against various biomolecules is very important in furthering the basic studies of cellular biology and pathology. One way to improve the binding properties of these probes is to have multiple binding domains or ligands on the surface of the probes. The more ligands, the greater is the chance of binding. One issue that comes up with such probes is whether, due to the multiple ligands, they will cause aggregation of target molecules within the cell. In order to show that new probes do not induce aggregation, researchers in the lab of Dr. Phil Santangelo at Georgia Tech compared the number of granules using a monovalent and a tetravalent probe in a cell plated at the same time and under the same biological conditions. The number of granules detected by each probe was recorded, and the researchers were interested to see if there were real differences between the numbers.

Using WSuRT, we want test the hypothesis of equality of the distributions and, subsequently, all theoretical moments at the significance level $\alpha = 0.05$.

```
monovalent =  [117   92   84 213   89   76   96 104 114 142 ...
               122 154 124   65 129   67 100 127   63   82 ...
               114   93 117   83   82   83 111   78   92   91];

tetravalent = [103   78 155 107 113   75   74   80 120 112 ...
               158   72   81 124 110   90   64   74 110 149 ...
                97   70 105   94 110   93 115 114 110 95];

[sumranks1, tstat, pval] = wsurt(monovalent, tetravalent, 0)

% sumranks1 = 925.5000
% tstat = 0.1553
% pval =  0.8737
```

The hypothesis that the two samples come from distributions with the same location is not rejected since the p-value is 0.8737. ✐

18.5 Kruskal–Wallis Test

The Kruskal–Wallis (KW) test is a generalization of the WSuRT. It is a non-parametric test used to compare three or more samples. It is used to test the null hypothesis that all populations have identical distribution functions against the alternative hypothesis that at least two of the samples differ only with respect to location (median), if at all.

The KW test is an analogue to the F-test used in one-way ANOVA. While ANOVA tests depend on the assumption that all populations under comparison are independent and normally distributed, the KW test places no distributional restriction, although the independence among the populations is required. Suppose that the data consist of k independent random samples with sample sizes $n_1, \ldots, n_k$. Let $n = n_1 + \cdots + n_k$.

Sample 1	$X_{11},$	$X_{12},$	$\ldots$	X_{1,n_1}
Sample 2	$X_{21},$	$X_{22},$	$\ldots$	X_{2,n_2}
$\vdots$	$\vdots$			
Sample $k-1$	$X_{k-1,1},$	$X_{k-1,2},$	$\ldots$	$X_{k-1,n_{k-1}}$
Sample k	$X_{k1},$	$X_{k2},$	$\ldots$	X_{k,n_k}

Under the null hypothesis, we can claim that all of the k samples are from a common population.

The expected sum of ranks for sample i, $\mathbb{E}(R_i)$, would be n_i times the expected rank for a single observation. That is, $n_i(n+1)/2$, and the variance can be calculated as $\mathrm{Var}\,(R_i) = n_i(n+1)(n-n_i)/12$. One way to test H_0 is to calculate $R_i = \sum_{j=1}^{n_i} r(X_{ij})$ – the total sum of ranks in sample i. The statistic

$$\sum_{i=1}^{k}\left[R_i - \frac{n_i(n+1)}{2}\right]^2 \tag{18.1}$$

will be large if the samples differ, so the idea is to reject H_0 if (18.1) is "too large." However, its distribution is quite messy, even for small samples, so we can use the normal approximation

$$\frac{R_i - \mathbb{E}(R_i)}{\sqrt{\mathrm{Var}\,(R_i)}} \overset{\text{appr}}{\approx} \mathcal{N}(0,1) \Rightarrow \sum_{i=1}^{k} \frac{(R_i - \mathbb{E}(R_i))^2}{\mathrm{Var}\,(R_i)} \overset{\text{appr}}{\approx} \chi^2_{k-1},$$

where the χ^2-statistic has only $k-1$ degrees of freedom due to the fact that only $k-1$ ranks are unique.

Based on this idea, Kruskal and Wallis (1952) proposed the test statistic

$$H' = \frac{1}{S^2}\left[\sum_{i=1}^{k} \frac{R_i^2}{n_i} - \frac{n(n+1)^2}{4}\right], \tag{18.2}$$

where

$$S^2 = \frac{1}{n-1}\left[\sum_{i=1}^{k}\sum_{j=1}^{n_i} r(X_{ij})^2 - \frac{n(n+1)^2}{4}\right].$$

If there are no ties in the data, S^2 becomes $n(n+1)/12$ and (18.2) simplifies to

$$H = \frac{12}{n(n+1)}\sum_{i=1}^{k} \frac{R_i^2}{n_i} - 3(n+1). \tag{18.3}$$

Kruskal and Wallis showed that this statistic has an approximate χ^2-distribution with $k-1$ degrees of freedom.

Correction for the ties in observations can be done in the following way. Let $c = 1 - \sum_{i=1}^{s}(t_i^3 - t_i)/(n^3 - n)$, where s is the number of different ranks and t_i is the number of observations that share ith rank. For example, in $[8,2,2,4,2,5,7,5,8,8]$, $s=5$ and t_is are $[3,1,2,1,3]$ indicating the number of tied observations for 2, 3, 5, 7, and 8. Then the corrected statistics is

$$H_c = H/c.$$

Note when there are no ties, $s = n$ and all $t_i = 1$ resulting in no correction, $c = 1$.

The MATLAB routine ✎ kruskalwallistest.m implements the KW test using a vector to represent the responses and another to identify the group from which the response came. It calculates H_c, the p-value for the test, and the average ranks for the groups. As an example, suppose we have the following responses from three treatment groups:

$$(1,3,4), \ (3,4,5), \ (4,4,4,6,5).$$

The code for testing the equality of locations of the three populations computes the corrected H as 3.8923 and a p-value of 0.1428. The average ranks in the three groups are 3.1667, 6, and 7.7.

```
data  =  [ 1 3 4   3 4 5   4 4 4 6 5 ];
belong = [ 1 1 1   2 2 2   3 3 3 3 3 ];
[p, H, aver] = kruskalwallistest(data, belong)
%p = 0.1428
%H = 3.8923
%aver  = 3.1667     6.0000     7.7000
```

Example 18.5. **Crop Yield.** The following data are from a classic agricultural experiment measuring crop yield in four different plots. For simplicity, we identify the treatment (plot) using the integers {1,2,3,4}. The null hypothesis (the treatment means are equal) is rejected with a p-value less than 0.0002. The average ranks for the four treatments are 21, 15.5, 26.6875, and 4.5714. In the context of this example, larger average rank is better and, given the significant test, the best treatment is 3, followed by 1, 2, and 4.

```
data= [83 91 94 89 89 96 91 92 90 84 91 90 81 83 84 83 ...
   88 91 89 101 100 91 93 96 95 94 81 78   82   81 77 79 81 80];
belong = [1 1 1 1 1 1 1 1 1 1 2 2 2 2 2 2 2 2 2 ...
              3 3 3 3 3 3 3 3 3 4 4 4 4 4 4 4 4];
[p, H, aver1] = kruskalwallistest(data, belong)
%p = 1.4451e-04
%H  =  20.3371
%aver1 = [21.0000   15.5000   26.6875   4.5714]
```

✎

Kruskal–Wallis Pairwise Comparisons. If the KW test detects treatment differences, we can determine if two particular treatment groups (e.g., i and j) are different at level α if

$$\left| \frac{R_i}{n_i} - \frac{R_j}{n_j} \right| > t_{n-k,1-\alpha/2} \sqrt{\frac{S^2(n-1-H')}{n-k} \cdot \left(\frac{1}{n_i} + \frac{1}{n_j} \right)}. \qquad (18.4)$$

Example 18.6. **Comparisons for Crop Treatments.** Since in Example 18.5 we found the four crop treatments significantly different, it would be natural to find out which ones seem better and which ones seem worse. In the table below, we compute the statistic

$$T = \frac{\left| \frac{R_i}{n_i} - \frac{R_j}{n_j} \right|}{\sqrt{\frac{S^2(n-1-H')}{n-k}\left(\frac{1}{n_i} + \frac{1}{n_j}\right)}}$$

for every combination of $1 \le i \ne j \le 4$ and compare it to $t_{30,0.975} = 2.042$.

(i,j)	1	2	3	4
1	0	1.856	1.859	5.169
2	1.856	0	3.570	3.363
3	1.859	3.570	0	6.626
4	5.169	3.363	6.626	0

This shows that the third treatment is the best, but not significantly different from the first treatment, which is second best. Treatment 2, which is third best, is not significantly different from treatment 1, but is different from treatments 4 and 3.

Confidence intervals for pairwise comparisons can be found using the function

```
% Kruskal-Wallis pairwise comparisons
kwpairwise(data, belong)
%
%      1.0000    2.0000   -0.5515    5.5000   11.5515
%      1.0000    3.0000  -11.9349   -5.6875    0.5599
%      1.0000    4.0000    9.9380   16.4286   22.9191
%      2.0000    3.0000  -17.5873  -11.1875   -4.7877
%      2.0000    4.0000    4.2912   10.9286   17.5659
%      3.0000    4.0000   15.2996   22.1161   28.9325
```

For example, the CI for difference of average ranks for treatments 1 and 2 is $[-0.5515, 11.5515]$. Since 0 belongs to this interval, the average ranks for treatments 1 and 2 are not significantly different, at the level of 0.05 that controls simultaneously all 6 comparisons.
🖉

MATLAB's built in function for KW test is ◀ [p t stats]=kruskalwallis(data, belong). The outputs are the p-value, ANOVA table, and a structure stats that can be used for multiple comparisons by multcompare similarly as in anovan.

Example 18.7. **Kruskal–Wallis versus ANOVA.** Kruskal–Wallis test does not assume normality but requires the group population distributions to be "similar in shape." To illustrate this point and differentiate between

ANOVA and Kruskal–Wallis test outcomes, we compiled a simple script
◀ kwvsanova.m where 30 observations are simulated from each, normal
$\mathcal{N}(1/2,1^2)$, gamma $\mathcal{G}a(1/4,1/2)$, and double exponential $\mathcal{DE}(1/2,\sqrt{2})$ distributions. All three distributions have mean $1/2$ and variance 1, but their
shapes differ. Given the three samples, we test ANOVA and Kruskal–Wallis
hypotheses and save the p-values. This is repeated $M = 20,000$ times on the
new random samples, and p-values for both tests are summarized as back-
to-back histograms, Figure 18.1. Under H_0, p-values are always uniform on
$(0,1)$, and this shows that ANOVA was not affected by nonnormality, as
long as the group sample means and variances were close. the Kruskal–
Wallis test, however, uses ranks and therefore is sensitive to the differences
between the corresponding sample quantiles, in particular the medians.
The median of gamma $\mathcal{G}a(1/4,1/2)$ distribution is not $1/2$, and the ranks
do not mix well among the samples. This causes Kruskal–Wallis' H statistic
to render significant more often than ANOVA's F statistic.

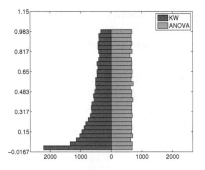

Fig. 18.1 Back-to-back histograms of p-values in Kruskal–Wallis (blue) and ANOVA
(green) tests where the locations of three populations are compared: normal $\mathcal{N}(1/2,1^2)$,
gamma $\mathcal{G}a(1/4,1/2)$, and double exponential $\mathcal{DE}(1/2,\sqrt{2})$.

18.6 Friedman's Test

Friedman's test is a nonparametric alternative to the repeated measures design from Section 11.6. It replaces the repeated measures design, or more
generally, the randomized block design, when the assumptions of normality are in question or when the variances vary from population to population. This test uses ranks of the data rather than their raw values to calculate
the test statistic. Due to the fact that Friedman's test does not make distri-

butional assumptions, it is not as powerful as the repeated measures design if the populations are indeed normal.

Milton Friedman published the first results for this test, which was eventually named after him. He received the Nobel Prize in economics in 1976, and one of the listed breakthrough publications was his article "The use of ranks to avoid the assumption of normality implicit in the analysis of variance," published in 1937 (Friedman, 1937).

Recall that the repeated measures design requires measurements for each subject (block) at each level of treatment. Let X_{ij} represent the experimental outcome of subject (block) i with treatment j, where $i = 1, \ldots, b$, and $j = 1, \ldots, k$.

Block	Treatment			
	1	2	$\ldots$	k
1	X_{11} X_{12}		$\ldots$	X_{1k}
2	X_{21} X_{22}		$\ldots$	X_{2k}
$\vdots$	$\vdots$ $\vdots$			$\vdots$
b	X_{b1} X_{b2}		$\ldots$	X_{bk}

Unlike the KW test where the ranks are assigned to all observations grouped together, here we assign ranks $\{1, 2, \ldots, k\}$ to each *row* in the table of observations. Thus the expected rank of any observation under H_0 is $(k + 1)/2$. We next sum all the ranks by columns (by treatments) to obtain $R_j = \sum_{i=1}^{b} r(X_{ij})$, $1 \le j \le k$. If H_0 is true, the expected value for R_j is $\mathbb{E}(R_j) = b(k + 1)/2$. The statistic

$$\sum_{j=1}^{k} \left(R_j - \frac{b(k + 1)}{2} \right)^2$$

is an intuitive formula to reveal treatment differences. It has expectation $bk(k^2 - 1)/12$ and variance $k^2 b(b - 1)(k - 1)(k + 1)^2/72$. Once normalized to

$$S = \frac{12}{bk(k + 1)} \sum_{j=1}^{k} \left(R_j - \frac{b(k + 1)}{2} \right)^2$$

$$= \frac{12}{bk(k + 1)} \sum_{j=1}^{k} R_j^2 - 3b(k + 1), \tag{18.5}$$

it has moments $\mathbb{E}(S) = k - 1$ and $\mathbb{V}\mathrm{ar}\,(S) = 2(k - 1)(b - 1)/b \approx 2(k - 1)$, which coincide with the first two moments of χ_{k-1}^2. Higher moments of S also approximate well those of χ_{k-1}^2 when b is large.

In the case of ties, a modification to S is suggested. Let $C = bk(k + 1)^2/4$ and $R^* = \sum_{i=1}^{b} \sum_{j=1}^{k} r(X_{ij})^2$. Then,

$$S' = \frac{k-1}{R^* - C} \left(\sum_{j=1}^{k} R_j^2 - bC \right) \tag{18.6}$$

is also approximately distributed as χ_{k-1}^2.

Although the Friedman statistic makes for a sensible, intuitive test, it turns out there is a better one to use. As an alternative to S (or S'), the test statistic

$$F = \frac{(b-1)S}{b(k-1) - S}$$

is approximately distributed as $F_{k-1,(b-1)(k-1)}$, and tests based on this approximation are generally superior to those based on chi-square approximation used for S. For details on the comparison between S and F, see Iman and Davenport (1980).

Example 18.8. **Cerebral Blood Flow and Metabolic Rate of Oxygen.** In the human brain, structures having high blood flow (e.g., gray matter) also tend to have increased metabolic rates, and structures containing low blood flow (e.g., white matter) have lower metabolic rates. This phenomenon is known as coupling, and it can be modeled by measuring cerebral blood flow (CBF) and comparing it with the variable cerebral metabolic rate of oxygen (CMR_{O_2}). This technique has been used to show how various brain regions differ in their metabolism characteristics and has been used as a fundamental basis for fMRI imaging. Fox and Raichle (1986) tested whether the measurements of CBF and CMR_{O_2} were dynamically coupled in the measurements they collected.

Three sets of stimulated-state measurements and one set of resting state measurements were acquired for each of the 9 subjects undergoing sensory stimulation. The 3 stimulation levels (S1, S2, S3) differed only in stimulus duration. The measured response to tactile stimulation for each subject was the regional blood flow ratio (contralateral/ipsilateral). Data are provided in the table below.

Subject	Resting	S1	S2	S3
1	0.82	1.14	1.05	1.12
2	1.05	1.45	1.38	1.42
3	1.10	1.45	1.49	1.58
4	0.90	1.05	1.22	1.14
5	1.02	1.27	1.27	1.39
6	1.03	1.25	1.23	1.20
7	1.04	1.30	1.33	1.29
8	0.98	1.30	1.42	1.34
9	1.04	1.29	1.23	1.24

We will conduct the repeated measures analysis in a nonparametric fashion and test the hypothesis that population distributions of regional blood flow ratios are the same. This would imply that the population means for the four groups are the same,

$$H_0 : \mu_{Resting} = \mu_{S1} = \mu_{S2} = \mu_{S3}.$$

To use the procedure ◀ friedmantest the input data should be formatted as a matrix in which the rows are blocks and the columns are treatments.

```
%cbfcmro2.m
cbf=[...
0.82 1.14 1.05 1.12;
1.05 1.45 1.38 1.42;
1.10 1.45 1.49 1.58;
0.90 1.05 1.22 1.14;
1.02 1.27 1.27 1.39;
1.03 1.25 1.23 1.20;
1.04 1.30 1.33 1.29;
0.98 1.30 1.42 1.34;
1.04 1.29 1.23 1.24];
%or cbf=xlsread ('cbfcmro2.xls');
[S F pS pF] = friedmantest(cbf)
%    S =16.4157
%    F =12.4076
%    pS =9.3179e-004
%    pF =4.2382e-005
```

The null hypothesis is rejected with respect to both S and F statistic. It is of interest to find which population locations differ.

Friedman's Pairwise Comparisons. If the p-value is small enough to warrant multiple comparisons of treatments, we consider two treatments i and j to be different at level α if

$$|R_i - R_j| > t_{(b-1)(k-1),1-\alpha/2} \sqrt{2 \cdot \frac{bR^* - \sum_{j=1}^{k} R_j^2}{(b-1)(k-1)}}. \qquad (18.7)$$

To conduct Friedman's pairwise comparisons on data from Example 18.8, we will use function ◀ friedmanpairwise.m which implements (18.7).

```
%cbfcmro2.m continued
out=friedmanpairwise(cbf)
%      1.0000    2.0000  -25.9653  -18.5000  -11.0347
%      1.0000    3.0000  -24.9653  -17.5000  -10.0347
%      1.0000    4.0000  -25.4653  -18.0000  -10.5347
%      2.0000    3.0000   -6.4653    1.0000    8.4653
%      2.0000    4.0000   -6.9653    0.5000    7.9653
```

```
%      3.0000     4.0000    -7.9653   -0.5000    6.9653
```

Since zero is not in the intervals for differences 1-2, 1-3, and 1-4, the Control treatment is significantly different from S1, S2, and S3 treatments. The intervals are all negative, meaning that the control measurements are significantly smaller. Yet, zero is in the intervals for differences 2-3, 2-4, and 3-4, and the locations for S1, S2, and S3 are not statistically different.

✏

18.7 Resampling Methods

> *... I was still a couple of miles above the clouds when it broke, and with such violence I fell to the ground that I found myself stunned, and in a hole nine fathoms under the grass, when I recovered, hardly knowing how to get out again. Looking down, I observed that I had on a pair of boots with exceptionally sturdy straps. Grasping them firmly, I pulled with all my might. Soon I had hoist myself to the top and stepped out on terra firma without further ado.*
>
> –Rudolph Raspe (1786, p. 22)

Resampling methods in statistics are computer-intensive procedures that use an observed sample to produce many surrogate samples subsequently used in the inference. Resampling can be applied in a broad range of statistical scenarios (estimation, testing, regression, experimental design, etc.), often providing answers of equal or higher quality compared to the traditional methods. This goes at the expense of computer time, but nowadays this hardly can be a serious objection.

18.7.1 The Jackknife

The jackknife method was introduced by Maurice Henry Quenouille in 1949 to correct for the bias of an estimator. In addition to bias reduction, the method is also used to estimate the variance of an estimator and to construct confidence intervals.

Recall that any estimator $\hat{\theta}$ is a function of the sample,

$$\hat{\theta} = f(X_1, X_2, \ldots, X_n).$$

Define n estimators $\hat{\theta}_{(-i)}$ re-computed the same way as $\hat{\theta}$, but with observation X_i omitted, that is, by using the remaining $n - 1$ observations, $(X_1, \ldots, X_{i-1}, X_{i+1}, \ldots, X_n)$. One gets exactly n such estimators, $\hat{\theta}_{(-1)}, \hat{\theta}_{(-2)}, \ldots, \hat{\theta}_{(-n)}$, one for each X_i omitted.

Define the pseudo-observation (pseudo-value) X_i^* as

$$X_i^* = n\hat{\theta} - (n-1)\hat{\theta}_{(-i)}$$

The motivation for this naming can be seen by considering the estimator $\hat{\theta} = \overline{X}$, for which the pseodo-observation X_i^* coincides with the observation X_i. Let $\hat{\theta}_{(.)}$ be the average of all $\hat{\theta}_{(-i)}$'s,

$$\hat{\theta}_{(.)} = \frac{1}{n}\sum_{i=1}^{n}\hat{\theta}_{(-i)}.$$

The jackknife estimator of θ is the average of pseudo-observations,

$$\hat{\theta}_J = \frac{1}{n}\sum_{i=1}^{n}X_i^* = \frac{1}{n}\sum_{i=1}^{n}\left[n\hat{\theta} - (n-1)\hat{\theta}_{(-i)}\right]$$

$$= n\hat{\theta} - \frac{n-1}{n}\sum_{i=1}^{n}\hat{\theta}_{(-i)} = n\hat{\theta} - (n-1)\hat{\theta}_{(.)}.$$

Bias b of an estimator $\hat{\theta}$ is estimated as

$$\hat{b} = \mathbb{E}\hat{\theta} - \theta = \frac{n-1}{n}\sum_{i=1}^{n}(\hat{\theta}_{(-i)} - \hat{\theta})$$

$$= (n-1)(\hat{\theta}_{(.)} - \hat{\theta}).$$

How does the jackknife estimator reduce the bias? Suppose that $\hat{\theta}$ is biased, that is, $\mathbb{E}\hat{\theta} = \theta + C/n$, where n is the sample size and C is a constant. Then the expectation of a pseudo-observation is

$$\mathbb{E}X_i^* = \mathbb{E}(n\hat{\theta} - (n-1)\hat{\theta}_{(-i)}) = n\left(\theta + \frac{C}{n}\right) - (n-1)\left(\theta + \frac{C}{n-1}\right) = \theta.$$

Therefore, if the bias is of order C/n, then $\mathbb{E}\hat{\theta}_J = \frac{1}{n}\sum_{i=1}^{n}\mathbb{E}X_i^* = \frac{1}{n}n\theta = \theta$, meaning the jackknife estimator $\hat{\theta}_J$ is unbiased.

The jackknife estimator of variance and jackknife confidence intervals were proposed by John Tukey in the late 1950s. Tukey also coined the term "jackknife" for the methodology. Quoting Tukey (1959) (also Brillinger, 1964),

... The procedure described here shares two characteristics with a Boy Scout Jack-knife:

(1) wide applicability to many kinds of problems,

(2) inferiority to special tools for those problems for which special tools have been designed and built.

Let s_{X^*} be an estimator of sample standard deviation for the pseudo-observations,

$$s_{X^*} = \sqrt{\frac{1}{n-1} \sum_{i=1}^{n} \left(X_i^* - \hat{\theta}_J \right)^2 }.$$

Since $\hat{\theta}_J$ is an average of pseudo-observations $\frac{1}{n} \sum_{i=1}^{n} X_i^*$, the estimator $s_{\hat{\theta}_J}$ is the usual estimator for sample standard deviation of the mean, $\frac{s_{X^*}}{\sqrt{n}}$.

Tukey postulated that the standard deviation of an estimator $\hat{\theta}$ can be well approximated by the standard deviation of its jackknife version,

$$s_{\hat{\theta}} \approx s_{\hat{\theta}_J} = \frac{s_{X^*}}{\sqrt{n}} = \sqrt{\frac{\sum_{i=1}^{n} \left(X_i^* - \hat{\theta}_J \right)^2}{n(n-1)}}.$$

In terms of $\hat{\theta}_{(-i)}$,

$$s_{\hat{\theta}_J} = \sqrt{\frac{n-1}{n} \sum_{i=1}^{n} \left(\hat{\theta}_{(-i)} - \hat{\theta}_{(\cdot)} \right)^2}, \quad \text{where} \quad \hat{\theta}_{(\cdot)} = \frac{1}{n} \sum_{i=1}^{n} \hat{\theta}_{(-i)}.$$

This leads to jackknife $(1 - \alpha)100\%$ CI for θ as

$$[\hat{\theta}_J - t_{n-1,1-\alpha/2} \cdot s_{\hat{\theta}_J}, \ \hat{\theta}_J + t_{n-1,1-\alpha/2} \cdot s_{\hat{\theta}_J}].$$

Here, $t_{n-1,1-\alpha/2}$ is $(1 - \alpha/2)$-quantile of a t-distribution with $n - 1$ degrees of freedom.

Example 18.9. **Spores of *Amanita Phalloides* Revisited.** Measurements in microns of 28 spores of *Amanita phalloides* have been discussed in Exercises 2.4 and 7.27. In Exercise 7.27, a $(1 - \alpha)100\%$ CI for population coefficient of variation (σ/μ) was found under assumption that data are coming from normal distribution. We will use jackknife to estimate the bias of sample CV, $s/\overline{X}$, and find the 95% jackknife CI for σ/μ.

9.2	8.8	9.1	10.1	8.5	8.4	9.3
8.7	9.7	9.9	8.4	8.6	8.0	9.5
8.8	8.1	8.3	9.0	8.2	8.6	9.0
8.7	9.1	9.2	7.9	8.6	9.0	9.1

```
%amanitajkn3.m, needs minusith.m
amanita =[...
9.2 8.8 9.1 10.1 8.5 8.4 9.3 8.7 9.7 9.9 8.4 8.6 8.0 9.5 ...
```

```
8.8 8.1 8.3 9.0 8.2 8.6 9.0 8.7 9.1 9.2 7.9 8.6 9.0 9.1];

%function handle for calculating CV statistics
coevar = @(x)  std(x)/mean(x);
%
n=length(amanita);
pseudo=[];    %collect pseudo-observations
coevarsj=[]; %collect estimators from jackknife samples
for i = 1:n
    jknsa = minusith(amanita, i); %i-th jackknife sample
    pseudoobs = n * coevar(amanita) - (n-1)*coevar(jknsa);
    pseudo =[pseudo pseudoobs];
    coevarsj=[coevarsj coevar(jknsa)];
end
%jackknife bias
bias=(n-1)*(mean(coevarsj)-coevar(amanita)  %-5.7487e-04
%two equivalent jackknife estimators of CV
pseudomean = mean(pseudo)           %0.0628
jackknifeset=coevar(amanita)-bias   %0.0628
pseudostd = std(pseudo)/sqrt(n)     %0.0081
%Jackknife CI
[pseudomean - tinv(0.975, n-1)*pseudostd,...
        pseudomean + tinv(0.975, n-1)*pseudostd]
%  0.0461    0.0795
%Compare to Miller 1991 CI for CV
    theta=coevar(amanita)           %0.0622
    [theta - norminv(0.975)*theta*sqrt(1/(n-1)*(1/2 + theta^2)),...
     theta + norminv(0.975)*theta*sqrt(1/(n-1)*(1/2 + theta^2))]
    %0.0456    0.0789
```

Note that $CV = s/\overline{X}$ is a biased estimator of σ/μ and that bias $\mathbb{E}(CV) - \sigma/\mu$ is estimated as -5.7487e-04; thus CV underestimates its population counterpart. A jackknife estimator of σ/μ for this data set is 0.0628 as compared to the directly calculated CV, which is 0.0622. A 95% jackknife CI for σ/μ is $[0.0461, 0.0795]$, which is very close to Miller's CI $[0.0456, 0.0789]$; see page 325.

✐

18.7.2 Bootstrap

Bootstrap is arguably the most popular resampling methodology, made systematic by Brad Efron, Professor at Stanford University. Before Efron's seminal paper (Efron, 1979a), which gave the theoretical foundations for bootstrap methodology, there were several important precursors, including, the work on jackknife by Quenouille and Tukey.

Among notable bootstrap evangelists was unconventional economist Julian Simon who in 1976 prophesied "that most everyday statistics eventually would be done the resampling way." His crusade for teaching statistics based on resampling at all levels was discounted by the statistical community until the work by Efron gave him legitimacy. An interesting account of bootstrap history is given in Hall (2003).

18.7.2.1 Bootstrap Sampling

To conduct bootstrap inference, one forms surrogate samples, called *bootstrap samples* or *bootstrap resamples*, by sampling with replacement from the original sample. The bootstrap samples are of the same size as the original sample. If the original sample is $X_1, X_2, \ldots, X_n$, then the bth bootstrap sample is denoted as $X_1^{*b}, X_2^{*b}, \ldots, X_n^{*b}$. Since the sampling is with replacement, some observations from the original sample may not be selected for a particular bootstrap sample, while some may be selected more than once.

Original sample	$X_1, X_2, \ldots, X_n$
Bootstrap samples	$X_1^{*1}, X_2^{*1}, \ldots, X_n^{*1}$
	$X_1^{*2}, X_2^{*2}, \ldots, X_n^{*2}$
	$\ldots$
	$X_1^{*B}, X_2^{*B}, \ldots, X_n^{*B}$

Here the number of bootstrap samples B is typically large. For example,

Original sample	$\{1, 4, 7, 2, 8\}$
Bootstrap samples	$\{8, 8, 1, 4, 2\}, \{1, 2, 4, 8, 7\}, \{2, 2, 1, 8, 4\},$
	$\{1, 8, 4, 1, 2\}, \{7, 4, 8, 4, 1\}, \{8, 1, 2, 2, 8\},$
	$\{2, 7, 1, 4, 8\}, \{1, 1, 4, 1, 2\}, \{4, 7, 7, 2, 7\},$ etc.

Next, we provide a simple MATLAB code for bootstrap sampling. The function ◀ bootsample.m takes a sample of size n as input, where each observation is p-dimensional and generates a bootstrap sample. The input (and the output) are $n \times p$ matrices, $n, p \geq 1$. MATLAB comes with an advanced function ◀ bootstr for more complex resampling approaches.

```
function vecout = bootsample(vecin)
% Bootstraping from the array "vecin" by random selecting the rows
% Usage
%    vecout = bootsample(vecin)
% Input
%    vecin - nxp data matrix
%    n - sample size
%    p - dimension of a single observation
% Output
%    vecout - a single bootstrap sample, size n x p
% Example
%    > bootsample([1 2; 2 3; 3 4; 4 5])
%    ans =
```

```
%       4    5
%       3    4
%       4    5
%       3    4
%
[n, p] = size(vecin);
selected_indices = floor(1+n.*(rand(1,n)));
vecout = vecin(selected_indices,:);
```

Assume that θ is the parameter of interest and that $\hat{\theta}$ is a statistic appropriate to make inference about θ. Recall that a statistic is a function of the sample, $\hat{\theta} = \hat{\theta}(X_1, X_2, \ldots, X_n)$. For only a few statistics $\hat{\theta}$, we can estimate their variability from the same sample. An example is the sample mean, $\hat{\theta} = \overline{X}$ for which the standard deviation is $s_{\overline{X}} = s/\sqrt{n}$. This result does not require normality and holds for any distribution of observations, under mild restrictions. However, for more complex estimators it is impossible to calculate their standard errors from a single sample, unless we know the distribution of observations. We will see later that even when the variability of an estimator can be found from the single sample, we may be interested in the sampling distribution of the estimator, which can be estimated by bootstrap.

If the sample $X_1, X_2, \ldots, X_n$ produces statistic $\hat{\theta}$ for estimating population parameter θ, then each of B bootstrap samples $X_1^{*b}, X_2^{*b}, \ldots, X_n^{*b}$, $b = 1, \ldots, B$, produces the counterpart statistic $\hat{\theta}_b^*$. When B is large, the ensemble of $\hat{\theta}_b^*$s approximates the sampling distribution of $\hat{\theta}$.

Original sample	$X_1, X_2, \ldots X_n$	$\longrightarrow \hat{\theta}$
Bootstrap samples	$X_1^{*1}, X_2^{*1}, \ldots, X_n^{*1}$	$\longrightarrow \hat{\theta}_1^*$
	$X_1^{*2}, X_2^{*2}, \ldots, X_n^{*2}$	$\longrightarrow \hat{\theta}_2^*$
	$\ldots$	
	$X_1^{*B}, X_2^{*B}, \ldots, X_n^{*B}$	$\longrightarrow \hat{\theta}_B^*$

For example, if the sampling distribution of $\hat{\theta}$ is $F(x) = \mathbb{P}(\hat{\theta} \leq x)$, then the bootstrap estimate of F is an empirical CDF,

$$F^*(x) = \frac{1}{B} \sum_{b=1}^{B} \mathbf{1}(\hat{\theta}_b^* \leq x),$$

as on page 27. Here the function $\mathbf{1}(A)$ is an indictor of A, equal to 1 if A is true, and to 0 if A is false.

One of the main uses of bootstrap is to estimate the standard error of the proposed statistic $\hat{\theta}$ based on a single sample and without any distributional assumptions about the underlying population. More generally, one obtains a bootstrap approximation of distribution for $\hat{\theta}$. With this approximation in hand, one can find not only the variability of sample estimators, but assess their bias, find confidence intervals, assess the hypotheses, etc.

Given B bootstrap replicates θ_b^*, the standard deviation of $\hat{\theta}$ is estimated as

$$s_{\hat{\theta}} = \sqrt{\frac{1}{B-1}\sum_{b=1}^{B}(\hat{\theta}_b^* - \overline{\theta^*})^2},$$

where $\overline{\theta^*}$ is the sample mean of $\hat{\theta}_b^*$s,

$$\overline{\theta^*} = \frac{1}{B}\sum_{b=1}^{B}\hat{\theta}_b^*.$$

The motivation for the rather curious name *bootstrap* is now apparent. To *pull oneself up by one's bootstraps* means to succeed without outside help. See also the original bootstrapper's account in the quote at the beginning of this section. In our context, this self-sufficiency refers to a single sample succeeding to produce an estimator of sampling distribution for any statistic calculated from that sample.

Another common use of bootstrap techniques is to assess the bias of an estimator. Recall (page 288) that an estimator $\hat{\theta}$ is unbiased for θ if $\mathbb{E}\hat{\theta} = \theta$, when the expectation is taken with respect to the sampling distribution of $\hat{\theta}$. Also, the bias of an estimator was defined as $b(\hat{\theta}) = \mathbb{E}(\hat{\theta}) - \theta$.

Since the bias depends on unobservable θ, it needs to be estimated. A bootstrap estimator of bias is defined as

$$\hat{b}(\hat{\theta}, \theta) = \overline{\theta^*} - \hat{\theta}.$$

The goal of bias estimation is to produce a bias-corrected estimator of θ,

$$\hat{\theta}_{bc} = \hat{\theta} - \hat{b}(\hat{\theta}, \theta) = 2\hat{\theta} - \overline{\theta^*}.$$

There are several approaches to calculating confidence intervals for θ by using bootstrap replicates. The simplest uses sample quantiles of $\hat{\theta}^*$. Let $\theta^*(\alpha/2)$ and $\theta^*(1 - \alpha/2)$ respectively denote $\alpha/2$ and $1 - \alpha/2$ sample quantiles for $\{\hat{\theta}_1^*, \hat{\theta}_2^*, \ldots, \hat{\theta}_B^*\}$. Then a $(1-\alpha) \times 100$-level bootstrap confidence interval for θ is

$$[\theta^*(\alpha/2),\ \theta^*(1 - \alpha/2)]$$

This is the most popular approach to bootstrap confidence intervals, because of its simplicity and natural appeal. A drawback of this method is that it requires at least an approximate symmetry of the sampling distribution of $\hat{\theta}$ around θ.

On the positive side, the quantile bootstrap confidence intervals are transformation invariant and they always lead to valid intervals. The latter property means that an interval will not go outside of the parameter domain (e.g., CI for the population proportion containing negative values). The transformation invariance property means that a confidence interval for parameter θ will be equal to the confidence interval calculated for $g(\theta)$ after applying g^{-1} on the interval endpoints, as long as the function g is monotone.

A centered bootstrap quantile method defines a $(1-\alpha)100\%$ CI as

$$[2\hat{\theta} - \theta^*(1 - \alpha/2),\ 2\hat{\theta} - \theta^*(\alpha/2)]$$

As $\hat{\theta}^* - \hat{\theta}$ is bootstrap approximation to $\hat{\theta} - \theta$, then

$$1 - \alpha = \mathbb{P}\left(\epsilon_L \le \hat{\theta} - \theta \le \epsilon_U\right) \approx \mathbb{P}\left(\epsilon_L \le \hat{\theta}^* - \hat{\theta} \le \epsilon_U\right).$$

From the first probability $\hat{\theta} - \epsilon_U \le \theta \le \hat{\theta} - \epsilon_L$ and from the second, $\epsilon_L \approx \theta^*(\alpha/2) - \hat{\theta}$ and $\epsilon_U \approx \theta^*(1 - \alpha/2) - \hat{\theta}$, leading to the centered bootstrap quantile interval.

The two quantile intervals differ when sampling distribution of $\hat{\theta}$ is skewed, and it is up to user to choose which tail should be more emphasized.

Bootstrap t-type confidence intervals derive their name from the fact that a t-like statistic $(\hat{\theta} - \theta)/SE(\hat{\theta})$ is bootstrapped. For example, if θ is the population mean μ, then statistic $t = (\overline{X} - \mu)/(s/\sqrt{n})$ is approximated by $t^* = (\overline{X}^* - \overline{X})/(s^*/\sqrt{n})$, where $\overline{X}^*$ and s^* are bootstrap sample counterparts of $\overline{X}$ and s. The CI for the mean is then $[\overline{X} + t^*(\alpha/2)s/\sqrt{n},\ \overline{X} + t^*(1 - \alpha/2)s/\sqrt{n}]$, where $t^*(\alpha)$ is αth quantile of $\{t_1^*, t_2^*, \ldots, t_B^*\}$.

Unlike the sample mean, most statistics do not have a formula for standard error and both $SE(\hat{\theta})$ and $SE(\hat{\theta}_b^*)$ need to be estimated by bootstrap as well. Then one can apply the jackknife or, alternatively, a bootstrap-within-bootstrap, that is, bootstrap of a bootstrapped sample to estimate $SE(\hat{\theta}_b^*)$ needed for t_b^*. In general form, a bootstrap t-type $(1-\alpha)100\%$ CI has the form

$$\left[\hat{\theta} + t^*(\alpha/2)\widehat{SE},\ \hat{\theta} + t^*(1 - \alpha/2)\widehat{SE}\right].$$

A nice feature of the bootstrap, particularly helpful if studentized bootstrap confidence intervals are used, is that it accounts for the skewness of the sampling distribution of the statistic. For some other approaches (ABC and BCA CIs), see Efron and Tibshirani (1993).

Example 18.10. **Anacapa Data Revisited.** In Example 17.14, concentrations of polychlorinated biphenyl (PCB) in yolk lipids of pelican eggs were checked for normality using Jarque–Bera test. The test statistic JB was approximated by chi-square distribution with 2 degrees of freedom and slow convergence of that approximation was discussed.

Here we revisit the problem and find bootstrap distribution of the JB statistic.

```
%anacapabci.m
jabe = @(sample) length(sample)/6 * (skewness(sample)^2 + ...
               (kurtosis(sample)-3)^2/4 );
anacapa=[452    184    %<60 observations deleted>  237  206   87];

thetahat = jabe(anacapa)    %9.5588
 B=20000;
 thetabs=[];
 for i = 1:B
   anacapab = bootsample(anacapa(:));
    thetabs=[thetabs jabe(anacapab)];
 end
%Quantile bootstrap CI
[quantile(thetabs,0.025)      quantile( thetabs,0.975)]   %0.1685   42.7039
% Centered bootstrap CI
[max(0,2*thetahat - quantile(thetabs,0.975) )  ...
      2*thetahat - quantile( thetabs,0.025)]              %0   18.9491
%Bootstrap-within-bootstrap t-type CI
B1=500; B2=1000;
tbi=[]; jabes=[];
for i=1:B1
    anacapab = bootsample(anacapa(:));
     samplej=[];
     for j=1:B2
        anacapabb = bootsample(anacapab(:));
        samplej=[samplej jabe(anacapabb)];
     end
     jbbar=mean(samplej); jbstd=std(samplej);
     tbi = [tbi (jabe(anacapab) - jbbar)/jbstd];
     jabes=[jabes jabe(anacapab)];
end
[ mean(jabes) + quantile(tbi, 0.025)*std(jabes) ...
   mean(jabes) + quantile(tbi, 0.975)*std(jabes) ] %1.4188   11.9744
```

Compare results of this example with results of Example 17.14 in Chapter 17. ✐

Example 18.11. **Polio 1954 Vaccine Trial.** In Chapter 1 we discussed the 1954 clinical trial of the Salk vaccine for preventing paralytic poliomyelitis. The results were summarized in the following table:

	Inoculated with vaccine	Inoculated with placebo
Total number of children inoculated	200,745	201,229
Number of cases of paralytic polio	33	115

The observed risk ratio is

$$RR = \frac{33/200745}{115/201229} = 0.2876,$$

meaning that the chance of disease in the vaccine group is about 29% of the chance of disease in the placebo group. In other words, a child in the placebo group is 3.5 times more likely to develop disease compared to a child from vaccine group. Was it possible that an ineffective vaccine produced such a result by chance? We provide an answer by a bootstrap analysis.

Assume that the presence of disease was coded by 1 and the absence by 0. Then the vaccine and control groups would correspond to samples of 33 ones and $200745 - 33$ zeros, and 115 ones and $201229 - 115$ zeros, respectively. The bootstrap sampling is conducted by the following commands:

```
vaccine = [ones(1, 33), zeros(1, 200745-33)]';
placebo = [ones(1, 115) zeros(1,201229-115)]';
rrs=[];
B=50000
for i = 1:B
    pvac  = sum(bootsample(vaccine))/200745;
    pplac = sum(bootsample(placebo))/201229;
    rr=pvac/pplac;
    rrs =[rrs rr];
end
```

From each group independent bootstrap samples are taken, and the bootstrapped risk ratio found and recorded. Figure 18.2 depicts a bootstrap distribution of the risk ratio for $B = 50,000$ bootstrap samples. The red bar indicates the position of observed risk ratio, while the green bars delimit the 95% bootstrap confidence interval calculated as $[0.1865, 0.4134]$. This interval is well separated from 1, thus based on data from this clinical trial, the vaccine significantly lowered the risk of polio. The detailed calculations are provided in m-file ◀ poliovaccine.m.
Note that by treating diseased in the two groups as independent binomials, the normal approximation of the log-risk ratio gives a similar result, as in Example 10.17.

In the following example we will use WinBUGS to produce bootstrap resamples. However, this is not a Bayesian bootstrap. A concept of Bayesian bootstrap was proposed by Rubin (1981); see Exercise 18.26.

Example 18.12. **Rubidium-to-Potassium Ratio.** In Exercise 9.19, measurements of a naturally occurring rubidium-to-potassium ratio (in hundreds of mEq of Ru to mEq of K) in 17 hospitalized patients were provided. A $(1 - \alpha)100\%$ CI for the population CV was to be found under the assumption of normality of population.

We will find this interval without any distributional assumptions using the bootstrap implemented via WinBUGS. The posterior distribution for CV would be identical to the bootstrap distribution, thus the 95% bootstrap

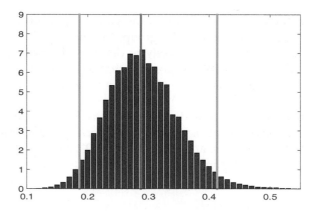

Fig. 18.2 Bootstrap distribution of risk ratio for the Polio Vaccine data. The red bar is the observed risk ratio, while the green bars delimit 95% bootstrap confidence interval for the risk ratio, $[0.1865, 0.4134]$.

confidence interval will be the 95% credible set in this approach. Note that WinBUGS does not use an explicit loop for bootstrap. The resampling is done automatically with each iteration of MCMC.

```
#Rubidium to Potassium Ratio
model{
   for (i in 1:N){
   p[i]  <- 1/N }  #discrete uniform on  1,2,...,N
  for( i in 1:N ){
     pick[i]  ~ dcat( p[ ] )
     yb[i] <- y[pick[i]]        }
  cv <- sd(yb[])/mean(yb[])
}
DATA
list(N=17,   y=c(0.028, 0.032, 0.031, 0.041, 0.028,
    0.039, 0.042, 0.036, 0.037, 0.029, 0.048, 0.037,
    0.037, 0.044, 0.039, 0.029, 0.038))
INITS
Just "gen inits"
```

Here is the summary of the WinBUGS output. As is evident from Figure 18.3, the posterior for CV is fairly symmetric. The posterior mean was found to be 0.1572 and the median 0.1575. The 95% credible set for cv is $[0.1125, 0.1996]$.

	mean	sd	MC error	val2.5pc	median	val97.5pc	start	sample
cv	0.1572	0.02232	7.112E-5	0.1125	0.1576	0.1996	1001	100000

See Exercise 18.21 for a traditional solution to this example and compare the confidence interval obtained there to the credible set produced by Win-BUGS.

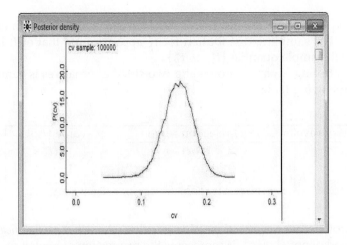

Fig. 18.3 Posterior density of CV equivalent to a bootstrap density.

18.7.3 Bootstrap Versions of Some Popular Tests

We will next discuss the bootstrap versions of three popular tests: testing for single mean, comparing two means, and one-way ANOVA. We also point out that there are various strategies for conducting a bootstrap sampling. Although, in principle, we follow the original sampling scheme when resampling the data, there is a range of valid resampling strategies in use. For example, in bootstrapping one-way ANOVA, one usually resamples each group separately (as described on page 923). An alternative option, among several, is to find group-residuals $y_{ij} - \overline{y}_i$, group them together, and then resample from that common group. Additional examples can be found in the excellent text by Kennett and Zacks (2004).

18.7.3.1 Testing the Mean

Let $X_1, \ldots, X_n$ be a sample from a population with finite mean μ. We are interested in testing $H_0 : \mu = \mu_0$. From the sample we find the studentized statistic

$$t = \frac{\overline{X} - \mu_0}{s/\sqrt{n}}.$$

We next generate B bootstrap resamples and find

$$t_b^* = \frac{\overline{X}_b^* - \overline{X}}{s_b^*/\sqrt{n}}, \quad b = 1, \ldots, B$$

where $\overline{X}_b^*$ and s_b^* are the mean and standard deviation of the bth bootstrap sample. The sampling distribution for t_b^* approximates that of t. Let $t^*(\alpha)$ be the αth sample quantile $\{t_1^*, \ldots, t_B^*\}$.

Then testing H_0 against one- and two-sided alternatives is summarized in the following table:

Alternative	α-level rejection region	p^*-value (ASL)
$\mu < \mu_0$	$t < t^*(\alpha)$	$\frac{1}{B}\sum_{b=1}^{B}\mathbf{1}(t_b^* < t)$
$\mu \neq \mu_0$	$t < t^*(\alpha/2)$ or $t > t^*(1-\alpha/2)$	$\frac{1}{B}\sum_{b=1}^{B}\mathbf{1}(|t_b^*| > |t|)$
$\mu > \mu_0$	$t > t^*(1-\alpha)$	$\frac{1}{B}\sum_{b=1}^{B}\mathbf{1}(t_b^* > t)$

Here the counterpart of p-value is called the *achieved significance level* (ASL) and denoted by p^*. It represents the proportion of bootstrap statistics t_b^* more extreme (in the sense of H_1) than t.

Example 18.13. **microRNA Regulation of Aortic Valve Disease Progression.** Aortic valve (AV) disease is known to occur preferentially on one side of the valve. The opposite sides of the valve are known to experience different hemodynamic conditions (i.e., oscillatory vs. laminar shear), which may influence the expression levels of mechano-sensitive miRNAs in AV endothelial cells. The study by Holliday et al. (2011) quantified and compared expression levels of miRNA-199a-3p expression in cells from both sides of the valve, exposed to either oscillatory or laminar shear. The null hypothesis for this experiment is that the cells exposed to oscillatory shear will show no difference in expression of miR-199a-3p from the cells exposed to laminar shear. The difference of the normalized expression levels in oscillatory-sheared samples and normalized expression levels in laminar-sheared samples $d_1, d_2, \ldots, d_9$ are given as

0.2442	0.0695	0.5447
0.8057	−0.4116	−0.5382
−0.1435	0.1743	0.1865

(a) Does an oscillatory shear induce a change in miR-199a-3p expression? Evidence in support of a change in the expression would require rejecting $H_0 : \delta = 0$, where $\delta = \mu_1 - \mu_2$ is the population parameter. Report the ASL in testing of H_0 against the one-sided alternative, $H_1 : \delta > 0$.

```
d=[ 0.2442 0.0695 0.5447 0.8057 ...
    -0.4116 -0.5382 -0.1435 0.1743 0.1865];
dbar = mean(d)               %0.1035
sd = std(d)                  %0.4270
t = dbar/(std(d)/sqrt(9))    %0.7272
p=1-tcdf(t, 8)               %0.2439
 %bootstrap
```

```
B=20000;
tbs =[];
for b=1:B
    db = randsample(d, length(d), true);
    tb = (mean(db) - dbar)/(std(db)/sqrt(9));
    tbs=[tbs tb];
end
sum(tbs > t)/B            %0.2421
```

Since the ASL was $p^* = 0.2421$, the null hypothesis was not rejected.

18.7.3.2 Comparing Two Means

Let $X_{11}, X_{12}, \ldots, X_{1n_1}$ and $X_{21}, X_{22}, \ldots, X_{2n_2}$ be samples from populations with finite means μ_1 and μ_2. We are interested in testing the hypothesis $H_0 : \mu_1 = \mu_2$ against one of the three possible alternatives. From the samples we compute the studentized statistic

$$t = \frac{\overline{X}_1 - \overline{X}_2}{\sqrt{s_1^2/n_1 + s_2^2/n_2}}.$$

If the samples were normal, this statistic would have a t-distribution. Without normality assumption, we will use bootstrap to approximate distribution of t under H_0; in fact we will need the quantiles of this distribution.

Since we have two independent samples, we can resample each group separately. From B bootstrap pairs of samples, we find

$$t_b^* = \frac{\overline{X}_{1,b}^* - \overline{X}_{2,b}^* - (\overline{X}_1 - \overline{X}_2)}{\sqrt{(s_{1,b}^*)^2/n_1 + (s_{2,b}^*)^2/n_2}}, \quad b = 1, \ldots, B.$$

Here, $\overline{X}_{1,b}^*$, $\overline{X}_{2,b}^*$, $(s_{1,b}^*)^2$, and $(s_{2,b}^*)^2$ are analogues of $\overline{X}_1$, $\overline{X}_2$, s_1^2, and s_2^2, calculated from the bth bootstrap resampled pair. Let, as before, $t^*(\alpha)$ denote the αth sample quantile of $\{t_1^*, \ldots, t_B^*\}$.

Then testing H_0 against one- or two-sided alternatives is summarized in the table, which is similar to the table in bootstrap testing for the single mean (μ replaced by $\mu_1 - \mu_2$ and μ_0 by 0).

Alternative	α-level rejection region	p^*-value
$\mu_1 < \mu_2$	$t < t^*(\alpha)$	$\frac{1}{B}\sum_{b=1}^B \mathbf{1}(t_b^* < t)$
$\mu_1 \neq \mu_2$	$t < t^*(\alpha/2)$ or $t > t^*(1-\alpha/2)$	$\frac{1}{B}\sum_{b=1}^B \mathbf{1}(|t_b^*| > |t|)$
$\mu_1 > \mu_2$	$t > t^*(1-\alpha)$	$\frac{1}{B}\sum_{b=1}^B \mathbf{1}(t_b^* > t)$

Note that one can propose a bootstrap CI for $\mu_1 - \mu_2$ as

$$
\left[\overline{X}_1 - \overline{X}_2 + t^*(\alpha/2)\sqrt{\frac{s_1^2}{n_1} + \frac{s_2^2}{n_2}}, \ \overline{X}_1 - \overline{X}_2 + t^*(1 - \alpha/2)\sqrt{\frac{s_1^2}{n_1} + \frac{s_2^2}{n_2}} \right].
$$

Example 18.14. **Microdamage in Bones Revisited.** In Example 10.5, we compared scores quantifying microdamage in bones for $n_1 = 13$ donors classified as young ($\leq$45 years old) and $n_2 = 17$ classified as old ($>$45 years old). We rejected H_0 in Bayesian testing, since the posterior probability of the alternative H_1 was found to be 0.9899. Here we repeat this two samples test using bootstrap methodology.

```
%Microdamage in bones revisited
x1=[0.790 0.944 0.958 1.011 0.714 0.256 0.406 ...
    0.135 0.316 1.264 1.410 1.160 0.179];
x2 =[1.137 0.601 1.029 1.264 1.183 1.856 1.899 ...
    0.486 0.813 1.327 1.325 2.012 1.026 1.130 0.605 0.870 0.820];
n1=length(x1); n2 = length(x2);
mx1 =mean(x1); mx2 = mean(x2);
 s1=std(x1); s2 = std(x2);
count=0;
 t = (mx1-mx2)/(sqrt(s1^2/n1 + s2^2/n2));
 ndf = (s1^2/n1 + s2^2/n2 )^2 /( (s1^2/n1)^2/(n1-1) + ...
    (s2^2/n2)^2 /(n2-1) )
%   Welch-Satterthwaite df = 26.4234
pval=tcdf(t, ndf)  %0.0096
% bootstrap
tstars =[];
B=100000;
for b = 1:B
   x1star = bootsample(x1');
   x2star = bootsample(x2');
   mxs1 = mean(x1star); mxs2 = mean(x2star);
   ss1 = std(x1star); ss2 = std(x2star);
   tstar = (mxs1 - mxs2 - (mx1 - mx2))/sqrt(ss1^2/n1 + ss2^2/n2);
   tstars = [tstars tstar];
   count = count + (tstar < t);       %for p*-value
end
count/B                             %p*-value 0.0083
quantile(tstars,  0.05)             %-1.6578
    %compare to   tinv(0.05, 26.4234) = -1.7046
reject = t < quantile(tstars,  0.05)  %0 no; 1 yes; here = 1
```

Figure 18.4 summarizes the output.

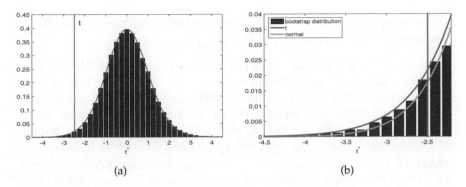

Fig. 18.4 (a) Bootstrap distribution of t^* statistic and superimposed t-distribution with Welch-Satterthwite df $= 26.4234$. The red bar is at the position $t = -2.4946$. (b) An enlargement of the left tail and comparison of bootstrap, t, and standard normal distributions.

18.7.3.3 Bootstrap ANOVA

Let k samples of size $n_1, n_2, \ldots, n_k$ result in sample means $\overline{X}_1, \ldots, \overline{X}_k$, and sample variances $s_1^2, \ldots, s_k^2$. Let the total sample size be $N = n_1 + n_2 + \cdots + n_k$, and let $\overline{X} = \sum_i n_i \overline{X}_i / N$ denote the grand mean.

We form the statistic

$$F = \frac{\frac{1}{k-1} \sum_{i=1}^{k} n_i (\overline{X}_i - \overline{X})^2}{\frac{1}{N-k} \sum_{i=1}^{k} (n_i - 1) s_i^2}.$$

This is a standard statistic for one way ANOVA, which under usual ANOVA assumptions (independence and normality of populations, constant variance) has an F-distribution with $k - 1, N - k$ degrees of freedom (see hint in Exercise 11.6).

Let $\overline{X}_{1,b}^*, \ldots, \overline{X}_{k,b}^*, (s_{1,b}^*)^2, \ldots, (s_{k,b}^*)^2$, and $\overline{X}_b^*, b = 1, \ldots, B$ be the bootstrap replicates of $\overline{X}_1, \ldots, \overline{X}_k, s_1^2, \ldots, s_k^2$, and $\overline{X}$, respectively. To obtain replicates, each of k original samples is bootstrapped separately. In general, for bootstrapping any statistical computation, we follow the original sampling scheme when resampling the data. Thus, each resample is a collection of k bootstrapped samples, for which we calculate

$$F_b^* = \frac{\frac{1}{k-1} \left[\sum_{i=1}^{k} n_i (\overline{X}_{i,b}^* - \overline{X}_i)^2 - N(\overline{X} - \overline{X}_b^*)^2 \right]}{\frac{1}{N-k} \sum_{i=1}^{k} (n_i - 1)(s_{i,b}^*)^2}, \quad b = 1, 2, \ldots, B. \quad (18.8)$$

As before, we utilize sample quantiles of $\{F_1^*, F_2^*, \ldots, F_B^*\}$ to make an inference. The ANOVA hypothesis $H_0 : \mu_1 = \mu_2 = \cdots = \mu_k$ is rejected at the level α when $F > F^*(1 - \alpha)$. This is because the F^* values are consistent with

H_0 since, by nature of resampling, the bootstrap samples are well mixed. The achieved significance level is $p^* = \sum_{b=1}^{B} \mathbf{1}(F_b^* > F)/B$.

Example 18.15. **Coagulation Times Bootstrapped.** In Example 11.1, we analyzed the coagulation times data by one-way ANOVA. Twenty-four animals were randomly allocated to 4 different diets, and blood coagulation times were measured for each animal. We found that population mean times were significantly different; the H_0 was decisively rejected with a p-value of 4.6585×10^{-5}. In this example, we repeat ANOVA analysis by bootstrap. MATLAB script ◀ coagulationboot.m performs bootstrap ANOVA. Figure 18.5 shows bootstrap distribution of statistic F^* from (18.8) as a normalized histogram and $F_{4-1,24-4}$ density superimposed. The F density assumes normality of data and equal variances, while the bootstrap distribution of F^* is slightly different.

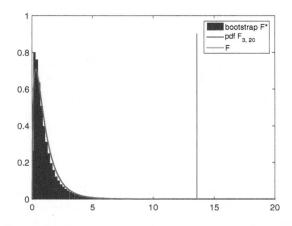

Fig. 18.5 Sampling distribution of bootstrap F^* statistic (histogram) with a $F_{4-1,24-4}$ density superimposed (in red) as a reference. The position of statistic F is marked by the green line.

Out of $B = 500{,}000$ bootstrap replicates, F^* exceeded F 248 times, leading to an ASL of $4.96e - 04$.

```
%coagulationboot.m
 times = [62, 60, 63, 59, ...
             63, 67, 71, 64, 65, 66, ...
             68, 66, 71, 67, 68, 68, ...
             56, 62, 60, 61, 63, 64, 63, 59];
  diets = [1 1 1 1   2 2 2 2 2 2   3 3 3 3 3 3  4 4 4 4 4 4 4 4];

ni =[4 6 6 8]; N=24; k=4;
barxi = [mean(times(diets==1))  mean(times(diets==2))  ...
   mean(times(diets==3))  mean(times(diets==4)) ]; %61 66 68 61
```

```
s2i = [var(times(diets==1)) var(times(diets==2)) ...
  var(times(diets==3)) var(times(diets==4)) ];  %3.3333  8. 2.8 6.8571
barx = sum(ni.*barxi)/N;  %64
Fnum= 1/(k-1) * sum( ni .* (barxi - barx).^2 );  %76
Fdenom = 1/(N-k) * sum( (ni - 1) .* s2i );  %5.6
F=Fnum/Fdenom;  %13.5714
%
B=500000;
Fstar=[];
for i= 1:B
select =  [randsample(1:4, 4, 'true') randsample(5:10, 6, 'true') ...
           randsample(11:16, 6, 'true') randsample(17:24, 8, 'true')];
timesstar = times(select);
barxistar = [mean(timesstar(diets==1)) mean(timesstar(diets==2)) ...
             mean(timesstar(diets==3)) mean(timesstar(diets==4)) ];
s2istar = [var(timesstar(diets==1)) var(timesstar(diets==2)) ...
           var(timesstar(diets==3)) var(timesstar(diets==4)) ];
barxstar = sum(ni.*barxistar)/N;
Fnumstar= 1/(k-1) *(sum(ni.*(barxistar-barxi).^2) - N*(barxstar-barx)^2);
Fdenomstar = 1/(N-k) * sum( (ni - 1) .* s2istar );
Fstar=[Fstar Fnumstar/Fdenomstar];
end
%achieved significance level p*
asl = 1- sum(F > Fstar)/B  %4.9600e-04
```

18.7.3.4 Bootstrapping Regression

There are two main approaches for bootstrapping regression data: random and fixed design.

Random Design Bootstrap. Random design bootstrapping is conceptually simple. From n observed $(k+1)$-tuples $(y_i, x_{1i}, x_{2i}, \ldots, x_{ki}), i = 1, \ldots, n$, bootstrap samples are formed by selecting n $(k+1)$-tuples with replacement.

Thus, we form B data sets

$$(y_{i*}, x_{1i*}, x_{2i*}, \ldots, x_{ki*})_b, \quad b = 1, \ldots, B,$$

where n indices i^* are sampled with replacement from $\{1, \ldots, n\}$.

For each regression

$$y_b = X_b \beta_b + \epsilon_b, \quad b = 1, \ldots, B,$$

we calculate relevant regression estimators in a standard manner, as in Chapter 14, which leads to their bootstrap distributions.

The term *random design* reflects the fact that design matrix X_b is not preserved in bootstrapping, but its rows are selected from the original design matrix X.

Fixed Design Bootstrap. In some situations it is desirable to preserve the design matrix. This may be in the case when sampling is informatively designed, when the index of covariates could be important, when the range of covariates is to be preserved, or in some other situations.

An alternative approach to bootstrap in this case is a fixed design or residual resampling. The design matrix remains fixed as the original and only bootstrap resamples of the response variable are taken. First, using the original sample, the fitted values $\hat{y}$ are found, $\hat{y} = X(X'X)^{-1}X'y$. Then, a bootstrap sample of responses is defined as

$$y_b = \hat{y} + e_b,$$

where e_b is a bootstrap sample of the residuals, $e = y - \hat{y}$.

Example 18.16. **Fat Data Bootstrapped.** In Section 14.6, we discussed multiple regression involving fat data, 🖳 fat.dat. The Brozek Index was regressed on 14 variables, and the detailed diagnostic analysis was implemented in 📄 fatregdiag.m. For this regression problem we employ both random and fixed design bootstrap approaches to assess distributions of some statistics.

As an illustration, we look at the bootstrap distribution of R^2. Figure 18.6 provides the random design bootstrap distribution for R^2 in panel (a), and for fixed design bootstrap distribution in panel (b). The bootstrap means, medians, and variances are 0.7627, 0.7635, and 0.0006013, for random design, and 0.7632, 0.7636, and 0.000381, for fixed design, respectively. For both designs we took $B = 50,000$ resamples. In Section 14.6, R^2 was found to be 0.7490, which is represented by the red bar in Figure 18.6.

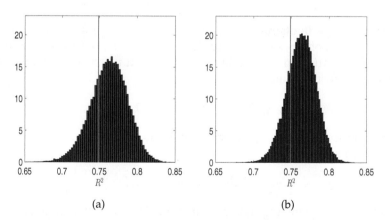

(a)   (b)

Fig. 18.6 Bootstrap distributions of R^2 for (a) random design, and (b) fixed design for fat data. The red bar in both panels is $R^2 = 0.7490$, found in the least squares regression.

Details can be found in 📄 fatregbootr.m and 📄 fatregbootf.m.

✐

18.7.4 Randomization and Permutation Tests

Suppose that in a statistical experiment the sample or samples are taken and a statistic S is constructed for testing a particular null hypothesis. The values of S extreme from the viewpoint of H_0 are critical for this hypothesis. The decision whether the observed value for S is extreme is made by looking at the distribution of S when H_0 is true. Sometimes such distribution is unknown or too complex to find.

Randomization methods that permute the original data can be used to approximate the null distribution of S. Given the sample, one forms permutations that are *consistent with the experimental design and H_0*, and for each permutation calculates the value of S. Such values of S are then used to estimate its empirical density. Finally, in usual way, one would find the ASL-like summary, often called a *permutation p-value*.

Permutations consistent with the null hypothesis depend on the problem. For example, in a two-sample problem we want to compare the means of two populations. The null hypothesis is $\mu_1 = \mu_2$. The permutations consistent with H_0 would be all permutations of a combined (concatenated) sample $X_{11}, \ldots, X_{1n_1}, X_{21} \ldots, X_{2n_2}$. Then, as the first permutation sample one takes first n_1 observations, and the second permutation sample is the rest.

As another example, suppose that a repeated measures design has observations that are triplets corresponding to three treatments, i.e., (X_{11}, X_{12}, X_{13}), $\ldots$, (X_{n1}, X_{n2}, X_{n3}), and that H_0 states that the three treatment means coincide, $\mu_1 = \mu_2 = \mu_3$. Then permutations consistent with this experimental design are random permutations among the triplets (X_{i1}, X_{i2}, X_{i3}), $i = 1, \ldots, n$ and a possible permutation might be

$$(X_{13}, X_{11}, X_{12})$$
$$(X_{21}, X_{23}, X_{22})$$
$$(X_{32}, X_{33}, X_{31})$$
$$\ldots$$
$$(X_{n2}, X_{n1}, X_{n3}).$$

Thus, depending on the design and H_0, consistent permutations can be quite different.

Example 18.17. **Monitoring Production Process.** Suppose that we are monitoring a production process for 11 hours and every hour record the temperature in °F, as a vector $X_1, X_2, \ldots, X_{11}$. Significant upward or downward trends could be critical for the process. We are interested in testing $H_0: X_i$'s are independent and identically distributed (no trend) versus H_1 : there is an overall increasing trend.

If the measurements are listed below:

176 175 164 175 168 160 179 181 207 189 205

Let $i = 1, 2, \ldots, 11$ be the hours. The statistic that is sensitive to trends is $f = \sum_{i=1}^{n} iX_i$. This cross-product (flux) exceeds $\frac{n(n+1)}{2}\overline{X}$ if the overall trend is increasing. Asymptotic normality can be applied and test devised, see, for example, Ross (2010, p. 690). We will not pursue that direction. Instead, we randomly permute $X_1, \ldots, X_{11}$ and obtain a permutation $X_1^*, \ldots, X_{11}^*$. For this permutation the flux statistic $f^* = \sum_{i=1}^{11} iX_i^*$ is calculated.

If this is repeated a large number of times, say for $B = 100,000$ permutations, the realizations for f^* will well approximate the sampling distribution of f, under H_0. The p^* value is calculated as in the bootstrap resampling tests, $p^* = \frac{1}{B}\sum_{b=1}^{B}\mathbf{1}(f_b^* \geq f)$. MATLAB code ◢ permute1.m calculates p^*-value as 0.008. The core of this code is the loop:

```
X = [176 175 164 175 168 160 179 181 207 189 205];
n = length(X);
i = 1:n;
f = i * X'    %12227
fps = [];
count = 0;
B = 100000;
for b = 1:B
    Xp = X(randperm(n));
    fp = i * Xp'; %permutation flux
    count = count + (fp >= f); %count fluxes >= f
    fps = [fps fp]; %save all fluxes
end
pstar = count/B    %0.0080
```

Thus, the hypothesis of no trend is rejected by this permutation test; the permutation p^*-value is 0.008. Figure 18.7 shows the original data in panel (a), and the permutation distribution for flux, with average flux (the green bar) and observed flux (the red bar) in panel (b).

✏

Remark. One of the reasons for relatively scarce use of resampling methodology is the large number of different approaches with many versions for the same task. We indicated that there are many kinds of bootstrap confidence intervals. Also the choice of statistic to replicate is quite liberal. For example, in testing the equality of two means, one can use t-like statistic, as in Example 1.3, or simply unstandardized difference $\overline{X}_1 - \overline{X}_2$.

How big should B be? Some researchers advise B in order of thousands irrespective of the size of original sample. Recommendations tied to the original sample size can be found in the literature, for example, $B = 40n$

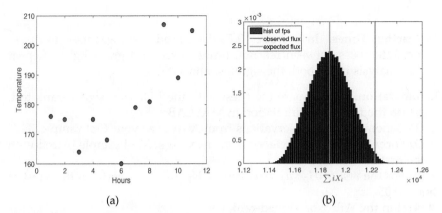

Fig. 18.7 (a) Process temperatures; (b) Permutation distribution for flux with average flux (green) and observed flux (red). The p^* value is 0.008.

or $B = n^2$. Such recommendations may not be feasible for large n (e.g., Example 10.17).

18.8 Exercises

18.1. **Friday the 13th.** The following data set is part of the larger study from Scanlon et al. (1993) titled "Is Friday the 13th bad for your health?" The data analysis in this paper addresses the issues of how superstitions regarding Friday the 13th affect human behavior. The authors reported and analyzed data on traffic accidents for Friday the 6th and Friday the 13th between October 1989 and November 1992. The data consist of the number of patients accepted in SWTRHA (South West Thames Regional Health Authority, London) hospital on the dates of Friday the 6th and Friday the 13th.

Year, month	Number of accidents Friday 6th	Friday 13th	Sign
1989, October	9	13	−
1990, July	6	12	−
1991, September	11	14	−
1991, December	11	10	+
1992, March	3	4	−
1992, November	5	12	−

Use the sign test at the level $\alpha = 10\%$ to test the hypothesis that the "Friday the 13th effect" is present. The m-file signtst.m could be applied.

18.2. **Reaction Times.** In Exercise 10.26 the paired t-test was used to assess the differences between reaction times to red and green lights. Repeat this analysis using both the sign test and WSiRT.

18.3. **Simulation.** To compare the t-test with the Wilcoxon signed-rank test, set up the following simulation in MATLAB:
(1) Generate $n = 20$ observations from $\mathcal{N}(0,1)$ as your first sample X.
(2) Form `Y = X + randn(size(x)) + 0.5` as your second sample paired with the first.
(3) For the test of $H_0 : \mu_1 = \mu_2$ versus $H_1 : \mu_1 < \mu_2$, perform the t-test at $\alpha = 0.05$.
(4) Run the Wilcoxon signed-rank test.
(5) Repeat this simulation 1,000 times and compare the powers of the tests by counting the number of times H_0 was rejected.

18.4. **Grippers.** Measurements of the left- and right-hand gripping strengths of 10 left-handed writers are recorded:

Person	1	2	3	4	5	6	7	8	9	10
Left hand (X)	140	90	125	130	95	121	85	97	131	110
Right hand (Y)	138	87	110	132	96	120	86	90	129	100

(a) Does the data provide strong evidence that people who write with their left hand have a greater gripping strength in their left hand than they do in their right hand? Use the Wilcoxon signed-rank test and $\alpha = 0.05$.
(b) Would you change your opinion on the significance if you used the paired t-test?

18.5. **Iodide and Serum Concentration of Thyroxine.** The effect of iodide administration on serum concentration of thyroxine (T_4) was investigated in Vagenakis et al. (1974). Twelve normal volunteers (9 male and 3 female) were given 190 mg iodide for 10 days. The measurement X is an average of T_4 in the last 3 days of administration, while Y is the mean value in three successive days after the administration stopped.

Subject	1	2	3	4	5	6	7	8	9	10	11	12
Iodide (X)	7.9	9.1	9.2	8.1	4.2	7.2	5.4	4.9	6.6	4.7	5.2	7.3
Control (Y)	10.2	10.2	11.5	8.0	6.6	7.4	7.7	7.2	8.2	6.2	6.0	8.7

Assume that the difference $D = X - Y$ has a symmetric distribution. Compare the p-values of WSiRT and the paired t-test in testing the hypothesis that the mean of control measurements exceed that of the iodide measurements.

18.6. **Weightlifters.** Blood lactate levels were determined in a group of amateur weightlifters following a competition of 10 repetitions of 5 different upper-body lifts, each at 70% of each lifter's single maximum ability. A sample of 17 randomly selected individuals was tested: 10 males and 7 females. The following table gives the blood lactate levels in female and male weightlifters, in units of mg/100 ml of blood:

Gender	N	Blood lactate
Female	7	7.9 8.2 8.7 12.3 12.5 16.7 20.2
Male	10	5.2 5.4 6.7 6.9 8.2 8.7 14.2 14.2 17.4 20.3

Test H_0 that blood lactate levels were not different between the two genders at the level $\alpha = 0.05$. Assume that the data, although consisting of measurements of continuous variables, are **not** distributed normally or variances are possibly heteroscedastic (nonequal variances). In other words, the traditional t-test may not be appropriate.

18.7. **Cartilage Thickness in Two Osteoarthritis Models.** Osteoarthritis (OA), characterized by gradual degradation of the cartilage extracellular matrix, articular cartilage degradation, and subchondral bone remodeling, is the most common degenerative joint disease in humans. One of the most common ways for researchers to study the progression of OA in a controlled manner is to use small animal models. In a study conducted in the laboratory of Robert Guldberg at the Georgia Institute of Technology, a group of 14 rats was randomly divided into two subgroups, and each subgroup was subjected to an induced form of OA. One of the subgroups ($n_1 = 7$) was subjected to a chemically induced form of OA via an intra-articular injection of monosodium iodoacetate (MIA), while the other subgroup ($n_2 = 7$) was subjected to a surgically induced form of OA by transecting the medial meniscus (MMT). In this study, all rats had OA induced by either MIA or MMT on the left knee, with the right knee serving as a contralateral control.

The objectives of this study were to quantify changes in cartilage thickness of a selected area of the medial tibial plateau, compare thickness values between the treatments and the controls, and determine if OA induced by MIA produced different results from OA induced by MMT. Cartilage thickness values were measured 3 weeks after the treatments using an ex vivo micro-CT scanner; the data are provided in the table below:

Rat MIA	Treated	Control	Rat MMT	Treated	Control
1	0.1334	0.2194	8	0.2569	0.2726
2	0.1214	0.1929	9	0.2101	0.2234
3	0.1276	0.1833	10	0.1852	0.2216
4	0.1152	0.1879	11	0.1798	0.1905
5	0.1047	0.2529	12	0.1049	0.1444
6	0.1312	0.2527	13	0.2649	0.2841
7	0.1222	0.2595	14	0.2383	0.2731

(a) Using the Wilcoxon signed-rank test, test the hypothesis that the difference between MIA-treated rats and its control thickness is significant. (b) Find the difference in thickness (treatment–control) for MIA and MMT, and test the hypothesis that they are the same. Apply a nonparametric version of a two-sample t-test on the differences. Conduct both tests at a 5% significance level.

18.8. **A Claim.** Professor Scott claims that 50% of his students in a big class get a final score of 90 or higher.

A suspicious student asks 17 randomly selected students from Professor Scott's class, and they report the following scores:

80 81 87 94 79 78 89 90 92 88 81 79 82 79 77 89 90

Test the hypothesis that Professor Scott's claim does not conform to the evidence, that is, that the 50th percentile (0.5-quantile, median) is different than 90. Use $\alpha = 0.05$.

18.9. **Claustrophobia.** Sixty subjects seeking treatment for claustrophobia are independently sorted into two groups, the first of size $n = 40$ and the second of size $m = 20$. The members of the first group each individually receive treatment A over a period of 15 weeks, while those of the second group receive treatment B. The investigators' directional hypothesis is that treatment A will prove to be more effective. At the end of the experimental treatment period, the subjects are individually placed in a series of claustrophobia test situations, knowing that their reactions to these situations are being recorded on videotape. Subsequently three clinical experts, uninvolved in the experimental treatment and not knowing which subject received which treatment, independently view the videotapes and rate each subject according to the degree of claustrophobic tendency shown in the test situations. Each subject is rated by the experts on a scale of 0 (no claustrophobia) to 10 (an extreme claustrophobia). The following tables show the average ratings for each subject in each of the two groups:

A
4.6 4.7 4.9 5.1 7.0 4.9
5.1 5.2 5.5 4.8 5.7 5.0
5.8 6.1 6.5 7.0 6.4 5.2
4.6 4.7 4.9 6.4 5.9 4.7
5.8 5.2 5.4 6.1 7.7 6.2
5.8 5.1 6.5 2.2 6.9 5.0
6.5 7.2 8.2 6.7

B
5.2 5.3 5.4 7.7 8.1 4.9
5.6 6.2 6.3 7.0 7.0 7.8
6.8 7.7 8.0 6.6 5.5 8.2
8.1 5.0

The investigators expected treatment A to prove more effective, and sure enough it is group A that appears to show the lower mean level of claustrophobic tendency in the test situations.

Using Wilcoxon's sum rank test, test the hypothesis H_0 that treatments A and B are of the same effectiveness, versus the alternative that treatment A is more effective.

18.10. **Nonparametric Stats with Raynaud's Phenomenon.** (a) Refer to Raynaud's phenomenon data from Exercise 11.20. Using Wilcoxon's signed-rank test, compare the responses (number of attacks) for drug/placebo effect. Ignore the Period variable. Compare this result with the result from a paired t-test.

(b) Compare the drug/placebo effect using Bayesian inference. Write a WinBUGS program that will read the drug/placebo info, take the difference between the measurements, d, and model it as normal, with noninformative priors on the mean mu and precision prec. Check if the credible set for mu contains 0 and draw the appropriate conclusion.

18.11. **Cry-babies.** Nurses at Northbay Healthcare participate in an Evidence Based Practice (EBP) program. The purpose of EBP is to research and implement improvements in nursing care, and then to gauge the success of these changes using statistical methods. In one study, they introduced changes in the way newborn babies are handled. The goal is to reduce pain experienced by the infants resulting from their vitamin K shot. Lawrence et al. (1993) provide data on 79 babies (infants) in a control group and 79 babies in the intervention group. The control group infants were handled using conventional methods. The intervention group infants were held by their mothers prior to, and during the administering of the shot, known as "kangaroo care." Kangaroo care seeks to provide restored closeness of the newborn with mother or father by placing the infant in direct skin-to-skin contact with one of them. This ensures physiological and psychological warmth and bonding.

File cry.mat contains two sequences cry.control and cry.kc of durations of crying after shot of K vitamin.

(a) Test the equality of means for the two groups using standard t-test, without any assumption on population variances. Report the p-value for the two-sided alternative.

(b) Using the Wilcoxon sum-rank Test procedure, repeat the test from (a). Compare the p-values and discuss?

18.12. **Cotinine and Nicotine Oxide.** Garrod et al. (1974) measured the nicotine metabolites, cotinine, and nicotine-1'-N-oxide in 24-hour urine collections from normal healthy male smokers and smokers affected with cancer of the urinary bladder. The data, also discussed by Daniel (1978), show the ratio of cotinine to nicotine-1'-N-oxide in the two groups of subjects.

	Ratio										
Cancer	5.0	8.3	6.7	3.0	2.5	12.5	2.4	5.5	5.2	21.3	5.1	1.6
patients	2.1	4.6	3.2	2.2	7.0	3.3	6.7	11.1	3.4	5.9	27.4	
Control	2.3	1.9	3.6	2.5	0.75	2.5	2.1	1.1	2.3	2.2	3.5	1.8
subjects	2.3	1.4	2.1	2.0	2.3	2.4	3.6	2.6	1.5			

(a) Using `qqplot` on cancer patients' data, argue that the normality condition for the ratio in this population may be violated.

(b) Using Wilcoxon's signed-rank procedure, test the hypothesis that the ratio in population of cancer patients is equal to 4 versus the two-sided alternative. Assume $\alpha = 0.05$.

(c) Using Wilcoxon's sum-rank procedure, test the hypothesis that the ratios in the two populations (cancer and control) are the same versus the alternative that the ratio in cancer population is significantly greater. Take $\alpha = 0.05$.

18.13. **Coagulation Times.** In Example 11.1 a standard one-way ANOVA gave p-value of 4.6585e-05. Repeat the test of equality of means using the Kruskal–Wallis procedure. If the test turns out significant, compare the means using the Kruskal–Wallis pairwise comparisons.

18.14. **Cryopreservation Preserving Circularity.** Cryopreservation is critical for bringing tissue engineered constructs from the benchtop to the clinic. In order to see if morphology of cells encapsulated in hydrogels was maintained after cryopreservation with respect to non-preserved controls, cell circularity per bead was measured in non-cryopreserved beads and in beads after cryopreservation treatment using either of 2 methods, A and B. The treatment A contains BSA (bovine serum albumin) in the cryopreservation medium while treatment B does not, in an effort to reduce use of animal-derived products for future clinical use. Fifteen beads were measured per experiment for each treatment.

Data set ▣ `cryopreservation.mat|xls` (Courtesy of Sambanis Lab, Georgia Tech) contains three columns: (1) circularity; (2) media, where 0 is control, 1 is treatment A, and 2 is treatment B; and (3) experiment number.

(a) Ignoring the experiment number (that is, assuming that experimental conditions were identical), compare the treatments A and B and the control. Use Kruskal–Wallis test. You can use ◢ `kruskalwallistest.m` or MATLAB's built in `kruskalwallis.m`.

(b) Using multiple comparisons for Kruskal–Wallis (◢ `kwpairwise.m`) discuss the circularity differences among the treatments.

18.15. **Honeybee Concentration Change.** In Exercise 11.11 we found that the ANOVA condition of equality of variances was violated and conducted Welch's test. A counterpart test that does not require normality and/or equality of variances is Kruskal–Wallis test. Show that ◢ `kruskalwallistest.m` indicates a significant difference between the batches (p-value 0.0486), while the one-way ANOVA is not significant (p-value 0.3854). Discuss this finding.

18.16. **Blocking by Rats.** In Example 11.7 the factor Procedure was found significant with a p-value of 8.7127e-04. Is this factor significant according to Friedman's test?

18.17. **Devices.** In an evaluation of a device performance, six engineers (A, B, C, D, E, and F) evaluated three devices (a, b, and c) in a randomized order. Their grades concern only the performance of the device and supposedly are not influenced by the brand name or similar exogenous information. Here are their rankings on a scale of 1 to 10:

	Device		
	a	b	c
A	7	8	9
B	6	10	7
Evaluator C	6	8	8
D	7	9	8
E	7	10	9
F	8	8	9

The null hypothesis is that the rankings for the three devices have the same distribution, and therefore the same means. The alternative is that the distributions are different, which is often interpreted as "locations or means are different."
(a) Test the null hypothesis using Friedman's test.
(b) If null hypothesis in (a) was rejected, find which means are different at the $\alpha = 0.05$ significance level.

18.18. **Differences in Differences.** Repeat Part (b) of Exercise 11.23 using Friedman's test and, if significant, do pairwise comparisons.

18.19. **Jackknifing MLE of Variance.** The MLE estimator for normal variance is $\hat{\sigma}^2 = \frac{1}{n}\sum_{i=1}^{n}(X_i - \overline{X})^2$. Show that the jackknife estimator of σ^2 is s^2.

18.20. **Beak-clapping.** Oppenheim (1968), also Hollander and Wolfe (1999), provide data (given in a data structure 🖴 beak.mat) on 25 chick embryos tested 12–40 h prior to hatching for light responsivity as measured by changes in the rate of occurrence of an overt behavioral response (beak-clapping). The rates of beak-clapping in the dark (beak.dark) and during the 1 min of light stimulation (beak.light) are provided as the two fields in the data structure file beak. Using bootstrap, test the hypothesis that the mean number of beak claps is significantly lower in the dark compared to that in the light.

18.21. **Rubidium to Potassium Ratio Revisited.** In Example 18.12 WinBUGS was used to produce bootstrap sample that was subsequently repeated using MCMC. Conduct the traditional bootstrap with $B = 100{,}000$ bootstrap resamples and compare the obtained 95% bootstrap confidence interval with the 95% credible set $[0.1125, 0.1996]$ from Example 18.12.

18.22. **High Ferritin Levels in Patients with Liver Disease.** Neghina et al. (2009) examined ferritin levels in patients with some form of liver disease. The authors looked at the links between mutant HFE genotypes (C282Y, H63D, and S65C) and iron overloads in patients with liver diseases.

Previous results suggested that the normal ferritin levels for males range from 12 to 300 μg/L. The ferritin levels (in μg/L) for the 15 male liver disease patients from the study are provided below:

$$
\begin{array}{ccccc}
1000 & 7837 & 920 & 620 & 280 \\
229 & 530 & 1320 & 123 & 112 \\
269 & 380 & 408 & 74 & 388
\end{array}
$$

(a) Using the bootstrap test for the mean, find whether men with liver disease have a significantly higher level of ferritin than the normal population, which has a mean ferritin level of 150 μg/L. Report the p^*-value.
(b) Repeat the preceding test after removing the extreme outlier (ferritin level of 7837 μg/L). Report the p^*-value. How does this affect the outcome of the test?
(c) Does a 90% bootstrap confidence interval for the mean contain 150?

18.23. **Sizes of Tumors in BC Patients.** Sharma et al. (2005) report sizes (in mm) of tumors found in 23 breast cancer patients.

```
sizes = [11 20 20 22 15  7 26  4 15 25 14  9 ...
         15 14  8 35 23 11 50 24 23 10  3];
```

(a) Follow steps (i)–(iii) to test the hypothesis that the population variance is $\sigma_0^2 = 80$ (in mm^2) versus the alternative that it is larger than 80.

   (i) Find sample variance s^2 for the original sample.
   (ii) For $B = 10{,}000$ bootstrap resamples, find sample variances $s_1^{*2}, s_2^{*2}, \ldots, s_B^{*2}$.
   (iii) Find the achieved significance level p^* as

$$
\frac{1}{B} \sum_{b=1}^{B} \mathbf{1} \left(\frac{s_b^{*2}}{s^2} > \frac{s^2}{\sigma_0^2} \right).
$$

(b) A $(1 - \alpha)100\%$ CI for the population variance can be calculated by either of the two formulas

$$
[s^{*2}(\alpha/2),\ s^{*2}(1 - \alpha/2)] \quad \text{or} \quad \left[\frac{s^4}{s^{*2}(1 - \alpha/2)},\ \frac{s^4}{s^{*2}(\alpha/2)} \right],
$$

where $s^{*2}(\alpha)$ is the αth quantile of $\{s_1^{*2}, s_2^{*2}, \ldots, s_B^{*2}\}$. For the dataset sizes find the 95% confidence intervals. Which interval is larger?

18.24. **Mandibles of Golden Jackals.** Higham et al. (1980) provide data on mandible length of the golden jackal (*Canis aureus*) from the British Museum of Natural History, London. The data consist of 10 measurements of mandible length on both male and female specimens and were collected as part of a study comparing prehistoric and modern canidae.

Male	120 107 110 116 114 111 113 117 114 112
Female	110 111 107 108 110 105 107 106 111 111

(a) Using the bootstrap two-sample test, assess the hypothesis that there is no significant difference between the mean lengths of mandible for male and female jackals.

(b) Modify the test when it is assumed that population variances are the same; that is, use the pooled standard deviation for studentized statistic.

18.25. **Ozone-Treated Rats.** Doksum and Sievers (1976) report data on a study designed to assess the effects of ozone on weight gain in rats. The experimental group consisted of 22 seventy-day-old rats kept in an ozone environment for 7 days. The control group consisted of 23 rats of the same age, and were kept in an ozone-free environment. Weight gain is measured in grams.

```
control =[41.0, 38.4, 24.4, 25.9, 21.9, 18.3, 13.1, 27.3,
          28.5,-16.9, 26.0, 17.4, 21.8, 15.4, 27.4, 19.2,
          22.4, 17.7, 26.0, 29.4, 21.4, 26.6, 22.7];
ozone   = [10.1,  6.1, 20.4,  7.3, 14.3, 15.5, -9.9,  6.8,
          28.2, 17.9, -9.0,-12.9, 14.0,  6.6, 12.1, 15.7,
          39.9,-15.9, 54.6,-14.7, 44.1, -9.0];
```

Find the bootstrap 95% CI for the difference between the population means.

18.26. **Bayesian Bootstrap.** A version of the bootstrap with Bayesian interpretation was proposed by Rubin (1981). In the standard bootstrap, each observation X_i from the sample $X_1, \ldots, X_n$ has a probability of $1/n$ to be selected.

In Bayesian bootstrapping, at each replication a discrete probability distribution $\mathbf{w} = \{w_1, \ldots, w_n\}$ on $\{1, 2, \ldots, n\}$ is generated and used to calculate bootstrap statistics (but not to generate bootstrap resample, see Exercise 18.27).

Operationally the distribution $\mathbf{w}$ is created by generating $n - 1$ uniform random variables on $[0,1]$ and ordering them. By this operation the interval $[0,1]$ is split in n intervals. The length of ith interval is taken as w_i. Now, any averaging in the process of calculating bootstrap estimates is weighted, with weights $\mathbf{w}$.

Using Bayesian bootstrap, estimate the variability of CV from Example 18.12.

Hint. Calculate $\overline{X}$ and s as $\sum_i w_i X_i$ and $\sqrt{\frac{n}{n-1} \sum_i w_i (X_i - \overline{X})^2}$.

18.27. **Bayesian Bootstrap – Correct and Incorrect.** In standard bootstrapping, observations are sampled with replacement. This implies that relative frequencies of observations in the resample follow (rescaled) multinomial distribution. In Bayesian bootstrap the multinomial distribution is replaced by the Dirichlet distribution.

Is there a difference between using weights taken from the Dirichlet distribution as (1) the probabilities when re-sampling from the original sample, as in

```
B=500
sample=[ 5      1     14    14    19     8    11    17 ...
     1      2     11    14     1     8     2     9    14 ...
    12     19     17];
n=length(sample);
resamples=[];
%
for b=1:B
    wi = diff([0  sort(rand(1,n-1))   1]);
    resample = randsample(sample,n,true,wi);
    %resampling with replacement according to the weights in wi
    resamples =[resamples; resample];
end
```

or, as (2) the weights of the observations in the original sample when calculating a bootstrapped statistic? Resampling with random probabilities **w**, as in the code above, is an incorrect way to do Bayesian bootstrap, while using **w** as the weights of elements in the original sample in calculating the statistic, as in Exercise 18.26, is correct.

Demonstrate using the sample from the code above that the variance of bootstrap distribution of the sample mean is substantially inflated when applying incorrect rather than the correct method.

18.28. **Honeybee Concentration Change Revisited.** In Exercise 11.11 the ANOVA condition of equality of variances was violated and we conducted Welch's test. In Exercise 12.15 we conducted the Kruskal–Wallis test. The Kruskal–Wallis test indicated significant difference between the batches (p-value of 0.0486) while Welch's test was not significant (p-value 0.3854). What are ASL's for bootstrap and permutation tests?

18.29. **Devices Revisited.** In Exercise 18.17 six engineers (A, B, C, D, E, and F) evaluated three devices (a, b, and c) in a randomized order.
The null hypothesis is that the rankings for the three devices have the same means. Conduct the test using the permutation method. Permute

the grades for a, b, and c for each evaluator independently and for each permutation find Friedman's statistics. Report the ASL.

18.30. **Law Schools Data.** Efron (1979b) provides data from 15 law schools in the United States concerning average scores on undergraduate GPA and the Law School Admission Test (LSAT).

School	LSAT	GPA	School	LSAT	GPA	School	LSAT	GPA
1	576	3.39	2	635	3.30	3	558	2.81
4	579	3.03	5	666	3.44	6	580	3.07
7	555	3.00	8	661	3.43	9	651	3.36
10	605	3.13	11	653	3.12	12	575	2.74
13	545	2.76	14	572	2.88	15	594	2.96

Estimate the intercept and slope in the regression of GPA on LSAT using
(a) random design bootstrap,
(b) fixed design bootstrap.
(c) Find the 95% bootstrap CIs for the two parameters under both designs.

MATLAB FILES AND DATA SETS USED IN THIS CHAPTER

http://statbook.gatech.edu/Ch18.NP/

amanitajkn.m, amanitajkn2.m, amanitajkn3.m, anacapa.m, anacapabci.m, bootsample.m, cbfcmro2.m, claustrophobia.m, coagulationboot.m, cryopreservation.m, ebola.m, fatregbootf.m, fatregbootr.m, fisherwsirt.m, flies.m, friedmanpairwise.m, friedmantest.m, grubbs.m, histn.m, kruskalwallistest.m, kwpairwise.m, kwvsanova.m, miammt.m, microRNA.m, microdamageboot.m, moonboot.m, muscariabci.m, NPdemo.m, npexamples.m, permute1.m, poliovaccine.m, quadeexample.m, quadetest.m, ranks.m, rubidium2.m, signtst.m, twins.m, walshnp.m, wsirt.m, wsirtcdf.m, wsirtexa.m, wsirtexact1.m, wsurt.m, wsurtcdf.m, wsurtexact.m

amanita28.dat, anacapa.dat|mat, beak.mat|xlsx, cbfcmro2.xls, cry.mat, cryopreservation.mat|xls

---

CHAPTER REFERENCES

Arbuthnot, J. (1710). An argument for divine providence. *Philosophical Transactions*, **27**, 186–190.

Bradley, J. V. (1968). *Distribution Free Statistical Tests*. Prentice Hall, Englewood Cliffs, NJ.

Brillinger, D. R. (1964). The asymptotic behaviour of TukeyấŹs general method of setting approximate confidence limits (the jackknife) when applied to maximum likelihood estimates. *Rev. Inst. Int. Stat.*, **32**, 3, 202–206.

Conover, W. J. (1999). *Practical Nonparametric Statistics*, 3rd ed. Wiley, New York.

Daniel, W. W. (1978). *Applied Nonparametric Statistics*. Houghton Mifflin, Boston.

Doksum, K. A. and Sievers, G. L. (1976). Plotting with confidence: graphical comparisons of two populations. *Biometrika*, **63**, 421–434.

Efron, B. (1979a). Bootstrap methods: another look at the jackknife. *Ann. Stat.*, **7**, 1–26.

Efron, B. (1979b). Computers and the theory of statistics: thinking the unthinkable. *SIAM Rev.*, **21**, 460–480.

Efron, B. and Tibshirani, R. J. (1993). *An Introduction to the Bootstrap*, CRC Press, Boca Raton, Fl.

Fox, P. T. and M. E. Raichle (1986). Focal physiological uncoupling of cerebral blood-flow and oxidative-metabolism during somatosensory stimulation in human subjects. *Proceed. Nat. Acad. Sci. USA*, **83**, 4, 1140–1144.

Friedman, M. (1937). The use of ranks to avoid the assumption of normality implicit in the analysis of variance. *J. Am. Stat. Assoc.*, **32**, 675–701.

Garrod, J. W., Jenner, P., Keysell, G. R., and Mikhael, B. R. (1974). Oxidative metabolism of nicotine by cigarette smokers with cancer of urinary bladder. *J. Nat. Cancer Inst.*, **52**, 1421–1924.

Hall, P. (2003). A short prehistory of bootstrap. *Statist. Sci.*, **18**, 2, 158–167.

Higham, C., Kijngam, A., and Manly, B. (1980). An analysis of prehistoric canid remains from Thailand. *J. Archaeol. Sci.*, **7**, 149–165.

Hollander, M. and Wolfe, D. A. (1999). *Nonparametric Statistical Methods*, 2nd ed. Wiley, Hoboken, NJ.

Holliday, C. J., Ankeny, R. F., Jo, H., and Nerem, R. M. (2011). Discovery of shear- and side-specific mRNAs and miRNAs in human aortic valvular endothelial cells. *Am. J. Physiol. Heart Circ. Physiol.*, **301**, 3, 856–867.

Iman, R. L. and Davenport, J. M. (1980). Approximations of the critical region of the Friedman statistic. *Commun. Stat. A.*, **9**, 571–595.

Kennett, S. R. and Zacks, S. (2004). *Modern Industrial Statistics*, Duxbury Press, Pacific Grove, CA.

Kruskal, W. H. and Wallis, W. A. (1952). Use of ranks in one-criterion variance analysis. *J. Am. Stat. Assoc.*, **47**, 583–621.

Kvam, P. and Vidakovic, B. (2007). *Nonparametric Statistics with Applications to Science and Engineering*. Wiley, Hoboken, NJ.

Lawrence, J., Alcock, D., McGrath, P., Kay, J., MacMurray, S. B., and Dulberg, C. (1993). The development of a tool to assess neonatal pain. *Neonatal Network*, **12**, 59–66.

Madansky, A. (1988). *Prescriptions for Working Statisticians*. Springer, New York.

Neghina, A., Anghel, A., Sporea, I., Popescu, A., Neghina, R., Collins, A., and Thorstensen, K. (2009). Mutant HFE genotype leads to significant iron overload in patients with liver diseases from western Romania. *J. Appl. Genet.*, **50**, 2, 173–176.

Oppenheim, R. W. (1968). Light responsivity in chick and duck embryos just prior to hatching. *Animal Behaviour*, **16**, 2-3, 276–280.

Quenouille, M. (1949). Approximate tests of correlation in time series. *J. Roy. Statist. Soc.*, *Ser. B*, **11**, 68–84.

Raspe, R. (1786). *The Singular Travels, Campaigns, and Adventures of Baron Munchausen.* Edition by John Carswell. Heritage Press, 1948. Riverton, WY.

Ross, S. M. (2010). *Introductory Statistics*, 3rd ed. Academic Press, Burlington, MA.

Rubin, D. B. (1981). The Bayesian bootstrap. *Ann. Statist.*, **9**, 1, 130–134.

Scanlon, T. J., Luben, R. N., Scanlon, F. L., and Singleton, N. (1993). Is Friday the 13th bad for your health? *Br. Med. J.*, **307**, 1584–1586.

Sharma, P., Sahni, N. S., Tibshirani, R., Skaane, P., Urdal, P., Berghagen, H., Jensen, M., Kristiansen, L., Moen, C., Sharma, P., Zaka, A., Arnes, J., Sauer, T., Akslen, L. A., Schlichting, E., Borresen-Dale, A.-L., and Lönneborg, A. (2005). Early detection of breast cancer based on gene-expression patterns in peripheral blood cells. *Breast Cancer Res.*, **7**, R634–R644.

Tukey, J. W. (1959). Approximate confidence limits for most estimates. Unpublished Manuscript.

Vagenakis, A. G., Rapoport, B., Azizi, F., Portnay, G. I., Braverman, L. E., and Ingbar, S. H. (1974). Hyperresponse to thyrothropin-releasing hormone acompanying small decreases in serum thyroid concentrations. *J. Clin. Invest.*, **54**, 913–918.

Walsh, J. E. (1962). *Handbook of Nonparametric Statistics I and II*. Van Nostrand, Princeton.

Wilcoxon, F. (1945). Individual comparisons by ranking methods. *Biometrics*, **1**, 80–83.

Chapter 19
Bayesian Inference Using Gibbs Sampling – BUGS Project

Beware: MCMC sampling can be dangerous!

– Disclaimer in WinBUGS User Manual

WHAT IS COVERED IN THIS CHAPTER

- Where to find WinBUGS, How to Install, Resources
- Step-by-step Example
- Built-in Functions and Common Distributions in BUGS
- MATBUGS: A MATLAB Interface to BUGS

19.1 Introduction

BUGS is a freely available software for constructing and evaluating Bayesian statistical models using simulation approaches based on the Markov chain Monte Carlo methodology.

BUGS and WINBUGS are distributed freely and are the result of many years of development by a team of statisticians and programmers at the Medical Research Council Biostatistics Unit in Cambridge, UK (BUGS and WinBUGS), and by a team at the University of Helsinki, Finland (Open-BUGS); see the project pages http://www.mrc-bsu.cam.ac.uk/software/bugs/ and http://www.openbugs.net.

Models are represented by a flexible language, and there is also a graphical feature, DOODLEBUGS, that allows users to specify their models as directed graphs. For complex models DOODLEBUGS can be very useful (Lunn et al., 2000). As of May 2017, the latest versions are WinBUGS 1.4.3 and OpenBUGS 3.2.3. A comprehensive overview of WinBUGS programming and applications can be found in Congdon (2005, 2006, 2010, 2014), Lunn et al. (2013), and Ntzoufras (2009).

19.2 Step-by-Step Session

We start this brief tutorial on WinBUGS with a simple regression example. Consider the model

$$
\begin{aligned}
y_i | \mu_i, \tau &\sim \mathcal{N}(\mu_i, \tau), \quad i = 1, \ldots, n, \\
\mu_i &= \alpha + \beta(x_i - \overline{x}), \\
\alpha &\sim \mathcal{N}(0, 10^{-4}), \\
\beta &\sim \mathcal{N}(0, 10^{-4}), \\
\tau &\sim \mathcal{G}a(0.001, 0.001).
\end{aligned}
$$

The normal distribution is parameterized by a *precision* parameter τ that is the reciprocal of the variance, $\tau = 1/\sigma^2$. Natural priors for precision parameters are gamma, and small values of the precision reflect the flatness (noninformativeness) of the priors. Assume that (x, y) pairs $(1,1)$, $(2,3)$, $(3,3)$, $(4,3)$, and $(5,5)$ are observed.

Estimators in classical, least-squares regression of y on $x - \overline{x}$ are given in the following MATLAB output:

```
y  = [1 3 3 3 5]'; %response
xx = [1 2 3 4 5]';
X  = [ones(size(xx)) xx-mean(xx)];
[b.b,b.int,res.res,res.int,stats] = regress(y,X);

b.b'
%    3.0000    0.8000
stats
%    0.8000  12.0000    0.0405    0.5333
```

Thus, the estimators are $\hat{\alpha} = \overline{y} = 3$, $\hat{\beta} = 0.8$, and $\hat{\tau} = 1/\hat{\sigma}^2 = 1/0.5333 = 1.875$.

What about Bayesian estimators? We will find the estimators by MCMC simulation, as empirical means of the simulated posterior distributions. Assume that the initial parameter values are $\alpha_0 = 0.1$, $\beta_0 = 0.6$, and $\tau = 1$. Start WinBUGS and input the following code in [**File > New**]:

```
# A simple regression
model{
```

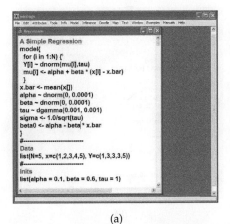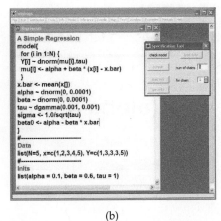

(a)   (b)

Fig. 19.1 (a) Opening WinBUGS front end with a simple regression task. The simple regression program is opened or typed in. (b) The front end after selecting **Specification** from the **Model** menu.

```
for (i in 1:N) {
 Y[i] ~ dnorm(mu[i],tau)
 mu[i] <- alpha + beta * (x[i] - x.bar)
 }
x.bar <- mean(x[])
alpha ~ dnorm(0, 0.0001)
beta ~ dnorm(0, 0.0001)
tau ~ dgamma(0.001, 0.001)
sigma <- 1.0/sqrt(tau)
}
#----------------------------
DATA
list(N=5, x=c(1,2,3,4,5), Y=c(1,3,3,3,5))
#----------------------------
INITS
list(alpha = 0.1, beta = 0.6, tau = 1)
```

Next, make sure that the cursor is somewhere within the scope of "model," that is, somewhere between the first open and the last closed curly bracket. Go to the **Model** menu and open **Specification**. The **Specification Tool** window will pop out (Fig. 19.1b). Next, press **check model** in the Specification Tool window. If the model is correct, the response on the lower left border of the window should be: **model is syntactically correct** (Fig. 19.2a). Next, data are read in. Highlight the "list" statement in the data part of your code (Fig. 19.2b). In the Specification Tool window, select **load data**. If the data are in the correct format, you should receive a response in the lower left corner of the WinBUGS window: **data loaded** (Fig. 19.3a). You will need to compile your model in order to activate the **inits** buttons.

Select **compile** in the Specification Tool window. The response should be: **model compiled** (Fig. 19.3b), and the **load inits** and **gen inits** but-

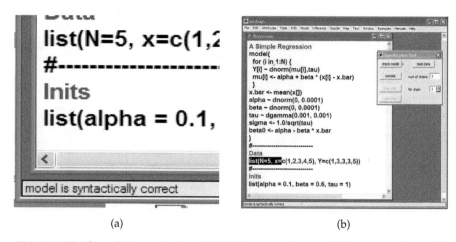

Fig. 19.2 (a) After selecting **check model**, if the syntax is correct, the response is **model is syntactically correct.** (b) Highlighting the list in the data prior to reading data in.

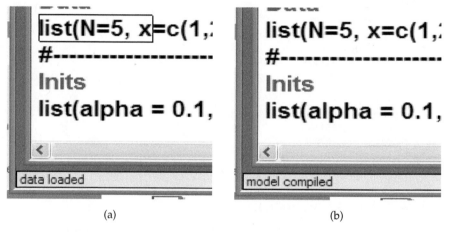

Fig. 19.3 WinBUGS' responses to (a) **load data** and (b) **compile** in the model specification tool.

tons become active. Finally, highlight the "list" statement in the initials part of your code, and in the Specification Tool window, select **load inits** (Fig. 19.4a). The response should be: **model is initialized** (Fig. 19.4b), and this completes the reading in of the model. If the response is **initial values loaded but this or another chain contains uninitialized variables,** click on the **gen inits** button. The response should be: **initial values generated, model initialized.**

Now you are ready to burn in some simulations and at the same time check if the program works. Recall that burning in the Markov chain model is necessary for the chain to "forget" the initialized parameter values. In

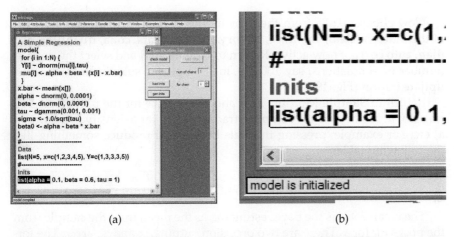

(a)   (b)

Fig. 19.4 (a) Highlighting the `list` to initialize the model. (b) WinBUGS confirms that the model (in fact a Markov chain) is initialized.

the **Model** menu, choose **Update...** and open **Update Tool** to check if your model updates (Fig. 19.5a).

From the **Inference** menu, open **Samples...**. A window titled **Sample Monitor Tool** will pop out (Fig. 19.5b). In the **node** subwindow, input the names of the variables you want to monitor. In this case, the variables are `alpha`, `beta`, and `tau`. If you correctly input the variable name, the **set** button becomes active and you should set the variable. Do this for all three variables of interest. In fact, `sigma` as a transformation of `tau` is available to be set as well.

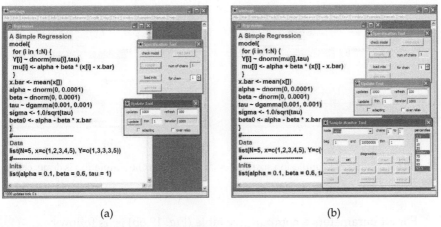

(a)   (b)

Fig. 19.5 WinBUGS' response to (a) **Update...** tool from the **Model** menu and (b) **Samples...** from the **Inference** menu.

Now choose `alpha` from the subwindow in **Sample Monitor Tool**. All of the buttons (**clear, set, trace, history, density, stats, coda, quantiles, bgr diag, auto cor**) are now active. Return to **Update Tool** and select the desired number of simulations, say 100,000, in the **updates** subwindow. Press the **update** button (Fig. 19.6a).

Return to **Sample Monitor Tool** and check **trace** for the part of the MC trace for α, **history** for the complete trace, **density** for a density estimator of α, etc. For example, pressing the **stats** button will produce something like the following table:

	mean	sd	MC error	val2.5pc	median	val97.5pc	start	sample
alpha	2.996	0.5583	0.001742	1.941	2.998	4.041	1001	100000

The mean 2.996 is the Bayes estimator, as the mean from the sample from the posterior for α. There are two precision outputs, `sd` and `MC error`. The former is an estimator of the standard deviation of the posterior and can be improved by increasing the sample size but not the number of simulations. The latter is the simulation error and can be improved by additional simulations. The 95% credible set (1.941, 4.041) is determined by `val2.5pc` and `val97.5pc`, which are the 0.025 and 0.975 (empirical) quantiles from the posterior. The empirical median of the posterior is given by `median`. The outputs `start` and `sample` show the starting index for the simulations (after burn-in) and the available number of simulations.

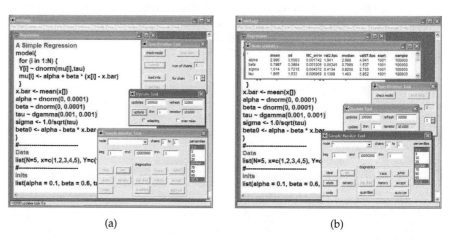

        (a)   (b)

Fig. 19.6 (a) Select the simulation size and update. (b) After the simulation is done, check the `stats` node.

For all parameters a comparative table (Fig. 19.6b) is as follows:

	mean	sd	MC error	val2.5pc	median	val97.5pc	start	sample
alpha	2.996	0.5583	0.001742	1.941	2.998	4.041	1001	100000
beta	0.7987	0.3884	0.001205	0.06345	0.7999	1.537	1001	100000
sigma	1.014	0.7215	0.004372	0.4134	0.8266	2.765	1001	100000
tau	1.865	1.533	0.006969	0.1308	1.463	5.852	1001	100000

We recall the least squares estimators from the beginning of this session: $\hat{\alpha} = 3$, $\hat{\beta} = 0.8$, and $\hat{\tau} = 1.875$, and note that their Bayesian counterparts are very close.

Densities (smoothed histograms) and traces for all parameters are given in Fig. 19.7.

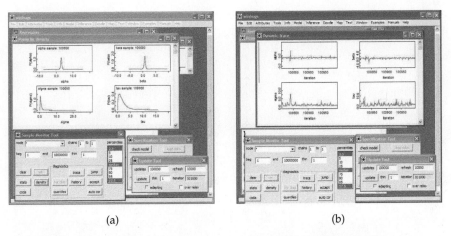

(a)   (b)

Fig. 19.7 Checking (a) density and (b) trace in the **Sample Monitor Tool**.

If you want to save the trace for α in a file and process it in MATLAB, select **coda**, and the data window will open with an information window as well. Keep the data window active and select **Save As** from the **File** menu. Save the αs in alphas.txt, where it will be ready to be imported into MATLAB. Later in this chapter we will discuss the direct interface between WinBUGS and MATLAB called MATBUGS.

19.3 Built-in Functions and Common Distributions in WinBUGS

This section contains two tables: one with the list of built-in functions and another with the list of available distributions.

A first-time WinBUGS user may be disappointed by the selection of built-in functions – the set is minimal but sufficient. The full list of distributions in WinBUGS can be found in **Manuals>OpenBUGS User Manual**.

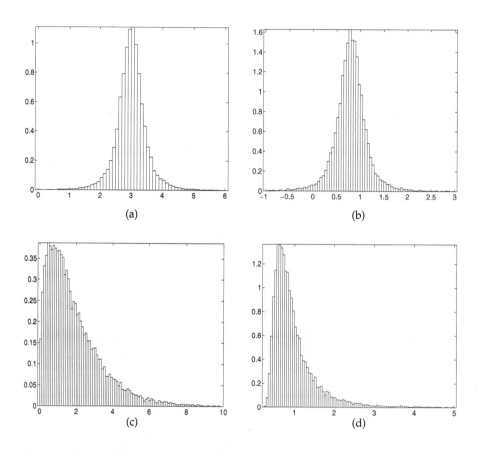

Fig. 19.8 Traces of the four parameters from a simple example: (a) α, (b) β, (c) τ, and (d) σ from WinBUGS. Data are plotted in MATLAB after being exported from WinBUGS.

WinBUGS also allows for the inclusion of distributions for which functions are not built in. Table 19.2 provides a list of important discrete and continuous distributions, with their syntax and parametrizations. WinBUGS has the capability to define custom distributions, both as a likelihood and as a prior, via the so-called zero-tricks (p. 353).

19.4 MATBUGS: A MATLAB Interface to WinBUGS

There is strong motivation to interface WinBUGS with MATLAB. Cutting and pasting results from WinBUGS is cumbersome if the simulation size is in millions or if the number of simulated parameters is large. Also, the data

Table 19.1 Built-in functions in WinBUGS

WinBUGS code	Function
abs(y)	$|y|$
cloglog(y)	$\ln(-\ln(1-y))$
cos(y)	$\cos(y)$
equals(y, z)	1 if $y = z$; 0 otherwise
exp(y)	$\exp(y)$
inprod(y, z)	$\sum_i y_i z_i$
inverse(y)	y^{-1} for symmetric positive–definite matrix y
log(y)	$\ln(y)$
logfact(y)	$\ln(y!)$
loggam(y)	$\ln(\Gamma(y))$
logit(y)	$\ln(y/(1-y))$
max(y, z)	y if $y > z$; y otherwise
mean(y)	$n^{-1}\sum_i y_i$, $n = dim(y)$
min(y, z)	y if $y < z$; z otherwise
phi(y)	standard normal CDF $\Phi(y)$
pow(y, z)	y^z
sin(y)	$\sin(y)$
sqrt(y)	$\sqrt{y}$
rank(v, s)	number of components of v less than or equal to v_s
ranked(v, s)	sth smallest component of v
round(y)	nearest integer to y
sd(v)	standard deviation of components of y ($n - 1$ in denom.)
step(y)	1 if $y \geq 0$; 0 otherwise
sum(y)	$\sum_i y_i$
trunc(y)	greatest integer less than or equal to y

manipulation and graphical capabilities in WinBUGS are quite rudimentary compared to MATLAB.

MATBUGS is a MATLAB program that communicates with WinBUGS. The program matbugs.m was written by Kevin Murphy and his team and can be found at: http://code.google.com/p/matbugs.

We now demonstrate how to solve Jeremy's IQ problem in MATLAB by calling WinBUGS. First we need to create a simple text file, say, jeremy.txt:

```
model{
for(i in 1 : N)
    {
        scores[i] ~ dnorm(theta, tau)
    }
 theta ~ dnorm(mu, xi)
```

and then run the MATLAB file:

```
dataStruct = struct( ...
    'N', 5, ...
    'tau',1/80,...
    'xi',1/120,...
    'mu',110,...
    'scores',[97 110 117 102 98]);
```

Table 19.2 Some important built-in distributions with WinBUGS names and their parameterizations.

Distribution	WinBUGS code	Density
Bernoulli	x ~ dbern(p)	$p^x(1-p)^{1-x}$, $x=0,1$; $0 \le p \le 1$
Binomial	x ~ dbin(p, n)	$\binom{n}{x}p^x(1-p)^{n-x}$, $x=0,\dots,n$; $0 \le p \le 1$
Categorical	x ~ dcat(p[])	$p[x]$, $x=1,2,\dots,\dim(p)$
Negative Binomial	x ~ dnegbin(p, r)	$\frac{(x+r-1)!}{x!(r-1)!}p^r(1-p)^x$, $x=0,1,\dots$; $0 \le p \le 1$
Poisson	x ~ dpois(lambda)	$\frac{\lambda^x}{x!}\exp\{-\lambda\}$, $x=0,1,2,\dots$; $\lambda > 0$
Beta	x ~ dbeta(a,b)	$\frac{1}{B(a,b)}x^{a-1}(1-x)^{b-1}$, $0 \le x \le 1$, $a,b > -1$
Chi-square	x ~ dchisqr(k)	$\frac{x^{k/1-1}\exp\{-x/2\}}{2^{k/2}\Gamma(k/2)}$, $x \ge 0, k > 0$
Double exponential	x ~ ddexp(mu, tau)	$\frac{\tau}{2}\exp\{-\tau|x-\mu|\}$, $x \in R, \tau > 0, \mu \in R$
Exponential	x ~ dexp(lambda)	$\lambda\exp\{-\lambda x\}$, $x \ge 0, \lambda \ge 0$
Flat	x ~ dflat()	constant; not a proper density
Gamma	x ~ dgamma(a, b)	$\frac{b^a x^{a-1}}{\Gamma(a)}\exp(-bx)$, $x,a,b > 0$
Normal	x ~ dnorm(mu, tau)	$\sqrt{\tau/(2\pi)}\exp\{-\frac{\tau}{2}(x-\mu)^2\}$, $x,\mu \in R, \tau > 0$
Pareto	x ~ dpar)alpha,c)	$\alpha c^\alpha x^{-(\alpha+1)}$, $x > c$
t	x ~ dt(mu, tau, k)	$\frac{\Gamma((k+1)/2)}{\Gamma(k/2)}\sqrt{\frac{\tau}{k\pi}}[1+\frac{\tau}{k}(x-\mu)^2]^{-(k+1)/2}$, $x \in \mathbb{R}, k \ge 2$
Uniform	x ~ dunif(a, b)	$\frac{1}{b-a}$, $a \le x \le b$
Weibull	x ~ dweib(v, lambda)	$v\lambda x^{v-1}\exp\{-\lambda x^v\}$, $x,v,\lambda > 0$,
Multinomial	x[] ~ dmulti(p[], N)	$\frac{(\sum_i x_i)!}{\prod_i x_i!}\prod_i p_i^{x_i}$, $\sum_i x_i = N$, $0 < p_i < 1$, $\sum_i p_i = 1$
Dirichlet	p[] ~ ddirch(alpha[])	$\frac{\Gamma(\sum_i \alpha_i)}{\prod_i \Gamma(\alpha_i)}\prod_i p_i^{\alpha_i-1}$, $0 < p_i < 1$, $\sum_i p_i = 1$
Multivariate normal	x[] ~ dmnorm(mu[], T[,])	$(2\pi)^{-d/2}|T|^{1/2}\exp\{-1/2(x-\mu)'T(x-\mu)\}$, $x \in R^d$
Multivariate t	x[] ~ dmt(mu[], T[,], k)	$\frac{\Gamma((k+d)/2)}{\Gamma(k/2)}\frac{|T|^{1/2}}{k^{d/2}\pi^{d/2}}\left[1+\frac{1}{k}(x-\mu)'T(x-\mu)\right]^{-(k+d)/2}$, $x \in \mathbb{R}^d, k \ge 2$
Wishart	x[,] ~ dwish(R[,], k)	$|R|^{k/2}|x|^{(k-p-1)/2}\exp\{-1/2Tr(Rx)\}$, x p.d.; $k > p-1$

```
initStruct = struct( ...
    'theta', 100 );

cd('C:\MyBugs\matbugs\')
[samples, stats] = matbugs(dataStruct, ...
fullfile(pwd, 'jeremy.txt'), ...
'init', initStruct, ...
        'nChains', 1, ...
'view', 0, ...
        'nburnin', 2000, ...
        'nsamples', 50000, ...
'thin', 1, ...
'monitorParams', {'theta'}, ...
        'Bugdir', 'C:/Program Files/BUGS');

baymean = mean(samples.theta)
frmean=mean(dataStruct.scores)

  figure(1)
  [p, x] = ksdensity(samples.theta);
  plot(x, p);
```

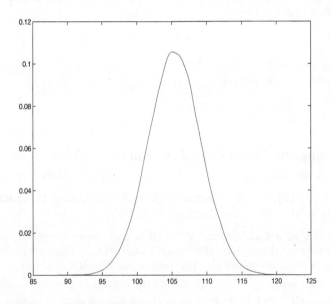

Fig. 19.9 Posterior for Jeremy's data set. Data are plotted in MATLAB after being exported from WinBUGS by MATBUGS.

19.5 Exercises

19.1. **A Coin and a Die.** The following WinBUGS code simulates flips of a coin. The outcome **H** is coded by 1 and **T** by 0. Mimic this code to simulate rolls of a fair die.

```
#coin
model{
flip ~ dcat(p.coin[])
coin <- flip - 1
}
DATA
list(p.coin=c(0.5, 0.5))
#just generate initials
```

19.2. **De Mere Paradox in WinBUGS.** In Exercise 3.6 (b) we examined de Mere's paradox: In playing a game with three fair dice, the sum 11 was advantageous to the sum 12.
(a) Using WinBUGS/OpenBUGS demonstrate that, in playing a game with 300 fair dice, the sum 1111 is advantageous to the sum 1112.
(b) Which of the two sums from (a) is more advantageous if the 300 dice are loaded, with probabilities 0.15, 0.15, 0.16, 0.2, 0.17, and 0.17, for sides $1, \ldots, 6$, respectively.
Hint:

```
 part of the code...
 for (i in 1:300) {
 dice[i] ~ dcat(p.dice[]);
 }
 is1111 <- equals(sum(dice[]),1111)
 is1112 <- equals(sum(dice[]),1112)
```

19.3. **Simulating the Probability of an Interval.** Consider an exponentially distributed random variable X, $X \sim \mathcal{E}\left(\frac{1}{10}\right)$, with density $f(x) = \frac{1}{10}\exp\{-x/10\}$, $x > 0$. Compute $\mathbb{P}(10 < X < 16)$ using (a) exact integration, (b) MATLAB's `expcdf`, and (c) WinBUGS.

19.4. **WinBUGS as a Calculator.** WinBUGS can approximate definite integrals, solve nonlinear equations, and even find values of definite integrals over random intervals. The following WinBUGS program finds an approximation to $\int_0^\pi \sin(x)dx$, solves the equation $y^5 - 2y = 0$, and finds the integral $\int_0^R z^3(1 - z^4)dz$, where R is a beta $\mathcal{B}e(2,2)$ random variable. Verify the following code and find the solution:

```
model{
F(x) <- sin(x)
int <- integral(F(x), 0, pi, 1.0E-6)
pi<- 3.141592659
```

```
y0 <- solution(F(y), 1,2, 1.0E-6)
F(y)  <- pow(y,5) - 2*y
zero <-  pow(y0, 5)-2*y0

randint <- integral(F(z), 0, randbound, 1.0E-6)
F(z) <- pow(z,3)*(1-pow(z,4))
randbound ~ dbeta(2,2)
}

NO DATA

INITS
     list(x=1, y=0, z=NA, randbound=0.5)
```

After model checking, one should go directly to compiling (no data to load in) and initializing the model. There is NO need to update the model, to go to the Inference tool, to set the variables for monitoring or to sample. One simply goes to the **Info** menu and checks **Node Info**. In the Node Info tool one specifies int for the approximation of an integral, y0 for the solution of an equation, zero for checking that y0 satisfies the equation (approximately), and randint for the value of a random interval.

| MATLAB AND WINBUGS FILES AND DATA SETS USED IN THIS CHAPTER |

http://statbook.gatech.edu/Ch19.WinBUGS/

simple.m

DeMere.odc, jeremy.odc, picktrick.odc, Regression1.odc, Regression2.odc, simulationd.odc

alpha.txt, beta.txt, sigma.txt, tau.txt

CHAPTER REFERENCES

Congdon, P. (2005). *Bayesian Models for Categorical Data*. Wiley, Hoboken, NJ.

Congdon, P. (2006). *Bayesian Statistical Modelling*, Second Edition. Wiley, Hoboken, NJ.

Congdon, P. (2010). *Hierarchical Bayesian Modelling*. Chapman & Hall/CRC, Boca Raton, FL.

Congdon, P. (2014). *Applied Bayesian Modelling*, Second Edition. Wiley, Hoboken, NJ.

Lunn, D., Jackson, C., Best, N., Thomas, A., and Spiegelhalter, D. (2013). *The BUGS Book*, CRC Press, Boca Raton, FL.

Lunn, D. J., Thomas, A., Best, N., and Spiegelhalter, D. (2000). WinBUGS – a Bayesian modelling framework: concepts, structure, and extensibility. *Stat. Comput.*, **10**, 325–337.

Ntzoufras, I. (2009). *Bayesian Modeling Using WinBUGS*. Wiley, Hoboken.

Index